SATELLITES OF JUPITER

SATELLITES OF JUPITER

Edited by
DAVID MORRISON

With the assistance of
MILDRED SHAPLEY MATTHEWS

With 47 collaborating authors

THE UNIVERSITY OF ARIZONA PRESS
TUCSON, ARIZONA

SPACE SCIENCE SERIES

Tom Gehrels, Space Sciences Consultant

PLANETS, STARS AND NEBULAE, STUDIED WITH PHOTOPOLARIMETRY, T. Gehrels Ed., 1974, 1133 pp.

JUPITER, T. Gehrels Ed., 1976, 1254 pp.

PLANETARY SATELLITES, J.A. Burns Ed., 1977, 598 pp.

PROTOSTARS AND PLANETS, T. Gehrels Ed., 1978, 756 pp.

ASTEROIDS, T. Gehrels Ed., 1979, 1181 pp.

COMETS, L.L. Wilkening, 1982, 766 pp.

THE SATELLITES OF JUPITER, D. Morrison Ed., 1982, 974 pp.

Back cover: This picture is a composite of two photographic plates (an overexposed one to show the satellites and an underexposed one to bring out some detail on Jupiter) taken in 1955 by G.P. Kuiper at the McDonald Observatory on the Struve Reflector.

THE UNIVERSITY OF ARIZONA PRESS

Copyright © 1982
The Arizona Board of Regents
All Rights Reserved

This book was set in 10/12 IBM MTSC Times Roman
Manufactured in the U.S.A.

Library of Congress Cataloging in Publication Data
Main entry under title:

Satellites of Jupiter.

    (Space Science Series)
    Includes index.
    1. Satellites—Jupiter. I. Morrison, David, 1940–
II. Matthews, Mildred Shapley. III. Series
QB 404.S34   523.4'5   81-13050

ISBN 0-8165-0762-7   AACR2

CONTENTS

COLLABORATING AUTHORS

BANERDT, W.B., *756*
CASSEN, P.M., *93*
CLARK, R.N., *174*
CRUIKSHANK, D.P., *129*
DAVIES, M.E., *911*
DEGEWIJ, J., *129*
ELSON, L.S., *756*
FANALE, F.P., *756, 872*
FINK, J.H., *340*
GAULT, D.E., *340*
GREELEY, R., *340*
GREENBERG, R., *65*
GURNIS, M., *237*
GUEST, J.E., *340*
HUNTEN, D.M., *782*
JEWITT, D.C., *44*
JOHNSON, T.V., *634, 756*
KIEFFER, S.W., *647*
KUMAR, S., *782*
LUCCHITTA, B.K., *435, 521*
MORRISON, D., *ix, 3*
OSTRO, S.J., *213*
PASSEY, Q.R., *379*
PEALE, S.J., *93*

PEARL, J.C., *724*
PILCHER, C.B., *807*
PLESCIA, J.B., *435*
POLLACK, J.B., *872*
REYNOLDS, R.T., *93*
SCHABER, G.G., *556*
SCHNEIDER, N.M., *598*
SHOEMAKER, E.M., *277, 379, 435*
SILL, G.T., *174*
SINTON, W.M., *724*
SISCOE, G.L., *846*
SODERBLOM, L.A., *521, 634*
STROBEL, D.F., *807*
SQUYRES, S.W., *435*
STROM, R.G., *237, 598*
SULLIVAN, J.D., *846*
THOMAS, P., *147*
VEVERKA, J., *147*
WILHELMS, D.E., *435*
WOLFE, R.F., *277*
WORONOW, A., *237*
ZELLNER, B.H., *129*
ZUREK, R.W., *756*

PREFACE

The 1970s were years of great progress in many sciences, but probably no discipline has been more fundamentally transformed than planetary studies. Stimulated by a flood of data from the deep space probes of the U.S. and U.S.S.R., astronomers, geologists, geophysicists, chemists, and plasma physicists have created this new interdisciplinary field. An essential conviction in this endeavor is that the planetary bodies, though individually diverse, can best be understood comparatively, as members of a family which have followed separate but related evolutionary paths in response to usually small differences in initial conditions or environmental factors. The Voyager explorations of the satellites of Jupiter and Saturn have brought over a dozen additional members of this family within our reach. The four large Galilean satellites of Jupiter, in particular, were shown to be comparable in beauty, variety, and scientific significance to the more familiar terrestrial bodies.

This book is the product of a unique moment in planetary exploration. Before the Voyager discoveries of March and July 1979, such a book could hardly have been conceived. Almost all the following chapters are based on data from these two spacecraft encounters with the Jovian system. Although constantly being enriched by Earth-based observations and by new theoretical and laboratory results, these data should stand until another spacecraft reaches the Jovian system, which unfortunately will not be before the late 1980s, when we hope that the Galileo Orbiter will carry out a two-year tour of Jupiter and its satellites. Until then, although much Voyager data remains undigested and there is unlimited scope for new ideas and synthesis, the period of initial discovery is complete.

Like other books of the University of Arizona Space Science Series, this volume is intended as both an introduction suitable for beginning graduate students and a reference for research workers. This is currently the only book to deal with the Jovian satellites both as individual worlds and as a system, and it benefits from the enthusiastic participation of many leading satellite scientists. Although publication has been delayed by nearly a year, some contributors have profited by the extra time to develop their topics more fully and to take advantage of new results from others. However, we regret that other authors who submitted their chapters within original deadlines have not been able to present their most recent work. It was probably too much to expect a definitive volume to be written within a year of the spacecraft encounter, but the book as a whole was benefited by the delays.

The meeting from which this book was derived, IAU Colloquium No. 57 "The Satellites of Jupiter," was held in Kailua-Kona, Hawaii, on May 13-16, 1980, and was sponsored by the Institute for Astronomy of the University of Hawaii with support from the International Astronomical Union, the Committee on Space Research, the Division for Planetary Sciences of the American Astronomical Society, the National Science Foundation, the National Aeronautics and Space Administration, and the Voyager Project. Approximately 250 scientists attended the 4-day meeting, at which 132 papers were given representing 199 authors. The Organizing Committee consisted of D. Morrison (Chair), K. Aksnes, I. Axford, A. Brahic, J. Burns, D. Cruikshank, A. Dollfus, F. Fanale, J. Guest, T. Johnson, H. Masursky, V. Moroz, H. Oya, C. Pilcher, G. Siscoe, E. Stone, and J. Veverka. The local organizing committee consisted of Morrison, Cruikshank, and Pilcher, assisted by D. Weiner and several staff members and graduate students at the University of Hawaii. The invited review papers at the colloquium are the basis for most of the chapters in this book, although all have been refereed and extensively revised after the meeting. Many of the contributed papers were published in a special issue of *Icarus* (November 1980), edited by J. Burns.

I cannot begin to thank the countless individuals who contributed to the meeting and the book. Instead, there are extensive acknowledgments at the end of the book. I will mention here only my associate, M. Matthews, without whose energetic and skillful efforts this book would not exist, and D. Weiner and M. Missback who contributed so much to the early organizational work. I also recognize that without the efforts of the Project Voyager science teams, led by E. Stone of the California Institute of Technology, the Jovian satellites would remain largely unknown to us. Finally, it is a pleasure to acknowledge financial contributions from T. Gehrels and the essential financial support received from the National Aeronautics and Space Administration, the Jet Propulsion Laboratory, the National Science Foundation, and the University of Hawaii. The University of Arizona Press is to be thanked for bringing about publication.

David Morrison

SATELLITES
OF JUPITER

1. INTRODUCTION TO THE SATELLITES OF JUPITER

DAVID MORRISON
University of Hawaii

The sixteen known satellites of Jupiter are a diverse and fascinating set of planetary bodies, categorized into four groups of four objects each: (1) the Galilean satellites, ranging in size from sublunar to larger than Mercury; (2) the smaller inner satellites, three of which have recently been discovered by Voyager; (3) the small prograde outer satellites; and (4) the small retrograde outer satellites. This introduction sketches the history of satellite studies, including the Pioneer and Voyager missions of the late 1970s. The topics dealt with in the chapters of this book are summarized, reviewing basic concepts and attempting to place each chapter in perspective.

The Jovian satellites provide the nearest analog within our solar system of the planetary system itself: diverse worlds orbiting a much larger primary, probably mainly derived from the condensation of a circumprimary nebula ~ 4.6 Gyr in the past. This miniature solar system has fascinated astronomers since Galileo's discovery in 1610 of the four largest Jovian satellites, but only recently have the Jovian moons been intensely investigated as unique planetary bodies that have formed and evolved through the same physical and chemical processes that produced the other planets and major satellites of our system. This book tells the story of the Jovian satellites as worlds, seen clearly for the first time in 1979 by the Voyager spacecraft, yet still new and mysterious.

This chapter, like the book, proceeds from the general to the specific, and from small satellites to large. It begins with the historical framework, leading to a summary of our knowledge of these objects in the mid 1970s, as represented in *Planetary Satellites,* University of Arizona Press (Burns 1977). I then outline the Pioneer and Voyager missions, in which four robot spacecraft reconnoitered and explored the Jovian system in the years 1973-1979, providing most of the data base for this book. From this perspective, we can then examine the elements of the satellite system: first, the rings that form its inner boundary, then Amalthea and the three newly discovered inner satellites, the system of small outer satellites that extends $> 10^6$ km from Jupiter, and finally the common dynamical processes that bind the system together and have provided critical boundary conditions on its evolution. Remote sensing from the Earth continues to provide most of our information on the mineralogical composition of satellite surfaces; groundbased radar also contributes to the probing of satellite regoliths.

Viewed from close range, the Galilean satellites reveal their individual identities. The universal discriminant of geologic history is the record of impact cratering, and three chapters discuss this link of the chronologies of all the solid-surface planets. We then look at the objects themselves, beginning with Callisto and Ganymede and moving inward to Europa and finally Io, progressing from old to young, from geologically dead to intensely dynamic. Io is the focus of the second half of the book, in chapters on its interior, surface, volcanic eruptions, unstable atmosphere generated by volcanism, the escape of neutrals and ions into the Jovian magnetosphere, and the sea of energetic particles in which all the Galilean satellites are bathed. Finally, the last chapter turns to the question of the origin and evolution of this miniature solar system.

I. THE SATELLITES: 1610 to 1974

Galileo Galilei's discovery of the four satellites that now bear his name was one of the most spectacular of early telescopic results, demonstrating that orbital motion can take place about a center other than the Earth. The actual discovery was not without controversy, since Galileo's contemporary, Simon Marius, claimed precedence; however, Galileo "published" first and no proof exists of Marius' earliest sightings, so credit is generally assigned to Galileo. Marius did suggest the names now in use: Io, Europa, Ganymede, and Callisto, all lovers of Zeus/Jupiter in Greco-Roman mythology.

Observations of positions of the Galilean satellites and timing of eclipses were important in the centuries after Galileo in such problems as determining the speed of light and establishing longitudes on the Earth, but it was not until the 19th century that physical observations became significant. Laplace and his successors developed a theory of motion for the satellites and used the resonant properties of their orbits to estimate their masses, while Barnard utilized the new refractors at Yerkes and Lick to measure their sizes. Although these measurements of mass and diameter had uncertainties on the order of 20%, they permitted the first estimate of bulk density, revealing that the inner satellites were apparently composed of denser materials than the outer. In the 1920s, Stebbins and Jacobsen (1928) obtained the first photoelectric measurements, determining the amplitude of the lightcurves and proving that all four satellites were in Jupiter-synchronous rotation. Also during the early 20th century, use of photographic techniques led to the discovery of the prograde group and retrograde group of small outer satellites.

Beginning about 1950, observers employing larger telescopes and modern techniques of photometry, polarimetry, and spectrophotometry began to examine the Jovian satellites. Colors were measured, color and albedo differences were noted, a spectroscopic search for atmospheres was carried out, and improved values were obtained for the satellite diameters and densities. Near-infrared photometry provided the first clue to surface composition; in a landmark abstract Kuiper (1957) suggested from limited data, never published, that Europa and Ganymede had water ice surfaces, and Moroz (1961) later came to the same conclusion from similar broadband infrared observations. Even at that time the anomalous nature of Io, with its high albedo, red color, and absence of ice could be noted. The rapid improvement of infrared detectors soon permitted thermal radiation from the Galilean satellites to be measured, and Murray et al. (1964) succeeded in measuring the cooling of Ganymede as it entered the shadow of Jupiter, demonstrating that the surface thermal conductivity was low, similar to that of the Moon.

In the early 1970s, studies of the large satellites in the outer solar system became increasingly important. The well-observed occultation of a star by Io (Taylor et al. 1972) yielded the first high-precision diameter for a satellite of Jupiter and set a stringent upper limit for an atmosphere; a year later similar occultation observations were made for Ganymede (Carlson et al. 1973). Pilcher et al. (1972) and Fink et al. (1973) demonstrated spectroscopically that H_2O ice was the dominant surface material on Europa and Ganymede but was apparently absent (to the 1% level) on Io, and the first measurements of eclipse heating and cooling curves led to an analysis of thermophysical surface properties of the Galilean satellites (Morrison et al. 1971; Morrison and Cruikshank 1973; Hansen 1973). The early theoretical work of Lewis (1971) was essential to understanding the interior structure of icy satellites. The most spectacular advance was the unexpected discovery in 1972 by Brown (1974) of sodium emission from Io. A number of observers quickly demonstrated the existence of a large cloud of neutral sodium near the satellite; surface sputtering by energetic particle bombardment was suggested by Matson et al. (1974) to explain its origin. These and other results set the stage for the first spacecraft observations of the Jovian system by Pioneer 10 in 1973.

At this point in the investigation of the Jupiter system, an international meeting devoted to the 33 known planetary satellites (excluding Earth's Moon) was held at Ithaca, New York, on 18-21 August 1974. The resulting book, *Planetary Satellites* (Burns 1977), provides a comprehensive picture of satellite knowledge before spacecraft exploration. A particularly detailed treatment of orbital dynamics, not repeated in this book, is found in Chapters 3-8 of *Planetary Satellites*. A number of other chapters are concerned primarily with the Jovian satellites, including chapters on photometry and polarimetry by Veverka (1977a,b) and Morrison and Morrison (1977), on radiometry by Morrison (1977), on spectrophotometry by Johnson and Pilcher (1977), on Io by Fanale et al. (1977), and on the thermal evolution and interior structure of the Galilean satellites by Consolmagno and Lewis (1977) and Fanale et al. (1977).

The reviews of photometry and polarimetry cited above remain the best sources of information on the disk-integrated properties of the surfaces of the Jovian satellites. Other chapters are more dated. Spectrophotometry has been much improved by new observations, many obtained at Mauna Kea in Hawaii by T. McCord and his colleagues, as discussed in Chapter 7 by Sill and Clark in this book. Radiometric studies of Io have been revolutionized by the discovery of active volcanism, discussed by Pearl and Sinton in Chapter 19. Fanale et al. (1977) provided an excellent summary of data then available on the surface and atmosphere of Io and introduced an ingenious hypothesis for its early thermal evolution, but this hypothesis was superceded when large-scale tidal heating and the consequent high level of volcanism were identified. The interior models of Consolmagno and Lewis (1977) also appear to require modification, to account both for tidal evolution and for solid-state convection as a heat transport mechanism in icy satellite mantles.

Many questions raised in 1974 about the Galilean satellites remain unanswered. A major curiosity on Io was the post-eclipse brightening phenomenon first identified by Binder and Cruikshank (1964); no such effect was seen by Voyager (Veverka et al. 1981a), and it is still unclear what, if anything, might cause sporadic short-term changes in albedo. The origin of the Io neutral sodium was hotly debated in 1974 and remains a puzzle now, although the sputtering model of Matson et al. (1974) is surely involved. The satellite's systematic longitudinal variations in disk-integrated color and albedo were a mystery, with both endogenic and exogenic hypotheses advanced; Voyager threw no light on this subject. The intriguing polarimetric asymmetry between the leading and trailing hemispheres of Callisto may be partly due to the cratering effects discussed in Chapter 10 by Shoemaker and Wolfe, but no full explanation has been worked out. Many of the problems that seemed dominant in 1974 are not emphasized in this book. The focus of our attention has shifted dramatically in the era of exploration by spacecraft, with the flood of new data drowning out old controversies and generating new ones.

II. PIONEER AND VOYAGER

The United States initiated direct exploration of the outer solar system through the Pioneer Jupiter project begun in 1969. Two spin-stabilized spacecraft, each having a mass of 258 kg and carrying 25 kg of scientific instruments, were built by TRW under the management of the NASA Ames Research Center. The Pioneers were launched by Atlas-Centaur rockets in 1972 and 1973, respectively. The Project Manager was C.F. Hall, and the Project Scientist was plasma physicist J.H. Wolfe; the scientific investigations are listed in Table 1.1.

The primary objectives of the Pioneer project were not so much to investigate the Jovian system as to demonstrate that spacecraft could be sent successfully to the outer solar system. Of particular concern were the potential hazards due to small debris in the asteroid belt and radiation damage from the Jovian magnetosphere. The scientific discoveries of the Pioneers at Jupiter were a valuable bonus beyond proving the accessibility of the outer solar system to direct exploration. For discussions of the Pioneer project and its results, see *Pioneer Odyssey, Encounter with a Giant* (Fimmel et al. 1974), and *Pioneer, First to Jupiter, Saturn, and Beyond* (Fimmel et al. 1980).

Primary emphasis in the Pioneer investigations was on *in situ* measurements of the particles and fields environment through which the craft passed, particularly within the Jovian magnetosphere. Because of the limitations of the spinning spacecraft and the low mass and power available for scientific instruments, very few measurements of the satellites themselves were made, although improved masses for all four Galilean satellites were derived from analysis of the spacecraft trajectories (Anderson et al. 1974), as well as some improvement in diameters for Callisto and Europa (Smith 1978). In addition, Pioneer 10 was targeted to provide a radio occultation of the spacecraft by Io, which revealed an ionosphere with peak electron concentration 6×10^4 cm^{-3} at 100 km altitude on the day side (Kliore et al. 1974,1975). An ultraviolet glow presumed to be Lyman α of hydrogen was also detected near the orbit of Io (Judge and Carlsen 1974), further supporting the idea of an extended atmosphere for the satellite.

The post-Pioneer view of the Galilean satellites is summarized in four chapters in *Jupiter*, University of Arizona Press (Gehrels 1976; Morrison and Burns 1976; Consolmagno and Lewis 1976; Judge et al. 1976; and Brown and Yung 1976). These chapters emphasize Io, its atmosphere and extended clouds of sodium and hydrogen, and its interaction with the magnetosphere of Jupiter.

TABLE 1.1

Pioneer Science Investigations
Project Scientist: J.H. Wolfe, NASA Ames

Investigation	Principal Investigator	Primary Objectives
Magnetic fields	E.J. Smith, JPL	Measurement of the magnetic field of Jupiter and determination of the structure of the magnetosphere.
Magnetic fields (Pioneer 11 only)	N.F. Ness, NASA Goddard	Measurement of the magnetic field of Jupiter and determination of the structure of the magnetosphere.
Plasma analyzer	J.H. Wolfe, NASA Ames	Measurement of low-energy electrons and ions, determination of the structure of the magnetosphere.
Charged particle composition	J.A. Simpson, U. Chicago	Determination of the number, energy, and composition of energetic charged particles in the Jovian magnetosphere.
Cosmic ray energy spectra	F.B. McDonald, NASA Goddard	Measurement of number and energy of very high energy charged particles in space.
Jovian charged particles	J.A. Van Allen, U. Iowa	Measurement of number and energy distribution of energetic charged particles and determination of magnetospheric structure.
Jovian trapped radiation	R. Walker Fillius, UC San Diego	Measurement of number and energy distribution of energetic charged particles and determination of magnetospheric structure.
Asteroid-meteoroid astronomy	R.K. Soberman, General Electric	Observation of solid particles (dust and larger) in the vicinity of the spacecraft.
Meteoroid detection	W.H. Kinard, NASA Langley	Detection of very small solid particles that strike the spacecraft.
Celestial mechanics	J.D. Anderson, JPL	Measurement of the masses of Jupiter and the Galilean satellites with high precision.
Ultraviolet photometry	D.L. Judge, U. Southern California	Measurement of ultraviolet emissions of the Jovian atmosphere and from circumsatellite gas clouds.
Imaging photopolarimetry	T. Gehrels, U. Arizona	Study of optical properties of Jupiter and zodiacal light; reconaissance imaging of planets and satellites.
Jovian infrared thermal structure	G. Münch, Caltech	Measurement of Jovian temperature and heat budget; determination of helium to hydrogen ratio.
S-Band occultation	A.J. Kliore, JPL	Probes of structure of Jovian atmosphere and ionosphere.

TABLE 1.2

Voyager Science Investigations
Project Scientist: E.C. Stone, Caltech

Investigation	Principal Investigator or Team Leader	Primary Objectives
Imaging science	B.A. Smith, U. Arizona	High resolution reconnaissance over large phase angles; measurement of atmospheric dynamics; determination of geologic structure of satellites; search for rings and new satellites.
Infrared radiation (IRIS)	R.A. Hanel, NASA Goddard	Determination of atmospheric composition, thermal structure, and dynamics; satellite surface composition and thermal properties.
Ultraviolet spectroscopy	A.L. Broadfoot, Kitt Peak Observatory	Measurement of upper atmospheric composition and structure; auroral processes; distribution of ions and neutral atoms in the Jovian system.
Photopolarimetry	C.F. Lillie/C.W. Hord, U. Colorado	Measurement of atmospheric aerosols; satellite surface texture and sodium cloud.
Planetary radio astronomy	J.W. Warwick, U. Colorado	Determination of polarization and spectra of radio frequency emissions; Io radio modulation process; plasma densities.
Magnetic fields	N.F. Ness, NASA Goddard	Measurement of the magnetic field of Jupiter, determination of the structure of the magnetosphere, investigation of the interactions of satellites, especially Io.
Plasma particles	H.S. Bridge, MIT	Measurement of magnetospheric ion and electron distribution; solar wind interaction with Jupiter; ions from satellites.
Plasma waves	F.L. Scarf, TRW	Measurement of plasma electron densities; wave-particle interactions; low-frequency wave emissions.
Low energy charged particles	S.M. Krimigis, Johns Hopkins U.	Measurement of the distribution, composition, and flow of energetic ions and electrons; satellite-energetic particle interactions.
Cosmic ray particles	R.E. Vogt, Caltech	Measurement of the distribution, composition, and flow of high energy trapped nuclei; energetic electron spectra.
Radio science	V.R. Eshleman, Stanford U.	Measurement of atmospheric and ionospheric structure, constituents, and dynamics; satellite masses.

The Pioneers served as pathfinders for a more ambitious NASA project, Voyager. Begun in 1972, Voyager was based on the sophisticated, three-axis-stabilized Mariner series of spacecraft developed at the Jet Propulsion Laboratory. Each of two 815 kg spacecraft supported 11 scientific investigations (listed in Table 1.2), including five remote-sensing instruments mounted on a stabilized, pointable scan platform. Under the guidance of Project Scientist E.C. Stone of the California Institute of Technology, the Voyagers carried out a comprehensive exploration of the Jovian system. When Voyager was conceived, the satellites were generally considered secondary objectives, but interest in the satellites grew, until at encounter the spectacular satellite images tended to steal the show. For an account of the Voyager project, see *Voyage to Jupiter* (Morrison and Samz 1980) and *Voyage to Saturn* (Morrison 1982).

The two Voyager spacecraft were launched by Titan-Centaur rockets in 1977 and arrived at Jupiter in March and July of 1979. Their trajectories through the satellite system were designed to be largely complementary. Voyager 1 was aimed for a close flyby of Io and a pass through the Io magnetic flux tube that links the satellite to Jupiter. It also provided moderately close flybys of Callisto, Ganymede, and Amalthea, but only distant views of Europa. Voyager 2 made its closest pass by Ganymede, improving substantially on the coverage by the first spacecraft, and also came much closer to Europa. Io, however, was not well placed for viewing by Voyager 2, although after the discovery of active volcanoes by Voyager 1 the second spacecraft was reprogrammed to permit a 10-hour "Io watch" from a range of $\sim 10^6$ km. Voyager 2 also obtained better coverage of the Jovian rings, including observations over a wide range of phase angles. Voyager 1 passed inside Io's orbit to permit direct measurements of the plasma torus associated with this satellite, which Voyager 2 did not. The satellite ranges and best imaging resolutions for each encounter are listed in Table 1.3.

Most of the data for this book were derived from the Voyager encounters. For the first time, it was possible to see the Jovian satellites as individuals; each new world emerged day by day as the spacecraft approached. For example: February 28—at a resolution of 200 km, albedo features prominent, especially on Ganymede, but no topographic features could be distinguished; March 1—sulfur [SIII] identified on the Io plasma torus; March 2—large circular features first distinguished on Callisto and Io, but those on Io appeared not to be impact craters, implying a geologically young surface; March 3—the "hoofprint" later identified as a vast pyroclastic deposit from the volcano Pele was photographed at 16 km resolution; March 4—discovery image of the ring of Jupiter; March 5—discovery of impact craters and grooved terrain on Ganymede; March 6—discovery of the Valhalla basin on Callisto; March 8—discovery image of volcanic eruptions on Io; March 11—identification of additional eruptions and of localized hot spots on Io. We

TABLE 1.3

Voyager Satellite Encounters

Satellite	Voyager 1		Voyager 2	
	Closest Approach (km)	Best Resolution (km/line-pair)	Closest Approach (km)	Best Resolution (km/line-pair)
Amalthea	420,000	8	558,000	11
Io	20,600	1	1,129,900	20
Europa	733,800	33	205,700	4
Ganymede	114,700	2	62,100	1
Callisto	126,400	2	214,900	4

TABLE 1.4

Satellite Discoveries

Year	Satellite	Discoverer	Country	V_0
1610	J1 Io	Galileo	Italy	5.0
1610	J2 Europa	Galileo	Italy	5.3
1610	J3 Ganymede	Galileo	Italy	4.6
1610	J4 Callisto	Galileo	Italy	5.6
1892	J5 Amalthea	Barnard	U.S.A.	14.1
1904/5	J6 Himalia	Perrine	U.S.A.	14.8
1904/5	J7 Elara	Perrine	U.S.A.	16.7
1908	J8 Pasiphae	Melotte	Britain	17.7
1914	J9 Sinope	Nicholson	U.S.A.	18.3
1938	J10 Lysithea	Nicholson	U.S.A.	18.4
1938	J11 Carme	Nicholson	U.S.A.	18.0
1951	J12 Ananke	Nicholson	U.S.A.	18.9
1974	J13 Leda	Kowal	U.S.A.	20.2
1979	J14 Adrastea	Jewitt et al.	U.S.A.	~ 17
1979/80	J15 Thebe	Synnott	U.S.A.	~ 16
1979/80	J16 Metis	Synnott	U.S.A.	~ 17

hope that some of the wonder and excitement of these discoveries have been preserved in the chapters of this book, in spite of the efforts of authors and editor to conform to the usual dry technical style.

The most significant single discovery of Voyager 1 at Jupiter was the volcanic activity on Io. Just before the encounter, an analysis of tidal heating by Peale et al. (1979) predicted that the interior of Io would be melted and surface activity was probable. During approach, the absence of impact craters appeared to support this prediction. However, direct observation of eight active volcanic plumes erupting to altitudes as high as 300 km, and the detection of hot spots and a possibly transient SO_2 atmosphere, clearly threw Io into a class by itself. The volcanoes provided a link among much of the accumulated disparate data on Io, such as the groundbased measurements of infrared outbursts, the anomalous dehydration of the surface and the unexpected concentration of sulfur and sulfur compounds there, and the apparent Ionian source of sulfur and oxygen ions to the Jovian plasma torus. More than half of this book deals directly or indirectly with Io volcanism.

Voyager observations of the other Galilean satellites were also highly productive. Europa appears enigmatic, with a relatively youthful ice surface and a remarkable geology dominated by tectonic effects. Ganymede and Callisto both were found to have crusts strong enough to preserve the cratering record, contradicting pre-Voyager predictions (e.g., Morrison and Burns 1976). Ganymede's geology shows evidence of tectonic activity, major alteration of terrain, and evolutionary variations in the strength of the crust; in surprising contrast, Callisto, the near-twin of Ganymede, shows no evidence of internal geologic activity. Voyager revealed the Galilean satellites as individuals, with much greater diversity than expected and therefore correspondingly greater interest as examples of planetary evolution. In the rest of this chapter, I summarize this diversity as reflected in the chapters of this book.

III. RINGS AND SMALL SATELLITES

The rings of Jupiter were discovered in 1979 by Voyager 1 (Smith et al. 1979a), quickly confirmed by groundbased astronomers (Becklin and Wynn-Williams 1979), and then reobserved with superior results by Voyager 2. Pioneer 11 had also detected the rings indirectly from a depletion of energetic particles near the planet (Acuña and Ness 1976), but the nature of this apparent absorber had not been ascertained. These Jovian rings were the third planetary ring system to be discovered, following that of Saturn (seen in 1610; recognized as rings in 1655) and of Uranus (discovered 1977). Of the three, the Jovian system is the least extensive, and it is almost surely an unstable phenomenon regenerated and controlled dynamically by adjacent small satellites, two of which were discovered by Voyager.

As described in Chapter 3 by Jewitt, the main ring of Jupiter is ~ 6000 km wide and extends to 1.80 R_J (R_J is equatorial Jovian radius, or 71,400 km), with maximum thickness 30 km. The ring is circular and equatorial, as well as we can determine from the Voyager images, and has a normal optical depth of only 3×10^{-5}, judging by surface brightness. Strong forward scattering suggests that the typical ring particle is small (~ 5 μm) and dark, essentially a mote of dust. The optical depth is too small to have been detected by the Voyager radio occultation (Tyler et al. 1981), also indicating that the particle sizes are small.

In the best available images, structure is apparent within the main ring, and in addition a fainter sheet extends inward toward the cloud tops. Jewitt argues for an out-of-plane component or halo of ring particles extending ~ 10^4 km above and below the equatorial plane. All these particles orbit Jupiter within the strong energetic particle belts of the magnetosphere, and they interact with these trapped electrons, protons, and ions (see e.g., Grün et al. 1980; Burns et al. 1980). In consequence, the ring particles acquire charge; also, sputtering can erode the larger lumps to generate the small particles observed. A picture of dynamic balance emerges, in which new particles are sputtered from unseen parent bodies while old particles are swept out of the system by electromagnetic and gravitational forces.

Closely connected with the rings, and possibly the source of many ring particles, are the innermost known satellites of Jupiter, J14 Adrastea (1979J1) and J16 Metis (1979J3). These two nearly coorbital objects were discovered by Jewitt et al. (1979) and Synnott (1981) orbiting at the outer edge of the rings, at 1.80 R_J, with periods of 7 hr, 5 min. Of the two, the orbit of Metis is better known; neither has been reliably identified in ground-based observations (Jewitt et al. 1981). Each is probably a dark object with a diameter of a few tens of km; for Metis, Synnott (1981) estimates 40 km. Their orbital interactions are not known but may resemble those of the two coorbital satellites of Saturn (e.g., Smith et al. 1981; Harrington and Seidelmann 1981). Other satellites could exist inside the rings, as in the Saturn system, but none is known, and their detection from Earth is not possible.

The next satellite outward from Adrastea and Metis is Amalthea, discovered by Barnard in 1892, the subject of Chapter 6 by Thomas and Veverka. Amalthea is 2.55 R_J from the planet and has a period of ~ 12 hours, slightly greater than a Jovian day. It was well observed by the Voyager cameras, which revealed a dark, red, irregular, and heavily cratered object with major diameter 270 km. Although small by the standards of the Jovian system, Amalthea is of considerable size, as large as a major asteroid and with a volume a thousand times greater than that of Phobos, which it superficially resembles. In size and shape, it seems similar to the Trojan asteroid Hektor.

Amalthea is heavily cratered, and its irregular outline (270 × 165 × 150 km) is suggestive of impact fragmentation (Veverka et al. 1981b). The albedo

is low (0.05–0.06) and the color redder than most other objects in the solar
system. Gradie et al. (1980) suggested that the visible layers contain sulfur
contaminants from Io, and that the surface has also been altered by interaction with the Jovian magnetosphere and micrometeoritic matter; carbona-
ceous-sulfur mixtures are the suggested dominant materials.

Brighter greenish spots associated with slopes on the positive relief features Lyctos and Ida may represent the underlying bedrock of Amalthea
revealed by downslope motion of the contaminated surface, but more likely
are simply older strata of the accreted sulfur and other material. Thomas and
Veverka argue in Chapter 6 that the available spectral data probably yield
little information on the bulk chemistry of Amalthea. Since the mass and
hence the density are also unknown, we can only speculate that this satellite,
and the other inner objects, are composed of refractory material able to
survive the heat emitted by Jupiter during early stages of evolution when it
was still undergoing accretion and gravitational collapse.

Between Amalthea and the Galilean satellites lies the newly found J15
Thebe (1979J2), discovered from the Voyager frames by Synnott (1980) and
imaged from the ground by Jewitt et al. (1982), but not discussed in this
book. Thebe has orbital radius 3.11 R_J, period 16 hr, 11 min, and diameter
~ 75 km; Jewitt et al. (1982) estimate its geometric albedo to be ~ 0.1.
Nothing else is known about this object, but we can speculate that the processes that affect Amalthea may be active here as well.

Chapter 5 by Cruikshank et al. skips over the Galilean satellites to the
outer satellites of Jupiter, divided into two groups on the basis of their orbits,
and perhaps their origins. The four objects in the prograde inner group (at
156-165 R_J) are Himalia (J6), Elara (J7), Lysithea (J10), and Leda (J13);
the retrograde outer group (291-333 R_J) consists of Pasiphae (J8), Sinope
(J9), Carme (J11), and Ananke (J12). None is brighter than visual mag 14.7,
and no spacecraft data exist; what we know of their physical properties
comes from painstaking work with large telescopes, primarily since the late
1970s.

Most of the observations (described in detail by Cruikshank 1977 and
Degewij et al. 1980a,b) are limited to the two brightest, Himalia and Elara,
from the inner prograde group. Himalia is ~ 185 km in diameter, Elara ~ 75
km; both objects are remarkably dark (geometric albedos 0.02–0.03) and
neutral in color, indistinguishable from the common C-type asteroids believed
to be of chemically primitive composition. Also, both have lightcurves of
substantial amplitude, suggesting irregular shapes. The rotation period of
Himalia is ~ 12 hr. Some still unverified observations indicate that members
of the retrograde group may be different physically from the prograde objects. These sparse physical data, and the irregular orbits and division into two
dynamical groups, would seem to indicate a capture origin for the outer
satellites, possibly including collision and fragmentation, but these ideas re-

TABLE 1.5

Satellite Orbits

Satellite	Orbital Radius		Period (day)	Eccentricity	Inclination (deg)
	$(10^3$ km)	R_J			
J14 Adrastea	128	1.80	0.295	~0.0	~0.0
J16 Metis	128	1.80	0.295	~0.0	~0.0
J5 Amalthea	181	2.55	0.489	0.003	0.4
J15 Thebe	221	3.11	0.675	~0.0	~0.0
J1 Io	422	5.95	1.769	0.004[a]	0.0
J2 Europa	671	9.47	3.551	0.000[a]	0.5
J3 Ganymede	1,070	15.1	7.155	0.001[a]	0.2
J4 Callisto	1,880	26.6	16.69	0.010	0.2
J13 Leda	11,110	156	240	0.146	26.7
J6 Himalia	11,470	161	251	0.158	27.6
J10 Lysithea	11,710	164	260	0.130	29.0
J7 Elara	11,740	165	260	0.207	24.8
J12 Ananke	20,700	291	617	0.17	147
J11 Carme	22,350	314	692	0.21	164
J8 Pasiphae	23,300	327	735	0.38	145
J9 Sinope	23,700	333	758	0.28	153

[a]Variable

main speculative. In any case, the connection between the irregular, outer satellites and the regular, inner satellites is tenuous.

IV. ORBITAL EVOLUTION AND TIDAL HEATING

The well-known Laplace resonance involving the orbital periods of Io, Europa, and Ganymede has long fascinated dynamicists, but only recently has it become apparent that the orbital motions can also exert dramatic control over the physical properties of the satellites. The resonance generates forces that alter the orbits; as first recognized by Peale et al. (1979) on the eve of the Voyager encounters, the consequence for Io is tidal heating resulting from noncircular motion ($e \simeq 0.004$) in the enormous gravitational field of Jupiter. The energy input depends on the resonant coupling; as the resonance has evolved, the heating has also changed with time. This orbital evolution is the subject of Chapter 3 by Greenberg.

Current observational estimates of the tidal heating of Io are on the order of 10^{14}W, about an order of magnitude greater than typical theoretical values. As noted by Yoder (1979), tidal dissipation of this magnitude not only extends our understanding of the interior of Jupiter (the dissipation function Q_J is involved because of the coupling between dissipation in Io and the counter-tides it raises on Jupiter), but implies a possibly unreasonably rapid orbital evolution for Io. Yoder and Peale (1981) have recently suggested that the heating may be episodic.

Greenberg examines two scenarios for the evolution of the resonance: one in which the system starts out of resonance and evolves into it as Io's orbit expands, locking first to Europa and then to Ganymede, and one in which the system began even deeper in resonance than it is today. These two scenarios make different predictions about the histories of the satellites. In the first, the heating rates increase with time, suggesting that the melting and outgassing of Io, and perhaps also the cracking of Europa, took place relatively recently. In the second, melting took place early for Ganymede, Europa, and Io, and the heating rates declined with time.

Several important questions remain in deciding the validity of these evolutionary models. An important parameter in assessing the degree to which the system is now in equilibrium is the amplitude of libration of the orbital longitude term ϕ, measured recently by Lieske (1980) but not independently verified. It would also be desirable to measure directly the lengthening of Io's orbital period caused by the tidal effects, but this has not been done. The tidal dissipation rates for both Jupiter and the satellites are dependent on interior structure, introducing numerical uncertainities into the calculations. More than ever before, the study of the satellites' physics and geology depends on the contributions of dynamicists to the understanding of the evolution of these bodies.

In Chapter 4, Cassen et al. utilize tidal dissipation theory to calculate the internal heat sources and resulting interior structures of the Galilean satellites, drawing upon their earlier calculations of tidal heating for Io (Peale et al. 1979) and Europa (Cassen et al. 1979,1980b). They analyze the response of a solid or liquid body to tidal stresses, noting that the current heating rate of Io depends on only one unknown, the effective coefficient of dissipation Q. A nominal value of Q yields heating rates of 2×10^{12} to 2×10^{13}W. In comparison, the energy necessary to produce the observed volcanic resurfacing rate (also discussed in Chapter 17) of Io has an estimated lower limit of $\sim 10^{12}$W, and the infrared measurements of hot spots (see Chapter 19) suggest a lower limit of 10^{13} to 10^{14} W, perhaps as good an agreement between theory and observation as can be expected. Cassen et al. calculate that Io has a molten interior with a relatively thin solid crust, although a variety of structures, including that of solid core and thin liquid mantle, are possible.

Recently Consolmagno (1981a,b) has also calculated the evolutionary history of Io, suggesting the presence of an FeS-rich core.

Europa, like Io, has a relatively high density. The tidal heating rate is of the order 10^{11} to 10^{12}W, depending on interior structure. This is substantially less than that on Io, but still significant in comparison to heating by radioactivity. Cassen et al. conclude that the water on Europa was melted early and some liquid could remain, although it is more probable that Europa is now frozen solid.

Ganymede and Callisto have much lower densities, and each has a bulk composition of $\sim 50\%$ H_2O. Following earlier work by Consolmagno and Lewis (1976), Cassen et al. calculate models for the interiors of these satellites and conclude that the mantles of both are solid ice, with radioactive heat from the core transported outward primarily by solid-state convection. They suggest that the surface differences of these "twin" satellites are probably due to Ganymede's larger radioactive source which provided the energy to keep its surface geologically active longer than Callisto's. A similar set of models for the satellite interiors was published by Schubert et al. (1981), who suggest that there has been little ice-rock differentiation on Callisto, while Ganymede probably fully differentiated through crustal foundering early in its history. All these models can be tested, to some degree, by the geological interpretations in later chapters.

V. SURFACE COMPOSITIONS AND PHYSICAL PROPERTIES

Earth-based observations still dominate the study of the Galilean satellites in the areas of high-resolution spectrophotometry and radar. In both techniques the analysis is based on the hemispheric-scale reflection of electromagnetic energy, and the spatial resolution is limited. However, no comparable instruments were carried to probe the satellites from close range, so the astronomical data reign supreme.

Most of what we know about the surface composition of the satellites has been derived from visible and infrared spectrophotometry, generally covering the wavelength region from 0.3 to 5 μm, with some recent contributions at shorter wavelengths. Typical spectral resolving powers are ~ 400, sufficient to distinguish the relatively broad features that occur in the reflection spectra of solid materials. In Chapter 7, Sill and Clark review the recent observations (presented in more detail in Clark and McCord 1980, Clark 1981, and McFadden et al. 1980) and interpret them with the aid of modern laboratory data. Their chapter is divided into two parts, the first dealing with anhydrous Io, the second with the other Galilean satellites, which all show the spectral signature of water ice.

D. MORRISON

TABLE 1.6

Satellite Physical Properties

Satellite	Radius (km)	Mass (10^{23} g)	Density (g cm^{-3})	Magnitude $V(1,0)$	Albedo (p_V)
J3 Ganymede	2631 ± 10	1490	1.93	− 2.1	0.4
J4 Callisto	2400 ± 10	1075	1.83	− 1.1	0.2
J1 Io	1815 ± 5	892	3.55	− 1.7	0.6
J2 Europa	1569 ± 10	487	3.04	− 1.4	0.6
J5 Amalthea	135×85×75 (± 5)	−	−	+ 7.4	0.05
J6 Himalia	90 ± 10	−	−	+ 8.1	0.03
J7 Elara	40 ± 5	−	−	+ 10.0	0.03
J15 Thebe	40 ± 5	−	−	∼+ 9	< 0.1
J16 Metis	20 ± 5	−	−	∼+ 10	< 0.1
J14 Adrastea	20 ± 5	−	−	∼+ 10	< 0.1
J8 Pasiphae	∼ 20	−	−	+ 11.0	−
J9 Sinope	∼ 15	−	−	+ 11.6	−
J11 Carme	∼ 15	−	−	+ 11.3	−
J10 Lysithea	∼ 10	−	−	+ 11.7	−
J12 Ananke	∼ 10	−	−	+ 12.2	−
J13 Leda	∼ 5	−	−	+ 13.5	−

Io's spectrum is dominated by sulfur and its compounds, particularly condensed SO_2, only recently identified by Fanale et al. (1979) and Smyth et al. (1979), on the basis of an infrared spectral feature at 4.08 μm discovered by Cruikshank et al. (1978) and Pollack et al. (1978). Chapter 7 presents extensive discussion of sulfur chemistry, emphasizing the complex behavior of the many allotropes of sulfur that can be produced by sequences of heating and cooling. All the varied colors (from white through yellow to orange, brown, or black) seen on the surface of Io (Soderblom et al. 1980; Clancy and Danielson 1981) can be reproduced qualitatively by sulfur allotropes, as first pointed out by Sagan (1979). SO_2 is a white frost; its presence can account not only for the 4.08-μm feature, but also for the relatively sharp decline in ultraviolet reflectance of Io which is not consistent with the properties of pure sulfur. Other colored compounds of sulfur may also be present, such as polymerized disulfur monoxide. The volcanoes of Io appear to be a reasonable source for this sulfur.

Other materials may also be present on Io; the sodium and potassium clouds that surround Io suggest that there must be a surface source for these elements, perhaps sputtering of alkali sulfides. No spectroscopic evidence for these compounds exists, however. There is much that we do not understand about the surface chemistry of Io. Several of the allotropes suggested to match colors on Io are unstable at Ionian temperatures and can only be made in the laboratory with some effort; how are they produced and maintained on Io? By observing color changes on Io could we monitor its surface temperatures? The idea that sulfur and sulfur compounds account for the colors on Io is an unproven hypothesis, highly compelling but still circumstantial, since we are making identifications based on broadband colors, not discrete spectral features.

The outer three Galilean satellites all have reflection spectra that are bland in the visible but dominated in the infrared by the prominent absorptions of water ice. Sill and Clark analyze the amount of ice present using the laboratory calibrations developed by Clark (1980), concluding that ice is $\geqslant$ 90 wt% for Europa, $\sim$ 90 wt% for Ganymede, and 30-90 wt% for Callisto. These percentages are substantially greater than those derived previously, e.g., by Pollack et al. (1978), suggesting that, possibly even on Callisto, the optical surface is nearly pure ice with other minerals present only as contaminants. The analysis in Chapter 7 also indicates that the surfaces physically consist of fine-grained frosts on larger ice blocks or crystals. Variations in albedo among Callisto, Ganymede, and Europa are presumably due to differing concentrations of the dark contaminant, which Sill and Clark suggest is spectrally similar to carbonaceous chondritic material or other minerals containing Fe^{3+}. If these minerals are exogenous—essentially meteoritic dust—the albedos are presumably an index of the average geologic ages of the surfaces.

In most respects Io is the most bizarre of the Galilean satellites, but in radar properties it is the most normal, reflecting microwaves in an essentially lunar fashion. In contrast, the 12.6-cm-wavelength radar properties of the three icy satellites are extraordinary, as described by Ostro in Chapter 8. (Observational data are presented in detail in Ostro et al. 1980.) Compared to the Moon and terrestrial planets, the radar reflectivities of Europa, Ganymede, and Callisto are enormous. Even more remarkable, the echo of a circularly polarized signal returns with the wrong sense of polarization, relative to any previously studied natural targets. As seen by radar, these three satellites represent a totally new kind of surface.

The anomalous radar properties of the satellites increase in proportion to the apparent importance of surface ice, as indicated by visible albedo or the spectral analysis presented in Chapter 7 by Sill and Clark, suggesting that the presence (or purity) of ice is associated with the remarkable radar behavior. Analysis of the radar echos permits approximate location of particularly anomalous scattering regions, and these tend to correlate with high albedo features on the Voyager images, again indicating the connection with ice. But what properties of ice actually cause this strange behavior?

Ostro discusses two physical models for the satellite surfaces. One requires that the surface be covered by deep hemispheric craters and have a refractive index substantially higher than that of ice. Ostro prefers the other model, proposed by Goldstein and Green (1980), which involves subsurface scattering from randomly oriented interfaces between two regolith components. One geological interpretation of such a model is that of veined mixtures of ice of different density.

Radar observations are sensitive to structural scales (10^{-2} to 10^2 m) not explored by any other technique. Thus, in spite of their low spatial resolution, these radar results make a unique contribution to the study of the satellite surfaces. However, attempts to distill useful geological information from them have just begun.

Other clues to the small-scale nature of the satellite regoliths are provided by visible and infrared photometry, topics not discussed in detail in this book. Thermal eclipse photometry is sensitive to the thermal properties of the topmost surface layer, usually no more than a few cm thick. The decade-old groundbased results for Callisto and Ganymede (Morrison and Cruikshank 1973; Hansen 1973) are still the best data base, but new Voyager data for the diurnal cooling of Callisto (Hanel et al. 1979) strongly support the earlier conclusion that the thermal conductivities of the satellite regoliths are very low. In Chapter 13, Shoemaker et al. use this result in modeling the thermal history of the crust of Ganymede.

Photometry from Voyager images and from the photopolarimeter also provide important extensions to earlier astronomical studies, profiting from both greatly extended phase angle coverage and high spatial resolution. Soderblum et al. (1980) have published preliminary photometry of Io; studies at higher spatial resolution were recently published by Clancy and Danielson (1981). Johnson et al. (1981) used similar techniques to produce multispectral mosaics of all four Galilean satellites. Squyres and Veverka (1981) have investigated the photometric behavior of individual areas on Callisto and Ganymede over phase angles from 10° to 124°, finding a typical normal reflectance of 0.18 for Callisto, and of 0.35 for the cratered terrain of Ganymede, and 0.44 for its grooved terrain (see also discussion in Chapter 13). On both objects, they find maximum reflectances, on brighter crater rims and ejecta deposits, to be as high as 0.7. Pang et al. (1981) have used photopolarimeter data for Io and Ganymede to determine geometric albedos and phase curves and to model the nature of the surface particles. They find that Io has a very steep phase curve, quite unlike the icy satellites, and they consider a number of surface models that might produce this surface optical behavior.

Finally, there is the question of ice migration on satellite surfaces by successive evaporation (or surface sputtering) and redeposition. Purves and Pilcher (1980) have published a treatment of this problem for all three icy satellites, while Squyres (1980a) has examined it specifically for Callisto and Ganymede. There is a consensus that significant amounts of water ice could be lost by evaporation over geologic time from the equatorial regions of the darker Galilean satellites, but not from temperate or polar areas; the effects of sputtering are more problematic. In Chapter 13, Shoemaker et al. also discuss the possible formation mechanisms of the faint polar "cap" of higher albedo material seen in Voyager images of Ganymede.

VI. IMPACT CRATERING

Cratering by meteoric impact is a fundamental process in molding planetary surfaces. The accretionary process itself involves innumerable impacts, although the associated heating of a protoplanet generally erases the record of this early period. Later, as the crust solidifies, the cratering record begins to be preserved, although erosion and volcanism may destroy the craters and set back the geological clock. Only on the most geologically active worlds, such as Earth and Io, are impact craters unusual; in more stable environments craters tend to be the most common major landform. A great deal can be learned from craters: relative age of surface units; the nature of the population of impacting bodies; past erosion rates; and bearing strengths of crustal materials and their variation with time. Craters can also reveal the surface stratigraphy, exposing windows to deeper layers. Three chapters in this book are devoted to the cratering process on the satellites of Jupiter.

The flux and population of incoming projectiles must be determined if crater counts are to be used to estimate ages. Before Voyager, it had not been known whether any cratered surfaces existed in the outer solar system, or what the origin of the impacting material might be. Chapter 9 by Woronow et al. and Chapter 10 by Shoemaker and Wolfe provide the first analyses, from two different perspectives, of the cratering flux at Jupiter.

Woronow et al. take the position that the cratered surfaces among both the terrestrial planets and Galilean satellites are not saturated; that is, they have not reached an equilibrium state in which old craters are destroyed as fast as new ones are created. In this circumstance, the size distribution and surface density of craters reflect the population of incoming bodies. In contrast, Shoemaker and Wolfe believe that the most densely cratered parts of Callisto and Ganymede are in an equilibrium state. However, this point is not critical to the validity of their chapter, in which they calculate the likely sources of projectiles and the expected distribution of impacts on satellite surfaces. Woronow et al. work from the satellite data to the properties of the projectiles, while Shoemaker and Wolfe reason in the opposite direction.

In Chapter 9, Woronow et al. describe the Voyager data on the cratered surfaces of Callisto and Ganymede and note that their large-crater populations differ markedly from those on the inner planets, which are all basically similar. Although conceding that part of this difference may be due to the icy surfaces of Ganymede and Callisto (e.g., as suggested by Parmentier and Head, 1981), they argue that the differences primarily reflect a different impacting population. Since the craters do not appear saturated on either satellite, they conclude that this population difference applies to the earliest history of the Jovian system, when the crusts first became strong enough to support topographic relief.

In the next chapter, Shoemaker and Wolfe calculate the expected impact rate for large-crater producing objects in an attempt to fix an absolute time scale for the Galilean satellite system. Since they work from the existing population of objects and extrapolate backward in time, their results are most applicable to the younger, less cratered surfaces. If any heavily cratered regions on Callisto or Ganymede date back to the era of accretion, they may of course represent the effects of a different population of debris.

Shoemaker and Wolfe argue that the prime source of impacts in the outer solar system is comets, and they undertake a detailed census of known objects, with corrections applied for incompleteness of discovery. The numerous long-period comets account for ~ 20% of the impacts, and the short-period comets, due to their frequent close passes by Jupiter, for a somewhat larger percentage. As many as half of the impacts may be due to outer solar system asteroids, although only two such objects are now known, Hidalgo and Chiron. Shoemaker and Wolfe argue that, in spite of the observational uncertainties, their calculated fluxes should be of the correct order of magnitude.

The Jovian gravity provides a strong concentration of the impacting flux close to the planet, so that similar crater densities translate into very different surface ages for the different satellites. In addition, Shoemaker and Wolfe find that the orbital motions of the satellites concentrate impacts on their leading hemispheres. The cratering rate at the apex is generally 10 to 20 times greater than at the antapex. Somewhat surprisingly, there is little observational evidence for a similar longitudinal gradient in the crater densities, an effect that Shoemaker and Wolfe attribute to the slower cooling of the crust in the leading, frequently impacted, hemisphere. That is, at different places the crust would have frozen during the same effective impact rate, since the less frequently impacted rear cooled first.

The cratering rates calculated in Chapter 10 suggest that the majority of craters on Callisto and Ganymede date from early solar system history when impacting fluxes were much higher. Europa, in contrast, is suggested to have retained craters for only a few times 10^7 yr.

In Chapter 11, Greeley et al. address the process of crater formation in icy satellites. All our previous experience has involved impacts into rocky planets such as the Earth and Moon, although recently some experimental and theoretical attention has been directed to crater formation in water- or ice-saturated soils such as may have existed on Mars.

Greeley et al. describe a series of cratering experiments carried out with the high-speed ballistic range at Ames Research Center, involving impacts into ice, ice-soil mixtures, and various analogs of the surface materials on Ganymede and Callisto. They found that some characteristic features observed on the Galilean satellites, such as pit craters, could be duplicated in several of their experiments. However, they caution against over-interpretation of their results. Scaling from the laboratory environment to impact features on real planetary surfaces is uncertain, and the size differences are enormous; the smallest craters that can be studied morphologically on Voyager images are tens of km across, while the largest experimental craters produced at Ames are $\sim 10^6$ times smaller in linear dimensions and perhaps $\sim 10^{17}$ times smaller in energy. Nevertheless, this series of experiments explores a new range of cratering conditions that may ultimately be important in understanding the processes that affect the geology of the icy Jovian satellites.

VII. CALLISTO AND GANYMEDE

The outer two Galilean satellites have the largest size (diameters $\sim$ 5000 km) and lowest densities ($\sim$ 1.9 g cm^{-3}); they are the first planetary bodies to be studied in detail that are $\sim$ 50% H$_2$O by mass. As argued originally by Lewis (1971), such bodies have low internal melting temperatures and should be fully differentiated, with rock and mud cores and water ice mantles and crusts. Voyager also revealed that the crusts of these two satellites are relatively old, preserving a geologic record that may extend back to the early solar system, although how far back is not clear. The geology of these two ice-and-rock worlds is the subject of Chapters 12 and 13.

Passey and Shoemaker, in Chapter 12, discuss the craters and basins and what these impact features can tell us about the evolution of the crusts of Callisto and Ganymede. Craters and their remnants are the dominant surface features on both; indeed, on Callisto there is little else for the geologist to study. Ganymede has a much more varied geology, as discussed in Chapter 13.

There is a wide variety of large craters on these satellites, including bowl-shaped craters, flat-floor craters, and craters with central peaks (all generally similar in morphology to the familiar lunar craters), as well as the more distinctive central-pit craters, basins, and giant multiring systems of ridges and furrows (such as Valhalla, the bullseye of Callisto). Most craters, in all size ranges down to the Voyager limit of $\sim$ 1 km, are highly flattened relative to those in silicate bodies, presumably by topographic relaxation due to flow in the icy crust. Pit craters are common for diameters $\geqslant$ 20 km, making this feature the most obvious characteristic of the cratered Galilean satellites.

Passey and Shoemaker introduce a new geologic nomenclature for the unusual features on Callisto and Ganymede that they interpret as remnants of old basins whose topography has been erased by crustal flow. There remains a circular feature of higher albedo, usually with a lower density of superposed craters than in the surrounding regions, which they call a crater palimpsest. Typically the palimpsest appears to include both the former crater and its ejecta blanket. On Ganymede, Passey and Shoemaker identify 22 palimpsests, and on Callisto 15, including the central regions of the Valhalla and Asgard ring structures. There are also transitional basin forms with some remaining relief that are called penepalimpsests; these occur on the younger grooved terrain of Ganymede.

The most impressive features on Callisto are the multiring structures; Valhalla appears to be the result of an impact that produced a 350-km diameter central crater with concentric ridges extending out to ~ 2000 km from the center; Asgard is about one third this size. The individual ridges in the multiring systems are irregular, ~ 15 km wide, spaced from tens to ~ 100 km apart, and form individual arcs hundreds of km in length. The ridge heights are not well determined but appear to be ~ 1 km. On Ganymede, a somewhat similar system of rings can be seen throughout the Galileo Regio of ancient cratered terrain. Only a portion of this system, which is comparable in size to Valhalla, remains; the central crater may have been destroyed by subsequent crustal evolution, although Passey and Shoemaker suggest as a candidate a palimpsest ~ 500 km in diameter. The rings of Ganymede are furrows (a few hundred m deep, with rims ~ 100 m high), rather than ridges as on Callisto. Other crater forms, notably many craters with brilliant ray systems (presumably made of relatively clear ice) and the mountain-rimmed basin Gilgamesh, are found on Ganymede but not on Callisto. The geology of crater and basin formation and evolution in an icy satellite is becoming an active research area, as shown by recent papers by McKinnon and Melosh (1980), Phillips and Malin (1980), Parmentier and Head (1981), Croft (1981), and Fink and Fletcher (1981).

Chapter 13 by Shoemaker et al. discusses in detail the geology of Ganymede, which unlike Callisto appears to have experienced extensive tectonic activity with break up and partial resurfacing of the ancient, heavily cratered crust. The old, Callisto-like surface is represented by large, polygonal, low-albedo areas such as Galileo Regio and Nicholson Regio. Shoemaker et al. describe the geologic record of Ganymede as most like that of the Moon and terrestrial planets, with surface units that bear a wide range of crater densities, indicating an extended period of surface evolution. Its diversity of geologic forms and the apparently long time base represented by its surface make Ganymede the most interesting Galilean satellite to many planetary geologists.

The ancient terrain on Ganymede is dark, though not as dark as that of Callisto (Squyres and Veverka 1981). The younger, higher albedo surfaces are classed together as grooved terrain, which is unique to Ganymede. These two major terrain types are lithologic rather than structural, representing composition differences that, according to Shoemaker et al., extend to a depth ⩾ 10 km. Apparently the grooved terrain, which covers ~ 60% of the surface, supplanted the ancient cratered terrain.

The grooved units on Ganymede are characterized by subparallel valleys with intervening ridges, commonly > 100 km long, spaced ~ 3 to 10 km apart. Topographic relief is generally 300 to 400 m and reaches a maximum of ~ 700 m. Slopes are gentle, averaging $\sim 5°$. Ejecta from craters formed in the grooved terrain share the high albedo of their surroundings, indicating that the albedo difference is not superficial. Mixed with the grooved terrain are smooth units that Shoemaker et al. argue were emplaced by flows of fluid material, presumably liquid water. From crater counts, it appears that the grooved terrains have not lost large craters by lithospheric flow as the older units have, indicating that the crust had thickened by the time the grooved terrain was formed.

Shoemaker et al. develop a model for the history of Ganymede in which the grooved terrain was emplaced through a sequence of faulting and subsidence coupled with flooding by water released from below. The margins of many grooves have the appearance of faults, and faults mark the boundaries between the grooved and ancient cratered terrain (Lucchitta 1980; Head et al. 1981). Shoemaker et al. argue that extensive faulting took place at a time of slight expansion of the core, a period they relate to the thermal evolution of the interior, a topic also discussed by Cassen et al. (1980a), Hsui and Toksoz (1980), McKinnon and Spencer (1981), Squyres (1980b,c,1981) and Squyres et al. (1981). Analysis of crater degradation suggests that the thickness of the lithosphere at the beginning of groove formation was ~ 35 km.

Chapter 13 also contains extensive discussion of the distinctive bright and dark ray craters of Ganymede and of the formation and evolution of the regolith. Shoemaker et al. argue that a thin surface layer of very low thermal conductivity was essential for the development of subsurface temperature gradients sufficient to produce the observed crater degradation by lithospheric flow. They calculate regolith thicknesses of tens of meters.

In Chapter 12, Passey and Shoemaker suggest a general pattern of crustal evolution applying to both Ganymede and Callisto. They conclude that the lithosphere was too thin during early heavy bombardment to sustain craters. The first part of the crust to harden would be near the antapex of motion, where heating by impact was least; Passey and Shoemaker believe that the oldest recognizable craters on both satellites are there. The crust then cooled progressively farther from the antapex. At any given place small craters were retained before large ones, and at any given time craters of any size were best retained where the crust was thickest. It also appears that Callisto always had a thicker crust than Ganymede, due to its smaller size (less radioactive heating per unit surface area) and lower impact rate due to the focusing effect of Jupiter's gravity.

On Ganymede, the youngest basin appears to be Gilgamesh, which has an age of 3.5 Gyr according to the cratering rates calculated by Shoemaker and Wolfe in Chapter 10. The lithosphere thickness required to support the topography of Gilgamesh is $\geq$ 300 km, suggesting that a thick crust had developed, at least at this latitude (60°N), by the same epoch that saw the final stages of mare volcanism on the Moon. The formation of the grooved terrain probably took place between 3.7 and 3.1 Gyr in the past, in response to a small expansion of the interior. It apparently formed first in the polar regions, where it occupies most of the surface, then in the trailing hemisphere, and finally in the leading hemisphere. Grooved terrain formation must have ceased either when the convecting layer disappeared or when convection was no longer able to displace the thickening ice crust. Since that time, the surface evolution of both Callisto and Ganymede has been primarily by cometary or meteoric impact, as outlined in Chapter 10 by Shoemaker and Wolfe. Taken together, Chapters 10, 12, and 13 present in considerable detail the geology of the icy satellites as developed during 1980-1981 by Shoemaker and his colleagues, and they will probably serve in the future as the baseline against which new developments are judged.

VIII. ENIGMATIC EUROPA

Of the Galilean satellites, Europa has the highest albedo and shows the strongest spectral signature of water ice, in spite of its rather high density of 3.0 g cm^{-3}, which suggests a basically rocky composition. Voyager 1 did not obtain very good images of this satellite, and even after the closer flyby by the second spacecraft, Europa remains the least well observed Galilean satellite, with best surface resolution $\sim$ 4 km. However, the available observations are sufficient to show that Europa has had a geologic history unlike any of its neighbor worlds; its relatively smooth youthful surface of ice is criss-crossed by a global network of light and dark linear features suggesting disruption of the surface by large-scale tectonic forces.

In Chapter 14, the geology of Europa is discussed by Lucchitta and Soderblom. Their analysis divides the surface into two basic units called plains and mottled terrain. Neither shows substantial relief or topography; they are distinguished instead by albedo features, with the mottled terrain covered by a dense population of dark spots and patches. Seen at low sun angles, the mottled terrain also appears hummocky, but the relief is apparently $\lesssim$ 100 m.

The most dramatic features of Europa are the dark and light bands that transect both terrain types and sometimes extend for thousands of km across the surface. The dark bands appear to trace a global system of cracks, and Lucchitta and Soderblom suggest that they were formed by filling in of gaps along fractures in the crust. Pieri (1980) and Finnerty et al. (1980) have argued that the lineament pattern represents surface cracking due to thermal evolution which caused expansion of the interior, while Helferstein and Parmentier (1980) consider the role of tidal deformation. In addition to the dark bands, there are numerous narrow, often curved, light ridges, which may have resulted from compression or dike-like filling of fractures by water that expanded on freezing. In some cases, the light ridges form central stripes in dark bands.

Europa is nearly free of craters; Lucchitta and Soderblom identify only five in the 10 to 30 km diameter range, suggesting that the age of the surface is $\sim 10^8$ yr, according to the flux rates presented in Chapter 10. In one area, ridges radiate from a crater; more generally, the fracturing of the surface may be partially controlled by impact cratering. The most striking geologic aspect of Europa is its apparent lack of a horizontally stratified crust. Instead of consisting of layers deposited one over the other, the surface appears to have been formed through disruption of a rather thick icy crust with subsequent intrusion of underlying material along nearly vertical fractures. There is also some evidence of small rotations of crustal plates. Lucchitta and Soderblom argue that the darker material was transported to the surface from a subjacent silicate lithosphere, probably no more than a few tens of km below the surface. Thus we may picture this strange object as a lunar-like core covered by a somewhat mobile ice crust repeatedly cracked by internally imposed stresses. It remains unclear whether the scarcity of large craters is due to topographic relaxation in the icy crust or to resurfacing events that flooded craters with fresh water. The role of internal tidal heating is also uncertain; the scenarios described in Chapter 3 permit the possibilities of both early or relatively recent maximum heating.

IX. GEOLOGY OF IO

Io would be a remarkable object even if its volcanism did not produce the spectacular volcanic plumes observed by Voyager. Decades of Earth-based observations had shown its redness, comparable to Mars and strikingly different from the other Galilean satellites. As revealed in spacecraft images, Io's surface is strikingly multihued, as discussed in Chapter 7 by Sill and Clark. Close views established that the morphology of surface features was also distinctive, with no impact craters, numerous dark calderas and surface flows, and surprisingly large mountains and steep scarps. In Chapter 15, Schaber describes the geology and presents a geologic map of the best-imaged 35% of the surface of Io, updating preliminary geologic analyses by Masursky et al. (1979), Carr et al. (1979), McCauley et al. (1979) and Schaber (1980).

Schaber divides the dominantly volcanic surface ($\sim 95\%$ is layered plains and vent flows) into a variety of units, including mountains, plains, flows, cones, and crater vents. The highest measured mountains rise 9 km above their surroundings, giving Io the greatest topographic relief of any Galilean satellite in spite of its presumably molten interior. The isolation and heavy erosion of these mountains suggest that they are relatively old. As pointed out earlier by Clow and Carr (1980), the steep scarps (up to nearly 2 km high) as well as the high mountains apparently exceed the strength limits of sulfur, and Schaber therefore argues that these features are dominantly silicate implying that the sulfur crust may be quite thin, at least over parts of Io. The morphology of the sulfur flows indicates a wide range of viscosity, and Schaber suggests that differing degrees of silicate admixture as well as different source temperatures may have contributed to this diversity. The volcanic vents themselves take the form of fissures, pit craters, and shield craters (concentrated strongly toward the equator). A total of 170 craters have been mapped on 35% of the satellite's surface, together with 151 lineaments and grabens indicative of tectonic processes. Erosion as well as deposition is evident, with the most obvious erosion at high southern latitudes.

The area of Io photographed in detail by Voyager and mapped by Schaber lies between $250°$ and $323°$ longitude and includes the active volcanoes Pele and Loki and their associated surface deposits. Generally, this is one of the reddest parts of Io, with the least indication of exposed SO_2 frost. Unfortunately, it will be difficult to extend such detailed analyses to the other parts of the satellite, where imaging resolution is poorer.

The nine major volcanic eruptions seen on Io are described in Chapters 16 and 18, and in Chapter 17 Johnson and Soderblom examine the implications of these eruptions for the geology of Io. In addition to the many surface flows mapped by Schaber, the giant pyroclastic plumes result in a deposition of sulfur and sulfur compounds over widespread surface areas. In the case of Pele, a surface area of $\sim 10^4$ km^2 changed dramatically in appearance during the four months between Voyager encounters (Smith et al. 1979*b*), demonstrating the importance of pyroclastic deposition on the surface.

In Chapter 17, Johnson and Soderblom extend their earlier work (Johnson et al. 1979) in an effort to quantify the effects of volcanic plumes in modifying the surface of Io. For the active plumes, they calculate a deposition rate of 10^{-1} to 10^{-2} cm yr^{-1}, averaged over the entire satellite. In this analysis they use the calculation by Collins (1981) that the solid material in the Loki plume is very small in size, 10^{-1} to 10^{-2} μm. A lower limit for the average deposition rate can also be derived from the absence of impact craters together with the expected crater production rate calculated in Chapter 10 by Shoemaker and Wolfe. The resulting deposition is $\geq 10^{-1}$ cm yr^{-1}. A third value is obtained from an estimate of the energy available from the tidal heat source, $\sim 10^{14}$ W (see Chapters 4 and 19), which can be satisfied by an average resurfacing rate by hot material of a few cm yr^{-1}.

If the typical deposition rate from both surface flows and plumes is $\sim 10^{-1}$ cm yr^{-1}, new surface will build up at a rate of ~ 1 km in 10^6 yr, and the entire mass of Io could be recirculated by volcanic processes in the age of the solar system. On the average, the surface we see must be very young: a few years for the optically active areas, and thousands to millions of years for even the largest visible landforms. It is interesting that there is stability on a 50 yr time scale in the hemispheric-scale variations with longitude of surface color and albedo of Io, as noted by Morrison et al. (1979). This preservation of large-scale structure suggests a deep-seated asymmetry in the volcanic sources that outlives individual eruptions. However, small photometric variations are seen on Io that apparently result from volcanic alteration of the surface on time scales of months to years, as recently documented by Lockwood et al. (1980).

X. THE VOLCANOES OF IO

The existence of volcanic plumes on Io was discovered by Morabito et al. (1979) and quickly became a subject of intense interest to the Voyager imaging team (e.g., Smith et al. 1979*a,b*; Strom et al. 1979). At the same time, volcanoes were identified as the source of anomalous thermal emission from hot spots, both by direct Voyager measurements (Hanel et al. 1979) and from groundbased observations (Matson et al. 1981; Sinton 1980). Chapters 16, 18, and 19 are devoted to discussion of these active eruptions and the associated hot spots.

In Chapter 16, Strom and Schneider review the Voyager observations of volcanic plumes, drawing in part on papers by Strom et al. (1979,1981). Eight plumes were measured by Voyager 1, either at the limb (against dark space) or superposed on the disk, where they appeared darker than the background. Four months later Plume 1 (Pele), the largest seen by Voyager 1, had ceased activity, while Plume 2 (Loki) had increased in height and apparently divided in two, issuing near opposite ends of an apparent fissure ~ 200 km long and tens of km wide. All the other plumes that could be reobserved by Voyager 2 were little changed, and no new eruptions were discovered. However, comparison of images taken four months apart near 335° longitude indicated that a major new ring-shaped surface marking had appeared surrounding the dark caldera-like feature Surt, suggesting that a large plume had been active there between the two spacecraft encounters.

The plumes seen by Voyager ranged in height from 60 to > 300 km, with corresponding velocities at the vent of 0.5-1.0 km s^{-1}. Two distinct plume morphologies were seen: regular, umbrella-shaped fountains typified by Plume 3 (Prometheus), and irregular plumes such as Plumes 2 and 9 (Loki). Plume 1 (Pele) differs from the other regular plumes in that it alone increases in brightness from bottom to top; the other plumes all show a monotonic decrease in brightness with altitude.

Strom and Schneider discuss ballistic models for the plumes that also consider the distinctive surface deposits associated with the active vents. The rather extensive data on Plume 3 (Prometheus) are well matched with a fairly simple model with constant ejection velocity 0.5 km s^{-1} and isotropic distribution of ejection directions from vertical down to a cut-off at 55°. The irregular plumes are less easily modeled and perhaps consist of the superposed ejecta from several closely-spaced vents. Strom and Schneider suggest that Plume 1 (Pele) with its distinctive bright envelope may indicate the presence of a shock front, possibly resulting from the greater eruptive volume at this site.

A major new analysis of the detailed dynamics and thermodynamics of terrestrial and Ionian volcanism is given by Kieffer in Chapter 18, in which she links some of the remarkable properties of Ionian plumes with terrestrial volcanic phenomena such as the explosive eruptions at Mount St. Helens and geothermal phenomena represented by geysers. (For instance, Kieffer calculates that the Old Faithful geyser in Yellowstone National Park, if transplanted to Io, would erupt a steam plume $\geqslant 38$ km high.) SO_2 was originally suggested as the major working fluid of the Io volcanoes by Smith et al. (1979c), while more recently Reynolds et al. (1980) have modeled volcanism dominated by high-temperature sulfur vapor heated by contact with silicate magmas below the surface. In this chapter Kieffer compares the terrestrial working fluids H_2O and CO_2 with the proposed Ionian fluids SO_2 and S; she emphasizes the many factors other than vapor chemistry that determine plume dynamics (such as the amount of entrained debris being carried by the vapor, and the shape of the volcanic system).

Kieffer develops a generalized dynamical model of a volcanic eruption that includes the geometry of the system as well as the thermodynamics and fluid dynamics of the working fluid. Conceptually, she separates the system into five regions: supply region, reservoir, conduit, surface crater, and plume, and discusses the dependence of plume dynamics on conditions in each subsurface region. The magmatic fluid is heated at depth in the supply region and driven to the surface from the reservoir, through a conduit and crater, into the plume. Because of the wide range of pressures between the reservoir and the atmosphere on Io, most fluids will undergo a series of phase changes dependent on the composition, initial temperature and pressure, and thermodynamic path. Kieffer concludes that sulfur at < 700 K cannot produce a plume, but that a wide variety of SO_2 reservoirs at > 400 K and sulfur reservoirs at > 700 K can produce plumes of the observed heights on Io.

The detailed plume structure depends on the relative pressures of the Ionian atmosphere and the fluid as it emerges at the surface. If the flow emerges through a narrow conduit, the flow velocity would be limited to the sound velocity (typically a few hundred m s^{-1}), and the sonic pressure (tens of bar) would generally exceed the cohesive strength postulated for the Ionian regolith. A steep-walled crater would be eroded, and the plume properties would depend on the depth and shape of the crater. If the crater is deep (~ 1 km), the flow pressure decreases to about ambient atmospheric pressure, and the flow in the plume would be smooth. The trajectories of the ejecta would be approximately ballistic, and umbrella-shaped plumes, such as Plume 3 (Prometheus), should result, with dynamics governed by a model such as that presented in Chapter 16 by Strom and Schneider. If the crater is shallow or if the eruption is through a fissure, the flow in the plume would be more complex because the fluid pressure would be greater than ambient as the fluid emerges at the surface. An irregular plume with complex internal shock and expansion waves will result; Plume 2 (Loki) may be an example.

In Chapter 18 Kieffer concludes that the observations cannot now distinguish between S and SO_2 volcanism, since both fluids can generate a wide variety of eruptive patterns starting from a deep sulfur source near 700 K. In both cases the plumes will contain both solid and fluid material, and it is plausible that both kinds of volcanoes are present and contribute to the surface deposits on Io.

The internal energy source of Io, presumably related to tidal heating, is indirectly responsible for the volcanic eruptions. But the escape of heat from the interior can also be measured directly. The average heat flux of 1 to 2 W m^{-2} would not be detectable if it were emitted uniformly from the surface, but much of the energy is channeled into volcanic vents and flows, which have a high enough temperature to be visible against the background thermal flux. These hot spots were detected individually by the Voyager IRIS (Infrared Inferometer Spectrometer), and their collective effect can be measured from Earth, as described in Chapter 19 by Pearl and Sinton.

Voyager 1 infrared observations of $\sim$ 30% of Io at resolutions between 700 and 70 km revealed eight regions of distinctly enhanced emission, with color temperatures $\sim$ 300 K; these tend to be associated with volcanic features, though not necessarily with plume eruptions. One of the strongest sources at the time of the Voyager 1 encounter was the so-called lava lake near Loki. Voyager also measured a temperature of $\sim$ 650 K emanating from a region just a few km in radius apparently located at the Plume 1 (Pele) vent. Groundbased data have revealed another type of event, the rapidly changing "outburst" apparently corresponding to a source at $\sim$ 600 K with a radius of tens of km.

The groundbased infrared observations described in Chapter 19 by Pearl and Sinton usually cannot distinguish individual sources, but they are a means of measuring the total heat flow through hot spots, and thus setting a current lower limiting value of the internal energy source. As first shown by Matson et al. (1981), even relatively old eclipse observations can now be recognized as affected by hot spot emission; they derived a flux of 2 ± 1 W m^{-2} for the period 1969-1972. Sinton (1981) used 1974-1979 observations to derive 1.8 ± 0.6 W m^{-2}, and most recently Morrison and Telesco (1980) found 1.5 ± 0.3 W m^{-2} in 1980. These values agree with each other and with the Voyager estimate obtained by summing the hot spots seen during the March flyby, of ~ 2 W m^{-2}. These observations strongly constrain the tidal evolution models discussed in Chapters 3 and 4, but unfortunately do not in themselves provide much insight into the volcanic processes through which the interior heat is released.

XI. THE ATMOSPHERE OF IO

The earliest evidence for a possible atmosphere on Io was from a surface condensation model for the apparent post-eclipse brightening (Binder and Cruikshank 1964), but it was not until 1973 that the Pioneer 10 radio occultation (Kliore et al. 1974) firmly established the existence of an ionosphere at Io, while Brown (1974) simultaneously discovered the extended sodium cloud that appeared to originate at the satellite. A neutral atmosphere (possibly transient) of SO_2 was found by Voyager, as well as the volcanic activity that appears to provide a plausible source for SO_2 and perhaps other volatiles. Thus, Io clearly possesses an atmosphere; the problem is to understand its stability and relationship to volcanic sources and to the Io plasma torus, which appears to be maintained primarily by sulfur and oxygen escaping from Io.

Pearl et al. (1979) first suggested that local buffering by cold traps might determine the quantity of available SO_2 in Io's atmosphere, and Kumar (1979,1980) estimated a mean SO_2 surface density of 10^{11} to 10^{12} cm^{-2}, noting that the amount could be in equilibrium with the vapor pressure of SO_2 ice at ~ 115 K as well as with the dusk-side ionosphere measured by Pioneer 11. (Near noon, where the temperature would be ~ 130 K, the equilibrium atmosphere would be $\sim 5 \times 10^{12}$ cm^{-3}, or ~ 0.2 cm-A, or $\sim 10^{-7}$ bar.) The primary observational difficulty with this equilibrium SO_2 atmosphere is the low abundance (~ 0.008 cm-A for the bright side disk average) indicated from IUE (International Ultraviolet Explorer) observations near 290 nm (Butterworth et al. 1980), although this interpretation has been questioned by Bertaux and Belton (1979), who argue that the IUE observations are compatible with an atmosphere nearer to the equilibrium value at Io's surface temperature. The question of the balance between atmosphere and surface SO_2 is further constrained by the absence of bright polar caps on Io that might be expected if the cooler polar regions acted as major traps for the gaseous SO_2.

The problems of the origin, evolution, and escape of Io's atmosphere are treated in detail in Chapters 20 and 21. In Chapter 20, Fanale et al. are particularly concerned with the atmosphere-surface interactions for SO_2, while Kumar and Hunten in Chapter 21 also discuss the photochemistry of the atmosphere and the interactions with the Io plasma torus. Both chapters present models that attempt to reconcile the available observations; Fanale et al. favor an SO_2-dominated atmosphere with regional cold trapping, while Kumar and Hunten suggest that O_2 may also play a major role.

Fanale et al. note that, given the importance of SO_2 frost on the surface, there must be a sizable SO_2 atmosphere on the day side of Io; the dominance of the plasma torus by sulfur and oxygen ions further demonstrates the major role of this gas. Io is like Mars in that the major atmospheric gas coexists in equilibrium with extensive surface deposits of solid phases. However, Fanale et al. point out that their models are limited by our current inability to distinguish various forms of condensed SO_2; adsorbed SO_2 may dominate the spectrum while free frost might be rare. Why does Io not have bright SO_2 polar caps analogous to the CO_2 caps on Mars? Fanale et al. argue that loss of SO_2 to the magnetosphere or radiation-induced darkening of the frost occurs on a time scale comparable to that needed to transport SO_2 to the poles, with cold trapping on bright regions at intermediate latitudes important in retarding the poleward migration. The poleward flux of SO_2 that they calculate is $\leqslant 10^{28}$ molecules s^{-1}, or $\leqslant 10^{-4}$ g cm^{-2} yr^{-1}. The mean dayside surface pressure of SO_2 in this model is 10^{-9} to 10^{-10} bar. A modification of the regional cold trapping idea, in which subsurface regolith cold trapping controls the surface pressure, has been recently developed by Matson and Nash (1981), who calculate that the SO_2 equilibrium surface pressure may be as low as $\sim 10^{-12}$ bar. Lee and Thomas (1980) and Ingersoll and Summers (1981) have further developed the theory of the dynamics of a thin atmosphere on Io.

An alternative reconciliation of the Voyager IRIS measurement of a large amount of SO_2 with the low IUE value is to assume that the atmosphere is highly variable, depending on the rate of volcanic activity. Possibly the SO_2 seen by IRIS was simply outflow from the nearby plume of Loki, rather than an equilibrium value. Unfortunately, the rate of outgassing from Loki cannot be determined independently with sufficient precision to test this hypothesis.

Kumar and Hunten, in Chapter 21, also conclude that SO_2 may be the major component of Io's atmosphere. But they note that the photochemistry of SO_2 could lead to a build up of O_2 until O_2 became the major gas on the night side and near the poles. The oxygen could then have a major role in retarding the flow of SO_2 away from the warm equatorial regions. In their model, both solar ultraviolet and energetic electrons produce ionization of the SO_2 to generate a day-side ionosphere consistent with the Pioneer 10 observations. Kumar and Hunten calculate a maximum SO_2 loss rate by photolysis of 10^{11} to 10^{12} cm^{-2} s^{-1}. Much of this material escapes from Io, at a rate corresponding to replenishment of the atmosphere and ionosphere every few weeks.

The neutral sodium cloud that surrounds Io also forms a kind of atmosphere, and its origin is considered in both Chapters 20 and 21. The most likely mechanism to release the sodium is still sputtering by energetic particles, first suggested by Matson et al. (1974). The discovery of abundant sulfur and oxygen in the plasma torus increases the significance of this process, since these ions are very efficient sputterers. However, surface sputtering only works if the atmosphere is tenuous ($< 10^{15}$ cm^{-2}), which is true only on the night side of Io in the models considered here. Unfortunately, the observations of the sodium cloud do not show an obvious day-night asymmetry (e.g., Matson et al. 1978). Possibly the sputtering takes place in the atmosphere or directly from volcanic plumes, processes discussed by Haff et al. (1981), but it is difficult to see how the required quantity of sodium can be present high above the surface.

Kumar and Hunten also briefly consider atmospheres on the other Galilean satellites. Very tenuous oxygen atmospheres ($\sim 10^{-12}$ bar) are possible or even likely (Yung and McElroy 1977), resulting from photodissociation of H_2O, but are well below present detection limits. An alternative model allows the possibility of larger atmospheres of O_2 for Callisto and Europa, but not for Ganymede.

XII. THE IO PLASMA TORUS

The satellites of Jupiter and the Jovian magnetosphere are closely related, as first indicated by the discovery that Io controls the strength of decametric radio bursts. The absorption of charged particles by the Galilean satellites was documented by the Pioneer flybys, and Voyager proved that the satellites are sources as well as sinks for the Jovian magnetospheric plasma. In particular, it now appears that Io-produced ions dominate the magnetosphere out to ~ 20 R$_J$, and that this prolific source of plasma is one of the major reasons for the striking differences between the Jovian and Saturnian magnetospheres. In this book, discussion of the inner magnetosphere is focused on what it can tell us about the satellites; the intended audience is primarily astronomers and geologists rather than plasma physicists. Remote sensing is covered in Chapter 22, and *in situ* measurements in Chapter 23. These topics are covered in much greater depth in many papers published in the special Voyager issue (September 1981) of the *Journal of Geophysical Research* and in the book *Physics of the Jovian Magnetosphere* (Dessler 1982).

Remote observations of the Io torus in visible and ultraviolet light have been made from the ground (e.g., Trafton 1975; Kupo et al. 1976; Pilcher and Morgan 1979,1980; Pilcher 1980), from Earth orbit (e.g., Strobel and Davis 1980), and from the Pioneer and Voyager spacecraft (e.g., Judge and Carlson 1974; Broadfoot et al. 1979; Sandel et al. 1979; Broadfoot et al. 1981). These results are described in Chapter 22 by Pilcher and Strobel. As of late 1980, emission had been measured from three neutrals (sodium, potassium, oxygen) and from ions of sulfur and oxygen. The neutrals are physically associated with Io, while the ions become trapped in the corotating Jovian field and quickly distribute themselves into a torus.

The best-observed neutral species is sodium, and the geometry of the sodium cloud has been mapped by a number of observers (e.g., Matson et al. 1978; Murcray and Goody 1978; Pilcher and Schemp 1979; Goldberg et al. 1980). The sodium atoms stream away from Io at substantial velocities (2 to 3 km s^{-1}), apparently emitted preferentially from the leading inner quadrant of Io. It is generally·believed that this localization of the source is related to the sputtering process, not to a concentration of sodium at these longitudes on Io's surface. The derived source magnitude is $\sim 10^{26}$ atoms s^{-1}. The lifetime of the sodium before it is ionized and swept up by the Jovian magnetic field is, from Voyager data, not more than a few hours. The ejection is not necessarily smooth, and Pilcher has photographed directional features that may be produced by time-variable collisional sweeping. Clearly much remains mysterious about this sodium cloud, first seen about a decade ago.

The oxygen and sulfur emissions are excited by electron collisions and therefore contain information about the exciting electrons that make up the bulk of the plasma. Pilcher and Strobel discuss the Voyager ultraviolet and groundbased visible observations in detail; a dynamic regime with sharply defined radial structure is revealed. The emissions are primarily from toroidal regions near the Jovian magnetic equator and the centrifugal equator. Azimuthal variations within the torus appear related to the Jovian magnetic longitude or local Jovian time rather than to the position of Io. Typical electron densities are $\sim 10^3$ cm^{-3}, while the plasma temperatures range from $\sim 10^4$ to $\sim 10^6$ K. All these parameters vary temporally as well as spatially, probably reflecting changes in the ion source at Io.

The remote observations of the plasma near Io are best interpreted along with direct measurements of ion composition, density, and temperature made by Voyager 1 as it crossed through the Io torus, coming within 4.9 R$_J$ of Jupiter. These results and their interpretation are described in Chapter 23 by Sullivan and Siscoe (see also, e.g., Dessler and Vasyliunas 1979; Bridge et al. 1979; Sullivan and Bagenal 1979; Dessler 1980; Bagenal et al. 1980; Bagenal and Sullivan 1981). The detectors on the spacecraft clearly showed several spatially distinct regions within what is broadly referred to as the Io plasma torus. The innermost part of the torus, inside the orbit of Io, is dominated by corotating S$^+$ ions with a thermal energy of $\sim 10^4$ K. Because it is relatively cool, this plasma collapses toward the centrifugal equator, where it attains maximum electron density $\sim 3 \times 10^3$ cm^{-3}.

Farther out, very near Io's orbit at 5.95 R$_J$ (and therefore presumably near the source of the plasma), there is a region of warm (several 10^4 K) ions of sulfur and oxygen with electron density $\sim 10^3$ cm^{-3}. In this region there are also a few ions with much higher energies; presumably these are the atoms most recently ionized and therefore not yet fully accelerated into the corotating plasma. Apparently the plasma diffuses outward from this source region; beyond ~ 8 R$_J$ departures from ridgid corotation begin, and the plasma density drops off substantially. At $\sim 9\ R_J$ the absorption of particles by Europa provides a convenient outer boundary to the Io torus.

A neutral atom of sulfur or oxygen escaping from Io enters the plasma torus when it loses one or more electrons. Within about one gyro period (10 hr), the ion is rigidly corotating. It then can take one of two routes. The majority of ions diffuse outward into a radially extended plasma disk, eventually exiting from the magnetosphere and joining with the solar wind. A smaller number of ions, primarily S^+, collapse inward to the centrifugal equator to form the dense, cool plasma ring inside the orbit of Io; from here the ion may ultimately precipitate into the atmosphere of Jupiter, where it can generate a polar aurora. The power radiated during these processes exceeds 10^{14} W. The total number of ions in the torus at any time is $\sim 5 \times 10^{34}$, and the typical residence time for a single ion is about one month.

Since Io is the source of much of the plasma in the Jovian magnetosphere, analysis of the plasma composition provides an indirect measure of the surface composition (or possibly the plume composition) of the satellite. In addition, the Io-produced ions, after acceleration by the Jovian magnetic field, provide the energetic impacts needed to sputter heavier atoms, such as sodium, from the surface. Thus, the interactions between Io and the magnetosphere are important in determining the chemical evolution of the surface and atmosphere of the satellite.

XIII. EVOLUTION OF THE SATELLITE SYSTEM

In the final chapter of this book, Pollack and Fanale summarize the main conclusions of previous authors and use the extensive data now available on the satellites to constrain theories for the origin and evolution of the system. No new data appear to require that we abandon the basic concept that the regular satellites condensed from a circum-Jovian nebula of substantial mass, or that in its early phases the proto-Jupiter was much larger and more luminous than it is today. Pollack and Fanale also discuss the origin of the irregular satellites, but I will discuss only the regular, inner objects here.

In Chapter 24, Pollack and Fanale assume that the Galilean satellites probably accreted from a disk of gas and dust that developed around Jupiter near the end of its period of rapid hydrodynamic collapse. As originally suggested by Pollack and Reynolds (1974) and further developed by Cameron and Pollack (1976), the high luminosity of Jupiter is thought to have precluded the condensation of H_2O in the region closer than $\sim 10^6$ km, creating a strong compositional gradient; H_2O became a major component of Ganymede and Callisto, but not of Io or Europa. We now realize that tidal heating and outgassing could also have contributed to the loss of H_2O from the inner two satellites, but the early compositional gradient in the system still seems the most likely cause for the two separate compositional groups of Galilean satellites. According to this theory, satellite formation should have proceeded rapidly, ending 10^5 to 10^6 yr after the end of the hydrodynamic collapse of Jupiter. This theory also predicts that Thebe, Amalthea, Metis, Adrastea, and of course the rings should be composed of refractory material, unless some of these represent captured bodies.

All four Galilean satellites are surely differentiated, as discussed by Cassen et al. in Chapter 4. Several other chapters discuss the subsequent evolution of the outer satellites; although Ganymede clearly remained active for longer than Callisto, the precise time scales are uncertain, and both had probably become geologically inactive by 4 Gyr ago.

Pollack and Fanale describe Io as an open system, in which volatiles have migrated to the surface under the influence of tidal heating and subsequently been depleted by atmospheric escape and surface sputtering (see also Pollack and Witteborn 1980). They calculate that even large initial quantities of H_2O, N_2, and CO_2 would by now have been lost, leaving S and SO_2 as the most active cosmically abundant volatiles still present in any quantity. Perhaps there once were volcanoes on Io powered by H_2O or CO_2, but at the present epoch sulfur and its compounds clearly dominate the surface of this satellite.

Acknowledgments. I thank all of the authors of this book for their excellent chapters, on which this introduction is based, and for many stimulating discussions over the 18 months in which this book was under preparation. This research was supported in part by grants from the National Aeronautics and Space Administration and the Voyager Project.

October 1981

REFERENCES

Acuña, M.H., and Ness, N.F. (1976). The main magnetic field of Jupiter. *J. Geophys. Res.* 81, 2917-2922.

Anderson, J.D., Null, G.W., and Wong, S.K. (1974). Gravity results from Pioneer 10 Doppler data. *J. Geophys. Res.* 79, 3661-3664.

Bagenal, F., and Sullivan, J.D. (1981). Direct plasma measurements in the Io torus and inner magnetosphere of Jupiter. *J. Geophys. Res.* 86, 8447-8466.

Bagenal, F., Sullivan, J.M., and Siscoe, G.L. (1980). Spatial distribution of plasma in the Io torus. *Geophys. Res. Letters* 7, 41-44.

Becklin, E.E., and Wynn-Williams, C.G. (1979). Detection of Jupiter's ring at 2.2 μm. *Nature* 279, 400-401.

Bertaux, J.L., and Belton, M.J.S. (1979). Evidence of SO_2 on Io from UV observations. *Nature* 282, 813-815.

Binder, A.B., and Cruikshank, D.P. (1964). Evidence for an atmosphere on Io. *Icarus* 3, 299-305.

Broadfoot, A.L., and the Voyager Ultraviolet Spectrometer Team (1979). Extreme ultraviolet observations from Voyager 1 encounter with Jupiter. *Science* 204, 979-982.

Broadfoot, A.L., and the Voyager Ultraviolet Spectrometer Team (1981). Overview of the Voyager Ultraviolet Spectrometry results through Jupiter encounter. *J. Geophys. Res.* 86, 8259-8284.

Bridge, H.S., and the Voyager Plasma Science Team (1979). Plasma observations near Jupiter: Initial results from Voyager 1. *Science* 204, 987-991.

Brown, R.A. (1974). Optical line emission from Io. In *Exploration of the Planetary System* (A. Woszcyzk and C. Iwariszewsha, Eds.), pp. 527-531. D. Reidel, Dordrecht, Holland.

Brown, R.A., and Yung, Y.L. (1976). Atmosphere and emissions of Io. In *Jupiter*, (T. Gehrels, Ed.), pp. 1102-1145. Univ. Arizona Press, Tucson.

Burns, J.A., Ed. (1977). *Planetary Satellites*, Univ. Arizona Press, Tucson.

Burns, J.A., Showalter, M.R., Cuzzi, J.N., and Pollack, J.B. (1980). Physical processes in Jupiter's ring: Clues to its origin by Jove! *Icarus* 44, 339-360.

Butterworth, P.S., Caldwell, J., Moore, V., Owen, T., Rivolo, A.R., and Lane, A.L. (1980). An upper limit to the global SO_2 abundance on Io. *Nature* 285, 308-309.

Cameron, A.G.W., and Pollack, J.B. (1976). On the origin of the solar system and of Jupiter and its satellites. In *Jupiter*, (T. Gehrels, Ed.), pp. 61-84. Univ. Arizona Press, Tucson.

Carlsen, R.W., Bhattacharyva, J.C., Smith, B.A., Johnson, T.V., Hidavat, B., Smith, S.A., Taylor, G.E., O'Leary, B.T., and Brinkman, R.T. (1973). An atmosphere on Ganymede from its occultation of SAO-186800 on 7 June 1972. *Science* 182, 53-55.

Carr, M.H., Masursky, H., Strom, R.G., and Terrile, R.J. (1979). Volcanic features of Io. *Nature* 280, 729-733.

Cassen, P., Peale, S.J., and Reynolds, R.T. (1980*a*). On the comparative evolution of Ganymede and Callisto. *Icarus* 41, 232-239.

Cassen, P., Reynolds, R.T., and Peale, S.J. (1979). Is there liquid water on Europa? *Geophys. Res. Letters* 6, 731-734.

Cassen, P., Reynolds, R.T., and Peale, S.J. (1980*b*). Tidal dissipation–A correction. *Geophys. Res. Letters* 7, 987-988.

Clancy, R.T., and Danielson, G.E. (1981). High resolution albedo measurements on Io from Voyager 1. *J. Geophys. Res.* 86, 8627-8634.

Clark, R.N. (1980). Ganymede, Europa, Callisto, and Saturn's rings: Compositional analysis from reflectance spectroscopy. *Icarus* 44, 388-409.

Clark, R.N. (1981). Water frost and ice: The near-infrared reflectance 0.65-2.5 μm. *J. Geophys. Res.* 86, 3087-3096.

Clark, R.N., and McCord, T.B. (1980). The Galilean satellites: New near-infrared reflectance measurements (0.65-2.5 μm) and a 0.325-5 μm summary. *Icarus* 41, 323-339.

Clow, G.D., and Carr, M.H. (1980). Stability of sulfur slopes on Io. *Icarus* 44, 268-279.

Collins, S.A. (1981). Spatial color variations in the volcanic plume at Loki, on Io. *J. Geophys. Res.* 86, 8621-8626.

Consolmagno, G.J. (1981*a*). An Io thermal model with intermittent volcanism. *Lunar Planet. Sci.* XII, 175-177 (abstract).

Consolmagno, G.J. (1981*b*). Io: Thermal models and chemical evolution. *Icarus* 47, 36-45.

Consolmagno, G.L., and Lewis, J.S. (1976). Structural and thermal models of icy Galilean satellites. In *Jupiter*, (T. Gehrels, Ed.), pp. 1035-1051. Univ. Arizona Press, Tucson.

Consolmagno, G.L., and Lewis, J.S. (1977). Preliminary thermal history models of icy satellites. In *Planetary Satellites*, (J. Burns, Ed.), pp. 492-500. Univ. Arizona Press, Tucson.

Croft. S.K. (1981). On the origin of pit craters. *Lunar Planet. Sci.* XII, 196-198 (abstract).

Cruikshank, D.P. (1977). Radii and albedos of four Trojan asteroids and Jovian satellites 6 and 7. *Icarus* 30, 224-230.

Cruikshank, D.P., Jones, T.J., and Pilcher, C.B. (1978). Absorption bands in the spectrum of Io. *Astrophys. J.* 225, L89-L92.

Degewij, J., Andersson, L.E., and Zellner, B. (1980*a*). Photometric properties of outer planetary satellites. *Icarus* 44, 520-540.

Degewij, J., Cruikshank, D.P., and Hartmann, W.K. (1980*b*). Near-infrared colorimetry of J6 Himalia and S9 Phoebe: A summary of 0.3 to 2.2 μm reflectances. *Icarus* 44, 541-547.

Dessler, A.J. (1980). Mass-injection rate from Io into the Io plasma torus. *Icarus* 44, 291-295.

Dessler, A.J., Ed. (1982). *Physics of the Jovian Magnetosphere*, Cambridge Univ. Press, Cambridge, U.K.

Dessler, A.J., and Vasyliunas, V.M. (1979). The magnetic anomaly model of the Jovian magnetosphere predictions for Voyager. *Geophys. Res. Letters* 6, 37-40.

Fanale, F.P., Brown, R.H., Cruikshank, D.P., and Clark, R.N. (1979). Significance of absorption features in Io's IR reflectance spectrum. *Nature* 280, 761-763.

Fanale, F., Johnson, T., and Matson, D. (1977). Io's surface and the histories of the Galilean satellites. In *Planetary Satellites*, (J. Burns, Ed.), pp. 379-405. Univ. Arizona Press, Tucson.

Fimmel, R.O., Swindell, W., and Burgess, E. (1974). *Pioneer Odyssey, Encounter with a Giant.* NASA SP-349, Washington, D.C.

Fimmel, R.O., Van Allen, J., and Burgess, E. (1980). *Pioneer, First to Jupiter, Saturn and Beyond.* NASA SP-446, Washington, D.C.

Fink, J.H., and Fletcher, R.C. (1981). Variations in thickness of Ganymede's lithosphere determined by spacings of lineations. *Lunar Planet. Sci.* XII, 277-278 (abstract).

Fink, U., Dekkers, N.H., and Larson, H.P. (1973). Infrared spectra of the Galilean satellites of Jupiter. *Astrophys. J.* 179, L155-L159.

Finnerty, A.A., Ransford, G.A., Pieri, D.C., and Collerson, K.D. (1980). Is Europa's surface cracking due to thermal evolution? *Nature* 289, 24-27.

Gehrels, T., Ed. (1976). *Jupiter.* Univ. Arizona Press, Tucson.

Goldberg, B.A., Makler, Y., Carlson, R.W., Johnson, T.V., and Matson, D.L. (1980). Io's sodium emission cloud and the Voyager 1 encounter. *Icarus* 44, 305-317.

Goldstein, R.M., and Green, R.R. (1980). Ganymede: Radar surface characteristics. *Science* 207, 179-180.

Gradie, J., Thomas, P., and Veverka, J. (1980). The surface composition of Amalthea. *Icarus* 44, 373-387.

Grün, E. Morfill, G., Schwehm, G., and Johnson, T.V. (1980). A model of the origin of the Jovian ring. *Icarus* 44, 326-338.

Haff, P.K., Watson, C.C., and Yung, Y.L. (1981). Sputter ejection of matter from Io. *J. Geophys. Res.* 86, 6933-6938.

Hanel, R., and the Voyager IRIS Team (1979). Infrared observations of the Jovian system from Voyager 1. *Science* 204, 972-976.

Hansen, O.L. (1973). Ten-micron eclipse observations of Io, Europa, and Ganymede. *Icarus* 18, 237-246.

Harrington, R.S., and Seidelmann, P.K. (1981). The dynamics of the Saturnian satellites 1980S1 and 1980S3. *Icarus* 47, 97-99.

Head, J.W., Allison, M.L., Parmentier, E.M., and Squyres, S. (1981). High-albedo terrain on Ganymede: Origin as flooded graben. *Lunar Planet. Sci.* XII, 418-420 (abstract).

Helferstein, P., and Parmentier, E.M. (1980). Fractures on Europa: Possible response of an ice crust to tidal deformation. *Proc. Lunar Planet. Sci. Conf.* 11, 1987-1998.

Hsui, A.T., and Toksoz, M.N. (1980). Thermal evolution of Ganymede and Callisto: Effects of solid-state convection and constraints from Voyager imagery. *Proc. Lunar Planet. Sci. Conf.* 11, 1957-1978.

Ingersoll, A., and Summers, M.F. (1981). A dynamically controlled atmosphere on Io. *Icarus.* In press.

Jewitt, D.C., Danielson, G.E., and Synnott, S.P. (1979) Discovery of a new Jupiter satellite. *Science* 206, 951.

Jewitt, D.C., Danielson, G.E., and Terrile, R.J. (1982). Ground-based observations of the Jovian ring and inner satellites. *Icarus* 48, 536-539.

Johnson, T.V., Cook, A.F., Sagan, C., and Soderblom, L.A. (1979). Volcanic resurfacing rates and implications for volatiles on Io. *Nature* 280, 746-750.

Johnson, T.V., and Pilcher, C.B. (1977). Satellite spectrophotometry and surface compositions. In *Planetary Satellites*, (J. Burns, Ed.), pp. 232-268. Univ. Arizona Press, Tucson.

Johnson, T.V., Soderblom, L.A., Mosher, J.A., Danielson, G.E., and Kupferman, P. (1981). Multispectral mosiacs of the Galilean satellites. *Lunar Planet. Sci.* XII, 509-510 (abstract).

Judge, D., and Carlson, R.W. (1974). Pioneer 10 observations of the ultraviolet glow in the vicinity of Jupiter. *Science* 183, 317-318.

Judge, D.L., Carlson, R.W., Wu, F.M., and Hartmann, U.G. (1976). Pioneer 10 and 11 ultraviolet photometer observations of the Jovian satellites. In *Jupiter* (T. Gehrels, Ed.), pp. 1068-1101. Univ. Arizona Press, Tucson.

Kliore, A., Cain, D.L., Fjeldbo, G., Seidel, B.L., and Rasool, S.I. (1974). Preliminary results of the atmospheres of Io and Jupiter from Pioneer 10 S-band occultation experiment. *Science* 183, 323-324.

Kliore, A.J., Fjeldbo, G., Seidel, B.L., Sweenham, D.N., Sesplankis, T.T., Woiceshyn, P.M., and Rasool, S.I. (1975). Atmosphere of Io from Pioneer 10 radio occultation measurements. *Icarus* 24, 407-410.

Kuiper, G.P. (1957). Infrared observations of planets and satellites. *Astron. J.* 62, 245 (abstract).

Kumar, S. (1979). The stability of an SO_2 atmosphere on Io. *Nature* 280, 758-761.

Kumar, S. (1980). A model of the SO_2 atmosphere and ionosphere of Io. *Geophys. Res. Letters* 7, 9-13.

Kupo, I., Mekler, Y., and Eviatar, A. (1976). Detection of ionized sulfur in the Jovian magnetosphere. *Astrophys. J.* 205, L51-L53.

Lee, S.W., and Thomas, P.C. (1980). Near-surface flow of volcanic gases on Io. *Icarus* 44, 280-290.

Lewis, J.S. (1971). Satellites of the outer planets: Their physical and chemical nature. *Icarus* 15, 174-185.

Lieske, J.H. (1980). Improved ephemerides of the Galilean satellites. *Astron. Astrophys.* 82, 340-348.

Lockwood, G.W., Lumme, K., and Thompson, D.T. (1980). The recent photometric variability of Io. *Icarus* 44, 240-248.

Lucchitta, B.K. (1980). Grooved terrain on Ganymede. *Icarus* 44, 481-501.

Masursky, H., Schaber, G.G., Soderblom, L.A., and Strom R.G. (1979). Preliminary geologic mapping of Io. *Nature* 280, 725-729.

Matson, D.L., Goldberg, B.A., Johnson, T.V., and Carlson, R.W. (1978). Images of Io's sodium cloud. *Science* 199, 531-533.

Matson, D.L., Johnson, T.V., and Fanale, F.P. (1974). Sodium D-line emission from Io: Sputtering and resonant scattering hypothesis. *Astrophys. J.* 192, L43-L46.

Matson, D.L., and Nash, D.B. (1981). Io's atmosphere: Pressure control by subsurface regolith coldtrapping. *Lunar Planet. Sci.* XII, 664-666 (abstract); also *J. Geophys. Res.* In press.

Matson, D.L., Ransford, G.A., and Johnson, T.V. (1981). Heat flow from Io (J1). *J. Geophys. Res.* 86, 1664-1672.

McCauley, J.F., Smith, B.A., and Soderblom, L.A. (1979). Erosional scarps on Io. *Nature* 280, 736-738.

McFadden, L.A., Bell, J.B., and McCord, T.B. (1980). Visible spectral reflectance measurements (0.33-1.1 μm) of the Galilean satellites at many orbital phase angles. *Icarus* 44, 410-430.

McKinnon, W.B., and Melosh, H.J. (1980). Evolution of planetary lithospheres: Evidence from multiringed structures on Ganymede and Callisto. *Icarus* 44, 454-471.

McKinnon, W.B., and Spencer, J. (1981). Tectonic deformation of Galileo Regio and limits to the planetary expansion of Ganymede. *Lunar Planet. Sci.* XII, 694-696 (abstract).

Morabito, J., Synnott, S.D., Kupferman, P.N., and Collins, S.A. (1979). Discovery of currently active extraterrestrial volcanism. *Science* 204, 972.

Moroz, V.I. (1961). On the infrared spectra of Jupiter and Saturn (0.9-2.5 μm). *Astron. Zh.* 38, 1080-1081 (trans. in *Soviet Astron-AJ* 5, 825-831).

Morrison, D. (1977). Radiometry of satellites and the rings of Saturn. In *Planetary Satellites,* (J. Burns, Ed.), pp. 269-330. Univ. Arizona Press, Tucson.

Morrison, D. (1982). *Voyage to Saturn.* NASA SP-451, Washington, D.C.

Morrison, D., and Burns, J.A. (1976). The Jovian satellites. In *Jupiter*, (T. Gehrels, Ed.), pp. 991-1034. Univ. Arizona Press, Tucson.

Morrison, D., and Cruikshank, D.P. (1973). Thermal properties of the Galilean satellites. *Icarus* 18, 224-236.

Morrison, D., Cruikshank, D.P., Murphy, R.E., Martin, T.F., Beery, J.G., and Shipley, J.P. (1971). Thermal inertia of Ganymede from 20-micron eclipse radiometry. *Astrophys. J.* 167, L107-L111.

Morrison, D., and Morrison, N.D. (1977). Photometry of the Galilean satellites. In *Planetary Satellites* (J. Burns, Ed.), pp. 363-378. Univ. Arizona Press, Tucson.

Morrison, D., Pieri, D., Veverka, J., and Johnson, T.V. (1979). Photometric evidence on long-term stability of albedo and color markings on Io. *Nature* 280, 753-755.

Morrison, D., and Samz, J. (1980). *Voyage to Jupiter.* NASA SP-439, Washington, D.C.

Morrison, D., and Telesco, C.M. (1980). Io: Observational constraints on internal energy and thermophysics of the surface. *Icarus* 44, 226-253.

Murcray, F.J., and Goody, R. (1978). Astronomical monochromatic imaging as applied to the Io sodium cloud. *Applied Optics* 17, 3117-3124.

Murray, B.C., Westphal, J.A., and Wildey, R.L. (1964). The eclipse cooling of Ganymede. *Astrophys. J.* 141, 1590-1592.

Ostro, S.J., Campbell, D.B., Pettengill, G.H., and Shapiro, I.I. (1980). Radar observations of the icy Galilean satellites. *Icarus* 44, 431-440.

Pang, K.D., Lumme, K., and Bowell, E. (1981). Interpretation of phase curves of Io and Ganymede: Nature of surface particles. *Lunar Planet. Sci.* XII, 799-801 (abstract).

Parmentier, E.M., and Head, J.W. (1981). Viscous relaxation of impact craters on icy planetary surfaces: Determination of viscosity variation with depth. *Icarus* 47, 100-111.

Peale, S.J., Cassen, P., and Reynolds, R.T. (1979). Melting of Io by tidal dissipation. *Science* 203, 892-894.

Pearl, J., Hanel, R., Kunde, V., Maguire, W., Kox, K., Gupta, S., Ponnamperuma, C., and Ranlin, F. (1979). Identification of gaseous SO_2 and new upper limits for other gases on Io. *Nature* 280, 755-758.

Phillips, R.J., and Malin, M.C. (1980). Ganymede: A relationship between thermal history and crater statistics. *Science* 210, 185-187.

Pieri, D.C. (1980). Lineament and polygon patterns on Europa. *Nature* 289, 17-21.

Pilcher, C.B. (1980). Images of Jupiter's sulfur ring. *Science* 207, 181-183.

Pilcher, C.B., and Morgan, J.S. (1979). Detection of singly ionized oxygen around Jupiter. *Science* 205, 297-298.

Pilcher, C.B., and Morgan, J.S. (1980). The distribution of SII emission around Jupiter. *Astrophys. J.* 238, 375-380.

Pilcher, C.B., Ridgway, S.R., and McCord, T.B. (1972). Galilean satellites: Identification of water frost. *Science* 178, 1087-1089.

Pilcher, C.B., and Schemp, W.V. (1979). Jovian sodium emission from region C_2. *Icarus* 38, 1-11.

Pollack, J.B., and Reynolds, R.T. (1974). Implications of Jupiter's early contraction history for the composition of the Galilean satellites. *Icarus* 21, 248-253.

Pollack, J.B., and Witteborn, F.C. (1980). Evolution of Io's volatile inventory. *Icarus* 44, 249-267.

Pollack, J.B., Witteborn, F.C., Erickson, E.F., Strecker, D.W., Baldwin, B.J., and Burch, T.E. (1978). Near-infrared spectra of the Galilean satellites: Observations and compositional implications. *Icarus* 36, 271-303.

Purves, N., and Pilcher, C.B. (1980). Thermal migration of water on the Galilean satellites. *Icarus* 43, 51-55.

Reynolds, R.T., Peale, S.J., and Cassen, P. (1980). Io: Energy constraints and plume volcanism. *Icarus* 44, 234-239.

Sagan, C. (1979). Sulfur flows on Io. *Nature* 280, 750-753.

Sandel, B.R., and the Voyager Ultraviolet Spectrometer Team (1979). Extreme ultraviolet observations from Voyager 2 encounter with Jupiter. *Science* 206, 962-966.

Schaber, G.G. (1980). The surface of Io: Geologic units, morphology, and tectonics. *Icarus* 43, 302-333.

Schubert, G., Stevenson, D.J., and Ellsworth, K. (1981). Internal structures of the Galilean satellites. *Icarus* 47, 46-59.

Sinton, W.M. (1980). Io's 5 μm variability. *Astrophys. J.* 235, L49-L51.

Sinton, W.M. (1981). The thermal emission spectrum of Io and a determination of the heat flux from its hot spots. *J. Geophys. Res.* 86, 3122-3128.

Smith, B.A., and the Voyager Imaging Team (1979a). The Jupiter system through the eyes of Voyager 1. *Science* 204, 951-972.

Smith, B.A., and the Voyager Imaging Team (1979b). The Galilean satellites and Jupiter: Voyager 2 imaging science results. *Science* 206, 927-950.

Smith, B.A., and the Voyager Imaging Team (1981). Encounter with Saturn: Voyager imaging science results. *Science* 212, 163-191.

Smith, B.A., Shoemaker, E.M., Kieffer, S.W., and Cook, A.F. (1979c). The role of SO_2 in volcanism on Io. *Nature* 280, 738-743.

Smith, P.H. (1978). Diameters of the Galilean satellites from Pioneer data. *Icarus* 35, 167-176.

Smythe, W.D., Nelson, R.M., and Nash, D.B. (1979). Spectral evidence for SO_2 frost or absorbate on Io's surface. *Nature* 280, 766-767.

Soderblom, L., Johnson, T., Morrison, D., Danielson, E., Smith, B., Veverka, J., Cook, A., Sagan, C., Kupferman, P., Pieri, D., Mosher, J., Avis., C., Gradie, J., and Clarey, T. (1980). Spectrophotometry of Io: Preliminary Voyager 1 results. *Geophys. Res. Letters* 7, 963-966.

Squyres, S.W. (1980a). Surface temperatures and retention of H_2O frost on Ganymede and Callisto. *Icarus* 44, 502-510.

Squyres, S.W. (1980b). Topographic domes on Ganymede: Ice volcanism or isostatic upwarping. *Icarus* 44, 472-480.

Squyres, S.W. (1980c). Volume changes in Ganymede and Callisto and the origin of grooved terrain. *Geophys. Res. Letters* 7, 593-596.

Squyres, S.W. (1981). The topography of Ganymede's grooved terrain. *Icarus* 46, 156-168.

Squyres, S.W., Parmentier, E.M., and Head, J.W. (1981). Origin of grooves on Ganymede. *Lunar Planet. Sci.* XII, 1031-1033 (abstract).

Squyres, S.W., and Veverka, J. (1981). Voyager photometry of surface features on Ganymede and Callisto. *Icarus* 46, 137-155.

Stebbins, J., and Jacobsen, T.S. (1928). Further photometric measures of Jupiter's satellites and Uranus, with tests for the solar constant. *Lick Obs. Bull.* 13, 180-195.

Strobel, D.F., and Davis, J. (1980). Properties of the Io plasma torus inferred from Voyager EUV data. *Astrophys. J.* 238, L49-L52.

Strom, R.G., Schneider, N.M., Terrile, R.J., Cook, A.F., and Hansen, C. (1981). Volcanic eruptions on Io. *J. Geophys. Res.* 86, 8593-8620.

Strom, R.G., Terrile, R.J., Masursky, H., and Hansen, C. (1979). Volcanic eruption plumes on Io. *Nature* 280, 733-736.

Sullivan, J.D., and Bagenal, F. (1979). In situ identification of various ionic species in Jupiter's magnetosphere. *Nature* 280, 798-799.

Synnott, S.P. (1980). 1979 J2: Discovery of a previously unknown Jovian satellite. *Science* 210, 786-788.

Synnott, S.P. (1981). 1979 J3: Discovery of a previously unknown satellite of Jupiter. *Science* 212, 1392.

Taylor, G.E. (1972). The determination of the diameter of Io from its occultation of β Scorpii B on May 14, 1971. *Icarus* 17, 202-208.

Trafton, L. (1975). Detection of a potassium cloud near Io. *Nature* 258, 690-692.

Tyler, G.L., Marouf, E.A., and Wood, G.E. (1981). Radio occultation of Jupiter's ring: Bounds on optical depth and particle size and a comparison with infrared and radio results. *J. Geophys. Res.* 86, 8699-8703.

Veverka, J. (1977a). Photometry of satellite surfaces. In *Planetary Satellites*, (J. Burns, Ed.), pp. 171-209. Univ. Arizona Press, Tucson.

Veverka, J. (1977*b*). Polarimetry of satellite surfaces. In *Planetary Satellites,* (J. Burns, Ed.), pp. 210-231. Univ. Arizona Press, Tucson.

Veverka, J., Simonelli, D., Thomas, P., Morrison, D., and Johnson, T.V. (1981*a*). Voyager search for posteclipse brightening on Io. *Icarus* 47, 60-74.

Veverka, J., Thomas, P., Davies, M., and Morrison, D. (1981*b*). Amalthea: Voyager imaging results. *J. Geophys. Res.* 86, 8675-8682.

Yoder, C.F. (1979). How tidal heating on Io drives the Galilean orbital resonance locks. *Nature* 279, 767-770.

Yoder, C.F., and Peale, S.J. (1981). The tides of Io. *Icarus* 47, 1-35.

Yung, Y.L., and McElroy, M.B. (1977). Stability of an oxygen atmosphere on Ganymede. *Icarus* 30, 97-103.

2. THE RINGS OF JUPITER

DAVID C. JEWITT
California Institute of Technology

The properties of the Jovian ring system are reviewed. The system has three morphologically distinct components. The bright ring is planar and circular. It merges into the faint sheet component at its inner edge. A broad halo of ring particles envelops the bright ring and the faint sheet, extending some 10^4 km above and below the plane of the bright ring, and apparently extending down to the atmosphere of the planet. The system is bounded on the outside by the small satellite Adrastea (1979 J1). The micrometer-sized ring particles observed at large phase angles may be generated by the erosion of parent bodies within the bright ring. The parent bodies probably lie in the size range 1 m to 1 km. Small particles may decay into the faint sheet under the action of Poynting-Robertson and other drag forces. Ring particles will maintain a net electric potential with respect to the ambient plasma, due to several charging processes in operation in the Jovian magnetosphere. Particles with diameters less than 0.4 μm may be lifted into the halo by an interaction with the Jovian magnetic field.

The Jovian ring system was discovered by the cameras of Voyager 1, in March of 1979, as the result of a deliberate search (Smith et al. 1979*a*). An earlier unrecognized detection of the ring had been made in 1974 by the Pioneer 11 spacecraft. The trajectory of Pioneer 11 caused it to pass through the charged-particle shadows of the ring and inner satellites (Fillius 1976). The particle shadow due to the absorption of high-energy protons and electrons by the ring was not immediately recognized in its true guise, although Acuña and Ness (1976) suggested that absorption by a ring was a possibility.

The single ring frame obtained from Voyager 1 enabled a preliminary determination of the maximum radius of the ring system, an estimate of the optical thickness of the bright ring, and an upper limit to the ring thickness. Using the information gained from the first spacecraft, Voyager 2 was programmed to obtain a more complete survey of the ring. In fact, 25 images were successfully acquired, both from the inbound leg of the trajectory (i.e., at small phase angles), and from the outbound leg (large phase angles).

I. THE VOYAGER PICTURES

Since almost all of the observations and deductions described in this chapter have been obtained by analysis of pictures acquired by Voyager, a brief description of the Voyager data set will be useful. Each picture or "frame" is subdivided into 800×800 resolution elements or pixels. Each pixel contains a data number (DN), which may take any integer value between 0 and 255. Under most circumstances, the DN is proportional to the surface brightness of the object that is imaged onto the relevant pixel. The image scale is given approximately by $6.9 \times 10^{-5} d$ km/pixel for the wide angle (WA) camera and by $9.3 \times 10^{-6} d$ km/pixel for the narrow angle (NA) camera, where d is the distance to the target, measured in kilometers. These image scales do not represent the effective resolutions of most frames, however, which are usually much poorer. Because of the low surface brightness of the ring, the exposure times needed were often very long (the eleven minute exposure given to Voyager 1 frame FDS 16368.19 is an extreme case). Most exposures were of either 15 or 96 s duration. In such long exposures the images generally become smeared out by drift in the pointing direction of the cameras. This motion is different from frame to frame, but smears of several tens of pixels are common. A smaller though still detectable effect is the parallax motion between the spacecraft and the ring, which typically produces one or two NA pixels of smear on ring images with small range.

The pictures of the ring were obtained during two brief time intervals, one just before encounter with Jupiter and the other just after. The pre-encounter images show the ring at small phase angles (typically about $10°$). The post-encounter images show the ring at phase angles of nearly $180°$, and were taken while the spacecraft was in the shadow of the planet so as to minimize the internal scattering of direct sunlight by the camera optics (Smith et al. 1979b).

II. THE BRIGHT RING

The brightest and most conspicuous component of the Jovian ring system will be referred to simply as "the bright ring" in this discussion. Its

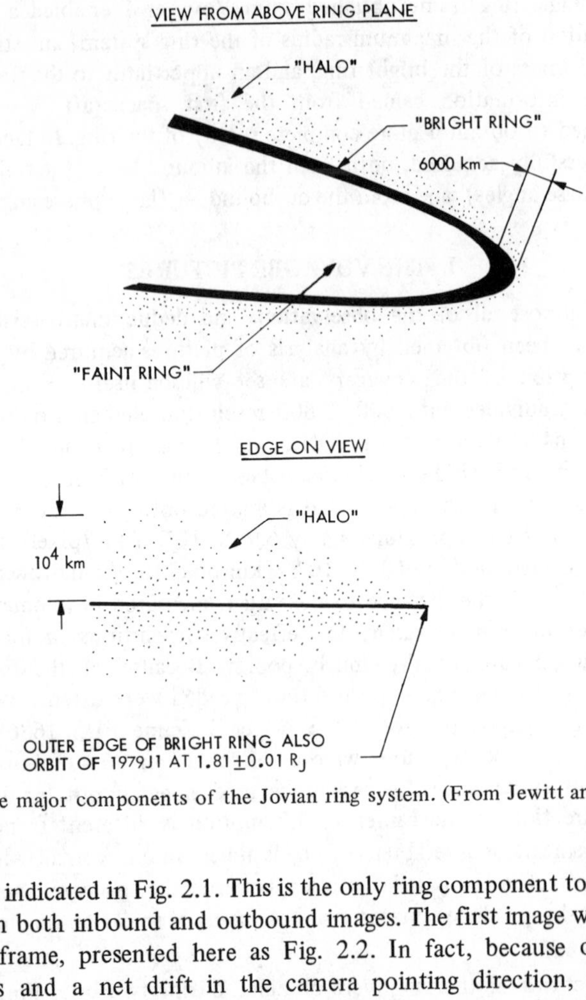

VIEW FROM ABOVE RING PLANE

EDGE ON VIEW

Fig. 2.1. The major components of the Jovian ring system. (From Jewitt and Danielson 1981.)

location is indicated in Fig. 2.1. This is the only ring component to have been observed in both inbound and outbound images. The first image was the ring discovery frame, presented here as Fig. 2.2. In fact, because of periodic oscillations and a net drift in the camera pointing direction, this frame actually contains six distinct images of the ring. Since the camera was approximately in the equatorial plane of the planet for the duration of the exposure, the ring was viewed edge-on. Direct measurement of the images on this frame yields an upper limit to the ring thickness of 30 km. By good fortune, the frame also records the outer edge of the ring, at 1.80 R_J with an estimated error of perhaps ± 0.01 R_J (T.C. Duxbury, JPL Inter-office Memo., 1979), where 1 R_J = 71400 km.

To search for any evidence of structure within the ring, brightness measurements were made along the images recorded in FDS 16368.19. The plots shown in Fig. 2.3 give the background DN value in the bottom trace, and the DN values along each of the six images of the ring in the upper six

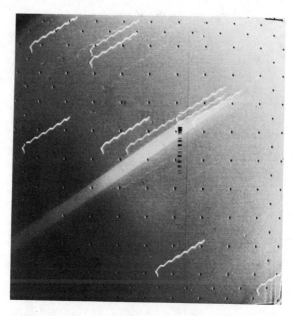

Fig. 2.2. Discovery frame of the Jovian ring, Voyager 1 narrow angle frame FDS 16368.19. Six adjacent images of the ring, each of about one minute effective exposure, extend from lower left to upper right. The variations in the pointing direction of the camera during this 672 s exposure, are evidenced by the "hairpin" star trails.

traces. Gaps in the traces correspond to reseaux and to field stars. The horizontal distance on the graphs corresponds to 256 pixels, or approximately 2900 km along the ring. The figure shows that the ring images are only 10 to 15 DN's above the background DN level at any point. Nevertheless, a gradual "bump" in the surface brightness is evident toward the right-hand side of the figure, centered at $\sim 1.79\ R_J$, and roughly $0.01\ R_J$ wide. An approximate contrast of 50% is observed at the bump in this backscattering image.

Further images of the ring were acquired by the second Voyager spacecraft in July of 1979 (Smith et al. 1979b). A mosaic of narrow angle frames proved to be too smeared to yield immediately useful information on the ring (though one of these frames very conveniently shows the new satellite Adrastea, 1979 J1). The inbound WA frames suffer less from smear, though they have unacceptably low signal-to-noise levels which prevent useful photometry. Figure 2.4 shows the WA frame FDS 20630.53. A considerable brightness gradient, due to the nearby fully illuminated disk of Jupiter, has been suppressed by special processing. The frame reveals the general appearance of the ring in backscattered light. It is very faint, narrow and

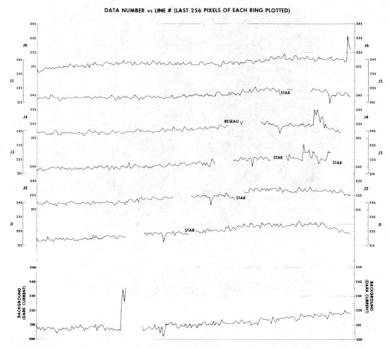

Fig. 2.3. Brightness plots along each of the six images of the Jovian ring recorded on FDS 16368.19 (Fig. 2.2). The horizontal scale represents linear distance along each ring image. The vertical scales for the six plots, separated for clarity, record the data number (DN) in each pixel along the ring. A plot of the background DN versus distance parallel to the ring is given at the botton. The ring plots exhibit a broad bump towards the right side of the figure, corresponding to a bright feature at ~ 1.79 R$_J$. Locations of stars and reseaux are indicated. (Plots by G.E. Danielson and V.L. Hall.)

frankly, unimpressive! Figure 2.5 also shows the ring in backscattered light, from a position $2°.5$ out of the ring plane.

Pictures of the ring taken while Voyager 2 was passing through the shadow of the planet show the ring in a more spectacular guise. Figure 2.6 is a composite of WA frames taken at this time. Figure 2.7 is a composite of NA frames. In these frames the bright ring is shown to have a width of approximately 6400 km. The ring extends from 1.72 R$_J$ to 1.81 R$_J$, each ±0.01 R$_J$, as determined from independent measurements of several Voyager frames (Jewitt and Danielson 1981). The ring appears to have only one definite surface feature—a bright annulus—though there is a general smooth variation of brightness over its surface. Both this variation and the annulus are evident in Fig. 2.8; the annulus corresponds in position with the bright bump shown in Fig. 2.3 and is presumed to be the same feature. On outbound

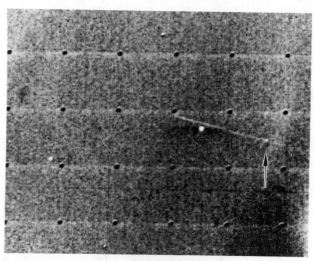

Fig. 2.4. The thin faint line on the right-hand side of the picture, frame FDS 20630.53, is the Jovian ring. The bright star below the ring is ρ Leonis. Satellite 1979J1 is indicated by an arrow. A strong brightness gradient due to glare from the day side of Jupiter has been removed from this image by digital processing.

frames, however, the bright annulus has a contrast of approximately 10%, significantly less than the 50% contrast evident from the inbound image. In a later section, it is argued that this is probably evidence for a population of larger bodies in the bright annulus.

A distinctive characteristic of the bright ring is that it has a very sharply delineated outer boundary. It decays over a distance of less than 0.01 R_J (possibly much less; the upper limit is set by image smear). At its inner edge, the bright ring merges into the faint sheet. The transition occurs over a radial distance of $\sim$ 0.02 R_J.

Since the bright ring is observed to have the same radius from several different azimuths, it seems reasonable to conclude that the ring radius is in fact independent of azimuth to within the errors of measurement (0.01 R_J). This imposes an upper limit of 0.003 to the eccentricity of the outer edge of the ring.

As seen from a single point in space, the phase angle varies from point to point around the ring. Variations of the surface brightness around the ring can be used to determine a phase curve, from which an estimate of the particle size is possible. Such measurements are presented in Fig. 2.9 (Jewitt and Danielson 1981). The measurements were taken from Voyager 2 frames covering the phase angle range from 174° to 176°. Each data point in the figure represents the mean brightness within an area chosen so as to average over radial brightness structures in the ring.

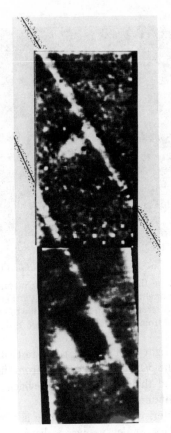

Fig. 2.5. Extreme enhancement of Voyager 2 frame FDS 20612.27 showing the very faint near and far arms of the ring. The picture was taken before encounter, with the spacecraft about 2°.5 out of the ring plane. The blotches result from the enhancement.

To deduce the properties of the ring particles from Fig. 2.9, several assumptions are necessary. It is assumed that the optical properties of the particles can be described in terms of Mie theory. The particles are assumed to be spheroidal, with a uniform refractive index. Fortunately, it is known that the forward scattering properties of Mie scatterers are only weakly dependent on particle shape and refractive index, so that these assumptions are plausible. Another assumption must be made concerning the size distribution of the particles. The simplest assumption is that the particles are monodispersed, i.e., they all have the same radius. The two lines plotted in Fig. 2.9 are for spheroidal Mie scatterers with scattering parameters 20 and 30 (the scattering parameter is defined to be the ratio of the circumference of the particle to the wavelength of the radiation). The scattering data were

Fig. 2.6. Mosaic of Voyager 2 wide angle frames, taken during solar eclipse. The Jovian limb is outlined by strongly forward-scattered light from its atmosphere.

Fig. 2.7. Mosaic of narrow angle Voyager 2 frames showing the sharp outer edge of the bright ring and the less distinct inner edge. The end frame appears blurred due to excessive smear motion.

taken from Gumprecht et al. (1952). It is evident from the figure that a particle diameter of between 3 and 5 μm is consistent with the observed forward scattering data. In the discussion which follows, a diameter of 5 μm is adopted. The important result is that the appearance of the bright ring at large phase angles is consistent with the presence of particles about ten (optical) wavelengths in diameter.

Measurements of the surface brightness of the ring have been used to estimate the normal optical depth through the relation

$$I = Fk\,\tau/\mu \qquad (1)$$

where I is the reflected surface brightness, F is the incident solar intensity, k is the geometric albedo of the individual ring particles, τ is the normal optical

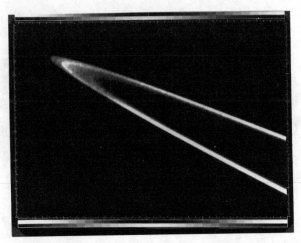

Fig. 2.8. The bright annulus within the bright ring at 1.79 R_J. The contrast in this rendition of FDS 20693.02 has been substantially enhanced. The ansa appears fainter than the arms of the bright ring due to the variation of the phase angle around the ring, and to the strongly forward-scattering ring particles.

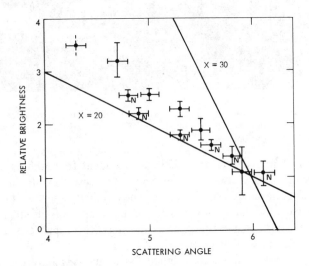

Fig. 2.9. Scattering diagram for the bright ring. The scattering angle, defined as 180° minus the phase angle, is plotted on the abscissa. The brightness measurements are normalized at 6° scattering angle. Points labeled 'N' are from the near arm of the ring. Straight lines were taken from the tables of Gumprecht et al. (1952). 'X' denotes the scattering parameter, equal to the particle circumference divided by the wavelength of the radiation (0.5 μm). (From Jewitt and Danielson 1981.)

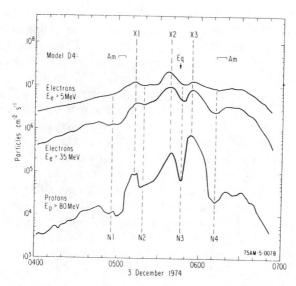

Fig. 2.10. Electron and proton fluxes measured by Poineer 11 close to periapsis. The minima at N1 and N4 are due to absorptions by Amalthea; those at N2 and N3 are due to the ring. 1979 J1 does not give a detectable absorption since its cross section is only a few percent of that of Amalthea or the ring. (From Fillius 1976.)

depth of the ring at the wavelength of observation, and μ is the sine of the angle of elevation of the spacecraft above the ring plane at the time of observation. Only small phase-angle images were used to estimate τ (Jewitt and Danielson 1981). The assumption was made that the ring particles could be accurately characterized by $k = 0.04$. This is tantamount to the assumption that the ring particles which backscatter are dark rocks, much larger than the wavelength of observation (about 0.5 μm), so that phase function effects are not strong. Under these assumptions, representative values of I lead to $\tau \sim 3 \times 10^{-5}$ for the bright ring. This value is an approximate average over the full radial extent of the bright ring.

An independent estimate of the optical depth of the bright ring is available from Pioneer 11 flux measurements of high-energy charged particles (Acuña and Ness 1976). The flux data are shown in Fig. 2.10 (taken from Fillius 1976). The pronounced dips in the particle fluxes correspond to absorptions by Amalthea and by the Jovian ring (though this was not realized at the time the observations were made). By assuming the proton drift rate at the ring to be the same as at Amalthea, Jewitt and Danielson (1980) used the flux measurements to infer the solid cross section of the ring to be 1.1 $\times$ 10^{11} m^2. A more refined calculation by Fillius (1980) gives a solid cross section of 2 $\times$ 10^{10} m^2, which, however, is likely to be a lower limit (W. Fillius, personal communication). If spread uniformly over the bright ring, between 1.72 and 1.81 R_J, these cross sections would give optical depths from 2×10^{-5} to 4×10^{-6}. These values are reasonably close to the optical value, 3×10^{-5}, and the agreement is fair in view of the many approximations made. The derived optical depths are consistent with upper limits imposed by Voyager IRIS observations at ~ 25 μm wavelength (Hanel et al. 1979). Since the high-energy charged particles recorded by Pioneer have typical absorption lengths of a few centimeters in rock (Ip 1980), it is likely that the solid cross section obtained from the charged particle absorption data of Pioneer 11 refers to ring particles much larger than the micrometer-sized ones that dominate the forward scatter.

Since the Voyager discovery of the rings, several Earth-based observations of the ring have now been secured. Within a few days the ring was detected at 2.2 μm by Becklin and Wynn-Williams (1979), who used the 2.2-m University of Hawaii telescope on Mauna Kea. Further studies (Neugebauer et al. 1980) have given a near infrared spectrum of the ring which precludes an icy composition. Optical wavelength images of the ring have been obtained (Smith and Reitsema 1980) using a CCD camera with the University of Arizona 1.5-m telescope. Continuation of such work (e.g., extension to polarization studies) may increase our knowledge of the ring in the near future.

III. THE FAINT SHEET AND THE HALO

The faint sheet appears as faint, continuous material in the space between the inner edge of the bright ring, 1.72 ± 0.01 R_J, and the surface of the planet. It has been detected only in large phase angle pictures, due to its low surface brightness (Fig. 2.11). In the approximate phase angle range 174° to 176°, it has $\sim 1/4$ the surface brightness of the bright ring, though there are variations with radial distance from Jupiter. A photometric analysis of the faint ring has not yet been attempted due to the complications imposed by the presence of the halo.

The halo component of the Jovian ring was not recognized for several months following the acquisition of the Voyager 2 images. There are at least three independent lines of evidence for its existence.

1. The shadow of the planet on the ring does not define a straight line. If the ring material were entirely in the plane of the Jovian equator, we would expect the (roughly cylindrical) shadow of the planet to intersect with it to produce a straight line cut tangential to the Jovian terminator. This is not observed. Instead, the shadow edge is curved, suggesting that material above the plane of the bright ring intercepts sunlight near the shadow edge.

2. Some faint material is seen apparently projected beyond the outer edge of the bright ring. The isophotes of this material do not describe ellipses

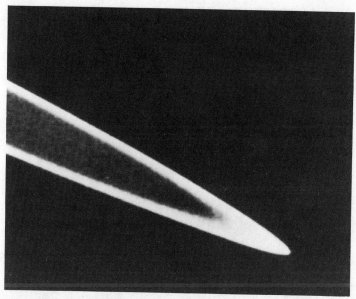

Fig. 2.11. Wide angle frame FDS 20693.02 taken by Voyager 2 through the clear filter. This picture has been processed to emphasize the faint ring; the granular appearance is an artifact.

which are concentric with those of the bright ring, but rather the isophotes jut in towards the bright ring at an acute angle. This behavior would not be expected of coplanar material extending beyond the bright ring.

3. The phase angle variation around the bright ring is such that some points on the near and the far arms possess the same phase angle. Photometry of the bright ring demonstrates that the far arm is consistently brighter than the near arm, when areas with the same phase angle are compared (see Fig. 2.9). This effect remains when various viewing-geometry related asymmetries are removed. A similar effect is seen in the faint ring. The simplest way to account for the observed anomaly is by the introduction of material into the line of sight to the far arm. This material, if optically thin, would add to the apparent brightness of the far arm.

Taken together, it is believed that these three arguments indicate beyond reasonable doubt the existence of out-of-plane material.

The halo has a dimension of about 10^4 km in a direction normal to the plane of the bright ring, as determined by direct measurement of FDS 20693.02. The halo thickness increases towards the planet; the images give the impression of a broad biconvex lens-like shape (see Fig. 2.12). The outer limit of the halo is very hard to define, but it can certainly be traced

Fig. 2.12. The Jovian ring halo, appearing in this picture as a broad, lens-like region enveloping most of the bright ring. The halo isophotes can be traced to just beyond the apparent tip of the bright ring. The black dots are reseaux. This is an extreme stretch of frame FDS 20693.02 shown in Fig. 2.11.

beyond the outer edge of the bright ring. There is a slight north-south asymmetry of the halo with respect to the bright ring. Measured from FDS 20693.02 this asymmetry is approximately 300 km, with the halo displaced to the south of the plane of the bright ring. The properties of the halo, and of the other two ring components, are summarized in Table 2.1.

IV. SATELLITE ADRASTEA (1979 J1)

The satellite 1979 J1, tentatively named for the nymph who suckled the infant Zeus, was discovered on frame FDS 20630.53 (see Fig. 2.4) (Jewitt et al. 1979). It appears as a 5th magnitude star in the plane of the Jovian ring. In Fig. 2.13 it is apparent that the satellite has trailed in a different direction and by a greater amount than have the field stars. Under the assumptions of zero orbital eccentricity and zero inclination, an orbit radius of 1.80 ±0.01 R_J was calculated from the first two frames on which the satellite was seen.

TABLE 2.1
Summary of Measurements of the Jovian Ring

	Bright Ring	Faint Ring	Halo
Outer Radius	$1.81\ R_J{}^a$	$1.72\ R_J$	$\sim 1.8\ R_J$
Inner Radius	$1.72\ R_J$	$1.0\ \ R_J$	$\sim 1.0\ R_J$
Thickness	< 30 km	?	$\gtrsim 10^4$ km
Normal Optical Depth	$\sim 3 \times 10^{-5}$	$\sim 7 \times 10^{-6}$	$\lesssim 5 \times 10^{-6\,b}$
Eccentricity	< 0.003	—	—

[a] $1\ R_J = 71400$ km.

[b] For V2 geometry.

Fig. 2.13. Narrow angle image FDS 20630.48 showing a trailed image of 1979 J1 projected against the ring. The field star just above the ring trails at a different angle and by a lesser amount than the satellite. Several other field stars are visible in the original image.

More recently the satellite has been identified on Voyager 1 frames as a silhouette against the bright face of the planet. From the Voyager 1 data an orbit computation gives the orbit radius to be 1.793 ±0.005 R_J (estimated error), though again under the assumptions of zero eccentricity and inclination (S.P. Synnott, personal communication, 1980). It may soon be possible to correlate the observations of 1979J1 made from both Voyagers. The improved time baseline would then permit a relaxation of the eccentricity and inclination assumptions, and a better orbit would result.

Limited information concerning the physical properties of 1979J1 is available. The strongly trailed NA image has an excess width of 2.5 pixels, suggesting a satellite diameter of $\sim 25 \pm 5$ km. From this quantity and the directly measured brightness, an albedo of ~ 0.05 is found, similar to that of Amalthea. This low albedo immediately precludes ice as the major surface constituent of 1979J1.

V. DISCUSSION

A discussion of the Jovian ring must attempt to account for the following.

1. The narrowness of the bright ring and its sharp outer and inner edges (the outer being sharper).
2. The narrow bright annulus seen at 1.79 R_J, within the bright ring.
3. The faint sheet.
4. The broad halo, including its slight north-south asymmetry with respect to the bright ring.
5. The overall photometric properties of the ring, including the very small optical depths and the deduced small particle sizes.
6. The close spatial relationship between the ring and 1979J1.

An important time scale for the Jovian ring is the Poynting-Robertson orbital decay time, given by

$$t_p \sim \frac{\xi a c^2}{F_\odot} \ \ell \, \mathrm{n} \, \frac{r_i}{r_f} \tag{2}$$

where ξ is the particle density, a is the particle radius, c is the speed of light, $F_\odot$ is the flux of solar radiation falling on the particle, and r_i and r_f are the initial and final orbital radii of the particle. A rock grain with $\xi = 3000$ kg m^{-3}, $a = 2.5 \times 10^{-6}$ m, and $F_\odot = 50$ W m^{-2} would take a time $t_p \sim 2.5 \times 10^5$ years to decay from $r_i = 1.8$ R_J to $r_f = 1.0$ R_J.

Jovian ring particles will also suffer from sputtering by magnetospheric particle impacts. The lifetime to sputtering, t_s, is given approximately by

$$t_s \sim \frac{a}{R} \tag{3}$$

where R is the surface sputtering loss rate (m/yr). Estimates of R for the Jovian environment vary considerably, but values of order 10^{-7} m/yr seem likely (Haff et al. 1980). For $a = 2.5 \times 10^{-6}$ m, we compute $t_s \sim 100$ yr. Both time scales are much shorter than the age of the solar system. We must either conclude that the ring is a very recent feature of the solar system, or that there exists a continuous source of ring particles. We assume the latter alternative to be true.

It is of interest to estimate the mass loss rate from the ring. Under the assumption that all particles are the same size, we may write the mass of the ring as

$$m \sim \xi \, aC \qquad (4)$$

where ξ is particle density, a is particle radius, and C is the total solid cross section of the ring as deduced from measurement of the optical depth. Dividing equation (4) by the Poynting-Robertson lifetime we obtain the mass loss rate

$$\frac{m}{t_p} \sim \frac{CF_\odot}{c^2} \frac{1}{\ell n(r_i/r_f)} \qquad (5)$$

which is independent of the particle size and density. Putting $C = 10^{11}$ m^2, $F_\odot = 50$ W m^{-2}, $c = 3 \times 10^8$ m s^{-1}, $r_i = 1.8\,R_J$, and $r_f = 1.0\,R_J$, we obtain $m/t_p \sim 10^{-4}$ kg s^{-1}. If instead we use the sputtering lifetime in the denominator of Eq. (5), the mass loss rate becomes $\sim 10^{-1}$ kg s^{-1}.

A plausible source of the micrometer-size ring particles, and one for which there is some evidence, is a population of large parent bodies within the bright ring. Such bodies would produce small particles by mutual collisions and by surface erosion due to the local micrometeoroid flux. The ring particle collision time is $t_c \sim 1/\Omega\tau$, where Ω is the Keplerian orbital motion appropriate for the bright ring and τ is its normal optical depth. Taking $\Omega = 2.5 \times 10^{-4}$ s^{-1} and $\tau \sim 3 \times 10^{-5}$, we find $t_c \sim 5$ yr. At each collision, material might chip from the larger bodies to maintain the small-particle population. The total production rate of small particles by this process is difficult to estimate, though 10^{-4} kg s^{-1} would seem to be a small amount to supply. In addition to mutual collisions between ring particles, the measurements from Pioneers 10 and 11 indicate that micrometeorites are able to supply the necessary small particle loss rates by more than four orders of magnitude.

Direct evidence for the parent body population is limited; we summarize it as follows.

1. The appearance of the ring at small phase angles cannot be matched by

the optical properties of 5-μm diameter particles. The ring is brighter in backscattered light than would be expected if it comprised only a collection of 5-μm particles. This may suggest the presence of more efficient backscatterers, possibly particles much larger than 5 μm.

2. Ground-based observations in the near infrared reveal a ring reflection spectrum that cannot be matched by 5-μm Mie spheres (Neugebauer et al. 1980). The spectrum closely resembles that of Amalthea, suggesting that large rocks in the ring are responsible for its appearance at small phase angles.

3. The bright annulus at 1.79 R_J, within the bright ring, has a contrast over the adjacent ring material of approximately 50% in backscattered light, but of only $\sim$ 10% in forward scattered light. In other words, the particles in the annulus have a different phase function from those in the adjacent bright ring. This suggests a different distribution of particle sizes in the annulus. The enhanced backscattering efficiency may result from the presence of larger bodies in the annulus.

4. The total solid cross section of the ring, as deduced from Pioneer measurements of charged particle flux, is comparable to the cross section deduced at optical wavelengths.

This strongly suggests that bodies larger than cm size are present. An empirical upper limit to the diameter of the parent bodies can be set at $\sim$1 km. Bodies much larger than this would be individually visible in the Voyager images. In the lifetime of the solar system, all bodies smaller than $\sim$ 1 m will have been removed from the bright ring by Poynting-Robertson drag. These simple arguments may be used to constrain the parent-body dimensions in the range of 1 m to 1 km, again assuming that the ring is 4.5×10^9 yr old.

The small particles generated from the parent bodies may become subject to orbital decay by Poynting-Robertson drag forces. Such particles would evolve from the bright ring into the faint ring over times of order t_p. Since we previously found that t_s is much smaller than t_p, we might expect to see a truncated inner edge to the faint ring, corresponding to the destruction of grains by sputtering. That such an edge is not observed may suggest that sputtered particles can re-stick on grain surfaces, so that there is an equilibrium grain size. Alternatively, the assumed sputtering rate, $R = 10^{-7}$ m/yr, may be inappropriate.

Very small particles generated from the parent bodies in the bright ring may populate the halo. A particle will undergo approximately $t_p/t_c \sim 10^5$ collisions while its orbit decays under the action of Poynting-Robertson drag. Any initial out-of-plane motion of the particle would be very quickly dampened by the collisions, in the absence of a continuous out-of-plane force acting on the particle. Since the halo is observed to extend down close to the

Jovian atmosphere and is not localized around the bright ring, the action of such a force seems certain. One plausible mechanism is outlined here.

Jovian ring particles orbit the planet deep within the magnetosphere. In general, each particle will possess a net potential with respect to the ambient plasma. Charging will result from the action of the photoelectric effect, of electron and proton sticking, of high-energy electron and proton induced secondary electron emission, and of other effects (Wyatt 1969).

A detailed calculation of the potential would require knowledge of the electron and proton energy spectra at the location of the ring, together with the detailed properties of the ring particles such as their composition and shape. However, to a first approximation, the charging due to electron and proton impacts may be taken to dominate the charging by the photoelectric effect and by secondary electron emission. Collisions between electrons and protons in the Jovian plasma tend to distribute energy equally between the two types of particles. Because of their low mass, the electrons achieve very much higher mean speeds than do protons. Hence the flux of electrons onto the surface of a particle is greater than the flux of protons. The resultant excess charging by electrons leads to a negative net potential on ring particles. The potential becomes more and more negative until the point is reached at which the electric field around a ring particle is strong enough to prevent further electrons from reaching it. The same field also acts to *enhance* the capture cross section for protons (Wickramasinghe 1967). In the Jovian plasma at 1.8 R_J, a grain potential on the order of -10 V seems likely (Consolmagno 1980).

The charged ring particles will interact with the Jovian magnetic field via the Lorentz force. Because the Jovian best-fit magnetic dipole is inclined to the spin axis of the planet by about 10° (Smith and Gulkis 1979), the Lorentz force does not lie entirely within the equatorial plane. As seen from a ring particle, the force oscillates through the ring plane with frequency $\Omega-\omega$, where Ω is the local orbital frequency and ω is the spin rate of the (corotational) magnetosphere. At the position of the bright ring, the oscillation has a period of 23 hr. Small particles will be lifted out of the plane into the halo. Larger particles will be held close to the equatorial plane by the component of Jovian gravity normal to the plane. The transition particle size between halo particles and those that remain in the ring plane may be found by comparing the gyration frequency, ω_g, with the orbital frequency. When $\omega_g \ll \Omega$, the motion is predominantly gravitational. When $\omega_g \gg \Omega$, a particle may make many gyrations in a single orbit, and the motion is dominated by the Lorentz force. The transition between the two regimes occurs when $\omega_g \sim \Omega$. We find

$$\frac{\omega_g}{\Omega} = \frac{3B\epsilon V}{\xi\Omega a^2} \qquad (6)$$

where B is the local magnetic flux density (Wb m^{-2}), ϵ is the permittivity of free space (F m^{-1}), and V is the potential on a particle of radius a and density ξ. We put $B \sim 10^{-4}$Wb m^{-2}, $\epsilon = 8.8 \times 10^{-12}$ F m^{-1}, $\xi \sim 3000$ kg m^{-3}, and $V = 10$ volts to find $\omega_g/\Omega = 3.5 \times 10^{-14}/a^2$. The left-hand side equals unity for a particle radius $a \sim 2 \times 10^{-7}$ m. This predicted transition size is consistent with the observed particles of $a \sim 2.5 \times 10^{-6}$ m in the bright ring.

By balancing the out-of-plane component of the Lorentz force against the normal component of Jovian gravity, we obtain the following expression for the height, Z, to which a charged particle may ascend from the plane

$$Z = \frac{B_x \cdot V\epsilon}{\xi a^2} \cdot \frac{(\Omega-\omega)r}{\omega(2\Omega-\omega)} \tag{7}$$

where B_x is the component of the magnetic flux density in the ring plane ($B_x \sim 10^{-5}$ WB m^{-2}). This equation assumes $Z/r \ll 1$ and assumes that the ring particles move independently. We find $Z \sim 2.7 \times 10^{-8} \, V/a^2$. For a potential V of 10 volts, $\sim 10^{-7}$ m diameter particles may ascend to $Z \sim 10^7$ m (the observed halo half-thickness) at distance 1.8 R$_J$ from Jupiter.

The observation that the symmetry plane of the halo is slightly offset from the plane of the bright ring may be explained by the poleward offset of the best-fit magnetic dipole from the center of Jupiter. The direction of the latter offset has not been determined unambiguously from radioastronomical (or other) measurements. Smith and Gulkis (1979) report a dipole latitude of $10° \pm 23°$, which is consistent with offsets in either hemisphere, and also with zero polar offset. If the charged grain interpretation of the halo is correct, the direction of halo offset shows the dipole offset to be towards the south of the Jovian equatorial plane. The observed halo displacement (~ 300 km) is likely to be an apparent average displacement of particles of many sizes, weighted by the particle size distribution and by the (size dependent) forward scattering intensities of the particles.

Finally, I briefly discuss the possible nature of the relationship between the ring and 1979J1, although present uncertainties in the measurements of the positions of both objects may render this discussion premature. The outer edge of the bright ring lies at 1.81 ± 0.01 R$_J$, as determined from small phase angle images. The satellite has an (assumed circular, uninclined) orbit radius of 1.793 ± 0.005 R$_J$. Dermott et al. (1980) have proposed that the bright ring of Jupiter is confined by a satellite embedded within it. The satellite would hold ring particles in stable horseshoe orbits about its Lagrangian equilibrium points. However, Dermott et al do not claim that 1979J1 is the satellite their process invokes. Indeed the horseshoe confinement mechanism requires and predicts a satellite at the center of the ring, at about 1.76 R$_J$, though no such object has been observed.

Work done by Goldreich and Tremaine (1979) shows the development of an edge due to gravitational perturbations by a satellite on a disk of particles. The edge develops because a ring particle that slowly moves towards a satellite gets into resonance with the satellite. The resulting radial excursions of the particle carry it repeatedly into and out of the ring until it undergoes a collision with another ring particle in the main body of the ring. At collision its velocity becomes reoriented and the particle is once again "absorbed" into the ring. The satellite and the ring effectively repel each other. Because of the need to have collisions, there is a lower limit to the optical depth of a ring in which this process may operate. The optical depth of the Jovian ring exceeds this limit (Goldreich and Tremaine 1980). If the process were perfectly efficient, there would be a standoff distance between the ring and the satellite (Goldreich and Tremaine 1979).

The relatively sharp inner edge of the bright ring similarly indicates that the ring is not spreading inwards. This may indicate the presence of a confining satellite at ~ 1.72 R_J, or it may simply indicate that the spreading time for the bright ring is much greater than its age. A search for a satellite at the inner edge of the ring has proved fruitless.

Acknowledgments. It is a pleasure to acknowledge the continued assistance of G.E. Danielson and P. Goldreich in this work. This work was supported by a contract from Jet Propulsion Laboratory, California Institute of Technology and a grant from the National Aeronautics and Space Administration.

July 1980

REFERENCES

Acuna, M. H., and Ness, N. F. (1976). The main magnetic field of Jupiter, *J. Geophys. Res.* 81, 2917-2922.

Becklin, E. E., and Wynn-Williams, C. G. (1979). Detection of Jupiter's ring at 2.2 μm. *Nature* 279, 400-401.

Consolmagno, G. J. (1980). Electromagnetic scattering lifetimes for dust in Jupiter's ring. *Nature* 285, 557-558.

Dermott, S. F., Murray, C. D., and Sinclair, A. T. (1980). The narrow rings of Jupiter, Saturn and Uranus. *Nature* 284, 309-313.

Fillius, W. (1976). The trapped radiation belts of Jupiter. In *Jupiter* (T. Gehrels, Ed.) pp. 896-927. Univ. of Arizona Press, Tucson.

Fillius, W. (1980). Trapped particle absorption by the rings of Jupiter and Saturn. IAU Colloquium 57. *The Satellites of Jupiter* (abstract 1-7).

Goldreich, P., and Tremaine, S. (1979). Towards a theory for the Uranian rings. *Nature* 277, 97-99.

Goldreich, P., and Tremaine, S. (1980). Disk-satellite interactions. Submitted to *Astrophys. J.*

Gumprecht, R. O., Sung, N. L., Chin, J. H., and Sliepcevich, C. M. 1952. Angular distribution of intensity of light scattered by large droplets of water. *J. Opt. Soc. of Amer.* 42, , 226-231.

Haff, P. K., Watson, C. C., and Tombrello, T. A. (1979). Ion erosion on the Galilean satellites of Jupiter. *Proc. Lunar Sci. Conf.* 10, 1685-1699.

Hanel, R., and the Voyager IRIS Team (1979). Infrared observations of the Jovian system from Voyager 2. *Science* 206, 952-956.

Ip, W. H. (1980). New progress in the physical studies of the planetary rings. *Space Sci. Rev.* 26, 97-109.

Jewitt, D. C., and Danielson, G. E. (1981). The Jovian ring. *J. Geophys. Res.* In press.

Jewitt, D. C., Danielson, G. E., and Synott, S. P. (1979). Discovery of a new Jupiter satellite. *Science* 206, 951.

Neugebauer, G., Becklin, E., Jewitt, D. C., Terrile, R.J., and Danielson, G.E. (1980). Spectra of the Jovian ring and Amalthea. Submitted to *Astron. J.*

Smith, B. A., and Reitsema, H. J. (1980). CCD observations of Jupiter's ring and Amalthea. IAU Coll. 57. *The Satellites of Jupiter* (abstract 2-4).

Smith, B. A., and the Voyager Imaging Team (1979a). The Jupiter system through the eyes of Voyager 1. *Science* 204, 951-972.

Smith, B. A., and the Voyager Imaging Team (1979b). The Galilean satellites and Jupiter: Voyager 2 imaging science results. *Science* 206, 927-950.

Smith, E. J. and Gulkis, S. (1979). The magnetic field of Jupiter: A comparison of radio astronomy and spacecraft observations. *Ann. Rev. Earth Planet. Sci.* 7, 385-415.

Wickramasinghe, M. C. (1967). *Interstellar Grains.* pp. 93-96. Chapman and Hall Ltd., London.

Wyatt, S. P. (1969). The electrostatic charge of interplanetary grains. *Planet. Space Sci.* 17, 155-171.

3. ORBITAL EVOLUTION OF THE GALILEAN SATELLITES

RICHARD GREENBERG
Planetary Science Institute

The orbital motions of the Galilean satellites exert dramatic control over their physical properties (most notably Io's) through tidal heating. In turn, tidal dissipation in the satellites, as well as in Jupiter, has governed the evolution of the orbits and, in particular, of the Laplace resonance. If the system started out of the resonance and evolved into it, forced eccentricities would have increased with time. Hence, the tidal melting of Io and the cracking of Europa's surface may have occurred relatively recently. This theory requires that Jupiter's tidal dissipation factor $Q_J \lesssim 2 \times 10^6$, a rather low value (high rate of tidal dissipation) compared with most models of Jovian interior processes. Alternatively, the system may have started even deeper in the resonance than it is today, a scenario which is consistent with larger values of Q_J. This model, with its correspondingly large initial forced eccentricities, would imply (1) that Io melted early and fast, and may have remained molten with only a thin solid skin until the present and (2) that the water mantles of both Europa and Ganymede remained largely molten for considerably longer than Callisto did, but later froze as their eccentricities and tidal heating decreased.

The orbital motion of the Galilean satellites helps govern their thermal evolution and hence their physical properties through the mechanism of tidal heating. Such control is dramatically demonstrated by volcanism on Io, apparently the result of runaway tidal heating (Peale et al. 1979) made possible by Io's relative proximity to Jupiter and by its substantial orbital eccentricity. Io's eccentricity is dominated by a forced component due to perturbations by Europa as part of a three-way orbital resonance involving Io, Europa, and Ganymede called the Laplace relation.

The causal relationship from orbital properties through tides to physical state also operates in reverse (Fig. 3.1); tidal heating in Io could have been a runaway effect because it would have induced melting and thus permitted greater response to the tidal potential and hence faster heat production. The reversibility goes further; tides govern orbital evolution. In this chapter I shall consider scenarios for the past evolution of the orbits of the Galilean satellites and, more specifically, evolution of the resonance. Some scenarios start with the system out of resonance, so initial forced eccentricities were small. I will also present an alternative evolutionary model, in which forced eccentricities were even greater in the past than now. I will identify key issues that need to be resolved in order to discriminate between these scenarios. Once we have constraints on the history of the resonance, we will be in a position to describe the past tidal heating and physical evolution of each satellite. Hence this chapter may provide a framework for understanding some of the observed physical properties, which are described and interpreted elsewhere in this book.

Because some readers may have minimal interest in the intricacies of celestial mechanics, I have structured this chapter so that the mathematical material is not absolutely essential to the thread of the logic; I do not treat the motions of other satellites or rotational dynamics since those subjects are covered in other books in this Space Science Series (Greenberg 1976; Peale 1977).

I. THE LAPLACE RESONANCE

The orbital periods of Io, Europa, and Ganymede are near a ratio of 1:2:4. There are many other examples in the solar system of pairs of satellites near the 1:2 commensurability. In any such case — or in any $j:(j+1)$ resonance, where j is any integer — the satellites' longitude of conjunction varies slowly. Hence their maximum mutual perturbations are repeated many

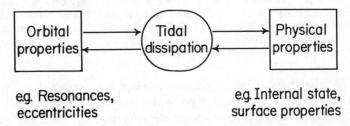

Fig. 3.1. Causal relationships in tidal evolution. Orbital properties govern tidal heating, which in turn modifies physical properties. The feedback effect of changed physical properties on tidal heating gives runaway melting of Io (Peale et al. 1979). Finally, tidal dissipation governs long-term evolution of orbits. Understanding of these relations can shed light on the past physical and orbital state of the system.

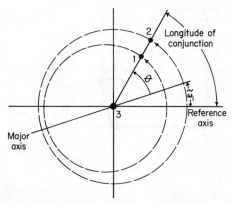

Fig. 3.2. System near a 2:1 commensurability. The longitude of conjunction of satellites (1) and (2) is given by $2\lambda_2 - \lambda_1$ and moves slowly, at rate $2n_2 - n_1$, where λ is a satellite's longitude and n is mean motion. This effect enhances perturbations that tend to maintain conjunction stable at an apse, such as the pericenter of the inner satellite's orbit ($\widetilde{\omega}_1$).

times near the same longitude, allowing a buildup of effects much greater than if conjunction occurred in a stochastic sequence of different longitudes where the effects would largely cancel one another.

If conjunction occurs at an angle θ in longitude from an inner satellite's pericenter (Fig. 3.2), the eccentricity e is increased at a rate

$$\dot{e} \propto \sin \theta. \tag{1}$$

This statement is rigorously true to lowest order in e, and also follows intuition; an extra pull outward on a satellite already moving out from pericenter to apocenter can only increase e. Similarly, the longitude of pericenter, $\widetilde{\omega}$, is varied according to

$$\dot{\widetilde{\omega}} \propto -\left(\frac{1}{e}\right)\cos \theta \tag{2}$$

where the inverse e describes the relative ease of rotating a nearly circular ellipse. The variation in e by expression (1) accelerates the variation of $\dot{\widetilde{\omega}}$, such that the nearest apse is pulled toward the point of conjunction. Thus the configuration with an apse at conjunction can be stable. To maintain this stable configuration, e must have just the right value so that $\dot{\widetilde{\omega}}$ matches the rate of motion of the longitude of conjunction. The closer the ratio of orbital periods is to exact resonance, the slower conjunction moves and the greater e must be according to expression (2). This value of e is called the forced eccentricity, e_f. If the system oscillates about equilibrium, the variation of θ is called libration and the variation of e is called the free or proper

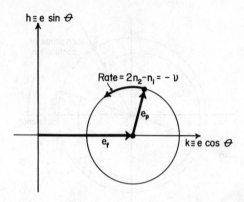

Fig. 3.3. Variation of eccentricity and θ (conjunction longitude minus pericenter longitude) during oscillation about a stable state for a typical case. In this (h,k) plane the true e is the vector sum of the forced eccentricity, e_f, and free or proper eccentricity, e_p. The closer the system is to exact commensurability the greater is $e_f : e_f \propto 1/\nu$. If $e_f > e_p$, θ librates about a fixed value as shown; if $e_p > e_f$, θ circulates through 360°.

eccentricity, e_p.[a] The relationship between these quantities is illustrated in Fig. 3.3.

Conjunction at a nearly fixed value of θ also varies the inner satellite's mean motion n according to

$$\dot{n} \propto e \sin \theta. \qquad (3)$$

The slow rate of migration of conjunction is $2n_2 - n_1$ (subscripts refer to satellites numbered from inner to outer). Thus, according to expression (3), conjunction is accelerated towards pericenter of the inner satellite, a stable configuration. For these examples I have considered the effects of resonance on the inner satellite of a pair. The outer satellite would be similarly affected with some sign reversals.

In the case of Io (subscript 1) and Europa (subscript 2), the conjunction longitude (given by $2\lambda_2 - \lambda_1$, where the λ's are each satellite's mean orbital longitude) is locked to Io's perijove and Europa's apojove:

---

[a]The choice of the word "proper" perhaps reflects a value judgment among celestial mechanicians. The free e is a constant of integration (just as the e of an unperturbed satellite is), while the forced e is not. In that sense the free e is more "proper." Traditionally only proper e has appeared in tables of orbital elements of the Galilean satellites (e.g., Greenberg 1976). This practice long hid from many workers the true e's of Io and Europa, which are dominated by forced e's, and caused the long delay in recognizing the true degree of tidal heating. I prefer the term "free e" to "proper e", but use the subscript p to distinguish it from the "forced e" (subscript f).

$$\theta_{11} \equiv 2\lambda_2 - \lambda_1 - \tilde{\omega}_1 \text{ librates about } 0°. \tag{4}$$

$$\theta_{12} \equiv 2\lambda_2 - \lambda_1 - \tilde{\omega}_2 \text{ librates about } 180°. \tag{5}$$

(In the notation defined here, the first subscript on θ identifies the pair of satellites whose conjunction is considered [1 for Io-Europa, 2 for Europa-Ganymede] and the second subscript indicates which satellite's perijove appears in the definition of a particular θ.) For Europa (2) and Ganymede (3), conjunction is locked to Europa's perijove but to neither apse of Ganymede:

$$\theta_{22} \equiv 2\lambda_3 - \lambda_2 - \tilde{\omega}_2 \text{ librates about } 0°. \tag{6}$$

$$\theta_{23} \equiv 2\lambda_3 - \lambda_2 - \tilde{\omega}_3 \text{ circulates through } 360°. \tag{7}$$

Combining (5) and (6) yields:

$$\phi \equiv \theta_{22} - \theta_{12}$$

$$= \lambda_1 - 3\lambda_2 - 2\lambda_3 \text{ librates about } 180°. \tag{8}$$

In fact the libration amplitude is remarkably small: $0°.066 \pm 0°.013$ with a period of ~ 6 yr according to Lieske's (1980) solution. Yoder (1979) suggests it may be even smaller. Kinematically, Eq. (8) means that whenever Europa and Ganymede are in conjunction, Io must be almost exactly $180°$ away (Fig. 3.4). Differentiation of Eq. (8) yields, on average over the libration,

$$n_1 - 2n_2 = n_2 - 2n_3 . \tag{9}$$

Both sides of this equation at present equal $0°.7395/$day. This number is much smaller than the mean motions (in Table 3.1, orbital periods are of the order of days), so each pair is very close to the exact 2:1 commensurability. In the past this quantity, which I call ν, may have been different; whether it was greater or smaller (implying smaller or greater forced eccentricities) depends on what theoretical scenario for long-term evolution one chooses to believe. (Later in this chapter, I will need to redefine ν slightly and introduce subscripts to differentiate between the left and right sides of Eq. (9); nevertheless, all ν's will retain their qualitative interpretation as a measure of distance of the system from exact commensurability.)

In Appendix A at the end of this chapter, I set up in more detail the equations of resonant interactions, which in essence are more complete versions of expressions (1), (2), and (3). I demonstrate mathematically some of the properties of the resonance that I discussed earlier, specifically the behavior

TABLE 3.1

The Galilean Satellites: Properties Relevant to Dynamics

	Io (1)	Europa (2)	Ganymede (3)	Callisto (4)
Semimajor axis, a (km)	4.22×10^5	6.71×10^5	1.07×10^6	1.88×10^6
Mean motion, n (deg/day)	203.4890	101.3747	50.3176	21.5711
Forced eccentricity, e_f	0.0041	0.0101	0.0006	—
Free eccentricity, e_p	$(1 \pm 2) \times 10^{-5}$	9×10^{-5}	0.0015	0.01
Mass, $\mu = m/m_J$ ($\times 10^{-5}$)	4.684 ± 0.022	2.523 ± 0.025	7.803 ± 0.030	5.661 ± 0.019

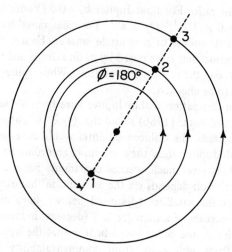

Fig. 3.4. The Laplace relation with $\phi = \lambda_1 - 3\lambda_2 - 2\lambda_3 \approx 180°$. When Europa and Ganymede are in conjunction ($\lambda_2 = \lambda_3$), as shown, Io is $180°$ away.

of eccentricities illustrated in Fig. 3.3 and the inverse dependence of forced eccentricity on ν. I also show that the behavior of ϕ can be described by a pendulum equation, Eq. (44), which explains the observed stability of ϕ at $180°$, seen in Fig. 3.4 and Eq. (8), as well as the observed relation among mean motions shown in Eq. (9). The derivation of the pendulum equation requires that ν have a value greater than or equal to its present value, a point that must be remembered when considering evolution models with initially small values of ν.

II. DISSIPATIVE EFFECTS ON ORBITS

Any satellite raises a tide on a planet. If the planet rotates faster than the satellite, as Jupiter does relative to the Galilean satellites, dissipation in the planet causes the tidal bulge to be oriented slightly ahead of the satellite. Thus the satellite gains orbital energy at the expense of the planet's rotational energy. Correspondingly, the satellite's mean motion decreases at a rate (Kaula 1964):

$$\frac{\dot{n}}{n} = -\frac{9}{2} k_J \left(\frac{R_J}{a}\right)^5 \frac{\mu n}{Q_J} \equiv -c \qquad (10)$$

where a is the satellite's semimajor axis, R_J is the planet's radius, Q_J is its tidal dissipation parameter (inversely proportional to dissipation rate), k_J is the "second degree Love number," which expresses the enhancement of the tidal potential due to distortion of the planet, and μ is again the satellite-planet mass ratio. For fluid Jupiter $k_J = 0.5$ (Yoder 1979). For tides raised by Io, $c \approx 6 \times 10^{-13} \ Q_J^{-1} \ \mathrm{s}^{-1}$. For tides raised by other Galilean satellites, c would be orders of magnitude smaller. Hence, throughout this chapter, I shall assume that tides raised by or on Europa and Ganymede have negligible effects on their orbital evolution. While quite plausible, this assumption needs to be checked.

Constraints on dissipation within Jupiter have been obtained by various means. Goldreich and Soter (1966) noted that $Q_J \gtrsim 10^5$; otherwise, over the age of the solar system, the Galilean satellites would have been pushed out much farther from Jupiter than they are now, according to Eq. (10). (The other inner satellites give much weaker constraints because of their small masses.) That conclusion depends on the assumption that other effects, such as tides raised on the satellites (discussed below), have not significantly counteracted the increase of semimajor axis (decrease in mean motion) given by Eq. (10). Goldreich and Soter noted further that the hypothesis that the system evolved from well away from commensurability into resonance requires $Q_J \lesssim 10^6$; otherwise there would have been insufficient change in mean motion ratios. Yoder's (1979) discussion of this hypothesis requires the same upper limit (see Sec. III).

Estimates of Q_J from models of tidal energy dissipation within Jupiter yield significantly higher values. From the mixing length theory of convection driven by internal heat sources in Jupiter, Goldreich and Nicholson (1977) estimated $Q_J \sim 5 \times 10^{13}$. Upward propagation of tidally driven atmospheric waves gives $Q_J \sim 5 \times 10^7$ according to an analysis by H.C. Houben and P.J. Gierasch (personal communication). There would probably be additional tidal dissipation in a rocky core. Dermott (1979) suggests a heating rate in the core equivalent to $Q_J \sim 10^6$. However, that rate is somewhat extreme for reasons clearly stated by Dermott. The Q value he assumed for the rocky core ($Q = 34$) is lower than one might estimate from comparison with the solid components of the Earth ($Q \gtrsim 100$). Even a value of $Q \sim 100$ requires the narrow assumption that the material be solid but just barely so; otherwise the rocky material would have a much larger Q. Furthermore, on cosmogonic grounds W. Hubbard (personal communication) expects Jupiter's core to contain a substantial water component, so the rocky mass would be much smaller than assumed by Dermott. Hubbard concludes that dissipation in Jupiter's core could not account for Q_J less than 10^7.

These estimates of Q_J based on physical models of Jupiter thus generally give values somewhat larger than are required for the hypothesis of capture into resonance. For this reason, in this chapter (Secs. IV and V) I propose alternative orbital evolution scenarios which are consistent with larger values of Q_J. The more developed of these alternative scenarios (Sec. V) assumes that the system was originally even closer to the exact commensurability (smaller ν's) than it is today. If the present best estimates of Q_J prove to be correct, a scenario of the latter type would be implied. On the other hand, physical modeling of Jupiter continues, motivated in part by these dynamical issues, and may ultimately yield a mechanism for smaller Q_J.

Another torque on Io's orbit is due to the electric current driven by the $v \times B$ electric field as the satellite moves through Jupiter's magnetosphere. For a given flux tube current I (in amp), the torque gives $\dot{n}_1$ equivalent to a tidal torque with $Q_J = 3 \times 10^{13}/I$ (Peale and Greenberg 1980). (The current through Io is about $2I$ due to contributions of two magnetospheric current loops.) The largest estimated value of I from Voyager data, 5×10^6 amp (Ness et al. 1979), is equivalent to $Q_J = 6 \times 10^6$. Kivelson et al. (1979) find I to be an order of magnitude smaller, and Ness' estimate of I has been revised downward to 3×10^6 amp (see Chapter 4 by Cassen et al.). Thus, at present any torque due to I must have an equivalent $Q_J > 10^7$. In my analysis in following sections, I include only tidal effects, but magnetic torque can be included by simply considering the equivalent Q_J.

Tides raised on the satellites by Jupiter must have brought them into synchronous rotation in times short compared with the age of the solar system (e.g., $\sim 10^4$ yr for Io, $\sim 10^6$ yr for Ganymede). However, tidal energy dissipation continues in each satellite due to its orbital eccentricity. The material is worked as, with each orbital period, (i) the distance to Jupiter and hence the amplitude of the tide oscillates and (ii) the orbital velocity and hence the orientation of the tide with respect to the satellite's body librates. These tides introduce the following variations (Kaula 1964):

$$\frac{\dot{n}}{n} = 7 Dce^2 \tag{11}$$

$$\dot{e} = -\frac{7}{3} Dce \tag{12}$$

where $D = (k/k_J)(R/R_J)^5 \mu^{-2}(Q_J/Q)$, with R, k and Q, the satellite's radius, Love number, and tidal dissipation parameter, respectively. As before, μ is the satellite-planet mass ratio.

The Love number k depends strongly on the rigidity of a body.[a] For a Moon-like Io with rigidity 6.5×10^{10} Pa, k would be 0.027. For a largely melted Io, the tidal response would be greater, with $k \sim 1$. The increased dissipation due to a bigger k in a melted body is at least partially offset by the fact that elastic dissipation occurs in only a small part of the total volume, namely the crust, according to the model introduced by Peale et al. (1979). The net enhancement of dissipation, relative to that in a solid body, can be

---

[a]As used here, k is the "second degree Love number", often written with a subscript 2. Because it is the only Love number used in this chapter, that subscript is not needed. Throughout this chapter numerical subscripts identify satellites unless otherwise stated.

given by a factor η (called f in Chapter 4 by Cassen et al.), which is a function of the fraction of volume melted. In general k/Q can be replaced in our equations by $\eta k/Q$ where k and Q are the values for the solid body case. For a model of equilibrium crustal thickness on Io, η may be ~ 13 (Peale et al. 1979).

While reading the following sections, bear in mind that terms proportional to c are due to tides raised on Jupiter by a satellite and those with Dc are due to tides raised on the satellite. D represents the ratio between the two. Where subscripts are not used, the satellite referenced is Io.

III. CAPTURE INTO RESONANCE

According to a hypothesis introduced by Goldreich (1965) and expanded upon by Sinclair (1975), the system of Galilean satellites began out of resonance and evolved into it. In this section I summarize the detailed analysis by Yoder (1979), using my own notation for consistency, and then I discuss the implications of his results. Some of Yoder's main results were also obtained independently by Lin and Papaloizou (1979). In the evolution scenario proposed by Yoder, Io moves outward from Jupiter by tidal evolution until the 2:1 commensurability with Europa is reached and the two satellites become locked in resonance. Then the two satellites move out together until the 2:1 commensurability between Europa and Ganymede is reached and the system becomes locked into the present resonance.

In order to study the long-term evolution of the system, the tidal effects discussed in the previous section must be incorporated into equations equivalent to those derived in Appendix A. In the mathematical analysis given in Appendix B, I show how the pendulum equation governing ϕ is modified (Eq. 51) and an additional term proportional to $\dot{\phi}$ is added to it (Eq. 56). The equation,

$$\dot{v}_{11} = -0.32c\, n_1 \left(1 - 13D\, e_{1f}{}^2\right) \tag{54}$$

for the evolution of v_{11} (Fig. 3.5) is also derived. This is the rate at which the system evolves toward or away from the exact 2:1 commensurability after the three-body resonance is established. According to Eq. (54), the evolution stops, i.e., the system is in equilibrium if $e_{1f}{}^2 = 1/(13D)$. That equilibrium is stable, because, if e_{1f} is too big, v_{11} increases. Since $e_{1f} \propto 1/v_{11}$, e_{1f} decreases and the system returns to equilibrium (Fig. 3.5).

Given the assumption that the system evolved from outside the resonance (large v_{11}, small e_{1f}) toward equilibrium, e_{1f} must now be less than its equilibrium value (curve A, Fig. 3.5). Thus we have

$$D \leqslant \frac{1}{13e_{1f}{}^2} = 4600. \tag{13}$$

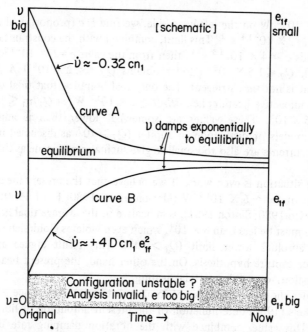

Fig. 3.5. The behavior of $\nu \equiv n_1 - 2n_2 = n_2 - 2n_3$ governed by Eq. (54). Curve A illustrates the scenario studied by Yoder (1979) and discussed in Sec. III. The system starts out of resonance (ν large and, correspondingly, forced e small). It evolves toward the exact commensurability ($\nu = 0$) according to $\dot{\nu} = -0.32\, c\, n_1$, but slows as the equilibrium condition $D = 1/(13e_{1f}^2)$ is approached. D, which contains the ratio Q_J/Q_1, must be less than the present value of $1/(13e_{1f}^2) = 4600$. Evolution from deeper in resonance (curve B) is discussed in Sec. V.

Recall from the definition of D (Sec. II) that, constant factors aside, D represents the ratio of the tidal parameters $Q_J/(Q_1/\eta_1)$. Thus by using Eq. (13) we may convert constraints on Q_1/η_1 to constraints on Q_J.

We do know something about Q_1/η_1, because Io apparently has undergone considerable tidal heating. Certainly the heating rate must be (and must have been for considerable time) greater than for the Moon, a nearly equal-sized, cold body (see Chapters 4 by Cassen et al. and 19 by Pearl and Sinton). The present radiogenic lunar heating rate is $\sim 5 \times 10^{11}$ W. Even the early radiogenic heating rate of 2×10^{12} W may not have been sufficient to melt the Moon, although it was probably nearly so (Peale et al. 1979). After the presumed runaway melting in Io, Peale et al. (1979) estimate tidal heating to continue at a rate of $\sim 2 \times 10^{13}$ W (equivalent to $Q_1/\eta_1 \approx 10$). The tidal heating rate, from the rate of orbital energy loss Eq. (11), is

$$\frac{dE}{dt} = \frac{7}{3} m_1 n_1^2 a_1^2 c\, D\, e_1^2 . \tag{14}$$

With 2×10^{13} W on the left-hand side, we find $D c$ (proportional to η_1/Q_1) must be $\gtrsim 2 \times 10^{-14}$ s^{-1}. This limit, combined with the constraint on D (Eq. 13), yields $c \geqslant 4 \times 10^{-18}$ s^{-1}; then from the value $c = 6 \times 10^{-13} Q_J^{-1}$ s^{-1} (Sec. II), $Q_J \leqslant 1.5 \times 10^5$. (Yoder obtains $Q_J \ll 2 \times 10^6$.) Actually the constraint is not that stringent. The only real bound is that tidal dissipation exceed radioactive heating; i.e., $dE/dt \gtrsim 2 \times 10^{12}$ W (or $Q_1/\eta_1 \lesssim 100$). Thus $Q_J \lesssim 1.5 \times 10^6$. These values are somewhat lower than is admitted by present models of dissipation in Jupiter ($Q_J > 10^7$ as discussed in Sec. II). Electric currents are also too small to give such a great torque as this scenario requires.

The situation is even worse if we believe that the present measured heat flux out of Io, $\sim 6 \times 10^{13}$ W (Matson et al. 1980; Pearl 1980; Morrison and Telesco 1980; Sinton 1981), is indicative of the average tidal heating rate. Then Q_J must be less than 5×10^4, which even violates Goldreich and Soter's (1966) absolute lower limit ($Q_J > 10^5$). This result argues against the resonance capture hypothesis. On the other hand, the present heat flux may be anomalous, or not due to tidal heating.

Yoder showed that the scenario of evolution into resonance is supported by Lieske's (1980) determination of the present amplitude of libration of ϕ, $0°066$. This value, combined with the libration damping rate due to the friction term (Eq. 56) in the pendulum equation, helps constrain the age of the resonance. The libration damps with an exponential time scale approximately twice the reciprocal of the coefficient of $\dot{\phi}$; to go from an amplitude of $180°$ (initial capture into libration) down to $0°066$ would take $\lesssim 5 \times 10^8$ yr according to Yoder. Thus the resonance is very young. The implications are that melting and volcanism on Io, and the cracking of Europa's surface, if due to tides, are recent phenomena. Actually, since the average dissipation rate in Io may have been considerably smaller than assumed by Yoder, the age of the resonance might be as great as 2.6×10^9 yr (S.J. Peale, personal communication), a value which still supports the capture hypothesis.

These results on the age of the resonance depend on Lieske's determination of the libration amplitude. Yoder notes that Lieske's neglect of some periodic effects may be significant. Even if Lieske's amplitude is correct, the resonance may not necessarily be young. Possibly the resonance is ancient, and the current libration is due to an impact by a stray body in the solar system (K. Blasius, personal communication). A 10-km body traveling at 20 km s^{-1} and impacting $\lesssim 10^8$ yr ago would have changed the velocity of any one of the three resonant satellites enough to account for a libration of $\sim 0°1$. (Such an impact would also have introduced rotational motion which would damp on a relatively short time scale.) Thus orbital libration is a questionable measure of the age of the resonance.

IV. RECENT OR EPISODIC MELTING OF IO

In this section I suggest modifications to the hypothesis of capture into resonance which may reconcile it with the larger values of Q_J ($\gtrsim 10^7$) indicated by physical models of Jupiter. I consider cases in which the resonance is very old, but the melting of Io was recent. Such recent melting may have been the first such melting event, or the most recent in a sequence of melt-refreeze cycles. These speculative models depend on the following notion: As ν evolves toward an equilibrium value dependent on the tidal parameter D (Fig. 3.5), the corresponding orbital eccentricities change. In turn, heating rates change and could induce sudden catastrophic physical changes (e.g., melting) which would drastically affect the value of D. Conceivably such a change in D, if recent, would throw the system out of equilibrium in such a way that the critical limit given in Eq. (13) would not hold. Q_J would no longer be required to be $\lesssim 2 \times 10^6$. I next give examples of how such a process might have worked.

Suppose Io was originally solid with $k_1 = 0.027$ and $Q_1 = 1000$. Peale et al. assumed that $Q_1 = 100$ for a solid Io, but the scenarios I discuss in this section require a larger value; 1000 is not out of the question. In this case with $Q_J = 10^7$, $D = 5000$. Suppose capture into resonance occurred early in the history of the solar system. According to Eq. (54) (remember that $e_{1f} \propto 1/\nu_{11}$), e_{1f} increases toward the equilibrium value 0.004 exponentially on a time scale of a few 10^9 yr, as in curve A of Fig. 3.5. As e_1 nears this value, the tidal heating rate in Io is $\sim 2 \times 10^{11}$ W. If the radiogenic heating ($\sim 10^{12}$ W) is already on the verge of melting Io, this amount of tidal heat could be enough to initiate runaway melting which, once it starts, might occur on a time scale of $\sim 10^8$ yr. The increase in Io's tidal heating (η_1 increasing to ~ 10) means a rapid increase in D by at least an order of magnitude. There is a corresponding shift in the equilibrium value of e_1. (In Fig. 3.5, the equilibrium line would suddenly move upward.) If this melting occurred recently (in the last few 10^8 yr), e_1 would not have had time to come (moving upward on curve B of Fig. 3.5 with an exponential time scale of 10^9 yr) to its new smaller equilibrium value. In this way D might be much greater than $1/(13e_{1f}^2)$ in contrast to the constraint of Eq. (13) imposed by Yoder's model; the need for $Q_J \lesssim 2 \times 10^6$ would no longer exist.

An episodic melting scenario might begin with the same initial conditions just described, but the system must be in resonance initially with e_1 almost as large as 0.004, so that the runaway melting is triggered immediately. D increases by an order of magnitude in $\sim 10^8$ yr, the melting time scale. Then e_{1f} decreases toward the correspondingly low new equilibrium value on a time scale $\sim 10^9$ yr. When e_1 gets small enough, I speculate that rapid runaway refreezing might occur: As the heating slows due to smaller e_1, the crust thickens, so k decreases and heating slows still further. D decreases to its original value and the cycle begins again. (This behavior can be visualized in terms of Fig. 3.5. First the system evolves along curve A, but the equilibrium line suddenly jumps upward. Then the system evolves upward on a curve like B toward this new equilibrium line until the equilibrium line suddenly jumps downward again, completing one cycle.) If a melting episode has occurred recently, this scenario also reconciles the present value of e_{1f}, plausible values of Q_J, and the observed volcanism. However, the episodic melting scenario remains speculative until more detailed models of the melting and refreezing process are analyzed (see Chapter 4 by Cassen et al.).

V. PRIMORDIAL DEEP RESONANCE

The scenarios discussed in Secs. III and IV, with the system originally further from exact commensurability than it is today, have various apparent shortcomings: Yoder's model requires small Q_J; the recent melting model requires special conditions; the episodic melting model is highly speculative. Even if that class of scenario is ultimately shown to be viable, we should consider an alternate hypothesis, specifically, that the system began even deeper in resonance than it is now, i.e., with all the ν's smaller and the forced e's larger than they are today (curve B, Fig. 3.5).

Peale and Greenberg (1980) used Eq. (54) to study the evolution from deeper in the resonance. As before, e_{1f} approaches its equilibrium value $(1/\sqrt{13D})$ but now from initially higher values. For various initial values of ν_{11}, and corresponding initial values of e_{1f}, we integrated Eq. (54) to find the values of Q_J and Q_1/η_1 that give the present e_{1f} after 4.6×10^9 yr.

The results are shown for three representative values of the initial e_{1f}, called e_{1_0} (Fig. 3.6). For e_{1_0} equal to the present value of e_{1f} (0.004), the system must have started and remained in equilibrium; thus, $D = 4600$, represented by the diagonal line in Fig. 3.6. For larger e_{1_0}, if Q_1/η_1 is very small, the system must have reached equilibrium very quickly, according to Eq. (54). Hence the same diagonal line pertains in this case. On the other hand, if Q_1/η_1 is large enough, the system stays well out of equilibrium throughout the evolution, so the evolution is independent of Q_J. Hence the loci in Fig. 3.6 become vertical.

We next compute the energy dissipated in Io during the evolution. Before equilibrium is approached $\dot{\nu}_{11} = 4D c n_1 e_1^2$ from Eq. (54). Dividing the energy variation, (Eq. 14) by $\dot{\nu}_{11}$ yields

$$\Delta E = \frac{1}{2} m_1 n_1 a_1^2 \Delta\nu_{11}. \tag{15}$$

So if $\Delta\nu_{11}$ changes from a small value to near its current value, $0°.9/day$, we find $\Delta E = 4 \times 10^{28}$ J, independent of the value of Q_1/η_1. After the system

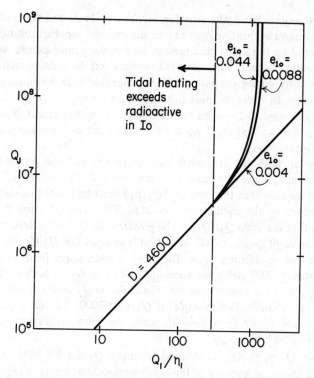

Fig.3.6. Evolution starting deeper in resonance (curve B, Fig. 3.5). At correspondingly large original values of e_{1f}, the solid curves show values of Q_J and Q_1/η_1 that lead to the present e_{1f}. A limit on Q_1/η_1 yields a limit on Q_J. (After Peale and Greenberg 1980.)

becomes close to equilibrium ($e_1 \sim 1/\sqrt{13D}$), the energy equation (Eq. 14) yields the following additional energy change over several billion years:

$$\Delta E = 10^{19} \text{ J} \times (\eta_1/Q_1)e_{1f}^2 = 3 \times 10^{31} \text{ J} \times (\eta_1/Q_1). \qquad (16)$$

In order for tidal heating to exceed total radiogenic heating (assumed chondritic, $\sim 1.3 \times 10^{29}$ J) Q_1/η_1 must be < 300. Peale and Greenberg (1980) took this as a necessary constraint to explain the great thermal activity on Io relative to the Moon. With this constraint on Q_1/η_1, we see from Fig. 3.6 that Q_J must be $< 5 \times 10^6$ or, alternatively, the electric current I must be $\gtrsim 6 \times 10^7$ amp. Since measured values of I are less than 6×10^7 amp (Sec. II), Peale and Greenberg concluded that the scenario of originally deep resonance requires $Q_J < 5 \times 10^6$.

On the other hand, the criterion invoked by Peale and Greenberg based on total integrated heating may not be the relevant one. Perhaps more to the point would be to require that present tidal heating *rates* exceed radioactive heating *rates,* because the observed activity on Io reflects energy input predominantly during the past $\lesssim 10^9$ yr rather than over the entire age of the solar system. In order to meet this constraint we require $Q_1/\eta_1 < 100$, the same requirement as I applied to the resonance capture model. Now reading from Fig. 3.6, we find $Q_J < 1.5 \times 10^6$, exactly the same limit as was implied by the capture model.

Why do both the resonance capture model and the primordial deep resonance model give the same constraint on Q_J? The former (cf. curve A, Fig. 3.5) requires that the ratio $Q_J/(Q_1/\eta_1)$ must be less than a certain fixed value defined by the equilibrium condition. The latter (cf. curve B, Fig. 3.5) requires that the ratio $Q_J/(Q_1/\eta_1)$ be greater than that same fixed value, but, for any value of $Q_1/\eta_1 < 300$, evolution is so rapid that $Q_J/(Q_1/\eta_1)$ must be very near the equilibrium value. Therefore, a given upper limit on Q_1/η_1 (as long as it is < 300) yields the same upper limit on Q_J for both models. Note that if Q_1/η_1 were greater than 300, the limits on Q_J would be very different for the two models. For example, if $Q_1/\eta_1 = 3000$, the resonance capture model would imply $Q_J < 5 \times 10^7$ while the primordial resonance model would place no limit on Q_J.

Since $Q_1/\eta_1 \lesssim 100$, both models imply $Q_J < 1.5 \times 10^6$. As I have discussed above, acceptance of the capture hypothesis seems to require that a mechanism be found to dissipate energy in Jupiter at such a rate. On the other hand, the primordial resonance scenario may work even if Q_J proves to be truly greater than 10^7, for the following reason. Any process that might have slowed the evolution from the early deep resonance toward equilibrium (curve B, Fig. 3.5) would imply that the system is presently further from equilibrium than computed by Peale and Greenberg, based on Eq. (54). In terms of Fig. 3.6, the turn-up of the Q_J versus Q_1/η_1 curves would occur further to the left. If the turn-up were shifted leftward in Fig. 3.6 by one order of magnitude, it would be possible to have $Q_1/\eta_1 < 100$ and $Q_J \sim 5 \times 10^7$, so the primordial resonance model would be viable. There are several factors that may have tended to slow the evolution compared with the rate implied by Eq. (54):

1. *Coefficient in Eq. 54.* Derivation of Eq. (54) (reduction of the nine governing equations to a pendulum equation) required the assumption that the circulation rate ν is much faster than the pendulum libration frequency. That restriction is satisfied now, albeit barely. It certainly is satisfied throughout the scenario studied by Yoder, with ν's greater than their present values. However, for the scenario considered here, the ν's were originally smaller; the restriction would surely not have been

satisfied. In order to investigate this scenario in which the system was once deeper in resonance, we need to consider the complete set of governing equations, including tidal terms. I have performed such a solution with the result that Eq. (54), which described the long-term evolution of Io's forced eccentricity and of v (Fig. 3.5), has been modified. The equilibrium solution $(e_{1f} = (13D)^{-1/2})$ remains unchanged, but the coefficient that was 0.32 in Eq. (54) is now a function $F(v)$. For $v > 0.8/\text{day}$, $F(v) = 0.32$, but for smaller values of v, $F(v)$ is smaller as shown in Fig. 3.7. So for evolution starting close to the exact commensurability, the evolution (curve B, Fig. 3.5) goes more slowly than Eq. (54) indicates.

2. *Variable Q_1/η_1.* The value of Q_1/η_1 may well have been larger in the past than at present. In the past, heating rates were probably several times what they are now. Radioactive elements were more abundant, and the higher forced eccentricities would have tended to make tidal heating greater as well. Such extra heat would decrease the thickness of the solid crust relative to the present thickness, according to the conductive equilibrium model of Peale et al. (1979). Hence there would be less volume in which elastic dissipation could occur and Q_1/η_1 would be larger. (The increase in the Love number k, which would tend to decrease Q_1/η_1, would be negligible.)

3. *Uncertainty about behavior in deep resonance.* No one has yet considered initial states with v's $\lesssim 0.1/\text{day}$. There is no basis for ruling out the possibility that evolution began that deep in resonance (the shaded zone in Fig. 3.5), or even on the other side of the exact

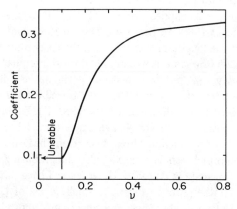

Fig. 3.7. The coefficient $F(v)$ replacing the coefficient 0.32 in Eq. (54). Only if $v \gtrsim 0.8/\text{day}$, is $F(v) = 0.32$. The lower values for small v represent a slower evolution than given by Eq. (54).

commensurability (negative ν's). Such cases have been neglected to date only because of analytical difficulties. When the ν's are small, forced e's are large, so the theory must be carried out to high orders in e. (In fact, Peale and Greenberg probably pushed the theory a bit too far when they considered the case $e_{1_0} = 0.044$ [Fig. 3.6].) If initial conditions were in the zone of small ν's (large forced e's) or negative ν's, evolution to the present value of ν would have been considerably delayed. The Q_J versus Q_1/η_1 loci in Fig. 3.6 might lie well leftward of the curves shown.

Are these three factors sufficient to reconcile the primordial resonance hypothesis with $Q_J > 10^7$? Factors (1) and (2) account for a substantial step towards that reconciliation; I find from preliminary calculations that a reasonable estimate of the magnitude of the variation in Q_1/η_1, combined with the correction in the coefficient of Eq. (54), shifts the turn-up in Fig. 3.6 leftward by half an order of magnitude. It seems quite plausible that factor (3) will contribute the remaining necessary shift, but the theory of the resonance must be extended to higher orders in eccentricities before we know for sure.

Further work is also needed on the model of variation of Q_1/η_1 with time (factor 2, above); I used the conductive equilibrium model in evaluating factor (2). On the other hand, there are complications that could significantly affect those calculations. If most of the heat escapes through local hot spots, as indicated by the observations of heat flux (see Chapter 19), the crust could be considerably thicker than given by the conductive equilibrium model. Deep cracks suggested by the observed local hot spots might serve as hinges to permit tidal flexing with less dissipation than Peale et al. (1979) computed from elastic bending. How such processes and others might have affected the variation of Q_1/η_1 with time remains to be studied.

If the value $Q_1/\eta_1 \sim 3$ implied by recent heat flow measurements (Matson et al. 1980; Pearl 1980; Morrison and Telesco 1980; Sinton 1981) is truly representative of the long-term average value, then according to Fig. 3.6 $Q_J < 5 \times 10^4$. This value (the same as was obtained for the resonance capture scenario) is unacceptably low, not only because it violates estimates of Q_J based on models of Jupiter, but because it violates as well the absolute lower limit ($Q_J = 10^5$) obtained by Goldreich and Soter (Sec. II). We ran into this same problem when considering the resonance capture model (Sec. III). In that context we could only conclude that the present heat flow must be anomalous. In the primordial resonance model also the problem goes away if we interpret the heat flow data in that way. However, as I have shown, in this model there may be another way to reconcile small values of Q_1/η_1 with reasonable values of Q_J. Conceivably, given our uncertainty about behavior in deep resonance (factor 3), the system may not yet have reached the equilibrium state. This suggestion would require that the system, having begun with small or negative values of ν ($< 0°.1/\text{day}$), retained such values until quite recently ($\lesssim 10^7$ years ago). Thus the heat flow measurements seem to require that something be special about the present time, either that the heating rate is anomalously high, or that the system has only recently come out of deep resonance.

The latter explanation could be tested by measurement of Io's secular acceleration (dn_1/dt). If the system is indeed well out of equilibrium at present, then according to Eq. (54) $D e_1^2 >> 1/13$. In that case dn_1/dt should be dominated by tides raised on Io (Eq. 11) rather than by tides raised on Jupiter (Eq. 10). In fact, direct determination of Io's orbital acceleration by de Sitter (1928; see also Brouwer and Clemence 1961b) based on the historical record of observations since the mid-1600s, yielded $\dot{n}_1/n_1 \approx 10^{-17}$ s^{-1}, exactly the rate that corresponds to $Q_1/\eta_1 \approx 3$, the value implied by the thermal measurements. De Sitter himself (1931) and Goldstein (1975) rejected de Sitter's determination because they felt that $\dot{n}_1$ should be negative; they assumed that tides on Jupiter should dominate over tides on Io. Now, with the primordial resonance scenario discussed in this chapter and with our raised consciousness of the importance of tides on Io, de Sitter's value does not seem necessarily unreasonable. On the other hand, there may be problems with de Sitter's determination due to neglected variations in the Earth's rotation, which served as the clock for timing historical data (Goldstein 1975).

The primordial resonance hypothesis implies that Io's tidal heating rate was greater earlier in the evolution, due to the large forced eccentricity. For a conservative example, if ν were once $0°.1/\text{day}$ and Q_1/η_1 were 1000, from Eq. (14) $\dot{E}_{1o} = 4 \times 10^{12}$ W, an order of magnitude greater than present radioactive rates, twice original radioactive rates, and certainly sufficient for melting according to the model of Peale et al. (1979).

In Europa, assuming $Q_2/\eta_2 = 100$ and using the present value of e_{2f}, Peale et al. (1979) find the dissipation rate 8×10^{10} W. If ν were $0°.1/\text{day}$ ($e_{2f} = 0.08$) early, the rate would have been 5×10^{12} W. If ν were once even smaller, as appears to be needed to reconcile the primordial resonance scenario with $Q_J > 10^7$ (factor [3] above), the tidal heating rate would have been even greater. Such heating would have helped delay the freezing of Europa's water mantle, a possible explanation for the peculiarly bland topography and absence of craters on Europa (see Chapter 14 by Lucchitta and Soderblom).

Cassen et al. (1980) compute the required tidal heating to keep Ganymede or Callisto melted, if they started out that way. The requirement is $\sim 8 \times 10^{13}$ W. For tides on Ganymede, that value corresponds to e_3 of 0.037 (assuming $Q_3/\eta_3 = 100$). (Cassen et al.'s argument that even an initial eccentricity of 0.1 would not have kept Ganymede melted long is irrelevant

TABLE 3.2

Constraints On Q_J Imposed by Two Scenarios

Scenario Q_J/η_1	Resonance Capture	Primordial Deep Resonance
< 100 (as volcanism implies)	$Q_J < 1.5 \times 10^6$	$Q_J < 1.5 \times 10^6$ or possibly no constraint
~ 3 (heat flux measurements)	$\begin{cases} Q_J > 10^5 \\ Q_J < 5 \times 10^4 \end{cases}$	$\begin{cases} Q_J > 10^5 \\ Q_J < 5 \times 10^4 \text{ or recent escape} \\ \text{from deep resonance} \end{cases}$

in this situation; they have ignored *forced* eccentricity because it was small in the case they considered.) For the resonance to have forced such a large eccentricity for Ganymede, ν would have had to have been $0°.02/\text{day}$. Such a deep resonance would be beyond the range of applicability of first order theory because $e_2 \sim 0.1$; again I emphasize the need to extend the theory to higher order in eccentricities. However, we can see that tidal heating in primordial deep resonance could have delayed freezing long enough to permit the apparent great tectonic activity (plate motions and grooving) on Ganymede relative to Callisto which had no such tidal heat source (see Chapter 13 by Shoemaker et al.).

VI. CONCLUSION

The implications of the resonance capture scenario and of the primordial deep resonance scenario are summarized in Table 3.2. If $Q_1/\eta_1 < 100$, as seems certain from observed volcanism, the first scenario requires $Q_J < 1.5 \times 10^6$. The second scenario will require that limit only if the plausible evolution-delaying factors described in the previous section fail to prove significant. Since $Q_J > 10^7$ according to present physical models of Jupiter, there is perhaps a slight bias in favor of the primordial deep resonance scenario.

If the value $Q_1/\eta_1 \sim 3$ shown by heat flow measurements is not anomalous, the resonance capture model places conflicting constraints on Q_J, which would seem to rule this scenario out completely. The primordial deep resonance model, on the other hand, may admit plausible Q_J values, if the system evolved out of deep resonance in the past ten million years. Again I think the primordial resonance model seems slightly favored. The primordial resonance model moreover gives early heating of Io, Europa and Ganymede, which can help explain their various properties and the contrast with the surface of nonresonant Callisto.

There is another reason for favoring the primordial resonance model. The large number of commensurabilities in the solar system suggests that planets and satellites may tend to form in resonance. (The newly announced fifteenth Jovian satellite, 1979 J2 [Synnott 1980] nearly fits the following relation with Io [J1] and Amalthea [J5] : $n_1 + n_{15} - n_5 = 0$. The dynamical significance, if any, needs consideration.) Most resonances cannot be explained by tidal evolution. We still do not understand in detail the mechanism that favors formation near resonances, but some of the same properties of resonances discussed in this chapter probably played important roles in the process of planet and satellite growth (Greenberg et al. 1978; Greenberg 1978).

Definitive selection of any particular model of formation and evolution of the Galilean satellites would be premature. Orbital evolution is intimately coupled with coeval physical properties. Therefore the evolution models I have discussed represent definition of and tentative manipulation of some of the pieces of the puzzle. These models give some examples of how the pieces can be fitted together, while other chapters in this book will define and arrange other portions of the puzzle. I anticipate with excitement that the pieces of the puzzle discussed here may prove ultimately to be arranged quite differently than I have suggested. In this way the complete picture of the Jovian satellites may emerge.

Acknowledgments. I thank K. Blasius, J. Burns, P. Cassen, C. Chapman, D. Davis, W. Hubbard, S. Peale, R. Reynolds and C. Yoder for sharing their ideas and insights with me, and A. Hostetler for her patient help with computations. My work on this chapter was principally supported by the Planetary Astronomy Program of the National Aeronautics and Space Administration, and in part by their Geophysics and Geochemistry program and the Galileo project. The Planetary Science Institute is a division of Science Applications, Inc.

October 1980

APPENDIX A

Derivation of the Pendulum Equation

The equations for variation of orbital elements are obtained by Fourier-expanding the gravitational potential at each satellite due to each other one (cf. Brouwer and Clemence 1961a). Most of the terms have short periods so their effects can be ignored. But I retain the terms of form $\cos \theta$ where θ's are the slowly varying (near the 2:1 commensurability) quantities defined by Eqs. (4) - (7) (see Sec. I above). I ignore terms higher than first order in e; the results are probably reasonable if all e's are < 0.1.[a] The variation equations for mean motions give

$$\dot{h}_1 = -3\mu_2 n_2{}^2\alpha^{-2} \left(C_1 e_1 \sin \theta_{11} + C_2 e_2 \sin \theta_{12}\right) \tag{17}$$

$$\dot{n}_2 = 6\mu_1 n_2{}^2 \left(C_1 e_1 \sin \theta_{11} + C_2 e_2 \sin \theta_{12}\right)$$
$$- 3\mu_3 n_3{}^2 \alpha^{-2} \left(C_1 e_2 \sin \theta_{22} + C_2 e_3 \sin \theta_{23}\right) \tag{18}$$

$$\dot{n}_3 = 6\mu_2 n_3{}^2 \left(C_1 e_2 \sin\theta_{22} + C_2 e_3 \sin \theta_{23}\right) \tag{19}$$

where μ_i is the ith satellite's mass expressed in units of Jupiter's mass, $\alpha \approx 0.63$ is the nearly constant ratio of semimajor axes, and the C's are functions of α with $C_1 \approx -1.18$ and $C_2 \approx 0.42$ (Yoder 1979). The variations of eccentricities and apsides are

$$\dot{e}_1 = -\mu_2 n_1 \alpha C_1 \sin \theta_{11} \tag{20}$$

$$\dot{\tilde{\omega}}_1 = \mu_2 n_1 \alpha C_1 / e_1 \cos \theta_{11} + \dot{\tilde{\omega}}_{1s} \tag{21}$$

$$\dot{e}_2 = -\mu_1 n_2 C_2 \sin \theta_{12} - \mu_3 n_2 \alpha C_1 \sin \theta_{22} \tag{22}$$

$$\dot{\tilde{\omega}}_2 = (\mu_1 n_2 C_2 / e_2) \cos \theta_{12} + (\mu_3 n_2 \alpha C_1 / e_2) \cos \theta_{22} + \dot{\tilde{\omega}}_{2s} \tag{23}$$

$$\dot{e}_3 = -\mu_2 n_3 C_2 \sin \theta_{23} \tag{24}$$

$$\dot{\tilde{\omega}}_3 = \mu_2 n_3 C_2 / e_3 \cos \theta_{23} + \dot{\tilde{\omega}}_{3s} \tag{25}$$

---

[a] A recent evaluation of higher order terms in e by Peale (personal communication), suggests that the first order theory is valid up to this limit, but not beyond it. Further work on this problem is needed and is being continued by Peale, Yoder and myself.

where the subscript s denotes variations due to secular (nonperiodic) terms in the distrubing function. Secular terms include Jupiter's oblateness well as secular components of the satellites' mutual disturbing potential. According to Chao (1976), $\dot{\tilde{\omega}}_{1s} = 0.16/\text{day}$, $\dot{\tilde{\omega}}_{2s} = 0.04/\text{day}$, and $\dot{\tilde{\omega}}_{3s} = 0.01/\text{day}$ (these values include a second-order correction due to the resonance terms). I assume here that these secular rates are constants while, in fact, they may have varied with time, as Jupiter's oblateness changed and the satellites' distances from Jupiter varied; the consequences need study.

These equations can be put into a more nearly linear form that avoids the inconvenience of a small quantity in the denominator by replacing the e's and $\tilde{\omega}$'s by h's and k's defined in the following way:

$$h_{11} \equiv e_1 \sin \theta_{11}, k_{11} \equiv e_1 \cos \theta_{11} \tag{26}$$

$$h_{12} \equiv e_2 \sin \theta_{12}, k_{12} \equiv e_2 \cos \theta_{12} \tag{27}$$

$$h_{22} \equiv e_2 \sin \theta_{22}, k_{22} \equiv e_2 \cos \theta_{22} \tag{28}$$

$$h_{23} \equiv e_3 \sin \theta_{23}, k_{23} \equiv e_3 \cos \theta_{23}. \tag{29}$$

Equations for variation of the h's and k's can replace the equations (20-25) for e's and $\tilde{\omega}$'s. We obtain these by differentiating Eqs. (19) - (22) and applying (13) - (18) yielding, for example,

$$\dot{h}_{11} = -\mu_2 n_1 \alpha C_1 - \nu_{11} k_{11} \tag{30}$$

$$\dot{k}_{11} = \nu_{11} h_{11} \tag{31}$$

where $\nu_{11} \equiv -(2n_2 - n_1 - \dot{\tilde{\omega}}_{1s}) \approx 0.9/\text{day}$. (This definition is one refinement of the quantity ν discussed after Eq. (9) in Sec. I. Subscript notation for ν's follows the example of θ's.) If ν_{11} is nearly constant, these equations are linear and the solution is

$$\begin{cases} h_{11} = -A_{11} \sin (\nu_{11}t + \Delta_{11}) & \tag{32} \\ k_{11} = A_{11} \cos (\nu_{11}t + \Delta_{11}) - \mu_2 \alpha C_1 n_1 / \nu_{11} & \tag{33} \end{cases}$$

where A_{11} and Δ_{11} are constants of integration.

Let us compare this solution with the behavior shown in Fig. 3.3. The (h_{11}, k_{11}) vector behaves just as the (h, k) vector does in Fig. 3.3. Its magnitude is Io's eccentricity. From this comparison A_{11} is identified as Io's free eccentricity (e_{1p}) and the second term in k_{11} is Io's forced eccentricity (e_{1f}). Note that e_{1f} is proportional to μ_2 but goes inversely as ν_{11}, i.e., e_{1f} would be greater if Europa were more massive or if the system were closer to the exact commensurability. From observations of the actual satellite motion we find $e_{1p} < e_{1f}$ (Table 3.1), consistent with θ_{11} librating about 0 (Eq. 4), as shown in Fig. 3.3.

A solution similar to Eqs. (32) and (33) is obtained for h_{23} and k_{23}, which describes the behavior of Ganymede's eccentricity and of θ_{23}. For Ganymede $e_{3p} > e_{3f}$ (Table 3.1) consistent with circulation of θ_{23} through $360°$ (Eq. 7).

For h_{22} and k_{22} the equations are more complicated:

$$\left\{ \begin{array}{ll} \dot{h}_{22} = -K_1 + K_2 \cos \phi - \nu_{22}k_{22} & (34) \\ \dot{k}_{22} = \qquad -K_2 \sin \phi + \nu_{22}h_{22} & (35) \end{array} \right.$$

where $K_1 = \mu_3 n_2 \alpha C_1 < 0$, $K_2 = -\mu_1 n_2 C_2 < 0$, $\nu_{22} = -(2n_3 - n_2 - \tilde{\dot{\omega}}_{2s}) \approx 0°.8/\text{day}$. Recall that $\phi \equiv \theta_{22} - \theta_{12} \equiv \lambda_1 - 3\lambda_2 - 2\lambda_3$. There is a similar set of equations for h_{12} and k_{12}:

$$\left\{ \begin{array}{ll} \dot{h}_{12} = K_2 - K_1 \cos \phi - \nu_{12}k_{12} & (36) \\ \dot{k}_{12} = \qquad -K_1 \sin \phi + \nu_{12}h_{12} & (37) \end{array} \right.$$

where $\nu_{12} = -(2n_2 - n_1 - \tilde{\dot{\omega}}_{2s})$.

The solution is straightforward assuming the ν's and ϕ to be nearly constant on the right sides of the $\dot{h}$ and $\dot{k}$ equations. This assumption, consistent with the present state of the system, permits the definition $\nu \equiv \nu_{12} = \nu_{22}$. For h_{22} and k_{22},

$$\left\{ \begin{array}{ll} h_{22} = (K_2/\nu) \sin \phi - A_{22} \sin (\nu t + \Delta_{22}) & (38) \\ \\ k_{22} = (K_2/\nu) \cos \phi + A_{22} \cos (\nu t + \Delta_{22}) - K_1/\nu & (39) \end{array} \right.$$

where A_{22} and Δ_{22} are integration constants. This solution describes behavior of the (h_{22}, k_{22}) vector whose magnitude is Europa's eccentricity e_2. Just as for the other (h, k) behavior described earlier, this solution vector is the sum of (i) a constant vector whose magnitude is identified as the forced eccentricity e_{2f} and (ii) a circulating vector of magnitude A_{22}, identified as the free eccentricity e_{2p}. We know $\phi \approx 180°$, so e_{2f} is $-(K_1 + K_2)/\nu$; just as for the other satellites, the forced eccentricity would be large near the exact commensurability (ν small). As shown in Table 3.1, $e_{2f} > e_{2p}$, so θ_{22} librates about 0, which is consistent with observation (see Eq. 6).

The solution for h_{12} and k_{12} is

$$
\begin{cases}
h_{12} = (K_1/v) \sin \phi - A_{12} \sin (vt + \Delta_{12}) & (40) \\
k_{12} = (K_1/v) \cos \phi + A_{12} \cos (vt + \Delta_{12}) + (K_2/v). & (41)
\end{cases}
$$

This solution is not independent of Eqs. (38) and (39) since the magnitude of both the (h_{12}, k_{12}) and the (h_{22}, k_{22}) vectors that equal e_2. To satisfy this constraint, $A_{12} = A_{22}$.

I next demonstrate the stability of ϕ at $180°$ following much the same argument at Laplace (1805) used. Differentiating Eq. (8) twice with respect to time yields

$$
\phi = \dot{n}_1 - 3\dot{n}_2 + 2\dot{n}_3 \tag{42}
$$

Substitution from Eqs. (17), (18), and (19) gives

$$
\phi = X_1 (C_1 h_{11} + C_2 h_{12}) + X_2 (C_1 h_{22} + C_2 h_{23}) \tag{43}
$$

where $X_1 \equiv -3\mu_2 n_2{}^2 \alpha^{-2} - 18\mu_1 n_2{}^2$ and $X_2 \equiv 9\mu_3 n_3{}^2 \alpha^{-2} + 12\mu_2 n_3{}^2$.

The solutions for the h's obtained above, e.g., Eqs. (38) and (40), can be inserted into Eq. (43). In the expressions for the h's, the oscillation rates (the v's) are fast compared to any significant change in ϕ, so we need insert only the mean values of the h's into Eq. (43), which yields

$$
\begin{aligned}
\phi &= [(X_1 C_2 K_1 + X_2 C_1 K_2)/v] \sin \phi \\
&= [0.036 \, (\text{deg/day})^3 /v] \sin \phi. \tag{44}
\end{aligned}
$$

Thus ϕ behaves like a pendulum, with stability at $180°$ and a libration frequency of $\sim 0°\!.2/\text{day}$.

Bear in mind that this derivation requires the v's to be considerably greater than $0°\!.2/\text{day}$, a constraint only barely satisfied at present with $v \approx 0°\!.9/\text{day}$. The constraint would not have been satisfied if, earlier in history, the system had been closer to the exact commensurability (smaller v's). Such a scenario is discussed in Sec. V.

APPENDIX B

Application and Modification of the Pendulum Equation

From the time the first commensurability is reached, θ_{11} and θ_{12} are slowly varying quantities, so Eqs. (30) and (31) for the behavior of e_1 apply. Adding terms for damping of e_1 due to tides on Io, we have

$$\dot{h}_{11} = -\mu_2 n_1 \alpha C_1 - \nu_{11} k_{11} - \frac{7}{3} Dc h_{11} \tag{45}$$

$$\dot{k}_{11} = \qquad\qquad + \nu_{11} h_{11} - \frac{7}{3} Dc k_{11}. \tag{46}$$

The solution is

$$h_{11} = -A_{11} \exp[-(7Dc/3)t] \sin (\nu_1 t + \Delta_{11}) - \frac{\mu_2 \alpha C_1 n_1 (7Dc/3)}{[(7Dc/3)^2 + \nu_{11}^2]} \tag{47}$$

$$k_{11} = A_{11} \exp[-(7Dc/3)t] \cos (\nu_1 t + \Delta_{11}) - \frac{\mu_2 \alpha C_1 n_1 \nu_1}{[(7Dc/3)^2 + \nu_{11}^2]} \tag{48}$$

Compare this (h_{11}, k_{11}) vector with the tide-free solution for the eccentricity behavior Eqs. (32) and (33), and with Fig. 3.3. The free eccentricity e_{1p} can again be identified as the coefficient of the cos and sin in the solution. We see that it is now damped exponentially. The forced eccentricity, e_{1f}, is the magnitude of the vector given by the second terms in Eqs. (47) and (48). To first order in Dc, e_{1f} is the same as in the tide-free case. An important difference between this solution and the tide-free solution is that the mean value of h_{11} is no longer zero. Rather it is given by the second term in Eq. (47), which is $(7/3) Dc\, e_{1f}/\nu_{11}$ to first order in Dc. This nonzero value contributes a term to both $\dot{n}_1$ and $\dot{n}_2$ via their dependence on h_{11} (cf. Eqs. (17) and (18), and recall $h_{11} \equiv e_1 \sin \theta_{11}$), in addition to the direction effects of Jupiter tides on $\dot{n}_1$. Tidal variation (denoted by subscript t) is thus

$$\dot{n}_{1t} = c\, n_1 (1 - 14D\, e_{1f}^2) \tag{49}$$

$$\dot{n}_{2t} = -10.3\, c\, n_1 D\, e_{1f}^2 \tag{50}$$

where known constants have been evaluated numerically.

As the three-body commensurability is approached, the pendulum equation applies. From Eq. (42), including tides, we obtain

$$\phi = A \sin \phi - c\, n_1 (1 - 45D\, e_{1f}^2) \tag{51}$$

where A is the same coefficient as in the tide-free pendulum equation (44), but there is now an additional tidal term. Equation (51) describes a pendulum with a small constant torque, which acts to offset equilibrium of ϕ slightly from $180°$. The equilibrium value about which ϕ librates is how

$$\sin \phi = c\, n_1\, (1 - 45D\, e_{1f}^2)/A. \tag{52}$$

A similar procedure to that which gave us the pendulum equation yields

$$\dot{\nu}_{11} = \dot{n}_1 - 2\dot{n}_2 = +0.68\, A \sin \phi - c\, n_1\, (1 - 34.6D\, e_{1f}^2). \tag{53}$$

Following Yoder (1979) we can take Eq. (52) as the average value of $\sin \phi$ and insert it into Eq. (53) to obtain

$$\dot{\nu}_{11} = -0.32\, c\, n_1\, (1 - 13D\, e_{1f}^2). \tag{54}$$

Equation (54) has important implications for the tidal evolution of the system as discussed in Sec. III.

Tides introduce an additional important term in the pendulum equation. Note that e_{1f} in Eq. (51) is not necessarily constant since it goes as ν_{11} which varies according to Eq. (53). The change in e_{1f}^2 is

$$\delta(e_{1f}^2) = \frac{\partial(e_{1f}^2)}{\partial \nu_{11}}\, \delta \nu_{11} = -\frac{2e_{1f}^2}{\nu_{11}}\, \delta \nu_{11}. \tag{55}$$

Comparing Eqs. (51) and (53) and ignoring the tidal terms, yields $\delta \nu_{11} \approx 0.68\, \dot{\phi}$. Hence the change in the tidal part of Eq. (51), to first order in tidal effects, is

$$-61\, D\, c\, e_{1f}^2 n_1 \nu_{11}^{-1}\, \dot{\phi}. \tag{56}$$

This term added to the right-hand side of Eq. (51) has the form of libration-damping friction in the pendulum.

October 1980

REFERENCES

Brouwer, D., and Clemence, G.M. (1961a). *Methods of Celestial Mechanics.* Academic Press, N.Y.

Brouwer, D., and Clemence, G.M. (1961b). Orbits and masses of planets and satellites. In *Planets and Satellites* (G.P. Kuiper, Ed.), p. 88. Univ. Chicago Press, Chicago.

Cassen, P., Peale, S.J., and Reynolds, R.T. (1980). On the comparative evolution of Ganymede and Callisto. *Icarus* 41, 232-239.

Chao, C.C. (1976). *A General Perturbation Method and its Application to the Motion of the Four Massive Satellites of Jupiter.* Ph.D. Dissertation, Univ. California at Los Angeles.

Dermott, S.F. (1979). Tidal dissipation in the solid cores of the major planets. *Icarus* 37, 310-321.

de Sitter, W. (1928) Orbital elements determining the longitudes of Jupiter's satellites, derived from observations. *Leiden Annals* 16, part 2.

de Sitter, W. (1931). Jupiter's Galilean satellites. *Mon. Not. Roy. Astron. Soc.* 91, 706-738.

Goldreich, P. (1965). An explanation of the frequent occurrence of commensurable mean motions in the solar system. *Mon. Not. Roy. Astron. Soc.* 130, 159–181.

Goldreich, P., and Nicholson, P.D. (1977). Turbulent viscosity and Jupiter's tidal Q. *Icarus* 30, 301-304.

Goldreich, P., and Soter, S. (1966). Q in the solar system. *Icarus* 5, 375-389.

Goldstein, S.J. (1975). On the secular change in the period of Io, 1668-1926. *Astron. J.* 80, 532-539.

Greenberg, R. (1976). The motions of satellites and asteroids: National probes of Jovian gravity. In *Jupiter* (T. Gehrels, Ed.)., pp. 122-132, Univ. Arizona Press, Tucson.

Greenberg, R. (1978). Orbital resonance in a dissipative medium. *Icarus* 33, 62-73.

Greenberg, R., Hartmann, W.K., Chapman, C.R., and Wacker, J.F. (1978). The accretion of planets from planetesimals. In *Protostars and Planets* (T. Gehrels, Ed.), pp. 599-622. Univ. Arizona Press, Tucson.

Kaula, W.M. (1964). Tidal dissipation by solid friction and the resulting orbital evolution. *Rev. Geophys.* 2, 661-685.

Kivelson, M.G. Slavin, J.A., Southwood, D.J., and Walker, R.J. (1979). Interaction of Jovian planet with I. *EOS* 60, 920 (abstract).

Laplace, P.S. (1805). *Mercanique Celeste* 4, Courcier, Paris. Translation by N. Bowditch reprinted (1966). Chelsea Publishing Co., N.Y.

Lieske, J.H. (1980). Improved ephemerides of the Galilean satellites. *Astron. Astrophys.* 82, 340-348.

Lin, D.N.C., and Papaloizou, J. (1979). On the structure of circumbinary accretion disks and the tidal evolution of commensurable satellites. *Mon. Not. Roy. Astron. Soc.* 188, 191-201.

Matson, D.K., Ransford, G.A., and Johnson, T.V. (1980). Heat flow from Io. *Lunar Planet. Sci.* XI, 686-687 (abstract).

Morrison, D., and Telesco, C.M. (1980). Io: Observational constraints on internal energy and thermophysics of the surface. *Icarus* 44, 226-233.

Ness, N.F., Acuna, M.H., Lepping, R.P., Buriaga, L.F., Behannon, K.W., and Neubauer, F.M. (1979). Magnetic field studies at Jupiter by Voyager 1: Preliminary results. *Science* 204, 982-986.

Peale, S.J. (1977). Rotation histories of the natural satellites. In *Planetary Satellites* (J.A. Burns, Ed.), pp. 87-112. Univ. Arizona Press, Tucson.

Peale, S.J., and Greenberg, R. (1980). On the Q of Jupiter. *Lunar Planet. Sci.* XI, 871-873.

Peale, S.J., Cassen, P., and Reynolds, R.T. (1979). Melting of Io by tidal dissipation. *Science* 203, 892-894.

Pearl, J.C. (1980). The thermal state of Io on March 5, 1979. IAU Coll. 57, The Satellites of Jupiter (abstract 4-1).

Sinclair, A.T. (1975). The orbital resonance amongst the Galilean satellites of Jupiter. *Mon. Not. Roy. Astron. Soc.* 171, 59-72.

Sinton, W.M. (1981). The thermal emission spectrum of Io and a determination of the heat flux from its hot spots. *J. Geophys. Res.* In press.

Synnott, S.P. (1980). 1979 J2: Discovery of a previously unknown satellite of Jupiter. *Science.* In press.

Yoder, C.F. (1979). How tidal heating in Io drives the Galilean orbital resonance locks. *Nature* 279, 767-770.

4. STRUCTURE AND THERMAL EVOLUTION OF THE GALILEAN SATELLITES

PATRICK M. CASSEN
Ames Research Center

STANTON J. PEALE
University of California, Santa Barbara

and

RAY T. REYNOLDS
Ames Research Center

The dynamic nature of Io's surface is probably due to intense heating by tidal dissipation, but this conclusion still admits a variety of models for the interior. The total energy flux passing through the surface is an important constraint on these models. Theoretical limits on this flux can be obtained from estimates of the mechanical properties of the interior based on comparisons with other planetary bodies, estimates of resurfacing rates, and considerations of orbital mechanics. Ganymede and Callisto are comparable in size and mean density, but present remarkably different appearances. Without introducing ad hoc assumptions, this difference is most plausibly attributed to a small difference in radioactive content, together with a steeply decreasing impact rate early in their histories. The water mantles of both satellites are probably completely solid, unless they are significantly contaminated by dissolved salts or ammonia. The major uncertainties in thermal models of these bodies are due to uncertainties in the creep properties of ice and the degree of contamination of the water mantles. Tidal heating undoubtedly has contributed significantly to Europa's thermal history, but whether it is enough to maintain liquid water depends on the viscosity of the ice, the history of the orbital resonances, and the quantity of impurities in the ice.

Theoretical models of planetary interiors and their evolutions proceed from the constraints imposed by the basic properties of mass, size, and surface appearance, to considerations of features such as surface composition, atmosphere, gravitational field structure, topography, and magnetic field, finally to be tested by *in situ* measurements of seismic and thermal properties, and the analysis and dating of surface material. Only for the Earth and Moon do we have data of sufficient sensitivity to permit the construction of detailed structural models, and rock samples that serve as probes of the deep chemical and thermal properties. Even in these cases many uncertainties remain. Nevertheless, the wealth of information that has been obtained from both groundbased and spacecraft observations has permitted the development of tentative models of the structures and evolutions of all of the terrestrial planets, albeit with varying degrees of confidence. With the extensive observations of the Galilean satellites, particularly in the 1970s, culminating in the spectacular results of the Voyager mission, these bodies have assumed their anticipated role as prime objects for the application of theories of planetary structure and evolution.

As has often been remarked, the Galilean system is in many ways an analog of the solar system, so theoretical studies of satellite formation and evolution are relevant to solar system studies in general. (See, for instance, Pollack and Reynolds [1974], Cameron and Pollack [1976], and Chapter 24 by Pollack and Fanale, for discussions of the theory of the condensation of material and satellite formation in the Jovian system as compared with that in the solar system.) No doubt the processes of melting, differentiation, subsolidus convection, and volcanism are important for the histories of the Galilean satellites, as they have been for the more thoroughly studied terrestrial planets. There is surely much to be learned by studying the consequences of these processes in new and, in some cases, surprisingly different contexts.

Based on their sizes, the Galilean satellites could be classified as terrestrial-type planets. But their location in the outer solar system and their proximity to Jupiter have had profound effects on their compositions and histories, which distinguish them from those of the inner planets. Europa, Ganymede, and Callisto formed in an environment that was cold enough for large quantities of H_2O ice to condense and be incorporated in their structures; their subsequent thermal evolutions have been determined to a large degree by the thermodynamic and mechanical properties of H_2O. Europa and Io are close enough to Jupiter so that tidal dissipation, maintained by the action of commensurate orbits, has been important for each.

We first discuss Io, for which intense internal heating, apparently derived from the dissipation of tidal energy, dominates all discussion of its interior. We then skip to Ganymede and Callisto, whose similar gross properties invite combined treatment. Europa is dealt with last because its evolution has been influenced by factors relevant to both Io and the outer pair. In each case, our approach is to describe models based on the simplest assumptions and then identify the most important complications that are likely to exist and explore their consequences.

I. IO

A. Internal Energy Source

Io is similar to the Moon in mass and radius. On the basis of solar abundances, both bodies would be composed primarily of silicates with the higher mean density of Io presumably due to a higher total iron content. The remarkable surface activity manifested in plumes, abundant calderas (Morabito et al. 1979; Smith et al. 1979a; Chapter 15 by Masursky et al.; Chapter 16 by Strom and Schneider), and the rapid obliteration of impact craters (Johnson et al. 1979 and Chapter 17 by Johnson and Soderblom) then imply an extraordinary internal energy source, regardless of the details of energy transport within the planet. The probability that typical abundances of radioactive elements (those characteristic of the Earth, meteorites, and the Moon) are insufficient to account for Io's properties is illustrated by the pre-Voyager thermal history calculations by Fanale et al. (1977), who obtained thermal profiles that could apply equally well to the Moon. The simple postulate that Io possesses an overabundance of radioactive elements is not refutable at this time, but this proposal would be purely *ad hoc*, and a more justifiable explanation exists.

It is most likely that Io is heated by the dissipation of tidal strain energy. The theory that describes this process goes back to the work of Darwin (1908), who was concerned with its effects on orbital evolution; it was developed by Kaula (1963, 1964), Kaula and Yoder (1976), and Peale and Cassen (1978), who studied its role in heating the Moon, and was applied to Io by Peale et al. (1979).

Because the gravitational field of a planet is not uniform across a satellite of finite size, the nominally spherical satellite is distorted by the field into a prolate spheroid with the long axis nearly aligned with the direction to the planet. This distortion increases with planetary mass, decreases with separation of planet and satellite, and is greater for larger satellites. Internal friction in a satellite rotating relative to its primary causes the tidal bulge to be misaligned in such a way that the resulting torque retards the relative rotation. This effect has caused the Galilean satellites to rotate synchronously with their orbital periods, as is also observed for our own Moon (e.g., Peale 1977).

If these satellites were in circular orbits, the tidal bulge would be fixed in the satellite and there would be no tidal flexing, and therefore no heating. However, the satellites are in eccentric orbits, and, because the orbital motion is not uniform whereas the rotation is uniform, Jupiter appears to move back and forth in the "sky" of the satellite, by an angle that increases as the orbital eccentricity increases. This causes the tidal bulge to increase and decrease in size as the separation between the bodies varies and to move back and forth on the satellite as it follows the motion of the planet in the satellite's sky. So, in an eccentric orbit tides flex the satellite and heat the interior even though the rotation is synchronous with the *average* orbital motion.

The gravitational field of Jupiter thus has periodic components in the reference frame of the satellite. Each periodic forcing term acts independently on the elastic satellite, provided the amplitudes of the distortions are small. The satellite responds as a harmonic oscillator forced by a sum of periodic functions, where the natural frequencies of the satellite oscillator are high compared with the important tidal frequencies. Like the harmonic oscillator, the satellite distorts in phase with the low-frequency tides, except for a small lag in the phase of the response due to the dissipation of the strain energy. When the dissipation is small, the satellite distortion is nearly the same as it would be in equilibrium with the instantaneous disturbing forces, except for the small phase lag. This simplifies the theory of tidal dissipation to a calculation of the equilibrium tidal stresses and strains, except for the phase lag of each periodic term in the response. The usual procedure is therefore to determine the equilibrium strains and strain rates as a sum of periodic terms, and to assign a phase lag to each argument relative to its corresponding forcing term. This phase lag is $1/Q$ rad where Q is the specific dissipation function (Munk and MacDonald 1960). The frequency dependence of Q is not important in the application to the Galilean satellites because the important terms all have the same period as the satellite orbital revolution. In addition to the value of Q, the rate of energy dissipation depends on the rigidity μ and density ρ as functions of position in the satellite. The distributions of μ, ρ, and Q define a model.

Barring unusual and *ad hoc* circumstances, the contribution of tidal dissipation to the thermal history of the Moon was found to be unimportant (Kaula 1964; Peale and Cassen 1978), and it is almost certainly unimportant for most bodies in the solar system; but special circumstances are responsible for its dominance of Io. First, Io is close to the massive Jupiter, which would be expected to produce a permanent tidal bulge on a homogeneous Io of about 8 km. But it is the strain energy associated with the variable tide that is subject to dissipation, and this variable part is due only to the slight eccentricity of Io's orbit, since Io is a synchronously rotating satellite. The eccentricity of 0.0041 is due almost entirely to perturbations by Europa, for the free eccentricity (the eccentricity Io's orbit would have in the absence of Europa) is less than 10^{-5} (Lieske 1980). Ordinarily the tidal dissipation in Io would cause the eccentricity to be damped very rapidly to a negligibly small value (e.g., Peale et al. 1979; Cassen et al. 1980a), but Io is in a resonance with Europa in which Io's eccentricity is forced to the value 0.0041. The mean orbital angular velocities of Io and Europa are nearly in the ratio of 2:1, with conjunctions between the two bodies always occurring at the pericenter of Io's orbit. It is this latter property that keeps Io's eccentricity at the given value (e.g., Peale et al. 1979). Even with the orbital resonances (which include Ganymede in a 2:1 resonance with Europa), dissipation in Io would still reduce the eccentricity and destroy the resonances in the process, were it not for a torque from a tide raised on Jupiter by Io, which tends to expand Io's orbit and drive the system deeper into the resonance (Yoder 1979).

Greenberg (see Chapter 3) investigates the consequences of the situation where the dissipation function for Jupiter Q_J is too large to result in a significant tidal torque from Jupiter. In this case, the resonance is hypothesized to be of primordial origin, and we are observing the system as it decays past the current configuration toward smaller eccentricities from an initial condition much deeper in the resonance. However, an excessively large Q of $\simeq 10^3$ for Io is required to prevent relaxation beyond the current configuration in 4.6×10^9 yr, if the observed thermal activity requires tidal dissipation to exceed expected heating from the decay of radioactive elements. Also, Yoder (personal communication, 1980) finds that the system cannot be arbitrarily deep in the resonance (larger eccentricities) without becoming unstable, so the eccentricity from which a primordial resonance could have relaxed is restricted to $e \lesssim 0.01$. Regardless of the history of the orbital resonances, their persistence maintains Io's orbital eccentricity and with it the dissipation of tidal energy within the body.

Peale and Cassen (1978) considered the heating of bodies of uniform density ρ, and both uniform and nonuniform rigidity μ. For a completely homogeneous body, the total rate of dissipation is given by their equation (31), which for our purposes can be written as:

$$\dot{E}_T = \frac{21}{2} k_2 \frac{GM_J^2 R_s^5 n e^2}{a^6 Q} \tag{1}$$

In this formula G is the gravitational constant; M_J is Jupiter's mass; R_s, n, e, and a are Io's radius, mean motion, orbital eccentricity, and orbit semimajor axis, respectively; Q is the dissipation function; and k_2 is the Love number of second degree for a homogeneous planet (Munk and MacDonald 1960), defined by

$$k_2 = \frac{3/2}{1 + (19\mu/2\rho g R_s)} \tag{2}$$

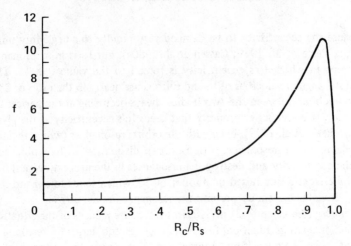

Fig. 4.1. The function $f(R_c/R_s)$ for Io $\mu = 6.5 \times 10^{11}$ dyne cm^{-2}, $\rho = 3.5$ g cm^{-3}, $g = $ 180 cm s^{-2}, R_s - 1.82 $\times$ 10^8 cm.

Here g is Io's surface gravity. The conditions under which Eq. (1) holds are that the satellite is homogeneous and incompressible; is far from its primary, in the sense that $a \gg R_s$; has a small eccentricity ($e \ll 1$); and that contributions from obliquity terms are negligible. Deviations from homogeneous density and incompressibility would be likely to require only small corrections (Kaula 1964), but a strongly varying rigidity is quite possible and has important consequences. In particular, a planet with a molten core ($\mu = 0$ in the core) experiences greater strain and, therefore, greater volumetric dissipation in the solid mantle, than an otherwise similar planet subject to the same tidal force. This effect can be taken into account in the formula for the total dissipation by multiplying the right-hand side of Eq. (1) by a factor f, which is a function of R_c/R_s (where R_c is the core radius) and $\mu/\rho g R_s$. The function f must be found by integrating the actual dissipation over the elastic shell, for various values of the core radius, subject to the condition that the shear stress vanishes on the core boundary (Peale and Cassen 1978). The function f (which includes the increase in k_2 as the elastic shell thickness decreases) is given in Fig. 4.1 for nominal values of the parameters given below. Although the dissipation per unit volume increases to a maximum at $R_c = R_s$, $f \to 0$ because the volume in which the dissipation occurs vanishes.

In applying Eq. (1) to Io, it is somewhat comforting to note that all of the quantities necessary for the calculation of $\dot{E}_T$ are well determined except for μ, Q, and f. The values of these, however, are constrained, although to varying degrees. Note from Fig. 4.1 that, typically, $f \lesssim 11$. If Io is indeed made mostly of silicate rocks, μ is probably in the range of 2×10^{11} to 7×10^{11} dyne cm^{-2}. The low end of this range corresponds to the value for rocks typical of the Earth's crust and the deep lunar interior; the high end corresponds to values of the Earth's rigidity at depths equivalent to 60 kbar pressure and to values appropriate to the outer layers of the Moon (Bullen 1975; Nakamura et al. 1976; Cheng and Toksoz 1978).

The correct value of Q is quite uncertain. Data summarized by Kaula (1968) show that Q for rock varies between 10^2 and 10^3 for a wide range of seismic frequencies, but data are sparse at tidal frequencies (cf. Burns 1977). Tidal dissipation in the Earth occurs mainly in the oceans, so the solid Earth Q is difficult to evaluate; Lambeck (1975) suggests $Q \sim 100$. The determination of an offset of the lunar spin axis from its mean Cassini position has indicated that Q for the Moon for very long-period oscillations may be as low as 10 to 20 (Yoder et al. 1978; Ferrari et al. 1980; Kappalo 1980). However, this surprising result may be due to the dissipation at a liquid-core-solid-mantle interface and not to dissipation in the solid material (Yoder et al. 1978). Perhaps our best guess comes from values derived for Mars from the secular acceleration of Phobos, for which $50 < Q < 150$ (Smith and Born 1976; Pollack 1977). For our purposes we shall use $Q = 100$.

Choosing nominal values of $\mu = 6.5 \times 10^{11}$ dyne cm^{-2} (the value for the outer part of the Moon) and $Q = 100$ yields $E_T = 1.7 \times 10^{12}$ W for an elastic, homogeneous Io, or as much as 1.9×10^{13} W for a "thin shell" Io (large molten core). These values should be compared with that which would be supplied by, say, lunar radioactive abundances, 6×10^{11} W. In the homogeneous case, the tidal energy would be deposited nonuniformly, with maximum at the center greater, by a factor of ~ 3, than the average. Based on a comparison with the Moon's present thermal state and energy sources, as inferred from data obtained by the Apollo program and thermal history models, Peale et al. (1979) argued that melting of Io's center was possible, and that this could trigger catastrophic melting of most of the planet. Thus the thin-shell model may apply to Io. Variations on this possibility are discussed below.

Although tidal heating is theoretically plausible, and capable of accounting for Io's large internal energy source, it has also been proposed that electrical heating is important (Ness et al. 1979; Drobyshevski 1979; Yanagisawa 1980). Indeed, the fact that the satellite exists in an intense electromagnetic environment is reflected in many Io-related phenomena. In particular, the magnetometer on Voyager 1 recorded a magnetic perturbation attributed to an electrical current of about 2.8×10^6 amp in each of two current loops induced by Io's motion through the Jovian magnetosphere (Acuna et al. 1981). Such currents were predicted by Piddington and Drake (1968) and by the Goldreich and Lynden-Bell (1969) theory of the Io-modulated decametric radiation.

The maximum electrical power available for heating Io, according to the

Voyager measurements, is about 10^{12} W (Ness et al. 1979), which could only be attained if all of the energy in the current loops were dissipated in Io. Because geologically abundant materials at Io's low surface temperatures are good insulators, it is usually proposed that the currents are conducted into Io through paths of relatively high conductivity, possibly associated with hot spots (e.g., Gold 1979). Colburn (1980), who performed a thorough study of the electrical heating mechanism, provides the first such analysis that takes into account the conducting ionosphere. Colburn concludes that:

1. The maximum heating by currents induced by Jupiter's variable magnetic field (as seen by Io) is negligible;
2. A conducting ionosphere such as that observed by Pioneer 10 (Kliore et al. 1974) effectively shunts transverse-magnetic mode (unipolar inductor) currents, thereby prohibiting them from passing through Io (cf. Dermott 1970);
3. Even if currents could be concentrated in local hot spots, temperature rise times would be too long to be compatible with the time scale of plume activity.

These conclusions pertain to even the most optimistic assumptions regarding the electrical conductivity of silicates; if sulfur is a dominant surface material (e.g., Sagan 1979), conductivities would be even less favorable for electrical heating. The calculations do not preclude significant electrical heating in the past if Jupiter's magnetic field was much larger than its present one.

It should also be mentioned that direct heating of Io's surface by energetic magnetospheric protons was considered by Witteborn et al. (1979) as a possible explanation of a transient infrared event observed by them prior to the Voyager encounters. (They also discussed what is probably, in retrospect, the correct explanation: the appearance of a temporary volcanic not spot.) Aside from the fact that energetic particle bombardment cannot account for the obviously endogenic surface features, the total energy flux, although substantial, is small compared with tidal or even radioactive sources (e.g., Pollack and Witteborn 1980).

Our discussion proceeds under the hypothesis that tidal dissipation has dominated Io's thermal evolution. The construction of detailed models will require knowledge of the real dissipation rate, which cannot be predicted accurately due to the uncertain values of μ and Q. Three lines of argument, each stemming from a different type of observation, can be brought to bear on the question of Io's dissipation rate. The conclusions are not all in agreement.

First Argument. The absence of visible impact craters on Io demands that the planet be resurfaced at a minimum rate of ~ 0.1 cm/yr (Johnson et al.

1979; see also Johnson and Soderblom, Chapter 17). No doubt the actual rate is considerably higher; statistical variations in both impact rates and removal processes were not taken into account in the above estimates. The numerous calderas and flow fronts revealed in Voyager images are evidence that resurfacing is accomplished largely through the transport of liquids to the surface. The energy flux density h through Io's surface is then (Reynolds et al. 1980)

$$h > (d\ell/dt) \times \Delta E \qquad (3)$$

where $d\ell/dt$ is the average resurfacing rate and ΔE is the energy released by a cubic centimeter of molten material upon solidification and cooling to Io's surface temperature. This must be regarded as an extreme lower bound, since the conducted flux has been completely ignored. If the mobilized material is sulfur, $h > 0.016$ W m^{-2}, which corresponds to a total flux of 6.7×10^{11} W. Since this flux density alone exceeds that expected from typical radioactive abundances, it attests to Io's exceptional energy source, but is otherwise a weak constraint. If resurfacing were accomplished by mobilization of silicates, $h > 0.18$ W m^{-2} (Reynolds et al. 1980), and the total power would exceed 7.5×10^{12} W.

Second Argument. This line of argument comes from observations of Io's present orbital configuration and from the theory of the origin of the orbital resonances. These resonances are most frequently referred to by the relation among the three orbital angular velocities $n_1 - 3n_2 + 2n_3 = 0$, known as the Laplace relation (subscripts 1, 2, and 3 refer to Io, Europa, and Ganymede, respectively). In contrast to the minimal resurfacing rate, Yoder's (1979) theory of the origin and evolution of the orbital resonances places an upper bound on the dissipation rate (cf. Chapter 3 by Greenberg). According to Yoder's theory the resonances were assembled from an originally nonresonant configuration by the differential tidal expansion of the orbits due to dissipative tides raised on Jupiter by the satellites. Io's orbit would expand more rapidly than the orbits of the more distant satellites and would thereby approach the configuration where $n_1 : n_2 \approx 2:1$. The two satellites are captured into this stable resonance so that during subsequent tidal expansion of Io's orbit, the commensurability of the mean orbital motions becomes more exact as Io transfers onto Europa part of the angular momentum it picks up from Jupiter.

Io and Europa cannot be driven arbitrarily close to the 2:1 commensurability, however. A condition of the resonance is that conjunctions of the satellites occur when Io is near its pericenter and Europa is near its apocenter. This means that the argument $\lambda_1 - 2\lambda_2 + \tilde{\omega}_1$, where λ_i are mean longitudes and $\tilde{\omega}_1$ is the longitude of the pericenter of Io's orbit, must librate about $0°$, and $\lambda_1 - 2\lambda_2 + \omega_2$ must librate $180°$ But it is observed that $n_1 - 2n_2 > 0$ ($d\lambda_i/dt = n_i$); this requires $\dot{\tilde{\omega}}_1 = \dot{\tilde{\omega}}_2 = -(n_1 - 2n_2) < 0$ to maintain the conditions of the resonance. (The dot indicates a time derivative.) The retrograde motions of the longitudes of the pericenters are induced by the resonant interaction and are inversely proportional to the respective orbital eccentricity, to lowest order in e. Hence, the values of e_1 and e_2 are determined by the value of $\dot{\tilde{\omega}}_1$, which in turn is determined by how close to the commensurability the system has been pushed by expansion of Io's orbit.

As the system gets closer to the commensurability that is, $n_1 - 2n_2$ gets smaller, the corresponding reduction in the magnitude of $\dot{\tilde{\omega}}_i$ means that both e_1 and e_2 are *forced* to larger values; but as e_1 increases, the tidal dissipation in Io increases and this dissipation opposes the growth in e_1 (and e_2). That is, tides raised by Io on Jupiter cause Io's orbit to expand, thereby driving the system deeper into the resonance and increasing e_1 and e_2, whereas tides raised on Io by Jupiter tend to decrease e_1 and e_2. As e_1 grows, the dissipation in Io (and less importantly in Europa) increases until the effects of dissipation in Jupiter and Io balance. At this point, e_1 and e_2 have reached equilibrium values and the system can approach no closer to the 2:1 commensurability, but continues to be pushed away from Jupiter. The attainment of the equilibrium configuration happens in a short time compared with that for significant orbital expansion.

As the orbits expand, a 2:1 commensurability between the mean motions of Europa and Ganymede is approached. Capture of Ganymede into the 2:1 resonance with Europa and simultaneous capture of the resonance variable $\phi = \lambda_1 - 3\lambda_2 + 2\lambda_3$ into libration about $180°$ is almost certain. So the Laplace relation is established with conjunctions between Europa and Ganymede required to be near Europa's pericenter, but anywhere in Ganymede's orbit. This added resonant interaction forces e_2 and consequently e_1 to larger values, and the high dissipation in Io especially causes the libration amplitude of ϕ to damp relatively rapidly. Coincident with this damping, e_1 and e_2 are driven to new equilibrium values. The current amplitude of libration of the Laplace angle ϕ is $0°.066$ (Lieske 1980), which is insignificantly different from 0 and may in fact be due to a perturbation whose period is close to that of the Laplace libration.

Thus the current value of e_1 is probably very close to the equilibrium value. The significance of this is that, in equilibrium, the ratio of dissipation in Jupiter to that in Io is determined (Yoder 1979). However, there is an absolute upper bound on the time-averaged dissipation in Jupiter, which is determined by the proximity of Io to Jupiter after 4.6×10^9 yr of tidal evolution, under the assumption that the Laplace relation is ancient (Goldreich and Soter 1966). Since the ratio of dissipations is fixed by the equilibrium value of e_1, the upper bound on Jupiter's dissipation translates to an upper bound on that in Io. This upper limit is $\sim 3.3 \times 10^{13}$ W, which corresponds to an average surface flux density of 0.79 W m^{-2}, and is only slightly larger than the nominal thin-shell model heat flux for $Q_1 = 100$.

It should be emphasized that given the firm upper bound on Jupiter's dissipation, this result depends only on e_1 having its equilibrium value (regardless of its prior evolution), and on the assumption that only tidal torques have significant effects on Io's orbit. If the evolution of the orbits has been as described by Yoder (1979), the former condition is almost certainly valid, since (1) the observed libration amplitude is very small, and may in fact be forced, and (2) the theoretical libration damping time is short.

If the present eccentricity is not an equilibrium value because the resonances are of primordial origin and are decaying, the dissipation would be necessarily less than that for the equilibrium configuration (Peale and Greenberg 1980). Otherwise, the system would have relaxed to smaller eccentricities in 4.6 X 10^9 yr than those observed. Thus the limit on Io's dissipation quoted above appears to be the greatest upper bound permitted by theories of tidally driven orbital evolution that do not require special conditions on the histories of the dissipation parameters. More dissipation might be possible, however, if Io's mechanical properties varied due to changes in internal structure on a time scale which is short compared to the orbital relaxation time scale (see below, and Chapter 3 by Greenberg).

The possibility that a magnetic torque acts on Io was discussed by Goldreich and Lynden-Bell (1969). However, its effect would be negligible for the values of electrical current inferred from the Voyager 1 measurements (Ness et al. 1979).

Third Argument. A third way of obtaining Io's heat flux is by direct observation of that flux in the infrared region of the spectrum, as first pointed out by Matson et al. (1981). The main problem with the technique is the separation of reflected radiation and that absorbed and reemitted, from the energy coming from the interior. Ordinarily this would be an almost impossible task, given the limited accuracy of Earth-based measurements, since the internal flux would be expected to be a small fraction of the absorbed incident solar flux. This contrasts with Jupiter's internally derived flux (Ingersoll et al. 1976), which is approximately equal to the absorbed solar radiation. However, the identification of hot spots on Io's surface, the possibility that most of the internal energy flux may be dynamically transported through the hot spots, the elimination of most of the solar contribution by observing Io when eclipsed, and the possibility of estimating effective hot-spot temperatures from narrow-band spectra in the near infrared offer a means of estimating the internal flux and perhaps monitoring its variability from groundbased observations. The estimate of the internal flux is based on the expression (Sinton 1980; Chapter 19 by Pearl and Sinton)

$$E(\lambda) = \sum_{i=1}^{n} A_i B_\lambda(T_i) + B_\lambda(T_b)\left(1 - \sum_{i=1}^{n} A_i\right) \qquad (4)$$

where $E(\lambda)$ is the measured emittance from Io per unit wavelength at wavelength λ; A_i is the fraction of the surface area in hot spots that are at temperature T_i; B_λ (T_i) is the Planck function; and T_b is a uniform background temperature in equilibrium with the incident sunlight or cooling during an eclipse. By determining $E(\lambda)$ at several wavelengths, a set of equations can be established and possibly solved for the $2n + 1$ unknowns, n areas A_i at temperatures T_i, and the temperature T_b. The flux density from the hot spots is $\sum_{i=1}^{n} A_i \, \sigma T_i^4$, which must be of internal origin if all the T_i are significantly larger than T_b.

Matson et al. (1981) use a slightly different procedure with a limited amount of infrared data at three wavelengths: 8.4 μm, 10.6 μm, and 21 μm. They ignore the possible contributions to the measured fluxes at T_b during eclipse and find for $n = 1$ a hot-spot temperature near 200 K with $A_1 \approx 1\%$ and a total flux from the hot spots of 2 ± 1 W m^{-2}.

With data from a sunlit Io at 2.2 μm, 3.8 μm, 4.8 μm, 11.1 μm, and 20 μm, Sinton was able to obtain convergent solutions for $A_1, A_2, T_1, T_2,$ and T_b that required some minor adjustments to the observational values of $E(\lambda)$ for one or two of the wavelengths. With no adjustment the solution diverged. These adjustments were within the observational errors. Sinton's preferred model solution yielded $T_1 \approx 600$ K, $T_2 \approx 300$ K, $T_b \approx 126$ K, $A_1 \approx 2.1 \times 10^{-5}$, $A_2 \approx 3.9 \times 10^{-3}$, and a flux density averaged over the surface of 1.8 ± 0.6 W m^{-2}, which agrees surprisingly well with the value of Matson et al.

The sets of data used by both Matson et al. and Sinton were not taken simultaneously and are not necessarily from the same hemisphere. This introduces the possibility, even probability, that the distribution of temperatures over the surface was unique for each observation, given the observed high variability of thermal activity (Smith et al. 1979). Using infrared observations obtained from the Voyager 1 spacecraft, Pearl (1980) has obtained a preliminary estimate of an internally derived flux of energy from Io of ~ 2 W m^{-2}, which is similar to the values obtained by Matson et al. (1981) and Sinton (1980).

A still more direct measurement of the hot spot emission from Io was made by Morrison and Telesco (1980). They observed emission during an eclipse with sufficiently broad spectral coverage (3 to 30 μm) to permit the determination of the excess emission independent of a specific hot spot model. They found a surface flux of 1.5 ± 0.3 W m^{-2}, corresponding to a total energy source of $6 \pm 1 \times 10^{13}$ W.

These four separately determined hot-spot flux densities (also discussed in Chapter 19 by Pearl and Sinton) correspond to a total energy flux of ~6 X 10^{13} W. Given the uncertainties in μ and Q, tidal dissipation in Io could possibly supply this flux, although it is about four times greater than in the nominal thin-shell model. However, the upper bound on Io's dissipation imposed by the lower bound on Q_J of 10^5 is also exceeded by a factor of two. This is a serious discrepancy whose resolution requires further study.

A possible explanation for the higher infrared fluxes reported might be that Io's heat output is variable on a variety of time scales, and that we are currently observing an anomalous "hot" period. Although this hypothesis is unattractive because it is difficult to test, one can imagine possible causes of such behavior. If Io's interior is at or near the melting point, large changes in its mechanical properties (such as rigidity and viscosity) could occur due to relatively small changes in thermal state. Since both heating and cooling rates depend on these properties, feedback mechanisms exist. Greenberg (see Chapter 3) speculates about the consequences of periodic changes between molten and solid states for Io's interior possibly induced by such feedback mechanisms.

The theoretical and observational estimates of Io's internal energy sources discussed in this section are summarized in Table 4.1. There, we list values for the total powers and average surface flux densities expected from radioactive sources and tidal dissipation models, the theoretical limit on internal power derived from orbital history arguments, and observational estimates based on infrared data.

B. Models of Io's Interior

The conclusion that Io is heated at a high rate by tidal dissipation still admits a variety of models for the interior. Peale et al. (1979) showed that runaway melting could leave the satellite with a very thin elastic shell over a molten interior, but it should be clear from the analysis of the shell model (Peale and Cassen 1978) that the essential characteristic of "melting" in this context is a drastic reduction of the rigidity. This might require only a small amount of partial melting. It is even conceivable that melting could be avoided if the creep viscosity of Io's interior decreased rapidly enough with temperature so that the tidally generated heat could be transported entirely by subsolidus convection (Schubert et al. 1980b). However, the interior of Io is certainly expected to be hotter than the lunar interior. It is not known whether the Moon is molten toward its center (Nakamura et al. 1976); it is known that seismic shear waves are strongly attenuated below a depth of about 900 km. Lunar thermal history models constrained by seismic, heat flow, and petrological data indicate that the interior could be as much as 400 K below the melting point of dry silicates, but the models cannot rule out melting temperatures either (Cassen et al. 1979a). The evidence for large

TABLE 4.1

Theoretical and Observational Estimates of Io's Internal Energy Source

Energy Sources		Total Power		Equivalent Surface Flux Density	
		$\mathrm{erg\,s^{-1}}$	W	$\mathrm{erg\,cm^{-2}\,s^{-1}}$	$\mathrm{Wm^{-2}}$
Radioactives (present day)	Chondritic	4.5×10^{18}	4.5×10^{11}	11	0.011
	Lunar	6.1×10^{18}	6.1×10^{11}	15	0.015
Tidal Dissipation	$Q = 100$ $\mu = 6.5 \times 10^{11}$ dyne cm^{-2} (thin shell)	1.9×10^{20}	1.9×10^{13}	460	0.46
	$Q = 50$ $\mu = 3 \times 10^{11}$ dyne cm^{-2} (thin shell)	6.6×10^{20}	6.6×10^{13}	1600	1.6
	Theoretical upper limit from orbital configuration	$<3.3 \times 10^{20}$	$<3.3 \times 10^{13}$	790	0.79
Infrared Observations	Matson et al. (1980)	$(8.3 \pm 4.2) \times 10^{20}$	$(8.3 \pm 4.2) \times 10^{13}$	2000 ± 1000	2 ± 1
	Sinton (1980)	$(7.5 \pm 2.5) \times 10^{20}$	$(7.5 \pm 2.5) \times 10^{13}$	1800 ± 600	1.8 ± 0.6
	Pearl (1980)	$\sim 9 \times 10^{20}$	$\sim 9 \times 10^{13}$	~ 2000	~ 2
	Morrison and Telesco (1980)	$(6 \pm 1) \times 10^{20}$	$(6 \pm 1) \times 10^{13}$	1500 ± 300	1.5 ± 0.3

amounts of sulfur on Io's surface implies extensive processing of the planet, which in turn would indicate either large-scale melting of the interior, or at least circulation of a large amount of material through molten zones near the surface.

Consolmagno and Lewis (1980) have developed a chemical-thermal evolution model of Io that provides for sulfur enrichment of the surface regions. They postulate an extensively heated Io with an initial composition similar to that of C2 chondritic meteorites. This model predicts the early loss of CO_2 followed by the escape of hydrogen and water from the near-surface regions where extensive oxidation would take place. Reaction with the deeper mantle material could then lead to the buildup of a sulfur crust. Continued heating over time would then lead to the production and loss of SO_2 gas and the maintenance of a mobile surface layer of sulfur.

It should be noted that it is not necessary that Io's interior be completely molten for a thin-shell model, with enhanced dissipation, to apply. A liquid or partially molten zone large enough to decouple an outer elastic shell is all that is required (Peale and Cassen 1978). This possibility leads to some interesting models of Io's interior. Consider, for instance, the thermal structure depicted in Fig. 4.2 as an illustrative example. The temperature rises steeply from the mean subsurface value of 100 K to the silicate solidus, attained at a depth of perhaps 35 km. The primary tidal heating occurs in this elastic shell, at a rate of $\sim 1.5 \times 10^{13}$ W (assuming $\mu = 6.5 \times 10^{11}$ dyne cm^{-2} and $Q = 100$). Below 35 km is a molten or partially molten shell of low rigidity, here chosen arbitrarily to be 45 km thick. Because the solidus temperature rises with pressure, it is possible that a solid core exists below the liquid layer. This would in fact be expected if the liquid were devoid of radioactive sources and were nondissipative, for it then could not maintain the heat flow corresponding to the conductive flux along the melting curve. (We assume that the solid-liquid boundary at a depth of 80 km is not a compositional boundary also.) The solid core could be heated by both retained radioactive elements and tidal dissipation. A crude estimate of the tidal heating is obtained from Eq. (1) using the solid-core radius instead of R_s; $\overset{\cdot}{E_T} \approx 1.3 \times 10^{12}$ W. (Eq. (1) does not give the exact dissipation for the core, of course, because the core's distortion is affected by the layers above it.) However, Io's mean density of 3.5 g cm^{-3} allows the possibility that a substantial Fe-FeS *inner* core exists. Since the Fe-FeS eutectic melting point at pressures expected in Io ($\lesssim 60$ kbar) is about 1000 K (Usselman 1975) well below typical silicate melting temperatures (≈ 1500 K), at least the outer part of the inner core would be molten. Thus the inner solid silicate zone would actually be a shell and experience slightly enhanced tidal heating also.

Clearly, the thermal structure of Io depends in subtle ways on its composition, as well as on the energy sources and transport mechanisms. Because much of Io is likely to be near melting temperatures, and both

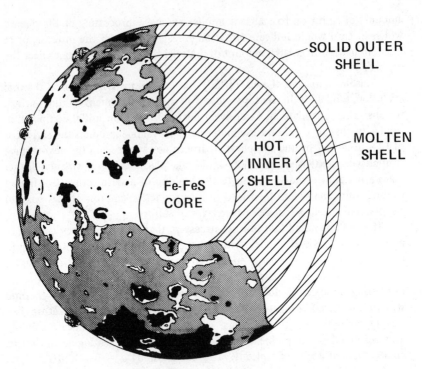

Fig. 4.2. An illustrative example of a model of Io's thermal structure. Hatched regions are solid, clear regions are molten or partially molten. Most of the tidal dissipation would occur in the thin outer solid shell, which would be subject to greater tidal deformation than the inner solid shell. Other models are possible; see text.

heating and cooling rates depend on mechanical properties near the melting curve that are either ill-defined or rapidly varying functions or both, it will be difficult to make detailed models with confidence. Even the thermal structure of the outermost layers is difficult to predict, because we do not know how much of the heat flow is delivered by dynamic processes, that is, by processes that involve the rapid transport of material, and how much is thermally conducted. Obviously, the thermal gradient will not correspond to that which would conduct the planetary heat flow below depths at which a significantly abundant constituent (S or SO_2, perhaps) becomes molten, and that could be shallow indeed.

The problem is further complicated by the fact that tidal dissipation is nonuniform. If, for example, the tidally heated region were a shell of homogeneous composition, mechanical properties, and thickness, the maximum heat deposition would occur at the poles (Peale and Cassen 1978). However, deviations from these special conditions could easily dominate the distribution of dissipation. It is not known if the observed concentration of active plumes toward low latitudes (Strom et al. 1979 and Chapter 16) reflects the heat source distribution, the stress distribution, or merely a statistical fluctuation of plume activity. The fact that calderas are uniformly distributed (cf. Chapter 15) over the planet seems to support the latter.

II. GANYMEDE AND CALLISTO

A. Structure

It has been known since the end of the last century, when the first useful measurements of the radii of the Galilean satellites were made, that the mean densities of Ganymede and Callisto were less than those of the inner two satellites, and probably less than the density of silicate rocks. (Their masses were known reasonably well from the effects of orbital perturbations.) This led to the suggestion that water, a cosmically abundant, light, condensable substance, was a major constituent of these satellites (e.g., Kuiper 1952; Urey 1952). The subsequent detection of H_2O absorption in their reflectance spectra (Pilcher et al. 1972; Fink et al. 1973; Lebofsky 1977) is consistent with the idea that their surfaces are covered to a great extent with frost or ice (cf. Chapter 7 by Sill and Clark).

Huaux (1951) was apparently the first to publish a detailed model of Callisto, based on the assumption that it was composed completely of H_2O, but Lewis (1971a,b) was the first to consider the structure and evolution of icy satellites in general. He emphasized that heating by radioactive elements in concentrations proportional to the rocky component, even though much less than for an inner solar system planet, could easily lead to melting and differentiation. This would be due not just to the presence of water, but also to the possible presence of NH_3, which could depress the melting point well below that of pure ice. Thus Lewis concluded that satellites like Ganymede and Callisto would have rocky cores, water mantles, and thin ice crusts, although he did not consider the effects of solid convection, which could result in completely frozen mantles (Reynolds and Cassen 1979). Lewis also noted that even if the outermost crust escaped primary differentiation, it would be unstable and founder, thereby bringing "clean" H_2O to the surface. So the surfaces should be primarily H_2O ice, perhaps contaminated by fine particulate foreign material or meteoritic debris, disturbed by impact-caused "light-colored craters and rays." This picture of the surfaces has been generally confirmed by the Voyager images (Smith et al. 1979a,b). Major questions that remain are the origin of Ganymede's grooved terrain, and differences from Callisto in general, and whether the H_2O mantles have indeed frozen.

Models of the icy satellites were further developed by Consolmagno and Lewis (1976, 1977), who explored the effects of progressive differentiation,

TABLE 4.2

Two Component Models (Ice Mantle — Rock Core) of Ganymede and
Callisto for Different Core Densities

	Ganymede			Callisto		
Radius (km)	2640			2420		
Mass (g)	1.48×10^{26}			1.08×10^{26}		
Mean density (g cm^{-3})	1.92			1.81		
Models	A	B	C[a]	A	B	C[a]
Core density (g cm^{-3})	3.0	3.3	3.5	3.0	3.3	3.5
Core radius (km)	1942	1826	1763	1695	1593	1542
Core mass fraction	0.623	0.574	0.548	0.571	0.526	0.505
Ice mantle mass fraction	0.377	0.426	0.452	0.429	0.474	0.495

[a]In model C, the cores of Ganymede and Callisto are assumed to have the mean density of
Io.

phase transitions in ice at high pressures, and different assumptions regarding
the state of hydration of the core. They showed that in the absence of
convection in the solid state, early differentiation was likely, even from a cold
(100 K) start and with no NH_3.

The major uncertainty in *structural* models of Ganymede and Callisto (as
opposed to *evolutionary* models) is the density of the rocky component.
("Rocky" here includes metallic iron.) Core constituents could have
uncompressed densities between those of low-density hydrated silicates and
denser iron-rich dry rocks. Io may be almost completely depleted in H_2O and
hence could represent the density of the nonvolatile fraction of the Jovian
nebula from which the satellites formed. On the other hand, if it is assumed
that Europa retained water of hydration upon formation, its mean density
could be representative of the hydrated cores of the outer satellites. Table 4.2
gives results for three models each of Ganymede and Callisto for core
densities of 3.0, 3.3, and 3.5 g cm^{-3}. These models were calculated using
temperature-dependent equations of state for the various phases of H_2O (see
Lupo and Lewis 1980) and constant values for the core densities. Differences
in the models due to different assumed temperature distributions are small.

For model C, in which all H_2O is assumed to be in the mantle, the
ice-mass fraction is the planetary H_2O mass fraction. The similarities between
these ratios for Ganymede and Callisto suggest that they approximate the
values within the Jovian nebula. According to Lewis and Prinn (1980), CO
existed in the solar nebula to low temperatures in preference to CH_4,
diminishing the amount of O available for forming H_2O. In this case, the

expected mass fraction of H_2O in the condensables is 0.36 from Cameron's (1973) cosmic abundances. If the carbon were primarily contained in CH_4 (as it would in an equilibrium condensation sequence) the expected mass fraction of H_2O would be 0.64. These values bracket the H_2O mass fraction in the nominal models C, for which it is assumed that the core contained no water. Perhaps the partitioning of carbon between CO and CH_4 determined the H_2O abundance in the Jovian nebula. The slight difference in calculated H_2O mass fractions between Ganymede and Callisto might reflect radial (or temporal) variations in the physical state of the nebula.

B. Thermal Evolution

Unlike Io, Ganymede and Callisto have probably not experienced extraordinary heating rates; their internal heat sources are believed to be similar in magnitude to those of the inner solar system planets. However, the abundance of water in those bodies permits rapid transport of heat at temperatures much lower than those characteristic of the interiors of the inner planets. This is due both to the low melting point of water, which promotes liquid convection, and the low creep viscosity of Ice-I, which allows efficient solid convection. Thus, understanding the evolutions of the icy satellites requires careful attention to the cooling mechanisms as well as the heat sources.

Consolmagno and Lewis (1976, 1977) followed the evolution of their models from an initially homogeneous mixture of rock and ice at 100 K, heated by radioactive elements and the release of gravitational energy as differentiation proceeded. (One case of heterogeneous accretion was considered for Callisto.) Heat was assumed to be rapidly removed by convection in a liquid layer, but to be transported only by conduction in regions below the melting point. Typically these models evolve to states in which they have extensive water mantles overlying silicate cores, capped by relatively thin ($\lesssim 250$ km) ice crusts. Fanale et al. (1977) also calculated thermal history models of these bodies, and obtained similar results, although compositional assumptions varied. Since all models were started from cold initial states, their history is primarily one of heating and expansion. The final ice crust thickness in these models depends on the surface heat flux, as determined mainly by the total radioactive abundance and the size of the satellite; higher abundances yield greater heat flows, and hence steeper thermal gradients and thinner crusts.

However, Reynolds and Cassen (1979) showed that the ice crusts were likely to be unstable to thermal convection. Their calculations indicated that the heat transported by convection would be sufficient to solidify the mantles in a time short compared with the age of the satellites. Also, Parmentier and Head (1979, 1980) emphasized that dense ice phases overlying water (as exhibited by some conductive thermal history models) would be unstable, as would a primitive, undifferentiated outer crust overlying pure water or ice (cf. Lewis 1971b). These factors would rapidly remove such density inversions and also enhance the outward flow of heat. The modeling of heat transport due to solid-state deformation is beset by difficulties beyond those encountered in conductive models, largely because the rheological properties of solids are complicated and sensitive to the state of the material. It therefore seems useful to study Ganymede and Callisto on a comparative basis, while avoiding estimates of yet undetermined quantities wherever possible.

The main thrust of such an effort must be to explain the difference in appearance, as revealed by Voyager, between the two satellites. Callisto's surface is apparently a primitive one, saturated with craters, whereas all areas of Ganymede appear to be younger, with extensive regions of presumably endogenous "grooved terrain" (Chapter 9 by Woronow and Strom; Chapter 12 by Passey and Shoemaker; Chapter 13 by Shoemaker and Plescia). A plausible explanation is that internal activity persisted on Ganymede until after the decline of an early, heavy bombardment, but did not so persist on Callisto. Why might this have happened? In an attempt to answer this question, we examine first the heat sources for the two satellites and then the cooling mechanisms, with the objective of discerning differences inferable from their known properties and positions in the Jovian system.

We consider three heat sources: accretion, the long-lived radioactive elements (^{238}U, ^{235}U, ^{232}Th, ^{40}K), and tidal dissipation. (Electromagnetic heating of these presumably ionosphereless satellites should probably not be dismissed out of hand, but it seems doubtful that it was important. We do not consider it here, except to note that it would favor heating of Ganymede.)

Accretion. Suppose that the satellites accreted homogeneously. The total gravitational energy per gram available is $3G\ M_s/5R_s$, or 2.24×10^{10} erg g^{-1} for Ganymede and 1.79×10^{10} erg g^{-1} for Callisto. Core formation would add about another 10%. In models of the growth of the Moon, Kaula (1979) found that typically, more than one half of this energy is retained. Using this as a guide, we expect that on the order of 10^{10} erg g^{-1} would be retained. Averaged over the mass of either satellite, this substantially exceeds the energy required to melt its mass of ice. It is therefore probable that extensive melting and differentiation occurred upon (the postulated homogeneous) accretion of both satellites (cf. Schubert et al. 1980b). If this was the case, both would have virtually the same temperature distribution soon after formation because of the strong thermal control provided by the liquid H_2O. That is, below a thin ice crust, the temperature would be constrained to an adiabat rising from the melting point of Ice I to the point where it intersected the melting curve of the dense phases of ice, which has a

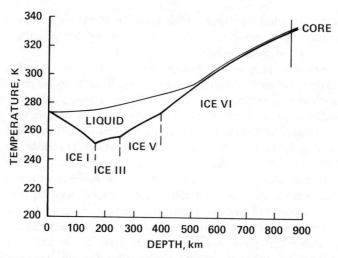

Fig. 4.3. The structure of Ganymede's mantle based on a model in which the satellite is composed of a rocky core with the density of Io, and an H_2O mantle. The heavy solid line is the mantle melting curve; the fine line is the expected postaccretional temperature profile for the mantle. The latter follows an adiabat through the melting point near the surface until it intersects the melting curve deep in the mantle.

steeper slope than the adiabat (see Fig. 4.3). Below this, the temperature would follow the melting curve, or even lie below it if the dense ice began to convect. Initial temperatures in the core probably would not exceed by much the maximum temperature in the mantle.

The initially stored energy could only be greater than that corresponding to the temperature distribution in Fig. 4.3 if rapid accretion occurred. For the purposes of a rough calculation, assume that the adiabatic temperature gradient in liquid water was nearly zero, and that the mantle was heated to some temperature $T_o > 273$ K. Since only a slightly superadiabatic gradient could be maintained in the water, the surface temperature would be almost T_o, and the mantle would cool at a rate determined by

$$M_m C_P \ (dT/dt) = \sigma T^4 \ 4\pi R_s^2 \tag{5}$$

where M_m is the mass of the liquid mantle; C_p is the specific heat of water; and σ is the Stefan-Boltzman constant. The time to cool the mantle to $T = 273$ K is thus less than

$$(M_m C_P / 12\pi R_s^2 \sigma T^3) \approx 2.8 \times 10^4 \text{ yr.} \tag{6}$$

Only if the accretion time were less than this could the water mantle attain temperatures appreciably exceeding the melting temperature. These arguments apply to both satellites.

If heterogeneous accretion occurred, with silicate cores forming first, one might expect the core of Ganymede to be significantly hotter than that of Callisto, since it is inferred to be about 50% more massive. In model C of Table 4.2, Ganymede's core is about the same size and mass as that of the Moon; moreover, its boundary temperature, if set by the melting point of ice, is only slightly (≈ 60 K) higher than that for the Moon. Therefore one might expect the evolution of such a core to resemble that of the Moon. Calculations of lunar evolution indicate that if the Moon is initially molten, liquid and subsequently solid convection rapidly cool it, resulting in a surface heat flow that can substantially exceed that corresponding to the radiogenic energy, for a period $< 5 \times 10^8$ yr (Cassen et al. 1979a; Schubert et al. 1980a). After that time, the surface heat flow is comparable to (but somewhat greater than) the radiogenic value. Therefore, in the extreme case in which Ganymede's core is initially molten but Callisto's core is not, Ganymede would experience an augmented heat flow for a period of $\leq 5 \times 10^8$ yr.

Long-lived Radioactive Elements. The fact that total radioactive heating is expected to be greater for Ganymede than for Callisto (because the former has a larger silicate core) was noted by Consolmagno and Lewis (1976) and Fanale et al. (1977). The ratio of surface heat flux densities in equilibrium with radioactive sources at any time is given by

$$(h_G/h_C) = (m_G R_C^2 / m_C R_G^2) \tag{7}$$

where m_G and m_C are the core masses. For model C, Ganymede's radiogenic surface heat flow is 1.26 times Callisto's. Cassen et al. (1980a) showed that this could lead to a delay in the cessation of surface activity on Ganymede relative to Callisto, for any process that depended on the maintenance of a minimum heat flow, of about 4×10^8 yr.

Tidal Dissipation. Could tidal dissipation have heated these bodies? If so, Ganymede would be preferentially heated because it is closer to Jupiter. Equation (1) gives an upper bound on the tidal heating if we use the rigidity of ice, $\mu = 4 \times 10^{10}$ dyne cm^{-2} (Proctor 1966) and the satellite's mean density in the formula for k_2 (Eq. 2). Substitution of these and the other appropriate values into Eq. (1) shows that present-day tidal dissipation is negligible for both satellites (Cassen et al. 1980a). But $\dot{E}_T$, could have been higher in the past due to a higher eccentricity. Ganymede's forced eccentricity is currently 6×10^{-4} and could never have exceeded 0.002, in view of the instability of the resonance for larger eccentricities (Yoder, personal communication, 1980). This means that any dissipation in Ganymede and Callisto must have been primarily due to their free eccentricities, which decrease with time, thereby causing a corresponding decrease in $\dot{E}_T$. Since the torques produced by Jupiter's tidal distortion are negligible for Ganymede and Callisto, $e(t)$ can be calculated by equating the dissipation rate to the loss of orbital energy for these bodies. Cassen (1980a) derives

$$e = e_0 \exp[(t - t_0)/\tau] \tag{8}$$

where subscript o refers to a reference time, and

$$\tau = M_s (GM_J/a) / (\dot{E}_T/e^2) . \tag{9}$$

Here M_s is the satellite mass, and the factor $(\dot{E}_T/e^2)$ is constant if the satellite's internal structure and material properties do not vary. Changes in the semimajor axis a are higher order in e and are negligible as long as $e \ll 1$. We can use Eq. (8) to calculate the past eccentricity of Ganymede, given $\dot{E}_T (t_0)$, subject to the conditions that $\dot{E}_T/e^2$ remained constant and e evolved to its present value. The quantity $\dot{E}_T(t)$ then follows directly; it is shown in Fig. 4.4 for the choices of $\dot{E}_T(0)$ equal to the primordial chondritic radioactive heating rate, and twice that value. The chondritic radioactive heating rate is also shown. High values of e_0 are required to produce these early dissipation rates: $e_0 = 0.087$ and 0.119, respectively, for the two values of $\dot{E}_T (0)$. Given such high eccentricities, tidal heating could have affected Ganymede's first 5×10^8 yr of evolution. The dissipation in Callisto, assuming the same material properties and eccentricities, would have been more than two orders of magnitude less.

Thus, there are three effects which could prolong the surface activity of Ganymede relative to that of Callisto. Radioactive heating and heterogenous accretion could be effective in this process for only the first 5×10^8 yr, whereas only the most optimistic tidal heating history would cause a difference for this length of time. We next consider the cooling mechanisms.

Figure 4.3 shows the expected postaccretional temperature profile for model C of Ganymede, and the melting curve of water. A minimum in the latter occurs at a depth of 165 km. (The corresponding depth in model C of Callisto would be about the same.) As the mantle freezes, the Ice-I layer thickens and the temperature in the liquid zone decreases, remaining on an adiabat essentially parallel to the one shown. Because the dense phases of ice are heavier then water and Ice I is lighter than water, and because the melting curve gradient exceeds the adiabatic gradient below the Ice I-III phase

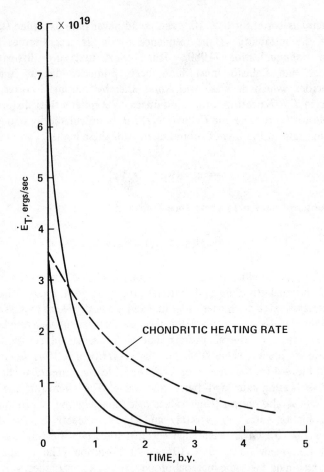

Fig. 4.4. The decay with time of tidal dissipation in Ganymede for initial values equal to the primordial chondritic radioactive heating rate, and twice that rate. Very high initial orbital eccentricities are required to attain these dissipation rates: $e = 0.087$ and 0.119. The calculations were constrained by the requirement that the eccentricity evolve to its present day value, and it was assumed that there were no changes in Ganymede's internal properties during the evolution. The curves should be regarded as rough upper limits on Ganymede's tidal dissipation.

boundary, the mantle freezes both from the bottom up and from the Ice-I layer down. The last water to freeze would be at the minimum in the melting curve.

Reynolds and Cassen (1979) calculated that solid convection, driven by the temperature differential maintained across the Ice-I layer as long as liquid remained, would produce a surface heat flux between 35 mW m^{-2} and 115 mW m^{-2}. The range of values is due to the uncertainty in the average temperature that would be attained by the convecting region, but the higher values are favored. The result depends on the assumption of a specific rheology law. Surface heat flows greater than ~ 55 mW m^{-2} would be capable of solidifying the water mantles of Ganymede or Callisto in $\leqslant 5 \times 10^8$ yr.

Conditions that could alter the conclusion of rapid mantle freezing are discussed below. But the main points we wish to emphasize here are the following:

1. The rate at which energy leaves the satellite is the rate at which it is transported through the Ice-I layer; if this layer is convecting, the energy flux density is determined by the temperature differential across it, and not by its thickness (Priestley 1954). Thus, as long as liquid remains in the mantle and the Ice-I layer is convecting, the planetary heat flow is independent of the internal energy sources, as well as such factors as convection in the dense phases of ice and the inhibition (or enhancement) of convection at phase transitions.

2. The heat flux densities through convecting Ice-I layers of Ganymede and Callisto would be essentially the same. The convective flux density is given by (Rossby 1969; Turcotte and Oxburgh 1967)

$$h_{conv} = k \ \Delta T/L \ b \ Ra\beta \tag{10}$$

where k is the thermal conductivity; ΔT is the temperature difference across the layer; L is the layer thickness; Ra is the Rayleigh number; and b (= 0.1 to 0.2) and β ($\approx 1/3$) are constants whose values depend on boundary conditions. The Rayleigh number is

$$Ra = \alpha g \Delta T L^3 \ / \ \kappa \nu \tag{11}$$

where α is the volumetric coefficient of thermal expansion; κ is the thermal diffusivity; and ν is the effective kinematic viscosity. Note that h_{conv} is almost independent of L, since $\beta \approx 1/3$. Following the initial differentiation, the material properties of the Ice-I layers of Ganymede and Callisto would be the same; also, until one of the mantles freezes, ΔT is almost the same for both satellites. Thus the only difference between the total heat flows from the satellites is contained in the factor $R_s^2 g\beta$, which gives a higher heat flow for Ganymede by a factor of 1.25. It should be noted that for model C, Ganymede's water mantle is 1.23 times more massive than Callisto's, and therefore contains this much more latent heat.

Because the initial cooling rates of the satellites are nearly the same, differences in their early rates of evolution should be due entirely to differences in the energy sources discussed previously. These differences all favor more energy for Ganymede, although the only one that is reasonably certain is the higher radioactive complement. None of the extra energy supplies for Ganymede should result in an internal evolution delayed by more than 5×10^8 yr relative to Callisto's. Thus, if Ganymede's surface features are the surviving manifestations of an early evolution that was common to both satellites, we surmise that they only escaped obliteration by virtue of the rapid decrease in the flux of impacting objects believed to have occurred in the first 5×10^8 yr of solar system history (Soderblom 1977).

It has been inferred from impact crater densities that even though Ganymede's surface is younger than Callisto's, both are a few 10^9 yr old (cf. Chapters 9, 10, and 12). A rapid evolution induced by efficient solid convection, as implied by the calculations of Reynolds and Cassen (1979), is consistent with this observation. On the other hand, surface modification may simply have ceased when a thick enough crust was established, since it would not be easy for underlying water to reach the surface (Parmentier and Head 1979). Can we be sure that no liquid water persists in either satellite? The answer is no; there are at least two factors that could alter the early freezing conclusion. The first is the possibility that the rheological behavior of ice in the situation under consideration is substantially different from that assumed by Reynolds and Cassen (1979). Because ice behaves as a non-Newtonian fluid, the viscosity is a function of local stress as well as temperature. Although considerable experimental data exist (cf. Weertman 1973; Hobbs 1974), it is difficult to predict the average stress level that will exist in a convecting system and, therefore, difficult to predict the effective viscosity. Reynolds and Cassen (1979) implicitly assumed a nonhydrostatic stress level of a few bars, which is typical for convection applicable to rocky planetary mantles, but may not be in the present context. Lower stress levels would result in higher effective viscosities, which in turn would reduce the convective heat flux. Numerical calculations such as those of Parmentier (1978) might be used to insure self-consistency.

A second factor that could negate the conclusion that Ganymede and Callisto have solid ice mantles would be the inclusion of NH_3, salts, or other soluble substances that would depress the melting point. All models since those of Lewis (1971a,b) have been restricted to two components, rock and pure water. As stressed by Lewis (1971a,b), the H_2O-NH_3 eutectic melting point is only 173 K. It may be that NH_3 does not occur in the Jovian planets because its condensation point is too low or because it never formed in the first place. Lewis and Prinn (1980) concluded that NH_3 would not be formed in the solar nebula. But even a small amount formed in the Jovian nebula could be important. Suppose that a water mantle contained 1% NH_3; freezing of all but 1/35 of the mantle would produce an NH_3-H_2O solution of eutectic composition (Lewis 1969), if all the NH_3 remained in solution. In the model C of Ganymede this solution would occupy a layer 25 km thick, if it remained at the bottom of the Ice-I layer. However, it should be noted that the density of a concentrated amonia solution is less than that of water, and may approach that of Ice I. It could then erupt on the surface, or reach higher (colder) levels where it would freeze anyway. Heavier salt solutions might descend through the layers of dense ice, particularly if they contain strong concentrations of the radioactive elements.

Up to this point we have avoided all discussion of specific mechanisms by which Ganymede's grooved terrain might have been produced, but it is to be hoped that considerations of thermal evolution can be brought to bear on this question. The grooved terrain could be the reflection of early convection (now defunct or greatly diminished), either liquid (Johnson 1978) or solid (Reynolds and Cassen 1979). For the reasons already discussed, the visible effects of convection would be expected to be produced for a longer time on Ganymede than on Callisto. The terrain could be the result of global scale tectonics associated with planetary expansion or contraction (Shoemaker and Passey 1979, and Chapter 12 by Passey and Shoemaker), tidal stresses (Parmentier and Head 1979), despinning (Burns 1976; Melosh and Dzurisin 1978), or polar wandering (McAdoo and Burns 1975). The following discussion is restricted to volume changes, since these are most directly related to thermal evolution.

Planetary volume changes are usually due to either differentiation or changes in thermal state. The former could be sudden; the latter are usually gradual. Differentiation of initially homogeneous ice-rock satellites results in a net expansion, as the less compressible rock descends to high pressure regimes and the ice expands on rising. Heating and cooling result in expansion and contraction, respectively, except for some phase transitions and the well-known anomalous behavior of water near its freezing point, at 1-bar pressure. Freezing of water to Ice I produces expansion; freezing to other phases of ice (and freezing of rock) results in contraction. Solid-solid phase transitions behave differently depending on the slope of the Claperon curve (e.g., Hobbs 1974, for ice). Clearly, the volume history of an icy satellite can be complicated.

If the heat of (homogeneous) accretion was sufficient to melt the water in the satellites, as postulated earlier, differentiation would have been essentially immediate for both. For Ganymede model C, there would have been a net expansion of about 70 km in radius, corresponding to a surface area increase of $\sim$ 5%, over the homogeneous planet dimensions.

Freezing of an initially liquid H_2O mantle (see Fig. 4.3) would reduce the mean radius by $\sim$ 13 km (1% decrease in area). Although the solidification of the outer H_2O layers to Ice-I results in a density decrease, the mean density increases due to the freezing of deeper layers to dense phases of ice. Continued cooling of the mantle would generally result in a further density increase, which produces relatively small changes in radius (of the order of several kilometers for changes in the mean temperature of 50 K). The magnitude (and even the sign in some special cases) of these changes depends on the temperature profile since the boundaries between the various phases of ice are both pressure- and temperature-dependent. Aside from differentiation, these changes are the result of mantle evolution only. The center of the silicate core would heat up at an initial rate of ~ 125 K in 10^8 yr, diminishing to 75 K per 10^8 yr after 1 billion years, based on heating by chrondritic radioactives. For a rock thermal expansion coefficient of 3×10^{-5} K^{-1} and a parabolic temperature distribution in the core, core heating would give a planetary expansion of about 3 km after 10^9 yr. During this time, outgassing of H_2O retained in the core as water of hydration (not included in model C) might increase the radius of the planet by as much as 20 km (relative surface area change = 1.5%), based on a serpentine core composition. It is clear that volume changes due to differentiation are considerably greater than those due to thermal effects (see Squyres 1980).

Is it possible that Ganymede differentiated later than Callisto, or that Callisto *never* differentiated? The former hypothesis would require different formation processes for the two satellites, because the arguments for greater heating rates for Ganymede all favor earlier differentiation for that satellite, even if its differentiation were delayed until after accretion. Callisto might never have differentiated if the satellites started out cold (which would require radiation of most of their accretional energy), and solid convection prevented melting in Callisto but not Ganymede. Although these alternatives cannot be dismissed, they are evidently *ad hoc*.

We conclude that, whatever the origin of Ganymede's surface features, their persistence was due to a prolonged thermal evolution of Ganymede relative to Callisto, probably caused by a higher radioactive content, perhaps supplemented by extra accretional energy (if accretion was heterogeneous) or tidal dissipation (if Ganymede had an initially large orbital eccentricity).

III. EUROPA

A. Structure

The mean density of Europa is 3.03 g cm^{-3}, which alone does not rule out a completely rocky satellite; but the presence of H_2O ice was established by infrared spectroscopy (Pilcher et al. 1972), and the Voyager images showed the satellite to be almost devoid of very dark terrain and substantial topographic relief (Smith 1979b; Chapter 14 by Lucchitta and Soderblom). The accepted interpretation of these observations is that the satellite is covered with at least several kilometers of H_2O, which conceal any underlying mountains or other silicate topography. Compositional models

(Consolmagno and Lewis 1976; Fanale et al. 1977; Cassen et al. 1979b) suggest that Europa contains an H_2O mass fraction of from 5% to 10%. (Note that if one assumes that the cores of Europa, Ganymede, and Callisto are compositionally similar, Europa's density of 3.03 g cm^{-3} is a lower limit on the core densities of the others.) A differentiated model, in which Europa's core has Io's mean density, has a core radius of 1443 km covered by an H_2O layer 122 km thick. The pressure at the core- mantle boundary would be $<$ 1.5 kbar, so Ice II would be present there only if the temperature were very low ($\lesssim 150$ K).

B. Thermal Evolution

The arguments given earlier with regard to satellite cooling by convection in an outer ice layer also apply to Europa. The heat transport rate per unit surface area is controlled essentially by the rheological properties of ice and the temperature drop across the convecting layer. For plausible values of the effective viscosity, and in the absence of internal heat sources exceeding typical present-day abundances of radioactive elements, Europa's mantle would freeze rapidly. But Europa's large silicate fraction and small size cause its surface heat flux density to be greater than Ganymede's, even though its total radioactive content is probably less than Callisto's. In addition, tidal dissipation could be quite important for Europa.

The total available accretional energy for Europa is equivalent to a temperature rise of about 1000 K g^{-1}. Therefore, early rock-ice differentiation is likely, even if only half of this energy is retained. Chrondritic abundances of radioactive elements relative to the silicate fraction would have produced $\sim 2.0 \times 10^{12}$ W at the time of origin, diminishing to 2.3×10^{11} W at the present time. At any time, the surface heat flux densities due to radioactives would be 1.5 times that of Ganymede, if the satellites were in approximate steady state with their radiogenic sources.

Unlike Ganymede and Callisto, Europa's orbital eccentricity attains an equilibrium value once the three-body resonance has been established. Therefore, tidal dissipation is not a self-limiting process, and can contribute importantly to Europa's thermal evolution. Using $\mu = 6.5 \times 10^{11}$ dyne cm^{-2}, and other values of the parameters appropriate for Europa, in Eq. (1), one finds $\dot{E}_T = 7.0 \times 10^{12}$ Q^{-1} W for an assumed homogeneous (rocky) Europa. For $Q = 100$, this heating would be small compared to the early chondritic heating rate, but would be about 30% of the present rate. But suppose that the satellite's H_2O mantle was mostly liquid, capped by a thin ice crust. The volumetric heating rate in the ice crust would be greater than in the rest of the satellite because it is mechanically decoupled from the core by the underlying water, and therefore undergoes greater deformation than otherwise. The core can be assumed to be approximately spherical, and its

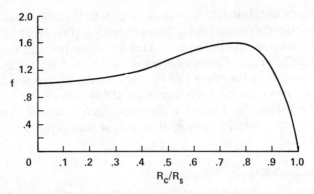

Fig. 4.5. The function $f(R_c/R_s)$ with a decoupled ice shell on a planetary body with Europa's mass and mean density: $\rho = 3.0$ g cm^{-3}, $g = 131$ cm s^{-2}, $R_s = 1.56 \times 10^8$ cm, $\mu = 4 \times 10^{10}$ dyne cm^{-2}.

effect will be simply to increase g compared with the value appropriate to a water core. Thus, in Eq. (1), we can use g and ρ for the whole planet and μ for ice; the result, multiplied by the correct value of f, gives $\dot{E}_T$ for the shell. Here it is important to realize that f is a function of $\mu/\rho g R_s$ as well as R_c/R_s, a point foolishly overlooked by Cassen et al. (1979b). The curve for $f(R_c/R_s)$, for $\mu = 4 \times 10^{10}$ dyne cm^{-2}, and other parameters appropriate to Europa ($\rho = 3.0$ cm^{-3}, $g = 131$ cm s^{-2}, and $R_s = 1.56 \times 10^8$ cm) is given in Fig. 4.5. It is seen that the maximum value of this function is much less than that for the f shown in Fig. 4.1 for a silicate body of roughly the same size. Since Cassen et al. (1979b) used the latter function, they overestimated the shell heating rate for Europa (see Cassen et al. 1980b).

For Europa's present orbital eccentricity, we find

$$\dot{E}_T = 9.8 \times 10^{13} \ f/Q \ \text{W}. \tag{12}$$

Since Europa's H_2O mantle is probably no more than about 125 km thick, $R_c/R_s > 0.92$. (R_c is now to be regarded as the radius of the bottom of the ice layer.) Therefore, from Fig. 4.5, $f \lesssim 1.1$. Hence, for the nominal $Q = 100$, $\dot{E}_T \leqslant 1.1 \times 10^{12}$ W. This maximum value is four to five times the expected present radioactive heating rate, but note that f, and therefore $\dot{E}_T$, drops steeply as R_c/R_s approaches unity. The thickness of an ice shell in conductive equilibrium with the maximum tidal dissipation (assumed uniformly distributed throughout the shell for the purpose of making a simple estimate) would be

$$L = (2k \ \Delta T \ 4\pi R_s^2/\dot{E}_T) \geqslant 35 \text{km}. \tag{13}$$

Here ΔT is again the difference between the mean subsurface temperature and the melting point of ice, assumed to be 173 K, and an average value of $k = 3.7 \times 10^5$ erg cm^{-1} s^{-1} K^{-1} was used. In order for such a shell to be stable against thermal convection, the Rayleigh number (Eq. 11) much not exceed its critical value. That value depends on boundary conditions, but is typically of the order of 10^3 (Chandrasekhar 1961). For average values of $\alpha = 10^{-4}$ K^{-1} and $\kappa = 2 \times 10^{-2}$ cm^2 s^{-1} the effective kinematic viscosity would have to be greater than 5×10^{18} cm^2 s^{-1} for stability against convection. Near the melting point, the kinematic viscosity of ice is in the range 10^{14} - 10^{16} cm^2 s^{-1} for stress levels of 1-10 bar (Weertman 1973). Thus it is likely that convection would occur in the ice shell considered here for Europa, as it is expected to in the mantles of Ganymede and Callisto. This would tend to freeze the ice layer, but at a slower rate than would occur in the outer pair of icy satellites because of the higher surface flux density produced by Europa's internal sources. Tidal heating in the shell could yield a maximum surface heat flux density of 36 mW m^{-2} (for $Q = 100$), slightly augmented by dissipation in Europa's core. Radiocactive sources could provide 65 mW m^{-2} initially, but would decrease with time to 7.5 mW m^{-2} at the present. The estimated convective flux density of 35-115 mW m^{-2} mentioned previously (Reynolds and Cassen 1979) applies to Europa as well as Ganymede and Callisto. Thus it is probable that freezing of Europa's ice mantle would have been delayed considerably, relative to the freezing of Ganymede and Callisto; it is even possible that liquid H$_2$O still exists within Europa. However, we emphasize that the latter possibility now seems much weaker than the conclusion reached by Cassen et al. (1979b) based on the use of an incorrect f function. Much better estimates than we presently have of the rheological properties and of the dissipation function Q, as well as the actual convective heat transfer rates, are required to make a firm prediction of whether liquid water could exist on Europa. Of course, the question of the role of contaminants is important there too, as it is for Ganymede and Callisto. Finally, the effect of dissipation in a thin liquid layer between the silicate core and the outer layer of ice has also been ignored.

We remind the reader that the present value of Europa's orbital eccentricity has been used in the above calculations of tidal dissipation. Therefore it has been implicitly assumed that the establishment of the three-body resonance is not recent, since it is the resonance-associated perturbations that maintain Europa's eccentricity. If the mantle were ever frozen, tidal dissipation could not remelt it (Cassen et al. 1979b).

The theoretical arguments concerning Europa's thermal evolution can be summarized as follows. There was enough accretional energy available to cause early differentiation. Radiogenic heat sources alone are expected to produce a higher surface flux density than on either Ganymede or Callisto. Tidal dissipation in a decoupled ice shell could exceed the present radiogenic heating rate if Europa's mantle were partly liquid when the orbital eccentricity attained its equilibrium value and if the convective heat transport in ice was near the lower end of the range of values estimated by Reynolds and Cassen (1979). Tidal dissipation would contribute significantly but *not* exceed radioactive heating if the mantle is frozen completely. This would be the case if: (1) the orbital resonance is of relatively recent origin; (2) a liquid H_2O region large enough to decouple the outer ice crust was never established, perhaps because of slow devolatilization of the satellite; or (3) convection in the outer ice layer was vigorous enough to freeze the mantle in spite of dissipation in the outer ice crust.

Europa's surface is covered with planetary scale (and smaller) linear features very similar to cracks (Chapter 14 by Lucchitta and Soderblom). They are quite different from the features comprising Ganymede's grooved terrain, but must also surely be related to some global process. The *average* cyclical tidal stresses in a Europa-type ice shell over a water layer would not exceed 8 bar. The tensile yield strength of ice at temperatures somewhat below melting is of the order of 10-20 bar (Glen 1975; Weeks and Assur 1969) and increases with decreasing temperature. Thus it is not apparent that tidal stresses could produce planetary scale tectonic features on Europa's surface. However, Helfenstein and Parmentier (1980) have examined the orientation of the linear features on Europa and conclude that their interpretation as fracture-produced is consistent with a deformation with a principal axis aligned toward Jupiter. The preferred axis suggests a tidally controlled origin for the fractures. The propagation of cracks initiated by meteorite impacts might produce the observed linear features in a floating ice layer. Indeed, some of the linear features appear to radiate from apparent impact craters (Chapter 14 by Lucchitta and Soderblom). However, even in this case the features would have to be widened by some process, perhaps by repeated tidal stresses operating over long periods of time.

Another possible explanation of the features is that they are caused by planetary expansion or contraction. Smith et al. (1979b) estimate that an area increase of 5% to 15% is necessary to account for them if they are tension cracks, based on length and width measurements from Voyager 2 images. They state that a 5% to 10% area increase would result from gradual outgassing of a 50-km-thick water mantle, but this is an overestimate. The correct value is closer to 1% or 2%, since release of this much water from hydrated silicates would increase the density of the source rock. Planetary expansion due to complete freezing of a 100 km Ice-I mantle would yield an area increase of about 1%, and core heating could provide less than 1% more. Malin (1980) cites evidence for both tension and compression. Finally, the absence of grooved terrain such as that found on Ganymede is in itself notable. Perhaps it formed but was removed by the bombardment of energetic particles (Smith et al. 1979b) or was covered by the evaporation and recondensation of underlying water (Cassen et al. 1979b).

The low topographic relief, the sparcity of identifiable impact features, and the high abundance of water frost on the surface of Europa would be plausible consequences of the existence of a thin ice crust over liquid water as suggested by Cassen et al. (1979*b*). However, our incomplete knowledge of the material properties that govern both tidal dissipation and convection in ice precludes, at present, an unequivocal explanation of the morphology of Europa's surface based on considerations of the several inputs to the satellite's thermal history.

Acknowledgments. We thank the many colleagues whose work and ideas have contributed to this review. In particular, G. Schubert and D.J. Stevenson stressed the importance of considering various configurations for Io's interior. Discussions with C.F. Yoder led to the realization that our first calculations of the dissipation rate in Europa's ice shell were incorrect. M. Carr, T.V. Johnson, L. Soderblom, and R. Strom gave us the benefits of their insights into interpretation of Voyager data. We express our appreciation to C. Alexander and A. Summers for their contributions to model calculations of the icy satellites. S.J.P. gratefully acknowledges support by the National Aeronautics and Space Administration and by NASA-Ames Research Center. During the preparation of this chapter, he was a Visiting Fellow at the Joint Institute for Laboratory Astrophysics, University of Colorado and National Bureau of Standards, Boulder, Colorado.

24 October 1980

REFERENCES

Acuña, M.H., Neubauer, F.M., and Ness, N.E. (1981). Standing Alfvén wave current system for Io: Voyager I observations. *J. Geophys. Res.* In press.

Bullen, K.E. (1975). *The Earth's Density*, p. 360. Wiley, New York.

Burns, J.A. (1976). Consequences of the tidal slowing of Mercury. *Icarus* 28, 453-458.

Burns, J.A. (1977). Orbital evolution. In *Planetary Satellites* (J.A. Burns, Ed.), pp. 113-156. Univ. Arizona Press, Tucson.

Cameron, A.G.W. (1973). Abundances of the elements in the solar system. *Space Sci. Rev.* 15, 121-146.

Cameron, A.G.W., and Pollack, J.B. (1976). On the origin of the solar system and of Jupiter and its satellites. In *Jupiter* (T. Gehrels, Ed.), pp. 61-84. Univ. Arizona Press, Tucson.

Cassen, P., Peale, S.J., and Reynolds, R.T. (1980*a*). On the comparative evolution of Ganymede and Callisto. *Icarus* 41, 232-239.

Cassen, P., Peale, S.J., and Reynolds, R.T. (1980*b*). Tidal dissipation in Europa: A correction. *Geophys. Res. Letters.* 7, 987-988.

Cassen, P., Reynolds, R.T., Graziani, F., Summers, A.L., McNellis, J., and Blalock, L. (1979*a*). Convection and lunar thermal history. *Phys. Earth Planet. Int.* 19, 183-196.

Cassen, P., Reynolds, R.T., and Peale, S.J. (1979*b*). Is there liquid water on Europa? *Geophys. Res. Letters* 6, 731-734.

Chandrasekhar, S. (1961). *Hydrodynamic and Hydromagnetic Stability*, Ch. 2. Clarendon Press, Oxford.

Cheng, C.H., and Toksoz, N. (1978). Tidal stresses in the Moon. *J. Geophys. Res.* 83, 845-853.

Colburn, D.S. (1980). Electromagnetic heating of Io. *J. Geophys. Res.* 85, 7257-7261.

Consolmagno, G.J., and Lewis, J.S. (1976). Structural and thermal models of icy Galilean satellites. In *Jupiter* (T. Gehrels, Ed.), pp. 1035-1051. Univ. Arizona Press, Tucson.

Consolmagno, G.J., and Lewis, J.S. (1977). Preliminary thermal history models of icy satellites. In *Planetary Satellites* (J.A. Burns, Ed.), pp. 492-500. Univ. Arizona Press, Tucson.

Consolmagno, G.J., and Lewis, J.S. (1980). The chemical thermal evolution of Io. IAU Coll. 57. The Satellites of Jupiter (abstract 7-11).

Darwin, G. (1908). Tidal friction and cosmogony, Vol. 2. *Scientific Papers,* Cambridge, University Press, New York.

Dermott, S.F. (1970). Modulation of Jupiter's decametric radio emission by Io. *Mon. Not. Roy. Astrom. Soc.* 149, 35-44.

Drobyshevski, E.M. (1979). Magnetic field of Jupiter and the volcanism and rotation of the Galilean satellites. *Nature* 282, 811-815.

Fanale, F., Johnson, T., and Matson, D. (1977). Io's surface and the histories of the Galilean satellites. In *Planetary Satellites* (J. Burns, Ed.), pp. 379-405. Univ. Arizona Press, Tucson.

Ferrari, A.J., Sinclair, W.S., Sjogren, W.L., Williams, J.G., and Yoder, C.F. (1980). Geophysical parameters of the Earth-Moon system. *J. Geophys. Res.* 85, 3939-3951.

Fink, U., Dekkers, N.H. and Larson, H.P. (1973). Infrared spectra of the Galilean satellites of Jupiter. *Astrophys. J.* 179, L155-L159.

Glen, J.W. (1975). *The Mechanics of Ice.* Cold Regions Science and Engineering Monograph 11-C2b, Cold Regions Research and Engineering Laboratory, Hanover, New Hampshire.

Gold, T. (1979). Electrical origin of the outbursts on Io. *Science* 206, 1071-1073.

Goldreich, P., and Soter, S. (1966). Q in the solar system. *Icarus* 5, 375-389.

Goldreich, P. and Lynden-Bell, D. (1969). Io, a Jovian unipolar inductor. *Astrophys. J.* 156, 59-78.

Helfenstein, P. and Parmentier, E.M. (1980). Fractures on Europa: Possible response of an ice crust to tidal deformation. *Proc. Lunar Planet. Sci. Conf.* 11. In press.

Hobbs, P. (1974). *Ice Physics.* Clarendon Press, Oxford.

Huaux, A. (1951). Sur un modèle de satellite en glace. *Bull. Acad. Roy. Sci. Belgique* 37, 534-539.

Ingersoll, A.P., Münch, G., Neugebauer, G., and Orton, G.S. (1976). Results of the infrared radiometer experiment on Pioneers 10 and 11. In *Jupiter* (T. Gehrels, Ed.), pp. 197-205. Univ. Arizona Press, Tucson.

Johnson, T.V. (1978). The Galilean satellites of Jupiter: Four worlds. *Ann. Rev. Planet. Sci.* 6, 93-125.

Johnson, T.V., Cook, II, A.F., Sagan, C., and Soderblom, L.A. (1979). Volcanic resurfacing rates and implications for volatiles on Io. *Nature* 280, 746-750.

Kappalo, R.J. (1980). *The Rotation of the Moon.* Ph.D. Dissertation, Massachusetts Inst. of Technology, Cambridge, Mass.

Kaula, W.M. (1963). Tidal dissipation in the Moon. *J. Geophys. Res.* 68, 4959-4965.

Kaula, W.M. (1964). Tidal dissipation by solid friction and the resulting orbital evolution. *Rev. Geophys.* 2, 661-685.

Kaula, W.M. (1968). *An Introduction to Planetary Physics,* p. 99. Wiley, New York.

Kaula, W.M. (1979). Thermal evolution of Earth and Moon growing by planetesimal impacts. *J. Geophys. Res.* 84, 999-1008.

Kaula, W.M., and Yoder, C.F. (1976). Lunar orbit evolution and tidal heating of the Moon. *Proc. Lunar Sci. Conf.* 7, 440-442.

Kliore, A., Cain, D.L., Fjeldbo, G., Seidel, B.L., and Rasool, S.I. (1974). Preliminary results of the atmospheres of Io and Jupiter from Pioneer 10 S-band occulation experiment. *Science* 183, 323-324.

Kuiper, G. (1952). *The Atmospheres of the Earth and Planets,* pp. 306-405. Univ. Chicago Press, Chicago.

Lambeck, K. (1975). Effects of tidal dissipation in the oceans on the Moon's orbit and the Earth's rotation. *J. Geophys. Res.* 80, 2917-2925.

Lebofsky, L.A. (1977). Callisto: Evidence for water frost. *Nature* 269, 785-787.

Lewis, J.S. (1969). The clouds of Jupiter and the NH_3-H_2O and NH_3-H_2S systems. *Icarus* 10, 365-378.

Lewis, J.S. (1971a). Satellites of the outer planets: Thermal models. *Science* 172, 1127-1128.

Lewis, J.S. (1971b). Satellites of the outer planets: Their physical and chemical nature. *Icarus* 15, 174-185.

Lewis, J.S. and Prinn, R.G. (1980). Kinetic inhibition of CO and N_2 reduction in the solar nebula. *Astrophys. J.* 238, 357-364.

Lieske, J. (1980). Improved ephemerides of the Galilean satellites of Jupiter. *Astron. Astrophys.* 82, 340-348.

Lupo, M.J., and Lewis, J.S. (1980). Mass-radius relationships in icy satellites. *Icarus* 40, 157-171.

Malin, M.C. (1980). Tectonics on Europa and Ganymede. *EOS* 61, 286 (abstract).

Matson, D.L., Ransford, G.A., and Johnson, T.V. (1981). Heat flow from Io (J1). *J. Geophys. Res.* In press.

McAdoo, D.C., and Burns, J.A. (1975). The Coprates trough assemblage: More evidence for Martian polar wander. *Earth Planet. Sci. Letters* 25, 347-354.

Melosh, H.J., and Dzurisin, D. (1978). Mercurian global tectonics: A consequence of tidal despinning? *Icarus* 35, 227-236.

Morabito, J. Synnott, S.P., Kupferman, P.N., and Collins, S.A. (1979). Discovery of currently active extraterrestrial volcanism. *Science* 204, 972.

Morrison, D. and Telesco, C.M. (1980). Io: Observational constraints on internal energy and thermophysics of the surface. *Icarus* 44, 226-233.

Munk, W.H., and MacDonald, G.J.F. (1960). *The Rotation of the Earth,* Ch. 5. Cambridge University Press, London.

Nakamura, Y., Latham, G.V., and Dorman, H.J. (1976). Seismic structure of the Moon. *Proc. Lunar Sci. Conf.* 7, 602-603.

Ness, N.F., Acuña, M.H., Lepping, R.P., Burlaga, L.F., Behannon, K.W., and Neubauer, F.M. (1979). Magnetic field studies at Jupiter by Voyager 1: Preliminary results. *Science* 208, 982-986.

Parmentier, E.M. (1978). A study of thermal convection in non-Newtonian fluids. *J. Fluid Mech.* 84, 1-11.

Parmentier, E.M., and Head, J.W. (1979). Internal processes affecting surfaces of low-density satellites: Ganymede and Callisto. *J. Geophys. Res.* 84, 6263-6276.

Parmentier, E.M., and Head, J.W. (1980). Some possible effects of solid state deformation on the thermal evolution of ice-silicate planetary bodies. *Lunar Planet. Sci.* 10, 2403-2420 (abstract).

Peale, S.J. (1977). Rotation histories of the natural satellites. In *Planetary Satellites* (J. Burns, Ed.), pp. 87-112. Univ. Arizona Press, Tucson.

Peale, S.J., and Cassen, P. (1978). Contribution of tidal dissipation to lunar thermal history. *Icarus* 36, 245-269.

Peale, S.J., Cassen, P., and Reynolds, R.T. (1979). Melting of Io by tidal dissipation. *Science* 203, 892-894.

Peale, S.J., and Greenberg, R.J. (1980). On the Q of Jupiter. *Lunar Planet. Sci.* 11, 871-873 (abstract).

Pearl, J.C. (1980). The thermal state of Io on March 5, 1979. IAU Coll. 57. *The Satellites of Jupiter* (abstract 4-1).

Piddington, J.H., and Drake, J.F. (1968). Electrodynamic effects of Jupiter's satellite Io. *Nature* 217, 935-937.

Pilcher, C.B., Ridgway, S.T., and McCord, T.B. (1972). Galilean satellites: Identification of water frost. *Science* 178, 1087-1089.

Pollack, J.B. (1977). Phobos and Deimos. In *Planetary Satellites* (J.A. Burns, Ed.), pp. 319-345. Univ. Arizona Press, Tucson.

Pollack, J.B., and Reynolds, R.T. (1974). Implications of Jupiter's early contraction history for the composition of the Galilean Satellites. *Icarus* 21, 248-253.

Pollack, J.B., and Witteborn, F.C. (1980). Evolution of Io's volatile inventory. *Icarus* 44, 249-267.

Priestley, C.H.B. (1954). Convection from a large horizontal surface. *Aust. J. Phys.* 7, 176-201.

Proctor, Jr. T.M., (1966). Low-temperature speed of sound in single crystal ice. *J. Acoust. Soc. Amer.* 39, 972-977.

Reynolds, R.T., and Cassen, P. (1979). On the internal structure of the major satellites of the outer planets. *Geophys. Res. Letters* 6, 121-124.

Reynolds, R.T., Peale, S.J., and Cassen, P. (1980). Io: Energy constraints and plume volcanism. *Icarus* 44, 234-239.

Rossby, H.T. (1969). A study of Benard convection with and without rotation. *J. Fluid Mech.* 36, 309-335.

Sagan, C. (1979) Sulphur flows on Io. *Nature* 280, 750-753.

Schubert, G., Stevenson, D., and Cassen, P. (1980a). Whole planet cooling and the radiogenic heat source contents of the Earth and Moon. *J. Geophys. Res.* 85, 2531-2538.

Schubert, G., Stevenson, D.J., and Ellsworth, K. (1980b). Internal structures of the Galilean satellites. Submitted to *Icarus*.

Shoemaker, E.M., and Passey, Q.R. (1979). Tectonic history of Ganymede. *EOS* 60, 869 (abstract).

Sinton, W.M. (1980). The thermal emission spectrum of Io and a determination of the heat flux from its hot spots. *J. Geophys. Res.* In press.

Smith, B.A., and the Voyager Imaging Team (1979a). The Jupiter system through the eyes of Voyager 1. *Science* 204, 951-972.

Smith, B.A., and the Voyager Imaging Team (1979b). The Galilean satellites and Jupiter: Voyager 2 imaging science results. *Science* 206, 927-950.

Smith, J.C. and Born, G.H. (1976). Secular acceleration of Phobos and Q of Mars. *Icarus* 27, 52-54.

Soderblom, L.A. (1977). Historical variations in the density and distribution of impacting debris in the inner solar system. In *Impact and Explosion Cratering* (D.J. Roddy, R.O. Pepin, and R.B. Merrill, Eds.) Pergamon, New York.

Squyres, S.W. (1980). Volume changes in Ganymede and Callisto and the origin of grooved terrain. *Geophys. Res. Letters.* 7, 593-596.

Strom, R.C., Terrile, R.J. Masursky, H., and Hansen, C. (1979). Volcanic eruption plumes on Io. *Nature* 280, 733-736.

Turcotte, D.L., and Oxburgh, E.R. (1967). Finite amplitude convective cells and continental drift. *J. Fluid Mech.* 28, 29-42.

Urey, H.C. (1952). *The Planets, Their Origin and Development,* p. 221. Yale Univ. Press, New Haven.

Usselman, T.M. (1975). Experimental approach to the state of the core: Part 1. The liquids relations of the Fe-rich portion of the Fe-Ni-S system from 30 to 100 kb. *Amer. J. Sci.* 275, 278-290.

Weeks, W.F., and Assur, A. (1969). Fracture of lake and sea ice. Research Report 269 of the Cold Regions Research and Engineering Laboratory, Hanover, New Hampshire.

Weertman, J. (1973). Creep of ice. In *Physics and Chemistry of Ice.* (E. Whalley, S.J. Jones, and L.W. Gold, Eds.), pp. 320-337. Roy. Soc. Canada, Ottawa.

Witteborn, F.C., Bregman, J.D., and Pollack, J.B. (1979). Io: An intense brightening near 5 micrometers. *Science* 203, 643-646.

Yanigisawa, M. (1980). Can electromagnetic induction current heat Io's interior effectively? *Lunar Planet. Sci.* 11, 1288-1290 (abstract).

Yoder, C.F. (1979). How tidal heating in Io drives the Galilean orbital resonance locks. *Nature* 279, 767-770.

Yoder, C.F., Sinclair, W.S., and Williams, J.G. (1978). The effects of dissipation in the Moon on the lunar physical librations. *Lunar Planet. Sci.* 9, 1292-1293 (abstract).

5. THE OUTER SATELLITES OF JUPITER

D. P. CRUIKSHANK
University of Hawaii

J. DEGEWIJ
Jet Propulsion Laboratory

and

B. H. ZELLNER
University of Arizona

The eight known irregular, outer satellites of Jupiter can be divided into two dynamically different groups. The physical properties of the satellites themselves are known only superficially because of their faintness. Of the prograde inner group, J6 (Himalia) is the best known, and present evidence shows that this 185-km diameter body is similar in size, surface composition, and shape to many common dark, C-type asteroids in the outer asteroid belt. J7 (Elara), though less than half the size of J6, is probably very similar, while the other two in the same dynamical group, J10 (Lysithea) and J13 (Leda), have not been studied. Of the outer retrograde group, preliminary optical spectrophotometry of J8 (Pasiphae) and J9 (Sinope) suggests that they may be compositionally different from J6 and J7, resembling dark asteroids (particularly some Trojans) of distinct red color. The remaining two in this group, J11 (Carme) and J12 (Ananke), have not been studied. Thus there is marginal evidence that the two groups are of separate compositional classes, corresponding to two compositional classes of Trojan asteroids and other asteroids in the main belt. The few data presently available do not yet answer such fundamental questions as the origin of the outer Jovian satellites and their mineralogical compositions.

Jupiter's family of satellites consists of three groups. The inner group of eight known bodies has nearly circular, coplanar orbits in the planet's equatorial plane, and includes Amalthea, 1979 J1 (Adrastea), 1979 J2, 1979 J3, and the Galilean satellites. The two outer groups consist of four known objects moving prograde in eccentric orbits inclined $\sim 27°$ to Jupiter's equator at a distance of about 164 R_J, and four known objects moving retrograde with inclinations $\sim 150°$ at a distance of ~ 322 R_J. An additional probable outer satellite, designated 1975 J1, was found by Kowal (1975) but subsequently lost and not recovered; observations were insufficient to determine if it belonged to the recognized prograde or retrograde group. Information on the orbital geometry, names, and discoveries of the eight confirmed outer satellites is given in Table 5.1, while Fig. 5.1 shows schematically the relative sizes of the semimajor orbital axes and the orbital inclinations; see also Anonymous (1975, 1976). A short review of these objects can be found in Morrison and Burns (1976).

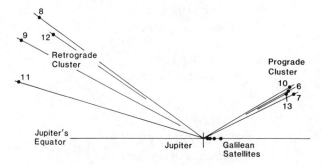

Fig. 5.1. Orbital distances and inclinations of the satellites of Jupiter shown schematically. These orbital planes are not actually aligned in three-dimensional space because perturbations cause a diverse range of orientations of the orbital planes of members of a given cluster. (From Pollack et al. 1979.)

The outer Jovian satellites are very faint, J6 (Himalia) being the brightest with a mean opposition visual magnitude[a] of 14.7, and consequently there is very little physical information available on the nature of these small bodies. There are, however, compelling reasons for the study of the outer satellites, particularly in light of current interest in the possible interrelation of

---

[a]The mean opposition visual magnitude, V_0, is computed with $V_0 = V(1,0) + 5 \log [a_p(a_p - 1)]$, in which a_p is the semimajor axis of the planet to which the satellite belongs. The absolute magnitude $V(1,0)$ is computed with $V(1,0) = B(1,0) - (B-V)$, where $B(1,0) = B - 5 \log r - 5 \log\Delta + 0.538 - 0.134|\alpha|^{0.714} - 7\beta$, for $|\alpha| < 7°$, and $B(1,0) = B - 5 \log r - 5 \log\Delta - |\alpha|\beta$, for $|\alpha| \geq 7°$. Here r is the distance between object and Sun in AU; Δ is the distance between object and Earth in AU; α is the phase angle; and the phase coefficient $\beta = 0.039$ mag/deg. This formula, described in more detail by Gehrels and Tedesco (1979), assumes that the outer satellites display an asteroidal opposition surge. The data for J6 (Himalia) (Fig. 5.6) show that this is indeed the case.

TABLE 5.1

The Outer Satellites of Jupiter — Orbital Data

Satellite Name[a]		Orbital Radius ($\times 10^3$ km)	(R_J)	Period (day)	Eccentricity	Inclination (deg)	Discoverer	Year	References
J13	Leda	11110	156	240	0.146	26.7	C.T. Kowal	1974	Kowal et al (1975)
J6	Himalia	11470	161	251	0.158	27.6	C.D. Perrine	1904/5	Perrine (1905a)
J10	Lysithea	11710	104	260	0.130	29.0	S.B. Nicholson	1938	Nicholson (1939)
J7	Elara	11740	165	260	0.207	24.8	C.D. Perrine	1904/5	Perrine (1905b)
J12	Ananke	20700	291	617	0.17	147	S.B. Nicholson	1951	Nicholson (1951)
J11	Carme	22350	314	692	0.21	164	S.B. Nicholson	1938	Nicholson (1939)
J8	Pasiphae	23300	327	735	0.38	145	P.J. Melotte	1908	Melotte (1908a, 1908b)
J9	Sinope	23700	333	758	0.28	153	S.B. Nicholson	1914	Nicholson (1914)

[a]Distant satellites with direct orbits have names ending in a, and those in retrograde orbits have names ending in e (Marsden 1975).

asteroids, comets, and planetary satellites. Plausible mechanisms that have emerged for the formation of the outer satellites include the following:

1. Capture and break up in the proto-Jovian nebula (Kuiper 1951, 1956);
2. Collision within Jupiter's gravitational sphere of influence (Colombo and Franklin 1971);
3. Capture through the Jovian Lagrangian points after dissipation of the proto-Jovian nebula (Bailey 1971a,b, 1972).

Kuiper (1951, 1956) proposed that as the proto-Jupiter lost mass during the later phases of condensation, small outer satellites were lost to free space because the planet's gravitational radius of action decreased in size. The gravitational sphere of influence is analogous to Kuiper's radius of action. If the gravitational sphere of influence μ is given by the relation

$$\mu \equiv \frac{M_p}{M_\odot + M_p} \tag{1}$$

where $M_\odot$ and M_p are the masses of the Sun and the planet, respectively, the radius of action R_A is given by

$$\frac{\log R_A}{a} = 0.318 \log \mu - 0.327 \tag{2}$$

where a is the distance of the planet from the Sun in AU. The practical limit for satellites in near-circular orbits is 0.5 R_A, but the outer Jovian group has slightly larger values (Kuiper 1956). Some of these shed objects were later captured as they encountered the gas-dust envelope of Jupiter on subsequent orbits around the Sun, dissipating their orbital energy through viscous interaction with the envelope. This capture could result either in direct or retrograde motion around Jupiter, depending on the geometry of the collision. Kuiper (1951) argued that because of the flatness of the proto-Jupiter nebula, low inclinations would be expected for the orbits of the recaptured satellites. Since the outer satellites clearly fall into two distinct groups, both with large inclinations, he proposed that only two captures occurred, with each mass dividing into several (at least four) parts, analogous to the break up of meteors in the Earth's atmosphere. Pollack et al. (1979) have considered fragmentation by gas drag in detail and reaffirm its importance and feasibility in the Jovian case. Once in orbit about Jupiter, the fragments made repeated encounters with the denser interior parts of the nebula, resulting in orbits of less eccentricity, until the present stable values were achieved. The outermost group of Jupiter's irregular satellites lies near the limit of gravitational stability corresponding to Jupiter's present mass (Kuiper 1956) and could not have been captured in its present location. Kuiper regarded this as further evidence that the planet's mass continued to decrease (with a consequent decrease in the radius of action) after capture of this group.

The possibility that both satellite groups originated from a single collision event within the gravitational sphere of influence of Jupiter was explored by Colombo and Franklin (1971). They examined the satellite orbits precessed back in time, finding that the two groups frequently come within interaction distance. Colombo and Franklin suggested that the two satellite groups resulted from the collision of an asteroid with a satellite, but found that collisions of two asteroids or two satellites were also plausible. Because this hypothesis, as does that of Kuiper, involves fragmentation by friction or collision, the implication is that numerous undiscovered pieces remain. The smallest debris have probably spiraled inward to Jupiter, but larger pieces with sizes on the order of a few kilometers may remain.

We note that a fragmentation history for the outer Jovian satellites would suggest that the individual satellites are irregular in shape. Both Kuiper (1951) and Colombo and Franklin (1971) suggested that the rotational light curves of the satellites might reveal irregular shapes by their relatively large amplitudes of several tenths of a magnitude. As we show below, the single object for which such measurements have been made, J6 (Himalia), has a light-curve amplitude of ~ 0.13 mag, probably resulting from an irregular shape.

The third mechanism for the origin of the outer satellites was proposed by Bailey (1971a,b, 1972) wherein the bodies were captured as a consequence of their motion through a Jovian Lagrangian point, and were independent of a viscous medium. The bodies, presumed originally to be in solar orbit, moved with nearly zero relative velocity through the L_2 point into the Jovian radius of action, but in order to affect capture the events would have had to occur when Jupiter was at parihelion or aphelion. Conceptual and analytical flaws in the Bailey hypothesis were pointed out by Davis (1974) and Heppenheimer (1975; see also Heppenheimer and Porco 1977), and as Greenberg (1976) has noted in his review, the hypothesis should be abandoned.

Gehrels (1977) has pointed out that the magnitude-frequency (hence the size-frequency) distribution of the outer satellites is unlike that of the main belt asteroids or Trojans, with the satellite population lacking the very large and very small members. Gehrels' own search with the Palomar Schmidt extended to photographic magnitude 21.2 (corresponding to visual magnitude ~ 20.6); if the geometric albedo of a hypothetical faint satellite is 0.03, the diameter corresponding to the limiting magnitude is ~ 13 km. Gehrels suggests that the absence of very small or large satellites supports the concept of capture in a viscous medium rather than from a purely gravitational interaction. Small pieces would spiral inward to Jupiter, while larger pieces would not have been affected by the viscous circum-Jovian cloud.

The outer satellites of Jupiter have not been studied from spacecraft, nor are they likely to be in the near future. Groundbased telescopic observations are therefore our only source of information on these objects. The techniques that have yielded our meager store of data are filter photometry, spectrophotometry, polarimetry, and infrared radiometry; we consider the results of each technique.

While the outer satellites have names recognized by the International Astronomical Union (Table 5.1), they are not in common use, probably because the objects themselves have not been commonly studied or referenced. While there is some sentiment (Kowal 1976) to leave unnamed the known outer satellites and those that may be discovered, the International Astronomical Union has named all but the most recently discovered satellites. In this review we shall normally use the nomenclature J6, J7, etc., for brevity.

I. PHOTOMETRY AND SPECTROPHOTOMETRY

Spectral Reflectance

The first physical information about objects in the solar system comes from photometric observations obtained over the widest possible wavelength range and over the maximum range in observing geometries. Photometry of the outer Jovian satellites is still in its infancy, but has provided most of the meager information we have on the nature of these bodies.

Measurements of the satellites in the UBV photometric system were the first to be obtained, but such data alone are not highly diagnostic of surface composition. The utility of photometry combined with other data, however, has been demonstrated by Bowell et al. (1978), who have used UBV, polarimetric, radiometric, and spectrophotometric data for more than 500 asteroids to develop a system of classification of these objects along compositional lines. In their scheme, seven directly observable optical parameters from the four observational techniques noted above are used to define C, S, M, U, and R-type asteroids. (The letter designations have a historical connection with early concepts of the compositions corresponding to observed spectral shapes.) In general terms, the C-type objects have low geometric albedo ($p_V \leqslant 0.065$) and flat spectral reflectances with no significant Fe^{2+} absorption band at 0.95 μm (pyroxene). The S-type objects have $0.065 \leqslant p_V \leqslant 0.23$ and spectral reflectances sloping upward toward the red, often showing a weak to moderate pyroxene absorption band. R objects have steeper reflectance slopes (redder) and/or deeper bands, and RD

TABLE 5.2

Wavelength and Bandpasses of Standard Filters in Micrometers

Filter	Central λ	FWHM[a] Bandpass
U	0.36	0.07
B	0.44	0.10
V	0.55	0.08
R	0.65	0.13
I	0.83	0.09
J	1.25	0.30
H	1.6	0.28
K	2.2	0.42

[a]Full width at half maximum of transmission curve.

objects (reddish and dark), a type introduced by Degewij (1978), are reddish but with low albedos (Degewij and van Houten 1979).

Photometry in filters other than *UBV*, particularly those in the red and near infrared, sheds additional light on the surface compositions of small solar system bodies, when laboratory comparisons with candidate mineral substances are available; recent progress has been made in observing satellites and asteroids at wavelengths throughout the range 0.3-4.2 μm. Data over the full range have not yet been obtained for any outer Jovian satellite because of the faintness of these bodies and the insensitivity of the instruments used, particularly in the near infrared. The wavelengths of the filters used for observations of the Jovian satellites are given in Table 5.2.

Degewij et al. (1980*a*) have published the results of a decade of photometric studies of J6, J7, and J8, with emphasis on J6 (Himalia), in the standard photometric filter bands. Data for J6 exist in the *UBVRIJHK* bands (Degewij et al. 1980*b*), for J7 (Elara) in *UBVRI*, and for J8 (Pasiphae) in *UBV*, with those for J6 having been obtained over a large range (11°) in phase. Further, data for J6 have been obtained with high time resolution in a study of the satellite's rotation (see below).

A *UBV* color diagram of distant asteroids (Hilda and Trojan groups) plus several faint outer planet satellites is shown in Fig. 5.2. In this figure J6, J7, and J8 fall within or near the field occupied by *C*-type asteroids, buy several other planetary satellites of apparently diverse surface composition do as well. For example, N1 (Triton) appears to have a rocky surface (Cruikshank et al. 1979) and a tenuous methane atmosphere (Cruikshank and Silvaggio 1979), S7 (Hyperion) has surface exposures of water ice (Cruikshank 1980) and has a high albedo (Cruikshank 1979), while S9 (Phoebe) is apparently a dark object of the *C* type (Degewij et al. 1980*b*). Thus, *UBV* photometry of the outer Jovian satellites merely tells us that these objects have reflectances that slope very slightly upward toward the red, as do many asteroids and other satellites.

Further information comes from a *VRI* color diagram (Fig. 5.3), which shows several Hilda asteroids, Trojan asteroids, and planetary satellites. Those near the top of the diagram represent the steepest slopes beyond the *R* wavelength toward the red, and have been designated the *RD* group. The *RD* group is defined by the color index $V - I \geqslant 0.9$, and geometric albedo in the *V* band $p_V \lesssim 0.05$. So far, the only planetary satellites found to have such steep red slopes are S8 (Iapetus) on its dark side (Cruikshank et al. 1981), and J8 and J9 (Smith et al. 1981). The *VRI* diagram is the most useful for separating bodies whose solid rocky (or dusty) surfaces are exposed, from those with covers of frozen volatiles, for it appears to distinguish among the *C,M*, and *RD* types (though only by association, since *C* and *M* types are not defined in terms of *R* and *I*).

Observations in the near-infrared *JHK* filter bands are also useful in distinguishing objects with rocky or dusty surfaces from those covered with frozen water or methane, but such observations of the faintest planetary satellites and asteroids are difficult to obtain because of limited instrumental

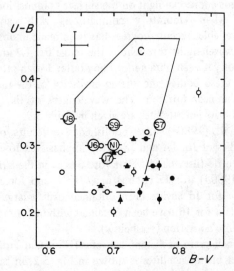

Fig. 5.2. *U-B* versus *B-V* colors for distant asteroids in the 153 Hilda group (○) and the Trojan cloud preceding (●) and following (▲) Jupiter, and some faint satellites. The data are from Table 1 in Degewij and van Houten (1979). Colors at red wavelengths are known for objects marked with bars: vertical bars indicate steep reddish *RD*-type spectra ($V-I \geqslant 0.9$ and $p_V \lesssim 0.05$; see Fig. 5.3) and horizontal bars indicate flat *C*-type spectra. The color of the Sun is at $U-B = 0.20$ and $B-V = 0.64$. The domain for *C* compositional types is drawn as defined by Bowell et al. (1978). The error bar represents the expected error in the mean of the measurements.

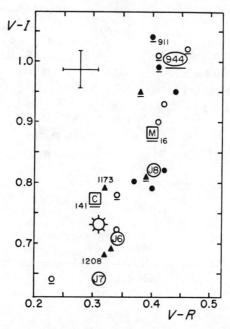

Fig. 5.3. *V-I* versus *V-R* colors for distant asteroids in the 153 Hilda group (○), the Trojan cloud preceding (●) and following (▲) Jupiter, and some faint satellites. The data are from Table 1 in Degewij and van Houten (1979). The underlined symbols represent synthetic colors obtained from spectra measured by Chapman and Gaffey (1979). The color of the Sun is *V-I* = 0.73 and *V-R* = 0.31. The error bar represents the expected error in the mean of the measurements.

sensitivity. Figure 5.4 shows a *JHK* color plot of some asteroids of the *S, U, C,* and *RD* types, plus many satellites having surfaces covered at least in part by water ice. J6 (Himalia) is the only Jovian satellite measured with sufficient precision to include it on this diagram; it clearly falls among the low-albedo objects and is distinct from those satellites having icy surfaces. Evidence for the low albedo of J6 comes from radiometric studies as well (see below).

The color diagrams (Figs. 5.3 and 5.4) show that J6 and J7 have very similar properties, but that J8, the only other body for which such data exist, is different in the sense of being redder. The results of preliminary spectrophotometric observations of J8 by Smith et al. (1981) corroborate the red color; their observations are shown in Fig. 5.5. The steep slope of J8 in the region 0.4-1.0 μm is comparable to that of the reddest asteroids (e.g., 624 Hektor, 1284 Latvia, 236 Honoria), the dark hemisphere of Iapetus (Cruikshank et al. 1981), the Moon (highlands), and Mercury (McCord and Clark 1979).

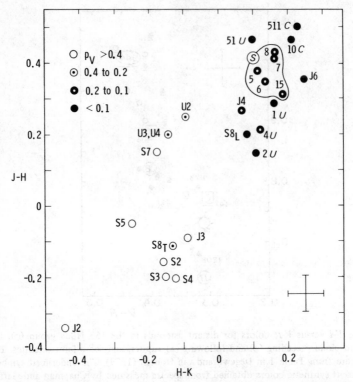

Fig. 5.4. *JHK* color diagram for various satellites and asteroids, with symbols indicative of geometric albedo (see key in upper left). A field of *S*-type asteroids, several *U*-type and *C*-type asteroids are shown. The addition of albedo to this color plot emphasizes that smaller bodies in the solar system divide into objects with dark stony surfaces and bright icy surfaces. From the outer main belt to Saturn, only J4 (Callisto) and the trailing side of S8 (Iapetus) have albedos between 0.1 and 0.4. Symbols: J2 = Europa, J3 = Ganymede, J4 = Callisto, S3 = Tethys, S4 = Dione, S5 = Rhea, J6 = Himalia, S7 = Hyperion, S8$_T$ = trailing side of Iapetus, S8$_L$ = leading side of Iapetus, U2 = Umbriel, U3 = Titania, U4 = Oberon. Cross in lower right corner represents error bars in measurements of J6 and (approximately) S7. (From Degewij et al. 1980*b*.)

While Degewij and van Houten (1979) had classified four Trojan asteroids (including 624 Hektor) and three Hilda asteroids as *RD*, there was no previous evidence that any of the outer Jovian satellites were similarly red. The significance of the Smith et al. (1981) work is that the similarity of some members of the Jovian system to at least some of the outermost asteroids may now be more clear. Smith et al. also observed four Trojans, finding that three of them have steep *RD*-type slopes comparable to that of J8. Other Trojans appear to have *C*-type reflectances (Degewij and van Houten 1979).

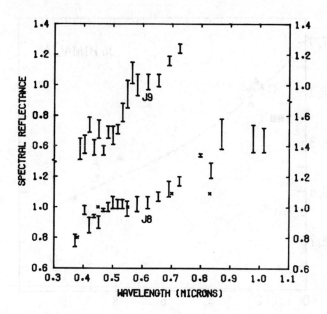

Fig. 5.5. Spectral reflectance of J8 and J9 from observations on a single night (each object) with the Hale 5−m telescope. (From Smith et al. 1981.)

The results of Smith et al. (1981) on the Trojans are in partial disagreement with those of Degewij and van Houten (1979); certain calibration uncertainties remain to be resolved.

Also shown in Fig. 5.5 is the spectral reflectance of J9 (Sinope). These observations reveal an extremely steep slope in the limited wavelength region covered. These and the data for J8, each of which was observed only once, require confirmation, and for the present time are regarded as preliminary. The reflectance shown for J9 does not compare with any other solid surface observed among the satellites or asteroids, and no interpretation is offered at this time.

The Solar Phase Function

Observations of J6 in the V-band over the maximum range observable from the Earth can be seen in Fig. 5.6 (Degewij et al. 1980a). The dashed line is the solar phase function for a typical asteroid; the scatter in the J6 data is partly caused by the rotation of the satellite. J6 shows the surge in brightness at opposition characteristic of bodies having fine-textured regoliths with very little internal light scattering. Thus the photometric data suggest that J6 has a surface analogous to that of a common asteroid of low albedo. There are no equivalent observations of the other outer Jovian satellites.

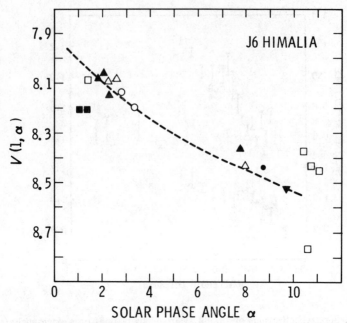

Fig. 5.6. Phase dependence of the V magnitude at unit distance to Sun and Earth for Jupiter's satellite J6 (Himalia) obtained during the apparitions of 1970 (O), 1972 (▲), 1975 (▼), 1976 (△), 1977-1978 (□), and 1979 (■). The data are given in Table III of Degewij et al. (1980*a*). The dashed line is the average phase relation for asteroids as defined by Gehrels and Tedesco (1979), and the scatter of the points can be explained by the rotational light curve (see Fig. 5.7).

Photometric Light Curves and Rotation

The rotation of J6 has been observed photometrically (Degewij et al. 1980*a*). The best data are shown in Fig. 5.7, in which a light curve with two distinct and differently shaped maxima and minima is seen, as in the case of many asteroids. The data in Fig. 5.7 were obtained on two consecutive nights and were combined on the basis of an assumed rotation period of 9.5 hr. Periods in the range 9.2-9.8 hr are possible with the limited data set, but 9.5 hr gives the best fit. More data would improve our confidence in this result. Except for J6, the photometric data are too sparse to derive a solar phase curve or to establish a rotation period. Preliminary observations of J7 by Degewij et al. (1980*a*) suggest a light curve amplitude of ~ 0.5 mag, but confirmation is clearly needed.

II. POLARIMETRY AND RADIOMETRY

A few polarimetric observations of J6 (Himalia) in the interval 1973–1976 have been published (Degewij et al. 1980*a*); the data show a negative branch in the polarization phase curve of depth ~−1.5%, consistent with more extensive data for other bodies of low surface albedo. There is no further information on the polarization of the outer satellites.

The radiometric technique of albedo and radius determination, consisting of observations of the thermal flux of solar system bodies without atmospheres (Morrison 1977), has been successfully applied to more than two hundred asteroids (Morrison and Zellner 1979) and numerous planetary satellites (Murphy et al. 1972; Morrison et al. 1975; Cruikshank 1977, 1979). Cruikshank (1977) observed to 20-μm thermal fluxed of J6 and J7 and concluded that their surface geometric albedos are on the order of 0.02 to 0.03, making them comparable to the low-albedo C-type asteroids. The thermal fluxes of the remaining outer Jovian satellites have so far not been detected, though the flux of J8 should be measurable if this satellite is also a low-albedo object. The others are probably too small for a measurement with the present level of detector sensitivity.

III. DISCUSSION OF THE OBSERVATIONS

The present evidence suggests that at least two outer Jovian satellites (J6 and J7) are comparable to C-type asteroids, and that two others (J8 and J9) are more nearly comparable to RD-type bodies. It is noteworthy that the apparent C-type satellites are both in the prograde cluster and the apparent RD-types are in the retrograde cluster. An important caveat is that only two of the four objects in each of two clusters have been studied, and in most

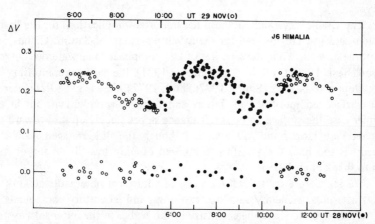

Fig. 5.7. V light curve of J6 (Himalia) by J. Degewij on 28 November UT 1976 (●) with the 1.5-m Catalina reflector and by B. Zellner on 29 November UT 1976 (○) with the 2.3-m Steward reflector of the University of Arizona Observatories. The points at the end are repeated from the beginning, and the lower part of the figure shows the measurements of the comparison star. Data are combined with an assumed period of 9.5 hr.

cases in only a very preliminary way. The limited data suggest that there may be a fundamental compositional difference between the two satellite groups.

An interesting hypothesis concerning the chemical nature of the *RD*-type bodies in the outer solar system has been proposed by Gradie and Veverka (1980). From their laboratory studies of the reflectances (both albedo and spectral reflectance) of meteorites and organic polymers, they suggest that the steep red reflectances and low geometric albedos of *RD* bodies might be explained by the presence of red, opaque organic compounds that are structurally similar to aromatic kerogen. They note that kerogen-like substances comprise the bulk of the organic matter in some carbonaceous chondrites, and propose that this material originated from non-biological processes in the early solar nebula. They simulated the reflectances of *RD*-type objects by making kerogen substances with various grain sizes, noting that contrary to the behavior of many minerals, the reflectance of the kerogens decreases with decreasing grain size. The kerogens were made by removing all solubles from coal tar; the residue is the substance most comparable to the opaque organic material found in the carbonaceous chondrites.

The proposal by Gradie and Veverka in many ways simplifies the picture of the formation of the objects in the outer solar system. Kerogens appear to form at lower temperatures than the carbonaceous materials in carbonaceous chondrites, which may explain why no *RD* asteroids have yet been found in the inner part of the asteroid zone. Gradie and Veverka suggest that the rocky material in cometary nuclei might be similar to *RD*-type bodies rather than to *C* types.

The Gradie-Veverka hypothesis has important implications for the Jovian system and the Trojans, and for the satellite systems of Saturn, Uranus, and Neptune as well. While the satellites of Uranus appear to be water ice covered (Cruikshank 1980; Cruikshank and Brown 1981), the particles comprising the rings appear to be dark (Sinton 1977; Smith 1977); they may be *RD* material. The surface compositions of Triton and Nereid are unknown, but in the former case the surface spectral reflectance slopes steeply upward toward the red. It may have some *RD* material, though the disk-averaged geometric albedo is too high for a surface composed of only low-albedo soil or rock (Cruikshank et al. 1979).

We see before us at least a decade of additional observational work on faint asteroids, planetary satellites and rings, and laboratory compounds, in the effort to understand the chemistry of the bodies of the outer solar system and their possible interrelationships. A major goal of the study of the outer Jovian satellites is to ascertain if some or all of them are captured asteroids or comets. There is a rather large diversity in mineralogy of the asteroids in the main belt, including the Hilda grouping (Chapman and Gaffey 1979; Zellner 1979; Degewij and van Houten 1979). However, the study of the physical

TABLE 5.3

The Outer Satellites of Jupiter — Physical Data

		Mean Visual Magnitude[a] (at opposition)	Diameter (km)	Inverse Mass[c] $(10^{-10} M_J)$
J6	Himalia	14.8	185	136
J7	Elara	16.8	75	9
J8	Pasiphae	17.1	50[b]	3
J9	Sinope	18.3	35[b]	1
J10	Lysithea	18.4	35[b]	1
J11	Carme	18.0	40[b]	1
J12	Ananke	18.9	30[b]	0.6
J13	Leda	20.2	15[b]	0.1

[a]Magnitudes for J6, J7, and J8 from Degewij et al. 1980a, others from estimates by E. Roemer on photographic plates (quoted by Andersson 1974).
[b]Assumes geometric albedo $p_V = 0.03$.
[c]Assumes mean density $\bar{\rho} = 2.6 \text{ g cm}^{-3}$.

parameters of distant asteroids and satellites with diameters between ~ 50 and 200 km beyond the Hilda grouping shows a surprising homogeneity in the optical reflection spectra and albedos; the differences between the apparent RD and C types among the asteroids and satellites are rather subtle, and in the case of the faintest bodies the uncertainties in the measurements of optical parameters make the classification into one type or another somewhat uncertain. There appear to be differences among these bodies along the lines already described, but the small sampling of objects and the large uncertainties in the data force us to be cautious in reaching conclusions on the origin and development of the small bodies of the outer solar system.

The fact that Jupiter strongly influences comets passing nearby (Kresák 1980; Rabe 1971) implies that at least some comets of the Jupiter group may have a genetic relationship with the Trojan clouds. Little can be said about the possible interrelation of Jovian satellites with comets from a compositional point of view, because there exists no reliable physical information about cometary nuclei. This is a topic of much interest, and some modest progress is being made. For example, observing techniques with two-dimensional detectors and special filters are being undertaken to map the gas and dust clouds around cometary nuclei at large solar distances. With special care, it may be possible to subtract the contribution of the comae so that some indication of the spectrum of the nuclear condensations can be derived. Likewise, the spectroscopic study of asteroids in search of comet-like emissions is of interest, particularly for asteroids presently occupying orbits suggestive of cometary origin. Degewij (1980) has made some preliminary observations of this type, but with negative results.

We have noted repeatedly the weakness of the data base upon which our studies of the physical properties of the outer Jovian satellites have been made. This is in large part due to the limited observing time on the very largest telescopes for the kinds of studies discussed here, and in part due to the lack of sensitivity of the detector systems required for diagnostic information on the extremely faint asteroids and satellites. Progress is being made in both areas, with more large telescopes becoming available for solar system studies and with steady progress in detector technology, particularly in the near infrared. We note also that much important information on outer satellites, asteroids, and comets can be obtained simply from repeated visual-band photometry of these objects over a long time base (years) and through the maximum range of viewing geometries. This will provide information on their shapes and rotation as well as the photometric properties of their surfaces. A weak outgassing, for example, will show up immediately. As the study of small bodies in the outer solar system progresses and gains momentum among planetary scientists, special telescopes and detectors (e.g., Weber 1978, 1979; Smith 1979) offer solutions to some of the problems peculiar to these interesting objects. The keystone position that the small bodies hold in the development of our understanding of the origin and evolution of the solar system merits this special attention.

Acknowledgments. D.P.C. acknowledges support from the National Aeronautics and Space Administration, and J.D. acknowledges support as a National Research Council Research Fellow at Jet Propulsion Laboratory. In part, this chapter presents the results of one phase of research carried out under a NASA Contract at the Jet Propulsion Laboratory, California Institute of Technology.

October 1980

REFERENCES

Andersson, L.E. (1974).*A Photometric Study of Pluto and Satellites of the Outer Planets.* Ph.D. Dissertation, Indiana University, Bloomington.
Anonymous (1975). Notes on Jupiter's moons. *Sky Tel.* 50, 380.
Anonymous (1976). All the outer satellites of Jupiter. *Sky Tel.* 51, 242--243.
Bailey, J.M. (1971*a*). Jupiter: Its captured satellites. *Science* 173, 812–813.
Bailey, J.M. (1971*b*). Origin of the outer satellites of Jupiter. *J. Geophys. Res.* 76, 7827–7832.
Bailey, J.M. (1972). Studies on planetary satellites. *Astron. J.* 77, 177–182.
Bowell, E., Chapman, C.R., Gradie, J.C., Morrison, D., and Zellner, B. (1978). Taxonomy of asteroids. *Icarus* 35, 315–335.
Chapman, C.R., and Gaffey, M.J. (1979). Reflectance spectra for 277 asteroids. In *Asteroids* (T. Gehrels, Ed.), pp. 655–678. Univ. Arizona Press, Tucson.
Colombo, G., and Franklin, F. (1971). On the formation of the outer satellite groups of Jupiter. *Icarus* 15, 186–189.

Cruikshank, D.P. (1977). Radii and albedos of four Trojan asteroids and Jovian satellites 6 and 7. *Icarus* 30, 224–230.

Cruikshank, D.P. (1979). The radius and albedo of Hyperion. *Icarus* 37, 307–309.

Cruikshank, D.P. (1980). Near-infrared studies of the satellites of Saturn and Uranus. *Icarus* 41, 246–258.

Cruikshank, D.P., and Brown, R.H. (1981). Uranian satellites: Water ice on Ariel and Umbrich. *Icarus.* In press.

Cruikshank, D.P., Bell, J.F., Beerman, C., and Rognstad, M. (1981). The dark side of Iapetus. Submitted to *Icarus.*

Cruikshank, D.P., and Silvaggio, P.M. (1979). Triton: A satellite with an atmosphere. *Astrophys. J.* 233, 1016–1020.

Cruikshank, D.P., Stockton, A.N., Dyck, H.M., Becklin, E.E., and Macy Jr., W. (1979). The diameter and reflectance of Triton. *Icarus* 40, 104–114.

Davis, D.R. (1974). Secular changes in Jovian eccentricity: Effect on the size of capture orbits. *J. Geophys. Res.* 79, 4442–4443.

Degewij, J. (1978). *Photometry of Faint Asteroids and Satellites.* Ph.D. Dissertation, Leiden University, Holland.

Degewij, J. (1980). Spectroscopy of faint asteroids, satellites, and comets. *Astron. J.* 85, 1403–1412.

Degewij, J., Andersson, L.E., and Zellner, B. (1980a). Photometric properties of outer planetary satellites. *Icarus.* 44, 520–540.

Degewij, J., Cruikshank, D.P., and Hartmann, W.K. (1980b). Near-infrared colorimetry of J6 Himalia and S9 Phoebe: A summary of 0.3–2.2 μm reflectances. *Icarus* 44, 541–547.

Degewij, J., and Van Houten, C.J. (1979). Distant asteroids and outer Jovian satellites. In *Asteroids* (T. Gehrels, Ed.), pp. 417–435. Univ. Arizona Press, Tucson.

Gehrels, T. (1977). Some interrelations of asteroids, Trojans and satellites. In *Comets, Asteroids, Meteorites: Interrelations, Evolution and Origins* (A. Delsemme, Ed.), pp. 323–325. Univ. Toledo, Toledo, Ohio.

Gehrels, T., and Tedesco, E.F. (1979). Minor planets and related objects. XXVII. Asteroid magnitudes and phase relations. *Astron. J.* 84, 1079–1087.

Gradie, J., and Veverka, J. (1980). The composition of the Trojan asteroids. *Nature* 283, 840–842.

Greenberg, R.J. (1976). The motions of satellites and asteroids: Natural probes of Jovian gravity. In *Jupiter* (T. Gehrels, Ed.), pp. 122–132. Univ. Arizona Press, Tucson.

Heppenheimer, T.A. (1975). On the presumed capture origin of Jupiter's outer satellites. *Icarus* 24, 172–180.

Heppenheimer, T.A., and Porco, C. (1977). New contributions to the problem of capture. *Icarus* 30, 385–401.

Kowal, C.T. (1975). Probable new satellite of Jupiter. *IAU Circ.* 2845.

Kowal, C.T. (1976). The case against names. *Icarus* 29, 513.

Kowal, C.T., Aksnes, K., Marsden, B.G., and Roemer, E. (1975). Thirteenth satellite of Jupiter. *Astron. J.* 80, 460–464.

Kresák, L. (1980). Dynamics, interrelations and evolution of the systems of asteroids and comets. *Moon Planets* 22, 83–98.

Kuiper, G.P. (1951) Satellites, comets, and interplanetary material. *Proc. Nat. Acad. Sci.* 39, 1153–1158.

Kuiper, G.P. (1956). On the origin of the satellites and the Trojans. In *Vistas in Astronomy* (A. Beer, Ed.), vol. 2, pp. 1631–1666. Pergamon Press, New York.

Marsden, B.G. (1975). Satellites of Jupiter. *IAU Circ.* 2846.

McCord, T.B., and Clark, R.N. (1979). The Mercury soil: Presence of Fe^{2+}. *J. Geophys. Res.* 84, 7664–7668.

Melotte, P.J. (1908a). Note on the discovery of a moving object near Jupiter (1908 CJ). *Mon. Not. Royal Astron. Soc.* 68, 373.

Melotte, P.J. (1908b). Report on the discovery of a satellite of Jupiter, meeting of March 25, 1908. *Jour. Brit. Astron. Assn.* 18, 231–232.

Morrison, D. (1977). Radiometry of satellites and the rings of Saturn. In *Planetary Satellites* (J. Burns, Ed.), pp. 269–301. Univ. Arizona Press, Tucson.

Morrison, D., and Burns, J.A. (1976). The Jovian satellites. In *Jupiter* (T. Gehrels, Ed.), pp. 991–1034. Univ. Arizona Press, Tucson.

Morrison, D., Jones, T.J., Cruikshank, D.P., and Murphy, R.E. (1975). The two faces of Iapetus. *Icarus* 24, 157–171.

Morrison, D., and Zellner, B. (1979). Polarimetry and radiometry of the asteroids. In *Asteroids* (T. Gehrels, Ed.), pp. 1090–1097. Univ. Arizona Press, Tucson.

Murphy, R.E., Cruikshank, D.P., and Morrison, D. (1972). Radii, albedos, and 20-micron brightness temperatures of Iapetus and Rhea. *Astrophys. J.* 177, L93–L96.

Nicholson, S.B. (1914). Discovery of the ninth satellite of Jupiter. *Publ. Astron. Soc. Pacific* 26, 197.

Nicholson, S.B. (1939). Discovery of the tenth and eleventh satellites of Jupiter and observations of these and other satellites. *Astron. J.* 48, 129–132.

Nicholson, S.B. (1951). An unidentified object near Jupiter, probably a new satellite. *Publ. Astron. Soc. Pacific.* 63, 297–299.

Perrine, C.D. (1905*a*). Discovery of a sixth satellite to Jupiter. *Publ. Astron. Soc. Pacific* 17, 22–23.

Perrine, C.D. (1905*b*). The seventh satellite of Jupiter. *Publ. Astron. Soc. Pacific* 17, 62–63.

Pollack, J.B., Burns, J.A., and Tauber, M.E. (1979). Gas drag in primordial circumplanetary envelopes: A mechanism for satellite capture. *Icarus* 37, 587–611.

Rabe, E. (1971). Trojans and comets of the Jupiter group. In *Physical Studies of Minor Planets* (T. Gehrels, Ed.), pp. 407–412. NASA SP–267, Washington, D.C.

Sinton, W.M. (1977). Uranus: The rings are black. *Science* 198, 503–504.

Smith, Bradford A. (1977). Uranus rings: An optical search. *Nature* 268, 32.

Smith, Bruce A. (1979). Ground-based electro-optical deep space surveillance system passes reviews. *Aviation Week Space Tech.*, Aug. 27, 48–53.

Smith, D.W., Johnson, P.E., and Shorthill, R.W. (1981). Spectrophotometry of J8, J9, and four Trojan asteroids from 0.32 to 1.05 μm. Submitted to *Icarus.*

Weber, R. (1978). The ground-based electro-optical detection of deep-space satellites. *SPIE* 143, 59–69.

Weber, R. (1979). Large-format Ebsicon for low-light-level satellite surveillance. *SPIE* 203, 6–11.

Zellner, B. (1979). Asteroid taxonomy and the distribution of the compositional types. In *Asteroids* (T. Gehrels, Ed.), pp. 783–806. Univ. Arizona Press, Tucson.

6. AMALTHEA

P. THOMAS

and

J. VEVERKA
Cornell University

Voyager images have revealed Amalthea to be an irregular object 270 × 165 × 150 km in size. The spin period is probably synchronous with the orbital period of 11.9 hr, with the long axis pointing toward Jupiter. The satellite's surface is heavily scarred by impact craters, the largest of which has a diameter of 90 km (comparable to the mean radius of the satellite). Amalthea is very dark (reflectance ∼ 5-6%) and very red, but isolated bright spots (reflectance up to 20%) occur. The spectrum of these bright spots is less red and may show an absorption feature near 0.6 μm. It is likely that the surface of Amalthea has been severely altered by its environment and by contamination from Io (especially by sulfur). It may, therefore, be very difficult to obtain definitive information on the composition of the intrinsic Amalthea material from remote sensing measurements.

I. DISCOVERY AND ORBIT

Amalthea was discovered by E.E. Barnard in September, 1892, using the Lick 0.9-m refractor. According to the discoverer, Amalthea is "much more difficult [to see] than the satellites of Mars" (Barnard 1892). No wonder that little was known about this object until the Voyager flyby in March 1979.

Observations by Barnard and others immediately after the discovery established the satellite's orbit to be approximately circular, with a period of 11.92 hours (Table 6.1). More recent positional observations are discussed by Sudbury (1969) and by Pascu (1977). Traditionally much of the interest in

TABLE 6.1

Orbits of Jupiter's Inner Satellites[a]

		Semimajor Axis (R_J)	Period (hr)	e	i (deg)
Ring	Inner Edge	1.72	6.60	–	–
	Outer Edge	1.81	7.12	–	–
Adrastea (1979 J1)		1.79	7.12	–	–
Amalthea		2.55	11.92	0.003	0.4
1979 J2		3.11	16.05	–	–

[a]Ring and satellite data from Smith et al. (1979*b*), Synnott (1980), and Jewitt et al. (1979). 1979 J3 is not included.

Amalthea's orbit has centered around the high rate of nodal regression (period = 0.4 yr), which provides important constraints on the J_2 and J_4 moments of Jupiter's gravity potential (Anderson 1976).

Barnard could never determine satisfactorily the apparent magnitude of his new satellite, but we now know (Sec. IV) that the mean opposition V magnitude is close to +14. Thus, viewed from Earth, Amalthea, which is never more than 1.55 R_J from the limb of the planet, is a hopelessly inconspicuous object. If, on the other hand, it were viewed from the cloud-tops of Jupiter, the story would be very different. Amalthea would be about 40 times brighter than the brightest planet and would rival the faintest Galilean satellite (Callisto) in apparent magnitude. It would move slowly east to west through the Jovian sky, completing one apparent revolution every six Jovian days. Because of its proximity to the planet, the satellite could be seen only within $\sim \pm 60°$ of the equator.

Since Amalthea is so difficult to observe from Earth, almost all of the material reviewed in this chapter is based on observations made by the Voyager spacecraft during the two encounters with Jupiter (Smith et al. 1979 *a,b*; Hanel et al. 1979). To our knowledge, only two non-spacecraft physical measurements of Amalthea have been published to date: a temperature determination by Rieke (1975) and *B, V* photometric measurements by Millis (1978), both of which are discussed below in the context of the more recent Voyager observations. The only pre-Voyager review of our knowledge of Amalthea is that contained in the general article on the satellites of Jupiter by Morrison and Burns (1976).

In addition to discovering a ring around the planet, the Voyagers'

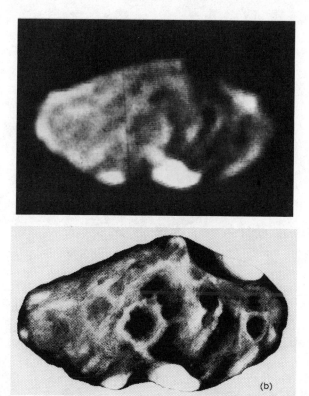

Fig. 6.1. (a) Image of the leading side of Amalthea; Jupiter is to the right, north is at top. Voyager image 16377.34, resolution 13 km/lp. (b) Sketch of leading side of Amalthea by J. Kolva based on entire image sequence 16377.32 to 16377.40.

discovered three small satellites near Amalthea (Jewitt et al. 1979; Synnott 1980; see also Table 6.1). While we still know very little about these bodies, they will be discussed briefly following the review of our current information on Amalthea, since the origins of all four objects may conceivably be related.

II. DIMENSIONS AND SHAPE

Voyager images of Amalthea (Fig. 6.1) have been used to deduce the silhouettes shown in Fig. 6.2 (Veverka et al. 1980). The three views correspond to: (a) looking along the orbital path at the leading face of Amalthea; (b) looking toward Jupiter along the line joining the center of the satellite and the center of the planet; and (c) looking down at the north pole. Silhouette (a) is the best determined; the outline viewed from the north pole is uncertain, but must approximate a diamond shape. Available Voyager data are consistent with the assumption that Amalthea's spin rate is synchronous with its orbital period (12 hr). According to Peale (1976) the time for tidal

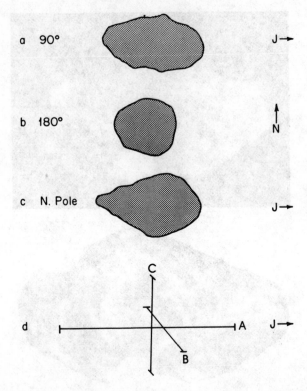

Fig. 6.2. Silhouettes of Amalthea and positions of measured axes. (a) View from leading side. (b) View from anti-Jupiter side. (c) View from above north pole. (d) Axial dimensions and positions. (From Veverka et al. 1980.)

damping of a non-synchronous spin is only on the order of 10^4 yr; thus synchronous rotation is fully expected.

Amalthea's irregular shape cannot be represented well by a triaxial ellipsoid. Veverka et al. (1980) describe the dimensions (Table 6.2) in the manner used for irregular sedimentary particles (cf. Krumbein 1941). The dimensions listed in Table 6.2 are the maximum dimensions along three mutually orthogonal axes, which need not pass through a common point. The longest axis, $2A$, points toward Jupiter; the intermediate axis, $2B$, is parallel to the orbital velocity vector; the shortest axis, $2C$, is orthogonal to the other two. The B axis is $\sim$ 100 km from the sub-Jupiter point on Amalthea; the C axis is $\sim$ 120 km from the sub-Jupiter point. The significant asymmetry of shape relative to the longest axis suggests that the rotation axis of the satellite is displaced toward the sub-Jupiter point, and does not lie halfway along the long axis.

These dimensions for Amalthea are consistent with an earlier estimate obtained by Rieke (1975) from Earth based infrared observations. Rieke

TABLE 6.2

Dimensions of Amalthea

	A	270 ± 10
Diameter	B^a	165 ± 15
of axes		
(km)	C^b	150 ± 10
Mean radius (km)		83 ± 10
Volume (km^3)		$2.4 \pm 0.5 \times 10^6$

[a]B axis is ~ 100 km from sub-Jupiter point.

[b]C axis is ~ 120 km from sub-Jupiter point.

derived a mean radius of 120 ±30 km. Near elongation, when Earth-based observations are usually made, one sees the A-C face of Amalthea – silhouette (a) in Fig. 6.2 – which corresponds to a mean radius of $\sim 100 \pm 10$ km.

The size of Amalthea and its degree of irregularity are comparable to those of the Trojan asteroid Hektor. Hartmann and Cruikshank (1980) find the longest and shortest dimensions of Hektor to be ~ 300 and 150 km (A/C = 2). The A/C ratio of 1.8 for Amalthea is slightly smaller than that of Hektor, but is larger than that for the satellites of Mars (~ 1.5). For another comparison, Eros, one of the more elongated asteroids, has an A/C ratio of 2.8 (Zellner 1976).

The A/C and A/B ratios fall well within the field of ratios typical of products of collisional fragmentation (Fujiwara et al. 1978). The sharp ridges and large concavities seen on this satellite, as well as its asymmetric shape, are all suggestive of fragmentation processes. The shaping of an object by cratering, spallation, and complete fragmentation can lead to extremely irregular shapes, provided that debris is not retained in such a manner as to smooth out the shape. Obviously, such rounding has not occurred on Amalthea. Even though the satellite has large craters, those 20 – 90 km in diameter could have excavated only a few percent of the satellite's volume. The implications of Amalthea's observed shape on the question of the satellite's effective internal strength are discussed in Sec. VIII below.

III. MAJOR SURFACE FEATURES

A sketch map of the known surface features of Amalthea is shown in Fig. 6.3. This map (Veverka et al. 1980) is based on all available Voyager images,

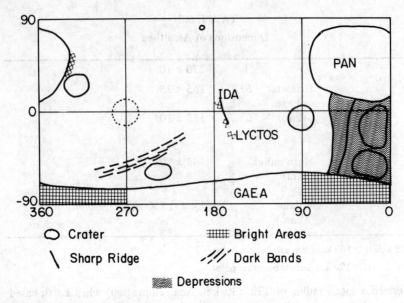

Fig. 6.3. Sketch map of Amalthea. (From Veverka et al. 1980.)

the best of which have a surface resolution of 8 km/line pair (1p). The coordinates used are very approximate and are planetocentric, spherical coordinates of points on the surface measured from the center of the long axis; longitude is measured west from the sub-Jupiter point.

At least two round depressions, probably impact craters, are clearly visible (Fig. 6.1, 6.3). Five other possible craters are noted on the map. The two largest craters have been named Pan and Gaea by the International Astronomical Union (IAU) (Fig. 6.3).

Near latitude $0° - 60°$ W between Pan and Gaea is a region of distinctive complex topography, consisting of a series of troughs and ridges, or possibly large, nested, noncircular depressions. These features have length scales of tens of kilometers and are well over 20 km wide. Their depths are unknown but could exceed several kilometers. The anti-Jupiter point lies near a ridge which extends some 40-50 km in length. The ends of the ridge are marked by bright patches, each roughly 15 km across, which in the IAU nomenclature are designated as Ida and Lyctos (Fig. 6.3). Bright areas also occur inside Gaea and on the eastern rim of Pan.

IV. PHYSICAL CHARACTERISTICS OF THE SURFACE

Photometric and Spectral Properties

The photometric properties of Amalthea derived from Voyager observations by Veverka et al. (1980) and from telescopic measurements by

TABLE 6.3

Photometric Properties of Amalthea

	Voyager clear filter[a]	Earth-based[b]	Derived[a]
Opposition magnitude	$V_o = +14.07$	$V = +14.1 \pm 0.2$ $B\text{-}V = +\ 1.5 \pm 0.3$	
Phase coefficient	$\beta = 0.042 \pm 0.004$ mag/deg		
Geometric albedo	$p = 0.056$		
Phase integral			$q \lesssim 0.3$
Bond albedo			$A \lesssim 0.015$

[a]From Veverka et al. (1980).
[b]From Millis (1978).

Millis (1978) are summarized in Table 6.3. The mean opposition magnitude of + 14.07 derived from the Voyager data (clear filter with effective wavelength $\sim 0.5\ \mu$m) agrees very well with Millis' value of $V = 14.1 \pm 0.2$ for Amalthea at western elongation. The Voyager data indicate that the brightness of the leading and trailing sides of the satellite is the same to within 0.1 mag. Veverka et al. (1980) adopt a mean normal reflectance of 0.056 as representative of Amalthea in the Voyager clear filter. The individual values are 0.054 for the trailing side and 0.059 for the leading side, values indistinguishable within the accuracy of the determination (± 0.010).

Voyager 1 obtained several images of Amalthea at a solar phase angle of only $0°\!.8$. Very little, if any, limb darkening is present in these images (Fig. 6.4), as is to be expected for such a dark surface which probably has the fairly intricate texture characteristic of regoliths. Hence Amalthea's geometric albedo can be taken to be equal to the normal reflectance of the surface.

The disk-integrated phase coefficient of Amalthea derived from observations between phase angles of $0°\!.8$ and $42°$ is 0.042 ± 0.004 mag/deg, a value consistent with that measured for other objects having rough, very dark surfaces. For example, Andersson (1974) found a value of 0.04 mag/deg between $2°$ and $8°$ for Himalia (J6), while Klaasen et al. (1979) quote values of 0.039 and 0.036 between $0°$ and $20°$ for Phobos and Deimos, respectively. According to Bowell and Lumme (1979), a value of 0.036 is typical for C asteroids, whose average geometric albedo is given by those authors as 0.05.

It is interesting that the phase curve published by Veverka et al. (1980) does not show any opposition effect, but this may be an artifact of having only five data points over the phase angle range in question. From the phase

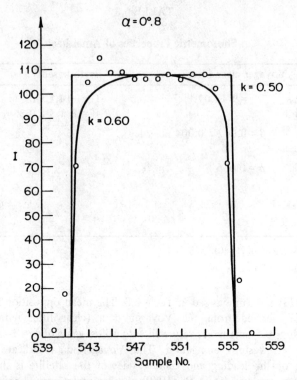

Fig. 6.4. Intensity along photometric equator of Amalthea viewed at phase angle of 0°.8, showing a minimal amount of limb darkening.

curve, Veverka et al. (1980) estimate that Amalthea's phase integral is less than 0.3. For comparison, Klaasen et al. (1979) found values of 0.27 and 0.32 for Phobos and Deimos, respectively. The conclusion is that Amalthea must be close to a blackbody, with a Bond albedo of less than 0.015.

The observations of Millis (1978) gave a *B-V* color for Amalthea of +1.5 ± 0.3, making Amalthea the reddest known object in the solar system. The Voyager data generally support this finding, although they suggest that Amalthea may be somewhat less red than found by Millis, and specifically less red than some areas on Io (Fig. 6.5). Amalthea is definitely much redder than the Trojan asteroids. From the compilation of Degewij and van Houten (1979), we find that a typical *B-V* color for Trojans is between 0.7 and 0.8, while typical geometric albedos are near 0.03. Amalthea is about twice as bright as a typical Trojan, and very much redder. The Voyager data suggest that the trailing side of Amalthea is very slightly redder than the leading side (Fig. 6.6).

The absolute normal reflectance of the leading side of Amalthea is shown as a function of wavelength in Fig. 6.7. Also shown are data for two of the remarkable bright spots mentioned in Sec. III. The bright spots are between

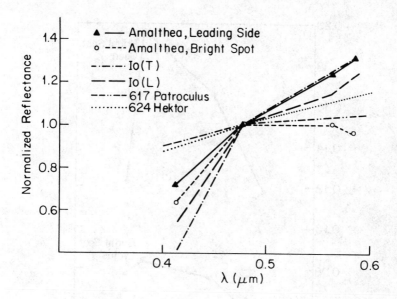

Fig. 6.5. Colors of Amalthea compared to leading and trailing sides of Io and to asteroids 617 Patroculus and 624 Hektor. Reflectances are normalized at 0.48 μm (clear filter). Triangles show leading side of Amalthea, circles bright spot Lyctos.

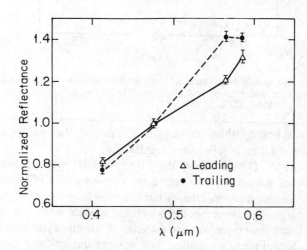

Fig. 6.6. Colors of leading and trailing sides of Amalthea. Reflectances normalized at 0.48 μm. Error bars are estimates of uncertainty and are not formal errors. (From Veverka et al. 1980.)

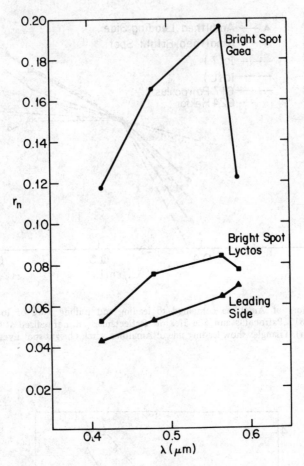

Fig. 6.7. Estimated normal reflectances of three areas on Amalthea: entire leading side, brightest part of crater Gaea, and bright spot Lyctos at anti-Jupiter area. (From Veverka et al. 1980.)

1.5 and 3 times brighter than the background material; they are less red and show evidence of a drop in reflectance near 0.6 μm.

Gradie et al. (1980) argue that the unusual spectrophotometric characteristics of Amalthea — the very dark, very red aspect of most of the surface, and the distinctive spectral signature of the bright spots — probably result from the severe environment in which the satellite exists (see Sec. X). They suggest that the original surface material of Amalthea has been altered strongly by charged particle radiation, high-velocity impacts, and contamination from Io. The most likely contaminant which has reddened the surface is a combination of sulfur allotropes and sulfur compounds. Gradie et al. (1980) have succeeded in reproducing the spectral reflectance curve of Amalthea by contaminating a carbonaceous chondrite analog with "brown sulfur," although they point out that such a simulation is certainly not unique in terms of the background material. In fact, they conclude that the available spectral reflectance data contain little information about the bulk composition of the satellite, because sulfur and sulfur compounds have probably masked the spectral characteristics of the intrinsic surface material.

The composition of the greenish bright spots remains an enigma. Commonly expected materials do not have deep absorptions near 0.6 μm; certain allotropes of sulfur do, and might predominate in these areas (cf. Sec. XI). Alternatively, as will be discussed at length below, these spots may represent regions of recently exposed material that may contain less sulfur-rich glass. Gradie et al. (1980) note that due to the very high impact velocities at Amalthea, glass may be an important constituent of the regolith. The production of sulfur-containing glass is an attractive possibility in explaining the colors seen on the satellite. Silica glass with small amounts of sulfur can be green or blue in color, whereas glass containing larger amounts of sulfur can be a deep brown.

Surface Temperature

A thermal infrared spectrum of Amalthea was obtained by the Voyager IRIS instruments (Hanel et al. 1979). The data indicate a blackbody of about 180 ± 5 K (J. Pearl, personal communication). This unexpectedly high disk-averaged temperature is higher than the Earth-based measurement of Rieke (1975), who found a 10 μm color temperature of 155 ± 15 K. A disk-averaged temperature of 180 K is difficult to reconcile with solar heating alone; simple solar heating models give a maximum surface temperature on Amalthea of about 164 K, even if the surface is assumed to have zero albedo (Fig. 6.8).

Detailed modeling of Amalthea's thermal state is certainly in order. Since the satellite orbits Jupiter synchronously every 12 hr, and one hemisphere suffers an eclipse every orbit, much of it radiates to a 125 K blackbody (Jupiter). In addition to solar radiation, charged particles and inductive Joule heating may provide energy. The latter is probably insignificant due to Amalthea's low velocity with respect to the lines of Jupiter's magnetic field. Protons and electrons certainly deposit energy within the surface layer, but it is not clear that they can account for the ~ 20 K discrepancy that seems to exist.

V. CRATER MORPHOLOGY

The two largest craters on Amalthea, Gaea with a diameter of 75 km and Pan with a diameter of 90 km, were formed in a very irregular surface and have very irregular rim crests. The floors of both craters are visible in some

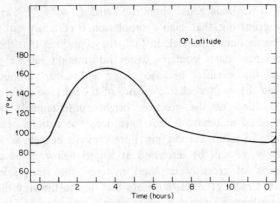

Fig. 6.8. Calculated surface temperature during one rotation of Amalthea for a point on the equator and on the anti-Jupiter side. Assumed parameters: conductivity = 8×10^1 erg cm^{-1} s^{-1} K^{-1}; density = 1.0 g cm^{-3}; specific heat = 4×10^6 erg g^{-1} K^{-1}; emissivity = 1; and albedo = 0. The maximum temperature is $\sim$165 K. ($t = 0$ corresponds to sunrise). Graph courtesy of D. Simonelli.

images, and crude estimates of the depths can be made. The depth of Pan is at least 8 km and probably twice this value; Gaea is at least 10 - 20 km deep. While these estimates are uncertain, there is no doubt that these two craters are significantly deeper than craters of a similar size on the Moon. From the data of Pike (1977), craters on the Moon having diameters between 75 and 90 km are typically less than 5 km deep. If simple gravity scaling of the diameter at the transition from bowl-shaped to more complex craters is applied (Pike et al. 1980; Hartmann 1971), then all craters that can form on Amalthea should be bowl-shaped.

The critical diameters at which craters change from bowl-shaped to flatter, more complex depressions is shown in Fig. 6.9 as a function of surface gravity; the data from Earth, Mars, Moon, and Mercury are from Pike et al. (1979). The anomalously low transition diameter of craters on Mars has been attributed to subsurface volatiles (Cintala and Mouginis-Mark 1980). In Fig. 6.9 Amalthea's position is plotted as a minimum because the transition diameter has probably not been observed. The apparent bowl shapes of even the largest craters on Amalthea are consistent with gravity scaling of the transition diameter of craters. Since the largest possible crater that can form on a small body of diameter D can have a diameter of only $0.5 - 0.6\ D$ (Housen et al. 1979; Thomas and Veverka 1979), Pan and Gaea are among the largest bowl-shaped craters one could expect to find anywhere in the solar system.

VI. CRATER PRODUCTION ON AMALTHEA

Shoemaker and Wolfe (see Chapter 10) estimate a crater production rate

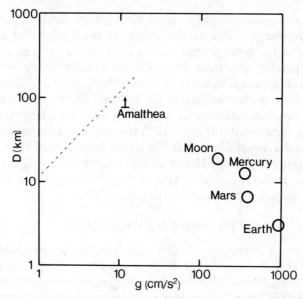

Fig. 6.9. Diameters of transition from simple to complex craters for Amalthea and other bodies. Data for the Moon, Mercury, Mars, and Earth are from Pike et al. (1979). Dashed line indicates the approximate limiting value of crater diameter for strength-dominated objects of density 3.0. The transition diameter on Amalthea is apparently greater than 90 km, and would be close to the estimated largest crater possible on the satellite.

in the Galilean system such that the rate at Callisto is only about one-half that on the Moon. Yet the large focusing effect of Jupiter's gravity greatly increases both the flux and velocity of particles as one goes from the orbit of Callisto (26.6 R_J) to that of Amalthea (2.6 R_J). If an average approach velocity relative to Jupiter of 8 km s^{-1} applies (cf. Smith et al. 1979b), impact velocities on Amalthea will be $\sim$ 38 km s^{-1}, compared to 14 km s^{-1} on Callisto, and the flux will be increased by a factor of $\sim$ 7.3. The actual enhancement in the production rate of craters will depend both on the relation between impact energy and crater diameter, and on the size distribution of impacting objects. The diameter-energy relation probably lies between $D \propto E^{0.33}$ and $D \propto E^{0.29}$ (Dence et al. 1977; Housen et al. 1979), or, in terms of velocity, between $D \propto V^{0.66}$ and $D \propto V^{0.58}$.

On Callisto, the number of craters per unit area varies approximately as D^{-2} to $D^{-2.5}$ (Smith et al. 1979b). Thus, increasing the impact velocity by a factor of 2.7 will increase the crater density at a particular diameter by 3 to 5 times. In combination with the higher flux, the rate at Amalthea increases by a total factor of $\sim$ 22 to 37 times relative to Callisto.

However, a given impact energy may produce a larger crater in ice (Callisto) than in rock (Amalthea) (Croft et al. 1979; Cintala et al. 1979).

Because we are considering large craters where gravity may be important, this enhancement may be much less than a factor of two (Cintala et al. 1979). We conclude that the crater production rate at Amalthea has been at least ten times and possibly forty times that at Callisto. It probably has been at least four times, and perhaps 20 times, as large as that on the Moon. Whatever the absolute crater production rate in the Galilean satellite system may have been, the surface of Callisto is nearly saturated with craters larger than 20 km in diameter (Smith et al. 1979b) ($\sim 10^{-4}$ craters/km^2). A production rate ten times this value would form about 100 craters larger than 20 km in diameter on Amalthea, and probably 10 over 50 km in diameter. Thus, the presence of at least two craters larger than 50 km in the available images of Amalthea is entirely consistent with the cratering record on Callisto.

VII. REGOLITH FORMATION

The intense bombardment that Amalthea has undergone must have produced abundant loose material for a regolith; in fact, it should have produced a highly crushed and fractured megaregolith similar to that in the highlands of the Moon. The disposition of the ejecta on Amalthea should be intermediate between that on the Moon ($g \sim 162$ cm s^{-2}) and on Deimos ($g \sim 0.2$ cm s^{-2}). Cratering on Amalthea occurs in a gravity of ~ 10 cm s^{-2} (assuming $\rho = 3$ gm cm^{-3}) and a velocity of escape into Jupiter orbit of ~ 100 m s^{-1}. Ejecta should be spread farther on Amalthea than on the Moon, though it is probable that only the largest craters send significant amounts of material into Jupiter orbit (cf. Housen et al. 1979; Cintala et al. 1978; McGetchin et al. 1973). The asteroidal models of Housen et al. (1979) suggest that only 10% of the ejecta from an asteroid 300 km in diameter would be lost. On this basis we would expect that somewhat more ejecta would be lost temporarily from Amalthea (mean diameter ~ 170 km). While some of this lost material may be swept up again later (Soter and Harris 1977; Dobrovolskis and Burns 1980), we expect that a substantial fraction of debris is retained as localized ejecta blankets; however, the low resolution of the Voyager images does not permit a test of this idea.

The total amount of ejecta produced on Amalthea may have been considerable if the large craters really are as bowl-shaped as they appear. The crater Gaea ($D = 75$ km), if approximated by a spherical segment of depth 14 km, has a volume of 2.8×10^4 km^3. Spread over the satellite, this volume could make a layer 280 m thick, and if confined to a quarter of the satellite, such a volume of material could produce a deposit one kilometer in thickness. Thus the few very large craters that we see on Amalthea should have generated a considerable amount of regolith. The models of asteroidal regoliths of Housen et al., if applied to Amalthea, suggest regolith depths could reach more than one km. Additionally, the extremely high density of craters larger than 20 km in diameter, that we infer to have formed on Amalthea (about one per 10^3 km^2), implies nearly complete excavation of one to two km of material. Because it appears to be difficult to lose a significant fraction of this material permanently from Amalthea, a layer of crater ejecta of such depth can be expected to have accumulated over the satellite's surface. However, it would be wrong to infer that the bulk of this layer has the fluffy texture and high porosity of the extreme lunar surface. Confining pressures at a depth of one km on Amalthea reach two bars, and much higher dynamic pressures were involved during the emplacement of these ejecta, some of which may have consisted of very coarse fragments. Crudely scaling from lunar and terrestrial analogy (Gault et al. 1963), we can expect that blocks as large as 300 meters across were produced during the formation of a crater as large as Gaea.

VIII. STRENGTH OF AMALTHEA

Two different, indirect means of estimating the strength of small satellites have been suggested: tidal distortion (Soter and Harris 1977) and survival following large impacts (Pollack et al. 1973). In this section we examine their application to Amalthea. Soter and Harris (1977) have suggested that the density of satellites of negligible strength could be determined if their shapes are distorted into Roche ellipsoids by planetary tides. This approach assumes that the satellite effectively cannot support shear stress; the ultimate shape is determined by the mean density and the tidal field.

This method clearly does not apply to Amalthea because the satellite's shape does not approximate an ellipsoid. Our best estimate of the actual volume of Amalthea is ~ 2.4 $\times$ 10^6 km^3 (Veverka et al. 1980). A circumscribed triaxial ellipsoid would have axes of at least 275, 180, and 165 km, and a total volume of 4.3 $\times$ 10^6 km^3 — nearly double the actual volume. Using the axes measured (as described in Sec. II) as the axes of a triaxial ellipsoid would give a volume of 3.7 $\times$ 10^6 km^3, still far different from the actual volume. It is basically impossible to approximate Amalthea's irregular shape by any triaxial ellipsoid. This fact should also be evident by simply considering the significant asymmetry along the A axis (Fig. 6.2c) and the straight-line ridge crests which are obvious in the images.

It should be emphasized that there are at least three ways of producing an ellipsoidal satellite under the influence of tides:

1. Distortion of a fluid body;
2. Distortion of an initially solid sphere into an ellipsoid;
3. Modification of an irregular shape into an ellipsoidal one by gradual erosion and sedimentation.

The first probably cannot apply to any satellite in the solar system. The second could result only in a small deviation from sphericity since the distortion would effectively be elastic (flow at the low temperature and small shear stress applicable to Amalthea would be negligible unless the satellite is made of ice). The third could hypothetically form a Roche ellipsoid (and could also apply to accretionary processes), but even if it does occur in nature it clearly has not progressed very far on Amalthea.

If Amalthea has more than zero shear strength two possibilities arise: it can be a single strong object, or it can be a relatively weak one reassembled by the agglomeration of smaller objects. The latter scenario itself has two end members: one in which the satellite is made of a few large fragments, the other in which it has the consistency of loose sand. The loose sand model could evolve toward an ellipsoidal shape under impact reworking. It clearly has had time to do so, as indicated by the large craters, but has not.

Could Amalthea be an agglomeration of a few large fragments, similar to the configuration suggested for the asteroid Hektor by Hartmann and Cruikshank (1980)? One possible version of this scenario would have one large fragment, $\sim$ 160 km across, on which crater Pan and possibly part of crater Gaea are superposed, and a smaller very angular fragment extending to the anti-Jupiter point (see Fig. 6.2). The smooth area between longitudes 180° and 270°, and also probably between 90° and 180°, would require considerable sedimentary filling between the two fragments, since one could not expect the original fragments to fit exactly at their interface. However, such an arrangement of fragments, or any other, is unlikely in the case of Amalthea for at least three reasons: a) the linear feature on the trailing side suggests structural continuity between the sub- and anti-Jupiter points; b) the ridge between Ida and Pan also suggests structural continuity across the satellite; and c) Lyctos and Ida are on a part of the satellite that is almost blade-shaped (Fig. 6.2) – reaccumulation of such a long splinter on end seems most unlikely. If Amalthea were composed of several smaller fragments, they would have reaccumulated into a more symmetrical object. To us, the available evidence suggests that Amalthea is a single, rigid object which has survived a long history of collisional evolution. This single-fragment view has strong implications for the strength and history of this satellite. An important question concerns the lifetime of an object the size of Almathea against catastrophic fragmentation by impact.

The energy involved in completely fragmenting objects has been investigated experimentally by Gault and Wedekind (1969), Fujiwara et al. (1977, and Hartmann 1978), among others. These small-scale experiments show that a progression from small impact craters to significant spallation to complete fragmentation occurs over an impact energy range of only a factor of ten. For example, impacts into basalt at energies of 10^6 ergs cm^{-3} (of target) produce craters, but impacts of 10^7 ergs cm^{-3} cause complete

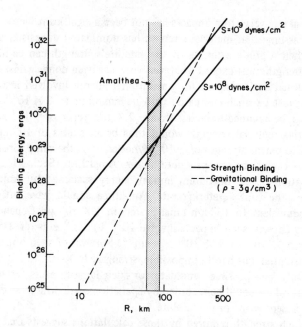

Fig. 6.10. Strength and gravitational binding energies for spherical objects as a function of radius. Gravitational strength is for a density of 3 gm cm^{-3}; material strengths are tensile strengths. Position of Amalthea is shown as 83 ± 10 km mean radius.

fragmentation. The impact energies needed for fragmentation usually are less than the binding energies calculated from tensile strength and target volume. Small samples of basalts tend to have tensile strengths in excess of 10^8 dynes cm^{-2}, and thus, calculated binding energies greater than 10^8 ergs cm^{-3}.

The strength binding energy of spherical objects can be expressed as $E_s = \dfrac{4\pi}{3} R^3 S$, where R is radius and S is tensile strength. The gravitational binding energy is

$$E_g = \frac{3G}{5} \left(\frac{4\pi}{3}\right)^2 \rho^2 R^5 = 10.5\, G\, \rho^2\, R^5. \tag{1}$$

These two quantities are plotted in Fig. 6.10 for objects with a density of 3 gm cm^{-3} and tensile strengths of 10^7 and 10^8 dynes cm^{-2} as a function of radius. Objects as small as Phobos with a radius of 12 km are entirely strength dominated (if composed even of very weak rock). Objects of a radius 500 km are gravitationally bound. Amalthea, if made of common rocks, is clearly in a transition region, as are many asteroids. The irregular shape implies that gravitational recovery from impacts has not been a significant factor.

If we assume that material strength has dominated its resistance to large impacts, then a crude estimate of the satellite's strength can be made. The material strength must exceed whatever work has been done against gravity in changing Amalthea's shape. The gravitational energy involved in modifying the shape must be small compared with its maximum ($\sim 2 \times 10^{29}$ ergs for ρ = 3) and can be assumed to be less than $\sim 2 \times 10^{28}$ ergs, or $\sim 10^7$ ergs cm^{-3}. The effective material strength would thus be in excess of 10^7 ergs cm^{-3}, equivalent to tensile strength of $\sim 10^8$ dynes cm^{-2} – the value for solid rock.

Estimates of crater production on Amalthea (Sec. VI) allow an approximation of the maximum impact energy sustained by Amalthea. At four times the lunar cratering rate, Amalthea should have suffered four impacts equivalent to 100-km lunar craters in 4×10^9 yr. Depending upon the scaling chosen, such impacts involve 10^{29} to 10^{30} ergs (cf. Dence et al. 1977), or 5×10^7 to 5×10^8 ergs cm^{-3}. Survival of such impacts would imply substantial, but hardly impossible, strengths (5×10^8 to 5×10^9 dynes cm^{-2}). Only one possible candidate for such large impacts (crater Pan) is visible, but others could be recorded by spallation scars, which are compatible with the present shape of the satellite.

The high strength required by these calculations suggests that Amalthea may not have survived in its present form for a full 4.5×10^9 yr. Its shape, in fact, practically demands that it be the end product of a process of collisional fragmentation from a larger body, and that this process be one in which the bulk of the fragmentation products were not reaccreted by the current satellite.

IX. SURFACE PROCESSES

Even if Amalthea's shape is maintained by material strength, gravity may be important in determining the nature of the damage associated with the formation of the largest craters. Strength dominated objects can show severe spallation and fracturing at points antipodal to large impacts. Large impacts on gravitationally dominated objects can produce massive disruption of the antipodal area by landslides, faulting and associated processes (Schultz and Gault 1975). The likely presence of a thick regolith on Amalthea and the nonnegligible escape velocity of ~ 100 m s^{-1} suggest that large craters could have produced severe antipodal effects in the regolith (if the highly irregular shape of Amalthea permits the focusing of seismic waves) in addition to possibly severe fracturing of the body near these craters. Massive movement of loose materials, such as landslides on large long regional slopes, may have occurred, but if so, it has not smoothed significantly the ridges and depressions.

Gaea and Pan are nearly antipodal to each other; Gaea is probably the younger feature. Antipodal effects associated with the impact that formed Gaea may well have smoothed off parts of Pan. Severe shaking from other impacts could cause regolith movement on the inner walls of Gaea. The walls of this crater may reach heights of 20 km near longitudes $0°$ and $180°$ and have slopes of $\sim 30°$, making these crater walls the highest scarps yet discovered in the solar system! The steepness of these slopes suggests that the potential for slumping is large, and the great height of the scarps means that massive landslides could occur. Watching part of the rim of Gaea slump would be spectacular; free fall time to the base of the scarp is on the order of ten minutes. The extremely complex topography between Gaea and Pan may be the side effects of the formation of these craters. Mobilization of a one to two km regolith by seismic waves and severe faulting of the solid body of Amalthea could form the rough topography with relief of the order of one to two km (see above).

Features analogous to the grooves on Phobos would not be visible in the Voyager images. The low resolutions (8 km/lp, at best) might allow detection of long linear features only a few kilometers across, but isolated objects would have to be at least ten km across to be seen. According to the model of Thomas et al. (1979), the width of Phobos-like grooves is limited to a few times the depth of the loose regolith. While one to two km of regolith may have been produced on Amalthea (see above), only a few hundred meters of this can be expected to be loose. Thus grooves on Amalthea, if they existed, would probably be less than one km across, and would not be conspicuous in the Voyager images.

X. ENVIRONMENT OF AMALTHEA AND CONTAMINATION OF THE SURFACE

At 2.55 R_J from Jupiter, Amalthea is not only deep within the gravity field of the planet but also in a very intense part of the magnetosphere. Typical impact velocities of micrometeoroids with Amalthea average about 38 km s^{-1}. Pioneers 10 and 11 found that Amalthea absorbs both keV and MeV protons and electrons (Simpson and McKibben 1976; McDonald and Trainor 1976).

Gradie et al. (1980) review the processes that may contaminate and alter the surface of Amalthea. Generally these processes can be divided into those involving charged particle bombardment and those of micrometeoritic impact. In the relatively benign inner solar system, micrometeoritic impacts are generally more efficient at darkening and reddening regoliths than is the solar wind (Matson et al. 1977). However, at Amalthea both the micrometeorite and charged particle environments are more severe. Proton energies are typically in the MeV range, rather than in the keV range, as in the solar wind (Divine 1979; Bridge et al. 1979). Uncharged micrometeorites that originate outside the Jovian system have average impact velocities on Amalthea of about 38 km s^{-1}; charged dust, much of which originates at Io (Johnson et al. 1980), impacts at an average of 60 km s^{-1}

Gradie et al. (1980) note that while proton bombardment probably causes some regolith darkening and reddening, the impacts of heavier ions and dust have far more drastic effects. The composition of heavy ions and dust diffusing away from Io is largely S, O, and Na (Krimigis et al. 1979). Charged dust impacting at 60 km s^{-1} probably produces abundant melt in the ejecta, providing the surface with a significant glass component whose composition reflects that of both the regolith and the impacting material. Although Amalthea has a low escape velocity ($\sim$ 100 m s^{-1}), most glassy ejecta may be recaptured even if initially lost to Jupiter orbit. Thus an Io-contaminated, glass-rich regolith could form on the satellite.

It is known that the addition of varying amounts of sulfur to silicate melts produces vividly colored glasses, the colors of which can range from the blue end of the spectrum to the red (Weyl 1959; Fuwa 1937). What causes the specific colors is poorly understood and no doubt complex. Fuwa believed that the amount and oxidation state of iron — which is controlled by the presence of sulfur — was the crucial factor in determining the color of the glass produced. Colors also depend on the cooling history of the glass (S. Ostro, personal communication, 1980).

Since we know so little about what goes on near the surface of Amalthea, and not much more about the conditions under which sulfur glasses of different colors can be produced, we cannot prove that such glasses are an important constituent of Amalthea's regolith. But certainly they are a possible constituent, along with elemental sulfur in various allotropic forms and probably even simple sulfur compounds. The relative amounts of these possible constituents cannot be determined from existing data.

XI. BRIGHT PATCHES

The material in the bright patches must differ in some way — composition, texture, or phase — from the rest of the surface. The presence of a probable dip in the spectrum of the bright material near 0.6 μm strongly suggests that the difference is not simply one of texture. By reducing the mean particle size of material one can often increase its albedo, but at the same time the contrast of absorption features decreases. In attacking the nature and origin of these bright spots, three important facts should be kept in mind:

1. The location of the spots on relatively steep slopes (topographic highs and crater walls);
2. The spectral dip near 0.6 μm;
3. The likely contamination of Amalthea's surface by material from Io.

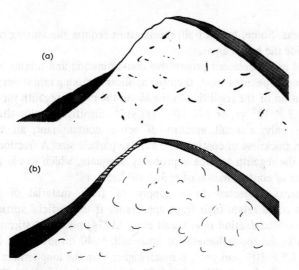

Fig. 6.11. (a) Schematic cross section of part of Amalthea showing bright areas as exposures of bedrock and dark red areas as regolith. (b) Schematic cross section showing bright areas covered by regolith, but more rapidly stirred, or otherwise distinct from the regolith on more gentle slopes.

Possible models of origin can be grouped into two very broad categories (Fig. 6.11). The first considers the bright spots to be exposures of a relatively fresher surface material. Most of Amalthea is covered by a dark red regolith whose spectrophotometric properties reflect an equilibrium result of the interaction of Amalthea's environment on pristine surface material. Downslope movement of this regolith — expected to be especially effective on steeper slopes — exposes fresher less modified (and less dark and less red) material. The fresher material could be "bedrock" or simply less weathered regolith — less weathered because it was until fairly recently buried and protected from the surface environment. A second extreme possibility is that both the dark, red areas and the bright spots owe their spectrophotometric characteristics to outside contamination. The difference would be that in the bright spots, as a result of possible regolith mixing by downslope movement, the contamination would be less developed (see below).

We exclude any schemes in which pervasive dark, red material represents the uncontaminated, pristine material of Amalthea for reasons discussed in Sec. IV, and elaborated upon by Gradie et al. (1980).

Mechanisms requiring that the bright areas be exposed bedrock are difficult to reconcile with their color (Fig. 6.5). Gradie et al. (1980) note that the absorption band near 0.6 μm is most compatible with allotropes of sulfur, strongly implying that external contamination is responsible for producing the characteristic spectral reflectance curves. Both categories of mechanisms require that the contamination rate exceed the meteoritic stirring rate outside the bright areas. Some, but not all, mechanisms require the stirring rates to dominate inside the bright areas.

If the red areas represent progressive contamination and mixing in areas with little downslope movement, then the applicable mixing rate is simply the rate of formation of the regolith. On the Moon the rate of regolith formation is ~ 10 m in 3×10^9 yr, or 3×10^{-7} cm yr^{-1}. Significant color alteration need involve only a small amount of actual contaminant, an amount dependent on thickness of coatings and average particle size. A fraction equal to one-tenth the regolith volume is probably adequate, which would require an average rate of contamination of $\sim 3 \times 10^{-8}$ cm yr^{-1}

Bright areas subjected to a supply of fresh material or altered contaminants could retain their fresh appearance if the surficial stirring rate exceeded the contamination rate. Gault et al. (1974) note that stirring rates are very high in the top millimeter of lunar soil, ~ 40 turnovers per 10^6 yr. This rate, $\sim 2.5 \times 10^{-5}$ cm yr^{-1}, is much higher than the long-term rate, and could help keep fresh material exposed. On Amalthea the effective rate could be much higher; 10^{-3} cm yr^{-1} is a good upper limit, equivalent to $\sim 10^{-4}$ cm yr^{-1} of contaminant. The minimum contamination thus is $\sim 3 \times 10^{-8}$ cm yr^{-1} for a lunar-like flux, and 3×10^{-6} cm yr^{-1} if the rate of regolith formation is 100 times that on the Moon. If the re-exposure rate is the important quantity, rather than bulk contamination of the regolith, then the maximum rate could be as high as 10^{-4} cm yr^{-1}.

Can this much material be supplied from Io? To obtain a lower limit on the rate of supply, we can simply convert the rates on Amalthea to rates on Io by the ratio of their surface areas, $\sim 3 \times 10^{-3}$. The higher rate on Amalthea, 3×10^{-6} cm yr^{-1}, is equivalent to 10^{-8} cm yr^{-1} on Io; the lower, 3×10^{-8}, is equivalent to 10^{-10} cm yr^{-1} on Io. Johnson et al. (1979) note that the supply of S to the Io torus corresponds to a surface loss on Io of $\sim 2 \times 10^{-6}$ cm yr^{-1}. If the minimum stirring rate applies on Amalthea, then perhaps only $\sim 10^{-4}$ of S lost from Io to the torus need coat Amalthea. If the higher stirring rates apply, then $\sim 1\%$ needs to be captured by Amalthea. If the highest surficial stirring rates apply, then 30% needs to be captured. We know of no calculations of the efficiency with which ions will be transferred from Io to Amalthea. In fact, we do not even know if the net effect of such incoming ions will be to coat or to erode. Furthermore, the transfer of neutral sulfur species could also be involved.

Slow downslope movement may effectively stir the regolith and supply fresh material or reequilibrated contaminants to the surface. Downslope movement on airless bodies may be driven by a variety of mechanisms:

1. Meteoritic stirring of the regolith (Soderblom 1970);
2. Seismic shaking induced by meteorite impacts (Houston et al. 1973);
3. Thermal cycling (Duennebier 1976).

Less effective mechanisms may include electrostatic effects and charged particle impacts. Passive mechanisms depend upon either momentum (e.g., landslides triggered by another mechanism) or flow (slow deformation of particles). The mechanisms that involve ballistic transport are more effective at lower surface gravity (g). Thermal effects may be independent of g. Slides require longer slopes at lower g, but therefore may on the average involve greater masses. Flow phenomena are dependent on surface gravity, temperature, material properties, and layer thickness. On Amalthea, g is ~ 10 cm s^{-2} and $T \sim 150$ K, conditions which clearly do not encourage rapid flow. Processes relating to impacts, and possibly to thermal cycling, are expected to be the most effective on Amalthea.

The high slopes that apparently exist in the crater Gaea suggest that sliding phenomena may be very effective in producing the associated bright spot. Such slides may be triggered by impacts or continuing downhill motion of ejecta, long after their initial deposition. Slopes on the walls of Ida and Lyctos are entirely unknown, but could also be high.

XII. ORIGIN OF AMALTHEA

Since the origin of our own Moon is still under debate a decade after the Apollo missions, it would be presumptuous for us to attempt to solve the origin of Amalthea, an object about which we know almost nothing. However, we have learned enough from Voyager to pose a few questions.

Our subsequent discussion is based on the premise that Amalthea as we now see it represents the remains of a larger body. How much larger we do not know, but a possible guess is that Amalthea was originally a roughly spherical object some 270 km across and that its present shape was fashioned by a continuum of impacts. The question of what happened to most of the mass in this scenario is not tractable. Some of the larger fragments could have ended in unstable orbits causing them either to crash into the planet or to be ejected out of the system. A more dramatic (but more unlikely) possibility is that Amalthea and the three small satellites recently discovered in its neighborhood (see Sec. I and Table 6.1) are all fragments of a single catastrophic collison. A major difficulty is that we do not know precisely how objects accrete, although schemes that can produce very nonspherical objects have been suggested (Hartmann and Cruikshank 1980). As noted earlier (Fig. 6.12), the size and shape of Amalthea are not anomalous in comparison with other small bodies at the distance of Jupiter from the Sun.

Is Amalthea a captured object, or did it form at roughly its present distance from Jupiter? If it was captured, one could expect a composition similar to that of the Trojan asteroids (cf. Gradie and Veverka 1980). Such an

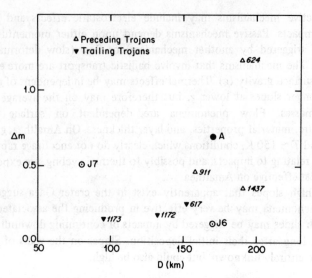

Fig. 6.12. Comparison of light curve amplitudes (shapes) of Amalthea, several Trojan asteroids, and of J6, Himalia, and J7, Elara. The Amalthea point is calculated; others are from data in Degewij and van Houten (1979). D is the diameter of the object, Δm the amplitude of the lightcurve in magnitudes.

object would have a low mean density, probably less than 2 gm cm^{-3}. Gradie et al. (1980) have shown that contaminating a carbon-rich material, probably similar to that found in the Trojans, with sulfur allotropes, can match the spectrum and albedo of Amalthea. We have also seen that Amalthea's size and shape are not at variance with such a possible origin. Unfortunately, capturing Amalthea into its present orbit seems to be highly unlikely (Pollack et al. 1979) because the time during which such a capture could take place is extremely short, even compared with that during which the postulated capture of the distant, irregular satellites might have occurred. Interactions with the Galilean satellites could also complicate such a capture process.

If Amalthea accreted near its present orbit, its composition was no doubt affected by the early high-luminosity phase of Jupiter. Cameron and Pollack (1976) estimate temperatures at Amalthea's distance in the circum-Jovian nebula at between 500 and 1400 K, depending on the nebular opacity. Temperatures near the upper end of the range would restrict Amalthea's composition to refractory oxides, nickel-iron, and possibly pyroxene. The mean density of such an object would be higher than 3 gm cm^{-3}. There appears to be no hope for testing the bulk composition of Amalthea, given the available color data. If the degree of surface contamination and alteration is as severe as Gradie et al. suspect, it may in fact be impossible to obtain any definitive information of the material that actually makes up Amalthea (below its contaminated rind) from spectral remote sensing measurements. The only hope lies in determining the satellite's mean density from a determination of its mass.

Acknowledgment. We are grateful to J. Gradie for helpful discussions, to J. Kolva for the drawing of Amalthea used in Fig. 6.1, and to D. Cruikshank, M. Davies, and D. Morrison for detailed comments on the first draft of this manuscript. This research was supported by the Planetary Geology Program of the National Aeronautics and Space Administration.

September 1980

REFERENCES

Anderson, J.O. (1976). The gravity field of Jupiter. In *Jupiter* (T. Gehrels, Ed.), pp. 113-121. Univ. of Arizona Press, Tucson.

Andersson, L.F. (1974). A photometric study of Pluto and satellites of the outer planets. Ph.D. Dissertation, Indiana Univ., Bloomington, Indiana.

Barnard, E.E. (1892). Discovery and observations of a fifth satellite to Jupiter, *Astron. J.* 12, 81-85.

Bowell, E., and Lumme, K. (1979). Colorimetry and magnitudes of asteroids. In *Asteroids.* (T. Gehrels, Ed.), pp. 132-169. Univ. of Arizona Press, Tucson.

Bridge, H.S., and the Voyager Plasma Science Team (1979). Plasma observations near Jupiter: Initial results from Voyager 1. *Science* 204, 987–991.

Cameron, A.G.W., and Pollack, J.B. (1976). On the origin of the solar system and of Jupiter and its satellites. In *Jupiter* (T. Gehrels, Ed.), pp. 61-84. Univ. of Arizona Press, Tucson.

Cintala, M.J., Head, J.W., and Veverka, J. (1978). Characteristics of the cratering process on small satellites and asteroids. *Proc. Lunar Sci. Conf.* 9, 3803-3830.

Cintala, M.J., and Mouginis-Mark, P.J. (1980). Martian crater depths: More evidence for subsurface volatiles? *Geophys. Res. Lett.* 7, 329-332.

Cintala, M.J., Parmentier, E.M., and Head, J.W. (1979). Characteristics of cratering process on icy bodies: Implications for outer planet satellites. *NASA TM* 80339, 179-181.

Croft, S.K., Kieffer, S.W., and Ahrens, T.J. (1979). Low velocity impact craters in ice and ice-saturated sand with implications for Martian crater count ages. *J. Geophys. Res.* 84, 8023-8032.

Degewij, J., and van Houten, C.J. (1979). Distant asteroids and outer Jovian satellites. In *Asteroids* (T. Gehrels, Ed.), pp. 391-416. Univ. of Arizona Press, Tucson.

Dence, M.R., Grieve, R.A.F., and Robertson, P.B. (1977). Terrestrial impact structures: Principal characteristics and energy considerations. In *Impact and Explosion Cratering* (D.J. Roddy, R.O. Pepin, and R.B. Merrill, Eds.), pp. 247-275. Pergamon Press, New York.

Divine, N. (1979). Distribution of energetic electrons and protons in Jupiter's radiation belts. *JPL IM* 3574-79-51.

Dobrovolskis, A., and Burns, J. (1980). Life near the Roche limit: Behavior of ejecta from satellites close to planets. *Icarus* 42, 422-441.

Duennebier, F. (1976). Thermal movement of the regolith. *Proc. Lunar Science Conf.* 7, 1073-1086.

Fujiwara, A., Kamimoto, G., and Tsukamato, A. (1977). Destruction of basaltic bodies by high velocity impacts. *Icarus* 31, 277-288.

Fujiwara, A., Kamimoto, G., and Tsukamato, A. (1978). Expected shape distribution of asteroids obtained from laboratory impact experiments. *Nature* 272, 602-603.

Fuwa, K. (1937). A study of the series of glasses coloured by sulfurous matters, II-III. *J. Soc. Chem. Ind. Japan* 40, Suppl. Bd. 64-65.

Gault, D.E., Horz, F., Brownlee, D.E., and Hartung, J.B. (1974). Mixing of the lunar regolith. *Proc. Lunar Sci. Conf.* 5, 2365-2386.

Gault, D.E., Shoemaker, E.M. and Moore, H.J. (1963). Spray ejected from the lunar surface by meteoroid impact. *NASA TN* D-1767.

Gault, D.E., and Wedekind, J.A. (1969). The destruction of tektites by micrometeoroid impact. *J. Geophys. Res.* 74, 6780-6794.

Gradie, J., Thomas, P., and Veverka, J. (1980). Surface composition of Amalthea. *Icarus* 44, 373-387.

Gradie, J., and Veverka, J. (1980). The composition of the Trojan asteroids. *Nature* 283, 840-842.

Hanel, R., and the Voyager IRIS Team (1979). Infrared observations of the Jovian system from Voyager 1. *Science* 204, 972-976.

Hartmann, W.K. (1971). Interplanet variations in scale of crater morphology, Earth, Mars, Moon. *Icarus* 17, 707-713.

Hartmann, W.K. (1978). Planet formation: Mechanism of early growth. *Icarus* 33, 50-61.

Hartmann, W.K., and Cruikshank, D.P. (1980). Hektor: The largest highly elongated asteroid. *Science* 207, 976-977.

Housen, K.R., Wilkening, L.L., Chapman, C.R., and Greenberg, R. (1979). Asteroidal regoliths. *Icarus* 40, 317-351.

Houston, W.N., Moriwaki, Y., and Chang, C.S. (1973). Downslope movement of lunar soil and rock caused by meteoroid impact. *Proc. Lunar Sci. Conf.* 4, 2425-2435.

Jewitt, D.C., Danielson, G.E., and Synnott, S.P. (1979). Discovery of a new Jupiter satellite. *Science* 206, 951.

Johnson, T.V., Cook, A.F. II, Sagan, C., and Soderblom, L.A. (1979). Volcanic resurfacing rates and implications for volatiles on Io. *Nature* 280, 746-750.

Klaasen, K.R., Duxbury, T.C., and Veverka, J. (1979). Photometry of Phobos and Deimos from Viking Orbiter images. *J. Geophys. Res.* 84, 8478-8486.

Krimigis, S.M., and the Voyager Plasma Science Team (1979). Hot plasma environment of Jupiter: Voyager 2 results. *Science* 206, 977-984.

Krumbein, W.C. (1941). Measurement and geologic significance of shape and roundness of sedimentary particles. *J. Sed. Pet.* 11, 64-72.

Matson, D.D., Johnson, T.V., and Veeder, G.J. (1977). Soil maturity and planetary regoliths: The Moon, Mercury, and the asteroids. *Proc. Lunar Sci. Conf.* 8, 1001-1011.

McDonald, F.B. and Trainor, J.H. (1976). Observations of energetic Jovian electrons and protons. In *Jupiter* (T. Gehrels, Ed.), pp. 961-987. Univ. Arizona Press, Tucson.

McGetchin, T.R., Settle, M., and Head, J.W. (1973). Radial thickness variation in impact ejecta: Implications for lunar basin deposits. *Earth Planet. Sci. Lett.* 20, 226-236.

Millis, R.L. (1978). Photoelectric photometry of J V. *Icarus* 33, 319-321.

Morrison, D., and Burns, J.A. (1976). The Jovian satellites. In *Jupiter* (T. Gehrels, Ed.), pp. 991-1034. Univ. of Arizona Press, Tucson.

Pascu, D. (1977). Astrometric techniques for the observation of planetary satellites. In *Planetary Satellites* (J. Burns, Ed.), pp. 63-86. Univ. of Arizona Press, Tucson.

Peale, S.J. (1976). Rotation histories of the natural satellites. In *Planetary Satellites* (J. Burns, Ed.), pp. 87-112. Univ. of Arizona Press, Tucson.

Pike, R.J. (1977). Size dependence of the shape of fresh impact craters on the Moon. In *Impact and Explosion Cratering*, (D.J. Roddy, R.O. Pepin, and R.B. Merrill, Eds.), pp. 489-509. Pergamon Press, New York.

Pike R.J., Roddy, D.J., and Arthur, D.W.G. (1980). Gravity and target strength: Controls on the morphologic transition from simple to complex impact craters. *NASA TM* 81776, 108-110.

Pollack, J.B., Burns, J.A., and Tauber, M.E. (1979). Gas drag in primordial circumplanetary envelopes: A mechanism for satellite capture. *Icarus* 37, 587-611.

Pollack, J.B., Veverka, J., Noland, M., Sagan, C., Duxbury, T.C., Acton, C.H., Born, G.H., Hartmann, W.K., and Smith, B.A. (1973). Mariner 9 television observations of Phobos and Deimos, 2. *J. Geophys. Res.* 78, 4313-4326.

Rieke, G.H. (1975). The temperature of Amalthea. *Icarus* 25, 333-334.

Schultz, P.H., and Gault, D.E. (1975). Seismic effects from major basin formations on the Moon and Mercury. *Moon* 12, 159-177.

Simpson, J.A., and McKibben, R.B. (1976). Dynamics of the Jovian magnetosphere and energetic particle radiation. In *Jupiter* (T. Gehrels, Ed.), pp. 738-766. Univ. of Arizona Press, Tucson.

Smith, B.A., and the Voyager Imaging Team (1979a). The Jupiter system through the eyes of Voyager 1. *Science* 204, 951-972.

Smith, B.A., and the Voyager Imaging Team (1979b). The Galilean satellites and Jupiter: Voyager 2 imaging results. *Science* 206, 927-950.

Soderblom, L.A. (1970). A model for small-impact erosion applied to the lunar surface. *J. Geophys. Res.* 75, 2655-2661.

Soter, S., and Harris, A. (1977). The equilibrium figures of Phobos and other small bodies. *Icarus* 30, 192-199.

Sudbury, P.V. (1969). The motion of Jupiter's fifth satellite. *Icarus* 10, 116-143.

Synnott, S.P. (1980). 1979 J2: Discovery of a previously unknown satellite of Jupiter. *Science* 210, 786-788.

Thomas, P., and Veverka, J. (1979). Grooves on asteroids: A prediction. *Icarus* 40, 394-405.

Thomas, P., Veverka, J., Bloom, A., and Duxbury, T. (1979). Grooves on Phobos: Their distribution, morphology and possible origin. *J. Geophys. Res.* 84, 8457-8477.

Veverka, J., Thomas, P., Morrison, D., and Davies, M.E. (1981). Amalthea. *J. Geophys. Res.* In press.

Weyl, W.A. (1959). *Coloured Glasses.* Dawson's of Pall Mall, London.

Zellner, B. (1976). Physical properties of asteroid 433 Eros. *Icarus* 28, 149-153.

7. COMPOSITION OF THE SURFACES OF THE GALILEAN SATELLITES

GODFREY T. SILL
University of Arizona

and

ROGER N. CLARK
University of Hawaii

The compositions of the surfaces of the Galilean satellites are dominated by condensed volatiles along with other components of variable content. Io's surface is dominated by sulfur and its compounds, particularly condensed sulfur dioxide; the exact form of the other sulfur components is not known. Elemental sulfur can exist as various colored allotropes, and exhibits the colors detected on Io's surface, from white through yellow to orange, brown, and black. Other colored sulfur components is not known. Elemental sulfur can exist as various A possible source of the Na and K in Io's torus is alkali sulfides sputtered from the surface. The other three satellites have more or less copious amounts of water ice. Of the spectroscopically detectable materials, water makes up ≥ 95 wt% of the surface of Europa, ~ 90 wt% of Ganymede, and 30 to 90 wt% of Callisto, on the assumption that the remaining contaminants have densities near three. The reflectance spectra of the other outer three satellites indicate a surface similar to fine-grained frost on larger ice blocks or ice crystals. The upper limit on bound water is estimated at 5 ± 5 wt% of the bulk surface. Since such a large amount of the surfaces of Europa, Ganymede, and Callisto is water ice, the remaining material constitutes only a fraction of the bulk surface weight and thus could be completely hydrated. The nonwater ice material on the outer three satellites appears to be spectrally like that of carbonaceous chondritic material or other minerals containing Fe^{3+}.

Most of what we know of the compositions of Io, Europa, Ganymede, and Callisto has come through measurements of the spectral and photometric properties of these surfaces. By comparing the spectral and photometric behavior of laboratory samples with that of the Galilean satellites, we may draw certain conclusions, some more firm than others, as to the chemical compositions of the satellites.

Regarding the photometry of the Galilean satellites, Johnson and Pilcher (1977) summed up the observations then available; more recently Clark and McCord (1980a) have brought us up to date. Measurements of spectral reflectivity have covered a broad range of wavelengths. In the ultraviolet, visible and very near infrared, Nelson and Hapke (1978) in the range $0.32 \mu m$ $- 0.86 \mu m$ and more recently McFadden et al. (1981) in the $0.325 \mu m -$ $1.1 \mu m$ region measured the changing reflectance of the Galilean satellites at various orbital and solar phase angles, finding fascinating variations. Pollack et al. (1978) measured the spectral reflectance of the leading and trailing hemispheres of the satellites in the $0.7 \mu m - 5 \mu m$ region and Clark and McCord (1980a) obtained high-precision data in the region $0.65 \mu m$ to $2.5 \mu m$.

The low reflectivity of the satellites in the ultraviolet, especially in the case of Io, was first noticed by Stebbins (1927) and confirmed by Harris (1961). The more diagnostic infrared data permitted Pilcher et al. (1972) and Fink et al. (1973) to identify H_2O ice in the spectra of Europa, Ganymede, and Callisto. The near infrared measurements at first seemed to reveal a blank Io until Cruikshank et al. (1978) and Pollack et al. (1978) discovered a strong feature near $4.08 \mu m$, which Fanale et al. (1979) and Smythe et al. (1979) identified as SO_2 frost. The current state of spectral observations of the satellites may be seen in Figs. 7.1−7.6. Figures 7.1 and 7.2 show visible and near infrared spectral reflectivities measured with high precision. Composite spectral reflectivities from $0.325 \mu m$ to $5 \mu m$ are illustrated in Figs. 7.3−7.6. The first part of this chapter treats Io and sulfur chemistry, the second the icy satellites and the spectral features of water ice mixed with other materials.

I. SULFUR CHEMISTRY ON IO

Pre-Voyager measurements of the reflectivity of Io led various observers to propose sulfur as a constituent of the surface of Io (Kuiper 1973; Wamsteker et al. 1974; Fanale et al. 1974; Nash and Fanale 1977; Nelson and Hapke 1978). Since the Voyager flybys Smith et al. (1979a, b), Consolmagno (1979), Sagan (1979), Soderblom et al. (1980), and Carr et al. (1979) have proposed models of Io volcanism that involve the presence of elemental sulfur (see also Chapters 16, 18, and 20).

The early conclusions that Io's surface might be covered with sulfur were based on Io's strong absorption in the near ultraviolet and lack of infrared

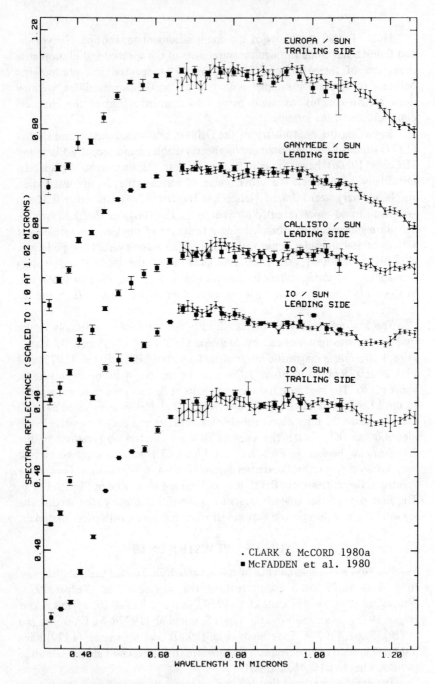

Fig. 7.1. The visible and very near-infrared spectra of the Galilean satellites. (From Clark and McCord 1980*a*.)

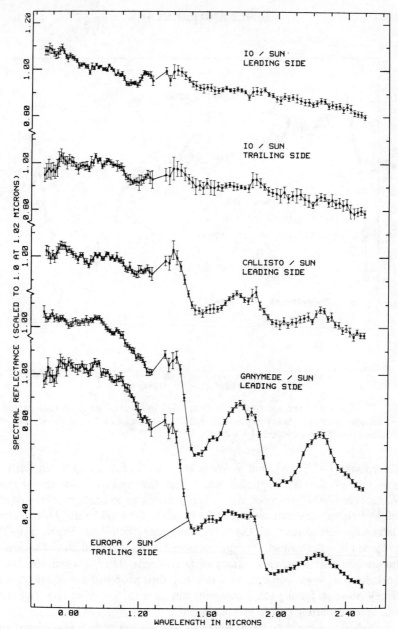

Fig. 7.2. The reflectance spectra of the Galilean satellites from 0.65 to 2.5 μm (from Clark and McCord 1980a). The geometric albedos at 1 μm are approximately 0.90 (Io – leading), 0.85 (Io – trailing), 0.85 (Europa – leading), 0.65 (Europa – trailing), 0.53 (Ganymede – leading), 0.45 (Ganymede – trailing), 0.23 (Callisto – leading), and 0.24 (Callisto – trailing). High precision reflectance spectra of this quality have not yet been published for the trailing sides of Callisto and Ganymede, and the leading side of Europa. See Pollack et al. (1978) for spectra of these objects of slightly lower resolution and precision. (From Clark and McCord 1980a.)

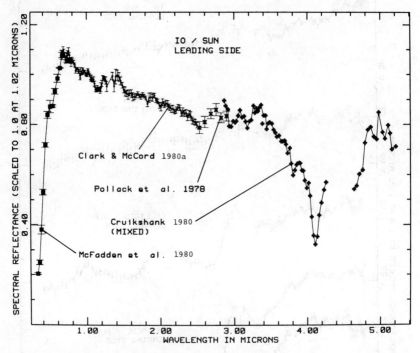

Fig. 7.3. The composite spectrum of Io from 0.325 to 5.3 μm. The Cruikshank data are averaged from spectra at many orbital phases; all others are of the leading side. (From Clark and McCord 1980a.)

absorption (1–2.5 μm), and as Wamsteker (1972) first pointed out, sulfur was one of the few materials which had the desired properties. Later Wamsteker (1974) proposed that a better match to Io could be obtained if 60% of elemental sulfur were mixed with 40% of the unidentified ultraviolet darkening component of the Saturnian ring. Nash and Fanale (1977) proposed that the elemental sulfur was mixed with various sulfates of sodium, magnesium, and iron, plus a trace of ferric oxide. They also simulated the Jovian radiation enrivonment by irradiating their proposed components with 5 keV protons, finding some darkening due to reduction of sulfate. H_2S gas was detected during proton bombardment.

Nelson and Hapke (1978) obtained new spectra of Io and observed an absorption edge near 0.33 μm that was compatible with the absorption spectrum of sulfur rapidly chilled by liquid nitrogen. The spectrum of this "glassy" sulfur is shown in Fig. 7.7 along with Nelson and Hapke's spectrum of Io. They proposed that the sulfur on the surface of Io could be due to outgassing of H_2S and subsequent photolytic dissociation into hydrogen and free sulfur, an idea originally proposed by Kuiper (1973) and Fanale et al.

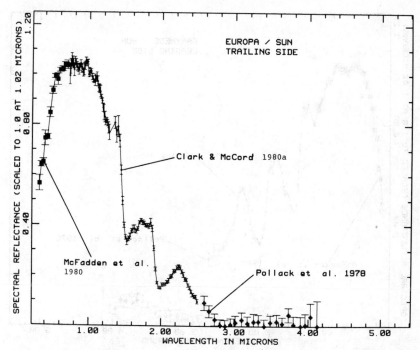

Fig. 7.4. The composite spectrum of the trailing side of Europa. Europa appears essentially black from 4 to 5 μm (Pollack et al. 1978). (From Clark and McCord 1980a.)

(1974). Such a process would necessarily be irreversible, inasmuch as the hydrogen would be lost from the surface of Io. Fanale et al. (1974) noted that elemental sulfur could also be produced by hydrogen reduction of sulfates.

Following the Voyager mission, Smith et al. (1979a) ascribed the various colors of Io to different allotropic forms of sulfur, a conclusion based on the previous spectroscopic studies of Io as well as on the color versatility of elemental sulfur (see Plate 1b in the color section). This opinion has been elaborated upon by others (e.g., Sagan 1979; Carr et al. 1979; Smith et al. 1979b; Soderblom et al. 1980).

Allotropic Forms of Sulfur

Allotropes of elements consist of different forms of the same element. The word "form" is deliberately vague, sometimes referring to the solid state, as with the two allotropes of carbon, graphite and diamond. Here the allotropes differ from each other by the way the carbon atoms are bonded to each other in the two crystalline phases. Some solid allotropes differ from each other only in the geometrical packing of their constituent molecules,

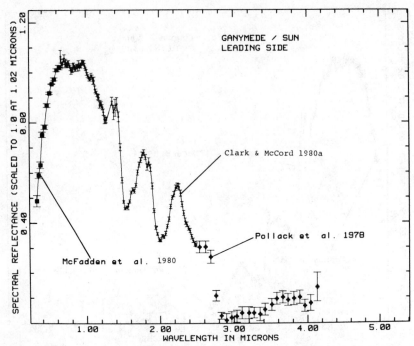

Fig. 7.5. The composite spectrum of the leading side of Ganymede. (From Clark and McCord 1980*a*.)

e.g., both red and white phosphorus contain P_4 molecules. Elements in the gaseous state can also exhibit allotropy; oxygen gas has two allotropes, ordinary oxygen (O_2) and ozone (O_3). It is safe to say that among all the elements sulfur exhibits the most allotropes, and in all states: solid, liquid, and gas.

The Molecular States of Elemental Sulfur

Sulfur has molecules ranging from the monatomic S to the polyatomic S_x, where x may equal many thousands. Some sulfur molecules of short length are cyclic, that is, exist as rings, S_8 and S_6 being common examples. Ring S_8 is called cycloocta sulfur. Other S molecules exist as simple chains, with ends not bonded together. Chain S_8 is called catenaocta sulfur. Catena sulfur species are mostly confined to the liquid and solid states. The cyclic species are found in all three states.

Gaseous Allotropes

The molecular species of sulfur vapor in equilibrium with liquid sulfur have been investigated by Berkowitz and coworkers (1963, 1964). Utilizing

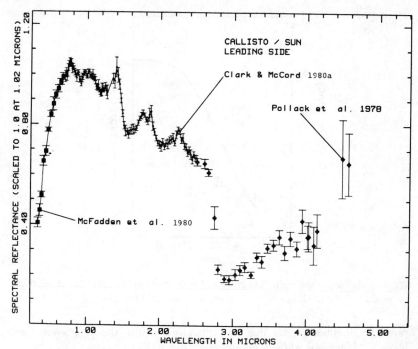

Fig. 7.6. The composite spectrum of the leading side of Callisto. (From Clark and McCord 1980*a*.)

thermodynamic data and their own measurements of vapor composition, Berkowitz and Marquart (1963) derived formulas for computing the equilibrium composition of sulfur vapor at pressures different from the vapor pressure. The general trend is that the percentage or mole fraction of the heavier molecules increases at higher pressures in accord with Le Chatelier's principle, which states that when a system in equilibrium is stressed the equilibrium shifts to relieve the stress. In the case of gaseous equilibria an increase in pressure favors that side of the chemical equilibrium with fewer molecules. Ran et al. (1973) examined the molecular composition of sulfur vapor at higher temperatures and arrived at conclusions that differed from those of Berkowitz. The molecular species in the vapor state include S, S_2, S_3, S_4, S_5, S_6, S_7, S_8, S_9, and S_{10}. The monatomic species is present only at high temperatures. S_8 is dominant from the melting point to 900 K, S_7 from 900 to 970, and S_2 from 970 to 1300, as can be seen in Fig. 7.8*b*. The ring structure of these molecules dominates at least from S_6 to S_{12} (Meyer et al. 1972).

Berkowitz and Chupka (1964) found that the vapor species in equilibria with solid allotropes were dominated by the solid phase. S_8 is dominant above orthorhombic (*a*) sulfur; rhombohedral (Aten) sulfur has a vapor

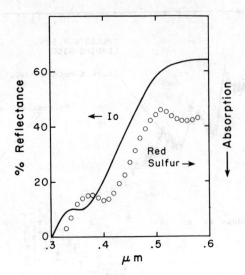

Fig. 7.7. Reflection spectrum of Io compared with an absorption spectrum of red sulfur, produced by quenching a film of boiling sulfur in liquid nitrogen.

pressure 50 times greater than orthorhombic sulfur, the main constituent being S_6. Polymeric sulfur produces vapor rich in S_7.

Liquid Allotropes

The remarkable properties of liquid sulfur are well known. Near the melting point (388 K) sulfur forms a yellow liquid of low viscosity (0.1 poise). Near 423 K the viscosity suddenly increases to 100 poise as the color darkens to a red-brown. Maximum viscosity of 600 poise occurs around 455 K and gradually declines as the color gets darker. At the boiling point, 718 K, the color is almost black and the viscosity has dropped to about one poise. In comparison, the viscosity of lava lakes in Hawaii is a few thousand poise (Yoder 1976, p. 190). At the other end of the scale, water and glycerin at 273 K have viscosities of 0.02 and 121 poise respectively.

At the melting point liquid sulfur contains mostly cycloocta S_8 molecules with some evidence for catena S_8 molecules (Meyer 1968), as well as S_6 and S_7 species. Near 423 K equilibrium polymerization of sulfur molecules occurs according to the following reactions:

$$cyclo-S_8 = catena-S_8 \tag{1}$$

$$(catena-S_8)_n + cyclo-S_8 = (catena-S_8)_{n+1}. \tag{2}$$

The number n approaches 10^5 at 450 K. At higher temperatures the process begins to reverse itself, but not in a simple sense. Mass spectrometric analysis (Berkowitz and Marquart 1963) and vapor density measurements (Ran et al. 1973) indicate that substantial quantities of other sulfur species, S_6 and S_7, must be present near the boiling point, and spectroscopic evidence (Meyer et al. 1972) indicates the presence of S_3 and S_4 species.

Solid Allotropes

The most common allotropic form of sulfur, stable at room temperature, is crystalline orthorhombic (α) sulfur, with 16 S_8 molecules per unit cell. This allotrope is stable up to 368 K where it reverts to monoclinic (β) sulfur. Liquid sulfur typically freezes as the monoclinic allotrope. The crystals are very long needles, often centimeters in length and millimeters wide. These large crystals are transparent, if pure, and permit long optical paths to be realized in the crystalline material. If the monoclinic crystals are cooled to room temperature, the mass of large monoclinic crystals revert in a few days to the orthorhombic phase. Since the orthorhombic phase is denser than the monoclinic phase, the growth of the microcyrstals of the orthorhombic phase is accompanied by cracking and crazing of the sulfur mass.

Another monoclinic phase exists, called γ sulfur, having an iridescent luster. It has been prepared from crystallization of an undercooled melt and has been reported in nature. γ sulfur transforms into monoclinic (β) and orthorhombic (α) in time.

Rhombohedral (ρ) sulfur, also called Aten or Engel sulfur, with dark yellow-red crystals, is distinctive in that its molecular composition is not cyclo–S_8, but rather cyclo–S_6. The material is made by the reaction of concentrated HCl with a solution of sodium thiosulfate near 0°C. The cyclo–S_6 is more reactive than cyclo–8_8. While it is unstable in the presence of impurities, as a pure crystal it is stable for weeks.

Beside the crystalline S_8 and S_6 allotropes there are also crystalline forms of catenasulfur, with S chain lengths of many thousands. This fibrous sulfur exhibits an X-ray pattern when stretched that is similar to that of stretched rubber (Meyer 1968).

Commonly catenasulfur is produced by quenching hot ($>$ 440 K) liquid sulfur in air, cold water, or liquid nitrogen. The resultant plastic material is a mixture of orthorhombic sulfur and catenasulfur. The orthorhombic sulfur may be removed by solution in CS_2; the plastic material remains insoluble. Without stretching and aligning the long polymers of plastic sulfur there is little order in the material and no X-ray pattern is discernible. In time the plastic material loses its elasticity and reverts to orthorhombic sulfur if kept at room temperature. This plastic material is usually dark brown to black, but changes to pale yellow when it reverts to the orthorhombic variety, provided there are no impurities in the original sulfur melt.

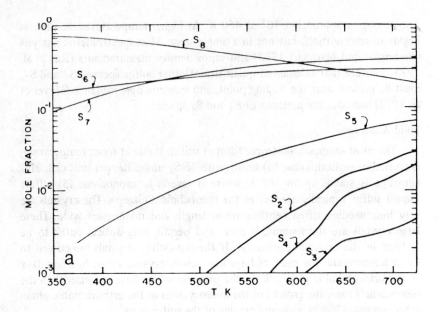

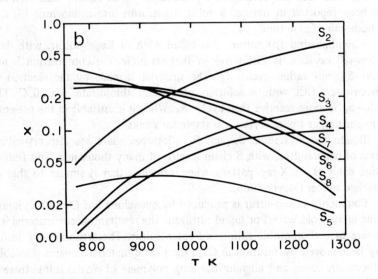

Fig. 7.8. (a) Sulfur species in the vapor state in equilibrium with liquid sulfur (from Berkowitz and Marquant 1963). (b) Sulfur species in the vapor state in equilibrium with liquid sulfur (from Ran et al. 1973).

Not only can heat cause S–S bond scission, leading to the formation of catenasulfur, but also radiation can destroy the cyclo–S_8 ring structure with the consequent production of polymers. Radford and Rice (1960) exposed orthorhombic S at liquid nitrogen temperatures to 40 megaroentgens of γ radiation from ^{60}Co. The pale sulfur turned a deep red or red-brown color. When warmed to room temperature the red sulfur explosively reverted to the normal yellow color. The material is stable if kept at low temperature. Electron paramagnetic resonance (EPR) studies of the red material gave no detectable signal indicating lack of unpaired electrons. If a cyclo–S_8 molecule is cleaved into a catena–S_8 chain, then at each end of the chain there should be a free electron, hence an EPR signal should have been detected. Meyer (1968) suggests that short chains of eight atoms may stabilize the unpaired electrons by resonance and produce a very broad EPR signal, which would be difficult to identify.

Unstable forms of sulfur, those produced by condensing sulfur vapor issuing from a hot furnace on to a liquid nitrogen cold finger, were examined by Radford and Rice (1960) and Chatelain and Buttet (1965). The arrangement of the experimental apparatus requires evaporation into a vacuum with the sulfur vapor passing through a heated collar and orifice, and immediate deposition onto the liquid N_2-cooled cold finger a few centimeters from the orifice. The cooling process occurs in a fraction of a second. The sulfur species trapped are apparently free radicals, since there are 10^{13} to 10^{16} unpaired electrons mg^{-1} (Chatelain and Buttet 1965). Radford and Rice (1960) reported that the sulfur condensed from a furnace temperature of 800-1200 K was purple in color with constant EPR signal. Below 800 K, the number of electron centers decreased with decreasing temperature. Olive-green S was deposited when the furnace temperature was 550-800 K, green S at temperatures 475-550, and yellow at 415-475. The purple sulfur is apparently the least stable; when heated to 195 K (dry ice) it changed to the olive color. The olive color was also produced when S vapor at 700 K was frozen out at 195 K. Cold deposition was also carried out on sulfur vapor subjected to electric discharge; the deposited sulfur had the reddish-brown color that was also characteristic of γ-irradiated sulfur.

MacKnight and Tobolsky (1965) measured the fraction of polymeric sulfur present in quenched liquid sulfur. Droplets of liquid sulfur quenched from a temperature > 432 K in ice water or liquid air produced 40% by weight polymeric catena S. Quenching of sulfur from 470 to 520 K in a dry ice–acetone bath produced a yellow translucent glassy sulfur, which was stable below 243 K. When heated to 263 K this sulfur first became elastic and later crystallized into the orthorhombic phase.

One of us (Sill) investigated the behavior of a 100-g mass of molten sulfur cooled in a vacuum maintained by a rotary vacuum pump. The first sample was heated in air just above the melting point (120°C) and placed in the vacuum system. Vaporization in the vacuum deposited a pale yellow film on the walls of the vacuum chamber, and after an hour the yellow-brown liquid froze into large monoclinic crystals of similar color. After 24 hr at room temperature the sulfur began to revert to orthorhombic yellow crystals, with audible cracking. In two days conversion to massive yellow sulfur was complete; pseudomorphs of the large monoclinic crystals were notable. The second sulfur sample was heated to 400°C in air and also placed in the vacuum system. Copious white clouds of condensing sulfur appeared in the chamber, and pale yellow sulfur deposits on the walls of the chamber accumulated and so heated the walls that yellow droplets of sulfur became noticeable. The intense color of the brownish-black liquid diminished slightly as it cooled. No solid crust formed on the surface of the liquid. After two hours the liquid began to crystallize into large brown monoclinic needles. The sulfur was transferred to cold storage (−30°C), and the brown monoclinic crystals faded in color, but no reversion to the orthorhombic phase occurred in two weeks. The sample was then kept at room temperature and after a day began to revert to the orthorhombic phase, destroying the transparency of the monoclinic crystals, finally producing a brownish-yellow mass. Meyer (1968) warns of contaminants even in laboratory grade sulfur of 99.99% purity. Besides Se, As, Te, H_2S, and "ash" already in the sulfur, heating in air can produce SO_2, O_2, and H_2O in the molten sulfur. Therefore the sulfur used in the above experiments is certainly not pure and produces color changes that Meyer refers to as irreversible. The sulfur molecules, particularly the catenasulfur, may have other impurity atoms terminating the long sulfur chains. The impurities prevent complete conversion to cyclo−S_8. Meyer et al. (1972) estimate that sulfur needs only 10^{-5} weight of hydrogen or 10^{-4} weight of hydrocarbon impurity to cause chains of polymeric S to be terminated by hydrogen atoms. Since we have no *a priori* knowledge of the purity of sulfur on Io, one might reasonably suppose that planetary processes may produce impure sulfur. Sulfur dioxide is one very likely impurity, as well as compounds of alkali metals (Fanale et al. 1974; Nash and Fanale 1977; Nash and Nelson 1979).

Spectra of Elemental Sulfur

Reflection spectra of sulfur at room temperature and at cooler temperatures have been published by Sill (1973), Wamsteker et al. (1974), Fanale et al. (1974), Nash and Fanale (1977) and Nelson and Hapke (1978). As pointed out in these papers the reflection spectra of sulfur vary with temperature and also with the size of the sulfur particles. Veverka et al. (1978a,b,1979) showed that in order to compare the laboratory reflection spectrum of sulfur with the geometric albedo of Io it is necessary to determine the wavelength dependence of the sample's photometric function.

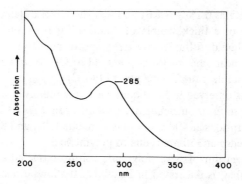

Fig. 7.9. Spectrum of orthorhombic sulfur at room temperature.

The reflection spectrum of a solid may need to be raised or lowered and the slope of an absorption feature to be varied in order to compare the laboratory reflection spectrum with the albedo of Io.

What is true of laboratory reflection spectra is even more so in regard to spectra taken in transmission. Nevertheless, it is instructive to compare the absorption spectra of thin films of sulfur in order to determine where characteristic absorptions in sulfur occur. An absorption spectrum (a plot of log I/I_o, where I is the measured intensity and I_o the intensity of the light without the sample) is a fair approximation to a reflection spectrum. A comparison of the absorption spectrum of red sulfur with the albedo of Io is presented in Fig. 7.7.

Meyer et al. (1972) have published a number of absorption spectra of sulfur allotropes, both in the pure state and in solution, at various temperatures. The spectrum of orthorhombic (a) sulfur at room temperature indicates the material is virtually opaque at 350 nm, but a thin film spectrum (Fig. 7.9) shows a relative absorption maximum centered at 385 nm, caused by the first allowed electronic transition of cyclo–S_8. The absorption edge, located near 350 nm, shifts with temperature 0.23 nm/K from 77 to 300 K. The shift is caused by the thermal population of the vibrational levels of the electronic ground state of cyclo–S_8. The temperature shift is observed in the reflection spectrum of cold sulfur and is responsible for sulfur turning white when cooled in liquid nitrogen.

The same temperature shift can be observed in liquid sulfur. Fig. 7.10 shows the absorption of a thin film of hot sulfur at two temperatures, 250°C (above the point of maximum viscosity) and at 500°C (above the boiling point). Meyer et al. (1972) describe this hot spectrum as due to at least four species of sulfur: thermally broadened cyclo–S_8, thermally broadened polycatena S, an absorption peak at 400 nm due to an electronic transition in S_3, and a similar absorption at 550 nm due to S_4. They compute that the liquid sulfur contains $0.1\% - 3\%$ S_3 and S_4. The effect of temperature on the absorption edge of a thick sample of liquid sulfur is seen in Fig. 7.11. The appearance of liquid sulfur follows the spectrum, varying from a straw-colored liquid near the melting point ($120°C$) to a virtually opaque brown-black liquid near the boiling point ($444°C$).

The features observed in hot liquid sulfur are preserved when a thin film of boiling liquid sulfur is quenched in liquid nitrogen. Fig. 7.12 shows this red sulfur. The absorptions at 400 and 550 nm are due to S_3 and S_4, respectively. The absorption edge at 350 nm is due to polymeric S.

Another spectrum of pure polymeric sulfur, extracted into glycerol, in which it is soluble, is illustrated in Fig. 7.13. It shows a relative absorption maximum at 360 nm that is not present in cyclo—S_8 or cyclo—S_6. Spectra of cyclo—S_8 and S_6 are seen in Fig. 7.14; again the pure species have been extracted by solvents, the S_8 in isopentane-methylcyclohexane and the S_6 in methanol. The extended slope of the S_6 curve indicates a deeper, redder color for S_6. The S_8 spectrum has a maximum near 285 nm, which was also observed in the spectrum of solid cyclo—S_8 (Fig. 7.9). The temperature effect can be clearly seen.

Sulfur and the Spectrum of Io

Do the colors of elemental sulfur correspond to the colors of Io? They do; there are white colors corresponding to cold orthorhombic sulfur, yellow characteristic of some rapidly cooled liquid sulfur, orange and red produced by quenched sulfur near the boiling point. Impure sulfur freezing into monoclinic crystals has a brownish-yellow or brown appearance. Cold, quenched monoclinic crystals could appear black, as well as pools of liquid sulfur at temperatures near 500 K. Brownish-red sulfur can be produced by irradiation of orthorhombic sulfur. Any color on Io can be matched by some stable or metastable sulfur allotrope. Whether the colored material on Io is

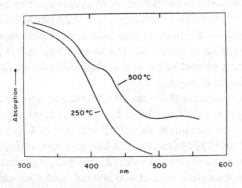

Fig. 7.10. Spectrum of liquid sulfur at two temperatures.

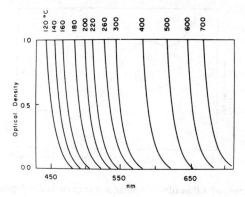

Fig. 7.11. The shift of the absorption edge of liquid sulfur with temperature.

indeed due to sulfur is not so easily answered. In order to make this judgment it is necessary to examine the reflectivity of sulfur and of Io spectroscopically, or at least with multicolor photometry, over all available wavelengths.

Spectroscopy of the whole disk of Io has been performed by various groups; we have reproduced in Fig. 7.7 a spectrum of Io by Nelson and Hapke (1978), and compared it with the absorption of quick-chilled red sulfur. The match between the two is good, but not definitive. The features ascribed to S_3 and S_4 at 400 and 550 nm are evident in spectra published by Nelson et al. (1981). Still, it must be remembered that here the spectrum of a single type of sulfur is compared to a multihued disk of Io. The Voyager spacecraft produced color images of the satellite at high resolution, and Soderblom et al. (1980) have been able to extract four-color photometry from the Voyager data. Figure 7.15 shows plots of this photometry for five differently colored areas on Io, corrected to normal reflectance. Also shown are the normal reflectances of laboratory samples of sulfur: the orange, red, and brown sulfur produced by heating liquid sulfur to various temperatures and quenching it at low temperatures. The general conclusion might be that the reflectivity of the colored sulfur species does not match exactly the colors observed on Io, especially in the near ultraviolet, where in most cases sulfur is too dark. Furthermore, the slope of the Io colors does not match that of the sulfur. Perhaps some adjustment, either in the photometric reductions or in the calibration of the laboratory spectra, may bring better agreement. Under the assumption that the spectra are accurate it is clear that either Io does not really correspond with sulfur colors produced by quick quenching of molten sulfur, or that the method of quenching is different on Io than in the laboratory. In the spectrum of red sulfur and dark liquid sulfur (Figs. 10 and 12 from Meyer et al. 1972), the absorption due to S_4 at 550 nm is clearly evident, but this same feature is not observed in the reflectance spectrum of

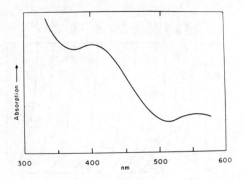

Fig. 7.12. Spectrum of a thin film of boiling sulfur quenched in liquid nitrogen. Color is red.

red or brown sulfur. The broad photometry of Io cannot show this feature, although the polar yellow-brown, the yellow-orange and red materials on Io seem to indicate a turning over in the reflectance at 550 nm. There are other ways of producing red, brownish, brownish-red, or yellow-brown colors of sulfur, and these involve the brown irreversible color of frozen sulfur caused by impurities as well as the red or brownish-red color produced by irradiating orthorhombic sulfur at low temperatures. If the 5 MeV protons measured by Voyager 1 (Vogt et al. 1979) in the vicinity of Io are as effective as gamma radiation is in producing red sulfur, then 1.5 yr exposure on the surface of Io is about equivalent to the 40 megaroentgens of γ irradiation used in the laboratory experiment. Furthermore, Meyer (1968) cautions that ultraviolet radiation is capable of S–S bond scission, and colored effects may be visible in elemental sulfur exposed to it.

While Sinton (1980*a*,*b*) has observed a hot spot (600 K) on Io that may correspond to a dark or black area (a pool of liquid sulfur?), most of the hot spots have temperatures nearer 300 K (Pearl et al. 1979; see also Chapter 19

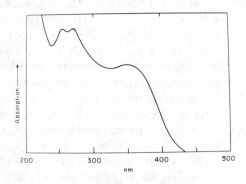

Fig. 7.13. Spectrum of catena sulfur dissolved in glycerol.

by Pearl and Sinton). Most of these warm spots are associated with dark areas. If the dark spots are pure elemental sulfur they should have reverted to the yellow color characteristic of orthorhombic sulfur at room temperature. All metastable phases of sulfur revert in a few days to the orthorhombic at this temperature (Meyer 1972). However, if impurities are present, the sulfur undergoes an irreversible color change, usually to brownish-yellow. The dark areas are usually associated with caldera floors, and it is interesting that the brown sulfur in Fig. 7.15 is enclosed in the area corresponding to the caldera floor material. However, this brown sulfur is darker than the yellow-brown color imparted to sulfur by impurities.

Compounds of Sulfur on Io

Sulfur dioxide frost has been identified on the surface of Io by its infrared absorption at 4.08 μm (2454 cm^{-1} combination band) (Fanale et al. 1979; Smythe et al. 1979; Hapke 1979). High resolution spectra of Io and of laboratory frosts and thin films of SO_2 have been reported by Fink et al. (1980). Besides the major absorption at 2454 cm^{-1} (4.075 μm), two very sharp satellite absorption lines at 2413 (4.144 μm) and 2434 cm^{-1} (4.108 μm) were evident both in the laboratory spectra of SO_2 and in the spectrum of Io, confirming the identification of SO_2 frost on Io. Nelson et al. (1981)

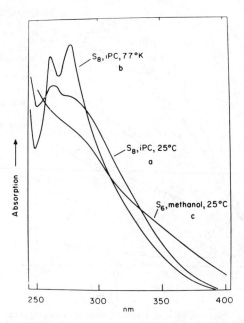

Fig. 7.14. Spectrum of cyclo–S_8 dissolved in isopentane-methylcyclohexane at two temperatures, and cyclo–S_6 dissolved in methanol.

have observed Io in the ultraviolet at numerous phase angles utilizing the International Ultraviolet Explorer (IUE). They have shown that the ultraviolet absorption of SO_2 near 330 nm is characteristic of the lower ultraviolet reflectivity of Io on one of its hemispheres where white areas are abundant, whereas the higher ultraviolet reflectivity of Io's other hemisphere (more characteristic of sulfur allotropes) corresponds to the yellow and red color of that hemisphere. The strength of the 4.08-μm band of SO_2 in the two hemispheres correlates with what is observed in the ultraviolet and visible. They caution that the white material on Io need not be pure SO_2, but only that SO_2 frost is more abundant in that material. Similarly, the yellow, orange, and reddish colors show a preponderance of features characteristic of sulfur, while the SO_2 features are still present, but muted.

Fanale et al. (1979) have suggested that SO_2 might be present on Io's surface dissolved or occluded in other surface materials, besides existing as a frost. Nash and Nelson (1979) suggested that SO_2, and perhaps H_2S as well, are adsorbed in the surface constituents of sulfur and alkali sulfides. Their

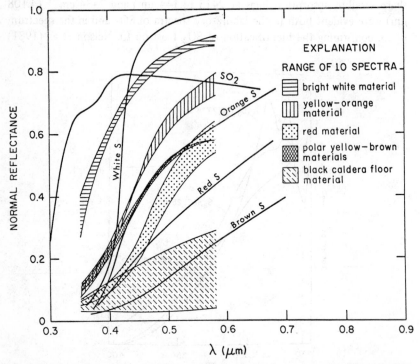

Fig. 7.15. Voyager colors of various colored areas on Io compared to the reflection spectra of sulfur. Colored species made by quenching liquid sulfur from progressively higher temperatures. (From Soderblom et al. 1980.)

laboratory spectra of Na_2S and $NaHS$ do indeed reveal the presence of "absorbed" H_2O, despite their efforts to dry the samples. The extent to which SO_2 and H_2S can be absorbed or adsorbed on to sulfur or other constituents remains to be shown, however. Their spectra of the alkali sulfides indicate that there are some relatively sharp features corresponding to H_2S and SO_2 vibrations near 3.9 and 4.1 μm, respectively. The alkali sulfides, as both Fanale et al. (1979) and Nash and Nelson (1979) point out, are attractive candidates for at least partial coverage of the surface of Io because of the presence of Na and K in the Io torus.

Smith et al. (1979) have proposed a model of volcanism on Io that utilizes SO_2 gas at $\sim$ 40 bar pressure and a temperature $\sim$ 400 K as the propelling mechanism. Isentropic expansion of the gas produces velocities of ejection near 1 km s^{-1}, characteristic of the plumes produced on Io. Peale et al. (1980) have constructed another model of plume formation in which the driving gas is not SO_2 but sulfur vapor, S_2. Consolmagno (1979) also proposed an S_2 driven volcanism. In Peale's model S_2 at temperature of 1000 to 1500 K with pressures of 8 to 24 bar was considered. These temperatures and pressures were sufficient to produce the proper ejection velocities. Molten silicate magma coming into contact with sulfur provides the necessary temperature. They postulate that SO_2 gas could be entrained in the S_2 vapor flow.

The mechanism of Peale provides for the formation of another sulfur compound besides SO_2; in contact with hot sulfur vapor SO_2 can produce a new oxide, disulfur monoxide, S_2O. Hapke (1979) has proposed S_2O as a constituent on the surface of Io, using an electric discharge mechanism. S_2O can also be produced under other conditions (Schenk and Steudel 1968). Thermodynamic data in the JANAF tables (Stull and Prophet 1971) can be used to calculate the equilibrium concentration of S_2O in a mixture of S_2 and SO_2 gases, according to the chemical reaction

$$3\,S_2 + 2\,SO_2 = 4\,S_2O. \tag{3}$$

At an initial S_2 pressure of 20 bar and SO_2 pressure of one bar, and at a temperature of 1000 K, the pressure of S_2O becomes 0.16 bar, and SO_2 falls to 0.92 bar when equilibrium is established. The ratio of S_2O/SO_2 therefore is 0.17. When S_2O gas is condensed onto a cold surface, irreversible polymerization occurs producing polysulfur oxide of general formula $(S_7O_2)_n$. The color of this oxide is red if pure, with other shades of orange and yellow possible depending on the amount of SO_2 that has been frozen out along with the S_2O polymer. The colors are shown in Table 7.1. Since the two materials that produce polymerized S_2O exist on Io, it is quite possible that S_2O could be formed in the initial hot gas mixture in the plumes of Io.

TABLE 7.1[a]

**Relationship Between the S_2O Content of a Gaseous S_2O/SO_2
Mixture and the Color of the Resulting Condensate**

S_2O (Mole %)	Color (77 K)
< 2	yellow
5−10	orange-yellow
20−30	orange-red
40−70	cherry-red
> 85	dark-red

[a]From Schenk and Steudel (1968).

Do the plumes show that this may be the case? The heart-shaped deposition area of the Pele plume shows an orange or reddish-orange color in the area surrounding the vent out to the region near maximum trajectory. This could correspond to condensed $(S_7O_2)_n$ with a small percentage of SO_2 cocondensed with the polymer. Farther out, the plume deposit has a yellow-white fringe, which could be richer in SO_2. In the area around the Prometheus plume, the yellow-white outer plume is quite conspicuous, whereas the orange deposit is of a smaller extent. This plume deposit would seem to be richer in SO_2 than the Pele deposit. If the SO_2 deposits on the surface of Io evaporate in time, then the older outer yellow-white ring deposits should darken with age, inasmuch as the SO_2 has much greater volatility than the S_2O polymer. There appear to be older relics of yellow-white deposits similar in diameter to the Prometheus ring, that are darker than the fresh Prometheus ring. The older rings may darken due to loss of SO_2 and consequent enrichment of the S_2O polymer.

In the reaction of S_2 with SO_2 to produce S_2O, we have chosen an excess of S_2 vapor in our calculations. If S_2 is the dominant constituent then coloring may be caused by S_2 as well as S_2O polymer. It is not easy to decide what form or color would be produced by the S_2 vapor as it condenses under adiabatic expansion. Hot liquid sulfur produces a condensed vapor that in a vacuum chamber looks white. As the sulfur cloud collects on the surface of the vessel it appears yellow-white. This sulfur appears like "flowers of sulfur"; the process that is used to produce flowers of sulfur is that of volatilization of liquid sulfur with the vapors collected on a cold substrate. The composition of flowers of sulfur is mostly orthorhombic S_8 and some catenasulfur. In order to produce the highly colored condensed sulfur species it is necessary to retain a high vacuum with immediate condensation on a cryogenic substrate. Apparently, higher pressures with long reaction times produce S_8 along with polymeric S. The free radicals characteristic of the purple, green, and yellow-green cryodeposits are destroyed in the sustained reaction at higher temperatures and pressures. If the S_2 can condense into liquid sulfur without going directly to the solid state, then it seems that the colors characteristic of quenched liquid sulfur would be formed. In this case one would surmise that the frozen droplets of catenapolysulfur are deposited close to the volcanic vents, and the lighter colors due to SO_2 and/or S_2O are carried farther in the gas phase to the periphery of the vent deposits.

Further study of the surface of Io is necessary to decide which of the various mechanisms proposed are real. Further laboratory studies of the reflection spectra of elemental sulfur with and without added impurities, as well as studies of polymerized sulfur oxide, are needed. High resolution spectroscopy of Io, from both the ground and spacecraft, will help to resolve the composition of Io's surface.

II. THE ICY GALILEAN SATELLITES

The infrared portion (0.7 to 5 μm) of the spectra of Europa, Ganymede, and Callisto includes well-defined absorptions that contain mineralogical information. These infrared absorptions when analyzed for band depth, shape, and position, along with the continuum shape in the visible and infrared, can yield compositional information about the surfaces. In order to best interpret the planetary reflectance spectra, we must have an adequate understanding of the spectral properties of minerals and mineral mixtures. Also, when possible, the planetary data should be equivalent in quality to the laboratory data used in any interpretations.

Only recently has the quality of planetary infrared data for some objects begun to approach that obtainable in the laboratory, and only recently has attention been focused on spectral properties of mixtures. Pollack et al. (1978) obtained data on the Galilean satellites from the Kuiper Airborne Observatory (KAO), which were compared to spectra of mixtures of water ice and other materials, computed using a simple multiple-scattering theory. They determined that the 2.9 μm absorption in Callisto's spectrum is due primarily to bound water, and that the fractional amounts of water ice cover on the trailing and leading sides of Ganymede and the leading side of Europa are $50 \pm 15\%$, $65 \pm 15\%$, and $\geqslant 85\%$, respectively. The trailing side of Europa has very asymmetric ice absorptions, which they attributed to magnetospheric particle bombardment with the fractional amount of water probably being comparable to that of the leading side. Given the resolution and precision of the KAO data, the use of a multiple scattering theory was probably adequate. For more detailed, higher resolution studies, present scattering theories, while adequate for predicting the positions of absorptions, may not be adequate for predicting band shape and continuum level and shape (see Emslie and Aronson 1973; Aronson and Emslie 1973). This difficulty is due to the extreme complexity of reflectance spectra of mineral crystals, which often have irregular, nonspherical shapes and are optically anisotropic. Orientation, compaction, grain size, and grain size distribution also affect the spectral properties, and thus the spectra measured in the laboratory are affected as well as computed spectra.

Although laboratory spectral reflectance data are more accurate than theoretical simulations, simulations are useful in understanding the reflectance spectrum of a planetary surface. Theoretical spectra allow systematic simulation of various possibilities that can be difficult to form in the laboratory, particularly when working with ices. For example, growing a frost of the desired grain size and thickness requires a good environmental chamber, skill, and practice. Due to the difficulty in measuring a nearly infinite variety of samples, some "arm waving" occurs in analyzing a reflectance spectrum. However, it should be understood that an exact match of a laboratory or theoretical spectrum to a reflectance spectrum of a surface whose composition is not known is not required in order to understand the surface composition. What is needed is an understanding of the spectral properties of materials so that the features in a planetary reflectance spectrum can be understood.

Clark (1980*a*, 1981*a,b*) measured the spectral reflectance properties of water frosts of different grain sizes, of water frost on ice blocks, and of water ice in mixtures with other minerals. These studies, combined with the higher precision reflectance data in the near infrared on the Galilean satellites (Figs. 7.1–7.6) presented by Clark and McCord (1980*a*) and by Pollack et al. (1978), have increased our understanding of the surfaces of the icy Galilean satellites.

As a result of the laboratory investigations by Clark (1980*a*,1981*a,b*) and many other studies of the spectral properties of mixtures (e.g., Singer 1981; Kieffer 1970; Nash and Conel 1974), it is apparent that the reflectance spectrum of a mixture is a highly complex, nonlinear combination of the spectra of the individual components. Thus simple additive combinations of the many reflectance spectra of pure materials to model a planetary surface may result in erroneous conclusions about the surface, if the materials are in an intimate mixture. Such additive models are valid only if the different materials are optically isolated.

Laboratory Studies of Water Frost, Ice, and Mixtures with Other Minerals

The spectral reflectance of optically thick water frost in the near infrared is shown in Fig. 7.16 for several different grain sizes. Water ice has overtone absorptions in the near infrared at 2.00, 1.65, 1.55, 1.25, 1.04, 0.90, and

0.81 μm (see Clark 1980a, 1981a). Since the absorption coefficient increases in an absorption band, different spectral wavelengths probe to different depths in a frost or ice layer. The number of scattering boundaries also affects the penetration depth. For instance, in the coarse grained frost in Fig. 7.16, the photons with wavelengths of 2.0 μm (at the bottom of the strong 2 μm absorption) penetrate to perhaps one mm, while photons at 0.80 μm (in a weak absorption) or in the continuum (e.g., 0.85 μm) penetrate to ~ 10 mm. Thus the apparent band depths may vary according to the grain size distribution or the microstructure of a block of ice, as illustrated in Fig. 7.17 (from Clark 1980a), which shows a fine-grained frost on a block of ice. The ice was milky in appearance, since the water froze with crystals in intimate contact, providing many internal scattering boundaries. The reflectance spectrum in the strong 1.5 and 2.0 μm absorptions was primarily influenced by the thin frost layer at the surface, since any photons that may have penetrated into the ice block were essentially totally absorbed. However, at

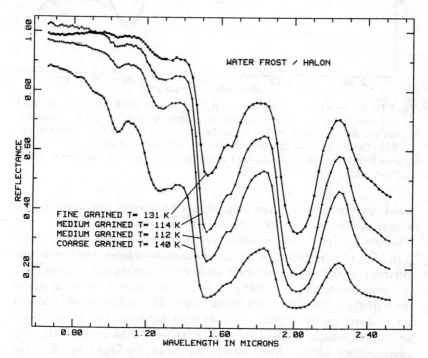

Fig. 7.16. The spectral reflectance of water frost of different grain sizes. Halon is a white reflectance standard. These data were taken with the same circular variable filter spectrometer as the Clark and McCord Galilean satellite data. The phase angle is 10° with normal incidence for the laboratory data in this figure and in Figs. 7.8 to 7.11. The grain sizes for the frosts (top to bottom) are: 50, 100, 200, and 400-2000 μm. (From Clark 1981a.)

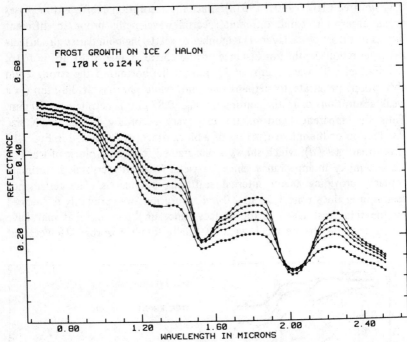

Fig. 7.17. The growth of a fine-grained frost layer on an ice block. The reflectance level rises with thicker frost depths. Note the asymmetric shape of the 1.5–μm absorption and the depth of the 1.04–μm absorption as compared to the frost spectra in Fig. 7.16. The thickness of the frost ranged from ～0.5 mm (bottom curve) to ～1 mm (top curve), and the frost grain size was less than a few tenths of a millimeter. (From Clark 1981a.)

wavelengths outside the strong absorptions (the apparent continuum and the weaker absorptions), the ice block has a strong effect on the spectrum. The weak absorptions become relatively strong compared to an optically thick frost. For clear ice with few or no scattering boundaries, visible and near infrared (e.g., 0.5–0.8 μm) radiation could probe to many meters depth.

Impurities in an ice or frost layer can radically change the appearance of the reflectance spectrum (Clark 1980a, 1981b). The addition of other mineral grains dusted on the surface of a frost layer decreases the 1.5 and 2.0 μm absorptions roughly in proportion to the fractional areal coverage (the only approximately additive trend seen in the studies by Clark), but since the surface of the frost is not a flat layer but a complex fairy castle structure, the mineral grains are somewhat mixed in the uppermost part of the surface and multiple scattering is drastically reduced at wavelengths where the mineral reflectance is not near unity. Because of reduced multiple scattering, the albedo of a surface may be lowered many times more than the fractional areal coverage in an additive model would indicate. Thus, mineral grains on a frost layer are detectable in very small quantities (e.g., fractional areal coverage $\lesssim 0.005$), if the mineral has suitable absorptions outside the major water ice features and the data are of sufficiently high quality – 0.5% precision (Clark 1980a, 1981b).

The effects of red cinder grains on a frost are shown in Fig. 7.18 (from Clark 1981c). Due to the penetrating power at different wavelengths, a thick frost layer ($\gtrsim 1$ mm) is required to mask another mineral at wavelengths short of ~ 1.4 μm, depending on the mineral reflectance features. Of the samples studied by Clark (1980a, 1981b), there are differences between the spectra of mineral grains dusted on a frost and the spectra of frost grown on a mineral sample (and containing the same type and size grains as that dusted on the frost). However, the spectra are qualitatively similar, and given an unknown spectrum of either condition, at present there is no way of determining which scattering condition obtains. One of the primary difficulties in determining quantitative parameters is that the grain size distribution is probably at least

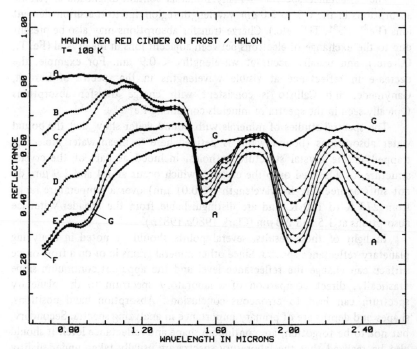

Fig. 7.18. The spectral reflectance of Mauna Kea red cinder grains ($\leqslant 125$ μm) on frost with fractional areal coverage of: (A) 0.0, (B) 0.05 ± 0.01, (C) 0.10 ± 0.03, (D) 0.3 ± 0.07, (E) 0.5 ± 0.1, and (F) 0.8 ± 0.1. Spectrum (G) is a pure Mauna Kea red cinder sample at a temperature of 189 K, without frost. (From Clark 1981b.)

slightly different for each sample – a very difficult set of parameters to measure.

Frost on a very dark surface (e.g., reflectance 6%) is easily seen, but a dark material intimately mixed with water ice could completely mask the water absorptions short of ~ 2.5 μm (Clark 1980a, 1981b). The higher overtones (0.80 μm to 1.25 μm) are masked more readily than the strong 1.5 and 2.0 μm absorptions. In fact, these weaker absorptions become suppressed with minerals dusted on frost. Even relatively high reflectivity (e.g., 80%) grains can greatly reduce the multiple scattering in a frost layer, suppressing the higher overtone absorptions. These effects change the apparent absorption band depth ratios of higher to lower overtones (e.g., 1.25/1.5 to 0.8/1.5) to a lower value. This is in contrast to the case of frost on ice, in which the apparent band depth ratio increases.

One important conclusion by Clark (1980a,1981b) is that the decrease in reflectance at visible wavelengths seen in the spectra of many icy planetary objects, including Europa, Ganymede, and Callisto, may be explained by water-mineral mixtures, mineral grains on frost, or frost on mineral grains.

The reflectance spectra of many minerals contain diagnostic electronic absorptions in the 0.3 to 3.0 μm wavelength region due to transition element ions (Fe^{2+}, Fe^{3+}, Ti^{3+}, etc.). Charge transfer absorptions may also be present due to the exchange of electrons between adjacent ions in the crystal (Fe, Ti, O, etc.), and usually occur at wavelengths < 0.9 μm. For example, the decrease in reflectance at visible wavelengths in the spectra of Europa, Ganymede, and Callisto is consistent with charge transfer absorptions typically seen in the spectra of minerals containing Fe^{3+}.

The spectral studies of minerals with bound water show that the bound water absorptions (bound water is defined here to mean water molecules trapped in the crystal structure in holes, included as part of the crystal structure, or adsorbed onto the crystal), which occur at 1.4 and 1.9 μm, do not shift appreciably in wavelength ($\lesssim 0.01$ μm) over a temperature range from 273 K to 150 K, and are distinguishable from the broader water ice absorptions at 1.5 and 2.0 μm (Clark 1980a,1981b).

In light of these results, several points should be noted in analyzing planetary reflectance spectra. Since other mineral grains in or on a frost or ice surface can change the reflectance level and the apparent continuum slope drastically, direct comparison of a laboratory spectrum to the planetary spectrum can lead to erroneous conclusions. Absorption band positions, shapes and depths are of primary importance in analyzing spectra. Secondary, but not to be forgotten, are continuum shape and reflectance level. It should also be realized that the laboratory spectra are usually taken under slightly different conditions from the planetary spectra. For instance, Earth-based spectra of the Galilean satellites are of the integral disk, and the laboratory spectra of Clark (1980a,1981c) were taken with normal incidence and an emission angle of 10° for the reflected light. Thus, if Callisto had exactly the same material over the entire disk and that same material were measured in the laboratory with the conditions noted above, the two spectra would not necessarily agree exactly; there would probably be some differences in the continuum level and shape and maybe in band depth. See Veverka et al. (1978a,b) for more details on these effects.

Interpretation of Spectra of the
Icy Galilean Satellites

Using the previously described laboratory studies, Clark (1980a, 1981c) analyzed the available high-precision reflectance spectra of Europa, Ganymede, and Callisto, as described by Clark and McCord (1980a). Since the most important features to analyze are the absorptions themselves, Clark first removed the apparent continuum from the telescopic and laboratory spectra by dividing the reflectance data by a cubic spline (defined in Clark 1980b). The spline was computed using data points defining the peaks between the apparent absorption features in the reflectance spectrum. The results of this part of the analysis are shown in Figs. 7.19 and 7.20 for Europa, Ganymede, Callisto, Saturn's rings, and several laboratory spectra. From the apparent solid water absorption band depths, Clark (1980a, 1981c) determined that the absorptions in the spectrum of Saturn's rings are characteristic of a medium-grained frost and that those for Europa, Ganymede, and Callisto are more characteristic of frost on ice spectra. The apparent band depths observed in spectra of frost, frost on ice, the icy Galilean satellites, and Saturn's rings are given in Table 7.2 for comparison. Inspection of the data in this table shows that all the band depths for Saturn's rings agree well with the medium fine frost, whereas the Galilean satellites all have 1.04 and 1.25 μm absorptions stronger than indicated by a frost spectrum matched to the 1.55- and 2.02-μm absorption band depths. As the next step in analyzing these spectra, Clark (1980a, 1981c) compared the satellite spectra with typical laboratory spectra. This is shown for Ganymede in Fig. 7.21, in which it is compared with a frost on ice spectrum. From the excellent agreement between the two spectra in the short wavelength side of the absorptions, there is no evidence for bound water on Ganymede. The frost-on-ice spectrum has a reflectance level of 0.8 at 0.8 μm, but the addition of ~ 2 wt% of dark mineral grains (such as carbonaceous chondrite grains) could easily lower the reflectance level to that on Ganymede with virtually no change in the apparent band depths of the frost-on-ice component.

A ratio of the laboratory reflectance to the Ganymede reflectance at each wavelength (note that the frost on ice spectrum is flat in the visible whereas Ganymede's spectrum decreases at shorter wavelengths) enhances the differences between the spectra and can indicate spectral properties of other

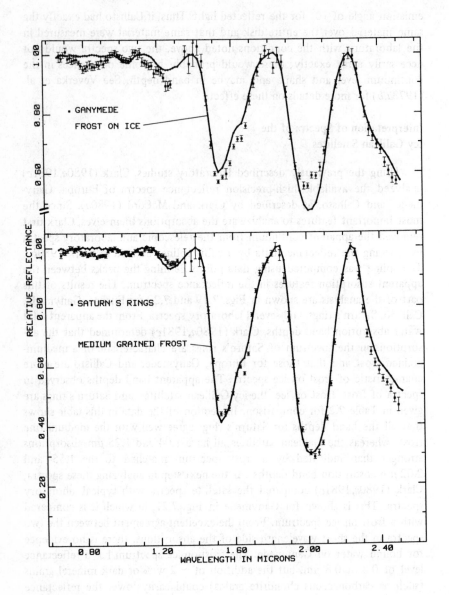

Fig. 7.19. The infrared spectra of the leading side of Ganymede and of Saturn's rings with cubic spline continua removed, plotted on similarly treated laboratory data, for comparison of apparent band depths. (From Clark 1981c.)

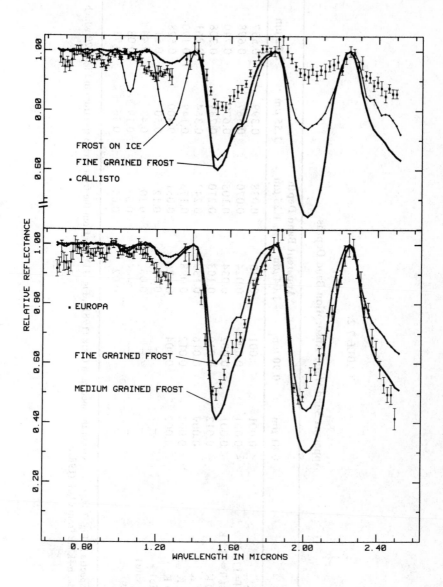

Fig. 7.20. The infrared spectra of Europa and Callisto with cubic spline continua removed. The fine and medium grained frost spectra are the top two spectra in Fig. 7.16 and were similarly treated. (From Clark 1981c.)

TABLE 7.2

Apparent Solid Water Absorption Band Depths

H$_2$O Solid[a]	0.81 μm	0.90 μm	Apparent Band Depth 1.04 μm	1.25 μm	1.55 μm	2.02 μm
a. Fine frost T=131 K	<0.001	<0.001	0.013	0.038	0.399	0.557
b. Medium fine frost T=114 K	<0.001	0.002	0.024	0.070	0.587	0.698
c. Medium coarse frost T=112 K	<0.001	0.002	0.034	0.105	0.670	0.740
d. Coarse frost T=140 K	0.012	0.019	0.103	0.210	0.737	0.708
e. Frost on ice T=115 K	0.009	0.016	0.137	0.247	0.364	0.261
f. Frost on ice T=110 K	0.004	0.012	0.098	0.173	0.149	0.190
g. Frost on ice T=110 K	0.002	0.004	0.026	0.051	0.400	0.528
h. Europa (trailing side)	–	–	0.04	0.12	0.50	0.51
i. Ganymede (leading side)	–	–	0.03	0.10	0.43	0.44
j. Callisto (leading side)	–	–	–	0.07	0.19	0.08
k. Saturn's rings	–	–	0.02	0.05	0.56	0.71

[a] a–d are from the frost spectra in Fig. 7.16; e–f are from the spectra in Clark (1981a, Fig. 11); g is from the frost on ice spectrum in Fig. 7.20 which is compared to Ganymede; h–k are from Clark (1981c).

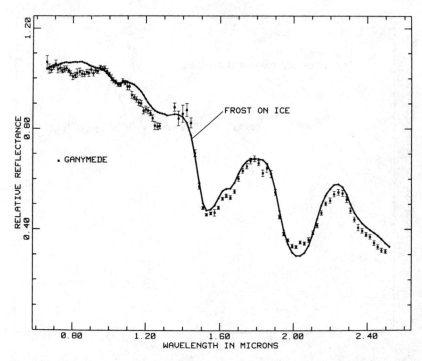

Fig. 7.21. The spectrum of the leading side of Ganymede compared to a frost on ice spectrum. The agreement is excellent. The frost on the ice has a particle size range from 30 to ~ 200 μm and a depth of ~ 1 mm. The ice block was ~ 17 mm deep. The intensities are scaled to unity at 1.02 μm. The frost on ice spectrum has a reflectance near 80% at 0.8 μm. (From Clark 1981c.)

components in or on the surface. This is shown in Fig. 7.22, although the result must be interpreted with caution. For instance, the peak at 2.0 μm is due to the slightly different depths of the 2.0-μm solid water absorptions. The important features in the spectrum are the already well-known decrease in visible wavelengths, the absorption at 0.85 μm, and the absorption at 1.15 μm. The 0.85-μm absorption and the decrease in reflectance at visible wavelenghts are typical of minerals containing Fe^{3+}. The cause of the 1.15-μm absorption is currently unknown. Clark (1980a,1981c) determined that the absorption is present in Io, Callisto, and possibly Europa, but not in Saturn's rings. Clark (1980a,1981c) determined that, of the spectroscopically detectable materials, the amount of water ice on Ganymede is ~ 90 wt%, considerably more than previously thought. The upper limit of bound water is 5 ± 5 wt% averaged over the surface.

Many approaches to the problem of the amount of water on the Galilean satellites have differed from Clark's. For example, Pilcher et al. (1972) and Pollack et al. (1978) determined the fractional amounts of frost cover on the

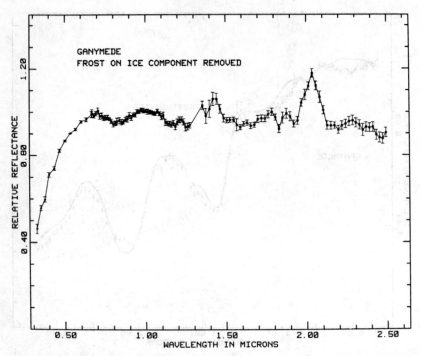

Fig. 7.22 The Ganymede spectrum divided by the frost on ice spectrum from Fig. 7.20, with the visible data included. The peak at 2.0 μm is due to a difference in the 2.0-μm water ice band depths.

surfaces by assuming a horizontally inhomogeneous situation in which a certain fraction of area has only water frost and the remaining area is spectrally bland (gray). Such an analysis can produce contradictory results depending on which absorption bands and what grain size of the frost are used for comparison. For example, the absorption band depths for the leading side of Ganymede can be compared to the frost band depths listed in Table 7.2. Since the 1.55-μm band depth in the Ganymede spectrum is similar to the fine-grained frost, one might conclude that there is 100% areal frost coverage on Ganymede. If the medium-coarse grained frost spectrum is compared with the Ganymede spectrum, one could conclude from the 1.55-μm band depths that $\sim$ 60% of the surface was covered by frost and $\sim$ 40% by a gray material, and if the 1.04-μm band depths of the same two spectra are compared, one would conclude that $\sim$ 100% of the surface is covered by frost. In practice, all absorption bands should be compared simultaneously, with the caution that different wavelengths probe to different depths in the surface. For example, the wavelengths in the 3.0-μm ice band probe to only a micrometer or so since the absorption coefficient of ice is extremely high, whereas the 1.04-μm wavelength region, where the ice absorption coefficient is very low, may probe to several centimeters (in a frost layer the penetration depth will most likely be controlled by scattering rather than absorption). The use of higher overtone ice absorption allows a better estimate of the bulk surface composition since light probes to deeper depths and increases the possibility of detecting other mineral absorptions.

The spectrum of the trailing side of Europa has water ice absorptions more asymmetric than any seen in the laboratory (Clark 1980a, 1981c), but there are indications that mixtures of ice and other minerals might cause such an asymmetry (Clark 1981a). In an analysis similar to that for Ganymede, Clark (1980a, 1981c) determined that 95 to 100 wt% of the spectroscopically detectable components on the surface of Europa is water ice.

The spectrum of Callisto is different from those of Europa and Ganymede in that the 1.5 and 2.0-μm water ice absorptions are much weaker (e.g., Pollack et al. 1978; Clark and McCord 1980a). Pollack et al. (1978) argued from the shape of the 3-μm water absorption that most of the water on Callisto is bound. Pollack et al. (1978) derived this conclusion by fitting the simulated spectra of water ice plus gray particles over the 1- to 5-μm region to the spectrum of Callisto. They also fit bound water (but did not include ice) absorptions to the Callisto spectrum in the 2.2- to 5-μm region. Since the bound water absorptions fit the 3-μm region and since none of the other simulations fit the Callisto spectrum well, they concluded that the 3-μm band was due mostly to bound water. However, they mostly ignored the presence of the 1.5- and 2.0-μm absorptions, saying that they may be due to only minor amounts of water ice. Lebofsky (1977) argued that the 3-μm band indicated water frost but could not rule out some bound water. Clark (1980a) pointed out potential problems in interpreting the 3-μm region, which is complicated by band saturation with very small amounts of water, and set a limit of bound water on Callisto of 5 ± 5 wt% since the 1.4- and 1.9-μm bound water absorptions were not detected. However, due to the low albedo of Callisto, the 1.5- and 2.0-μm ice bands may be suppressed (Clark 1980a,1981c). Using an optically isolated two-component (frost + gray) model, and the depth of the 1.5- and 2.0-μm absorptions in the spectra measured by Clark and McCord (1980a), Clark (1981c) derived an areal frost coverage of ≥ 45% for Callisto. The 1.25-μm frost absorption indicates ~ 90% areal coverage. However, the 3-μm band is stronger than indicated by areal frost coverage in optically isolated patches, and the higher overtone water ice absorptions are apparently stronger than those at longer wavelengths (e.g., the apparent 1.5-μm absorption is stronger than that at 2.0 μm). It is a physical impossibility for the real higher overtones to be stronger than the lower overtones. However, this is the case for the *apparent* absorptions in the frost on ice spectra (Clark 1980a,1981a). The wings of the strong absorptions overlap with the result of washing out the absorptions, leaving only weak apparent absorption bands. The continuum would be steeply sloped downward with increasing wavelength for a pure frost-on-ice spectrum, so in the case of Callisto (and probably to a lesser extent Europa) the continuum can be modified by other minerals if the minerals are mixed with the ice (Clark 1980a,1981c). From the laboratory data of Clark (1981a,b), Clark (1980b) determined an upper limit of ~ 90 wt% water ice on Callisto using a model in which dark particles are well mixed in the ice surface. From the indications of at least some dark particles mixed with the ice, the areal coverage of 45 to 90% and density of about 3 for the dark particles, Clark (1981c) determined a lower limit of ~ 30 wt% for the amount of water ice on the surface of Callisto.

III. DISCUSSION

The most recent studies of the icy Galilean satellites indicate that there is more water on their surfaces than previously thought. The other components in the surface have spectra that appear similar to those of minerals containing Fe^{3+} (Clark 1980a,1981c). Many carbonaceous chondritic meteorites show spectra similar to the indicated nonwater component (Clark 1980a, 1981c); however, this similarity is not proof of carbonaceous chondritic material on the satellites since other combinations of materials appear spectrally similar. Some sulfur allotropes may have the necessary spectral shape in the visible wavelength region. The exact spectral shape of the nonice component in the visible is not known due to the complications of the water frost in that region.

There has been some controversy over the presence of bound water in the nonice component on Callisto and Ganymede. The 5 ± 5 wt% bound water averaged over the surface as determined by Clark (1981c) does not contradict the work of others (e.g., Pollack et al. 1978). In the case of Ganymede, Clark (1981c) determined the nonice component as ~ 10 wt% of the surface. With a bulk average of 5 ± 5 wt% bound water in the surface, the nonice component contains 50 ± 50 wt% bound water. In the case of Callisto, Clark (1981c) gave a lower limit of 30 wt% water ice or 60 wt% for the nonice component. Thus, this limit gives ~ 8 ± 8 wt% bound water in the nonice component. The upper limit of 90 wt% water ice on Callisto given by Clark (1981c) implies 50 ± 50 wt% bound water in the nonice component. Thus no hydrated minerals could be excluded by Clark (1981c). The fact that Pollack et al. (1978) attributed the 3-μm band mostly to bound water may mean that there is a small amount of dust near the surface (top micrometers), and there may not be a discrepancy with the results of Clark (1981c). Clark et al. (1980) presented new data for the trailing side of Callisto that show a 1.04-μm ice absorption with a band depth of 2.3%, indicating that the upper limits determined by Clark (1981c) (e.g., ~ 90 wt%) for the water ice on Callisto may be the more correct value.

For Ganymede, the data from the Voyager imaging experiment combined with the spectral evidence for 90 wt% water have implications for the compositions of the different units. Clark (1981a) showed that a surface composed of a mixture of ice and dark material can appear as dark or darker than the dark material alone. Thus Clark (1980a,1981c) argued that: the dark ancient terrain on Ganymede may contain as much water near the surface as on Callisto (e.g., as much as 90 wt%); the younger grooved terrain contains more water ($\gtrsim$ 90 wt%); and the bright rayed craters have only trace amounts of other material ($\lesssim$ 2%). The thin polar caps (see Smith et al. 1979, Fig. 12) on Ganymede are typical of a thin frost layer $\leqslant$ 0.1 mm thick (Clark 1980a,1981c).

Due to the present limited understanding of reflectance spectra of pure and mixed materials, the conclusions by Clark (1980a,1981c) and others may need revision. The present conclusions refer to the spectroscopically detectable materials. The present infrared data (2–3% precision), while approaching the quality of laboratory data ($\lesssim$ 0.5% precision), need improvement, and the spectral properties of mixtures need much more investigation. If we are to understand more precise spectra of these objects, the laboratory studies need to be expanded. Such studies have important applications not only for the Galilean satellites but also for the satellites of Saturn, Uranus, and Neptune, and for comets, Mars, and even remote sensing of the Earth.

Infrared detector technology has advanced to the point of obtaining very high-precision reflectance spectra of the Galilean satellites. From new laboratory studies (e.g., Clark 1980a, 1981a,b; Smythe 1979; Singer 1981) and from high-precision telescopic data of Clark and McCord (1980a), it is evident that many diagnostic features become apparent in ice-mineral mixtures at the 1% to 5% level, implying the need for the highest quality data. Clark et al. (1980) have obtained data on the Galilean satellites with better than 1% precision; the limit to data quality is now the stellar calibrations used in deriving the reflectance spectra. Interpretation of these data is current in progress and was not available for this chapter. Infrared monitoring of Io in the 5- to 20-μm region from Earth will be useful for monitoring volcanic eruptions, as indicated by Sinton (1980a,1981c) and, combined with monitoring of the 4-μm SO_2 absorption, should provide a basis for a better understanding of the surface of Io. The next few years before Galileo arrives at Jupiter will probably provide significant advancement in understanding of the compositions of the surfaces of the Galilean satellites.

Acknowledgments. This research was supported by grants from the National Aeronautics and Space Administration.

REFERENCES

Aronson, J.R. and Emslie, A.G. (1973). Spectral reflectance and emittance of particulate materials. 2: Application and results. *Applied Optics* 12, 2573–2584.

Berkowitz, J. and Chupka, W.A. (1964). Vaporization processes involving sulfur. *J. Chem. Phys.* 40, 287–295.

Berkowitz, J. and Marquart, J.R. (1963). Equilibrium composition of sulfur vapor. *J. Chem. Phys.* 39, 275–283.

Carr, M.H., Masursky, H., Strom, R.G. and Terrile, R.J. (1979). Volcanic features of Io. *Nature* 280, 729–733.

Chatelain, Y.A. and Buttet, J. (1965). Electron paramagnetic resonance studies of unstable sulfur forms. In *Elemental Sulfur* (B. Meyer, Ed.), p. 208–215. Interscience, New York.

Clark, R.N. (1980a). Spectroscopic Studies of Water and Water/Regolith Mixtures on Planetary Surfaces at Low Temperatures. Ph.D. Dissertation, M.I.T., Cambridge.

Clark, R.N. (1980b). A large scale interactive one dimensional array processing system. *Publ. Astron. Soc. Pacific* 92, 221–224.

Clark, R.N. (1981a). Water frost and ice: The near-infrared reflectance 0.65–2.5 μm. *J. Geophys. Res.* In press.

Clark, R.N. (1981b). The spectral reflectance of water-mineral mixtures at low temperatures. *J. Geophys. Res.* In press.

Clark, R.N. (1981c). Ganymede, Europa, Callisto, and Saturn's rings: Compositional analysis from reflectance spectroscopy. *Icarus* 44, 388–409.

Clark, R.N. and McCord, T.B. (1980a). The Galilean satellites: New near-infrared reflectance measurements (0.65–2.5 μm) and a 0.325–5 μm summary. *Icarus* 41, 323–339.

Clark, R.N. and McCord, T.B. (1980b). The rings of Saturn: New near-infrared reflectance measurements and a 0.326–4.08 μm summary. *Icarus* 43, 161–168.

Clark, R.N., Singer, R.B., Owensby, P.D. and Fanale, F.P. (1980). Galilean satellites: High precision near-infrared spectrophotometry (0.65–2.5 μm) of the leading and trailing sides. *Bull. Amer. Astron. Soc.* 12, 713–714.

Consolmagno, G.J. (1979). Sulfur volcanoes on Io. *Science* 205, 397–398.

Cruikshank, D.P. (1980). Infrared spectrum of Io, 2.8–5.3 μm. *Icarus* 41, 240–245.

Cruikshank, D.P., Jones, T.J., and Pilcher, C.B. (1978). Absorption bands in the spectrum of Io. *Astrophys. J.* 225, L89–L92.

Emslie, A.G. and Aronson, J.R. (1973). Spectral reflectance and emittance of particulate materials. 1: Theory. *Applied Optics* 12, 2563–2572.

Fanale, F.P., Brown, R.H., Cruikshank, D.P., and Clark, R.N. (1979). Significance of absorption features in Io's IR reflectance spectrum. *Nature* 280, 761–763.

Fanale, F.P., Johnson, T.V., and Matson, D.L. (1974). Io: A surface evaporate deposit? *Science* 186, 922–925.

Fink, U., Dekkers, N.H., and Larson, H.P. (1973). Infrared spectra of the Galilean satellites of Jupiter. *Astrophys. J.* 179, L155–L159.

Fink, U., Larson, H.P., Lebofsky, L.A., Feierberg, M., and Smith, H. (1980). High resolution spectrum of Io from 2–4 μm, IAU Coll. 57. *The Satellites of Jupiter* (abstract 4–15).

Hapke, B. (1979). Io's surface and environs: A magmatic-volatile model. *Geophys. Res. Letters* 6, 799–802.

Harris, D.L. (1961). Photometry and colorimetry of planets and satellites. In *Planets and Satellites* (G.P. Kuiper and B.M. Middlehurst, Eds.), pp. 272–343. Univ. Chicago Press, Chicago.

Johnson, T.V. and Pilcher, C.B. (1977). Satellite spectrophotometry and surface compositions. In *Planetary Satellites* (J. Burns, Ed.), pp. 232–268. Univ. Arizona Press, Tucson.

Kieffer, H.H. (1970). Spectral reflectance of CO_2–H_2O frosts. *J. Geophys. Res.* 75, 501–509.

Kuiper, G.P. (1973). Comments on the Galilean satellites. *Comm. Lunar Planet. Lab.* 10, 28–34.

Lebofsky, L. (1977). Identification of water frost on Callisto. *Nature* 269, 785–787.

MacKnight, W.J. and Tobolsky, A.V. (1965). Properties of polymeric sulfur. In *Elemental Sulfur* (B. Meyers, Ed.), pp. 101–104. Interscience, New York.

McFadden, L.A., Bell, J., and McCord, T.B. (1981). Visible spectral reflectance measurements 0.3–1.1 μm of the Galilean satellites at many orbital phase angles 1977–1978. *Icarus* 44, 410–430.

Meyer, B. (1968). Elemental sulfur. In *Inorganic Sulfur Chemistry* (G. Nickless, Ed.), pp. 241–258. Elsevier, Amsterdam.

Meyer, B., Gouterman, M., Jensen, D., Oommen, T.V., Spitzer, K., and Stroyer-Hansen, T. (1972). The spectrum of sulfur and its allotropes. In *Sulfur Research Trends* (D.J. Miller and T.K. Wieniorski, Eds.), pp. 53–71. Amer. Chem. Soc., Washington.

Nash, D.B. and Conel, J.F. (1974). Spectral reflectance systematics for mixtures of powdered hypersthene, labradorite and ilmenite. *J. Geophys. Res.* 79, 1615–1621.

Nash, D.B. and Fanale, F.P. (1977). Io's surface composition based on reflectance spectra of sulfur/salt mixtures and proton irradiation experiments. *Icarus* 31, 40–80.

Nash, D.B. and Nelson, R.M. (1979). Spectral evidence for sublimates and adsorbates on Io. *Nature* 280, 763–766.

Nelson, R.M. and Hapke, B.W. (1978). Spectral reflectivities of the Galilean satellites and Titan 0.32 to 0.86 micrometers. *Icarus* 36, 304–329.

Nelson, R.M., Lane, A.L., Matson, D.L., Fanale, F.P., Nash, D.B. and Johnson, T.V. (1981). Io: Longitudinal distribution of SO_2 frost. Submitted to *Science*.

Peale, S.J., Reynolds, R.T., and Cassen, P.M. (1980). Sulfur vapor and sulfur dioxide models of Io's plumes. IAU Coll. 57. *The Satellites of Jupiter* (abstract 3–9).

Pearl, J., Hanel, R., Kunde, V., Maguire, W., Fox, K., Gupta, S., Ponnamperuma, C., and Raulin, F. (1979). Identification of gaseous SO_2 and new upper limits for other gases on Io. *Nature* 280, 755–758.

Pilcher, C.B., Ridgway, S.T., and McCord, T.B. (1972). Galilean satellites: Identification of water frost. *Science* 178, 1087–1089.

Pollack, J.B., Witteborn, F.C., Edwin, F.E., Strecker, D.W., Baldwin, J.B., and Bunch, B.E. (1978). Near-infrared spectra of the Galilean satellites: Observations and implications. *Icarus* 36, 271–303.

Radford, H.E. and Rice, F.O. (1960). Green and purple sulfur: Electron spin resonance studies. *J. Chem. Phys.* 33, 774–776.

Ran, H., Kutty, T.R., and Guedes de Varvalho, J.R. (1973). Thermodynamics of sulphur vapour. *J. Chem. Thermo.* 5, 833–844.

Sagan, C. (1979). Sulfur flows on Io. *Nature* 280, 750–753.

Schenk, P.W. and Steudel, R. (1968). Oxides of sulfur. In *Inorganic Sulphur Chemistry* (G. Nickless, Ed.), pp. 367–418. Elsevier, Amsterdam.

Sill, G.T. (1973). Reflection spectra of solids of planetary interest. *Comm. Lunar Planet. Lab.* 10, 1–7.

Singer, R.B. (1981). Near-infrared spectral reflectance of mineral mixtures: Systematic combinations of pyroxenes, olivine, and iron oxides. Submitted to *J. Geophys. Res.*

Sinton, W.M. (1980a). Io's 5 μm variability. *Astrophys. J.* 235, L49–L51.

Sinton, W.M. (1980b). Io: Are vapor explosions responsible for the 5 μm outbursts? *Icarus* 43, 56–64.

Sinton, W.M. (1981). The thermal emission spectrum of Io and a determination of the heat flux from its hot spots. *J. Geophys. Res.* In press.

Smith, B.A., Shoemaker, E.M., Kieffer, S.W., and Cook, A.F. (1979a). The role of SO_2 in volcanism on Io. *Nature* 280, 738–746.

Smith, B.A. and the Voyager Imaging Team (1979b). The Jupiter system through the eyes of Voyager 1. *Science* 204, 951-971.

Smythe, W.D. (1979). The Detectability of Clathrate Hydrates in the Outer Solar System. Ph.D. Dissertation. Univ. of California, Los Angeles.

Smythe, W.D., Nelson, R.M., and Nash, D.B. (1979). Spectral evidence for SO_2 frost or adsorbate on Io's surface. *Nature* 280, 766.

Soderblom, L., Johnson, T., Morrison, D., Danielson, E., Smith, B., Veverka, J., Cook, A., Sagan, C., Kupfermap, P., Pieri, D., Mosher, J., Avis, C., Gradie, J., and Clancy, T. (1980). Spectrophotometry of Io: Preliminary Voyager I results. *Geophys. Res. Letters* 7, 963–966.

Stebbins, J. (1927). The light variations of the satellites of Jupiter and their application to measures of the solar constant. *Lick Obs. Bull.* 13, 1–11.

Stull, D.R. and Prophet, H. (1971). *JANAF Thermochemical Tables*. Nat. Bur. Standards, Washington.

Veverka, J., Goguen, J., Yang, S., and Elliot, J.L. (1978a). Near-opposition limb darkening of solids of planetary interest. *Icarus* 33, 368–379.

Veverka, J., Goguen, J., Yang, S., and Elliot, J.L. (1978b). How to compare the surface of Io to laboratory samples. *Icarus* 34, 63–67.

Veverka, J., Goguen, J., Yang, S., and Elliot, J.L. (1979). On matching the spectrum of Io: Variations in the photometric properties of sulfur-containing mixtures. *Icarus* 37, 249–255.

Vogt, R.E. and the Voyager Cosmic Ray Team (1979). Voyager 1; Energetic ions and electrons in the Jovian magnetosphere. *Science* 204, 1003–1007.

Wamsteker, W. (1972). Narrow-band photometry of the Galilean satellites. *Comm. Lunar Planet. Lab.* 9, 171–177.

Wamsteker, W., Kroes, R.L., and Fountain, J. A. (1974). On the surface composition of Io. *Icarus* 23, 417–424.

Yoder, H.S. (1976). *Generation of Basaltic Magma*. Nat. Acad. Sci., Washington.

8. RADAR PROPERTIES OF EUROPA, GANYMEDE, AND CALLISTO

STEVEN J. OSTRO

Cornell University

Europa, Ganymede, and Callisto have extremely unusual 12.6-cm-wavelength radar properties. Their geometric albedos are enormous compared to those of the Moon and inner planets. The circular polarization ratio, of echo power received in the transmitted rotational sense (i.e., handedness) to that received in the opposite sense, exceeds unity for each satellite. This "circular polarization inversion" has not been encountered in radar studies of other solar system targets. The circular polarization ratios for Europa and Ganymede are indistinguishable, and considerably larger than that for Callisto. The albedos increase logarithmically in the order: Callisto, Ganymede, Europa. Significant albedo and/or polarization features are common in the echo power spectra and, in a few cases, the source of a radar scattering anomaly can be identified tentatively in Voyager images. The magnitude of the circular polarization inversion is extremely sensitive to wavelength in the range from 3.5 cm to 70 cm. If the unusual radar echoes arise from external surface reflections, the satellite surfaces must have a refractive index substantially higher than that of solid water ice and must be covered with nearly hemispherical craters. A geologically more attractive interpretation of the radar results postulates subsurface scattering from randomly oriented interfaces between two regolith components. In this random-facet model, the ratio of the components' refractive indices must be small (~ 1.2). Therefore, the components may have identical chemical compositions and slightly different densities, or vice versa. If the lower-index component exists as planar veins ~ 10 cm thick, the wavelength dependence of the satellites' radar properties is easily explained.

The presence of ice on the surfaces of Europa, Ganymede, and Callisto has endowed these bodies with highly unusual radar properties. In fact, the primary reaction to the initial measurement of radar albedos and polarization ratios (Campbell et al. 1978) of these icy satellites was incredulity. Since then, extensive observations have confirmed initial impressions that these bodies comprise a unique class of planetary radar targets. Within this class, Europa, Ganymede, and Callisto have distinctly individual properties, suggesting that differences between the satellites' morphology at centimeter-to-decameter scales may be as striking as the differences at much larger scales revealed in Voyager images.

Interpretation of the radar results in terms of surface characteristics has been more difficult for the icy Galilean satellites than for other planetary radar targets. For the Moon and the inner planets, several sound theoretical models can relate radar properties (e.g., albedo, polarization, angular scattering law) to the structural configuration of the surface. These theories cannot be used to interpret the radar echoes from Europa, Ganymede, and Callisto, simply because the scattering of electromagnetic waves from these objects is fundamentally unlike scattering from any other natural or man-made surface so far studied with radar.

The most perplexing aspect of the satellites' radar behavior involves the way they backscatter circularly polarized waves. As such waves propagate, the electric field vector rotates either as a right-handed screw or as a left-handed screw. The handedness is reversed upon reflection from a mirror or a smooth sphere. For example, a right-handed circularly polarized wave transmitted at a smooth sphere will produce an echo whose polarization is left-handed circular.

The Moon and the inner planets scatter radar waves like moderately smooth spheres, and most of the echo power has the rotational sense opposite to that transmitted. The relatively small amount of power received in the same rotational sense as transmitted can be attributed to multiple scattering or to backscattering from some rough component of the surface, such as wavelength-sized rocks. One might imagine a surface which is so rough that the polarization of the incident wave is completely randomized. In this limit, there would be as much echo power in the transmitted sense of circular polarization as in the opposite sense. Such an echo would be unpolarized.

It is convenient to define a circular polarization ratio μ_C as the ratio of echo power received in the same sense of circular polarization as transmitted to that received in the opposite sense. For the terrestrial planets, the circular polarization ratio is much less than unity. However, for Europa, Ganymede, and Callisto, the circular polarization ratio is actually greater than unity. In other words, the transmitted handedness is largely preserved. Compared to all previous radar experience, the circular polarization is inverted.

In addition to this startling result, the satellites' radar geometric albedos are enormous. Europa constitutes the extreme case, with a backscattering efficiency about thirty times greater than that of the Moon. Campbell et al. (1977) first pointed out that the radar geometric albedo, the visible-wavelength geometric albedo, and the areal water-frost coverage all increase by satellite in the order: Callisto, Ganymede, Europa.

The anomalous radar properties appear correlated with the presence of water ice on the satellite surfaces, but the detailed reflecting mechanism which gives rise to the observed behavior is hardly obvious. Although the radar data may well contain valuable physical information, geomorphological interpretation of the radar results certainly cannot precede a basic theoretical comprehension of the anomalous radar scattering. Therefore, efforts to interpret the radar results for Europa, Ganymede, and Callisto have been devoted primarily to explaining the circular polarization inversion.

During the late 1970s, a competent, first-order understanding of the scattering process has emerged. Radar echoes from the terrestrial planets are dominated by reflections from the surface *per se*. However, Ostro and Pettengill (1978) showed that surface reflections can produce a circular polarization inversion only if the surface has a geologically unlikely geometry. Goldstein and Green (1980) showed that subsurface scattering from a two-component regolith can invert the circular polarization if certain plausible electrical and geometric constraints are satisfied.

This chapter summarizes the radar properties of Europa, Ganymede, and Callisto and documents our comprehension of these properties. In the first section I describe the radar techniques, review the observational results, and discuss salient aspects of the radar data set. The second section focuses on theoretical interpretation of the satellites' anomalous radar properties. I discuss the physics of the circular polarization inversion and suggest several refinements of the Goldstein-Green scattering model, arriving at a tentative description of the small-scale geomorphology of the icy Galilean satellites.

I. OBSERVATIONAL RESULTS

Most radar investigations of the satellites, both from the National Astronomy and Ionosphere Center's Arecibo Observatory and from the Jet Propulsion Laboratory's Goldstone Tracking Station, have been simple continuous-wave observations. A highly monochromatic, unmodulated wave is transmitted for a duration corresponding to the round trip travel time of light to the target. The Doppler-shifted echo is then received for a similar duration. Since each illuminated part of the rotating target backscatters power at a particular Doppler shift, the received echo is dispersed in frequency, and each point on the received power spectrum is proportional to the brightness of a portion of the visible disk viewed through a slit oriented parallel to the target's rotation axis.

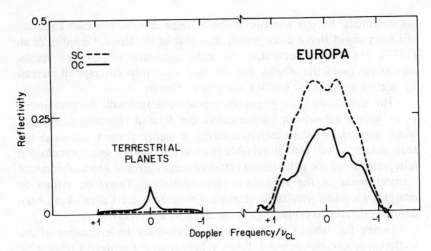

Fig. 8.1. Typical 12.6-cm-wavelength radar echo spectra for the terrestrial planets (left) compared to echo spectra for Europa obtained by Ostro et al. (1980). The abscissa is in units of the center-to-limb Doppler frequency, $\nu_{CL} = 4\pi a/\lambda P$, where a is radius, λ is wavelength, and P is rotation period. Spectral edges correspond to echoes from equatorial regions on the target limbs. The geometric albedo α equals the sum of the areas under the SC and OC curves (see text), is ~ 0.025 for the terrestrial planets, and is ~ 0.65 for Europa. The circular polarization ratio μ_C equals the ratio of SC reflectivity to OC reflectivity, is ~ 0.1 for the terrestrial planets, and is ~ 1.6 for Europa.

In the more valuable radar experiments, power spectra are obtained in two orthogonal polarizations, permitting estimation of the geometric albedo, $\alpha = (\hat{\sigma}_{SC} + \hat{\sigma}_{OC})/4 = (\hat{\sigma}_{OL} + \hat{\sigma}_{SL})/4$, and either the circular polarization ratio, $\mu_C = \hat{\sigma}_{SC}/\hat{\sigma}_{OC}$, or the linear polarization ratio, $\mu_L = \hat{\sigma}_{OL}/\hat{\sigma}_{SL}$. Here O and S denote polarization senses "orthogonal to" and "same as" the transmitted polarization. The normalized radar cross section $\hat{\sigma}$ is equal to $\sigma/\pi a^2$, where σ is the radar cross section and a is the target radius. The polarization ratios are defined so that each equals zero for the echo from an infinite plane mirror or a perfectly smooth sphere, and unity for a completely unpolarized echo.

Comparison with the Inner Planets

The left side of Fig. 8.1 shows OC and SC spectra which are representative of 12.6-cm-wavelength radar echoes from the Moon, Mercury, Mars, and Venus. The abscissa is in units of the target's center-to-limb bandwidth, $\nu_{CL} = 4\pi a/\lambda P$, where λ is the wavelength and P is the rotation period. The spectral edges correspond to echoes from equatorial regions on the target limbs. The curves are normalized so that the geometric albedo equals the sum of the areas under the SC and OC curves. The circular polarization ratio μ_C is simply the ratio of these areas. For the terrestrial planets, $\alpha \simeq 0.025$ and $\mu_C \simeq 0.1$. As discussed by Pettengill (1968, 1978), the radar echo from the inner planets can be separated into two portions. The stronger quasispecular portion has low polarization ratios, is due to scattering from a smooth component of the surface, and dominates the echo from the subradar regions. The weaker diffuse portion is largely unpolarized, is due to scattering from rough surface components, and dominates the echo far from the subradar point.

The right side of Fig. 8.1 shows 12.6-cm echo power spectra obtained for Europa by Ostro et al. (1980). The corresponding curves for Ganymede and Callisto have lower amplitudes but similar shapes. Clearly, the radar echo from the icy satellites is not dominated by a quasispecular component. Instead, the scattering is diffuse and is conveniently modeled by an angular law of the form $\cos^m \theta$. Here θ is the angle of incidence, measured between the normal to the surface and the radar line of sight. The satellites' mean scattering laws are intermediate between a Lambert law ($m = 2$) and the law for a uniformly bright disk ($m = 1$).

Diffuse scattering laws require structure that is very rough at a scale at least as large as the observing wavelength. For the icy Galilean satellites, critical constraints on the specific type of roughness are provided by the satellites' circular polarization ratios and, to a lesser extent, by their radar albedos. As Fig. 8.1 illustrates, the albedos of the icy satellites are huge. Even more striking is the fact that the circular polarization ratio exceeds unity for all three satellites. As noted earlier, this anomalous polarization inversion has not been encountered in radar studies of nonicy planetary objects, whose circular polarization ratios are much less than unity.

The Radar Data Set

Extensive 12.6-cm-wavelength observations at the Arecibo Observatory have yielded dual-polarization echo power spectra at an average of eight widely separated values of rotational phase for each satellite. (See Ostro et al. [1980] for spectra and detailed tabulation of results corresponding to particular observations.) Table 8.1 summarizes weighted mean values of the radar geometric albedo, the circular and linear polarization ratios, and the scattering-law exponents for Europa, Ganymede, and Callisto. Data at 3.5 cm and 70 cm are sparse, but sufficient to indicate that the anomalous radar properties depend on wavelength. As will be discussed in the next section, this wavelength dependence sets limits on the scale of structure responsible for those properties. Extremes in the satellites' disk-integrated radar properties are represented by Ganymede's 3.5-cm circular polarization ratio ($\mu_C = 2.0 \pm 0.3$), and Europa's 12.6-cm geometric albedo ($\alpha = 0.65 \pm 0.16$).

The inter-satellite differences in 12.6-cm radar properties are intriguing.

TABLE 8.1

Radar Properties of the Icy Galilean Satellites[a]

| | Europa | | Ganymede | | Callisto |
	12.6 cm[b]	70 cm[c]	12.6 cm[b]	3.5 cm[d]	12.6 cm[b]
α	0.65±0.16	<0.3	0.38±0.10	0.30±0.07	0.16±0.04
μ_C	1.56±0.11	<1.0	1.55±0.06	2.0 ±0.03	1.19±0.06
μ_L	0.47±0.07		0.47±0.08		0.55±0.10
m	1.73±0.08		1.46±0.04		1.43±0.05
m_{OC}/m_{SC}	1.08±0.10[e]		1.21±0.06[e]		1.13±0.09[e]

[a]Weighted mean values of the radar geometric albedo α, the polarization ratios μ_C and μ_L, and the scattering-law exponent m are listed. Here m is estimated from a spectrum formed by summing SC and OC spectra, or SL and OL spectra.

[b]The 12.6-cm results are from Ostro et al. (1980).

[c]The 70-cm values are from preliminary analyses of observations by Campbell and Ostro (1981).

[d]The 3.5-cm results are from Goldstein and Green (1980).

[e]The ratio, m_{OC}/m_{SC}, of single-polarization estimates of the scattering-law exponent, is discussed in the text.

The radar albedo increases logarithmically from Callisto to Ganymede to Europa, mimicking the well-known relation among visible-wavelength albedos. The circular polarization ratios of Ganymede and Europa are virtually identical, and much greater than Callisto's. The scattering-law exponents of Ganymede and Callisto are indistinguishable and significantly smaller than Europa's, suggesting that Europa is more limb-darkened and/or has brighter polar regions than Ganymede or Callisto. (Either situation would produce a narrower echo spectrum, and thus a larger apparent scattering-law exponent, for Europa.) For Ganymede, the scattering-law exponents estimated from OC spectra are larger than those estimated from corresponding SC spectra, suggesting that the circular polarization ratio increases toward the limbs or decreases toward the poles, or both. Finally, the linear polarization ratio of each satellite is ~ 0.5.

Radar Features on the Satellites

For each satellite, the rms fluctuations of measurements of geometric albedo and circular polarization ratio at particular values of rotational phase Φ are within 10% of the corresponding weighted mean values. There is limited evidence (Ostro et al. 1980) for correlation between the radar- and visible-wavelength light curves, although uncertainties in radar system calibration make this conclusion tentative. No clear dependence on rotational phase is apparent among estimates of the scattering-law exponent or of the circular polarization ratio. Nevertheless, Callisto's dual-polarization spectra (Fig. 8.2) suggest that substantial surface areas with $\mu_C \approx 1$ exist on the trailing side but not on the leading side. Since the anomalous polarization inversion is presumably related to the presence of surface ice, the association of larger values of μ_C with the visually darker hemisphere is curious.

On all three satellites, localized radar scattering anomalies are suggested by statistically significant "features" in the spectra. Occasionally, features in OC and SC spectra are correlated, indicating regions of anomalous reflectivity. These albedo features are more common than polarization features, such as the high-μ_C feature produced by a dip (located by ↑) in the OC spectrum of Europa obtained on 23 Nov. 77 (Fig. 8.3). This particular feature could lie anywhere along the spectral "salient" shown as the left-hand solid curve drawn over the map of Europa in Fig. 8.4. In this figure, the strip between the dashed salients corresponds to a high-μ_C spectral band in the echo from Europa obtained on 27 Nov. 77 (Fig. 8.3). The overlapping of these salients suggests tentative identification of the dark regions on Europa near longitude 246° as a source of extreme polarization inversion. Unfortunately, since radar echoes from the satellites have never been resolved simultaneously in delay and Doppler, the current data preclude definitive correlation between radar features and Voyager images. (Consider the possible sources of the high-albedo feature, located by ↓, in the Europa spectrum obtained on 23 Nov. 77, corresponding to the right-hand solid curve in Fig. 8.4.)

Observations by Ostro et al. (1980) of Ganymede at rotational phases 138° and 168° suggest that Galileo Regio has a relatively low radar reflectivity (Fig. 8.5). This feature, the largest single piece of undisturbed heavily cratered terrain on Ganymede, lies in the northern hemisphere between longitudes ~ 90° and ~ 180°. The radar spectra obtained at $\Phi =$ 168° (i.e., with the west longitude of the central meridian equal to 168°) have low relative amplitudes at negative Doppler frequencies, suggesting that areas east of longitude 168° have a lower albedo than areas west of that longitude. During the observation at rotational phase 138°, Galileo Regio was nearly centered about the central meridian. The echo spectra for this observation are centrally depressed, probably due to the low radar reflectivity of Galileo Regio. At this particular rotational phase, the scattering-law exponent deduced from the echo spectral shape ($m = 1.05 \pm 0.08$) is significantly lower than Ganymede's weighted mean value ($m = 1.46 \pm 0.04$). In other words, Ganymede looks less radar limb-darkened at $\Phi = 138°$ than at other rotational phases. Tentative identification of Galileo Regio as a region of relatively low radar reflectivity is consistent with the apparent correlation of the disk-integrated radar albedo with areal water-frost coverage.

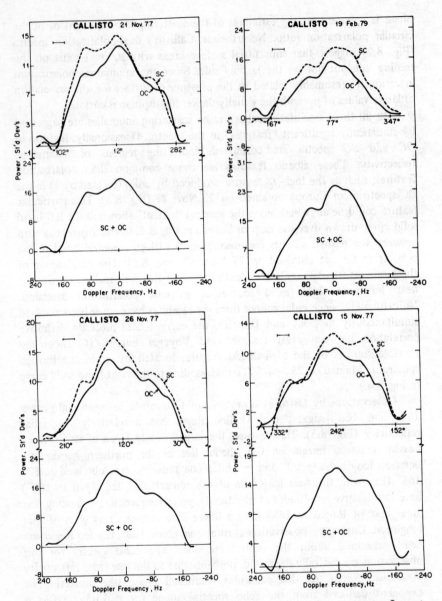

Fig. 8.2. Callisto echo spectra obtained by Ostro et al. (1980) for orbital phase Φ equal to $12°$, $77°$, $120°$, and $242°$. The $SC + OC$ spectra were obtained by summing SC and OC spectra. Due to the satellites' synchronous rotation, Φ also represents the rotational phase and is approximately equal to the longitude of the central meridian. Longitudes of the approaching and receding limbs are indicated. Note that $\mu_C \approx 1$ for substantial areas on the trailing side ($180° \leqslant \Phi \leqslant 360°$), but $\mu_C > 1$ for the leading side ($0° \leqslant \Phi \leqslant 180°$).

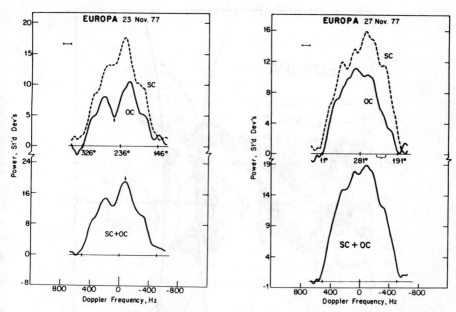

Fig. 8.3. Europa spectra obtained by Ostro et al. (1980). The dip (↓) in the OC spectrum for 23 Nov. 77 produces $\mu_C > 2$ at a certain frequency. The left-hand solid curve seen in Fig. 8.4 indicates positions on Europa possibly contributing echo at this frequency. Within the bracketed spectral band in the 27 Nov. 77 spectrum, $\mu_C \geqslant 2$ and received power in each polarization at least four times as large as the rms receiver noise. This band corresponds to the strip enclosed by dashed curves in Fig. 8.4. The high albedo feature (↓) in the 23 Nov. 77 spectrum corresponds to the right-hand solid curve in Fig. 8.4.

II. THEORETICAL INTERPRETATIONS

In this section, the physics of the anomalous circular polarization inversion is examined in some detail. As we have seen, for Europa, Ganymede, and Callisto, the radar scattering must be dominated by some process which returns more power in the transmitted sense of circular polarization than in the opposite sense. All single-reflection backscattering reverses the rotational sense of incident circularly polarized waves, so at least two reflections must be involved. Scattering involving surface areas whose dimensions or radii of curvature are not large compared to the observing wavelength will destroy the phase coherence and the degree of polarization of the transmitted wave. Thus the backscattering process probably involves surface areas (facets) that are large and quite flat at centimeter-to-decimeter scales, and a geometric optics approximation to the scattering is justified.

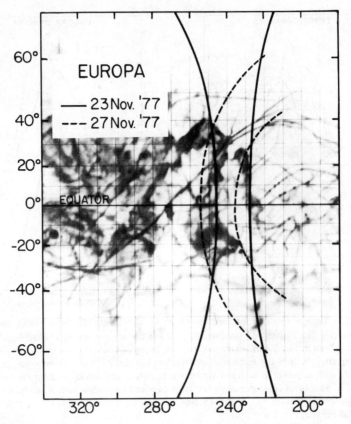

Fig. 8.4. Spectral salients (loci of constant Doppler frequency) corresponding to spectral features in radar observations of Europa (Fig. 8.3). The large, dark, north-south regions near longitude $246°$ may be the source of echo with $\mu_C > 2$. (See text.)

External Scattering: Double Reflections from Hemispherical Craters

Ostro and Pettengill (1978) assumed that the satellites' high geometric albedos required the Fresnel power reflection coefficient of the surface material to be large. Water ice, with a refractive index n equal to 1.8, was the only positively identified chemical species on the satellites. However, in order to satisfy the observed albedos, it was necessary to invoke a dense matrix of ice plus some higher-index material – possibly silicates, with $n \sim 2.4$ (Campbell and Ulrichs 1969).

Noting that first-order backscattering yields $\mu_C \simeq 0$, while high-order multiple external scattering from randomly oriented facets rapidly depolarizes the transmitted wave ($\mu_C \to 1$), Ostro and Pettengill suggested second-order reflections as the primary source of echo. (Here, "external"

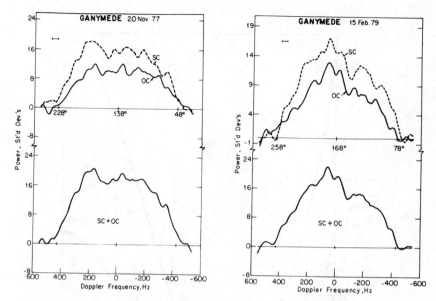

Fig. 8.5. Ganymede spectra obtained by Ostro et al. (1980). Echo from Galileo Regio would be received at central (low positive and low negative) Doppler frequencies during observation of Ganymede at a rotational phase $\Phi = 138°$ (left), but almost exclusively at negative Doppler frequencies during observation at $\Phi = 168°$ (right). Galileo Regio is tentatively identified as having a lower radar reflectivity than Ganymede's average surface.

reflection from the boundary between two media denotes reflection from the medium with the higher refractive index.) Two successive reflections from a pair of facets will produce backscatter if, and only if, the facet normals are perpendicular to each other and lie in a plane containing the radar line of sight. If this geometrical constraint is satisfied, the backscattered intensity and polarization will be a strong function of the angle of incidence θ_1 for the first reflection. (Because of the geometry, the angles of incidence for the two reflections are complementary and interchangeable.) As shown in Fig. 8.6, the circular polarization is inverted ($\mu_C > 1$) when both angles of incidence are between the Brewster angle, $\theta_B = \tan^{-1} n$, and its complement. The maximum value of μ_C is reached at $\theta_1 = \theta_2 = 45°$ and is an increasing function of refractive index. However, the reflected intensity is proportional to the product W of the power reflection coefficients for the two reflections. This product, plotted on an arbitrary scale in Fig. 8.6, is considerably greater for a single reflection at normal incidence to a facet than for 45° double-reflection backscatter. If the effects of facet capture area ($\sim \cos \theta$) and gain (i.e., directivity) are considered, the inequality $W(0°) \gg (45°)$ becomes even more pronounced. Thus an ensemble of randomly oriented

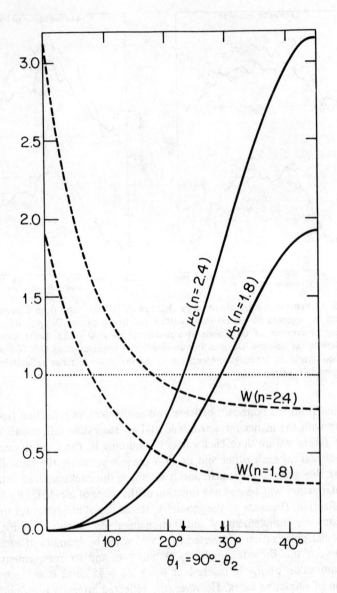

Fig. 8.6. The circular polarization ratio μ_C and the scaled product W of the Fresnel power reflection coefficients for double-external-reflection backscatter from facets with refractive index n equal to 1.8 (ice) and 2.4 (silicate). The figure is symmetrical about $\theta_1 = \theta_2 = 45°$. The ratio μ_C exceeds unity when the angles of incidence for the two reflections are between the Brewster angle and its complement (designated by arrows).

facet pairs (or grooves, or corner reflectors) will backscatter much more power with $\mu_C < 1$ than with $\mu_C > 1$.

On the other hand, hemispherical craters geometrically favor 45° double-reflection backscatter for all orientations and can produce a net circular polarization inversion ($\mu_C > 1$). Single-reflection echoes will be seen from the part of the crater farthest from the radar, while double-reflection echoes will arise from an annular region of the crater wall inclined 45° to the line of sight. In the Ostro-Pettengill scattering model, the surfaces of the postulated icy craters are endowed with flat facets much larger than the observing wavelength, in accordance with the geometric optics approximation. Results of their scattering calculations indicate that both μ_C and α decrease as the facet size decreases. The desirable radar polarization properties of hemispherical craters will probably hold for crater diameters $\gtrsim$ 100 m, but probably not for diameters $\lesssim$ 1 m. Implicit in the geometric optics approximation is the constraint that scattering interfaces be smooth at scales $\sim \lambda/10$. Hence the craters must be smooth to a scale of several millimeters (to satisfy the 3.5-cm results) as well as nearly hemispherical (to maximize the relative probability of 45° double-reflection backscatter).

Although meteoritic impact, volcanism, or thermal ablation are plausible sources of craters with diameters below the Voyager resolution limit, craters with very smooth surfaces and nearly hemispherical shapes do not seem highly plausible. Thus one can infer from the work of Ostro and Pettengill that external surface reflections are an unlikely explanation for the anomalous radar properties of the icy satellites.

The Random-Facet Model

Goldstein and Green (1980; hereafter G&G) proposed a different explanation for the circular polarization inversion, postulating that the upper few meters of the icy surface are "crazed and fissured and covered by jagged ice boulders." They suggested that the radar echo is due to high-order multiple scattering from subsurface, randomly oriented, ice-vacuum interfaces. In their model, the polarization inversion arises from the prominent role played by grazing reflections (i.e., reflections at high angles of incidence), and by total internal reflections in particular.

G&G use a probabilistic, ray-tracing approach to calculate the radar properties of a layer of randomly oriented facets (i.e., ice-to-vacuum and vacuum-to-ice interfaces). Individual photons are followed through many encounters with facets that are distributed within a slab whose depth is H photon mean free paths. The Fresnel reflection coefficients determine the relative probabilities associated with possible outcomes of photon-facet encounters. G&G monitor each photon's location, propagation direction, and polarization state until it leaves the slab. Using Monte Carlo techniques (and about 3000 trial photons), they estimate the net circular polarization ratio for the echo from the slab. The free parameters in their model are the slab thickness H and the refractive index of the ice. The scattering facets are assumed to be sufficiently large, flat, and smooth for ray optics to be valid.

Several constraints on regolith morphology are implicit in the G&G probabilistic scattering calculations. Following each photon-facet encounter, all information about the randomly chosen orientation and location of the facet is discarded. Thus G&G assume implicitly that (1) the spatial configurations of any two ice-vacuum interfaces are completely decoupled, and that (2) photons are unlikely to encounter any facet more than once. Since the facets are statistically isolated, they must be spatially separated, and the length of the photon mean free path must be larger than the dimensions of the facets. To satisfy the 12.6 cm observations, the scattering facets and the photon mean free path must be at least $\sim$ one meter in scale. Since $H = 6$ for the model reported by G&G, randomly oriented ice-vacuum interfaces must be distributed within a layer many meters thick. Since the facets' orientations are uncorrelated, nearly parallel ice-vacuum interfaces (i.e., fissures) are rare. Furthermore, nearly horizontal ice-vacuum interfaces must be as common as nearly vertical ice-vacuum interfaces.

Total Internal Reflection

Consider double-internal-reflection backscatter from empty cavities in ice. If a circularly polarized wave traveling through ice with refractive index n meets an ice-vacuum interface at an angle of incidence θ_i that is larger than the critical angle, $\theta_C = \sin^{-1} n^{-1}$, all the incident energy will be reflected, and the rotational handedness of the incident wave will be preserved, yielding $\mu_C > 1$. However, circularity is not preserved for total internal reflection. As discussed by Born and Wolf (1975), total internal reflection rotates the phases of the electric field components perpendicular and parallel to the plane of incidence by different amounts, $\delta_\perp (\theta_i, n)$ and $\delta_\parallel (\theta_i, n)$, resulting in the insertion of a net phase difference, $\delta = \delta_\perp - \delta_\parallel$. Since this angle is nonzero for $\theta_C < \theta_i < 90°$, the reflected wave is elliptically polarized. The maximum departure from circularity, corresponding to insertion of a phase, δ_{max}, occurs at $\theta_i = \theta_{\delta max}$ (Fig. 8.7). If the plane of incidence for successive reflections is constant, as must be the case for double-reflection backscatter, the total phase insertion is simply the sum $\Sigma_i \delta (\theta_i, n)$. If $n = 2^{1/2}$ as in Fig. 8.8a, the critical angle is 45°. Since the phase insertion approaches zero as the angle of incidence approaches the critical angle, the net phase insertion for two consecutive reflections at $\theta = 45°$ approaches zero, and the net circular polarization ratio approaches infinity (cf. Figs. 8.8a and 8.6). For $n \neq 2^{1/2}$, μ_C reaches finite maxima at the critical angle and its complement. Figure 8.8b shows the net value of μ_C expected from an ensemble of double internal reflections from right-angle facet pairs, whose orientations with respect to the

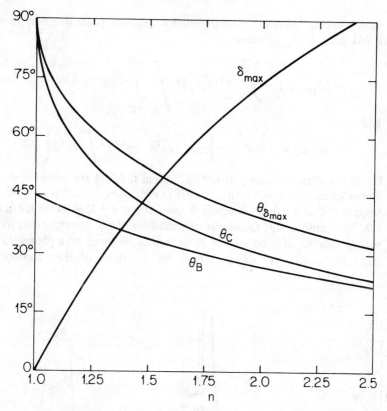

Fig. 8.7. The functions discussed by Born and Wolf (1975) as applied to internal reflections within a medium of refractive index, $n \geqslant 1$. They are: the Brewster angle $\theta_B = \tan^{-1} n^{-1}$; the critical angle $\theta_C = \sin^{-1} n^{-1}$; the maximum phase insertion for total internal reflection $\delta_{max} = 2 \tan^{-1} [(n-n^{-1})/2]$; and the angle of incidence $\theta\delta_{max} = \sin^{-1} [2/(n^2 + 1)]^{1/2}$, which yields δ_{max}.

radar line of sight are uniformly distributed over the interval $0 \leqslant \theta_1 \leqslant 90°$. Clearly, double-reflection backscatter from vacuum facets in ice with $n \simeq 2^{1/2}$ could produce the observed polarization inversion.

Multiple Total Internal Reflection

In the model postulated by G&G, the radar echo results from high-order multiple scattering, and the location and orientation of any interface are assumed to be independent of the locations and orientations of all other boundaries. Thus the planes of incidence are totally uncorrelated, and the cumulative effect of many successive random phase insertions is no longer a simple summation. Rather, it can be modeled as a one-dimensional random walk involving N steps of random length δ_i. The rms phase insertion for such a walk is $\Delta = N^{1/2}\delta_{rms}$, where

$$\delta^2_{rms}(n) = \int_{\theta_C}^{\pi/2} \delta^2(\theta, n)d\theta \bigg/ \int_{\theta_C}^{\pi/2} d\theta \qquad (1)$$

and

$$\delta(\theta, n) = 2\tan^{-1}\left[\cos\theta\,(\sin^2\theta - n^{-2})^{1/2}\bigg/\sin^2\theta\right]. \qquad (2)$$

Definition of the circular polarization ratio in terms of the components of a Stokes vector (Ostro and Pettengill 1978) yields $\mu_C = (1 + \cos\Delta)(1 - \cos\Delta)^{-1}$. Since $\Delta \to 0$ as $n \to 1$, μ_C increases dramatically as $n \to 1$, as illustrated in Fig. 8.9. The curve in this figure was calculated for $N = 8$, corresponding to the average number of total internal reflections experienced by a photon in the the G&G model. The physical basis for the shape of their numerically

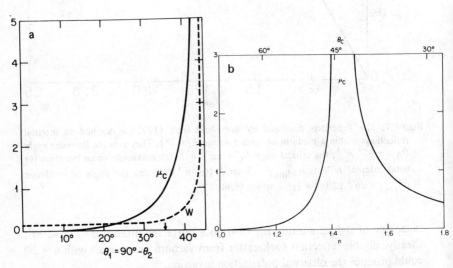

Fig. 8.8. (a) The circular polarization ratio μ_C for double-internal-reflection backscatter from vacuum facets within a medium with refractive index $n = 2\frac{1}{2}$. Also shown is the scaled product W of the Fresnel power reflection coefficients $R(\theta_1, n)$ and $R(90°-\theta_1, n)$ for the two successive reflections. The arrow locates the Brewster angle θ_B. Note differences between these curves and those in Fig. 8.6. (b) The net value of μ_C for double-internal-reflections from an ensemble of randomly oriented vacuum facet pairs in a medium with refractive index n. Angles of incidence $\theta_1 = 90°-\theta_2$ are uniformly distributed over the range $0° \leqslant \theta_1 \leqslant 90°$. The contribution at each value of θ_1 is weighted by W. The upper abscissa indicates values of the critical angle $\theta_C = \sin^{-1} n^{-1}$.

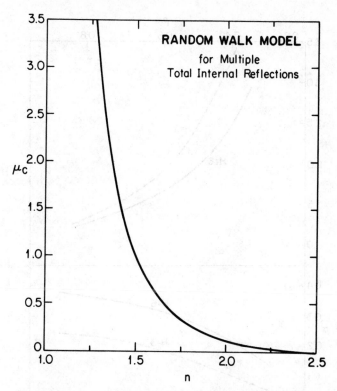

Fig. 8.9. The circular polarization ratio μ_C for eight successive total internal reflections from orientationally uncorrelated vacuum facets inside a medium with refractive index n, modeled as a one-dimensional random walk. Differences between this curve and that calculated by G&G (the solid line in the upper part of Fig. 8.10) are partially due to an extra $\sin \theta$ factor in the G&G model. (See text.)

calculated curve, reproduced here as the solid line in the upper part of Fig. 8.10, is substantiated by the above analysis.

For any encounter of a photon with an interface in the G&G model, the normal $\hat{n}$ to the boundary is chosen randomly from a distribution of directions which is isotropic with respect to the photon propagation direction $\hat{k}$. Such a distribution weights angles of incidence $\theta_i \equiv \cos^{-1}(\hat{n} \cdot \hat{k})$ as $\sin \theta_i$, favoring grazing angles of incidence. (This is important because a circularly polarized photon does not have its polarization seriously altered by a grazing reflection.) However, a facet's capture cross section actually varies as $\cos \theta_i$, diminishing the relative probability of grazing interactions. Therefore G&G have overestimated slightly their model's ability to invert the circular polarization.

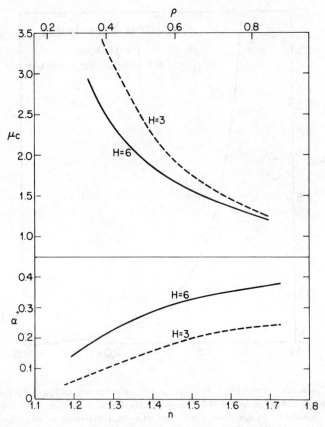

Fig. 8.10. The circular polarization ratio μ_C and the geometric albedo α for the G&G ice-vacuum random-facet model plotted against the ice refractive index n and the ice density ρ, for two values of the thickness H of the scattering layer. The curves are based on calculations by Goldstein (personal communication, 1980).

The High Radar Geometric Albedos

If the thickness H of the random-facet layer is reduced, the number of total internal reflections per photon decreases and the net circular polarization ratio is enhanced. Unfortunately this welcome effect is accompanied by a simultaneous decrease in the radar geometric albedo (Fig. 8.10). One can compensate for a low model albedo by including unmodeled sources of echo power, such as scattering from dielectric boundaries, sharp corners, linear edges, rough surfaces, and small particles, with radii of curvature which are not much larger than the shortest observing wavelength (3.5 cm). However, since such contributions are characterized by circular polarization ratios between zero and unity, they will degrade the net value of

μ_C. Figure 8.11 illustrates this effect for a pair of three-component scattering models. In these ternary diagrams, the fraction of echo power due to a component with $\mu_C = \mu_*$ is unity at the vertex so labeled, is zero for points on the side opposite that vertex, and increases linearly from that side. Contours of constant net μ_C in an echo due to contributions from three components are plotted for values measured for Ganymede at 12.6 cm and 3.5 cm.

Using Figs. 8.10 and 8.11, one can evaluate the ability of the G&G random-facet model to satisfy the observational results. The G&G model with slab depth H equal to six photon mean free paths can yield Ganymede's 3.5-cm circular polarization ratio ($\mu_C = 2.0 \pm 0.3$) and geometric albedo ($\alpha = 0.30 \pm 0.07$) if the ice in the ice-plus-vacuum regolith has a refractive index equal to ~ 1.4. (Since the refractive index of solid ice is 1.8, ice with an index equal to 1.4 would have to be very porous and might better be described as rigid snow.) However, since unmodeled sources of power will degrade the model circular polarization ratio, the model value of μ_C must exceed the net value of μ_C. Suppose the model value of μ_C is 4.0, corresponding to an ice refractive index of ~ 1.15. In the lower ternary diagram in Fig. 8.11, the contour labeled $\mu_C = 2.00$ designates proportions of each component in those three-component models that yield a net circular polarization ratio of 2.00. For example, if 56% of the echo were due to a G&G component with $\mu_C = 4.0$ and the balance were due to a component with $\mu_C = 1$, the net circular polarization ratio would be 2.00. (The net albedo would be $1/0.56 \simeq 1.8$ times larger than the model albedo.) Any three-component models within the tiny triangle bounded below by the line labeled $\mu_C = 2.00$ could satisfy the Ganymede results. However, if more than 17% of the echo were due to a component with $\mu_C = 0$ and the balance due to a G&G component with $\mu_C \leqslant 4.0$, the net polarization ratio would be less than 2.00 and Ganymede's 3.5-cm circular polarization ratio could not be satisfied. The random-facet model clearly cannot tolerate competition from unmodeled scattering processes with low circular polarization ratios.

A Tenuous Upper Layer

Since first-order backscattering always reverses the handedness of the transmitted wave, elimination of all first-order backscattering from the surface might improve the random-facet model. If the regolith were electrically matched to free space by a very tenuous upper layer a few centimeters thick, surface reflections would be largely curtailed. A layer, the density of which increases gradually over a distance $\gtrsim 10$ cm, would provide a particularly effective impedance match (Simpson 1976). In this context, it is interesting to note that results of infrared eclipse radiometry of the satellites apparently exclude homogeneous thermal models (see review by Morrison 1977). On the other hand, those data were easily fitted by a model

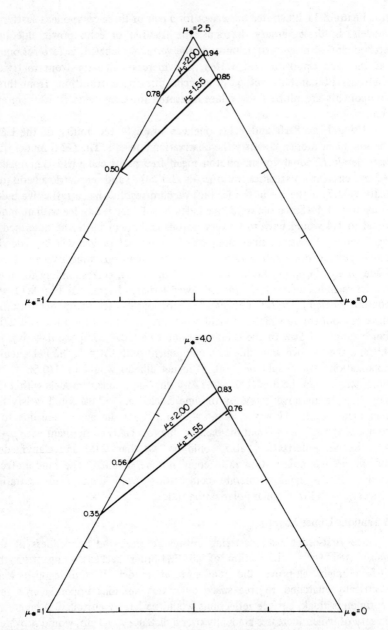

Fig. 8.11. Ternary diagrams of the net circular polarization ratios for three-component scattering models. The components have individual circular polarization ratios μ_* equal to 0, 1, and either 2.5 (top) or 4.0 (bottom). The fraction of echo power due to a component with ratio μ_* is unity at the vertex so labeled, is zero on the opposite side, and increases linearly as the distance from that side. Contours of constant net μ_C are plotted for values measured at two wavelengths for Ganymede (see Table 8.1).

postulating a low-thermal-conductivity layer several millimeters thick above a higher-conductivity lower layer. It may be worthwhile to investigate models with a thicker upper layer characterized by a gentle conductivity gradient.

It is difficult for the G&G model to satisfy simultaneously a large albedo and a large circular polarization ratio. (The general overestimation of μ_C in the G&G calculations, discussed earlier, aggravates this problem.) Europa's enormous 12.6-cm albedo ($\alpha = 0.65 \pm 0.16$) cannot be satisfied unless (1) the thickness of the scattering slab is much greater than six photon mean free paths, or (2) large unmodeled contributions of echo with $\mu_C = 1$ are postulated. For either alternative, survival of Europa's 12.6-cm circular polarization ratio requires the regolith ice to have a very low refractive index. The curves in Figs. 8.10 and 8.11 suggest that the random-facet model probably cannot satisfy the Europa results with refractive indices larger than ~ 1.2. However, ice with a low refractive index necessarily has a low density (see the upper abscissa in Fig. 8.10) and little mechanical strength. Extremely tenuous ice may be incapable of supporting empty cavities far below the surface. The following interpretation of the random-facet scattering model overcomes this difficulty.

A Two-Component Regolith

The parameter n is merely the ratio of the refractive indices of two regolith components, neither of which need be vacuum. The two components may have similar chemical compositions and slightly different densities (e.g., snow and solid ice), or similar densities and slightly different compositions (e.g., clean ice and dirty ice). The regolith can thus contain regions with different refractive indices but with similar mechanical strengths, obviating objections to deep, empty spaces in low-density ice. There are no constraints on the bulk physical properties of the regolith or on the depth dependence of these properties. If such a regolith is electrically matched to free space (e.g., by a very tenuous upper layer ~ 10 cm thick), multiple grazing reflections will be the dominant source of echo power. The polarization will be inverted if the regolith is interspersed with large, smooth, flat, randomly oriented interfaces between regions of similar but unequal refractive index. Since n must be small, H should be large, and the lower-index component should be interspersed in the regolith to a depth of several tens of meters.

The Morphology of a Two-Component Regolith

Constraints on regolith morphology can be inferred from the wavelength dependence of the circular polarization ratio (Table 8.1). For Ganymede, μ_C is higher at 3.5 cm than at 12.6 cm. For Europa, the 70-cm albedo is much less than the 12.6-cm albedo, and the circular polarization is apparently not inverted. Although the precise forms of $\mu_C(\lambda)$ and $\alpha(\lambda)$ for a given target cannot be determined from available data, it seems that some component of regolith structure must have a characteristic dimension between one centimeter and one meter.

A variety of regolith morphologies might satisfy this constraint. If the scattering facets appear much larger, smoother, and flatter at 12.6 cm than at 70 cm, the geometric optics assumption on which the random-facet model rests would be less valid at the longer wavelength.

A different explanation of the wavelength dependence involves total internal reflection. Consider the well-known phenomenon of "frustrated" total internal reflection (Kapany 1967). Suppose that the regolith contains many randomly oriented flat fissures. (Flat fissures require considerable spatial and orientational correlation between scattering facets, contradicting G&G's assumption that any two facets are statistically isolated.) Let the fissures have a thickness t and be filled with a secondary component whose refractive index is slightly less than that of the more prevalent primary component. Photons traveling through the primary component may experience total internal reflection when they meet one of the low-index veins at grazing angles of incidence, as described earlier. However, total internal reflection will occur only if the vein is sufficiently thick. If the vein is too thin, total internal reflection will be frustrated and the photon will merely be transmitted. For an angle of incidence halfway between the critical angle θ_C and $90°$, the power reflection coefficient R is nearly unity if $t > \lambda$. However, $R \ll 1$ if $t < \lambda/10$. If the planar veins are ~ 10 cm thick, total internal reflection will be more efficient at 3.5 cm than at 12.6 cm. At much longer wavelengths, R will plummet and both μ_C and α will decrease dramatically. The 70-cm radar observations of Europa reveal precisely this trend.

Flat, randomly oriented veins might result from meteoritic impact, volcanism, or thermal stresses. Since the surfaces of Europa, Ganymede, and Callisto have had very different geologic histories, the postulated veined regolith might be polygenetic. If such a regolith exists, regions where the randomly oriented veins meet the surface might resemble superficially the "patterned ground" so common to terrestrial periglacial regions (Embleton and King 1975).

Compositional Effects

Differences in the radar properties of Europa, Ganymede, and Callisto might be due to variations in (1) the thickness t of low-index veins, (2) the optical depth (proportional to H) of the scattering layer, (3) the relative refractive index of the two regolith components, or (4) the extinction index (i.e., the imaginary part of the complex refractive index; see Muller 1969) of either regolith component. Microwaves are attenuated much more rapidly in silicates than in ice because, at temperatures of the satellite surfaces, the extinction index of silicates is several orders of magnitude larger than that of ice (Campbell and Ulrichs 1969; Whalley and Labbé 1969; Von Hippel 1954). Furthermore, the efficiency of total internal reflection decreases as the extinction coefficient of the reflecting medium increases. Thus, for Callisto, substantial amounts of silicates in the regolith might cause rapid attenuation of the subsurface radar wave, upsetting the conditions necessary for high values of the geometric albedo and the circular polarization ratio.

CONCLUSIONS

Multiple total internal reflection from randomly oriented subsurface facets can explain the anomalous circular polarization inversion in the radar echoes from Europa, Ganymede, and Callisto. A variety of two-component regoliths can probably satisfy the most important geometrical and statistical constraints in the G&G scattering model. If the lower-index component exists as planar veins ~ 10 cm thick, the wavelength sensitivity of the circular polarization ratio is easily explained. This veined regolith model, when fully parameterized, might serve as a framework for understanding the differences among the satellites in albedo and circular polarization ratio, as well as the features in the satellites' radar echo spectra.

Many questions remain about interpretation of the radar results, but we seem to be pointed in a sensible direction. Radar observations at 3.5 cm are needed to elucidate the wavelength dependence of albedo and circular polarization ratio for each satellite. The G&G scattering model should be used to investigate effects of vein thickness and extinction indices of each regolith component on albedo and circular polarization ratio as functions of wavelength. Model values of the linear polarization ratio should be calculated and compared to the 12.6-cm observational results ($\mu_L \approx 0.5$ for each satellite). Finally, geologic processes that might produce appropriate subsurface structure on Europa, Ganymede, and Callisto should be explored.

Acknowledgments. I thank R. M. Goldstein for valuable discussions and for providing unpublished results of his Monte Carlo calculations. I am indebted to C. R. Chapman for his comments on the original manuscript. I am grateful to D. Baker and D. Jones for programming assistance, to J. Veverka and S. Squyres for helpful discussions, and to M. Roth and A. Alberga-Martin for secretarial assistance. This research was supported by the National Astronomy and Ionosphere Center, operated by Cornell University with support by the National Science Foundation and the National Aeronautics and Space Administration.

July 1980

REFERENCES

Born, M. and Wolf, E. (1975). *Principles of Optics,* pp. 47-51. Pergamon Press, New York.

Campbell, D.B., Chandler, J.F., Ostro, S.J., Pettengill, G.H., and Shapiro, I.I. (1978). Galilean satellites: 1976 radar results. *Icarus* 34, 254-267.

Campbell, D.B., Chandler, J.F., Pettengill, G.H., and Shapiro, I.I. (1977). Galilean satellites of Jupiter: 12.6 cm radar observations. *Science* 196, 650-653.

Campbell, D.B., and Ostro, S.J. (1981). Europa: 70 cm radar observations. In preparation.

Campbell, M.J., and Ulrichs, J. (1969). The electrical properties of rock and their significance for lunar radar observations. *J. Geophys. Res.* 74, 5867-5881.

Embleton, C., and King, C.A.M. (1975). *Periglacial Geomorphology,* pp. 67-95. Wiley and Sons, New York.

Goldstein, R.M., and Green, R.R. (1980). Ganymede: Radar surface characteristics. *Science* 207, 179-180.

Kapany, N.S. (1967). *Fiber Optics.* pp. 5-49. Academic Press, New York.

Morrison, D. (1977). Radiometry of satellites and the rings of Saturn. In *Planetary Satellites* (J.A. Burns, Ed.), pp. 269-301. Univ. Arizona Press, Tucson.

Muller, R.H. (1969). Definitions and conventions in ellipsometry. *Surface Science* 16, 14-33.

Ostro, S.J., Campbell, D.B., Pettengill, G.H., and Shapiro, I.I. (1980). Radar observations on the icy. Galilean satellites. *Icarus.* 44, 431-440.

Ostro, S.J., and Pettengill, G.H. (1978). Icy craters on the Galilean satellites? *Icarus* 34, 268-279.

Pettengill, G.H. (1968). Radar studies of the planets. In *Radar Astronomy* (J.V. Evans, and T. Hagfors, Eds.), pp. 275-321. McGraw-Hill, New York.

Pettengill, G.H. (1978). Physical properties of the planets and satellites from radar observations. *Ann. Rev. Astron. Astrophys.* 16, 265-292.

Simpson, R.A. (1976). Reflection and transmission from interfaces involving graded dielectrics with applications to planetary radar astronomy. *Proc. IEEE AP* 24, 17-24.

Von Hipple, A.R. (1954) *Dielectrics and Waves,* pp. 26-37. M.I.T. Press, Cambridge, Mass.

Whalley, E., and Labbe, H.J. (1969). Optical spectra of rotationally disordered crystals. III. Infrared spectra of the sound waves. *J. Chem Phys.* 51, 3120-3127.

9. INTERPRETING THE CRATERING RECORD: MERCURY TO GANYMEDE AND CALLISTO

ALEX WORONOW, ROBERT G. STROM AND MICHAEL GURNIS
University of Arizona

The large-crater populations on the terrestrial planets differ markedly from those on the Galilean satellites. On the densely cratered terrains of Mercury, Mars, and the Moon, the crater populations > 8 km diameter) bear fundamental similarities to one another. These similarities encompass both the shape of their crater curves (i.e., their size-frequency distribution functions) and their overall crater densities. We agrue that these similarities reflect the attributes of the ancient impacting population and are not, as sometimes hypothesized, the result of crater saturation. The densely cratered terrains of Ganymede and Callisto, and the grooved terrains of Ganymede, have some similarities with each other but bear few common attributes with those on the terrestrial planets. Some of the differences between the crater populations on Ganymede and Callisto apparently result from differences in the physical states of their icy crusts during and soon after the period of late heavy bombardment. The emplacement of the grooved terrain clearly altered the young crater population on Ganymede, and some evidence exists that a similar alteration of the most ancient crater population also occurred. We are unable to reconcile the observed crater populations on Ganymede and Callisto (modified by any of the proposed mechanisms) with an initial population similar to that of the terrestrial plants, suggesting that a different reservoir of impacting bodies cratered surfaces in the inner and outer solar system. Devolving from this interpretation are constraints on the origins of both of the impacting populations and strong cautions about extrapolating time scales derived from the lunar cratering record to the Galilean satellites.

Analyses of the size-frequency distributions and densities of crater populations have aided interpretations in many fields peripherial to the studies of the populations *per se*. Examples of such successful applications include the correlation of bistatic-radar echo broading with crater densities and size-distribution functions, the correlation of the 25-km-deep change in p-wave velocity within the Moon with crustal fracturing due to impact cratering, the correlation of lunar reflection spectra with the impact-glass content of the soil, and most recently, the recognition that the infrared speckle interferometry of Ganymede and Callisto resolves the bright-rimmed craters on their surfaces.

Many studies, however, utilize crater statistics to probe directly the histories of planetary surfaces. Relative age dating of surfaces is one such common application; namely, of two surfaces with significantly different densities of superposed craters, the one having the lowest crater density is the youngest. This is often the only dating technique available until we have an adequate number of returned samples to derive an absolute time scale. Even with extensive returned and dated samples, many geologic units on the Moon remain unsampled, particularly the older units; their absolute dates are inferred by their crater densities as compared to those of the sampled surfaces. These applications are relatively straightforward, but do not begin to tap the vast amount of information that can be gleaned from crater statistics.

Each geologic process leaves its distinctive imprint on the preexisting crater population. Volcanic eruptions preferentially obliterate the smaller members of the population by inundation, and the viscous relaxation of craters in the ices of Ganymede and Callisto would likewise obliterate the larger craters. However, the recognition of these signatures rests on our ability to determine what the preexisting population looked like. Some divergence of opinion exists on this matter even for terrestrial planets, and it would be out of the question to make any *a priori* assumptions about the primitive size-frequency distribution function for the Galilean satellites.

In this chapter we will offer a first analysis of what the Galilean satellites' crater production function is (and is not), and some interpretations that devolve from these conclusions. To begin, we need a basic premise upon which to build; the one we use has only recently been recognized as more in concert with observations than its alternative: *the larger-crater population of the lunar highlands is not at saturation density* (Woronow 1977*b*, 1978; Strom 1979; Neukum and Dietzel 1971). The ramifications of that are great: the ancient history of the highlands is there to be read; the imprint of the size-frequency distribution of the impacting bodies may be recovered from the observed crater size-frequency distribution; and, most important to our purpose, the impact histories of different terrains and different planets can be meaningfully compared to that preserved in the ancient lunar highlands. In fact, many recently developed lines of evidence support the hypothesis that on some terrains of all of the heavily cratered bodies thus far studied (Moon, Mercury, Mars, Ganymede, Callisto) the observed large-crater populations (craters > 8 km diameter) bear a reasonable likeness to their production functions. Admittedly, the strengths of the arguments vary considerably from planet to planet. Yet if this working hypothesis proves basically correct, we can extract more information from the observed crater records than has previously been considered possible. We will show the following.

1. Even the very densely cratered lunar highlands still retain considerable information about their production function.
2. Remarkable similarities exist among the cratering histories of all of the terrestrial planets, both in terms of their production functions and of their total crater densities.
3. The Galilean satellites seem to have experienced quite a different impact history from that of the terrestrial planets.

The amalgamation of these three deductions places strong constraints on the origins of the impacting bodies recorded both in the inner solar system and in the Jovian system, but before interpreting the cratering histories of either the Galilean satellites or the terrestrial planets, we must establish the framework for those interpretations. This framework is dependent on the basic concept that saturation of lunar highlands can no longer be regarded as the best hypothesis, at least for large craters.

I. THE ISSUE OF SATURATION

We begin by briefly recounting some of the most commonly expressed views of saturation, as applied to the lunar highlands (see Woronow 1977b for a complete discussion). The basic belief was that the highlands have such a high crater density that each new crater (and its ejecta deposit) obliterates craters and pieces of craters so that, on the average, the total crater density remains unchanged (i.e., the crater density is in a steady state). If this were the case for the lunar highlands, all information about the size-frequency distribution of the crater production function is irrevocably lost as is all information about the size-frequency distribution of the impacting bodies and all information on the cumulative intensity of the meteorite bombardment experienced by the highlands. By accepting the existence of saturation we accept the idea that a principal era of lunar history is forever beyond our direct study; such a large concession should not be made without a complete understanding of the relevant facts and models.

Perhaps the first interpretations that the lunar highlands are saturated arose from the marked contrast in crater densities between the maria and the highlands. This large crater-density difference easily sways an observer's intuition toward the idea of a saturated lunar-highlands surface. To bolster this initial intuitive impression, one needs to presume only that the highlands have been exposed to meteoritic bombardment through the time of accretion, a presumption widely held until post-Apollo times. Add to these two ideas the theoretical studies which predicted that for a variety of production functions (although not all) the saturation crater population should follow the power-law equation

$$dN = kD^{-3} dD \qquad (1)$$

where N is the number of craters, D is the crater diameter, k is a constant of proportionality, and the exponent is commonly called the slope index. Early studies reported finding just this result, and while it was considered whether this could be a coincidence, most workers believed otherwise and concluded that the highlands are saturated. Subsequently, numerous models were produced to explain how saturation comes about through the physical and stochastical processes of cratering and ejecta blanketing. These models were developed to *explain* saturation, but instead they were widely regarded as proof of its reality.

No distinctions were made between processes effecting small craters, like those observed in great profusion by the Ranger and Lunar Orbiter spacecraft, and those effecting the sparser larger craters. Concepts and experiments developed with relevance to the small craters formed in powdered regolith were infused into the interpretations of the large-crater populations formed in bedrock. The laboratory experiments reported by Moore et al. (1974) and Gault (1970) utilized production functions that may be relevant to small-crater populations (particularly if secondary craters are included as part of the population), but bear no similarity to the production functions that must have driven the larger crater population. Yet those results are frequently extrapolated to larger diameters where different processes dominate the obliteration of craters.

Some theoretical determinations of the crater density at which saturation occurs devolve from geometric considerations, e.g., the densest packing of nonoverlapping craters. Although such arrangements of craters are often referred to as representing the maximum number of craters that could be placed on a surface; this is clearly not true because craters not only overlap but even nest one inside the other, still retaining their identity, and are included in most statistical analyses. Only if one were interested in the maximum number of nonoverlapping craters would geometric packing be relevant to a study of crater densities, but then most existing crater counts would not.

The concept of a lunar highlands saturated at large diameters grew without ever being meaningfully challenged; let us now challenge the propositions mentioned above. First, we can not counter intuitive impressions that the highlands are so densely cratered that they must be saturated since impressions stand outside the realm of scientific test. The once tenable theory that the lunar highlands date back to the time of accretion is no longer widely held by scientists who have studied the melting history of the Moon (e.g., see Taylor 1975). Studies of lunar samples suggest that the Moon underwent complete crustal melting after the accretional bombardment had essentially terminated. This being the case, all remnants of the accretional surface were obliterated, and the surface did not start reaccumulating craters until the crust resolidified ~ 4.4 Gyr ago. Therefore, no need exists for presuming the highlands to be saturated because of the accretion process.

The analytical models of saturation all predicted a size-number distribution at saturation identical to that given by Eq. (1). However, those results are relevant to a particular crater population only if the original production function had a size-number distribution whose slope index was substantially more negative than −3. If the production function had a slope index more positive than −3, it will not produce a stable saturation population; instead the population slope index will diverge to larger and larger values with time (see Fig. 9.1 for the effects of different production functions as crater densities increase). Therefore the analytical models would be consistent with the highlands being at saturation if and only if the production function can be demonstrated to have had a slope index that was significantly more negative than −3, and if and only if the highlands' crater population can be shown to now have a −3 slope index.

The asteroidal population has commonly been cited both as being the source of highlands' craters and as satisfying the necessary production size-frequency distribution function (Hartmann 1971). The production function does not, in fact, meet the requirement of having an exponent significantly less than −3 if the source of the highlands' bombardment was the asteroids. Although many early studies of asteroid sizes contended that the asteroids followed a −3 distribution function, those studies often employed correction factors to adjust the data for possible observational losses. To a degree the sizes of some of the corrections were themselves adjusted to bring the observed distribution into accord with the theories of fragmentation which predicted a −3 distribution (e.g., Marcus 1965). However, even if the asteroidal sizes did follow a −3 slope index, the imprint they would leave at saturation would not be a −3 population, but one of much larger slope index (see the B curves in Fig. 9.1). This is, however, a moot issue as the asteroid population is now recognized as not following a −3 distribution, or any simple power law (Chapman 1974; Woronow 1977a; Zellner 1979), and so we can not call upon the asteroid population to justify the idea that the highlands are saturated. This, too, matters little because we now recognize, with statistical confidence levels exceeding 99.99%, that the large lunar craters on the highlands do not follow a −3 distribution. Therefore, the analytical models are not relevant to the actual large-crater

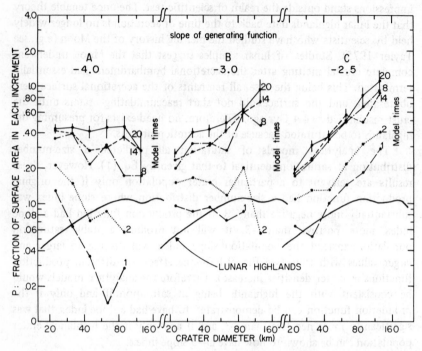

Fig. 9.1. Monte Carlo simulations of various production functions showing their signatures as the crater densities increase. Marked deviations from the production functions do not occur until crater densities vastly exceed those found on either the terrestrial planets or the Galilean satellites. Notice also, that the -3 saturation curve is only obtained for the case where the production function has an exponent much more negative than -3. For the other cases, the observed crater population has an index that becomes progressively larger (less negative) with continuing bombardment. This size-density plot and most others in this chapter are the "relative plots" recommended by the Crater Analysis Techniques Work Group (1979), except that $P = 3.66 R$. On such plots a -3 distribution is a horizontal line. (From Woronow 1977b.)

population on the lunar highlands (although they may be appropriate for much smaller craters).

Actually, for both the highlands' crater population and for the asteroids, significant departures from a -3 slope index have been known for a long time (e.g., McDonald 1931), but in both instances these were attributed to incompleteness in the data rather than to real departures. Neither distribution follows a simple power law. Although the nonpower-law behavior of the asteroids has now gained wide acceptance, the nonpower-law nature of the highlands' crater curve seems slow to achieve similar status. Figure 9.2 illustrates the actual size-density distribution of the lunar highlands, showing

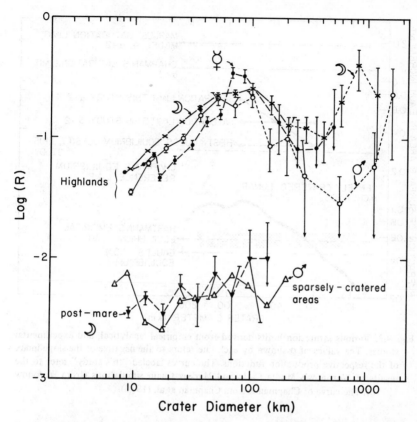

Fig. 9.2. The crater size-density distributions of the terrestrial planets. The distributions of the heavily cratered terrains are remarkably similar in shape and density. The lunar maria, however, have a different crater population. The Martian sparsely-cratered region is from Hartmann (1973). ☽ Moon, ☿ Mercury, ♂ Mars.

that it is not a −3 power law distribution. The lunar curve in Fig. 9.2 is of the average frontside highlands in the diameter range of 8 to ∼ 100 km. Although some regions show variations about this average curve, those variations are all readily ascribed to known geologic processes; no region of the highlands has a crater population so radically different from this curve as to be commensurate with a −3 power law.

As for the models developed to explain the physical processes that lead to saturation, most of them violate observable constraints. For instance, when Marcus (1970) presented eight possible models, his preferred one required that each new crater totally obliterates all craters within three crater radii. Clearly, no such thing occurs or else each relatively fresh highlands' crater would sit neatly in the center of an annulus of obliteration. Many other models were even less realistic (see Woronow 1977a, for a review). In any

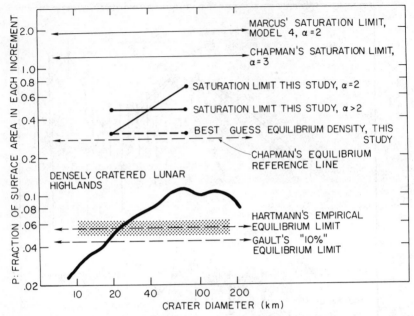

Fig. 9.3. Various saturation limits derived from empirical, analytical, and experimental studies. The values of α shown by each line refers to the negative of the slope index of the respective production functions. The curves labeled "this study" refer to the results of Woronow's Monte Carlo and Markov Chain simulations (from Woronow 1977b). The curve of Chapman is from Chapman et al. (1970).

case, models designed to exploit a hypothesis can not be taken as proof of the validity of the initial hypothesis. Figure 9.3 illustrates the breadth of possible saturation limits derivable from analytical, experimental, simulation, and observational evidence. Most of these are taken directly from the literature as they were presented; the Marcus data were recalculated using a more reasonable value for the effects of ejecta blanketing. Some of the limits actually suggest that the lunar highlands exceed the saturation limit, but most of the analytical studies and simulations do not support the hypothesis that the highlands are densely cratered enough to have reached saturation.

The final line of reasoning, suggesting that the lunar highlands may be saturated at large diameters, came from the similarity of the crater densities on the heavily cratered terrains on the Moon, Mercury, and Mars. While all three bodies do have surprisingly similar crater densities, the complexity of the size-density curves does not comply with the analytical and simulation predictions of simple power-law relationships at saturation. Too much large-scale structure occurs in the curves to suppose that they represent a final state of evolution that no longer retains any information on the size-frequency distribution of the production functions. We should expect much blander curves if these populations were actually at their saturation densities. Other explanations for the similarities of the terrestrial planets' curves are more plausible. The similarities may be merely due to the fact that the cumulative fluxes experienced by each of the bodies, after resolidification of their crusts, were nearly the same. This concept is discussed in the summary of terrestrial-planet cratering given in Sec. IV.

In summary, the concept that the lunar highlands are saturated with large craters was (1) developed from data sets that were not as good as those now available, (2) extrapolated from observations of and experiments pertaining to much smaller craters, and (3) bolstered by theoretical studies that were not appropriate to what we now know to be the actual crater size-frequency distribution function. Many new insights, based on new data sets, new simulations, and new theoretical treatments, supplant the earlier works. Among the new evidence supporting the hypothesis that the large-crater population of the lunar highlands is not saturated are the theoretical studies of Neukum and Dietzel (1971), the reanalysis of previous theoretical studies by Woronow (1977b), the simulations of Woronow (1977b, 1978), and the crater-population analyses done by Strom (1977).

Woronow's (1977b) Monte Carlo computer simulations of cratering, wherein each crater could obliterate an underlying one only by overlap, showed that crater densities far in excess of those observed for the larger craters of the lunar highlands were theoretically attainable. Although this result was in agreement with some of the more elaborate analytical models (e.g., Model 4 of Marcus [1970], adjusted for more realistic effects of ejecta blanketing), these simulations had a small dynamic range of crater diameters, spanning a factor of 16, and did not allow for any crater obliteration by ejecta. In a subsequent simulation, using a Monte Carlo driven Markov Chain computer code, Woronow (1978) extended the dynamic range to cover all craters and basins > 8 km diameter, and included the effects of ejecta on crater obliteration. This and his preceeding simulations were the first that were able to examine production functions with slope indexes more positive than a −3. The results of this simulation did not differ significantly from the earlier one in the principal finding that the lunar highlands are far from their saturation density and that the observed crater size-frequency distribution function basically represents the production function (Fig. 9.4). The best estimate of the saturation limits predicted by these simulations is included in Fig. 9.3. Figure 9.5 shows the amount of crater obliteration that has occurred on the highlands due to crater and ejecta overlap, assuming the observed lunar curve to be the approximate production function as well. Woronow (1978) found that although a sizeable fraction of all large craters have been obliterated (perhaps as much as 40 %), the obliteration process is such that it tends to obliterate craters of a given diameter in close proportion to their

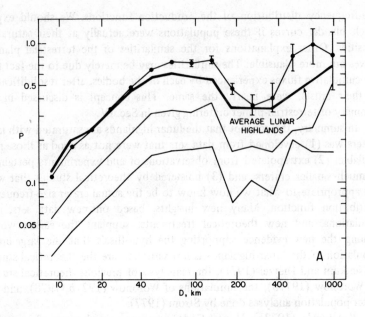

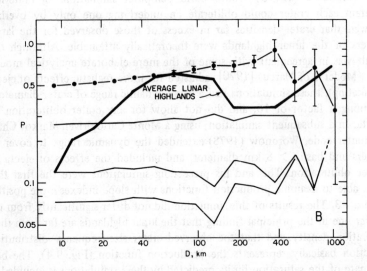

Fig. 9.4. (A) Markov Chain simulations which take into account ejecta-blanket obliteration of smaller craters and show that the observed crater population on the lunar highlands must represent its production function quite closely. (B) Simulations showing it could not have been the result of a −3 production function. The simulation results are shown at various times as the crater densities approach that of the lunar highlands. (From Woronow 1978, the ordinate values P can be converted to relative values by multiplying by 3.66.)

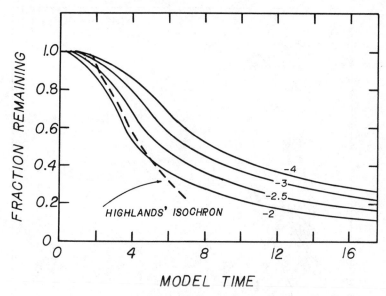

Fig. 9.5. Curves displaying the fraction of all craters produced that remain visible as a function of arbitrary model time and of the slope index of the production function. The lunar highlands have a slope index of approximately −2.3 over the diameter range 8-100 km. The expected crater loss in this range is ∼ 30-40 %. (From Woronow 1978; the ordinate values can be converted to relative values by multiplying by 3.66.)

production rate. This means that the overall size-frequency curve of the crater population does not differ radically from that of the production function even as the crater densities increase to values greater than those that occur on the lunar highlands.

The final line of recent evidence that demonstrates that the highlands are not saturated comes from a study of Orientale done by Strom (1979). He found that the size-frequency distribution of the craters superimposed on the Orientale basin and its ejecta blanket is essentially the same as the size-frequency distribution of the heavily cratered highlands (Fig. 9.6). Of course the actual crater density of the post-Orientale crater population is but a small fraction of that of the highlands population, and could not be anywhere near saturation. That the two curves are nearly parallel strongly argues that the highlands crater population retains the imprint of the production function and is not close to the saturation limit.

The above discussion has repeatedly used the term "saturation" to imply a condition of maximum crater density on a surface where the only processes that can obliterate craters are the subsequent overlaps of craters and of their ejecta blankets. These processes most likely dominate large-crater obliteration on the lunar highlands. However, on some other planetary surfaces such as those of Mars, Ganymede, and Callisto, other geologic processes may serve to

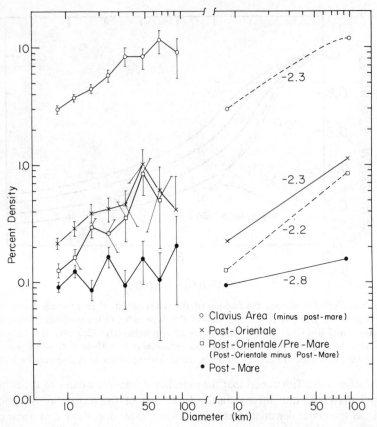

Fig. 9.6. Lunar crater diameter-density distribution of the craters superposed on the
Orientale basin and ejecta blanket, compared to the Clavius region in the highlands
and the best straight-line fits to the data shown on the left. The post-Orientale
production curve is virtually identical to the highlands curve, suggesting that the
highlands curve basically represents its production function. Present density can be
converted to relative values by multiplying by 0.0366. (From Strom 1977, with
modifications.)

limit the crater densities. Examples of such processes would be aeolian
erosion and deposition on Mars, and perhaps viscous relaxation on Ganymede
and Callisto. When other processes do dominate, we will refer to the ultimate
crater density as representing an "equilibrium" condition. Both equilibrium
and saturation represent steady-state conditions.

The above appraisal of our understanding of the cratering process, of
crater saturation, and of the observed cratering record of the lunar highlands
leads to the following as the most reasonable working hypotheses.

1. The lunar highlands are not saturated at crater diameters larger than at
    least 8 km.

2. The observed size-frequency distribution function for the large craters on the lunar highlands is essentially identical to the production function which generated them.

3. Neither the observed size-density function for the larger craters nor their production functions follow a simple power-law relationship.

Two subsidiary findings are: (i) extrapolating information on saturation conditions obtained from small craters to these larger craters is not valid; and (ii) extrapolating the larger-crater curve to much smaller diameters by using a simple power law to represent the highlands population grossly misrepresents the highlands data and consequently does not provide a meaningful reference line at the small-crater diameters.

Obviously if the lunar highlands are not saturated with craters > 8 km diameter, the mare surfaces also are not saturated at comparable crater diameters. Therefore, the post-mare crater population reflects its production function in this size range, and this crater population has a production function significantly different from that of the highlands population over the same diameter range (see Figs. 9.2 and 9.6). In the highlands, the population slope index is ∼ −2.0 m the diameter range 8-50 km, whereas the post-mare population slope index is ∼ −2.8 over the same diameter range. A Chi-squared test indicates that the two populations are different at the 99% confidence level. Therefore the Moon has been impacted by at least two populations of objects, one responsible for the period of late heavy bombardment and another responsible for the period of crater formation primarily after mare emplacement. Some evidence for these two populations comes from an entirely independent source. Although still an issue of debate, based primarily on the abundance patterns of certain siderophile elements (Ir, Au, Re, Ni, etc.) in the soils and breccias of the highlands, Morgan et al. (1977) believe that the impacting bodies responsible for the period of late heavy bombardment were chemically distinct from known meteorites. Because meteorites apparently reflect the composition of the majority of asteroids, the asteroids could not be the source of the ancient population; the source population must be extinct. The post-mare crater density is believed to be consistent with the expected integrated flux of asteroidal impacts (Shoemaker et al. 1979). If this is the case, then the highlands primarily sampled a population with both a different size-frequency distribution and a different chemistry than that responsible for the post-mare impacts.

With these perspectives on lunar cratering, we will now examine the cratering records of Mercury, Mars, Ganymede, and Callisto. As will become apparent, many close similarities exist among the terrestrial planets that are not shared by the Galilean satellites. Both the similarities and differences reveal the geologic processes and the histories of the impacting populations that affected the individual bodies.

II. THE CRATERING RECORD OF MERCURY

Like the Moon, Mercury has had no atmosphere to interact with its cratered surface. Aeolian and aqueous erosion and deposition have always been nonexistent. Unlike the Moon, however, Mercury has developed extensive intercrater plains. These plains, probably predominately of volcanic origin, were emplaced over a considerable portion of the early bombardment history of Mercury (Strom 1977, 1979; Malin 1976). Because of their pervasiveness, it is not possible to establish a cratering curve for Mercury exactly comparable to that for the lunar highlands. The lunar and Mercurian curves shown in Fig. 9.2 differ in two important respects. First, the Mercurian crater population has a different average slope index for diameters between ~ 15 and 70 km. This difference implies a relative depletion of successively smaller-diameter craters on Mercury and probably reflects the effects of intercrater plains emplacement. Volcanic flooding preferentially obliterates small craters, causing just such a departure from what must have been a production function very similar to that for the lunar highlands (Strom 1977).

The second feature in the Mercurian curve that has no counterpart in the curve of the lunar highlands is a knee at ~ 15 km diameter. In some regions of the densely cratered lunar farside, however, a similar knee does appear (Woronow et al. 1979). Although the source of the additional craters on the Moon that constitute this knee is not yet fully resolved, it appears that Orientale-basin secondaries may be the predominant contributors. The Orientale impact preferentially sprayed the farside with its ejecta, but the frontside highlands are apparently relatively free of large basin secondaries. Because Mercurian secondary craters are larger than their lunar counterparts, the knee in the Mercurian crater curve may be the result of basin secondaries as well. All things considered, the Mercurian densely cratered terrains and the nearside lunar highlands probably had very similar size-density curves—probably even more similar than the current Mercurian highlands data suggest.

The post-Caloris crater curve is virtually identical in both crater density and size-frequency distribution to that of the lunar post-Orientale curve (see Fig. 20 of Strom 1979) which, in turn, is the same as the lunar highlands curve. This too suggests that the ancient production function was very similar, perhaps identical, for both bodies. Furthermore, the most densely cratered terrains on both bodies have similar total crater densities.

To date, a younger crater population similar to the lunar post-mare population has not been recognized on Mercury (Strom 1979). Perhaps either it never reached Mercury in numbers large enough to leave a recognizable signature, or the youngest surfaces on Mercury were formed earlier than the lunar maria, when the objects responsible for late heavy bombardment of the highlands still dominated.

III. THE CRATERING RECORD OF MARS

Compared to the Moon and Mercury, Mars has the greatest potential for geologic processes complicating its cratering record. Aeolian erosion and deposition could obscure the ancient cratering record, as could possible rainfall and ground ice. Yet recent studies of regional Martian crater populations have shown that Mars, in some locales, still retains a cratering record remarkably similar to that of some locales on both the Moon and Mercury. In the southern cratered terrain, if one carefully excludes smooth plains units with few superposed craters and with wrinkle ridges, the size-density distribution on the truly ancient surfaces is similar to that of the lunar highlands in both crater density and size-frequency distribution function (Fig. 9.2). Therefore, although the Martian surface has undeniably been affected by aqueous and aeolian processes, some regions have retained, for many eons, their initial crater population.

The similarity of the cratering records of Mars and the Moon extends further. The northern plains of Mars record an impacting population with a size-density distribution markedly different from that of the ancient terrains, yet almost identical to that found on the lunar maria (Fig. 9.2). Thus both the ancient terrain and the northern plains on Mars mimic, in overall crater density and in their size-frequency distribution, the equivalent terrains on the Moon.

IV. SYNTHESIS OF THE TERRESTRIAL PLANETS' CRATERING RECORDS

In summarizing the facts now known about the cratering records of the terrestrial planets, we are first confronted by the great similarities they exhibit. The most ancient, most heavily cratered terrains of the Moon, Mercury, and Mars record a complex-shaped crater curve that differs little from one planet to the next in either density or shape. A similar relationship holds for the crater populations of the lunar maria and the Martian northern plains, but their size-frequency distribution is different from that of the ancient terrains and from the post-Orientale pre-maria population. These data require that two populations of objects have impacted the terrestrial planets: one responsible for the period of late heavy bombardment early in the history of Mars, the Moon, and Mercury, and probably the Earth and Venus as well, and another primarily responsible for the period of crater formation after mare formation on the Moon and plains formation on Mars. These two populations may represent two separate and distinct origins of the impacting bodies or they may represent one population which evolved with time through mutual collision. If the younger population is missing from Mercury or the chemical evidence of Morgan et al. (1974) proves valid, then most likely the two are separate and distinct populations.

Four principal origins have been suggested for the objects responsible for the period of late heavy bombardment in the inner solar system.

1. They may have originated from the asteroid belt through mutual collisional processes.
2. They may be bodies left over from the final accretion of the planets themselves.
3. They may be the remains of planetoids disrupted by gravitational forces due to a close approach to a larger planet (Wetherill 1975).
4. They may be bodies individually deflected into the inner solar system from beyond Jupiter (see Chapter 10 by Shoemaker and Wolfe).

The contribution from comets is not known, but their present impact rate is under intensive study (Chapter 10, by Shoemaker and Wolfe). However, if the impacting bodies originated in the asteroid belt, the similarities in crater densities among the terrestrial planets would be a grand coincidence. Because the impact fluxes would be different for each planet (Hartmann et al. 1980), the melting and resolidification of each planet's crust would have had to be timed to compensate exactly for its impact flux. This is not a likely scenario. The viability of the second alternative is also questionable. Wetherill (1975) calculates that essentially all the accretional bodies would be swept up prior to the resolidification of the lunar crust. Few accretional bodies would remain to cause the late heavy bombardment unless they were stored in inclined orbits. Under the latter circumstance, and using the thermal analyses of Hostetler and Drake (1980) suggesting that the crusts of the Moon and terrestrial planets all melted and subsequently solidified at about the same time ($\sim 10^8$ yr after planetary formation), we could account for the similarity of the cumulative impact flux recorded on all the terrestrial planets. If the impact bodies came from a gravitationally disrupted body (Wetherill 1975), they could imprint all of the terrestrial planets about equally, presuming all of the planets had solidified crusts prior to this event. This scenario seems to require fewer coincidences and to be plausible from an orbital mechanics standpoint. Finally, if Shoemaker and Wolfe's (Chapter 10) hypothesis were true, there should be a close correspondence between the size-frequency distribution of craters on the terrestrial planets and the Galilean satellites; we shall show this does not occur.

Alternatives two and three are both consistent with the chemical evidence of Morgan et al. (1977) that the objects responsible for the late heavy bombardment on the Moon were chemically different from present day meteorites (and asteroids).

V. CALLISTO: CRATER CHARACTERISTICS AND STATISTICS

The surface of Callisto is dominated by impact craters as large as 600 km in diameter (Smith et al. 1979*a*, *b*; Passey and Shoemaker, Chapter 12). No extensive areas of smooth plains or other types of nonimpact-related topography have been recognized. In this sense Callisto is the most extensively cratered, but not necessarily the most densely cratered, body of its size so far discovered in the solar system. The morphologies of the craters in the size range 10-100 km, seen on the highest resolution pictures (3-4 km/line pair), are remarkably similar to those on the inner planets despite their formation in icy material. However, at resolutions of 3-4 km, inner rim terracing or the character of ejecta blankets is not discernable. Many of Callisto's craters, unlike those on the inner planets, have central pits, probably due to a response of an icy target material to the impact process (Greeley et al., Chapter 11; Passey and Shoemaker, Chapter 12). As on the inner planets, the floor structures of the "fresher" craters vary with diameter (Fig. 9.7). Smaller craters appear deeper and more bowl shaped, while larger ones appear shallower with relatively flat floors. Older, more degraded craters have floors that appear to be almost at the same level as the surrounding terrain. However, where not disrupted by the subsequent impacts, the rims of most of these craters will appear fairly sharp and well preserved (Fig. 9.8), suggesting that, at least on Callisto, relaxation of the icy crust primarily affects the excavation cavity and largely leaves the uppermost part of the rim intact. Therefore crater obliteration solely by relaxation in an icy crust may not be very effective on Callisto in this size range.

The most heavily cratered regions on Ganymede show a peculiar type of structure termed a palimpsest (Smith et al. 1979*b*; Passey and Shoemaker, Chapter 12). These structures are circular bright areas with barely discernable rims about half the diameter of the bright area. They occur in the diameter range ~ 30-80 km and are probably impact scars formed by relaxation of the thin icy crust early in the planet's history when the crust was in a more plastic state. On Callisto these features are extremely rare or absent; no clear cases of circular regions of low crater density, indicative of large relaxation sites, have been found. The paucity of these features on Callisto suggests that the observed crater population formed primarily at a time when Callisto's crust was rigid enough to preserve its structure.

Valhalla and Asgard, the two largest basins on Callisto (600 and 400 km diameter) have a morphology very different from that of similar sized basins on the inner planets. They are characterized by rimless, bright, level, circular areas surrounded by closely spaced ridges extending outward beyond two basin diameters. This type of structure probably results from the response of the icy crust to very large impacts. The diameters measured for these basins are those coinciding with the brightest circular regions in their centers, which probably correspond to the excavation craters.

Crater measurements on Callisto in the size range 10-100 km were made from Voyager 2 images taken at quarter phase illumination with a resolution

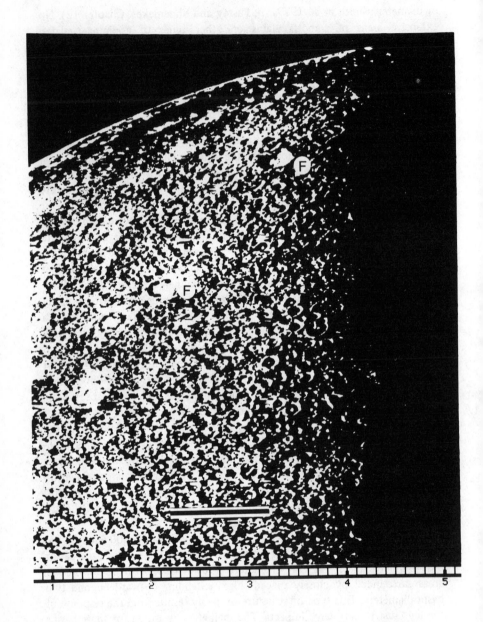

Fig. 9.7. The heavily cratered surface of Callisto showing the paucity of large craters relative to smaller ones. The letter F indicates relatively fresh craters, which are remarkably similar in morphology to lunar and Mercurian fresh craters. Black scale bar represents 200 km. South is at the top. (FDS 20617.41)

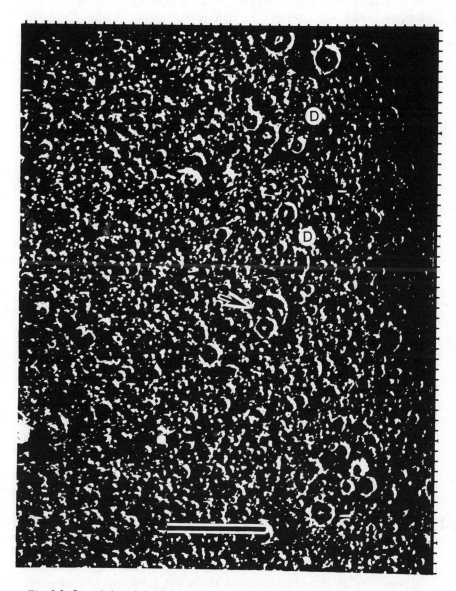

Fig. 9.8. Several degraded craters, indicated by *D*, on Callisto. They have been rather heavily impacted but are still recognizable. The black arrow points to a crater with its floor at about the same level as the surroundings but still exhibiting a well-defined, sharp rim. There is an absence of large palimpsests on Callisto, suggesting that most of the observed craters formed in a relatively rigid crust. Black scale bar represents 200 km. South is at the top. (FDS 20617.09)

of ~ 4 km. These images comprise a mosaic covering about 34 % of the satellite, which was amenable to crater measurements. Craters $\gtrsim$ 10 km diameter were measured out to 29° from the terminator, and craters $\gtrsim$ 30 km diameter out to higher sun angles (see Fig. 9.9). Although the crater measurements extended to smaller diameters, only for craters > 10 km diameter are the statistics judged reliably complete. Craters $\gtrsim$ 100 km diameter were measured out to 29° from the terminator, and craters $\gtrsim$ 30 km to larger diameters and improve the statistics at these sizes. Although about 74 % of Callisto was searched for craters $\gtrsim$ 100 km, only 5 large craters and basins were found.

The craters were divided into two broad categories: fresh and degraded. Fresh craters are those with relatively sharp, well-preserved rims, while degraded ones have disrupted rims and usually have floors more nearly at the level of the surrounding terrain. Degraded craters occur most frequently at the larger diameters ($\geqslant$ 45 km), which may reflect, at least in part, enhanced degradation by relaxation at these diameters. The data are given in Table 9.1. In no case can the observed populations be justifiably characterized by a simple power law.

VI. GANYMEDE: CRATER CHARACTERISTICS AND STATISTICS

The craters on Ganymede appear to have a larger spectrum of morphological characteristics than those on Calisto, probably due in part to the higher resolution coverage (~ 1 km/line pair) and in part to the more complex surface history of Ganymede. In general Ganymede's craters show a range of freshness from sharp rimmed with rays and well-developed ejecta blankets to very low-rimmed, flat-floored with no discernable ejecta deposits. Central pits appear to be abundant, but central peaks are also common. The ejecta blankets on some of the freshest craters have sharp, elevated, and somewhat lobate termini (Fig. 9.10), in some aspects similar to the rampart craters on Mars. Secondary impact craters are common around the large, fresh, and rayed craters. Aiso, as on Callisto, the small fresh craters are deep and bowl-shaped while the large ones are shallow with flat to somewhat domed floors (Fig. 9.10). The floors of some of the more subdued craters are at about the same level as the surrounding terrain. In most cases the rims are low and sometimes discontinuous, but still relatively sharp. Crater palimpsests are relatively common in the most heavily cratered regions of Ganymede, but appear to be less abundant on the grooved terrain. These structures have been described in the previous section and are shown in Fig. 9.11. Their concentration in the heavily cratered regions suggests that when these craters formed, the crust was generally in a more plastic and easily deformable state than it was for the later impacts into the grooved terrain. No obvious circular areas of low crater density, but lacking the albedo characteristics of

TABLE 9.1

Crater Counts on Ganymede and Callisto

Diameter Range (km)	Callisto Number of Craters	Callisto Area (10^6 km^2)	Ganymede Cratered Terrain Number of Craters	Cratered Terrain Area (10^6 km^2)	Grooved Terrain Number of Craters	Grooved Terrain Area (10^6 km^2)
4 – 5.7			2231	2.4		
5.7 – 8.0			1000	2.4		
8.0 – 11.3			559	3.4		
11.3 – 16.0	1311	4.7	291	3.4	216	7.8
16.0 – 22.6	696	4.7	154	3.4	195	13
22.6 – 32.0	346	4.7	106	3.4	123	13
32.0 – 45.3	211	6.1	71	3.4	79	13
45.3 – 64.0	92	6.1	20	3.4	43	13
64.0 – 90.5	22	6.1	6	3.4	20	13
90.5 – 128	7	6.1	1	3.4	2	13
128 – 256	2	55			1	13
256 – 724	3	55			1	13

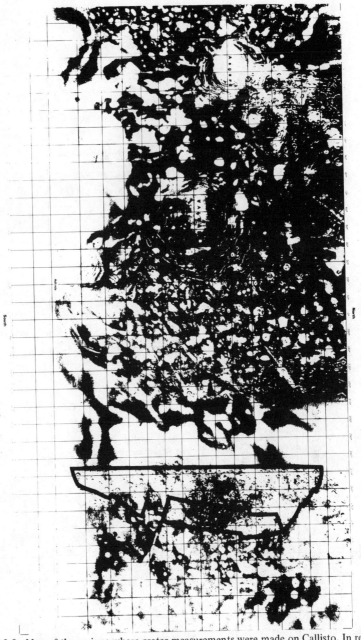

Fig. 9.9. Map of the regions where crater measurements were made on Callisto. In region A the measurements were completed down to 10 km; in region B the measurements were completed down to 30 km diameter. The rest of the area was measured for craters $\gtrsim 100$ km.

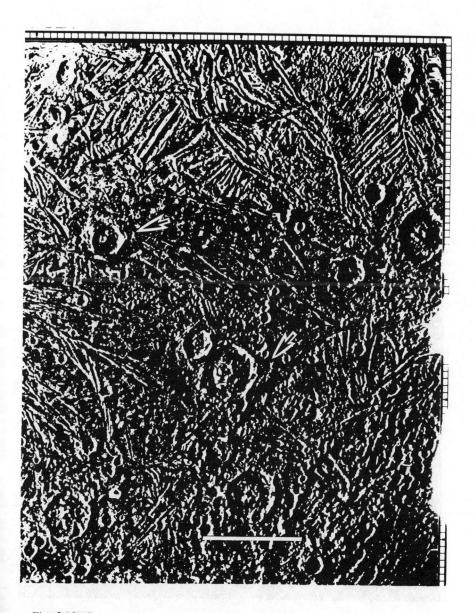

Fig. 9.10. Grooved terrain on Ganymede showing several fresh craters with ejecta blankets (black arrows) somewhat similar to Martian rampart craters. The overall morphology of these fresh craters is remarkably similar to that of lunar craters. Preliminary measurements indicate depths between 0.5 and 1 km, which are only slightly less than for lunar craters of similar size. White scale bar represents 100 km. North is at the top. (FDS 20639.05)

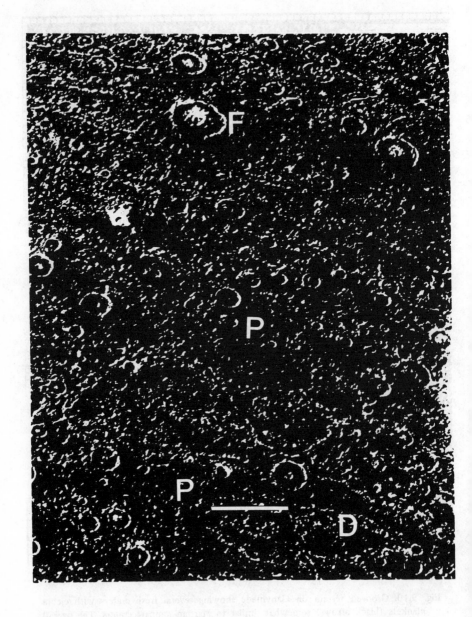

Fig. 9.11. Region of heavily cratered terrain on Ganymede. The *F* and *D* indicate fresh and degraded craters respectively, and *P* indicates palimpsests. White scale bar represents 100 km. North is at the top. (FDS 20636.59)

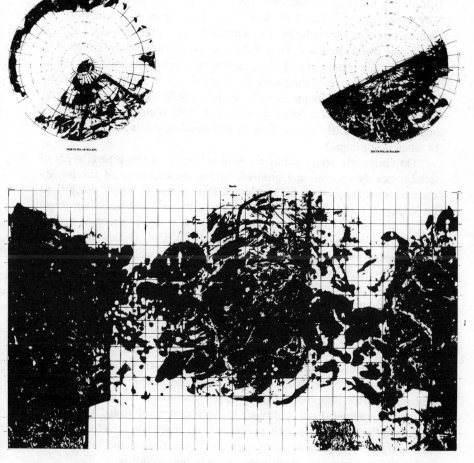

Fig. 9.12. Map of the regions where crater measurements were made on Ganymede. *C* indicates heavily cratered terrain, *G*, grooved terrain.

palimpsests, have been found. Apparently few, if any, palimpsests have progressed beyond easy recognition. A complete description and analysis of crater morphology is outside the scope of this chapter; however, the general observations of crater morphology noted on Callisto and Ganymede are consistent with the hypothesis that although relaxation of the icy substrate has been important in degrading craters, it has not been very effective at obliterating them.

These conclusions are antithetic to those of Passey and Shoemaker (Chapter 12). They contend that regions of low crater density flag sites of "cryptopalimpsests." However, low crater densities in themselves do not demonstrate the past existence of a palimpsest—many other processes of local resurfacing could cause that. Only if the low crater densities were confined to clearly circular regions would the proposal of cryptopalimpsests be viable. Furthermore, if cryptopalimpsests do exist, a gap in morphologies exists between the cryptopalimpsests and the clear albedo imprints of the crater palimpsests; intermediate forms do not occur. More likely the palimpsests we now see are essentially all that have formed since the crusts solidified enough to retain major impacts.

On Ganymede crater diameters were measured on two general types of terrain: heavily cratered and grooved. Crater measurements on the heavily cratered terrain encompass the two largest areas viewed at highest resolutions by Voyagers 1 and 2; Galileo Regio (2.4×10^6 km^2) and Nicholson Regio (1.0×10^6 km^2). Together these areas comprise $\sim 4\%$ of the total surface area of Ganymede (see Fig. 9.12). The largest area, Galileo Regio, was photographed by Voyager 2 at a resolution of ~ 1 km, while Nicholson Regio was photographed by Voyager 1 at a resolution of ~ 2.8 km. Galileo Regio is a dark circular area characterized by a relatively high crater density and broadly arcuate rimmed troughs. The morophogies, spacings (~ 50 km), and overall geometry of the troughs somewhat mimic the concentric ridges associated with the largest basins on Callisto (e.g., Valhalla), and they have been interpreted as resulting from a similar process, i.e., a major impact into an icy crust (Smith et al. 1979b). The center of the Ganymede ring system is near 30°S and 180°W, but no central bright area occurs at this location. Perhaps both the formation of the younger grooved terrains and the process of impact cratering have destroyed it. Several rimmed troughs found on Nicholson Regio may have similar origins. As discussed later, the formation of these rings may have been partly responsible for altering the shape of the crater diameter-frequency distribution curve in their vicinities.

The grooved terrain on Ganymede is a complex combination of transecting, elongate to polygonal equant segments (cf. Shoemaker and Plescia, Chapter 13). Some segments are complexly grooved, others have a very regular grid-like pattern of intersecting ridges, and still others are relatively smooth and featureless. Crater densities on individual segments vary widely and suggest a rather wide range in ages. In order to obtain more statistically reliable crater size-frequency distributions, all types of grooved terrain were combined.

The regions of grooved terrain on which craters were measured using both Voyager 1 and Voyager 2 images is shown in Fig. 9.12. Not included in the studied areas were regions of heavily cratered terrain. Craters > 16 km diameter were measured over an area of 1.07×10^7 km^2, craters > 11.3 km diameter over an area of 5.82×10^6 km^2, and craters > 8 km diameter over an area of 2.1×10^6 km^2. The data are given in Table 9.1. As on Callisto, the crater populations of Ganymede can not be represented by any simple power law.

Although crater measurements extended to smaller diameters, only for craters > 8 km diameter are the statistics reliably complete. Areas with heavy concentrations of secondary impact craters were not included in the statistics on the grooved terrain, but their complete exclusion is probably an unrealistic assumption. Therefore, below a diameter of ~ 10 km, the data may include a significant number of secondaries, particularly on the heavily cratered terrain. As on Callisto, the craters were divided into two broad categories: fresh and degraded. Although the same criteria used on Callisto were followed in assigning craters to these categories on Ganymede, the higher resolution pictures on Ganymede allowed greater certainties in classifications. Therefore the two categories on each satellite may be only approximately equivalent.

The diameters assigned to the palimpsests are those of the degraded rims surrounding the central, flat-floored regions. The larger high-albedo areas surrounding these central regions have surface topography much like that of their surroundings, and so the excavation cavity did not extend over the entire bright area. The edges of the bright areas are relatively sharp, compared with normal ejecta deposits, suggesting that they did not result directly from ejecta blanketing. Plausible explanations of these extended bright regions involve either disrupting or annealing of the ice. Mechanisms which may have contributed are (1) seismic disruption (e.g., small-scale fracturing and crystal deformations) and (2) local heating, both mechanisms as a direct result of a large impact. The outer limit of the bright zone might be either the point at which the shock wave decayed below the threshold value for the disruption of the ice or where temperatures were raised just enough to allow annealing of preexistent small-scale fracture. Alternatively, annealing of the ice over the area where it had been mobilized to fill and relax the excavation crater could be the cause.

VII. PRELIMINARY INTERPRETATION OF THE CRATER STATISTICS

Because of uncertainties in such fundamental parameters as the source and composition of the impacting bodies, the reaction of icy targets to hypervelocity impacts, and the diversity and vigor of crater degradation and removal processes, the interpretations presented here must be considered preliminary. Many of the uncertainties are currently under investigation, and the result of those investigations will provide feedback to analyses such as these.

Several striking features are immediately apparent in the crater size-density curves of Ganymede and Callisto, and in the appearances of their surfaces when compared to the Moon and Mercury (Fig. 9.13, 9.14, 9.15). The curves for the heavily cratered terrains on Ganymede and Callisto differ markedly in both shape and crater density. If the impacting bodies originate outside the Jovian system, Ganymede, being deeper in Jupiter's gravity field, should have experienced about twice the impact-flux rate of Callisto (Smith et al. 1979b; Shoemaker and Wolf, Chapter 10). However, the most heavily cratered and therefore oldest terrain on Ganymede is actually less densely cratered than the most cratered terrain on Callisto (Fig. 9.14). Although the two curves are nearly parallel between ~ 50 and 130 km diameter, at smaller diameters considerable differences occur. Finally, neither of the crater populations on Ganymede or Callisto resembles those encountered on the terrestrial planets. This is not only apparent from the crater curves but also from the striking visual differences in the cratered surfaces among Callisto, Mercury and the Moon as shown in Fig. 9.13. Immediately apparent from these similar-scale photographs is that Callisto has a very strong deficiency of large craters and basins compared to both the Moon and Mercury.

The observed overall crater density on the most heavily cratered terrain on Ganymede is down by a factor of ~ 3 compared to Callisto, indicating that the oldest surfaces of Ganymede began recording their crater population at a later time than did Callisto. This might be due to a large-scale (perhaps global) diameter-independent resurfacing event on Ganymede, or Ganymede simply may have developed a rigid crust capable of retaining large craters later than did Callisto. These two possibilities are both consistent with thermal history models by Cassen et al. (1980 and their Chapter 4) and by Parmentier and Head (1979); these models suggest that tidal heating, a higher radioactive content, greater accretional heating or some combination of these factors may have delayed the freezing of Ganymede's icy mantle and kept it thermally active longer than Callisto's was. In any event, this process of "resurfacing" must have taken place later on Ganymede, but still during the period of late heavy bombardment (cf. Shoemaker and Plescia, Chapter 13).

If at least a portion of the crater curves in Figs. 9.14 and 9.15 can be demonstrated to represent the production function, then a meaningful comparison of these Jovian satellites' crater populations to those found in the inner solar system can be made, and will allow speculations as to the nature and signature of degradational and obliterational processes on these icy bodies, and also place constraints on the origins of the objects responsible for bombarding both the inner and outer solar system.

From our previous discussion of the terrestrial planet cratering record, clearly neither Ganymede nor Callisto is saturated with craters at the diameters considered in this study. On Callisto at diameters of ≥ 30 km, the

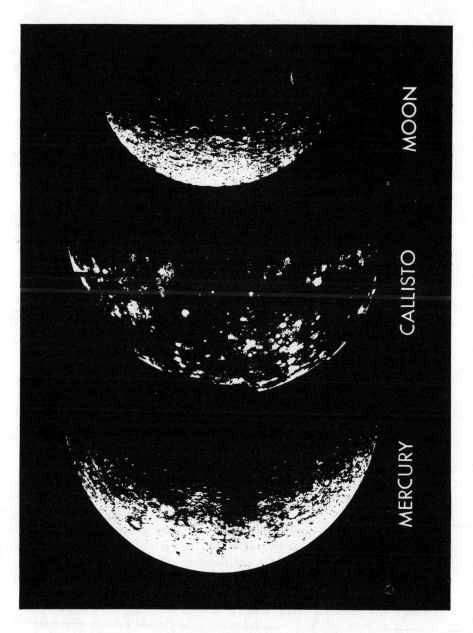

Fig. 9.13. Scaled comparison of the Moon, Mercury and Callisto seen under similar lighting conditions, but not similar resolutions. The visual comparison vividly demonstrates that the crater size-frequency distribution on Callisto is very different from that on Mercury and the Moon. Callisto has a great deficiency of large craters compared to the Moon or Mercury.

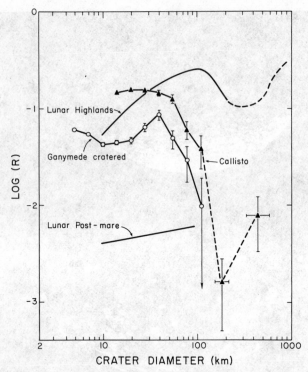

Fig. 9.14. Curves for the crater populations measured on the heavily cratered terrains of both Ganymede and Callisto, with the lunar curve for reference. The differences between the Moon and Callisto and Ganymede are much greater than are the similarities. The Ganymede and Callisto curves are similar for diameters $\gtrsim 50$ km, but differ substantially at smaller diameters.

crater density is equal to or much less than that of the lunar highlands (Fig. 9.14), and the Mercurian and Martian highlands as well. At diameters > 10 km, the crater density in the heavily cratered terrain on Ganymede is also well below that of the lunar highlands (Fig. 9.15). As shown in Figs. 9.1 and 9.4, these densities are far below the saturation limit for a wide range of population distribution functions, and therefore the shapes of the curves are not the result of saturation. However, at least over some range of the diameters equilibrium effects may be present (i.e., obliteration by crater relaxation or other nonimpact-related processes).

Figure 9.14 shows that the crater curve on Callisto consists of at least two parts. In the diameter range of $\sim$ 10-50 km the curve is practically horizontal on the relative plot, which is equivalent to a -3 distribution function on a differential plot, i.e., it follows (Eq. (1). Between diameters of $\sim$ 50 and 100 km, the nearly -5 slope index indicates a very rapid decrease in

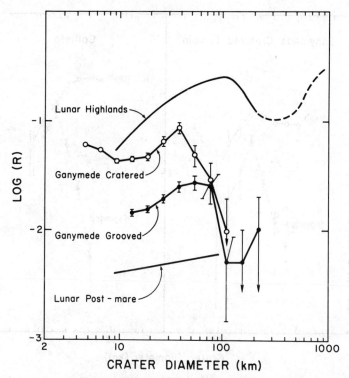

Fig. 9.15. Comparison of the crater populations on Ganymede's grooved and heavily cratered terrains, with the lunar curve for reference. The grooved terrain is similar in slope to the heavily cratered terrain for diameters > 30 km, but at smaller diameters a progressive loss of smaller craters has increased the slope index of the grooved terrain compared to that of the heavily cratered terrain. Grooved terrains which were measured are of at least two ages, one being about as densely cratered as the heavily cratered terrain and the other much less so.

number of craters with increasing diameter. Although the data are sparse, at diameters ≳ 200 km the rate of decrease may slow down. On Ganymede the size-frequency distributions can also be divided into two parts (Fig. 9.15). In the size range of ~ 10-40 km, the curves for both the heavily cratered regions and the grooved terrain slope downward to the left with a slope index of ~ −2. If both Callisto and Ganymede were impacted by the same population of objects, evidently a diameter-dependent obliteration of craters occurred in this size range on Ganymede but perhaps not on Callisto. Examining the curves for the fresh and degraded craters on the two bodies suggests the manner in which this obliterative process operated. The size-density curves of the degraded craters on both bodies show a high degree of diameter dependence (Fig. 9.16); the smaller the crater diameter, the smaller the proportion of degraded craters. If this is a valid observation and not a

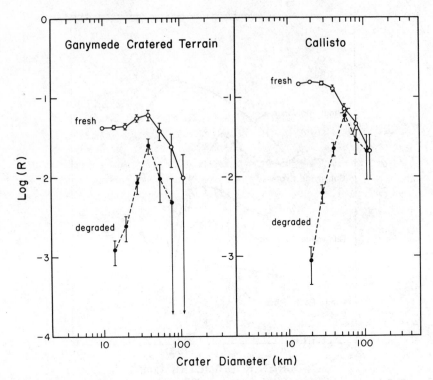

Fig. 9.16. Curves for the fresh and degraded craters on both Ganymede and Callisto. The degraded craters represent only a small fraction of the small craters, but a substantial proportion of the larger ones. Both satellites have similar curves.

problem in recognizing small, degraded craters, it means that the process is dominantly one of crater obliteration for the smallest craters, progressing toward one of degradation for the larger craters. On the terrestrial planets one process that behaves this way is crater and ejecta overlap. Small craters are most often either left unscathed or totally obliterated by each subsequent impact while larger craters are more often partially overlapped and thereby degraded. Although crater overlap may contribute to the observed degree of crater obliteration, we do not believe it to be the dominant process because of the generally low crater densities, particularly on the grooved terrain.

Comparison of the crater curves for the cratered and grooved terrains (Fig. 9.15) shows that the grooved-terrain curve bends downward more steeply than the cratered-terrain curve at diameters $\lesssim 70$ km. This suggests that the recent processes of crater removal on the grooved terrain has been more efficient than any on the cratered terrain. Apparently the formation of grooved terrain (i.e., emplacement of new ice) preferentially obliterated smaller craters (perhaps up to 70 km diameter), while only degrading larger craters. Figure 9.17 shows an area of grooved terrain where the formation of new ice has destroyed a large portion of the rims of several craters; smaller ones, of course, would have been completely obliterated. The preferential obliteration of small craters on the cratered terrain may have been the result of an ancient episode of grooved-terrain formation (now hidden by the recratering), which may have been associated with the resurfacing or late crustal freezing mentioned earlier. Alternatively, the formation of the arcuate troughs may have been responsible for the loss of the smaller craters. Regional variations in these processes could have caused some of the crater-density variations discussed by Passey and Shoemaker in Chapter 12.

At diameters $\lesssim$ 10 km the curve for the heavily cratered terrain on Ganymede turns up slightly. Without comparable diameter coverage on Callisto, we cannot ascertain whether this is indicative of the primary cratering population. However, secondary craters of sufficient size and perhaps of sufficient abundance to account for this upturn do occur on Ganymede. Although at small diameters on Callisto the curve may not record the actual production function, it does set an upper limit of ~ -3 for its slope index. If an ancient episode of obliteration, similar to that proposed for Ganymede, operated on Callisto as well, then the production function could be even more negative than -3. In any case, this observation provides the firs evidence that the production function was far different from the -2.3 observed on the terrestrial planets over the equivalent diameter range.

All terrains on both Ganymede and Callisto show a decrease in slope index for crater diameters $\gtrsim$ 50 km. Two plausible explanations for this decrease are (1) a great deal of crater obliteration due to crater relaxation in the ice crust, the vigor of the process increasing with crater size (Parmentier et al. 1980), or (2) a curve basically representing a production population with a relative deficiency of impacts in this crater size range compared to those on the terrestrial planets. We tend to favor the latter explanation for the following reasons. Figures 9.14, 9.15, and 9.16 show that four different crater curves, representing vastly different crater densities, on different terrains and even different satellites, all possess this same steep-slope (~ -4.7) distribution function. Furthermore, as pointed out earlier, even though older craters $\lesssim$ 100 km diameter have been degraded, i.e., flat floors at about the level of the surrounding terrain, their rim sharpness is more or less preserved, suggesting that relaxation is not very effective at totally obliterating craters. If the paucity of craters in this diameter range was solely due to obliteration by relaxation, one would expect a very different distribution function between, for example, the fresh craters preserved over long periods of time in rigid ice, and the degraded craters perhaps formed at a time when the ice was more able to flow. Therefore the observed large variations in crater densities, but similarities in slopes among the many different terrains, ages,

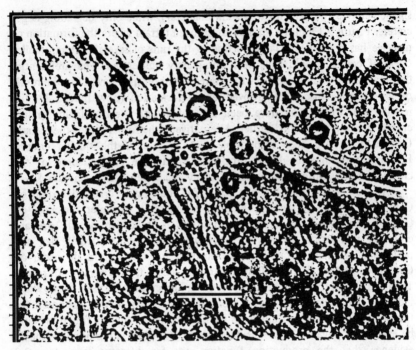

Fig. 9.17. A portion of the rims of several large craters (black arrows) removed by the formation of a segment of grooved terrain on Ganymede. Similar new ice formation must have totally obliterated many smaller craters. This process probably accounts for the overall lower crater density on grooved terrain than on cratered terrain and the greater diameter-dependent obliteration of the smaller craters shown in Fig. 9.13. Black scale bar represents 50 km. North is at the top. (FDS 20637.38)

degradational classes, and even satellites, argue against this curve in this diameter range being shaped by equilibrium. Furthermore, the absence of palimpsests on Callisto suggests that the presently observed crater population formed when the icy crust was rigid enough to retain the craters basically intact, with a minimum of obliteration due to viscous relaxation of the ice. The conclusion is that the data for craters $\gtrsim$ 50 km diameter basically represent their production function.

Three important consequences devolve from these interpretations.

1. Because of the similarity of the curves for the degraded and the fresh craters (over the range 30-130 km), the process degrading them must be nearly diameter-independent.

2. Although significant proportions of craters have been degraded, the process that degrades them is not too effective in totally obliterating them. If the degradational process is crater relaxation, it effectively stops

at a stress level still high enough to leave a residual topography that is identifiable as a crater. Observational evidence for this has already been presented.

3. The production function in the Jupiter region was vastly different from that in the inner solar system. We speculate, at this early stage, that Ganymede and Callisto may principally record a population of trans-Jupiter bodies that never penetrated the inner solar system in numbers great enough to have left a recognizable signature there.

At crater diameters $\gtrsim 150$ km the data are too sparse to be certain of the shape of the curves, and to interpret them would certainly be treading on thin ice. However, from the Callisto data set, one might speculate that the steep downward trend is reversed beyond ~ 250 km diameter, i.e., large impacts are slightly more abundant on Callisto than might be expected from the frequency of the immediately smaller impacts.

The above arguments were constructed to support the proposition that the crater populations of Ganymede and Callisto have not sustained significant crater losses due to relaxation of the icy crust. Although many analyses of relaxation times on these satellites suggest that craters could be obliterated on time scales short compared to the surface exposure ages (e.g., Johnson and McGetchin 1973; Parmentier et al. 1980), to date these studies have of necessity greatly simplified a difficult problem; for example, they often represent the craters as monofrequency sine waves although the craters are clearly far more complex structures. The high-frequency components of the craters, such as their sharp inner and outer rims and their central structures, will last many times longer than the crater bowls themselves. In the case of the palimpsests, for example, although their bowls are apparently nearly fully relaxed, under low enough illumination angles the remnants of the rim can still be easily recognized; in fact that is how we are able to assign a particular diameter to those structures.

Although the reality of complete crater relaxation must be considered an open question from the point of view of physical modeling, we recognize how readily the shape of the crater curves lend themselves to interpretations of crater obliteration by relaxation. The similarity of the curves of the grooved and cratered terrains beyond diameters of 60 km suggest abutment against an equilibrium limit. Yet we must be aware that this similarity spans only two diameter bins (64 to 128 km), with large associated error bars. A segment of the Callisto crater curve also nearly parallels those of Ganymede at the large diameters, again suggesting attainment of equilibrium, but at a greater density than on Ganymede, perhaps because the relaxation process is slower on Callisto. However, a significant upturn occurs in the curves of both Callisto and Ganymede's grooved terrain at still larger diameters; this is not in obvious concord with an interpretation of equilibrium conditions.

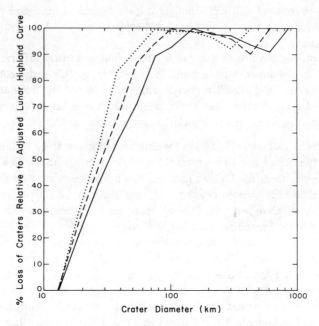

Fig. 9.18. A graph showing the per cent loss of craters on Callisto as a function of crater diameter. If we assume that the same population that impacted the lunar highlands impacted Callisto, a large proportion of craters would necessarily have been obliterated to obtain the present Callisto population. The solid line assumes that equal-sized craters result from equally energetic impacts into icy and rocky material; dashed line assumes a factor 1.4 enlargement, and the dotted line a factor of 2 enlargement, in ice compared to rock.

We do not believe that these curves truly represent attainment of equilibrium, but even if we accept such a conclusion, we can still make strong arguments that the crater production function recorded on the Galilean satellites was not the same one recorded in the inner solar system. We will do this by *reducto ad absurdum*. Let us assume that the lunar highlands' distribution was initially impressed on the ancient terrain of Callisto. We adjust the total crater flux at Callisto so that the minimum amount of crater obliteration would be required in order ultimately to evolve the observed Callisto curve from the lunar highlands' production function. This is done by assuming that no craters in the 8-11 km diameter range were obliterated on Callisto and none were obliterated at any diameter on the lunar highlands. If, for each diameter bin, we now calculate the percentage of all craters that would have to be obliterated in order to evolve the Callisto curve, we obtain the solid line in Fig. 9.18. Such a calculation is conservative not only because it assumes that no 8-km craters were obliterated, but also because it assumes that craters on the Moon and Callisto follow the same scaling relationship. If, as Boyce (1979) has suggested, craters in ice will be larger than their equivalents formed in rocky substrates, the degree of obliteration required to develop the observed Callisto curve from that of the Moon is also greater. The dotted and dashed lines in Fig. 9.18 show cases for enlargement factors of 1.4 and 2 for craters on Callisto. But returning to the most conservative case, namely no enlargement of Callisto's craters over the lunar craters, we see from the graph that more than 90 % of all craters > 80 km diameter would have to have been obliterated by relaxation. This is far more obliteration than is consistent with the appearance of the surface. If so many of these large impact structures and their ejecta deposits were totally obliterated by relaxation, many would still have left easily recognizable scars in the form of circular areas depleted in craters. Only if essentially all of the large bodies impacted very early in the cratering history would their crater scars now all be recratered to densities indistinguishable from the average background, but accepting such a scenario would immediately concede that the cratering histories of the terrestrial planets and of the Galilean satellites were different.

One may contend that low crater-density scars do exist, the best example being the Valhalla basin. However, our statistics include Valhalla and all degraded basins with measurable diameters. Therefore, these are not yet obliterated by relaxation and count among the observed craters and not among the low crater-density scars; no such scars have been found. We are forced to accept, therefore, that we can not subject Callisto to the lunar highlands' cratering history, add the phenomenon of crater relaxation, and develop a surface such as the one actually observed. We conclude that the crater production functions were different for the inner solar system and the Jovian region. Because we can not simultaneously match the lunar and Galilean satellites' impacting populations in terms of both their size-frequency distribution and their temporal size variations (if any), we have no reason to believe that the impacting bodies share a common origin and therefore cannot use the dated lunar surfaces to recover the flux histories or the terrain dates on the Galilean satellites.

VIII. SUMMARY

Several lines of evidence suggest that the crater curves on Callisto and Ganymede in the diameter range ~ 50-130 km basically represent the production population. In the diameter range ~ 10-50 km Ganymede has apparently suffered crater losses on both the grooved and cratered terrains, extending to 70 km on the grooved terrain; similar crater obliteration may or may not have occured on Callisto. In the case of the grooved terrain the crater loss was probably the result of the formation of new ice over some indeterminate period of time. The overall crater density is significantly less on the grooved terrain than on the cratered terrain, indicating that it is younger.

This is consistent with observed stratigraphic and transectional relationships between the two types of terrain. The loss of small craters in the heavily cratered terrain may have been due to the formation of ancient grooved terrain, which is now unrecognizable because of heavy recratering or the formation of the numerous troughs. The overall crater density on the most heavily cratered (oldest) terrain on Ganymede is significantly less than that on Callisto, indicating that either widespread resurfacing has taken place on Ganymede, or its icy crust became rigid enough to retain craters at a later time than Callisto's. In any event, the responsible mechanism must have acted during the period of intense bombardment.

Finally, the population of objects responsible for the period of intense bombardment at Jupiter appears to have had a very different size-frequency distribution from those responsible for the period of heavy bombardment on the terrestrial planets. The population of objects in the Jovian region may have been primarily a population of trans-Jupiter bodies that never penetrated the inner solar system in numbers great enough to leave a recognizable signature. It seems unlikely that this population was largely derived from the asteroid belt (Smith et al. 1979b). Furthermore, if we are dealing with two very different populations it is very unlikely that the impact flux history at Jupiter was similar to that in the inner solar system. Deciding among hypotheses regarding the origin of the objects responsible for the period of heavy bombardment is not yet possible, but the observed cratering records on Callisto and Ganymede suggest two broad alternatives for the origin of the impacting bodies.

First, if the objects responsible for heavy bombardment in the inner solar system were remnants left over from the accretion of the planets, this cannot also be the case at Jupiter, unless the objects there had very different diameter-frequency distributions. However, a major argument against an accretional origin is the short lifetime (halflife 30-70 $\times$ 10^6 yr) of such objects (Wetherill 1975).

A second alternative concerns a model proposed by Wetherill (1975) for the origin of the objects responsible for the period of heavy bombardment in the inner solar system. This model involves perturbation of Uranus- and Neptune-crossing objects into Saturn- and then Jupiter-crossing orbits, which eventually evolve into orbits similar to short-period comets. A single, randomly timed breakup of a large body ($\sim$ 10^{23} g) due to a close encounter with the Earth or Venus would then produce the objects responsible for the period of late heavy bombardment of the terrestrial planets. If this mechanism operated, one might expect that the population of the crossing objects would differ from those in the inner solar system and that the impact rate would be quite different as well (but very similar among terrestrial planets). Whether or not the surface of Callisto and Ganymede record the impact of such objects is not known. All that can be said at this time is that any model for the origin of the objects responsible for the periods of intense bombardment at Jupiter and in the inner solar system must account for the differences in the crater populations in these two regions.

Acknowledgments. We thank D. Morrison and S. Weidenschilling for their reviews and comments. We also thank E. Shoemaker for useful discussions and comments. This research was supported by a grant from the National Aeronautics and Space Administration.

Note added in proof: A recent Monte Carlo study demonstrated that Callisto's crater population could not have been produced by the lunar highlands impacting population and then altered by relaxation. The computer simulations (Woronow and Strom *Geophys. Res. Lett.*, in press, 1981) imposed a lunar highlands production function on a surface and allowed the oldest craters to relax until the size-density curve of Callisto's craters was matched. The resultant simulated surface had abundant "cryptopalimpsests" delineated by obvious expanses of lightly cratered terrains and dense clusters of small craters between the cryptopalimpsests. This clumpy areal distribution of craters is unlike that observed on Callisto. They draw two conclusions from this: (1) the impacting bodies in the Callisto region intrinsically lacked a portion of the larger members which impacted the terrestrial planets, and (2) crater obliteration by relaxation can not be as extensive as often presumed. Because different populations of distinct origins affected the Moon and Jovian systems, time scales derived from cratering fluxes on the Moon can not be translated to Ganymede or Callisto, nor can thermal and viscosity histories be constructed for those bodies by assuming a lunar-like crater production function.

REFERENCES

Boyce, J.M. (1979). Diameter enlargement effects on crater populations resulting from impacts into ice. *NASA TM* 80339, 119-120 (abstract).

Cassen, P.S., Peale, S.J., and Reynolds, R.T. (1980). On the comparative evolution of Ganymede and Callisto. *Icarus* 41, 232-239.

Chapman, C.R. (1974). Asteroid size distribution: Implications for the origin of stony-iron meteorites. *Geophys. Res. Letters* 1, 341-344.

Chapman, C.R., Mosher, J.A., and Simmons, G. (1970). Lunar cratering and erosion from Orbiter 5 photographs. *J. Geophys. Res.* 72, 1445-1466.

Crater Analysis Techniques Work Group (1979). Standard techniques for presentation and analysis of crater size-frequency data. *Icarus* 37, 467-474.

Gault, D.E. (1970). Saturation and equilibrium conditions for impact cratering on the lunar surface—criteria and implications. *Radio Sci.* 5, 273-291.

Hartmann, W.K. (1971). Martian cratering, II. Asteroid impact history. *Icarus* 15, 396-409.

Hartmann, W.K. (1973). Martian cratering, 4. Mariner 9 initial analysis of cratering chronology. *J. Geophys. Res.* 78, 4096-4116.

Hartmann, W.K., Strom, R.G., Weidenschilling, S., Blasius, K., Woronow, A., Dence, M., Grieve, R., Soderblom, L.A., Diaz, J., Chapman, C.R., Shoemaker, E.M. (1980). *Basaltic Volcanism Study Project.* Ch. 8. Pergamon Press, NY. In Press.

Hostetler, C.J., and Drake, M.J. (1980). On the early global melting of the terrestrial planets. *Proc. Lunar Planet. Sci. Conf.* 11. In Press.

Johnson, J.V., and McGetchin, T.R. (1973). Topography on satellite surfaces and the shape of asteroids. *Icarus* 18, 612-620.

MacDonald, T.L. (1931). Studies in lunar statistics. *J. Brit. Astron. Assoc.* 41, 288-290.

Malin, M.C. (1976). Observations of intercrater plains on Mercury. *Geophys. Res. Letters* 3, 581-584.

Marcus, A.H. (1965). Positive stability laws and the mass distribution of planetesimals. *Icarus* 4, 267-272.

Marcus, A.H. (1970). Comparison of equilibrium size distributions for lunar craters. *J. Geophys. Res.* 75, 4977-4984.

Moore, H.J., Lugn, R.V. and Newman, E.B. (1974). Some morphometric properties of experimentally cratered surfaces. *J. Res. U.S. Geol. Survey* 2, 279-288.

Morgan, J.W., Ganapathy, R., Higuchi, H., and Anders, E. (1977). *Meteoritic Material on the Moon.* pp. 659-689. NASA SP-370, Washington, D.C.

Neukum, G., and Dietzel, H. (1971). On the development of the crater population on the Moon with time under meteoroid and solar wind bombardment. *Earth Planet. Sci. Lett.* 12, 50-66.

Parmentier, E.M., Allison, M.L., Cintala, M.J., and Head, J.W. (1980). Viscous degradation of impact craters on icy satellite surfaces. *Lunar Planet. Sci.* 11, 857-859 (abstract).

Parmentier, E.M., and Head, J.W. (1979). Internal processes affecting surfaces of low-density satellites: Ganymede and Callisto. *J. Geophys. Res.* 84, 6263-6276.

Shoemaker, E.M., Williams, J.G., Helin, E.F. and Wolfe, R.F. (1979). Earth-crossing asteroids: Orbital classes, population, and fluctuation of population in late geologic history *NASA TM* 80339 3-5 (abstract).

Smith, B.A. and the Voyager Imaging Team. (1979a). The Jupiter system through the eyes of Voyager 1. *Science* 204, 951-972.

Smith, B.A., and the Voyager Imaging Team (1979b). The Galilean satellites and Jupiter: Voyager 2 imaging science results. *Science* 206, 927-950.

Strom, R.G. (1977). Origin and relative age of lunar and Mercurian intercrater plains. *Phys. Earth Planet. Interiors* 15, 156-171.

Strom, R.G. (1979). Mercury: A post-Mariner 10 assessment. *Space Sci. Rev.* 24, 3-70.

Taylor, S.R. (1975. *Lunar Science: A Post-Apollo View.* Pergamon Press, N.Y.

Wetherill, G.W. (1975). Late heavy bombardment of the Moon and terrestrial planets, *Proc. Lunar Sci. Conf.* 6, 1539-1561.

Woronow, A. (1977a). A size-frequency study of large Martian craters. *J. Geophys. Res.* 82, 5807-5820.

Woronow, A. (1977b) Crater saturation and equilibrium: A Monte Carlo simulation. *J. Geophys. Res.* 82, 2447-2451.

Woronow, A. (1978). A general cratering-history model and its implications for the lunar highlands. *Icarus* 34, 76-88.

Woronow, A., Strom, R.G., and Rains, E. (1979). Effects of the Orientale impact on the pre-existing crater populations. *NASA TM 80339,* 152-153 (abstract).

Zellner, B. (1979). Asteroid taxonomy and the distribution of the compositional types. In *Asteroids* (T. Gehrels, Ed.), pp. 783-806. Univ. Arizona Press, Tucson.

10. CRATERING TIME SCALES FOR THE GALILEAN SATELLITES

EUGENE M. SHOEMAKER

and

RUTH F. WOLFE
U.S. Geological Survey

Craters on the Galilean satellites of Jupiter have been produced in the recent geologic past primarily by the impact of active and extinct comet nuclei. Both comets of short period, sometimes called the Jupiter-family comets, and long period comets are represented in the flux of active comets in the vicinity of the Galilean satellites. Although the number of Jupiter-family comets is relatively small, their flux in the neighborhood of Jupiter is relatively high because of their short periods and because of their low velocities at Jupiter crossing, permitting strong concentration of the flux by Jupiter's gravity field. The much more numerous long period comets, including the nearly parabolic and so-called parabolic comets, probably account for about one-eighth to one-third of the present production of impact craters. Planet-crossing asteroids, consisting chiefly of recently extinct short period comets, may be responsible for roughly half of the present production of impact craters. Because of the high orbital velocities of the Galilean satellites, the present cratering rate at the apex of motion is about twice the mean rate for each satellite, and the cratering rate at the antapex is ~ 5 to 20 times less than the mean rate. The majority of craters on Callisto and on Ganymede probably date from a period of heavy bombardment early in solar system history. Impact craters tentatively identified on Europa, however, were formed late in geologic time.

In order to interpret the geologic record preserved on the surfaces of the Galilean satellites, it is crucial to understand the rates at which impact craters have been produced on their surfaces. If the cratering rates are known or, more precisely, if the history of bombardment is known, then ages can be assigned to various surface features, the rates of surface processes can be estimated or, in the case of Io, limits can be set on these rates, and important clues concerning the evolution of each satellite can be deciphered. At our present level of understanding of the Galilean satellites, even order-of-magnitude estimates of the cratering rates are useful. We would like to know, for example, whether the sparsely cratered surface of Europa is relatively old, say several billion years, or comparatively young – less than 100 million years. The answer profoundly affects our understanding of the present thermal structure of Europa. Or, we might ask, when was the grooved terrain of Ganymede formed? Different answers to this question lead to different models of tectonic mechanisms.

This chapter is an inquiry into the available observations and theory that can be brought to bear to assess the cratering history of the Galilean satellites. The larger solid bodies that collide with the satellites at the present time are primarily active and extinct nuclei of comets. Relatively abundant high-quality information is available on the orbits of active comets, and we can calculate the probability of their collision with the Galilean satellites with fairly good accuracy. On the other hand, very little reliable information is available on the physical properties of the comet nuclei that we need to know in order to calculate cratering rates. This difficulty can be partly overcome with bounds that may be set by the cratering record of the Earth. When these bounds are introduced, it is found that roughly half the present production of large impact craters on the Galilean satellites may be due to collision with objects of asteroidal appearance on Jupiter-crossing orbits. We interpret such bodies to be chiefly extinct comet nuclei. Unfortunately, our estimate of the population of these bodies depends on the discovery of a single object, 944 Hidalgo, although other evidence can be used to set an upper limit to the population.

Our goal here is to estimate the present cratering rate for each satellite within the correct order of magnitude. We will then attempt to extend the cratering rates back into the geologic past on the basis of evidence from the Earth-Moon system. Beyond ~ 3 Gyr this extension is highly model-dependent and uncertain. We will adopt what seems to us the most likely model and one that yields a cratering history for the Galilean satellites consistent with the present knowledge of the time scale of the solar system.

In many of the calculations reported here we will carry one or two more "significant" figures than are warranted by the observations. The purpose of this is to allow for comparison of the results for each satellite and for the different classes of impacting bodies. The relative rates of cratering on the

Galilean satellites can be determined with much greater confidence than the absolute rates. Similarly, the gradient in cratering rate across the surface of each satellite is fairly insensitive to the principal sources of error in estimation of the present total cratering rate or the variation of cratering with time. We will deliberately eschew assigning formal confidence limits to our various estimates, as we believe these to be misleading; they tend to convey a false sense of precision. It is fairly easy to derive these limits for certain parts of the problem, but for other, critical parts such limits would be utterly meaningless. Some of the essential observations needed are simply lacking and we are forced to rely on indirect arguments.

I. COLLISIONS WITH SHORT PERIOD COMETS

By the end of 1978, 113 objects classified as periodic comets had been discovered over the course of two centuries of astronomical observation (Marsden 1979). A comet is now designated as periodic if the period of its osculating orbit is less than 200 yr. Among the 113 objects, 72 had been observed at two or more appearances and 41 only once; the majority of periodic comets observed on only one appearance have been recently discovered, but 28 periodic comets discovered earlier are now lost. Eighty-six percent of the periodic comets have periods < 20 yr; these are referred to here as the Jupiter-family comets. The mean period of the Jupiter-family comets is 7.9 yr, and they are all in prograde orbits. Because of their frequent passages through perihelion, where comets are most active and generally brightest and most easily detected, discovery of the Jupiter-family comets tends to be much more complete than for other comets. The remaining periodic comets, 16 in all, have a mean period of 81 yr; a quarter of these have retrograde orbits, including the famous P/Halley. For purposes of computing cratering rates, we will group these remaining objects with the long period comets.

Magnitude and Size Distribution of Short Period Comet Nuclei

A systematic program of astrometric observations of comets was initiated at Flagstaff in 1957 by E. Roemer. Her observations of photographic magnitudes of comet nuclei, obtained from plates taken for astrometry, constitute a foundation for studies of the size-frequency distribution of comet nuclei and the populations of various classes of comets to given nuclear magnitudes or size limits. Precise positions and photographic nuclear magnitudes were obtained for 42 periodic comets and 24 nearly parabolic comets (Roemer 1965; Roemer and Lloyd 1966; Roemer et al. 1966). The reported magnitudes are the most homogeneous set of photometric data available for a large number of comet nuclei, with the possible exception of a later set of nuclear magnitudes obtained by Roemer, which have not yet been fully published.

Absolute blue magnitudes and sizes of the nuclei of the periodic comets can be estimated by a method described by Roemer (1966). Whenever possible, the estimates are based on observations made when the nuclei are stellar in appearance; preference generally is given to magnitudes obtained at maximum observed heliocentric distances, when the comets normally are accompanied by the least observable comae. In most cases, only faint traces of a coma were seen in the distant observations, and in some cases, none at all. Following Roemer (1966, 1968), we will first treat the effect of the coma on the magnitude of the nucleus as negligible when the nucleus is stellar in appearance. The observed radiation from the nucleus is thus assumed to be almost entirely sunlight reflected from the solid surface of the nucleus. This assumption is not likely to be correct, but it greatly simplifies the discussion. We shall show that, even when a comet is at large heliocentric distances and the nucleus is nearly stellar in appearance, the observed magnitude of the nucleus probably is strongly affected by unresolved coma. A correction will be made for the contamination of the observations by the coma as a later step. Absolute magnitudes will first be derived on the basis that the comet nuclei have solid surfaces with photometric properties like those of the asteroids; preliminary dimensions of the comet nuclei will then be calculated on the basis of plausible assumptions about the geometric albedos of comet nuclei.

The cumulative frequency distribution of absolute blue magnitudes for the nuclei for 42 periodic comets observed by Roemer, calculated with the above assumptions, is illustrated in Fig. 10.1. All of these objects are Jupiter-family comets. They constitute 55% of the then-known Jupiter-family comets and 71% of the Jupiter-family comets that had been observed on more than one appearance. Nearly all of the bright Jupiter-family comets of relatively small perihelion distance that are not lost are represented in the statistics. From Fig. 10.1 it may be seen that the cumulative magnitude distribution tends to follow an exponential law up to magnitude 16; if the comet nuclei all had the same albedo, their size distribution would follow a power law of the form shown in the figure. At absolute magnitudes higher than 16, the cumulative frequency drops below the extrapolated exponential curve. The observed drop-off almost certainly is due to incompleteness of discovery of faint comets and is the normal effect observed in population statistics when the threshold of detection is approached.

There is a weak negative correlation between absolute magnitude of the nucleus and perihelion distance of the observed Jupiter-family comets. This correlation has the effect of increasing slightly the proportion of bright to faint observed comet nuclei relative to the true proportion in the comet population. Allowance has been made for this bias in drawing the exponential curve. The brightest nucleus falls well above the exponential distribution. This object is P/Schwassmann-Wachmann 1, which is the most distant known

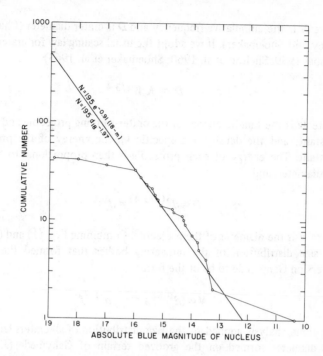

Fig. 10.1. Cumulative frequency distribution of absolute nuclear magnitude for 42 short period comets. Absolute blue magnitudes of comet nuclei are based on apparent nuclear magnitudes published by Roemer (1965), Roemer and Lloyd (1966) and Roemer et al. (1966).

short period comet; its current perihelion distance q is 5.45 AU. When account is taken of the distribution of Jupiter-family comets by perihelion distance, it will be seen that the brightest nucleus should have an absolute magnitude about equal to that observed for P/Schwassmann-Wachmann 1.

The population index of the exponential function fitted to the magnitude frequency distribution of the Jupiter-family comet nuclei is -0.9_1 [a]. This is equivalent to a slope (exponent) for the power law distribution of diameters of the comet nuclei of -1.9_7. If young impact craters on the Galilean satellites have been formed primarily by Jupiter-family comets, there should be a direct relationship between this slope and the slope of the size-frequency distribution of the diameters of the young craters. The diameters of bright ray craters larger than 25 km diameter on Ganymede follow the distribution

$$N \propto D^{-2.2} \tag{1}$$

[a]Subscripted figures are used in this chapter to indicate a digit that lies at or just beyond the threshold of significance.

where N is the cumulative frequency and D is crater diameter (Chapter 12 by Passey and Shoemaker). If we adopt the usual scaling law for crater diameters (Chabai 1959; Shelton et al. 1960; Shoemaker et al. 1963)

$$D = K W^{1/3.4} \qquad (2)$$

where W is the kinetic energy of the crater-forming projectile and K a scaling constant, and the density and specific kinetic energy of the projectile are constant. The energy of each projectile is then proportional to the cube of the diameter, and

$$D \propto d^{3(1/3.4)} = d^{0.88} \qquad (3)$$

where d is the diameter of the projectile. Combining Eqs. (1) and (3), we find the size distribution of the impacting bodies that formed the bright ray craters on Ganymede to be of the form

$$N \propto (d^{0.88})^{-2.2} = d^{-1.94}. \qquad (4)$$

The same analysis applied to the size distribution of all craters larger than 10 km diameter formed on the grooved terrain of Ganymede (J. B. Plescia, personal communication) yields $N \propto d^{-2.08}$. The derived slopes for the size distribution of impacting projectiles, -1.9_4 and -2.0_8, agree (within the errors of estimation) with the inferred slope of -1.9_7 for periodic comet nuclei.

A cumulative number of ~ 195 comet nuclei equal to or brighter than absolute blue magnitude 18 is obtained by extrapolation of the exponential function fitted to the absolute magnitudes derived from Roemer's observations (Fig. 10.1). This is the number that corresponds to the observed sample of Jupiter-family comets. As the observed sample constitutes 55% of the discovered Jupiter-family comets, the population of comet nuclei to magnitude 18 corresponding to the set of all discovered Jupiter-family comets is $\sim 195/0.55 = 355$.

Distribution of Perihelion Distances of Short Period Comets

Most observed periodic comets have perihelion distances less than 2 AU, but a few, such as P/Schwassmann-Wachmann 1, P/Oterma at $q = 3.39$ AU, and P/van Houten at $q = 3.96$ AU, have been found at much greater distances. From the conditions of capture of short period comets by Jupiter perturbations of intermediate and long period comets (Everhart 1972, 1973, 1977; Kazimirchak-Polonskaya 1972; Delsemme 1973), and from the continued perturbations of their orbits by Jupiter, there are strong reasons to expect that Jupiter-family comets are relatively uniformly distributed in q out to the orbit of Jupiter. As shown by a deliberate search for and discovery of distant comets by T. Gehrels with the 1.2-m Schmidt telescope at Palomar Mountain, the small number of discovered objects with $q > 2.0$ AU is simply a result of observational selection. The activity of short period comets generally is greatly diminished at heliocentric distances greater than 2 AU, and the nucleus also becomes too faint to be detected with the small aperture telescopes normally used by comet searchers. In order to obtain the number of bodies crossing Jupiter's orbit, it is necessary to make a correction for these selection effects.

From the frequency distribution of q for Jupiter-family comets discovered through 1965 (Fig. 10.2), it may be seen that the cumulative number Σn is a nearly linear function of q between 1.1 and 1.7 AU. Below $q = 1.1$ AU the number lies above the fitted line. This indicates that the differential frequency per unit increment of q below 1.1 AU is less than in the range 1.1 AU $\leq q \leq 1.7$ AU, a fact for which there is not yet an adequate explanation. Possibly the low abundance of comets with $q < 1.1$ AU is due to disruption

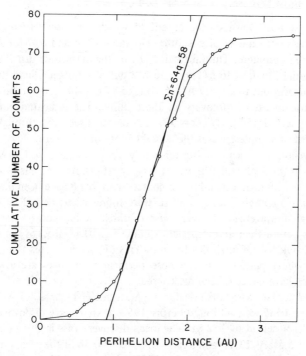

Fig. 10.2. Cumulative frequency distribution of perihelion distance for short period comets discovered through 1965. The distribution of q (perihelion distance) is essentially linear between 1.1 and 1.7 AU; in this interval it follows the line represented by the equation $\Sigma n = 64q - 58$.

or exceptionally rapid ablation during very close passes to the Sun. Above $q = 1.7$ AU, $\Sigma\, n$ falls below the fitted line, as expected for incomplete discovery. Extrapolating the fitted line to 5.527 AU, the maximum aphelion of Jupiter, yields a predicted number of ~ 296 Jupiter-family comets occasionally or frequently crossing Jupiter's orbit that correspond with 77 comets discovered. It should be noted that a number of very short period comets have aphelion distances currently inside Jupiter's orbit, but only one, P/Encke, is safe from Jupiter encounter. Deducting P/Encke from the count of known Jupiter crossers discovered through 1965, we find that the ratio of Jupiter-crossing to discovered short period comets is $\sim 296/76 = 3.9$. Hence the estimated total number of Jupiter-crossing short period comets to magnitude 18 is $\sim 3.9 \times 355 = 1.3_8 \times 10^3$. Many periodic comets with $q > 5.527$ AU are strongly perturbed by Jupiter and can collide with the Galilean satellites, but, as will be shown, these do not need to be counted in the method used here to estimate collision rates.

Completeness of Discovery of Short Period Comets

We must now consider the extent to which the discovery of Jupiter-family comets with nuclei of absolute magnitude $\leqslant 16$ and $q \leqslant 1.7$ AU may be considered complete. The fairly close fit of the magnitude distribution to an exponential function to $B(1,0) = 16$ (32 out of 42 objects) and the linear distribution of q out to $q = 1.7$ AU (51 out of 77 objects) suggest, but do not prove, completeness of discovery to these limits. The indicated number to these limits is $(32/55\%)(51/77) = 39$, for comets discovered through 1965. The best test of completeness is the actual history of discovery.

The majority of known Jupiter-family comets with $q < 1.5$ AU were discovered in the 18th and 19th centuries (Fig. 10.3). As time has progressed, more comets with higher q and fewer comets with low q have been discovered (Fig. 10.3). No comets with $q < 1$ AU were found in the period 1960-1978, even though the average discovery rate of short period comets was higher during this period than at any previous time (Fig. 10.4). Most of the recently discovered objects are intrinsically faint. This history indicates that most of the bright short period comets whose perihelion distances are low at the present time have, indeed, been discovered.

The 39 comets with intrinsically brightest nuclei and $q \leqslant 1.7$ AU discovered through 1965 were all found before 1952, at an average discovery rate of 0.26/yr over a period of 152 yr. The mean discovery rate in this early period amounts to 6 every 23 yr, and a ratio of $39/67 = 0.58$ of the total number of Jupiter-family comets found through 1952. In contrast, only two comets out of 25 Jupiter-family comets discovered from 1953 through 1975 (the last year for which nuclear magnitudes obtained by Roemer have been published) fall within the range of $B(1,0)$ and q for which discovery is provisionally taken to be complete. One of these comets, P/Kojima = 1970 XII [$B(1,0)$ = 13.7, q = 1.63], was perturbed into an orbit with $q < 1.7$ AU by encounter with Jupiter just prior to the perihelion passage on which it was discovered (Marsden 1971). It has since been perturbed by a second Jupiter encounter to an orbit with q = 2.40 AU. We do not count it among the comets with q currently less than 1.7 AU, as it was last observed in an orbit with higher q. The second comet, P/Kohoutek = 1975 III [$B(1,0)$ = 14.1, q = 1.57], is counted among the objects with current $q < 1.7$ AU. The drop in discovery rate since 1952 to 1/6 the former mean rate suggests that the number of objects remaining to be found is roughly 1/6 the prior average number remaining. Discovery of objects through 1952 to $B(1,0)$ = 16 and q = 1.7 AU, therefore, may be more than 5/6 complete.

Greater than 5/6 completeness of search to blue magnitude 16 for short period comets can be compared with the discovery of main belt asteroids, which is probably complete to the same degree at about $B(1,0)$ = 14. The limit magnitudes fainter by two of approximate completeness for short period comets is attributable chiefly to the fact that the integrated brightness of the coma, as a low q periodic comet approaches perihelion, is typically several magnitudes brighter than the magnitude of the nucleus. Hence a comet is discovered because the bright coma attracts notice. Moreover, observations of comets are much more consistently reported by astronomical observers than are observations of asteroids.

In addition to the discovery of P/Kohoutek, the fact that 17 short period comets are lost suggests that not quite all bright Earth-approaching short

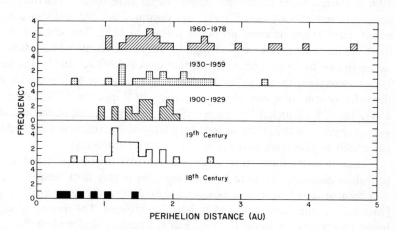

Fig. 10.3. Frequency distributions of perihelion distance for short period comets classified on the basis of time at which they were discovered. With the passage of time there is a progressive shift in mean q and in the range of q for discovered short period comets. This trend suggests that discovery of bright short period comets with currently small perihelion distance is now relatively complete.

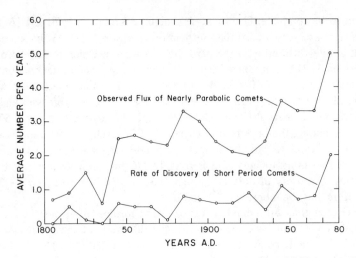

Fig. 10.4. Ten-year average discovery rates for comets for the 19th and 20th centuries. A surge of discovery of short period comets has occurred in the last decade, primarily as a result of increased coverage of the sky by large-aperture, wide-field telescopes. Many of the newly discovered short period comets are intrinsically very faint. The recent surge of discovery of nearly parabolic comets is due chiefly to increased surveillance of the southern sky.

period comets may have been found. Among the lost comets are 4 bright objects of small perihelion distances discovered in the late 18th and early 19th centuries. Some of the lost objects might have become inactive since discovery, and at least one, P/Biela, was disrupted in 1846 and then disappeared after its last observed perihelion passage in 1852. The orbits of other short period comets have been changed by Jupiter encounters into orbits of high q and are no longer observable (Kazimirchak-Polonskaya 1972). The best known example is that of P/Lexell, which might even have been ejected from the solar system by a very close encounter with Jupiter in 1779 (Leverrier 1848, 1857; Kazimirchak-Polonskaya 1972). However, a few of the lost objects probably are still intact, at least detectably active, still have $q \leqslant 1.7$ AU, and merely require more intensive search for their rediscovery.

Thus it seems likely that a few comets with bright nuclei and small perihelion distances remain to be discovered. Some may have escaped attention because recent perihelion passages occurred relatively near superior conjunction, a circumstance that makes detection difficult. The number of such bright objects may be equal to about half the number that are lost. We will tentatively set the number remaining to be found at 5 for Jupiter-family comets to $B(1,0) = 16$ and $q = 1.7$ AU. This is $\sim 1/8$ of the number discovered through 1965. About the same number may have been lost by disruption, extinction or near extinction of activity, and by increase of q beyond

1.7 AU by encounter with Jupiter. The estimated present total number to the limit of completeness, therefore, is not changed, and the total estimated population of short period comets presently crossing the orbit of Jupiter remains 1.4×10^3 to nuclear magnitude 18.

As a check on the total estimated population of short period Jupiter-crossing comets, let us assume that P/Schwassmann-Wachmann 1 is the brightest such comet. The equation

$$N = \mathcal{L} e^{-0.91(18 - m)} \tag{5}$$

is adopted as the form of the integral magnitude-frequency distribution, where $\mathcal{L}$ is the number at $m = 18$. By substituting $m = 10.45$ for P/Schwassmann-Wachmann 1 at $N = 1$, we find $\mathcal{L} = 0.96 \times 10^3$. This is within the estimated population of 1.4×10^3 to $m = 18$, but discovery of another short period comet of the same magnitude or brighter than P/Schwassmann-Wachmann 1 would suggest a higher population. It is tentatively concluded that P/Schwassmann-Wachmann 1 is the brightest and largest Jupiter-family comet and that the total population to magnitude 18 is about 1.4×10^3.

Diameters and Masses of Cometary Nuclei

In order to estimate diameters and masses of cometary nuclei, it is necessary to make plausible assumptions about their albedos and densities. No direct determinations of albedo have been obtained for comet nuclei, although a variety of indirect estimates have been published (Whipple 1978). C. R. Chapman obtained a spectrum of the nucleus of the very weakly active short period comet P/Arend-Riguax that resembles the reflection spectrum of RD-type asteroids (Degewij and van Houten 1979). The asteroid 944 Hidalgo, which has an orbit like that of some Jupiter-family comets and is almost certainly an extinct periodic comet nucleus, also has *UBV* and *VRI* colors similar to those of RD-type objects (Degewij and van Houten 1979). Radiometrically determined visual geometric albedos for three RD-type asteroids (including two Trojans and a Hilda) range from 0.030 to 0.035 (Degewij and van Houten 1979); $B-V$ for these asteroids ranges from 0.70 to 0.77, indicating blue geometric albedos, p_B, of 0.028 to 0.031. These values are within the range of albedos found by Whipple (1978) by comparison of the nongravitational forces acting on selected periodic comets with the magnitudes of these comets at maximum observed solar distances.

Although much higher albedos commonly have been assumed for comets in the past (see review by Whipple 1978), it appears entirely plausible that most periodic comets are as dark as RD- or, perhaps, C-type asteroids. Comets are known to be the sources of abundant dust, which is evidently embedded in a matrix of ice in the nucleus and liberated as the ice sublimates during approach to perihelion. It is generally thought that certain dust particles collected from the stratosphere (Brownlee et al. 1976, 1977) are actual samples of this dust (Brownlee 1978). These particles turn out to be similar in composition to primitive carbonaceous meteorites. If, as a periodic comet recedes from the Sun, a lag deposit of these particles is left on the surface of the nucleus, the surface of the nucleus may resemble photometrically a primitive carbonaceous meteorite or an optically similar C-type asteroid. The mean p_B for C-type asteroids is 0.034 (Morrison 1977). The peculiar spectral reflectance of RD objects may be due to an abundant clay mineral like that found in C1 and C2 meteorites (Degewij and van Houten 1979), which is a likely major component of cometary dust. Hence, a somewhat lower mean p_B of 0.030 for observed RD objects is also reasonable for periodic comet nuclei.

Roemer's estimated photographic blue magnitudes are taken to be very close to B photoelectric magnitudes that she used to calibrate her magnitude estimates. Adopting a mean value of 0.030 for p_B and the blue magnitude for the Sun of $- 26.14$ (Gehrels et al. 1964), we have

$$\log d_{18} = -0.5 \log p_B - 0.359 \tag{6}$$

where d_{18} is the diameter of the comet nucleus in km at $B(1,0) = 18$. Substituting $p_B = 0.030$ yields a diameter of 2.53 km at $B(1,0) = 18$.

Delsemme's (1977) heuristic model for the composition of comets is a useful starting point for estimating the mean density of comet nuclei. From consideration of the cosmic abundances of the elements and the observed depletion of the lightest elements in comets, Delsemme obtained 0.9 for the mass ratio of the anhydrous silicate (plus iron oxide) fraction to ice in comet nuclei. He finds this ratio to be consistent with the observed liberation of dust and gas during recent perihelion passages of five long period comets for which there are adequate data (Delsemme 1980). If we assume, (1) that the silicate fraction has $\sim$ 10% water of crystallization, as appropriate for the clay fraction of carbonaceous meteorites, (2) that the density of the rocky fraction is comparable to the mean density of C1 meteorites (2.33 g cm^{-3}), and (3) that the ice is predominantly H_2O ice, as deduced by Delsemme, we get a mean grain density for comet nuclei of $\sim 1.3_3$ g cm^{-3}. Allowing 15% porosity for gently compacted nuclei composed of accreted grains gives a bulk density of 1.2 g cm^{-3}. Selective loss of ice may have occurred in periodic comets, which could lead to higher densities, particularly for relatively inactive comets. If the residue of nonvolatile silicates is very porous, on the other hand, a weakly active short period comet might be less dense than a fresh icy comet. Whipple (1978) suggests that the density of comet nuclei may be in the range 1.0 g cm^{-3} to 1.5 g cm^{-3}. Here, we adopt 1.2 g cm^{-3} as a reasonable mean. The mass M_{18} of an absolute blue magnitude 18 comet nucleus ($d = 2.53$ km) then is $(1.2)(\pi/6)(2.53 \times 10^5)^3$ g $= 1.0 \times 10^{16}$ g.

Crater Scaling

To obtain crater diameters formed by impacting comet nuclei, we next evaluate the constant K in Eq. (2). For a nuclear crater formed in alluvium at the Nevada Test Site by explosion at a scaled depth appropriate for comparison with most impact craters, K_n = 0.074 km/(kiloton TNT equivalent)$^{1/3.4}$ (Shoemaker et al. 1963). As we will be concerned chiefly with impact of an icy or low density projectile into an icy satellite crust, the scaled depth for such an impact is taken to be comparable to the scaled depth for impact of a rocky projectile into a rocky crust. At crater diameters that will be considered here, it is assumed that cratering efficiency (as measured by diameter) is independent of the mechanical properties of the target but does depend on the acceleration of gravity at the surface of the satellite, on the density of the target material, and on the degree of gravitational collapse of the early formed crater. The presence of central peaks and terraced walls in craters on Ganymede suggests that nearly all craters ≥ 10 km in diameter formed on the icy Galilean satellites have collapsed. When account is taken of these factors, Eq. (2) becomes

$$D = s_g s_\rho \, c_f K_n \, W^{1/3.4} \tag{7}$$

where
$$s_g = \left(\frac{g_e}{g_s}\right)^{1/6} = \left(\frac{R_s^2 \, g_e}{G M_s}\right)^{1/6} \tag{8}$$

is the gravity scaling factor (Gault and Wedekind 1977), g_e is the surface gravity on Earth, g_s is the surface gravity on the satellite, R_s is the mean radius of the satellite, M_s is the mass of the satellite, and G is the universal gravitational constant is

$$s_\rho = \left(\frac{\rho_a}{\rho_s}\right)^{1/3.4} \tag{9}$$

The density scaling factor, ρ_a is the density of alluvium at the Nevada Test Site, ρ_s is the density of the satellite's crust, and it is assumed that the energy required to form a large impact crater of a given diameter is proportional to the target density; finally, c_f is the collapse factor, nominally taken to be 1.3 from studies of collapsed lunar craters (Shoemaker 1977a; Shoemaker et al. 1979).

The kinetic energy of an impacting magnitude 18 comet nucleus, in kilotons TNT equivalent, is given by

$$W = \frac{M_{18} \, v_i^2}{2 \times 4.185 \times 10^{19}} \, \text{kt} \tag{10}$$

where M_{18} is in g, and v_i, the impact velocity, is in cm s^{-1}. Combining Eqs. (7), (8), (9), and (10), we find the diameter D_{18} of the crater produced by this impact is

$$D_{18} = \left(\frac{R_s^2 g_e}{GM_s}\right)^{1/6}\left(\frac{\rho_a}{\rho_s}\right)^{1/3.4} \frac{c_f K_n M_{18}^{1/3.4} v_i^{2/3.4}}{8.37 \times 10^{19}} \text{km.} \qquad (11)$$

The following parameters will be used in Eq. (11)
$g_e = 9.80 \times 10^{-3}$ km s^{-2},
$\rho_a/\rho_s = 2.0$ for the icy Galilean satellites,
$\rho_a/\rho_s = 1.0$ for Io,
$c_f = 1.3$,
$K_n = 0.074$ km/(kt TNT equivalent)$^{1/3.4}$,
$M_{18} = 1.0 \times 10^{16}$ g,

and the following values will be used for R$_s$ (Davies and Katayama 1980) and GM_s (S. P. Synnott and J. K. Campbell, personal communication, 1980) for the Galilean satellites, in km and km^3 s^{-2}, respectively:

	Io	Europa	Ganymede	Callisto
R_s	1816	1563	2638	2410
GM_s	5962	3196	9891	7178

Equation (11) then reduces to:

	Io	Europa	Ganymede	Callisto	(12)
$D_{18} =$	(7.8,	10.2,	10.0,	10.3)	$v_i^{2/3.4}$ km

where v_i is given in km s^{-1}.

Collision Probabilities and Cratering Rates

Impact velocities for comet nuclei and their collision probabilities with the Galilean satellites can be obtained with the aid of Öpik's (1951) equations with appropriate modifications (see Appendix A). In applying Öpik's equations to the calculation of probabilities of comet impact and cratering rates on the Galilean satellites, we adopt a quite different point of view from that of Öpik in his original application of these equations to collision lifetimes of comets. The dynamical lifetimes of individual comets are very much less than those derived by Öpik from the reciprocals of the collision probabilities, because the orbit of nearly every periodic comet undergoes fairly rapid changes as a consequence of both distant and near encounters with Jupiter. In a typical case, a Jupiter-family comet is ejected from the solar system as a

result of these encounters in $\sim 10^5$ yr (see discussion of planet-crossing asteroids in Sec. III). This is many orders of magnitude less than the reciprocal of
the probability of collision of a Jupiter-family comet with one of the Galilean
satellites.

Here we will assume that the orbit of each comet is frequently changed
by Jupiter encounters but that the statistical distribution of orbital elements
for the ensemble of short period comets is relatively stable or in steady state.
We further assume that the observed orbits, at a given time, are a representative sample, for orbits with $q \lesssim 1.7$ AU. The sample is a "snapshot," in other
words, of the steady state orbit distribution. Thus, while most comets are
ejected from the solar system and never collide with Jupiter or its satellites,
the mean collision probability obtained from a given snapshot of orbits, when
extended over a short time, does give the true collision rate. The rate so
obtained is strictly valid only for the instant of the snapshot. We must depend
on other evidence to determine how steady that rate may be over a long
period of time.

In the derivation of his equations, it was assumed by Öpik that the
argument of perihelion ω of a comet (or other small body) with respect to
the ascending node of its orbit on the plane of the planet is, over time,
uniformly distributed between $0°$ and $360°$. When we apply these equations
in the snapshot mode, the equivalent assumption is that the instantaneous
distribution of ω for the ensemble of orbits is uniform between $0°$ and $360°$.
This assumption can be readily tested by observation; if the distribution
deviates from uniformity by an amount greater than that expected by chance,
then a correction must be made to the calculated mean probability of collision.

The distribution of ω (with respect to the ascending node on the orbit of
Jupiter) for the 97 discovered short period comets is illustrated in Fig. 10.5.
All values of ω are for the last observed perihelion passage. It can be seen that
the distribution is very strongly bimodal, with one mode near $0°$ and the
other near $180°$ (see also Porter 1963). This striking bimodality is chiefly a
consequence of the fact that most short period comets have one node close to
the orbit of Jupiter and also have aphelion distances comparable to the semimajor axis of Jupiter. The proximity of one node to Jupiter's orbit allows
many close encounters, which have been responsible, in turn, for the evolution of the orbit to short period. During this evolution, a short period comet
tends to be hinged on this node.

The distribution of heliocentric radii to the nodes of 83 observed short
period comets that currently cross the orbit of Jupiter has a very strong peak,
centered on the semimajor axis of Jupiter (Fig. 10.6). This peak corresponds
to the nodes near aphelion. A second peak at 1.5 AU corresponds to the
nodes near perihelion. For comparison, we have calculated the frequency
distribution of radii to the nodes for the same 83 orbits for a perfectly

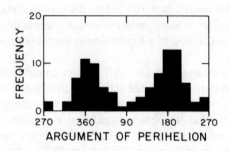

Fig. 10.5. Frequency distribution of the argument of perihelion (with respect to the ascending node on the orbit plane of Jupiter) for 97 short period comets discovered through 1978. The distribution is strongly bimodal; modes are centered near $\omega = 90^{\circ}$ and $\omega = 180^{\circ}$.

uniform distribution of ω (Fig. 10.6). The ratio of the observed frequency in the peak centered at the orbit of Jupiter to the frequency predicted for uniform distribution of ω represents the correction that must be applied to obtain the true mean collision probability. This ratio $\sigma = 2.09$.

In order to obtain an estimate of the long-term average cratering rate on the Galilean satellites, we use the time-averaged eccentricity of Jupiter to determine which orbits are Jupiter-crossing and to obtain the crossing fractions (see Appendix A). The constants used in our solutions of Öpik's equations are:

$M_{\odot} = 1.9891 \times 10^{30}$ kg (mass of the Sun),

$a_0 = 5.2027$ AU (semimajor axis of the orbit of Jupiter),

$e_0 = 0.0464$ (time-average eccentricity of Jupiter) (Wolfe and Shoemaker 1981),

$M_0/M_{\odot} = 9.5461 \times 10^{-4}$ (ratio of mass of Jupiter to mass of Sun) (Null 1976; M_0 does not include the masses of the satellites),

$v_0 = 13.058$ km s^{-1} (orbital velocity of Jupiter).

| Io | Europa | Ganymede | Callisto |

$r_s = 4.216 \times 10^5$ km, 6.709×10^5 km, 1.070×10^6 km, 1.880×10^6 km

(semimajor axes of the orbits of the Galilean satellites).

Orbital elements of the Jupiter-family comets are taken from Marsden (1979) and, in each case, the orbit for the last observed perihelion passage is used. Inclinations to the ecliptic given by Marsden are first transformed to inclinations to the orbital plane of Jupiter. The frequency distribution of the transformed inclinations and the distribution of eccentricities are shown in Fig. 10.7.

For each Galilean satellite, the asymptotic encounter velocity with Jupiter's sphere of influence U, the encounter velocity at the sphere of influence of the satellite U_c, the enhancement of the flux at the orbital radius of the satellite F, the collision velocity v_i, the collision probability P_c, the diameter of the crater produced by impact of a magnitude 18 nucleus D_{18}, and the equivalent rate of production of ≥ 10-km diameter craters Γ_i, are computed for orbital elements corresponding to each comet that currently crosses the orbit of Jupiter. The equivalent cratering rate to 10-km crater diameter is obtained from

$$\Gamma_i = \frac{\sigma P_c}{4\pi R_s^2} \left(\frac{D_{18}}{10 \text{ km}}\right)^{2.2} \tag{13}$$

$\bar{P}_c$ and $\bar{\Gamma}$ are obtained by averaging P_c and Γ_i calculated for each Jupiter-crossing short period comet.

Multiplying $\bar{\Gamma}$ by $1.3_8 \times 10^3$, the estimated population of Jupiter-family comets to magnitude 18, would yield a first estimate of the total rate of cratering by impact of this class of comets. The set of orbital elements corresponding to the observed Jupiter-family comets, however, is not representative of the entire class of short period comets. The majority of Jupiter-family comets have $q > 1.7$ AU, whereas only a relatively small number of orbits of this type is included in the set of observed orbits. An improved estimate of the mean equivalent 10-km rate of cratering by impact of Jupiter-family comets is obtained as follows: (1) $\bar{\Gamma}$ is multiplied by 355, the population of comet nuclei to magnitude 18 corresponding to the set of all observed Jupi-

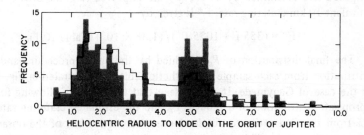

Fig. 10.6. Frequency distribution of heliocentric radii to the nodes on the orbit plane of Jupiter for 83 short period comets that currently cross the orbit of Jupiter (shaded bars). Because the distribution of ω is nonuniform, the distribution of the radii to the nodes is strongly bimodal; one mode is centered on the semimajor axis of Jupiter at 5.2 AU. Shown with the open histogram is the frequency distribution of the radii to the nodes for a population of orbits with the same distribution of a and e observed for Jupiter-crossing short period comets but with a perfectly uniform distribution of ω.

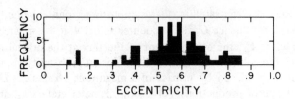

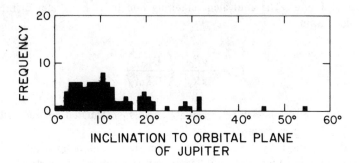

Fig. 10.7. Frequency distribution of eccentricity and frequency distribution of inclination to the orbit plane of Jupiter for 97 short period comets discovered through 1978.

ter-family comets; (2) for the 1025 objects corresponding to undiscovered short period comets with $q > 1.7$ AU, a set of "synthetic" orbital elements is generated (see Appendix B); a new mean equivalent 10-km cratering rate $\bar{\Gamma}'$ is computed and multiplied by 1025; (3) the improved estimate of the mean equivalent 10-km cratering rate, $\bar{\Gamma}*$ is given by

$$\bar{\Gamma}* = (355 \, \bar{\Gamma} + 1025 \, \bar{\Gamma}')/(1.38 \times 10^3) \text{ (Table 10.1)}.$$

The final distribution of P_c obtained by the above procedure and the contribution from each sample of synthetic orbits are illustrated in Fig. 10.8 for the case of Ganymede. It can be seen that the effect of allowing for an appropriate distribution of q reduces $\bar{P}_c$ and the estimated cratering rate by 11% from the first estimate obtained using only the orbits of the observed short period comets. This reduction arises because the observed orbits are biased toward cases where the aphelion is close to the orbit of Jupiter and the collision probability is anomalously high. The collision probability is reduced also because the mean period of the synthetic orbits is substantially higher than the mean period of the observed orbits. Not all orbits, either in the observed set or the synthetic set, are currently Jupiter-crossing. This circumstance is allowed for by including the cases of no crossing ($P_c = 0, \Gamma_i = 0$) in the calculated averages $\bar{\Gamma}$ and $\bar{\Gamma}'$.

TABLE 10.1

**Short-Period Comet Collision Parameters and Cratering Rates to
10-km Diameter on the Galilean Satellites[a]**

	Io	Europa	Ganymede	Callisto
$\bar{U}^2$	0.115	0.116	0.118	0.121
δ	46.9	42.8	20.9	15.6
$\bar{F}$	152.8	98.0	60.1	34.2
$\bar{v}_i$, km s^{-1}	25.0	20.0	16.2	12.6
$\bar{D}_{18}$, km	52.1	59.2	51.6	45.6
$2.09 \times \bar{P}_c$, $\times 10^{-11}$yr^{-1}	10.2_7	4.8_6	9.0_7	4.5_6
$\bar{\Gamma}$, $\times 10^{-17}$km^{-2}yr^{-1}	9.9	8.5	4.2_5	2.0_6
$\bar{\Gamma}^*$, $\times 10^{-17}$km^{-2}yr^{-1}	9.4	7.9	3.8	1.7_5

[a]All values of collision parameters shown are weighted means for observed plus synthetic orbits of short period comets. See below for definitions of table entries:

$\bar{U}^2$ = mean square of asymptotic encounter velocity at sphere of influence of Jupiter, weighted by P_c. (U is in units of Jupiter's mean orbital velocity.)

δ = ratio of cratering rate at apex of satellite's motion to cratering rate at antapex.

$\bar{F}$ = mean enhancement of the flux of comets by the gravitational field of Jupiter at orbital radius of satellite, weighted by P_c.

$\bar{v}_i$ = mean impact velocity, weighted by P_c.

$\bar{D}_{18}$ = mean diameter of crater produced by impact of a blue magnitude 18 comet nucleus, weighted by P_c.

$2.09 \times \bar{P}_c$ = mean collision probability corrected for nonuniform distribution of the arguments of perihelion of the short period comet orbits.

$\bar{\Gamma}$ = estimated mean rate of production of craters 10 km in diameter and larger per unit body of population to absolute blue magnitude 18, based on observed short period comet orbits only.

$\bar{\Gamma}^*$ = improved estimated mean rate of production of craters 10 km in diameter and larger per unit body of population to absolute blue magnitude 18, based on observed and synthetic short period comet orbits.

It should be noted that the periods of many synthetic orbits for short-period comets with $q > 1.7$ AU do not fall within the limit of 20 yr, which was used at the outset to separate the observed Jupiter-family comets from other periodic comets. The harmonic mean semimajor axis for the various samples of synthetic orbits and the corresponding periods are shown in Table 10.2. It can be seen that many synthetic orbits, in fact, have periods that are intermediate between those of the observed Jupiter-family comets and those of typical long period comets. Comets with these intermediate periods are interpreted by us to be part of a group of intermediate period comets that are

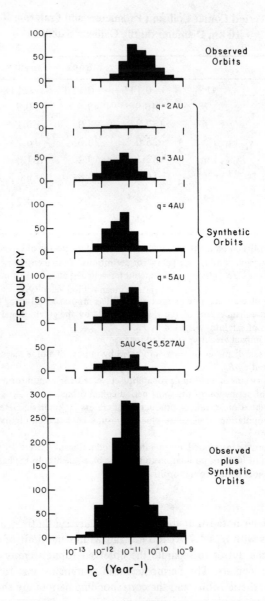

Fig. 10.8. Frequency distribution of probability of collision with Ganymede, P_c, for all Jupiter-family comets to absolute blue nuclear magnitude 18.

TABLE 10.2

Summary of Short Period Comet Orbits Used for Improved Estimate of Cratering Rates on Galilean Satellites

	Number of Orbits	Harmonic Mean a	Period Corresponding to Harmonic Mean a	Mean F (weighted by P_c)
Observed orbits	355	3.780	7.351	11.46
Synthetic orbits				
1.7 AU $< q \leqslant 2$ AU	28	4.166	8.50	10.24
2.0 AU $< q \leqslant 3$ AU	253	4.870	10.75	13.95
3.0 AU $< q \leqslant 4$ AU	291	6.378	16.11	26.09
4.0 AU $< q \leqslant 5$ AU	300	9.674	30.09	92.63
5.0 AU $< q \leqslant 5.527$ AU	153	15.916	63.50	260.15
Weighted mean	1380	7.153	21.01	60.10

inferred by Delsemme (1973) to be the precursors of most observed short period comets. A small fraction of the synthetic orbits turns out to have periods close to commensurabilities with the periods of Jupiter or Saturn, which is an artifact of the procedure by which the synthetic orbits were generated. A more realistic procedure would eliminate commensurable orbits by reassigning noncommensurable periods to them, but this refinement would not significantly change the derived values of $\bar{\Gamma}'$.

A somewhat unanticipated result obtained from the synthetic orbits is that the weighted mean enhancement factor F for the flux of comets at the orbital radius of each satellite is much higher than that found for the observed orbits (Table 10.2). This occurs because synthetic orbits corresponding to certain Jupiter-family comets of low inclination and low eccentricity, or with a semimajor axis near that of Jupiter, are included in the sample of synthetic orbits and have very low encounter velocities with Jupiter. As the collision probabilities for these orbits are also high (3×10^{-10} to 3×10^{-9} yr^{-1}), the mean F, weighted by P_c, is strongly influenced by a small number of orbits with very low encounter velocities. The interesting outcome from including the low encounter velocity cases is that the increase in mean F very nearly balances the effect of increase in period in the calculated collision probabilities for the synthetic orbits.

Total cratering rates for Jupiter-family comets can be obtained from the improved mean equivalent 10-km diameter cratering rates shown in Table 10.1 by multiplying $\bar{\Gamma}^*$ by the population of Jupiter-family comets to $B(1,0)$

= 18, nominally $1.3_8 \times 10^2$. We shall show, however, that our provisional estimate of the population to absolute magnitude 18, as calculated for the bare nucleus, may be too high, owing to contamination of the observed nuclear magnitudes by unresolved coma. Therefore, calculation of the total cratering rate for Jupiter-family comets is deferred until a corrected estimate of the population is made.

Gradients in Cratering Rates across the Surfaces of the Satellites

No account is taken of the orbital velocity of each satellite around Jupiter in our modifications of Öpik's equations. Some collisions are head-on and others are overtaking encounters, with respect to the motion of the satellite in its orbit. The contributions of the orbital velocity of the satellite to $\bar{P}_c$ and to mean v_i average very nearly to 0. However, as the orbital velocity of each Galilean satellite is a substantial fraction of U_c, there must be a significant gradient in the flux and in the mean impact velocity of comet nuclei from the leading point to the trailing point of the satellite.

The leading and trailing points remain nearly stationary with respect to the surface of each satellite. For a given encounter velocity U_c, the ratio γ_ϱ of the flux at the center of the leading hemisphere to the average flux on the satellite is approximately

$$\gamma_\varrho = \frac{2/3\ U_c + S_\varrho + S_0/\sqrt{2}}{2/3\ (U_c^2 + S_s^2)^{1/2}} \left(\frac{1 + S_s^2/U_\varrho^2}{1 + S_s^2/U_c^2} \right) \tag{14a}$$

and the ratio γ_t of the flux at the center of the trailing hemisphere to the average flux is

$$\gamma_t = \frac{2/3\ U_c + S_t - S_0/\sqrt{2}}{2/3\ (U_c^2 + S_s^2)^{1/2}} \left(\frac{1 + S_s^2/U_t^2}{1 + S_s^2/U_c^2} \right) \tag{14b}$$

where S_0 is the dimensionless escape velocity from Jupiter at the orbital radius of the satellite, S_s is the dimensionless escape velocity from the satellite, and

$$U_\varrho = [(2/3\ U_c + S_0/\sqrt{2})^2 + 5/9\ U_c^2]^{1/2}$$

$$U_t = [(2/3\ U_c - S_0/\sqrt{2})^2 + 5/9\ U_c^2]^{1/2}$$

$$S_\varrho = \frac{\sqrt{5}\ S_s U_c}{3\ U_\varrho} \tag{14c}$$

$$S_t = \frac{\sqrt{5}\ S_s U_c}{3\ U_t}.$$

The derivation of Eqs. 14a and b is given in Appendix C. Impact velocities at the leading and trailing points are

$$v_{\varrho} = v_0 \, (U_{\varrho}^2 + S_s^2)^{1/2} \tag{15a}$$

at the leading point, and

$$v_t = v_0 \, (U_t^2 + S_s^2)^{1/2} \tag{15b}$$

at the trailing point. The ratio δ of the cratering rate at the leading point to the cratering rate at the trailing point is

$$\delta = \frac{\overline{\gamma_{\varrho}}}{\overline{\gamma_t}} \left(\frac{\overline{v_{\varrho}}}{\overline{v_t}} \right)^{2.2} \tag{16}$$

where $\overline{\gamma_{\varrho}}$ and $\overline{\gamma_t}$ are the mean flux ratios weighted by P_c and $\overline{v_{\varrho}}$ and $\overline{v_t}$ are the mean impact velocities, weighted by P_c, for all cases of U_c that result in impact.

As S_0^2 is the major component of U_c^2 for short period comets that encounter any of the Galilean satellites, δ is much greater than 1 and increases with decreasing orbital radius. Account must be taken, therefore, of the position on the satellite in computing the cratering rate. The cratering rate, at any position on the satellite, is given by

$$\Gamma \, (\beta) = \overline{\Gamma} \left[1 + \frac{(\delta - 1)}{(\delta + 1)} \cos \beta \right] \tag{17}$$

where $\Gamma \, (\beta)$ is the cratering rate at β, $\overline{\Gamma}$ is the mean cratering rate on the satellite, and β is the great circle distance from the mean position of the apex of the satellite's motion. The solution for δ for impact of short period comets on each Galilean satellite is listed in Table 10.1.

II. COLLISIONS WITH LONG PERIOD COMETS

A total of 545 long period comets had been discovered through 1978 (Marsden 1979); osculating elliptical orbits have been computed for 162 of these comets and osculating hyperbolic orbits for 98 comets. No hyperbolic osculating orbit has been found, however, that can be shown with confidence to have originated from an orbit that was hyperbolic at great distance from the region of the planets. For the remaining 285 comets, only parabolic orbits have been reported. For the vast majority of these remaining comets, the observations (mostly pre-20th century) are insufficient to compute precise

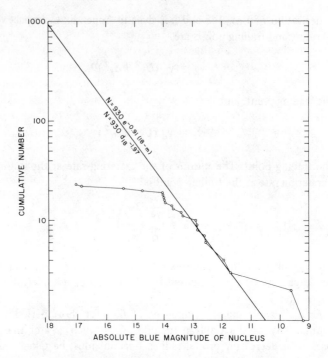

Fig. 10.9. Cumulative frequency distribution of absolute nuclear magnitudes for 23 long period comets. Absolute blue magnitudes of comet nuclei are based on apparent nuclear magnitudes published by Roemer (1965), Roemer and Lloyd (1966) and Roemer et al. (1966).

orbits. For purposes of calculating cratering rates, we will add here the 16 periodic comets with periods between 20 and 200 yr to the list of long period comets.

Magnitude Distribution of Long Period Comet Nuclei

The cumulative frequency distribution of absolute blue nuclear magnitudes for 23 long period comets observed by Roemer (1965), Roemer and Lloyd (1966), and Roemer et al. (1966) is shown in Fig. 10.9. As in the case of the periodic comets, there is a weak negative correlation between absolute nuclear magnitude and q, which is an observational selection effect. The two brightest nuclei shown in Fig. 10.9, with absolute blue magnitudes of 9.2 and 9.6, are those of comets 1957 VI Wirtanen ($q = 4.447$) and 1962 VIII Humason ($q = 2.133$). If these two bright comets are neglected, the correlation between absolute nuclear magnitude and q disappears. Therefore, we have neglected these two comets in fitting a theoretical cumulative frequency function to the observations.

The number of absolute nuclear magnitudes obtained from the Flagstaff observations of Roemer and her colleagues is too small to derive a reliable magnitude-frequency distribution solely from the data for long period comets. We assume that the cumulative frequency is an exponential function of the absolute magnitude, as found for the short period comets. For the exponential function, shown in Fig. 10.9, we used the population index of -0.9_1, adopted for the short period comets, and fitted the curve to the observed distribution below $B(1,0) = 13$ (exclusive of comets 1957 VI and 1962 VIII). This is the magnitude range over which the observations are likely to be most representative of the population of long period comets. At magnitude 14, the observed cumulative frequency drops $\sim 25\%$ below the fitted curve; at higher magnitudes the observed frequency rapidly becomes very incomplete.

The lower magnitude threshold of apparent completeness for observed long period comets, compared with the threshold of apparent completeness for short period comets, is related directly to the number and variety of opportunities for discovery. Discovery of a comet with an absolute nuclear magnitude near the threshold of completeness depends upon many factors that determine whether it will be bright enough to be detected and whether it will be favorably located in the sky so as to be found by comet searchers (Everhart 1967a). In the history of astronomical observations, the number of appearances of each long period comet discovered is essentially one, and there was only one set of observing conditions. For each relatively faint short period comet, on the other hand, there have been numerous appearances, some favorable and others unfavorable for discovery or detection of the comet. With a wide variation of observing conditions presented from one appearance to the next, the probability of discovery is proportional to the fraction of appearances in which the observing conditions are favorable; the larger the number of appearances, the greater the chances of discovery. In the time since 1840, when the discovery rate for short period comets approached the modern rate (Fig. 10.4), the average number of appearances of the short period comets is approximately the time elapsed since 1840 divided by the mean period, $\sim 140 \text{ yr}/8 \text{ yr} \simeq 18$. This factor of 18 corresponds to a difference in magnitude threshold of completeness of $(\ln 18)/0.91 = 3.2$, very close to the observed difference in threshold for short and long period comets. The relative completeness of observation for long period comets does not drop quite as rapidly below the threshold as does the relative completeness of short period comets, probably because long period comets of a given nuclear magnitude tend to be more active than the short period ones.

The fitted frequency function for the long period comets (Fig. 10.9) gives a cumulative frequency of ~ 930 comets to absolute blue magnitude 18. This is the number of comets to magnitude 18 corresponding to the comets actually observed that are estimated to have passed perihelion in the period of

Roemer's observations at Flagstaff (April, 1957 through October, 1965 = 8.58 yr). The ratio of cumulative frequency of comets at magnitude 18 to the observed comets is $\sim 930/25 = 37.2$. The number of perihelion passages per year is $25/8.58$ yr $= 2.9$ yr^{-1} for comets observed by the Roemer team; this is close to the average number of 3.3 perihelion passages per year for all observed long period comets in the decades 1950-1960 and 1960-1970 (Fig. 10.4). The number of long period comets to magnitude 18 passing perihelion each year that corresponds to the flux of observed long period comets is $\sim 37.2 \times 3.3 = 123$.

Distribution of Perihelion Distance of Long Period Comets

Just as in the case of the short period comets, we must estimate the distribution of perihelion distance q for long period comets in order to determine the flux of these objects in the vicinity of Jupiter. The perihelion distances of most discovered long period comets are less than 1 AU, and the cumulative frequency curve of q tends to level off above $q \simeq 1.1$ AU. Below $q = 1.1$ AU, the observed cumulative frequency is a fairly accurately linear function of q down to $q = 0.5$ AU. The slope of this function, normalized to the cumulative frequency at 1.0 AU, has been stable for long period comets discovered over the course of time (Fig. 10.10). As time has progressed and the discovery of comets has become more complete, the range over which the frequency of q for discovered objects is linear has progressively increased. The upper limit of the linear part of the distribution is 1.0 AU for comets discovered before 1800; it is 1.1 AU for comets discovered in the 19th century and 1.2 AU for comets discovered in the 20th century. As surveillance of the sky has improved with time, the relative completeness of search for comets has been pushed to progressively higher q. The leveling off of the cumulative frequency of q above $q \simeq 1.0$ to 1.2 AU, in other words, is purely an observational selection effect.

Öort (1950) and Weissman (1977) have shown that stellar perturbation of the Öort Cloud of long period comets should lead to a linear distribution of q in the region of the planets. Moreover, from a detailed investigation of the observational selection effects, Everhart (1967a,b) found that the observed distribution of q between about 1 AU and 4 AU was consistent with a linear cumulative distribution of q out to 4 AU for the long period comet population. From theory, we expect that the linear distribution extends out at least to 20 AU (Weissman 1977). Below $q = 0.5$ AU, the differential frequency of q is less than expected from the linear distribution (Fig. 10.11), as was found for periodic comets below $q = 1.1$ AU. However we should be able to determine the number of long period comets out to the maximum aphelion of Jupiter (5.527 AU) by extrapolation from the observed number at 1 AU.

One method of extrapolation is to use the function

$$\Sigma n = 408\,q - 81 \qquad (18)$$

where Σn is the cumulative number of comets at q, fitted to the linear part of the observed distribution of q for all long period comets (Fig. 10.11). This function is fitted over the range $0.5\ \mathrm{AU} \leqslant q \leqslant 1.1\ \mathrm{AU}$. Owing to observational selection, however, which favors the discovery of very low q comets that brighten dramatically as they approach the Sun, the slope of this fitted line is demonstrably too low to represent the long period comet population. A better curve, which allows for most of the recognized selection effects, is obtained by fitting Everhart's (1967b) solution for the intrinsic q distribution to the observations (Fig. 10.11). The Everhart distribution (solid dots) can be

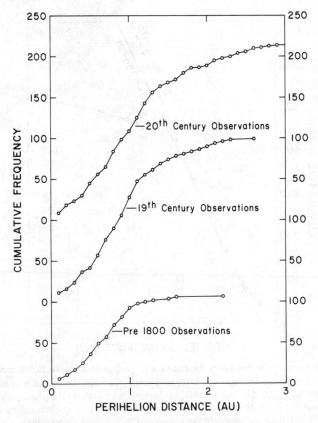

Fig. 10.10. Cumulative frequency distributions of perihelion distance for long period comets classified on the basis of the time at which they were discovered.

matched fairly closely to the observed distribution (open circles) over the range 0.2 AU $\leqslant q \leqslant$ 1.0 AU. Above $q = 1.3$ AU, the fitted Everhart distribution becomes linear and is represented by the equation

$$\Sigma n = 500\,q - 175. \tag{19}$$

The difference in Σn obtained from Eqs. (18) and (19) at $q = 5.527$ AU is 16% of the preferred number obtained from Eq. (19). Only a modest error would be introduced in extrapolating the comet flux to Jupiter's orbit by neglecting observational selection below $q = 1.1$ AU.

Using Eq. (19), we prodict the number of Jupiter crossers ($q \leqslant 5.527$ AU) to be ~ 2590, and the ratio of predicted to discovered Jupiter-crossing long period comets is $\sim 2590/540 = 4.7_9$. To obtain the number of Jupiter-crossing comets to $B(1,0) = 18$ that pass perihelion each year, we multiply the number corresponding to the discovered comets by 4.7_9, $123 \times 4.7_9 \simeq 590$. This is the predicted annual flux through perihelion of Jupiter-crossing

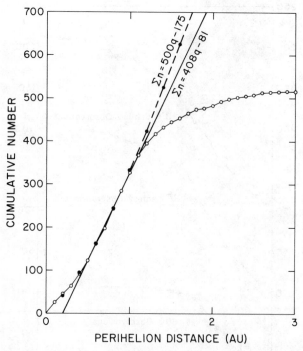

Fig. 10.11. Cumulative frequency distribution of perihelion distance for all long period comets discovered through 1978 (open circles). The distribution of q is essentially linear between 0.5 AU and 1.1 AU; in this interval it follows the line represented by the equation $\Sigma n = 408q - 81$. The intrinsic distribution of q obtained by Everhart (1967b), fitted to the observed frequencies of long period comets, is shown by solid dots. Everhart's distribution takes account of observational selection effects; this distribution is linear above $q = 1.3$ AU and follows the line represented by the equation $\Sigma n = 500q - 175$.

comets to absolute blue nuclear magnitude 18 that corresponds to the observed annual flux through perihelion at $q \leqslant 1.0$ AU of comets to blue nuclear magnitude 13.

Completeness of Discovery of Long Period Comets

To obtain the total annual flux of long period Jupiter crosssers to $B(1,0)$ = 18, we must determine the extent to which discovery of comets to $B(1,0)$ = 13 and $q \leqslant 1.0$ AU can be considered complete. Because there generally is opportunity for discovery on only one perihelion passage, a number of conditions work against complete discovery, even of bright long period comets with small perihelion distance. First, comets which pass perihelion near superior conjunction tend to be difficult to detect. This condition has long been recognized and is referred to by Everhart (1967a) as the Holetschek effect. Second, because the position of perihelion of long period comets is distributed fairly uniformly over the sky, the past discovery of comets has been biased by the nonuniform geographic latitude distribution of comet observers. Most astronomers, both professional and amateur, have been located in the northern hemisphere; thus there has been unequal surveillance of the northern and southern skies. As a consequence many comets have been missed that passed perihelion in the southern sky and remained in positions unfavorable for discovery by northern hemisphere observers (Everhart 1967a, b). This loss is reflected (Figs. 10.12 and 10.13) in the frequency distribution of the argument of perihelion of the discovered long period comets (see Appendix D).

In addition to the Holetschek effect and the loss due to poor coverage of the southern sky, there is a third important observational selection effect, which has not been fully analyzed in previous work. This is a strong seasonal effect in the discovery of long period comets (Porter 1963) that must be taken into account in estimating their flux. The seasonal effect is reflected (Fig. 10.14) in the frequency distribution of the ecliptic longitude of perihelion L for long period comets (see Appendix D).

From Everhart's (1967b) analysis of the Holetschek effect, the theoretical deficiency of discovered comets due to this effect alone is found to be 27.8%; considering only this effect, we find the relative completeness of discovery of long period comets is 72.2%. We may estimate the deficiencies or relative completenesses of discovery due to the past nonuniform latitude distribution of comet observers and to the seasonal selection effects directly from the observed distributions of ω and L. If we assume that the maximum (obtained from the 60° running mean) in each of these distributions represents complete discovery, the relative completeness for 238 long period comets discovered in the 20th century is 77.2%, from the ω distribution, and the relative completeness for all long period comets is 56%, from the L

distribution. We use the ω distribution for comets discovered in the 20th century because the ω distribution varies with time, and we wish to calculate the correction to the flux of observed comets from 1957 to 1965 that is used as the datum from which the flux in the neighborhood of Jupiter is obtained. As each of the three main selection effects is largely but not entirely independent, the total completeness of discovery [to $B(1,0) = 18$ and $q = 1.0$ AU] can be approximately estimated from the product of the relative completenesses for each effect, $0.722 \times 0.722 \times 0.56 = 31.2\%$. Our estimate of the total annual flux past perihelion of Jupiter-crossing comets to $B(1,0) = 18$, then, is $\sim 590/0.312 \simeq 1890$.

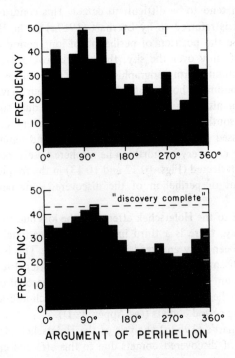

Fig. 10.12. Frequency distribution of the argument of perihelion (with respect to the ascending node on the orbit plane of Jupiter) for long period comets discovered through 1978. Upper histogram shows observed frequency per $20°$ intervals of ω; lower histogram shows $60°$ running mean frequency per $20°$ intervals of ω. The low frequency in the range $180° < \omega < 360°$ is due primarily to incomplete surveillance of the southern sky. The dashed line labeled "discovery complete" is the predicted frequency distribution of ω for discovered long period comets, if the southern sky had been covered as thoroughly as the northern sky, and if there were no selection effect due to perihelion passage near superior conjunction.

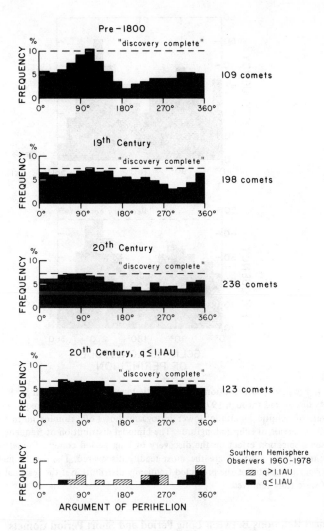

Fig. 10.13. Frequency distributions of the argument of perihelion (with respect to the ascending node on the orbit plane of Jupiter) for long period comets classified by time of discovery. The dashed lines labeled "discovery complete" are the predicted frequency distributions of ω, if the southern sky had been covered as completely as the northern sky and if there were no selection effect due to perihelion passages near conjunction with the Sun. Coverage of the southern sky and completeness of discovery increase with time. Because of improved coverage of the southern sky, comets with $q \leqslant 1.1$ AU discovered in the 20th century are only moderately deficient in the interval $180° < \omega < 360°$. Southern hemisphere observers have been particularly active in the last two decades; their contribution in this period is shown in the histogram at the bottom of the figure.

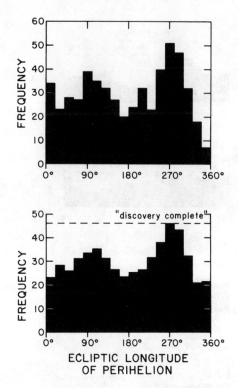

Fig. 10.14. Frequency distribution of the ecliptic longitude of perihelion for long period
comets discovered through 1978. Upper histogram shows observed frequency per 20°
intervals of ecliptic longitude; lower histogram shows 60° running mean frequency
per 20° intervals of ecliptic longitude. The bimodal distribution of frequency reflects
a seasonal selection effect on the discovery of long period comets. Comets that pass
perihelion at the solstices are the most readily discovered. The dashed line labeled
"discovery complete" is the predicted frequency distribution if the seasonal selection
effect were fully removed.

Dynamical Relations Between Long Period and Short Period Comets

As a check on the flux of long period comets at Jupiter's orbit, it is of
interest to compare our estimate of ~ 1890 annual perihelion passages inside
of 5.527 AU with the theoretical flux required to maintain the population of
short period comets in equilibrium. Delsemme (1973) estimated that a total
of 2400 ± 1200 comets passed perihelion each year in the range 4 AU
$< q < 6$ AU, about 2000 of which are periodic comets ($a < 30$ AU) and
~ 400 are long period comets ($a > 30$ AU). He concluded that this flux was
sufficient to maintain the observed number of active short period comets
through capture of long period comets into short period orbits by Jupiter
perturbations (Everhart 1972). At first glance this flux appears similar to that

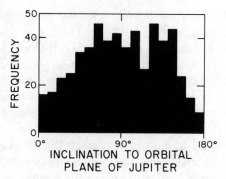

Fig. 10.15 Frequency distribution of inclination to the orbit plane of Jupiter for long period comets discovered through 1978.

determined by us, but Delsemme's estimate refers to relatively bright comets (of unspecified magnitude) and is, in fact, much higher than ours. When the perturbations of comet orbits by Saturn, Uranus, and Neptune are appropriately accounted for, however, the flux of long period comets required to maintain the short period comet population probably is less than that posited by Delsemme and may be consistent with our estimate.

From a Monte Carlo study of the perturbations of long period comet orbits by Jupiter, Everhart (1972) found that 90% of the captures to short periods occurred for orbits with initial q in the range $4 \, \mathrm{AU} < q < 6 \, \mathrm{AU}$ and with initial $i < 9°$. In this range, the average probability of capture to short period orbits, by Jupiter acting alone, is $2.5_4 \times 10^{-4}$ per perihelion passage. Our predicted number of perihelion passages per year in the range $4 \, \mathrm{AU} < q < 6 \, \mathrm{AU}$ [for comets to $B(1,0) = 18$], can be obtained from Eq. (19), with the use of our estimate of 1890 passages at $q = 5.527 \, \mathrm{AU}$. The incremental frequency of q per AU is 500, and the cumulative frequency at 5.527 AU is $\simeq 2590$; scaling the cumulative frequency to the ~ 1890 annual perihelion passages at 5.527 AU, we get $2 \times 500 \times 1890/2590 \simeq 730$ annual perihelion passages in the range $4 \, \mathrm{AU} < q < 6 \, \mathrm{AU}$. The observed distribution of i for discovered long period comets is illustrated in Fig. 10.15; the fraction of total frequency to $i = 9°$ is 2.8×10^{-2} (see Appendix E for dicussion of the observed distribution of i). Thus our predicted number of perihelion passages of long period comets in Everhart's capture region is $\sim 730 \times 2.8 \times 10^{-2} = 21 \, \mathrm{yr}^{-1}$ to $B(1,0) = 18$. The supply of short period comets to $B(1,0) = 18$, from the perturbations by Jupiter alone, is the flux in the capture region times the capture probability divided by the fraction of captures in the capture region, $21 \, \mathrm{yr}^{-1} \times 2.5_4 \times 10^{-4}/0.9 = 6 \times 10^{-3} \, \mathrm{yr}^{-1}$.

Kresák and Pittich (1978), who adopt a uniform space density and a random distribution of orbital inclination of long period comets, estimate the flux of these comets in Everhart's capture region to be $0.09 \, \mathrm{yr}^{-1}$. As in Delsemme's study, nuclear magnitudes are not specified for this flux, but we can estimate that it roughly corresponds to the flux at blue magnitude 14; the flux at $B(1,0) = 18$, then, would be 0.09 exp 0.91 × 4 yr^{-1} ≃ 3 yr^{-1}, almost an order of magnitude below our calculated flux. It should be noted, moreover, that Kresák and Pittich interpret the small number of faint comets observed to be a reflection of the intrinsic magnitude distribution rather than a consequence of observational selection; hence, their estimate of the flux in the capture region is more than 2 orders of magnitude below ours.

The groundwork needed from which an approximate estimate can be made of the total supply of short period comets from the combined perturbations of Jupiter, Saturn, Uranus and Neptune has also been provided by Everhart (1977). A comet that starts on a nearly parabolic orbit with 30 AU $< q <$ 34 AU can have both q and a decreased by successive perturbation by Neptune until q reaches the range of 15 to 21 AU, where Uranus will dominate the perturbations. Uranus, in turn, can hand the comet down to the range of control of Saturn, and Saturn can pass the comet on to Jupiter. Similarly, a comet may start with a nearly parabolic orbit with q in the range of control of Uranus or of Saturn, and the comet may be passed to the control of Jupiter, where a small fraction of the orbits are transformed to short period.

The efficiency of transfer of comets into short period orbits of $q < 2.5$ AU, starting with nearly parabolic orbits in the capture region of each major planet, was determined by Everhart by a Monte Carlo random walk method (Table 10.3). If it is assumed that the q distribution of long period comets is uniform to 34 AU, as suggested by the study of Weissman (1977), the relative

TABLE 10.3

Relative Yield of Short Period Comets from Planetary Perturbations of Nearly Parabolic Comets

Perturbing Planet	Capture Region Studied[a] (AU)	Relative Number of Perihelion Passages in Capture Region[b]	Efficiency of Capture to Short Period Orbits[a]	Relative Yield of Short Period Comets
Neptune	30 − 34	4	1.6 x 10^{-4}	0.6 x 10^{-3}
Uranus	15 − 21	6	2.2 to 6 x 10^{-4}	~ 2 x 10^{-3}
Saturn	7.6 − 10.7	3.1	1.4 x 10^{-3}	4.4 x 10^{-3}
Jupiter	5.2 − 5.9	0.7	7.8 x 10^{-3}	5.4 x 10^{-3}
TOTAL				12.4 x 10^{-3}

[a]From Everhart (1977).
[b]Based on a uniform distribution of q to 34 AU.

yield of short period comets from each capture region can be calculated as shown in Table 10.3. The numerical experiments by Everhart utilized an inclination of 6°, taken as the median for the region of efficient capture; the extent of the capture region in q and i is not fully defined. Here we will assume that the proportions of the yields from the capture regions of the planets given in Table 10.3 are roughly applicable to the regions of 90% capture, as defined in Everhart's (1972) more detailed study for Jupiter. From Table 10.3 it may be seen that $\sim 7/12.4 = 56\%$ of the total yield is from the capture regions of Saturn, Uranus, and Neptune, and that $\sim 44\%$ is from direct capture of long period comets by Jupiter. Using this proportion, we find that the total supply of short period comets is $\sim 6 \times 10^{-3}/0.44 = 1.4 \times 10^{-2}$ yr^{-1}.

Everhart's studies of the capture of comets to short period refer to comets on orbits that would permit their detection, nominally at $q < 2.5$ AU. If the observable short period comets are in steady state, then their average lifetime of activity is given by their number divided by the rate of supply (provided that the lifetime of activity is much less than the dynamical lifetime). The estimated number of short period comets to $B(1,0) = 18$ with $q < 2.5$ AU is 4.6_7 $(64 \times 2.5 - 58) \simeq 480$, and the estimated active lifetime is $480/1.4 \times 10^{-2}$ yr$^{-1} \simeq 3.4 \times 10^4$ yr. This lifetime is an order of magnitude greater than the lifetimes of short period comets estimated by Delsemme (1973), who adopts 200 ± 100 revolutions, or equivalently, 1400 ± 700 yr as the estimated mean lifetime for comets with periods less than 13 yr. Using the mean period obtained by us for Jupiter-family comets to $q = 2.5$ AU, our estimated lifetime is equivalent to ~ 4000 revolutions. This is somewhat greater than the minimum number of revolutions for P/Encke in its present short period orbit, found by Whipple and Hamid (1951) from meteor-stream evidence, and less than the maximum active lifetime of $\sim 10,000$ revolutions estimated for comets by Whipple (1962).

Two considerations suggest that the rather long mean lifetime derived here may be plausible. First, we include comets with q out to 2.5 AU, where losses by ablation are reduced, compared to losses for the typical discovered short period comet at a median q of ~ 1.5 AU. Second, our estimates of the diameters of comet nuclei suggest that their masses may be somewhat larger than is commonly supposed. If this is true, they may be expected to have somewhat longer mean lifetimes than have been calculated from previous models. We shall return to these estimates of size later. Finally, we note that the dimming of long period comets, as used by Delsemme, does not provide a satisfactory basis for estimating lifetimes of short period comets. As we have shown, the magnitude threshold for discovery of long period comets is 2 to 3 magnitudes brighter than the threshold for short period comets. Therefore the short period comets will have longer lifetimes than suggested by the number of revolutions required for dimming of long period comets to the magnitude limit for discovery.

Cratering Rates from Impact of Long Period Comets

The cratering rates from impact of long period comets on the Galilean satellites are obtained by methods closely analogous to those used for short period comets. For our initial calculations long period comet nuclei are assumed to have albedos and densities comparable to those of short period comets. The albedos are for the inactive nuclei observed *after* perihelion passage. Velocities and probabilities of impact are obtained from Öpik's equations, modified as indicated in Appendix A. For long period comets, Öpik's solutions for the encounter velocity with the sphere of influence of Jupiter U, the radial component of this velocity U_x, and the crossing fraction f, can be rewritten as functions of q:

$$U^2 = 3 - \frac{a_0(1-e)}{q} - 2\sqrt{\frac{q}{a_0}}\,(1+e)\,\cos i + \frac{4}{9}\,e_0{}^2 \qquad (20a)$$

$$U_x{}^2 = 2 - \frac{a_0(1-e)}{q} - \frac{q}{a_0}(1+e) + \frac{4}{9}\,e_0{}^2 \qquad (20b)$$

$$f = \frac{1}{\pi}\,\cos^{-1} \frac{\dfrac{q}{a_0} - (1 - e_0{}^2)}{\dfrac{e_0 q}{a_0}} \;,\quad \frac{q}{a_0} > 1 - e_0\,. \qquad (20c)$$

Orbital elements for the long period comets are taken from Marsden (1979), and inclinations to the ecliptic are then transformed to inclinations to the orbital plane of Jupiter. The distribution of transformed inclinations is illustrated in Fig. 10.15.

As was done for short period comets, Γ_i and other relevant impact parameters were calculated for the orbits of all observed long period comets and for a set of synthetic orbits with $q > 1$ AU. The relative proportions of observed and synthetic orbits are given in Table 10.4. Synthetic orbits for each interval of q were derived by selecting all observed orbits with q less than the lower bound of the interval and assigning a new value of q chosen at random within the interval. The elements e and transformed i for each selected orbit were preserved. The impact parameters found for the sum of observed and synthetic orbits are given in Table 10.5.

III. COLLISIONS WITH PLANET-CROSSING ASTEROIDS

The median time required for Jupiter to transform a nearly parabolic orbit ot a short period orbit ($q < 2.5$ AU) is 2.0×10^5 yr, when the initial perihelion is in the range 5.2 AU to 5.9 AU and the inclination is $6°$ (Everhart

TABLE 10.4

Relative Proportions of Observed and Synthetic Orbits

Observed orbits	394
Synthetic orbits	
1 AU $< q \leqslant$ 2 AU	252
2 AU $< q \leqslant$ 3 AU	339
3 AU $< q \leqslant$ 4 AU	357
4 AU $< q \leqslant$ 5 AU	356
5 AU $< q \leqslant$ 5.527 AU	192
Total	1890

1977). This process of capture, which is basically a series of random or nearly random perturbations, is reversible; the mean time for transformation of the resultant short period orbit back to a parabola or to a slightly hyperbolic orbit, at which point the comet escapes from the solar system, must also be close to 2.0×10^5 yr. More numerical experiments are needed to define accurately the mean dynamical lifetime for observed short period comets, but, meanwhile, we will adopt Everhart's time to capture as the best available estimate. As the mean time to ejection is an order of magnitude greater than our estimate of the average lifetime for cometary activity, we must assess the fate of a dying comet.

Does the average short period comet simply ablate to nothing, or is there something left? If the nucleus of every comet had a nonvolatile rocky core (cf. Sekanina 1971), for example, then there would be $\sim 2.0 \times 10^5/3.4 \times 10^4 \simeq 6$ such inactive cores left on Jupiter-crossing orbits for every active comet. Alternatively, there may be coarse rocky debris distributed through the icy nucleus of a typical comet, which is left as a residual aggregate of rubble at the end of cometary activity. It is also possible that a comet may become armored with rocky debris and cease its activity. In any of these cases, we should look for bodies of asterodial appearance that represent extinct comets.

The Population of Jupiter-crossing Asteroids

Precisely one asteroid, 944 Hidalgo, has been discovered on a Jupiter-crossing orbit. This body currently comes to perihelion at 2.01 AU and reaches aphelion at 9.71 AU; its period is 14.2 yr. In these respects its orbit is rather similar to the orbits of a number of Jupiter-family comets, such as P/Wild 1 or P/Vaisala 1. However the inclination of Hidalgo, 42°, is higher than that typical of short period comets and is exceeded only by the inclination of P/Tuttle. Because of its relatively long period and high inclination, the probability of encounter with Jupiter for Hidalgo is considerably lower than average for short period comets, a circumstance that probably has prolonged its survival in the solar system.

In addition to Hidalgo, it has been suggested by Öpik (1963), Wetherill (1976), and Shoemaker and Helin (1977), among others, that many asteroids that cross the orbits of the terrestrial planets are extinct comets. Until their orbits are changed by encounters with the terrestrial planets, these latter bodies presently are safe from encounter with Jupiter. Their dynamical lifetimes are 2 to 5 orders of magnitude greater than the mean lifetimes of short

TABLE 10.5

Long Period Comet Collision Parameters and Cratering Rates to 10 km Diameter on the Galilean Satellites[a]

	Io	Europa	Ganymede	Callisto
$\bar{U}^2$	2.51	2.71	2.86	3.01
δ	14.2	10.2	6.4_3	4.43
$\bar{F}$	3.96	2.73	2.0_6	1.51
$\bar{v}_i$, km s^{-1}	32.16	28.63	26.5	24.91
$\bar{D}_{18}$, km	60.4	72.8	68.4	67.3
$\bar{P}_c \times 10^{-11}$ yr^{-1}	1.9_9	1.1_2	2.5_7	1.7_6
$\bar{\Gamma}$, $\times 10^{-17}$ km^{-2} yr^{-1}	1.4	1.7	1.3	1.0_5
$\bar{\Gamma}^*$, $\times 10^{-17}$ km^{-2} yr^{-1}	2.5	2.9	2.0	1.6_4

[a]All values of collision parameters shown are weighted means for observed plus synthetic orbits of long period comets. See below for definitions of table entries:

$\bar{U}^2$ = mean square of asymptotic encounter velocity at sphere of influence of Jupiter, weighted by P_c. (U is in units of Jupiter's mean orbital velocity.)

δ = ratio of cratering rate at apex of satellite's motion to cratering rate at antapex.

$\bar{F}$ = mean enhancement of the flux of comets by the gravitational field of Jupiter at orbital radius of satellite, weighted by P_c.

$\bar{v}_i$ = mean impact velocity, weighted by P_c.

$\bar{D}_{18}$ = mean diameter of crater produced by impact of a blue magnitude 18 comet nucleus, weighted by P_c.

$\bar{P}_c$ = mean collision probability per perihelion passage.

$\bar{\Gamma}$ = estimated mean rate of production of craters 10 km in diameter and larger per unit body of population to absolute blue magnitude 18, based on observed long period comet orbits only.

$\bar{\Gamma}^*$ = improved estimated mean rate of production of craters 10 km in diameter and larger per unit body of population to absolute blue magnitude 18, based on observed and synthetic long period comet orbits.

period comets. Therefore extinct comets can accumulate, to some extent, in the region of the terrestrial planets, if there are mechanisms by which they can be injected into these "safe" orbits. P/Encke is an active comet that has arrived in such a safe orbit, apparently as a consequence of the nongravitational forces arising from the ablation of volatiles from its surface (Sekanina 1971). Some extinct comets also may be captured in safe orbits by encounter with Earth or Venus. Known Earth-crossing asteroids all have absolute blue magnitudes fainter than 14.8; they are well within the magnitude and probable size range of short period comet nuclei. The largest of them have *UBV* color ratios in the color domain of *C*-type asteroids, which we suggest may be appropriate for comet nuclei.

Hidalgo has an absolute blue magnitude of 11.58 (Bowell et al. 1979), which places it near the brighter end of the magnitude range for short period comet nuclei. Only P/Schwassmann-Wachmann 1 is known to be brighter. Thus it is reasonable to suppose not only that Hidalgo is an extinct comet, but also that it is the brightest surviving object of this class. Hidalgo never becomes brighter than apparent *B* magnitude 15.5. This is fainter than the range of apparent magnitudes in which the discovery of main belt asteroids is fully complete. Moreover, the mean opposition magnitude of Hidalgo is much fainter than 15.5. Its discovery by W. Baade in 1920 is a testimony to the alertness of the discoverer (who thought it might be a comet). If Hidalgo is the brightest extinct comet and if the magnitude distribution and distribution of orbital elements of Jupiter-crossing extinct comets resemble those of active short period comets, then we would not expect that any others should necessarily have been found.

Some Jupiter-crossing extinct comets should come much closer to Earth than Hidalgo, of course, and we must take these into account. Assuming that Hidalgo is the brightest object and that the magnitude distribution is given by Eq. (5), we have a total population of $\sim$ 320 to $B(1,0) = 18$, and a population of $\sim$ 700 to $V(1,0) = 18$. From the distribution of the perihelion distance (Fig. 10.2), we find that $\sim$ 60 objects to $V(1,0) = 18$ have $q \leqslant 1.3$ AU. The discovery of Earth-approaching asteroids ($q \leqslant 1.3$ AU), presumed to have about the same magnitude distribution as the Jupiter-crossing extinct comets, is only $\sim$ 2% complete to $V(1,0) = 18$ (Shoemaker et al. 1979). Thus we would predict that, on average, only $\sim$ 2% $\times$ 60 $\simeq$ 1 Jupiter-crossing extinct comet that approaches Earth should have been discovered to date; there is only an $\sim$ 50% chance of this prediction being fulfilled. From the lack of discovery of Jupiter-crossing extinct comets that approach Earth, we conclude that the population to $B(1,0) = 18$ probably is not much greater than 320.

Here we will adopt 320 as the most probable population of the Jupiter-crossing asteroids, even though the estimate is supported only by a statistic of one. If this estimate is correct, a vigorous campaign of search for new Earth-approaching asteroids should begin to turn up some more Jupiter crossers. We will further assume that the orbits of Jupiter-crossing asteroids resemble the orbits of short period comets and that the impact parameters given in Table 10.1 are appropriate for Jupiter-crossing asteroids.

Supply of Jupiter Crossers by Deflection of
Planet-crossing Asteroids on Small Orbits

In addition to asteroids resulting from the demise of active comets, Jupiter-crossing asteroids are produced by deflection of other planet-crossing asteroids during encounters with the terrestrial planets. Many of these objects probably are extinct comets to begin with, but some probably are derived from the asteroid belt. About half of the bodies crossing the orbits of the terrestrial planets are ultimately transferred to Jupiter-crossing orbits. The equilibrium number of Jupiter crossers N_j derived from this source is given approximately by

$$N_j \simeq 1/2 \frac{\ell_j}{\ell_s} N_s \qquad (21)$$

where ℓ_j is the mean dynamical lifetime of Jupiter-crossing asteroids, ℓ_s is the mean dynamical lifetime of the planet-crossing asteroids in the source region, and N_s is the equilibrium population in the source region. Adopting $\ell_j \simeq 2 \times 10^5$ yr, $\ell_s \simeq 3 \times 10^7$ for Earth-crossing asteroids (Wetherill 1976), and $N_s \simeq 800$ to $B(1,0) = 18$ for Earth crossers (Shoemaker et al. 1979), we get $N_j \simeq 2$ to $B(1,0) = 18$. A comparably small yield of Jupiter-crossing asteroids is found for deflection of Mars crossers. The total yield from the asteroids crossing the orbits of the terrestrial planets is on the order of 1% of the estimated number of Jupiter-crossing asteroids; this small number does not need to be treated separately from the objects derived directly from the demise of short period comets.

IV. PRESENT AND PAST CRATERING RATES ON
THE GALILEAN SATELLITES

Limits from the Cratering History of the Earth

To obtain a best estimate for the cratering rates on the Galilean satellites, we will first test our provisional estimates of the populations of active and extinct comets against the cratering record of the Earth. The production of craters on Earth by impact of active and extinct comets can be calculated readily from Öpik's equations, using the orbits from Marsden's (1979) catalogue and our provisional estimates of the physical properties and populations of the various classes of comets (Table 10.6). The mass, radius, and mean orbital elements of the Earth are introduced at the appropriate places in these equations. At the orbit of the Earth, there is no need to consider synthetic orbits with $q > 1$ AU. The short period comet orbits are adopted as representative of the orbits of the extinct comets. From Table 10.6, it may be seen that long period comets are responsible for $\sim 98\%$ of crater production by comet impact on Earth. This is due primarily to the fact that the orbits of only a small fraction of short period comets overlap the orbit of the Earth.

The total cratering rate by comet impact on Earth given in Table 10.6 is about four times higher than the average Phanerozoic cratering rate on Earth found by Shoemaker (1977a) and nearly six times higher than the rate estimated by Grieve and Dence (1979). We take this as evidence that our provisional estimates of the sizes and masses of the active comet nuclei probably are too high, or, equivalently, that our estimates of the flux of active comets to a given nuclear size are high. These estimates may be too high primarily because of the contribution of the unresolved coma to the nuclear magnitudes observed by Roemer and her colleagues. It is well known that observed magnitudes of comets depend on the aperture and focal length of the telescope used (see Bobrovnikoff 1941); the greater the focal length, the fainter is the observed magnitude. This relationship is well understood to arise from the degree to which the nuclear region is resolved from the extended coma.

A detailed evaluation of the contribution of the coma to Roemer's observations of nuclear magnitude can be made with the aid of her later observations at telescopes of various focal length, but this evaluation is beyond the

TABLE 10.6

Cratering Rates on Earth Calculated from the Provisional Estimated Populations of Jupiter-Crossing Active and Extinct Comets to 2.53 km Diameter of Nucleus[a]

	Provisional Estimate of Jupiter-crossing Population to 2.53 km Diameter of Nucleus	Production of Craters $\geqslant 10$ km Diameter on Earth[a] $(10^{-14} \text{ km}^{-2} \text{ yr}^{-1})$
Long period comets	1890	7.92
Short period comets	1380	0.16
Extinct comets[b]	320	0.04
Total	3590	8.12

[a]These estimated cratering rates are used only to correct the provisional estimated populations of the active comets. They do not indicate the actual cratering rates on Earth.

[b]Jupiter-crossing asteroids are here inferred to be extinct comets.

scope of this chapter. Here we will simply apply a correction to the estimated flux of active comets on the basis of the cratering record on the Earth. This procedure has the advantage that possible errors in our estimates of the mean geometric albedo and mean density of comet nuclei, and also possible errors in the crater scaling relationships used, tend to be corrected for. In other words, we will scale the cratering record of the Galilean satellites to the cratering record of the Earth. To do so, however, we must first allow for the contribution of Earth-crossing asteroids to the impact cratering on Earth.

The present production of craters $\geqslant 10$ km diameter by impact of Earth-crossing asteroids on Earth was estimated by Shoemaker et al. (1979) as 2.3 $\pm 1.1 \times 10^{-14}$ km^{-2} yr^{-1}. The central value of this estimate is nearly identical to the estimated Phanerozoic production of craters of $2.2 \pm 1.1 \times 10^{-14}$ km^{-2} yr^{-1} (Shoemaker 1977a). As a reasonable approximate upper bound on the cratering rate by impact of active comets, we might take either the one standard deviation uncertainty for the asteroid cratering rate or the uncertainty in the cratering record. Here we adopt 1.0×10^{-14} km^{-2} as an appropriate upper bound to the cratering rate to 10 km by comet impact. This implies that the provisional estimate of the long period comet flux should be reduced by a factor of $\sim 8.1 \times 10^{-14}$ km^{-2} yr^{-1}/1.0×10^{-14} km^{-2} yr^{-1} = 8.1.

It is possible that long period comet nuclei have a much higher average albedo than we have estimated for the short period comets. In this case, the correction would apply chiefly to long period comets. We cannot reduce the estimated sizes of long period comet nuclei very much, however, without running into a serious problem in explaining the population of short period comets. If we assume that the correction factor of 8.1 is required solely because of the contribution of the coma to the observed nuclear magnitudes of active comets, on the other hand, it should then apply both to short and to long period comets; but it does not apply to the estimated population of Jupiter-crossing asteroids. Thus we would reduce the estimated population of short period comets and the annual flux of long period comet nuclei $\geqslant 2.35$ km diameter by a factor of 8.1, to ~ 170 and ~ 230, respectively, and retain the estimated population of 320 for extinct comets (Jupiter-crossing asteroids). To the same size limit, the estimated number of extinct comets now becomes about twice the number of active short period comets.

Given our estimate of the mean active lifetime of short period comets, and a ratio of this lifetime to their dynamical lifetime $\sim 1/6$, it can be seen that the revised estimate of their population could not be reduced much further without running into a problem. If the population of active comets to 2.53 km diameter were more than a factor of ~ 3 less than the estimated 170, there would not be enough active comets to maintain the estimated population of the extinct comets. Indeed, the limit must be reached well before a factor of 3.

Suppose, for example, that $\sim 50\%$ of the mass of a new comet is composed of ice, following Delsemme (1977), with a density of ~ 0.93 g cm^{-3}.

Suppose, further, that half of this ice already has been lost by ablation from the average observed short period comet. If the rocky component of the comet has a density of $\sim$ 2.3 g cm^{-3}, as we suggest, and if it is largely preserved during ablation (in a core, for instance), then about 55% of the volume of the average observed short period comet is ice. About 45% of the volume would be left at extinction, if all the ice were ablated, and the extinct comet would have $\sim$ 75% of the diameter of the active comet. Upon their demise, active comets would then yield only $(0.75)^{1.97} = 55\%$ of the number of extinct comets to the same size limit. Under this example, the population of short period comet nuclei to 2.53 km diameter could not be $\lesssim$ 100 and still supply a steady population of 320 extinct comets. If much rocky material is lost during ablation as cometary dust, moreover, the minimum population of active comets must be appreciably higher than 100. Since abundant dust is, in fact, observed to be lost to the tails of active comets, we conclude that the estimated population of $\sim$ 170 short period comets and a corresponding annual flux of $\sim$ 230 long period comets, to 2.53 km diameter of the nucleus, probably are close to reasonable lower bounds. The lower bounds set by the requirement to supply the estimated number of extinct comets, in other words, are roughly comparable to the upper bounds set by the cratering record on Earth.

The average magnitude correction for the coma that should be applied to the observed nuclear magnitudes of Roemer, in order to derive diameters of the nuclei, apparently is $\sim$ ln 8.1/0.91 = 2.3. This is consistent with the correction estimated by Sekanina (1976). If our initial estimate of the mean albedo of the nuclei is approximately correct, light reflected from the solid surfaces of the nuclei contributed, on average, only $\sim$ 12% of the observed radiation from which Roemer estimated nuclear magnitudes. The rest was from unresolved comae. This leads to a reduction of the estimated sizes of the nuclei by a factor of $\sim (8.1)^{1/1.97} \simeq 2.9$. Thus our revised estimate of the diameter of the nucleus at absolute blue magnitude 18 (derived from Roemer's observations of nuclear magnitudes on the 1.0-m Ritchey-Chrétien telescope at Flagstaff) is 2.53 km/2.9 = 0.9 km. Another way that this result can be viewed is that the average effective blue geometric albedo of the active comet nuclei (including unresolved coma) is $\sim$ 0.25, rather than the value of 0.030 that was initially adopted.

Estimated Present Cratering Rates on the Galilean Satellites

With revised estimates of the populations and flux of comet nuclei to $D = 2.53$ km, the impact parameters of Tables 10.1 and 10.5 can now be utilized to calculate the present cratering rates on the Galilean satellites, as shown in Table 10.7. The total cratering rates for craters $\geqslant$ 10 km diameter range from 1.2×10^{-14} km^{-2} yr^{-1} on Callisto, to 5.2×10^{-14} km^{-2} yr^{-1} on Io. Active comets account for the major fraction of present crater production on Callisto, but, owing to the gravitational focusing of the flux, extinct comets probably dominate the present production of craters on Europa and Io. The impact of long period comets produces about one-third of the craters formed on Callisto but only about one-seventh to one-eighth of the craters on Europa and Io. The present cratering rate found for Ganymede is similar to the rate estimated for the Earth from the Phanerozoic cratering record; the rate on Callisto is about half the rate on Earth, and the rate on Io about twice the present cratering rate on Earth.

From the proportions of the estimated population of extinct comets and the revised population of short period comets and flux of long period comets, we can determine the ratio δ of the cratering rate at the apex to the cratering rate at the antapex of motion for each of the Galilean satellites. The values of $\bar{\gamma}_\varrho$, $\bar{\gamma}_t$, $\bar{v}_\varrho$, and v_t, weighted according to the satellite-wide mean rates of

Present Rates of Production of Craters $\geqslant 10$ km Diameter on the Galilean Satellites

	[a]$N*$	[b]$\bar{\Gamma}*$ $(10^{-17}$ km^{-2} yr$^{-1})$	[c]$C_\rho = N* \times \bar{\Gamma}*$ $(10^{-14}$ km^{-2} yr$^{-1})$
Io			
Extinct comets	320	9.4	3.0
Short period comets	170	9.4	1.6
Long period comets	233	2.5	0.6
Total	723	21.3	5.2
Europa			
Extinct comets	320	7.90	2.5
Short period comets	170	7.90	1.3
Long period comets	233	2.90	0.7
Total	723	18.70	4.5
Ganymede			
Extinct comets	320	3.80	1.2
Short period comets	170	3.80	0.6
Long period comets	233	2.00	0.5
Total	723	9.60	2.3
Callisto			
Extinct comets	320	1.75	0.55
Short period comets	170	1.75	0.30
Long period comets	233	1.64	0.38
Total	723	5.14	1.2_3

[a]$N*$ is the best estimate of the population of short-period and extinct comet nuclei to $D \geqslant 2.53$ km, and the annual flux of long-period comet nuclei to the same size limit.

[b]$\bar{\Gamma}*$ is the cratering rate per unit body of the population (from Table 10.1 for extinct and short period comets and from Table 10.5 for long period comets).

[c]C_p is the best estimate of present rate of production of craters $\geqslant 10$ km diameter.

TABLE 10.8

**Ratios of the Cratering Rate at the Apex to the Cratering Rate at the
Antapex of Orbital Motion for the Galilean Satellites**

	[a]$\bar{\gamma}_\varrho$	[b]$\bar{\gamma}_t$	[c]$\bar{v}_\varrho$ (km s^{-1})	[d]$\bar{v}_t$ (km s^{-1})	[e]δ
Io	2.06	0.136	39.7	19.5	38.2
Europa	2.03	0.162	32.4	16.3	30.6
Ganymede	1.99	0.304	27.1	14.3	14.9
Callisto	1.90	0.409	23.0	13.1	9.6

[a]$\bar{\gamma}_\varrho$ is the weighted mean ratio of the flux of impacting bodies at the apex of orbital
motion to the satellite-wide mean flux.

[b]$\bar{\gamma}_t$ is the weighted mean ratio of the flux of impacting bodies at the antapex of orbital
motion to the satellite-wide mean flux.

[c]$\bar{v}_\varrho$ is the weighted mean impact velocity at the apex of orbital motion.

[d]$\bar{v}_t$ is the weighted mean impact velocity at the antapex of orbital motion.

[e]δ is the ratio of cratering rate at the apex to the cratering rate at the antapex, obtained
from $\delta = (\bar{\gamma}_\varrho/\bar{\gamma}_t)(\bar{v}_\varrho/\bar{v}_t)^{2.2/1.7}$.

cratering by each class of cometary object, are given in Table 10.8. Again, the
extinct comets are presumed to have orbits like those of the active short
period comets. As shown in Table 10.8, δ ranges from 9.6 on Callisto to 38.2
on Io. The derived values of δ are independent of crater size, so long as the
distribution of orbits of the impacting bodies does not vary with the mass of
the bodies. For small particles, whose orbits are influenced by the Poynting-
Robertson effect or radiation pressure, δ must be determined for the orbits
appropriate to these particles.

Variation of Cratering Rate with Time

Cratering time scales for the Galilean satellites can be constructed, if
sufficient inferences can be drawn about the history of the cratering rates.
Here we are forced to depend on the cratering history of the Earth-Moon
system, which is the only part of the solar system where the cratering history
is even approximately calibrated by various methods of absolute age determination. In the recent geologic past (Phanerozoic) the cratering rates on the
Earth and on the Moon have been closely coupled to the cratering rates on
the Galilean satellites. From the arguments just presented, we concluded that
about one-third of the recent craters on Earth ≥ 10 km diameter were produced by long period comet nuclei, which also cross the orbit of Jupiter. The

long period comets are linked dynamically to the short period comets and to the extinct comets, which account for the majority of craters on the Galilean satellites. Any fluctuation in the past flux of long period comets must have been reflected in corresponding fluctuations of the short period and extinct comet flux.

The relationship of Earth-crossing asteroids to the long period comets is somewhat more remote and is still the subject of some debate. From the evidence summarized by Shoemaker et al. (1979), it appears likely that the majority of Earth-crossing asteroids are extinct comet nuclei, derived ultimately from long period comets. A significant fraction of these objects must have been derived from the main asteroid belt, however, and the exact contribution from the two sources is not known. Even if half of the Earth-crossing asteroids were derived from the asteroid belt, about two-thirds of the recent impact craters on Earth were produced by the combined impact of active and extinct comets. Hence, fluctuations of the cratering rate in the Earth-Moon system in the recent past probably were related chiefly to changes in the flux of long period comets and should have been strongly reflected in fluctuations in the cratering rates on the Galilean satellites.

As we trace the history of cratering back to the period of late heavy bombardment of the Moon, the dynamical linkage between the bodies impacting on the Moon at that time and the bodies colliding with the Galilean satellites becomes less clear. It has been suggested that the objects striking the Moon during late heaby bombardment were other satellites of the Earth (Gilbert 1893), Earth planetesimals in long-lifetime heliocentric orbits (Wetherill 1977), fragments of a large tidally disrupted planetesimal derived from the region of Neptune or Uranus (Wetherill 1975), or asteroidal bodies or collision fragments of asteroids derived from the main asteroid belt (Chapman 1976; Shoemaker 1977b). The distribution of craters on Mars and on Mercury suggests that the late heavy bombardment of each of these planets was contemporaneous with the late heavy bombardment of the Moon (Soderblom 1977); the projectiles striking each of these bodies during late heavy bombardment probably were related and probably were following heliocentric orbits (Chapter 9 by Woronow et al.).

It can be shown from Monte Carlo numerical experiments (Wetherill and Williams 1968) that, even if the orbits of the projectiles have only moderate initial eccentricity, objects encountering any one of the terrestrial planets tend to be diffused into orbits that will cross the other terrestrial planets. Hence it is likely that the late heavy bombardment history of the Moon and Earth is, indeed, related to heavy bombardment of each of the other terrestrial planets. About half of the present Earth crossers, in fact, will be deflected to Jupiter-crossing orbits. This does not necessarily mean, however, that late heavy bombardment of the Moon was accompanied by a corresponding late heavy bombardment of the Galilean satellites. Because the lifetimes of most Jupiter crossers are very short compared to the lifetimes of terrestrial planet crossers, the rates of cratering on the Galilean satellites due to impact of bodies originating from the inner solar system (e.g., objects from the main asteroid belt or Earth planetesimals) is two to three orders of magnitude less than rates of cratering on the Moon or terrestrial planets.

If the late heavy bombardment of the Moon was produced mainly by planetesimals arriving from the outer solar system, on the other hand, the dynamical circumstances were very similar to those found for the comets at the present time. Late arriving Neptune and Uranus planetesimals (presumed to be directly related to and, in part, physically identical with the long period comets) probably became active, just like present day comets, as they reached small perihelion distances. It is likely that many were transformed into Earth-crossing asteroids through capture by Jupiter to short period orbits and subsequent escape to small safe orbits by the action of nongravitational forces. Therefore late heavy bombardment of the Galilean satellites by such planetesimals would have been accompanied by simultaneous late heavy bombardment of the Moon and the terrestrial planets.

The crater record of Ganymede and Callisto revealed by the Voyager television images suggests to us that these satellites were subjected to late heavy bombardment; this bombardment probably was due to late arriving Neptune and Uranus planetesimals (cf. Wetherill 1975). If so, Neptune and Uranus planetesimals may have been primarily responsible for the late heavy bombardment of the terrestrial planets. While other scenarios are tenable, we will adopt this as a working hypothesis and model the history of cratering of the Galilean satellites to follow the fluctuation of cratering rates recorded in the Earth-Moon system. One of the consequences of this hypothesis is that any record of impact of Jupiter planetesimals, or of objects displaced from the outer asteroid belt by strong Jupiter perturbations (Kaula and Bigeleisen 1975) very early in the history of the Galilean satellites, would have been obliterated during later bombardment by the Neptune and Uranus planetesimals.

Comparison of the Phanerozoic record of cratering on Earth with the long-term record of cratering on the Moon indicates that the cratering rate down to 10-km diameter craters in the Earth-Moon system was about twice as high in the last 0.5 Gyr, as during the preceding $\sim$ 3 Gyr (Shoemaker et al. 1979). For the Galilean satellites, we extend the calculated present cratering rate back to 0.5 Gyr; from 0.5 to 3.0 Gyr we assume that the cratering rate was just half the present rate. Prior to 3.0 Gyr an exponentially decaying component of the cratering rate is introduced to simulate the late heavy bombardment. From the record on the Moon, the half-life for decay of this component is taken to be 0.1 Gyr (Shoemaker 1972; Soderblom and Boyce 1972; Neukum et al. 1975). The exponentially decaying component is scaled so that, when it is added to the constant component of the cratering rate, the integrated crater production for various isotopically dated geologic units on the Moon is accurately reproduced. Key units through which the curve of integrated crater production is constrained to pass are the Fra Mauro Formation (3.9 Gyr), lavas at the Apollo 11 landing site in the Mare Tranquillitatis (3.65 Gyr) and lavas at the Apollo 12 landing site in the Oceanus Procellarum (3.3 Gyr). (See Papanastassiou and Wasserburg 1973 for a review of the relevant $^{87}Sr/^{87}Rb$ ages.)

For Ganymede, which is posited to have a cratering history close to that of both the Earth and the Moon, the equation derived for crater production on a surface formed at time t prior to 0.5 Gyr is

$$\int_0^t C = \left[5.7 + \frac{(1.15 \times 10^{-8})t}{yr} + \frac{R_o}{\lambda} \, exp\lambda(t - 3.3 \times 10^9 \, yr)\right] 10^{-6} km^{-2} \quad (22)$$

where $\int_0^t C$ is the satellite-wide mean crater production to 10 km diameter, $\lambda = (\ln 2)/10^8$ yr is the decay constant of the exponential component of the cratering rate, and $R_o \simeq 2.6_3 \times 10^{-8} \, yr^{-1}$, the exponential component of the cratering rate at $t = 3.3$ Gyr. The integrated crater production for surfaces of a given age on the other Galilean satellites is proportional to the crater production on Ganymede. Relative to Ganymede the constants of proportionality are 0.52 for Callisto, 1.96 for Europa, and 2.26 for Io. Curves showing $\int C$ as a function of time for each of the Galilean satellites are given in Fig. 10.16.

The steady proportionality of the cratering rates on the Galilean satellites through time depends strictly on a steady distribution of orbital elements of the impacting bodies with time. It is inherent in our model of late heavy bombardment that the distribution of long and short period orbits of Neptune and Uranus planetesimals will resemble, to a certain degree, the present distribution of the orbits of the comets. The number of very long period orbits of large inclination (whose inclinations have been randomized by stellar perturbations of the Oort Cloud) will be a greatly reduced fraction of the total population, however. The effect of this is to decrease slightly the cratering rate on Callisto and increase slightly the cratering rates on Europa and Io, relative to the cratering rate on Ganymede, during late heavy bombardment. Similarly there are small changes of δ. In our adopted model, δ is slightly higher for all satellites during late heavy bombardment.

The cratering rates we have deduced for the post-heavy bombardment period probably should be regarded as near the upper limits. They cannot be increased much without introducing a significant discrepancy between the cratering predicted by astronomical observation of Earth-crossing comets and asteroids and the record of cratering on the Earth and the Moon. The lower limits depend on the single discovery of 944 Hidalgo and its observed magnitude and color (and inferred geometric albedo). If Hidalgo should prove to be an anomaly in the magnitude distribution of Jupiter-crossing asteroids, then

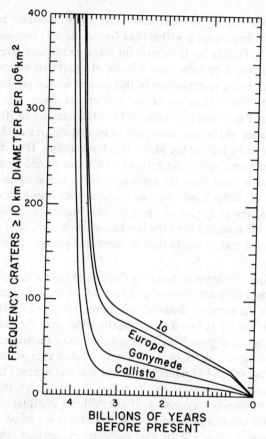

Fig. 10.16 Cratering history of the Galilean satellites. The ordinate represents the mean number of craters with diameter ≥ 10 km that have been produced in each 10^6 km^2 since the time shown on the abscissa.

the cratering rates might be less than we have calculated for the Galilean satellites. The ages obtained by comparison of observed crater densities with Eq. (22) or Fig. 10.16, therefore, are near the lower limit of the ages; the true ages might be somewhat higher.

Consistency of Derived Cratering Time Scales with the Cratering Record of the Icy Galilean Satellites

Because of the dependence on the observations of Hidalgo for setting an upper limit to crater ages, it is desirable to see whether or not the observations of the Galilean satellites might provide an independent upper bound. The oldest surface preserved in the Galilean satellite system evidently occurs near the antapex of Callisto (Passey and Shoemaker, Chapter 12). From Eq.

(22), the model 10-km crater retention age for this surface is found to be 4.4_4 Gyr. This derived age is within 0.11 Gyr of the 4.5_5 Gyr age obtained by Patterson et al. (1955) for the Earth; the crater production given by Eq. (21) cannot be decreased by more than a factor of about two without making the derived 10-km crater retention age of this ancient surface greater than the age of the Earth. (For discussion of the concept of crater retention age, see Chapter 12 by Passey and Shoemaker.) The crater density to 10 km diameter, near the antapex of Callisto, turns out to be slightly greater than the steady state density at 10 km for the Moon (cf. Trask 1966). Thus it is likely that the calculated crater age at the antapex of Callisto should be set even closer to the Patterson limit than the age suggested by the observed 10-km crater density. On the other hand, the true cratering rate may have decayed more rapidly during the early part of heavy bombardment than is indicated by the 0.1 Gyr half-life modeled from the late heavy bombardment of the Moon; the oldest preserved craters would then be somewhat younger than the ages given by Eq. 22.

The oldest preserved surfaces on Ganymede have a model 10-km crater retention age calculated from Eq. 22 of 3.9_5 Gyr, very near the time of formation of the Imbrium Basin on the Moon (Chapter 12). The Asgard ring structure on Callisto is found to be slightly older than the oldest craters on Ganymede, and the great Valhalla ring structure of Callisto and the Galileo Regio ring structure of Ganymede are somewhat younger and apparently close to the age of the Imbrium Basin of the Moon (Chapter 12).

In contrast to these ancient features, the crater retention age of the surface of Europa is very young. If the features interpreted to be impact craters by Lucchitta and Soderblom (Chapter 14) are taken to be the only preserved impact craters $\geqslant 10$ km diameter, in the area covered by high-resolution Voyager 2 images, then the observed crater density to 10 km diameter is equivalent to a satellite-wide mean density of $\sim 1.5 \times 10^{-6}$ km^{-2}. From the cratering rate given in Table 10.7, we obtain a crater retention age corresponding to this density of $\sim 30 \times 10^6$ yr. Even if allowance is made for additional craters near the threshold of recognition and for plausible over-estimation of the cratering rate, it appears unlikely that the true 10-km crater retention age of the surface observed on Europa is > 1 to 2×10^8 yr. Processes that have resurfaced Europa late in geologic time or that remove or greatly flatten craters at a very high rate are clearly implied by the low observed crater density. One such process could be rapid loss by cold viscous flow.

Cassen et al. (1979) suggested that, owing to tidal heating, a thin crust of ice floating on a layer of water might be present on Europa. The implications of this model for crater lifetimes can be evaluated by using a technique of rough analysis of crater collapse presented by Passey and Shoemaker (Chapter 12). Cassen et al. (Chapter 4) showed that if tidal dissipation is occurring in a thin ice shell, the thickness of the shell is $\geqslant 35$ km. The temperature difference across the shell is taken to be 173 K, and the thermal gradient is on the order of $\leqslant 5$ K km^{-1}; the temperature at a depth corresponding to the radius of a 10 km crater is $\leqslant 125$ K (at equatorial latitudes). From the viscosity relations for ice presented by Passey and Shoemaker, this corresponds to an effective viscosity of $\geqslant 10^{25}$ poise. At this viscosity, at a depth equal to the radius of the crater, more than 10^{11} yr would be required for a newly formed 10 km crater to flatten to the point at which it would not be recognized. A crater twice as large would flatten in $\sim 2 \times 10^8$ yr, but recognizable remnants of the rim probably would remain. Craters significantly > 20 km diameter would tend to disappear either by cold flow or prompt collapse, if formed in a 35 km ice crust floating on water. We cannot account, however, for the low abundance of 10 km craters. Other processes, perhaps connected with some global event (cf. Chapter 14 by Lucchitta and Soderblom; Yoder 1979), must have produced extensive resurfacing of the satellite late in geologic time.

Acknowledgments. We thank E. Roemer for sharing much of her data in advance of publication and for many courtesies over the past 17 years. We also thank S. J. Weidenschilling, A. Woronow, and C. R. Chapman for many constructive comments in their review of this chapter. This investigation has been partly supported by the Voyager Project and Planetary Geology Program of the National Aeronautics and Space Administration.

APPENDIX A

Equations for Probability of Collision of Small Bodies with the Galilean Satellites

Fairly accurate solutions for the probabilities of collision of comet nuclei with the Galilean satellites can be obtained by appropriate modification of equations derived by Öpik (1951) for the case of collision of a small body with a planet. These equations are:

$$P = \frac{\tau^2 \, U f}{\pi \sin i \, |U_x| \, a^{3/2}} \tag{23}$$

$$U^2 = 3 - \frac{1}{A} - 2\sqrt{A(1 - e^2)} \cos i + \frac{4}{9} e_0^2 \tag{24}$$

$$U_x^2 = 2 - \frac{1}{A} - A(1 - e^2) + \frac{4}{9} e_0^2 \tag{25}$$

$$\tau^2 = \left(\frac{R}{a_0}\right)^2 \left(1 + \frac{S^2}{U^2}\right) \tag{26}$$

$$S^2 = \frac{2a_0 M}{R M_\odot} \qquad (27)$$

$$f = \frac{1}{\pi} \cos^{-1} \frac{A(1-e)-(1-e_0{}^2)}{e_0 A(1-e)}, A(1-e) > 1-e_0 \left.\begin{array}{c} \\ \\ \\ \\ \\ \end{array}\right\}$$

$$= \frac{1}{\pi} \cos^{-1} \frac{(1-e_0{}^2)-A(1+e)}{e_0 A(1+e)}, A(1+e) < 1+e_0 \qquad (28)$$

where P is the probability of collision with the planet per year, τ is the dimensionless capture radius of the planet (ratio of capture radius to the planet's semimajor axis), U is the dimensionless encounter velocity at the sphere of influence of the planet (ratio of the encounter velocity to the mean orbital velocity of the planet), U_x is the radial (with respect to the Sun) component of the encounter velocity, f is the crossing fraction for cases of partial overlap of the orbit of the small body and the orbit of the planet, i is the inclination of the orbit of the small body to the orbital plane of the planet, A is the ratio of the semimajor axis of the small body orbit to the semimajor axis of the orbit of the planet, e is the eccentricity of the small body orbit, e_0 is the eccentricity of the orbit of the planet, S is the dimensionless escape velocity from the surface of the planet (ratio of escape velocity to the mean orbital velocity of the planet), R is the radius of the planet, a_0 is the semimajor axis of the orbit of the planet, M is the mass of the planet, and $M_\odot$ is the mass of the Sun.

Three modifications to Öpik's equations must be made in order to apply them to the collison of comets with the satellites of Jupiter. First, a term must be added to Eq. (23) to account for the enhancement of the flux of comets in the neighborhood of a satellite due to the gravitational field of Jupiter. This enhancement F is given by

$$F = 1 + \frac{S_0{}^2}{U^2} \left.\begin{array}{c} \\ \\ \\ \\ \\ \\ \end{array}\right\}$$

$$S_0{}^2 = \left(\frac{2a_0}{r_s}\right) \frac{M_0}{M_\odot} \qquad (29)$$

where S_0 is the dimensionless escape velocity from Jupiter at the semimajor axis of the satellite (ratio of escape velocity to Jupiter's orbital velocity), r_s is the semimajor axis of the orbit of the satellite, and M_0 is the mass of Jupiter.

Second, the acceleration of the comets by Jupiter must be accounted for in calculating the capture cross-section of the satellite. The encounter velocity at the sphere of influence of the satellite is

$$U_c = \sqrt{U^2 + S_0{}^2} \tag{30}$$

and the capture radius of the satellite is

$$\tau_c = \frac{R_s}{a_0} \left(1 + \frac{S_s{}^2}{U_c{}^2}\right)^{\frac{1}{2}} \tag{31}$$

where S_s is the escape velocity from the surface of the satellite. S_s is obtained from Eq. (26) by substitution of R_s for R and M_s, the mass of the satellite, for M_0. The expression for the collision probability of a comet with the satellite is

$$P_c = \frac{F\tau_c{}^2 \, U f}{\pi \sin i \, |U_x| \, a^{3/2}} \, . \tag{32}$$

Third, the acceleration by Jupiter must be taken into account in calculating v_i, the impact velocity on the satellite. The mean impact velocity is given by

$$v_i = v_0 \sqrt{U_c{}^2 + S_s{}^2} = v_0 \sqrt{U^2 + S_0{}^2 + S_s{}^2} \tag{33}$$

where v_0 is the orbital velocity of Jupiter.

APPENDIX B

Derivation of Synthetic Orbits for Short Period Comets

The set of synthetic orbital elements for short period comets with $q > 1.7$ AU is derived from the observed orbits of Jupiter-family comets by the following method: for each interval, 1.7 AU $< q \leqslant 2$ AU, 2 AU $< q \leqslant 3$ AU, 3 AU $< q \leqslant 4$ AU, 4 AU $< q \leqslant 5$ AU, 5.0 AU $< q \leqslant 5.527$ AU, the orbits of all observed Jupiter-family comets with q less than the lower bound of the interval are selected; these selected orbits are modified by raising q for each orbit, while preserving the original i. A value of q is selected at random, within the limits of the interval, and assigned to each orbit to be modified. From the studies of Everhart (1972) and Kazimirchak-Polonskaya (1972) it was found that, for Jupiter-family comets, substantial changes of q occur in a small number of Jupiter encounters. Under these conditions the encounter velocity U or a quantity T,

$$T = \frac{1}{2}\left(3 + \frac{4}{9}e_0{}^2 - U^2\right) \tag{34}$$

called the Tisserand invariant, is approximately conserved. We may also write

$$e = \frac{a - q}{a} \tag{35}$$

$$T = \frac{a_0}{2a} + \left[\frac{2q}{a_0} \left(1 - \frac{q}{2a} \right) \right]^{\frac{1}{2}} \cos i \tag{36}$$

$$\frac{1}{a} = \frac{2}{a_0} [T - \cos^2 i \, Q^2 - \cos i \, (\cos^2 i \, Q^4 + 2Q - 2TQ^2)^{\frac{1}{2}}] \tag{37}$$

where $Q = q/a_0$. In cases where $(\cos^2 i \, Q^4 + 2Q - 2 \, T \, Q^2) < 0$, this term is set equal to 0. This latter circumstance arises when the original orbit is unusually small and is not Jupiter-crossing. Holding T and i constant, we obtain new values of a and e from Eqs. (35) and (37) that correspond to the new randomly assigned values of q. All orbits with negative a (or $e > 1$), which represent hyperbolic orbits, are rejected. The remaining modified orbits are then taken as a sample of the orbits of the undiscovered comets.

The number Δn of synthetic orbits to be determined for each interval is given by

$$\Delta n = 4.67 \, [(\Sigma n_i - \Sigma n_{i-1})_{\text{predicted}} - (\Sigma n_i - \Sigma n_{i-1})_{\text{observed}}] \tag{38}$$

where Σn_i is Σn at $q = 2, 3, 4, 5$, $(\Sigma n_i - \Sigma n_{i-1})_{\text{predicted}}$ is the incremental frequency per unit q derived from the linear function fitted over $1.1 \text{ AU} \leqslant q \leqslant 1.7 \text{ AU}$, shown in Fig. 10.2

$$\Sigma n = 64q - 58 \tag{39}$$

and $(\Sigma n_i - \Sigma n_{i-1})_{\text{observed}}$ is the observed incremental frequency per unit q of discovered Jupiter-family comets. In the interval $5.0 \text{ AU} < q \leqslant 5.527 \text{ AU}$, Σn is obtained from

$$\Delta n = 4.67 [(\Sigma n_{5.527} - \Sigma n_5)_{\text{predicted}} - (\Sigma n_{5.527} - \Sigma n_5)_{\text{observed}}] \cdot \tag{40}$$

At the suggestion of Weidenschilling, we also derived synthetic orbits by letting i vary as well as a and e. For each interval of q, synthetic orbits were derived from the observed orbits of Jupiter-family comets by holding T constant and selecting q randomly as described above; i for each synthetic orbit was selected randomly from the observed i distribution of the short period comets. If the orbit thus derived was hyperbolic or if it failed to overlap the orbit of Jupiter (where the time averaged eccentricity of Jupiter was used for e_0), another value of i was chosen randomly. Where necessary, this procedure was repeated until a bound Jupiter-crossing orbit was found, up to a maximum of 20 trials. If no bound Jupiter-crossing orbit was obtained after 20 trials, the synthetic orbit was rejected. The use of a randomly selected value of i introduces a rather large statistical scatter in the mean impact parameters calculated from the synthetic orbits. Therefore we repeated the derivation of synthetic orbits for each interval of q 20 times. A total of $\sim$ 5000 orbits was derived, in order to obtain statistically stable results. The final calculated values of $\bar{P}_c$, $\bar{\Gamma}^*$, and δ, utilizing this large sample of synthetic orbits, are within a few percent of the values given in Table 10.1.

APPENDIX C

Variation of Flux of Impacting Bodies as a Function of Distance From the Apex of Orbital Motion on the Galilean Satellites

To obtain the variation of cratering rate as a function of distance from the apex of motion of a satellite we first derive the mean vertical component of the encounter velocity $\bar{U}_\perp$ onto a small massless sphere located on the orbit of the satellite but stationary with respect to Jupiter. As a first approximation, we take the directional distribution of the encounter velocity U_c in space near the sphere to be isotropic. This is an excellent approximation for the long period comets in the vicinity of Jupiter. The approximation is less good for the short period comets, as there is a deficiency in frequency of directions of U_c at high angles to the orbital plane of Jupiter. This deficiency arises from the inclination distribution of the short period comets, which is concentrated toward the orbit plane of Jupiter (Fig. 10.7). The deficiency has only a minor effect, however, on the computation of the mean vertical component $\bar{U}_\perp$ when the vertical to the surface of the satellite lies close to Jupiter's orbital plane.

As shown by Shoemaker (1962), for an isotropic flux of impacting bodies, both on a massless sphere and on a sphere with a spherically symmetric gravitational field, the frequency distribution of the elevation angle of impact is

$$\mathrm{d}f = \sin 2\epsilon \, \mathrm{d}\epsilon \tag{41}$$

where $\mathrm{d}f$ is the differential frequency of ϵ and ϵ is the elevation angle from plane tangent to sphere. Now

$$U_\perp = \sin \epsilon \, U_c \tag{42}$$

and

$$df_\perp = U_\perp \, df = U_\perp \sin 2\epsilon \, d\epsilon = U_c \sin \epsilon \sin 2\epsilon \, d\epsilon \qquad (43)$$

where $df_\perp$ is the differential frequency of $U_\perp$. Integration of Eq. (43) yields

$$U_\perp = U_c \int_0^{\pi/2} \sin \epsilon \sin 2\epsilon \, d\epsilon = \frac{2}{3} U_c. \qquad (44)$$

For the case of isotropic distribution of the direction of U_c, in other words, the mean vertical component of the encounter velocity $\bar{U}_\perp$ is precisely 2/3 U_c.

For a gravitating satellite moving relative to the frame of reference in which the distribution of U_c is isotropic, the flux of impacting bodies is no longer isotropic. The flux is enhanced at the apex of motion of the satellite and depressed at the antapex. The ratio of the flux of impacting bodies at any one place on a moving satellite to the flux at any other place is given, to good approximation, by the ratio of the mean vertical components of the respective impact velocities multiplied by the ratio of the respective capture cross-sections.

At the apex, the mean vertical component of velocity of the impacting projectiles is, to good approximation, the sum of $\bar{U}_\perp$, an increment to the velocity component at the apex due to acceleration of the projectiles in the gravitational field of the satellite S_ϱ, and the orbital speed of the satellite, $S_0/\sqrt{2}$. The ratio ξ_ϱ of this mean vertical component at the apex to the satellite-wide mean vertical component of impact velocity is

$$\xi_\varrho = \frac{\bar{U}_\perp + S_\varrho + S_0/\sqrt{2}}{2/3 \, (U_c^2 + S_s^2)^{1/2}} = \frac{2/3 \, U_c + S_\varrho + S_0/\sqrt{2}}{2/3 \, (U_c^2 + S_s^2)^{1/2}} \qquad (45)$$

where the satellite-wide mean impact velocity is $(U_c^2 + S_s^2)^{1/2}$, S_s is the escape velocity from the satellite, and the satellite-wide mean vertical component of the impact velocity is just 2/3 the mean impact velocity. The satellite-wide mean vertical component of the impact velocity is equivalent to the mean vertical component of the impact velocity at 90° from the apex, where orbital motion does not add to nor subtract from the vertical component. Similarly, the ratio ξ_t of the mean vertical component of the impact velocity at the antapex to the satellite-wide mean vertical component is given by

$$\xi_t = \frac{2/3 \, U_c + S_t - S_0 \sqrt{2}}{2/3 \, (U_c^2 + S_s^2)^{1/2}} \, , \quad 2/3 \, U_c + S_t - S_0/\sqrt{2} > 0 \qquad (46)$$

where S_t is the increment to the vertical component of velocity at the antapex due to acceleration in the gravitational field of the satellite.

The capture radius τ_β is dependent on the encounter velocity

$$\tau_\beta = \frac{R_s}{a_0} \left(\frac{1 + S_s^2}{U_\beta^2} \right)^{1/2} \tag{47}$$

where R_s/a_0 is the normalized radius of the satellite, and U_β is the asymptotic encounter velocity at the gravitational sphere of influence of the satellite for impact at a given distance β from the apex. Therefore the ratio $\tau_\varrho^2 / \tau_c^2$, where τ_ϱ is the capture radius for impact at the apex and τ_c is the satellite-wide mean capture radius, is given by

$$\frac{\tau_\varrho^2}{\tau_c^2} = \frac{(R_s/a_0)^2 \, (1 + S_s^2 / U_\varrho^2)}{(R_s/a_0)^2 \, (1 + S_s^2 / U_c^2)} = \frac{(1 + S_s^2 / U_\varrho)}{(1 + S_s^2 / U_c)} \tag{48}$$

where U_ϱ is the asymptotic encounter velocity for impact at the apex and U_c is the asymptotic encounter velocity at $\beta \quad 90°$, which is equal to the satellite-wide mean asymptotic encounter velocity. The product of ξ_ϱ and $\tau_\varrho^2 / \tau_c^2$ gives the ratio γ_ϱ of the flux of impacting bodies at the apex to the satellite-wide mean flux

$$\gamma_\varrho = \xi_\varrho \frac{\tau_\varrho^2}{\tau_c^2} = \frac{2/3 \, U_c + S_\varrho + S_0/\sqrt{2}}{2/3 \, (U_c^2 + S_s^2)^{1/2}} \left(\frac{1 + S_s^2 / U_\varrho^2}{1 + S_s^2 / U_c^2} \right) . \tag{49}$$

Similarly the ratio γ_T of the flux of impacting bodies at the antapex to the satellite-wide mean flux is

$$\gamma_t = \frac{2/3 \, U_c + S_t - S_0 \sqrt{2}}{2/3 \, (U_c^2 + S_s^2)^{1/2}} \left(\frac{1 + S_s^2 / U_t^2}{1 + S_s^2 / U_c^2} \right) \tag{50}$$

where U_t is the asymptotic encounter velocity for impact at the antapex.

To solve for γ_ϱ and γ_t, we derive U_ϱ, U_t, S_ϱ and S_t as functions of U_c. Both U_ϱ and U_t are simply the vector sums of the mean horizontal and mean vertical components of the asymptotic encounter velocities. For U_ϱ, the mean vertical component is $2/3 \, U_c + S_0/\sqrt{2}$, and the mean horizontal component is $[U_c^2 - (2/3 \, U_c)^2]^{1/2} = \sqrt{5/3} \, U_c$; hence U_ϱ is given by

$$U_\varrho = \left[(\tfrac{2}{3} U_c + S_0/\sqrt{2})^2 + (\sqrt{5/3} \, U_c)^2 \right]^{1/2} . \tag{51}$$

Similarly U_t is given by

$$U_t = \left[(\tfrac{2}{3} U_c - S_0/\sqrt{2})^2 + (\sqrt{5/3}\ U_c)^2 \right]^{1/2}. \qquad (52)$$

S_ϱ and S_t are defined as the vertical components of the velocity increment S_s, which is produced by acceleration in the gravitational field of the satellite. From similar triangles, the ratio S_ϱ/S_s is

$$\frac{S_\varrho}{S_s} = \frac{\sqrt{5/3}\ U_c}{U_\varrho}\ , \quad S_\varrho = \frac{\sqrt{5}\ S_s\ U_c}{3\ U_\varrho} \qquad (53)$$

and the ratio S_t/S_s is

$$\frac{S_t}{S_s} = \frac{\sqrt{5/3}\ U_c}{U_t}\ , \quad S_t = \frac{\sqrt{5}\ S_s\ U_c}{3\ U_t} \qquad (54)$$

As $U_c = (U^2 + S_0{}^2)^{1/2}$, where U is the asymptotic encounter velocity with the sphere of influence of Jupiter, it may be seen that U_c can never be less than S_0. Hence $2/3\ U_c - S_0/\sqrt{2}$ can never be less than $2/3 - 1/\sqrt{2} = -0.0404$. It is easy to show, for the Galilean satellites, that as $U_c \to S_0$, S_t is always greater than $+0.0404$. Therefore, the condition given for Eq. (46) is always satisfied.

APPENDIX D

Latitudinal and Seasonal Selection Effects in the Discovery of Long Period Comets

Many long period comets that reached perihelion in the southern sky have been missed as a result of the unequal distribution of comet observers between the northern and southern hemispheres. This selection effect is revealed by the frequency distribution of ω. Where $0° < \omega < 180°$, perihelion is located at northern ecliptic latitudes, and where $180° < \omega < 360°$, perihelion is at southern ecliptic latitudes. From Fig. 10.12 it may be seen that, for all discovered long period comets, the running mean frequency of ω near $\omega = 270°$ is only about 55% of the frequency near $\omega = 90°$. Hence, if the intrinsic distribution of ω is uniform, $\sim 45\%$ of the comets with ω near $270°$ that became bright enough near perihelion to have been discovered, had the southern sky been as carefully searched as the northern sky, were missed.

Everhart (1967b) has shown, for a smaller sample of orbits, that the observed shape of the ω distribution can be approximately predicted from the conditions of observation, if we assume the intrinsic distribution of ω is uniform. The history of discovery provides an independent confirmation that the intrinsic distribution of ω is indeed nearly uniform (Fig. 10.13). When comets are classified by the time of discovery, it is found that the distribution of ω becomes more uniform for comets discovered at later times. Specifically, the deficiency of frequency near $\omega = 270°$ is greatly reduced for comets with $q \leqslant 1.1$ AU that were discovered in the 20th century. This is due to improved coverage of the southern sky, particularly since 1965. The contributions of southern hemisphere observers for the period 1960 to 1978 is illustrated at the bottom of Fig. 10.13. In the 1970s, the amateur astronomer W. A. Bradfield of Dernancourt, Australia, has single-handedly discovered eight long period comets, seven of which came to perihelion at southern ecliptic latitudes. In the frequency distribution of ω for comets with $q \leqslant 1.1$ AU that were discovered in the 20th century, it can be seen that there is a residual deficiency of discovered objects with ω near $0°$ and near $180°$. This deficiency reflects the Holetschek effect; comets which pass perihelion near superior conjunction must have ω near $0°$ or $180°$.

A seasonal selection effect in the discovery of long period comets is revealed by the frequency distribution of ecliptic longitude L. As may be seen in Fig. 10.14, the distribution of ecliptic longitude is strongly bimodal; one mode is centered a little east of $90°$ longitude, a region in the middle of the night sky during the northern hemisphere summer; the strongest mode is centered on $270°$ longitude, which is the middle of the night sky at the nothern hemisphere winter solstice. This alignment of the modes of L with respect to the present orientation of the Earth's axis is unlikely to be due to chance, although the strong mode centered on the winter night sky was recognized by Van Flandern (1978) from a selected set of 92 very long period comets and used by him as supporting evidence for the hypothesis of relatively recent disruption of a planet.

The cause of the observed selection of L probably is complex and may include both human and astronomical factors. The strongest mode, at $L = 270°$, clearly is due in large part to the fact that the largest range of ecliptic latitudes is visible at winter solstice, from mid-geographic latitudes in the northern hemisphere, where most comet observers have been located (Porter 1963). Moreover, the winter nights are long, and conditions are favorable for astrometric observation of newly discovered objects. It is the superior conditions in winter, which favor this essential follow-up astrometry needed to derive accurate orbits, leading to the single strong mode of L in the winter sky for the orbits selected by Van Flandern. The other, summer sky mode, observed in the distribution of L for all long period comets, probably is related to the habits of comet observers, and to the prolonged opportunity to search the post-sunset and pre-dawn sky near the sun, where comets are most often found. In addition, weather patterns in the northern hemisphere probably result in some of the best average seeing conditions in the period around summer solstice.

APPENDIX E

Interpretation of the Observed Distribution of
Inclination of Long Period Comets

If the poles to the orbit planes of long period comets were uniformly distributed over the celestial sphere, the differential frequency of the inclination of these orbits would be proportional to the sine of the inclination (Porter 1963). It has long been recognized that the frequency of inclination of observed long period comets (Fig. 10.14) approximately follows a sine function but differs from this function by an amount greater than expected by chance. There is a deficiency of observed frequency near 90°, or, conversely, there is an excess of frequency near 0° and 180° relative to a sine function. Observational selection may account for part of the departure from an isotropic distribution of poles to the orbit plane.

Prolonged perturbation of the Oort Cloud of comets by stars passing near the Sun tends to randomize the orbits of the *very* long period comets and would be expected to produce a nearly isotropic distribution of the orbit plane poles for comets that have never returned to the region of the planets. Many of the observed long period comets, however, have been subjected to perturbations by the planets (Porter 1963). Comets on low inclination orbits, and, to a lesser extent, on retrograde orbits with inclinations near 180° have the highest probability of being perturbed into orbits of progressively shorter periods (Everhart 1972). Hence the flux of observable long period comets on low inclination orbits and on retrograde orbits near 180° inclination is enhanced relative to the flux of observable comets on inclinations near 90°. This selective perturbation ultimately yields the so-called periodic comets, which have predominantly low inclination orbits. In the calculation of equilibrium between short period and long period comets, the comets on orbits that have already partly evolved toward short period constitute an important fraction of the long period comets passing through Everhart's capture region ($i < 9°$). These partly evolved orbits are represented in the observed frequency distribution of inclinations of long period comets. Thus the observed distributions of i for all long period comets is preferred over the theoretical distribution for *very* long period comets, for use in our calculation of the equilibrium between short and long period comets.

January 1981

REFERENCES

Bobronikoff, N.T. (1941). Investigations of the brightness of comets. *Contrib. Perkins Obs.* 15.

Bowell, E., Gehrels, T., and Zellner, B. (1979). Magnitudes, colors, types and adopted diameters of the asteroids. In *Asteroids* (T. Gehrels, Ed.), pp. 417–435. Univ. Ari-Press, Tucson.

Brownlee, D.E. (1978). Microparticle studies by sampling techniques. In *Cosmic Dust* (J.A.M. McDonnell, Ed.), pp. 295–336. John Wilen & Sons, New York.

Brownlee, D.E., Horz, F., Tomandl, D.A., and Hodge, P.W. (1976). Physical properties of interplanetary grains. In *The Study of Comets*, NASA Special Paper 393, pp. 462–482.

Brownlee, D.E., Rajan, R.S., and Tomandl, D.A. (1977). A chemical and textural comparison between carbonaceous chondrites and interplanetary dust. In *Comets Asteroids Meteorites* (A.H. Delsemme, Ed.), pp. 137–141. Univ. Toledo, Toledo, Ohio.

Cassen, P.M., Reynolds, R.J., and Peale, S.J. (1979). Is there liquid water on Europa? *Geophys. Res. Letters* 6, 731–734.

Chabai, A.J. (1959). Cratering scaling laws for desert alluvium. *Sandia Corporation Report* SC-4391 (RR).

Chapman, C.R. (1976). Chronology of terrestrial planet evolution: The evidence from Mercury. *Icarus* 28, 523–536.

Davies, M.E. and Katayama, F.Y. (1980). Coordinates of features on the Galilean satellites. *J. Geophys. Res.* In press.

Degewij, J. and Van Houten, C.J. (1979). Distant asteroids and outer Jovian satellites. In *Asteroids* (T. Gehrels, Ed.), pp. 417–435. Univ. Arizona Press, Tucson.

Delsemme, A.H. (1973). Origin of the short-period comets. *Astron. Astrophys.* 29, 377–381.

Delsemme, A.H. (1977). The pristine nature of comets. In *Comets Asteroids Meteorites* (A.H. Delsemme, Ed.), pp. 1–13. Univ. Toledo, Toledo, Ohio.

Delsemme, A.H. (1980). Is the rock-to-ice ratio of comets the same as that of Ganymede and Callisto? IAU Coll. 57. *The Satellites of Jupiter* (abstract 7–13).

Everhart, E. (1967a). Comet discoveries and observational selection. *Astron. J.* 72, 716–726.

Everhart, E. (1967b). Intrinsic distributions of cometary perihelia and magnitudes. *Astron. J.* 72, 1001–1011.

Everhart, E. (1972). The origin of short-period comets. *Astrophys. Lett.* 10, 131–135.

Everhart, E. (1973). Examination of several ideas of comet origins. *Astron. J.* 78, 329–337.

Everhart, E. (1977). The evolution of comet orbits as perturbed by Uranus and Neptune. In *Comets Asteroids Meteorites* (A.H. Delsemme, Ed.), pp. 99–104. Univ. Toledo, Toledo, Ohio.

Gault, P.E. and Wedekind, J.A. (1977). Experimental hypervelocity impact into quartz sand. II. Effects of gravitational acceleration. In *Impact and Explosion Cratering* (D.J. Roddy, R.O. Pepin, and R.B. Merrill, Eds.), pp. 1231–1244. Pergamon Press, New York.

Gehrels, T., Coffeen, T., and Owings, D. (1964). Wavelength dependence of polarization III: The lunar surface. *Astron. J.* 69, 826–852.

Gilbert, G.K. (1893). The Moon's face: A study of the origin of its features. *Philosophical Soc. of Wash. Bull.* 12, 241–292.

Grieve, R.A.F. and Dence, M.R. (1979). The terrestrial cratering record II: The crater production rate. *Icarus* 38, 230–242.

Kaula, W.M. and Bigeleisen, P.E. (1975). Early scattering by Jupiter and its collision effects in the territorial zone. *Icarus* 25, 18–33.

Kazimirchak-Polonskaya, E.I. (1972). The major planets as powerful transformers of cometary orbits. In *The Motion, Evolution of Orbits, and Origin of Comets* (G.S. Chebotarey and E.I. Kazimirchak-Polonskaya, Eds.), pp. 373–397. D. Reidel, Dordrecht.

Kresák, L., and Pittich, E.M. (1978). The intrinsic number density of active long-period comets in the inner solar system. *Bull. Astron. Inst. Czechosl.* 29, 299–309.

Leverrier, U.J.J. (1848). Mémoire sur la comète périodique de 1770. *Compt. Rend. Acad. Sci. Paris.* 26, 465–469.

Leverrier, U.J.J. (1857). Thèorie de la comète périodique de 1770. *Ann. Obs. Paris Mem.* 3, 203–270.

Marsden, B.G. (1971). Reports on progress in astronomy. *Quarterly J. Roy. Astro. Soc.* 12, 244–273.

Marsden, B.G. (1979). *Catalogue of Cometary Orbits.* Smithsonian Astrophys. Obs. Cambridge, Massachusetts.

Morrison, D. (1977). Asteroid sizes and albedos. *Icarus* 31, 185–220.

Neukum, G., Konig, B., Fechtig, H., and Storzer, D. (1975). Cratering in the Earth-Moon system: Consequences for age determination by crater counting. *Proc. Lunar Sci. Conf.* 6, 2597–2620.

Null, G.W. (1976). Gravity fields of Jupiter and its satellites from Pioneer 10 and 11 tracking data. *Astron. J.* 81, 1153–1161.

Oort, J.H. (1950). The structure of the cloud of comets surrounding the solar system, .. and a hypothesis concerning its origin. *Bull. Astr. Inst. Netherlands* 11, 91–110.

Opik, E.J. (1951). Collision probabilities with the planets and distribution of .. interplanetary matter. *Proc. Roy. Irish Acad.* 54A, 165–199.

Opik, E.J. (1963). The stray bodies in the solar system. Part 1. Survival of cometary nuclei and the asteroids. *Advan. Astron. Astrophys.* 2, 219–262.

Papanastassiou, D.A., and Wasserburg, G.J. (1973). Rb-Sr ages and initial strontium in basalts from Apollo 15. *Earth Plan. Sci. Lett.* 17, 324–337.

Patterson, C.C., Tilton, G.R., and Ingrham, M.G. (1955). Age of the earth. *Science* 121, 69–75.

Porter, J.G. (1963). The statistics of comet orbits. In *The Solar System,* Vol. IV: *The Moon, Meteorites, and Comets* (G.P. Kuiper and B.M. Middlehurst, Eds.), pp. 550–572. Univ. Chicago Press, Chicago.

Roemer, E. (1965). Observations of comets and minor planets. *Astron. J.* 70, 397–402.

Roemer, E. (1966). The dimensions of cometary nuclei. *Mém. Royal Sci. Soc. of Liège,* 5th Series 12, 23–28.

Roemer, E. (1968). Dimensions of the nuclei of periodic and near-parabolic comets (abstract). *Astron. J.* 73, 533.

Roemer, E. and Lloyd, R.E. (1966). Observations of comets, minor planets, and satellites. *Astron. J.* 71, 443–457.

Roemer, E., Thomas, M., and Lloyd, R.E. (1966). Observations of comets, minor planets, and Jupiter VIII. *Astron. J.* 71, 591–601.

Sekanina, Z. (1971). A core-mantle model for cometary nuclei and asteroids of possible cometary origin. In *Physical Studies of Minor Planets* (T. Gehrels, Ed.), pp. 423–426. NASA-SP 267, Washington.

Sekanina, Z. (1976). A continuing controversy: Has the cometary nucleus been resolved? In *The Study of Comets* (B. Donn, M. Mumma, W. Jackson, M. A'Hearn, and R. Harrington, Eds.), pp. 537–585. NASA-SP 393.

Shelton, A.V., Nordyke, M.D., and Goeckermann, R.H. (1960). A nuclear explosive cratering experiment. *Univ. Calif. Lawrence Livermore Laboratory* Report 5766.

Shoemaker, E.M. (1962). Interpretation of lunar craters. In *Physics and Astronomy of the Moon* (Z. Kopal, Ed.), pp. 283–359. Academic Press, London.

Shoemaker, E.M. (1972). Cratering history and early evolution of the Moon. *Lunar Sci.* 3, 696–698.

Shoemaker, E.M. (1977a). Astronomically observable crater-forming projectiles. In *Impact and Explosion Cratering: Planetary and Terrestrial Implications* (D.J. Roddy, R.O. Pepin, and R.B. Merrill, Eds.), pp. 617–628. Pergamon Press, New York.

Shoemaker, E.M. (1977b). Why study impact craters? In *Impact and Explosion Cratering: Planetary and Terrestrial Implications* (D.J. Roddy, R.O. Pepin, and R.B. Merrill, Eds.), pp. 1–9. Pergamon Press, New York.

Shoemaker, E.M., Hackman, R.J., and Eggleton, R.E. (1963). Interplanetary correlation of geologic time. *Adv. Astronaut. Sci.* 8, 70–89.

Shoemaker, E.M. and Helin, E.F. (1977). Populations of planet-crossing asteroids and the relation of Apollo objects to main-belt asteroids and comets. In *Comets Asteroids Meteorites* (A.H. Delsemme, Ed.), pp. 297–300. Univ. Toledo, Toledo, Ohio.

Shoemaker, E.M., Williams, J.G., Helin, E.F., and Wolfe, R.F. (1979). Earth-crossing asteroids: Orbital classes, collision rates with Earth, and origin. In *Asteroids* (T. Gehrels, Ed.), pp. 253–282. Univ. Arizona Press, Tucson.

Soderblom, L.A. (1977). Historical variations in the density and distribution of impacting debris in the inner solar system: Evidence from planetary imaging. In *Impact and Explosion Cratering: Planetary and Terrestrial Implications* (D.J. Roddy, R.O. Pepin, and R.B. Merrill, Eds.), pp. 629–633. Pergamon Press, New York.

Soderblom, L.A. and Boyce, J.M. (1972). Relative ages of some near-side and far-side Terra plains based on Apollo 16 metric photography. *NASA Special Paper* 315, pp. 29-3–29-6.

Trask, N.J. (1966). Size and spatial distribution of craters estimated from the Ranger photographs in Mare Tranquillitatis. Ranger VIII and IX, Part II: Experimenters' analyses and interpretations. *Jet Propulsion Lab. Tech. Report* 32–800.

Van Flandern, T.C. (1978). A former asteroidal planet as the origin of comets. *Icarus* 36, 51–74.

Weissman, P.R. (1977). Initial energy and perihelion distributions of Oort-cloud comets. In *Comets Asteroids Meteorites* (A.H. Delsemme, Ed.), pp. 87–91. Univ. Toledo, Toledo, Ohio.

Wetherill, G.W. (1975). Late heavy bombardment of the Moon and terrestrial planets. *Proc. Lunar Sci. Conf.* 6, 1539–1561.

Wetherill, G.W. (1976). Where do the meteorites come from: A reevaluation of the Earth-crossing Apollo objects as sources of stone meteorites. *Geochim. Cosmochim. Acta* 40, 1297–1317.

Wetherill, G.W. (1977). Evolution of the Earth's planetesimal swarm subsequent to the formation of the Earth and Moon. *Proc. Lunar Sci. Conf.* 8, 1–16.

Wetherill, G.W., and Williams, J.G. (1968). Evaluation of the Apollo asteroids as sources of stone meteorites. *J. Geophys. Res.* 73, 635–648.

Whipple, F.L. (1962). On the distribution of semimajor axes among comet orbits. *Astron. J.* 67, 1–9.

Whipple, F.L. (1978). Comets. In *Cosmic Dust* (J.A.M. McDonnell, Ed.), pp. 1–73. John Wiley & Sons, New York.

Whipple, F.L., and Hamid, S. (1951). On the origin of the Taurid meteor streams. *Helwan Obs. Bull.* 41, 1–29.

Yoder, C.F. (1979). How tidal heating in Io drives the Galilean orbital resonance locks. *Nature* 279, 767–770.

11. EXPERIMENTAL SIMULATION OF IMPACT CRATERING ON ICY SATELLITES

RONALD GREELEY, JONATHAN H. FINK
Arizona State University

DONALD E. GAULT
Murphys Center for Planetology

and

JOHN E. GUEST
University of London Observatory

Voyager images of Ganymede and Callisto reveal impact craters morphologically distinct from those on the terrestrial planets. To simulate cratering processes that might have occurred on icy satellites, we performed a series of 102 laboratory impact experiments involving a wide range of target materials. The first group consisted of 36 impacts into homogeneous clay slurries (appropriate in some respects for scaling the rheological properties of icy planetary surfaces to laboratory sized experiments) in which impact energies were varied from 5×10^6 to 1×10^{10} erg, target yield strengths ranged from 100 to 38 Pa, and apparent viscosities ranged from 8 to 200 Pa s. The following morphologic progression was observed in targets with constant rheological properties as impact energy was increased: bowl-shaped craters, flat-floored craters, central peak craters with high relief, central peaks with little relief, craters with no relief. Crater diameters also increased steadily as energies were raised. A similar sequence was seen for experiments in which impact energy was held constant but target viscosity and strength *progressively decreased. These experiments suggest that the physical properties of the target media (strength, viscosity) relative to the gravitationally induced stresses (lithostatic, hydrostatic) determine the final crater morphology. In particular, crater palimpsests (large, circular, high-albedo features with low relief on Ganymede and Callisto) could form by prompt collapse of large central peak craters formed in low target strength (fluidized?) materials. Furthermore, ages estimated from crater size-frequency distributions that include these large craters may give values that are too high. Other experiments involved layered targets including clay substrates overlain by ice, unbonded or weakly bonded sand-sized particles, and water with or without a surface layer of highly viscous oil (η = 0.2 and 30 Pa s). Central pit craters formed in a few cases for which the surface layer had a critical thickness: a crater formed almost entirely within the surface layer; a narrow central peak rose up through the floor of the crater; and finally the peak receded, leaving a central depression. Other models for the origin of central pits suggested by the experiments involve the loss of volatiles and the formation of centrally located secondary craters caused by the fallback of ejecta.*

Even before the Voyager results became available, there was speculation that impact craters on the icy Galilean satellites would be markedly different from craters formed on dominantly silicate planets (for instance, Johnson and McGetchin 1973). The basic consideration was that the craters would be reduced topographically by the cold viscous flow of the ice in which they formed; there was, in fact, some concern that the icy satellites might show little topographic relief. Although this concern was for the most part unfounded, Voyager images of Ganymede and Callisto, both considered to be composed in large part of water, reveal impact craters that are indeed different in morphology from craters of similar sizes on the inner planets. Passey and Shoemaker (Chapter 12) provide a detailed morphological classification of craters on the icy satellites. They observe that larger craters are relatively shallow and degraded whereas smaller craters are relatively deep and appear to be better preserved (i.e., sharper rim crests), as would be expected from viscous deformation of icy materials.

Because impact craters are so dominant on the Moon, Mars, and Mercury, a great deal of attention has been given to their morphology. Various relationships, such as depth to diameter ratios, are well documented as described in introductory books on planetology (see, e.g., Mutch 1972; Short 1975). Of particular importance are the presence and shape of the cavity, form of the central peak and rings (if present), wall terraces, and characteristics of the ejecta deposits. The morphology of impact craters and related features is the result of three primary factors:

1. Properties of the impacting body, including mass, composition, velocity, and angle of impact;

2. Properties of the target, including composition ·(ice, water, or silicate materials, alone or in various mixtures) and complexity (homogeneous versus layered);

3. Degree of modification by external processes (e.g., superposed impacts), internal processes (e.g., viscous deformation), or a combination. The degree of modification is also a function of time.

Thus, the observed morphology of impact craters can provide information on these parameters, if the various functional relationships are known. The effect of these parameters on impact crater formation and geometry can be studied both theoretically and experimentally. The theoretical approach has advantages of testing conditions unattainable in the laboratory (i.e., impact velocity, scale), but the behavior of icy materials at high shock stresses is poorly known, and the theoretical results are generally difficult, if not impossible, to verify or substantiate by experimental observations. Physical experiments enable the isolation of individual variables to determine their role in the impact process. Such experiments are often visually gratifying, but scaling to planetary-sized events is difficult to achieve.

We have chosen the experimental approach, first to complement ongoing theoretical studies of other investigators, and second to help visualize the processes of impact crater formation and degradation for viscous planetary surfaces. Ultimately, the understanding of cratering processes on icy bodies will be derived by syntheses of theoretical studies, large- and small-scale impact experiments, and analyses of spacecraft data. Such an understanding will provide clues to the thermal history and surface evolution of the satellites, as is currently under investigation by Phillips and Malin (1980) and Passey and Shoemaker (Chapter 12).

I. PREVIOUS STUDIES

In this section we review studies relevant to craters on the icy satellites and provide the background for the impact cratering experiments. Johnson and McGetchin (1973) first considered the effect that the viscous relaxation of icy crusts on Ganymede and Callisto would have on the preservation of impact craters and predicted that all topography, including craters, would be essentially flat within $< 10^6$ yr of time of formation. A later study by Reynolds and Cassen (1979) used more appropriate temperature values to assess material properties and predicted that the crusts would be more rigid than predicted earlier and that the surface features would be preserved, a prediction borne out by the Voyager results. Parmentier and Head (1979) have further refined calculations of rates of viscous deformation for the surfaces of Ganymede and Callisto and speculate on the residence times of features including impact craters as a function of size. Passey and Shoemaker (Chapter 12) have applied the general equations for the viscous flow of craters derived by Scott (1967) to their analyses of impact craters on Ganymede and Callisto. Other theoretical studies include those by McKinnon (1978), who considered the plastic failure of craters formed in various target materials, and by Cintala et al. (1979, 1980), who calculated the partitioning of energy for impacts into icy materials and predicted the effect on crater size relative to comparable impacts into silicate targets. In most of the models of viscous deformation the results are based on lithospheres that are composed essentially of pure ice, although the investigators recognize that other substances such as silicates and possibly salts are present (indicated by spectral data and albedo contrasts). Such mixtures can have a significant effect on the rates of viscous relaxation, but because specific values for possible components are not known, no one has yet attempted to model complex mixed lithosphere cases.

Impact cratering experiments date from long before the space age. Green (1965) reviews various experiments beginning with Sir Robert Hooke who, in 1665, described impacts of bullets into mixtures of clay and water. One of the foremost experimental facilities for studying impact craters in the planetary context is the NASA-Ames Vertical Ballistic Range (VBR), established in the mid-1960s; this facility and the rationale for cratering simulations are described by Gault et al. (1968). Through the mid-1970s the facility was used to study the fundamental cratering process (Gault and Heitowit 1963; Gault and Moore 1964; Quaide et al. 1965; Oberbeck 1971, 1973; Gault et al. 1968; Gault 1970, 1973; Gault and Wedekind 1978), the nature of impact-generated regolith (Oberbeck 1975; Oberbeck and Quaide 1968), and to determine various morphological relationships among target properties and impact conditions (Gault et al. 1963; Quaide and Oberbeck 1968; Oberbeck 1970). Experimental cratering and computational modeling of cratering were also carried out at other institutions, including substantial efforts sponsored by the defense agencies in the U.S. and Canada. Field studies of natural impact craters by the U.S. Geological Survey and other institutions provided important constraints for these studies. The various investigators were drawn together at a symposium in 1976; the proceedings volume from this meeting (Roddy et al. 1977) contains seventy-eight papers on impact cratering and remains a standard reference.

Concurrent with these experiments, photogeological analyses of planetary surfaces showed that crater geometry could be related to multilayered targets at small scales (craters up to a few kilometers in diameter; Quaide and Oberbeck 1968) and possibly large scales (Head 1976). Studies of the effects of target characteristics on crater morphology include those of Cintala (1977) and Cintala et al. (1977), who analyzed such processes as central-peak formation. Viking orbiter images of Mars revealed impact crater ejecta deposits that are distinct from those on the Moon and Mercury, leading to the interpretation that the volatiles on Mars are somehow responsible for the flow-like form (Carr et al. 1976). An exploratory series of experiments was carried out by Gault and Greeley (1978) to assess the effects of viscous targets on crater morphology. These experiments were followed by a more rigorous series in which a generalized model of laboratory impact cratering in viscous materials was defined and the implications for Martian impact craters were discussed (Greeley et al. 1980). Martian multilobed craters have particular relevance to craters on icy satellites. Two explanations for the flow-like character of the Martian craters have been advanced: first, that they result from fluidization of the ejecta by entrainment of entrapped volatiles including liquid water (e.g., Carr et al. 1977); second, that they result from a sorting process involving the Martian atmosphere (Schultz and Gault 1979). Although a third possibility involves a combination of the first two, the question remains as to which might dominate the process of ejecta emplacement. The icy satellites afford an excellent opportunity to test the two cases because they lack atmospheres but presumably contain volatiles in the lithospheres. Thus if the craters lack Martian-like ejecta deposits, the atmospheric model would be enhanced. Unfortunately, the resolution of the Voyager images is less than the threshold of detectability. It is interesting to note that on Mars, Mariner 9 images did not show clearly the ejecta morphology, although there were suggestions of flow-like forms (Head and Roth 1976); it was not until the better resolution of Viking images was available that the details were observable. In a similar fashion, some high resolution Voyager images show a few dozen craters on Ganymede that appear to have flow-like forms (R.G. Strom, personal communication; Horner and Greeley 1981), suggesting that volatile entrainment is the important mechanism. Definitive testing, however, must await the higher resolution images anticipated from the Galileo mission. Thus, although Mars and the icy satellites may be worlds apart in many of their characteristics, certain aspects of impact cratering involving volatiles may be common.

II. EXPERIMENTAL FACILITY AND PROCEDURE

The experiments described below were carried out at the NASA-Ames Research Center Vertical Ballistic Range (VBR) Facility. This facility (Gault et al. 1968) consists of a vacuum chamber 2.5 m in diameter by 3 m high that is straddled by an A-frame on which light-gas guns (Charters and Curtis 1962; Curtis 1964) and powder guns can be mounted. The A-frame can be rotated into firing positions into the chamber at angles from horizontal to vertical in $15°$ increments. Projectiles of different composition (and density) such as steel, aluminum, or glass, in diameters ranging from 1.5 mm to 25.4 mm, can be launched at speeds up to 7.5 km s^{-1}. The chamber can be evacuated to $\sim$ 13.3 Pa pressure. Various viewing ports in the chamber enable observation and photodocumentation of impact events. High-speed 16-mm motion

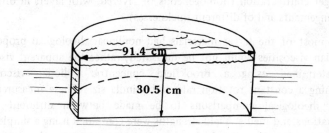

Fig. 11.1. Quarter-space target bucket. Impact point was aimed immediately inside the plexiglass window, permitting observation of subsurface cratering processes in near cross section.

picture cameras capable of recording up to 10^6 frames per second enable the sequence of cratering events to be studied in detail.

A wide range of target configurations is possible within the facility chamber. Our experiments involved three types of containers: 1) large bucket, consisting of a container 1 m in diameter by 0.3 m deep set in a false floor of the chamber so that when the bucket is filled with target material it forms a continuous surface flush with the floor; 2) small bucket, a container of identical configuration as the large bucket, but 0.6 m in diameter by 0.2 m deep; and 3) a "quarter-space" target bucket, consisting of a container of the same size as the large bucket, but cut in half and fitted with a plexiglass window (Fig. 11.1). Experiments using the quarter-space bucket involved transparent target materials that were impacted by projectiles fired within a centimeter of the bucket window in order to observe the formation and modification of the transient cavity and related features effectively in cross section as a function of time. The technique of observing targets in cross section has been used successfully by Gault et al. (1968) involving colored sand as target materials, but our experiments are the first to use transparent viscous target materials.

The experiments are subdivided into various series. Each series consists of a given target and bucket configuration that is impacted over a range of energies and impact angles. Controlled parameters that were analyzed systematically include:

1. Target density, viscosity and yield strength (see Greeley et al. 1980, for a discussion of the relationships between viscosity and yield strength appropriate for these experiments);
2. Impact projectile composition and size;
3. Impact velocity;

4.   Impact angle;
5.   Target configuration (homogeneous or layered, with layers in different arrangements and of different thicknesses).

Because most of the target materials had nonlinear rheological properties, Newtonian viscosities could not be measured; rather the apparent viscosity was determined using a Brookfield concentric cylinder viscometer maintaining a constant rotation rate and spindle size. These measurements allowed rheological comparisons to be made between different target configurations and media. Yield strengths were computed using a simple cone penetrometer. Density was calculated by weighing a known volume of target material. Density, yield strength and apparent viscosity of clay slurry targets could be varied by altering the slurry concentration; however, these parameters were not independent of one another. Thus, in our discussion, targets that are relatively stiff also have the highest viscosities and densities (Table 11.1).

After each shot, the crater geometry was described and measurements were made of such features as crater diameter, extent of ejecta deposits, and diameter of any central mound or pit. Subsequent analyses of the high-speed motion pictures (we used 400 frames per second throughout the experiments) enabled measurements of transient features such as maximum size of the transient cavity, maximum height of the central mound, and time when the mound formed. For some shots calculations were made of the proportion of kinetic impact energy that was converted into the potential energy of a transient central peak. Measurements and parameters for a given experiment or series of experiments were later combined in various dimensionless groups using the Buckingham Pi Theorem (Bird et al. 1960). This dimensional analysis enabled certain relationships to be assessed which could be extrapolated to full-size (i.e., planetary) cases.

III. CRATERING EXPERIMENTS

Very little experimental work has been conducted on cratering in ice or ice-rich materials, although some engineering studies and experiments have involved explosion cratering in ice as a potential method for clearing rivers and harbors. Some of the results may have application to impact cratering on the Galilean satellites (Gaffney 1980), but the use of explosion craters as analogs to impact craters requires a specific scaled depth of burst (Oberbeck 1971, 1977; Melosh 1980), and the engineering experiments may not be completely appropriate. Other experiments have involved small-scale impacts into ice and ice-saturated sand (Croft et al. 1979), which may provide some data appropriate to the problem, but scaling of the strengths of the materials is problematic. Almost all other considerations of impact cratering in ice,

TABLE 11.1

Impacts into Water-Based Clay Slurry Targets (Series I) [a]

Run No.	Shot No.	Density (kg m^{-3})	Apparent Viscosity (x 0.1 Pa s)	Yield Strength (Pa)	Impact Energy (joules)	Peak Diameter (m)	Crater Diameter (m)	Plume Deposit Diameter (m)	Transient Central Peak Height[b] (m)	Peak Potential Energy (joules)	Final Crater Depth (m)	Final Peak Height (m)
1	800404	1515	12.8	197	41	0.089	0.178	0.330	0.046	0.031	0.006	—
2	800405	1539	14.4	197	56	0.127	0.279	0.457	0.053	0.065	—	—
3	800413	1534	4.5	180	59	—	0.254	0.356	0.083	0.121	—	—
4	800501	1740	43.2	424	50	0.140	0.279	0.457	0.011	0.004	0.019	—
5	800502	1740	43.2	424	31	0.127	0.254	0.330	0.013	0.001	0.019	—
6	800503	1740	40.0	487	8	0.076	0.191	0.279	0.000	—	0.025	—
7	800504	1740	40.0	487	72	0.140	0.305	0.406	0.026	0.009	—	0.038
8	800505	1740	48.0	487	99	0.152	0.318	0.406	0.026	0.009	—	0.032
9	800506	1740	48.0	487	67	0.203	0.394	0.457	0.046	0.053	—	0.025
10	800507	1740	46.0	487	176	0.241	0.457	0.610	0.061	0.173	—	0.013
11	800508	1740	46.0	487	371	0.305	0.483	0.711	0.090	0.555	—	0.006
12	800511	1740	43.2	423	372	0.241	0.508	0.660	0.086	0.490	—	—
13	800512	1740	43.2	423	487	0.343	0.483	0.686	0.112	1.08	—	—
14	800514	1717	51.2	486	720	0.381	0.610	0.813	0.123	1.23	—	—
15	800515	1661	64.0	488	0.9	0.102	0.165	0.216	0.000	—	—	—
16	800519	1606	8.0	152	49	—	0.305	0.445	0.018	0.374	0.019	—
17	800520	1606	8.0	152	22	—	—	0.356	0.105	0.259	—	—
18	800521	1600	8.8	152	7	—	—	—	0.072	0.060	—	—
19	800522	1600	8.8	152	0.5	—	—	0.229	0.042	0.010	—	—
20	800523	1600	8.8	152	70	—	—	0.508	0.114	0.350	—	—
21	800524	1630	9.6	166	85	—	—	0.533	0.141	0.741	—	—
22	800525	1630	9.6	166	102	—	—	0.584	0.143	0.715	—	—
23	800526	1630	9.6	166	108	—	0.356	0.635	0.147	0.847	—	—
24	800528	1630	9.6	166	91	—	0.533	—	0.171	1.77	—	—

TABLE 11.1 (Continued)

Impacts into Water-Based Clay Slurry Targets (Series I) [a]

Run No.	Shot No.	Density (kg m^{-3})	Apparent Viscosity (× 0.1 Pa s)	Yield Strength (Pa)	Impact Energy (joules)	Peak Diameter (m)	Crater Diameter (m)	Plume Deposit Diameter (m)	Transient Central Peak Height[b] (m)	Peak Potential Energy (joules)	Final Crater Depth (m)	Final Peak Height (m)
25	800529	1630	9.6	166	309	—	—	—	0.237	4.30	—	—
26	800531	1630	9.6	166	511	—	—	—	0.255	6.01	—	—
27	800533	1600	9.9	166	618	—	—	—	0.285	7.44	—	—
28	800537	1600	130.0	3815	39	—	0.203	—	0	—	0.089	—
29	800540	1600	130.0	3815	750	—	0.400	—	0	—	0.107	—
30	800541	1600	130.0	3815	105	—	0.224	—	0	—	0.080	—
31	800542	1700	200.0	3815	356	—	0.380	—	0	—	0.085	—
32	800546	1700	200.0	3815	765	—	0.460	—	0	—	0.115	—
33	800547	1700	170.0	3815	930	—	0.435	—	0	—	0.110	—
34	800548	1700	170.0	3815	1050	—	0.470	—	0	—	0.106	—
35	800549	1720	80.0	2657	91	—	—	—	0	—	—	—
36	800550	1760	70.0	2047	740	0.178	0.483	0.559	0.029	0.005	0.035	—

[a]Peak diameter, crater diameter, plume deposit diameter, final crater depth, and final central peak height all measured after run completed.
[b]Transient central peak heights measured from high speed motion pictures.

ice-water, and ice-slush have been theoretical (e.g., Boyce 1979; Cintala et al. 1980; Austin et al. 1980).

The selection of target materials to simulate planetary-scale impacts in the laboratory is difficult, particularly as related to studying cratering processes and crater morphologies. The general problem of scaling laboratory impact experiments has been addressed by many investigators, including White (1971), Killian and Germain (1977) and Holsapple and Schmidt (1979). When scaled to laboratory-sized experiments, actual rock is far too strong to be used in cratering experiments. Cohesionless silica sand is commonly used to simulate rocky targets, as discussed by Chabai (1977), Gault and Wedekind (1977), and Schmidt and Holsapple (1980). Ice in the cold environment of Ganymede is very strong, comparable to rock. Thus sand could also be used to simulate the nonviscous aspects of cratering in icy targets. Dry sand is not, however, appropriate to simulate the viscous aspects of cratering into materials that may involve flow, such as melted ice, slushy ejecta, penetration through a solid crust into low-viscosity mantles, or to simulate the long-term viscous relaxation of craters. For these types of simulations, we selected various viscous materials whose properties could be scaled approximately to the planetary environment.

Comparisons among the various target media and likely planetary surface materials are made especially difficult by their different mechanical behaviors. For example, dry, crushed walnut shells behave like a cohesionless sand and are characterized by a dimensionless internal friction angle. Water-saturated shell particles, however, have cohesive strength (due to capillarity) and an effective viscosity, as well as the internal friction angle. Clay slurries have both a yield strength and a linear plastic viscosity. Solid ice (like basalt or granite) has compressive, tensile and shear strengths, each measured by different types of tests; yet it also has nonlinear viscous properties. Massive debris flows, such as those that might form during a large impact event, may be described by a yield strength and viscosity, as with mud slurries. Simultaneous scaling of all of these parameters is generally impossible in the laboratory; the parameters that we monitored, including density, yield strength, and plastic viscosity, are considered of prime importance and are readily amenable to laboratory measurements.

A. Homogeneous Viscous Targets (Series I)

This series of experiments was conducted to determine the effect that target rheology has on impact crater mechanics and morphology. Targets consisted of homogeneous clay-water slurries whose yield strength and viscosity could be varied. These slurries have a Bingham rheology; that is, when subjected to high stresses, they flow, but if the applied load is less than the yield strength, they remain rigid.

Analysis of 36 shots in this series (Table 11.1) leads to a model of crater formation that involves an oscillating central mound (Greeley et al. 1980). The impact process may be broken down into several stages (Fig. 11.2):

1. Formation and deposition of an ejecta plume;
2. Excavation of a crater bowl;
3. Collapse of crater bowl and rise of a central mound;
4. Fall of the central mound and outward flow of a surge wave;
5. Relaxation of the final crater form.

Depending upon the impact conditions and target properties, stages 3 and 4 may be repeated or a central mound may not form at all. The experiments indicated that each of these stages, and hence the final crater form, could be influenced by the target properties and impact energy (i.e., crater size).

Figure 11.3 shows profiles obtained from experiments in which the impact conditions were held constant but the viscosity and yield strength of the target were varied by altering the concentrations of the clay slurries. For high viscosities and strengths, bowl-shaped craters formed. As the viscosities and strengths were reduced, the crater diameters increased and the depths decreased. Central peaks began to form in an intermediate range of viscosities. For very fluid slurries, one or more central peaks formed but no topography was preserved. The transient cavity was defined by a rough surface texture resulting in part from outgassing of the zone of impact; evidently the impact caused release of some volatiles from the target. In many of these shots, *in situ* measurements of the viscosity of the target showed a decrease, also possibly related to release of volatiles upon impact (Greeley et al. 1980).

A similar sequence of crater forms was obtained by keeping the target properties constant while increasing crater dimensions (i.e., impact energy; Fig. 11.4 *a, b*). For relatively low-energy shots, bowl-shaped craters formed. With increasing energy a transition occured from bowl-shaped to flat-floored to central-peak craters. At higher energies, transient central peaks were higher, but the final peak relief was less. For the highest energy shots, no crater morphology remained. Use of stiffer slurries produced a similar morphologic progression, but the transitions from one crater type to another occurred for larger diameter craters (i.e., larger impact energies; Fig. 11.4*b*).

Most studies of impacts into icy satellites have assumed that the observed diameter-dependent changes of crater morphology are due to relaxation processes, and that the initial crater forms were constant for a given terrain. Our experiments suggest that the variety of crater types on Ganymede and Callisto may owe as much or more to initial target properties and impact conditions as to secondary modification by slow viscous creep. Impact into a planetary surface composed of ice or a mixture of ice and rock would cause comminution and melting of material in the impact zone. The resulting slurry of rock, ice and water might well behave like terrestrial debris flows that have

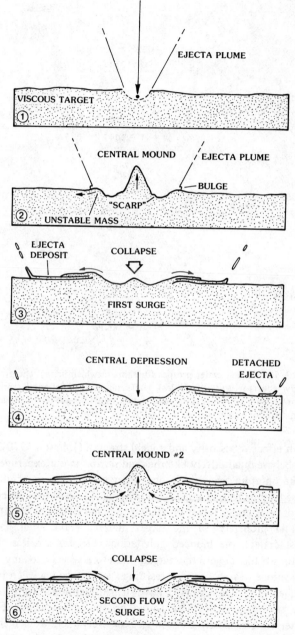

Fig. 11.2. Diagram of impact into viscous medium showing development of oscillating central mound and subsequent generation of ejecta flow surges. (From Greeley et al. 1980.)

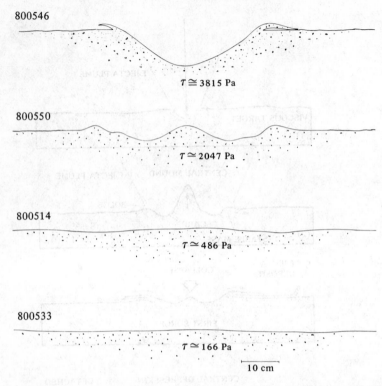

Fig. 11.3. Post-impact crater profiles for shots in which impact energy was maintained nearly constant (~ 700 joules) while target yield strength varied. Progression followed from bowl-shaped to central-peak to palimpsest-type craters with decreasing strength. Shot numbers given on the left (see Tables).

both an effective viscosity and a yield strength (Johnson 1970). This material would behave qualitatively like the mud slurries in our experiments.

The morphologic transitions observed as the crater diameters were increased by increasing impact energy (Fig. 11.4*a,b*) may be compared with the variety of crater forms on Ganymede, and both can be attributed to the solid-fluid properties of the target material (Fink et al. 1980). For the smallest craters, the induced gravitational stresses are less than the yield strength of the target; the target, therefore, reacts to any post-cratering modifications as a solid, resisting any relaxing deformations, and consequently retains craters as bowl-shaped features in form similar to those in dry sand targets or those found on the lunar surface. For the largest diameter craters, gravity-induced stresses everywhere exceed the target yield strength, causing post-cratering deformations that resemble those occurring in strengthless materials (e.g., water) where no topography is preserved. Intermediate to these two extremes, post-cratering deformations occur, but

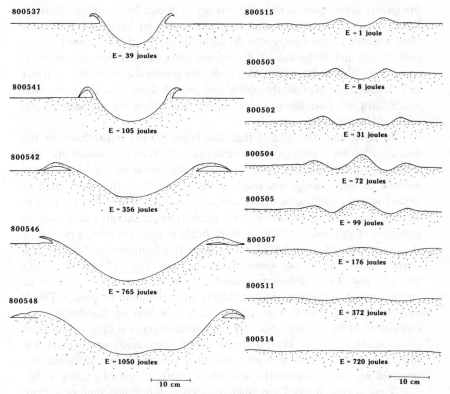

Fig. 11.4. Post-impact crater profiles for shots in which target properties were maintained constant while impact energies were increased to produce larger craters. (a) Relatively weak targets ($\tau \simeq 450$ Pa; $\eta \simeq 4.5$ Pa s); progression followed from bowl-shaped to high-relief central-peak to low-relief central-peak to palimpsest-type craters with increasing impact energy. (b) Relatively strong targets ($\tau \simeq 3800$ Pa; $\eta \simeq 13-20$ Pa s); only bowl-shaped craters formed within this energy range (i.e., small craters). Shot numbers given on the left (see Tables).

finite strength of the target material preserves morphological records of the impact. If this behavior were to occur on the surface of Ganymede, craters formed by the largest impacts might be immediately erased while craters formed by contemporaneous smaller bombardments in the same vicinity might be well-preserved. Subsequent impacts into areas previously hit (and left in a semifluidized state) by large bolides might form shallower craters than those in nearby areas hit by fewer recent large impacts. This might explain the observations (Passey and Shoemaker, Chapter 12) of totally degraded palimpsests having populations of well-preserved secondary craters.

The changes seen with decreasing viscosity (Fig. 11.3) can be explained in part as a viscous dissipation effect. The higher the viscosity of the target, the greater the proportion of kinetic energy of impact lost by viscous heating. Hence, whereas impacts into water generally convert between 4 and 8% of their available kinetic energy into the potential energy of the transient central peak (Gault and Wedekind 1978), impacts into mud slurries had between 0 and 2% conversion. On a planetary scale, the greater the proportion of water and fluidized material in the ejecta and impact slurry, the more the final crater form will resemble an impact into a strengthless material and the less relief will be preserved.

To determine the effects that the target bucket might have on the formation of the central mound, a subseries of shots was performed involving quarter-space experiments. A target bucket was cut in half and a plexiglass window installed; water was used as the target. Projectiles were fired to impact within a centimeter of the window. In effect, it was thus possible to observe the cratering process in cross section; use of various tracers such as streamers of dye and neutrally buoyant beads enabled the relative motion of the target material to be traced during the crater formation and degradation stages. Analysis of the high-speed motion pictures (Fig. 11.5) shows: (1) In the first few frames following impact, i.e., within ~ 1/200 s, a wave of bubbles passes across the window radially from the point of impact. This is interpreted as resulting from passage and reflection of the shock wave. Calculation of the shock pressure for the impact energies in these experiments gives a maximum of 65 kbar; the wave decays very rapidly and is much less than 1 bar by the time it reaches the floor of the bucket. Moreover, the reflected wave would encounter and interact with the growing cavity within $<$ 0.5 ms, a time interval less than the camera interframe time of 2.5 ms. More importantly, the embryonic cavity continues to grow to its maximum size for some tens of ms after the wave encounter. Thus any wave reflection effect is insignificant, at least in relation to the formation of the central mound. (2) After the transient cavity reaches its maximum size, a central mound is formed by the flow of target material from both below the cavity and from the sides. (3) A surface wave is generated from the transient cavity rim that travels toward the confining wall of the bucket and is reflected; however, it does not return to the cavity until *after* the central mound develops and therefore is not involved in the formation of the central mound. For impacts in which the craters are small relative to the bucket, the reflected surface wave does not reach the zone of impact until after the second oscillation of the central mound. We thus conclude that the oscillating central mound can form independently of the confining target bucket, and represents a significant post-cratering process for planetary applications.

B. Ice over Viscous Targets (Series II)

Analyses in which impact crater target strengths and other properties are

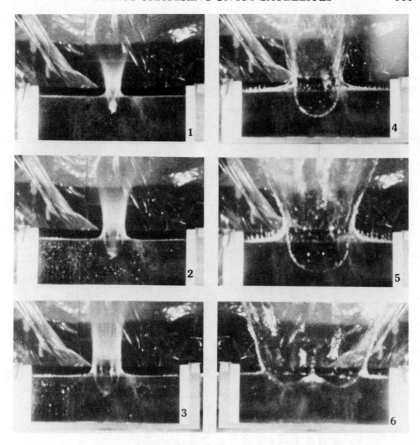

Fig. 11.5. Sequence of photographs showing impact into water in quarter-space bucket. In (1), a wave front of bubbles, seen passing across window, is interpreted to have resulted from the passage of an initial shock wave. In (6), the crater bowl collapses to form a central peak, with material flowing into the crater from all directions (sides and bottom).

scaled to full-size planetary events have shown that rocky materials are much too strong for use in simulations (Chabai 1977; Gault and Wedekind 1978). In an exploratory mode, the next series of experiments (Table 11.2) involved multilayered targets of an ice crust overlying viscous clay slurries of the same type used in the homogeneous target experiments (Series I). In some cases the crust consisted of relatively pure water ice up to a few centimeters thick; in other cases the crust consisted of frozen water and clay. In three runs the ice layer was overlain by a surface layer of unbonded sand-sized particles.

Relatively low-energy impacts simply punctured the crust and penetrated the underlying viscous material. With increasing energies, the crust was progressively more heavily fractured with segments being uplifted by the

TABLE 11.2

Impacts into Clay Slurries with Ice Surface Layers (Series II)

Run No.	Shot No.	Ice Layer (m)	Shell Layer (m)	Mud Density (kg m⁻³)	Mud Viscosity (x 0.1 Pa s)	Mud Yield Strength (x 0.1 Pa)	Impact Energy (joules)	Peak Diameter (m)	Crater Diameter (m)	Plume Deposit Diameter (m)	Crater Depth (m)
1	800419	?[b]	—[c]	1543	4.5	?	62	—	0.133	—	—
2	800420	0.006	—	1543	21.6	?	97	0.102	—	0.254	—
3	800421	?	—	1543	21.6	?	95	—	—	—	—
4	800423	—	—	1630	20.8	17.5	65	0.102	0.178	0.406	0.025
5	800424	0.013	—	1630	20.8	17.5	60	—	—	—	—
6	800425	0.025	—	1630	20.8	17.5	505	0.152	0.229	0.356	—
7	800426	0.008	0.018	1680	20.8	19.9	450	0.089	0.318	0.508	0.025
8	800427	0.018	0.018	1680	20.8	19.9	472	0.127	0.343	0.216	0.019
9	800428	0.018	0.006	1680	57.6	22.3	459	0.222	0.356	0.368	—

[a]Shell layers placed on top of ice layers in some shots.
[b]Question marks indicate measurements not taken.
[c]Dashes indicate features not present.

Fig. 11.6. Shot 800426. The target consisted of viscous clay overlain with ~20 mm of ice and an upper regolith layer a few mm thick of sand-sized shell particles. An impact of a 6.4–mm pyrex projectile at 1.76 km s^{-1} fractured the ice and produced a flat-floored crater. A central secondary crater resulted from fallback of a blob of clay ejecta held together by surface tension.

rising central mound of slurry. Crater dimensions generally increased with increasing impact energy and decreasing crustal thickness (Table 11.2). In some experiments, the central mound produced a detached segment of material, similar to that portrayed by Murray (1980), which fell back to the target forming a small secondary crater within the larger primary crater (Fig. 11.6). Although superficially resembling a central pit crater, this experiment is not considered an analog because the detached segment was a mass of material held together by surface tension. When scaled to planetary sizes, this feature would be dispersed and probably would not behave as a coherent mass capable of producing a clearly defined secondary crater. Because of the general problem of inappropriate scaling of material strengths, experiments involving natural ice were not continued. Quartz sand and other loose or weakly bonded particles have been used for more than a decade to simulate large-scale impacts into rocky materials, and this medium was selected for the next series of experiments.

C. Unbonded Particles over Viscous Targets (Series III)

Multilayered targets involving cratering into loose, incompetent materials overlying a more competent substrate were carried out to simulate impacts into lunar regoliths of varying thicknesses by Oberbeck and Quaide (1968). With increasing thickness of the incompetent layer, all other parameters (such as impact energy) held constant, they found that the geometry of the crater changed from a complex concentric form to a flat-floored crater to a simple bowl shape. We wished to determine the crater morphology for impacts involving an upper incompetent layer overlying a viscous medium of lower relative strength, which is the opposite case of the Oberbeck and Quaide experiments. Various thicknesses of sand or sand-size ($\sim 300~\mu$m) shell particles were emplaced on viscous clays as indicated in Table 11.3 (shell particles were used for most of the experiments because they floated on the viscous clay, in contrast to sand which tended to sink, particularly in the more fluid substrates). The overlying particles formed a layer that was less cohesive but had a greater shear strength than the viscous substrate. As shown in Fig. 11.7, impacts into targets with a thin upper layer produced a relatively shallow crater (the floor having deformed rapidly by viscous flow) with an extensive ejecta deposit. As in the earlier experiments with homogenous viscous targets, a central mound developed within the clay. With increasing thickness of the upper layer, the diameters of the crater, the ejecta deposit and the central mound all decreased (Fig. 11.7), reflecting the greater strength and more effective stress wave attenuation in the overlying material. For very thick surface layers, the transient cavity did not penetrate to the viscous substrate, and a bowl-shaped, lunar-type crater formed (e.g., Fig. 11.7, shot 800410). For a given impact energy there was a critical layer thickness for which the transient cavity barely penetrated into the viscous substrate. In these cases a narrow central mound rose up through the bowl-shaped crater in the surface layer (Fig. 11.8a) and then collapsed, leaving a very small disturbed zone. In a subseries of experiments, a water-saturated layer of shell particles was placed beneath the dry surface layer (Fig. 11.8b). This wet layer restricted the dispersal of the central mound and produced craters with a concentric central pit, analogous to those found commonly in the icy satellites (Fig. 11.9) as well as in a few terrestrial impact craters (Fig. 11.10).

D. Weakly Bonded Particles over Viscous Targets (Series IV)

In order to test an intermediate strength between natural ice and the relatively weak unbonded particles for the upper layer, a series of experiments was conducted in which the upper layer consisted of quartz sand weakly bonded with epoxy (Table 11.4). These layers were made in thicknesses of 6.4, 12.8, 19.0 and 25.4 mm and emplaced on the viscous clay substrates in the large target bucket. Low-energy impacts simply punctured the crust and penetrated the clay, similar to impacts into the natural ice-over-clay targets. With increasing impact energies, progressively larger

TABLE 11.3

Impacts into Clay Slurries with Surface Layers of Dry Granular Shell Particles (Series III)

Run No.	Shot No.	Dry Shell Layer (m)	Wet Shell Layer (m)	Mud Density (kg m^{-3})	Mud Viscosity (x 0.1 Pa s)	Mud Yield Strength (Pa)	Impact Energy (joules)	Peak Diameter (m)	Crater Diameter (m)	Plume Deposit Diameter (m)	Crater Depth (m)
1	800405	–	–	1539	14.4	197	56	0.127	0.279	0.457	–
2	800406	0.006	–	1572	15.2	197	50	0.165	0.216	0.356	–
3	800407	0.006	–	1572	15.2	197	54	0.152	0.254	0.406	–
4	800408	0.018	–	1549	19.2	216	53	0.127	0.191	0.305	0.016
5	800409	0.036	–	1535	12.8	131	52	0.076	0.114	0.178	0.019
6	800410	0.060	–	1535	11.5	148	54	–	0.140	–	0.013
7	800411	0.036	–	1535	11.5	148	56	0.064	0.140	–	–
8	800412	–	0.036	1535	14.4	148	59	0.057	0.165	0.114	–
9	800413	–	–	1534	4.5	161	59	–	0.254	0.356	–
10	800414	0.5	–	1534	4.8	109	56	–	0.229	0.305	0.013
11	800415	1.3	–	1534	4.6	101	50	–	0.171	0.254	0.032
12	800416	2.5	–	1534	4.6	101	50	0.089	0.178	0.076	0.019
13	800417	2.5	–	1534	4.6	101	61	–	0.165	–	–
14	800418	–	2.5	1543	7.2	109	60	–	0.229	0.406	–
15	800516	0.6	0.6	1661	64.0	788	32	0.076	0.229	0.127	0.038
16	800517	1.8	0.6	1661	51.2	486	60	0.076	0.229	0.178	0.032
17	800518	1.3	1.7	1661	51.2	486	51	0.064	0.216	0.165	0.038
18	800534	2.2	0.6	1600	9.9	166	35	0.051	0.178	–	–
19	800535	1.8	0.6	1600	9.9	152	55	0.089	0.165	–	0.019

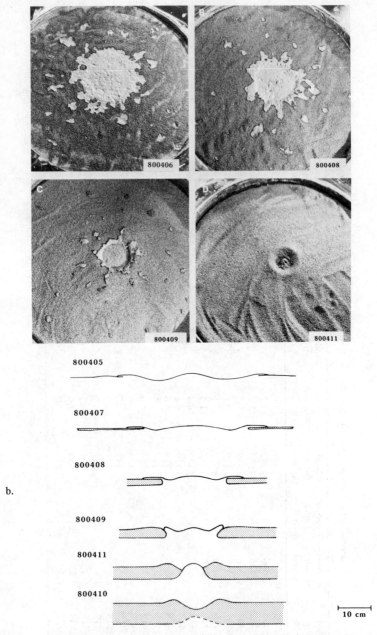

Fig. 11.7. (a) Overhead views and (b) profiles showing effect of multilayered target on crater diameter (shot numbers given on the left). Targets consist of an upper layer of sand-size shell particles (relatively strong, shown in stipple pattern) and a substrate of viscous clay (relatively weak); as the thickness of the upper, competent layer increases, the effective crater diameter decreases (see Table 11.3). In this set of experiments, impact energy was held relatively constant.

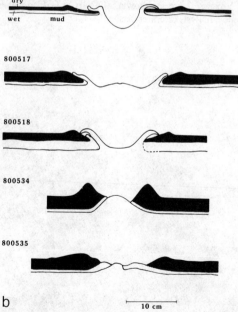

Fig. 11.8. (a) High-speed motion picture frame showing impact into multilayered target (shell particles over viscous clay), showing development of small central mound (Shot 800535). Post-impact crater was similar to that in shot 800411 (Fig. 11.7a). (b) Profiles for various shots involving wet and dry shell layers overlying mud substrate (shot numbers given on the left).

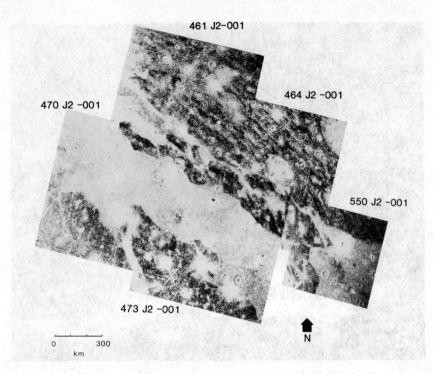

461 J2-001

464 J2 -001

470 J2 -001

550 J2 -001

473 J2 -001

0   300
km

N

Fig. 11.9. Mosaic of Voyager 2 images showing the part of Galileo Regio on Ganymede used in the frequency analysis of crater classes.

Fig. 11.10. Serra da Canghala structure in central Brazil, considered to be a central-pit-type impact crater formed in water-saturated sediments. Outer diameter is 12 km. (Photograph courtesy of J.F. McHone.)

TABLE 11.4

Impacts into Clay Slurries overlain by Weakly Bonded Sand Surface Layers (Series IVa)

Run No.	Shot No.	Layer Thickness (m)	Layer Compressive Strength (x 10³ Pa)	Mud Approx. Density (kg m⁻³)	Mud Viscosity (x 0.1 Pa s)	Mud Yield Approx. Strength (x 0.1 Pa)	Impact Energy (joules)	Crater Diameter (m)	First Ring Diameter[a] (m)	Second Ring Diameter[a] (m)
1	800642	0.018	63	1650	27.2	7	560	0.07	0.20	–
2	800643	0.013	63	1650	27.2	7	490	0.235	0.35	0.51
3	800644	0.006	285	1650	27.2	7	480	0.35	0.45	–
4	800645	0.006	118	1650	27.2	7	95	0.12	0.20	–
5	800646	0.013	118	1650	27.2	7	190	0.14	0.32	–
6	800647	0.013	118	1600	12.8	5	700	0.16	0.35	–
7	800701	0.006	125	1600	12.5	5	34	0.06	0.13	–
8	800702	0.006	125	1600	12.5	5	89	0.14	0.23	–
9	800703	0.013	125	1600	14.1	5	258	0.16	–	–
0	800704	0.006	50	1600	14.1	5	125	0.21	–	–
1	800705	0.006	50	1600	14.1	5	8	0.014	–	–
2	800706	0.006	50	1600	14.1	5	129	0.36	–	–

[a]Diameters of concentric ring fractures measured after runs completed.

Fig. 11.11. Ejection stage of impact into an upper, weakly bonded layer of quartz sand
    underlain by viscous clay, showing directed, explosive-like ejecta plume (shot
    800702).

craters formed and the complexity of the fracture patterns in the crust
increased. Both radial and concentric fractures developed. Analyses of the
motion pictures show that in these experiments the angle of the ejecta plume
is relatively steep and has the appearance of explosive venting (Fig. 11.11).
Venting may result from release of volatiles from the clay, as was noted in the
earlier homogeneous viscous clay experiments. The plume shape may also be
restricted by upturned plates of the surface crust.

    To assess the transient subsurface processes for impacts into
multilayered targets, the quarter-space bucket was filled with water overlain
by a layer of silicon oil having a greater viscosity than water. Analyses of
high-speed motion pictures enabled assessment of the growth of the central
mound from the upper oil layer and the water substrate. In addition to
varying the thickness of the oil layer, the impact energies, and the impact
angles, oils of two viscosities were used (Table 11.5). Although not directly
correlated with material strength, the greater viscosity of both oils relative to
the water had the effect of simulating a stronger material overlying a fluid.

    Results of these experiments qualitatively resembled those for weakly

TABLE 11.5

Impacts into Silicon Oil Layers Overlying
Water in "Quarter-Space" Target Bucket

Run No.	Shot No.	Oil Viscosity (x 0.1 Pa s)	Impact Angle[a] (deg)	Approx. Impact Energy (joules)	Oil Layer Thickness (m)
1	800611	2	90	3.0	0.007
2	800612	2	90	3.0	0.010
3	800613	2	90	3.0	0.013
4	800614	2	90	3.0	0.017
5	800615	2	90	4.3	0.002
6	800616	2	90	27.1	0.002
7	800617	2	45	3.3	0.010
8	800618	2	45	3.3	0.010
9	800619	2	30	3.3	0.010
10	800621	2	30	3.3	0.010
11	800622	2	90	3.3	0.010
12	800623	30	90	3.3	0.010
13	800624	30	90	32.2	0.010
14	800625	30	90	4.1	0.020
15	800626	30	90	3.3	0.020
16	800627	30	90	3.8	0.020
17	800628	30	90	3.0	0.025
18	800629	30	90	3.2	0.025
19	800630	30	90	3.0	0.025
20	800631	30	90	4.2	0.025
21	800632	30	90	10.9	0.025
22	800633	30	90	29.9	0.025
23	800634	30	90	190.6	0.025

[a]Impact angle measured from the horizontal (vertical impact = 90°).

bonded shell particle layers overlying mud substrates (Series III). For very thin surface layers, an oscillating central mound of water formed, essentially unaffected by the oil layer (Fig. 11.12). For relatively thick surface layers of oil, the projectile did not penetrate through to the underlying water and a bowl-shaped crater formed, entirely within the surface layer. For a critical thickness of the oil surface layer, a bowl-shaped crater formed which barely intersected the oil-water interface, followed by a narrow plume of water that rose through the surface crater bowl and squirted upward. Part of this plume detached and eventually fell back into the crater, but most of it receded directly, leaving a small depression in the center of the initial crater bowl (Fig. 11.13). This pit-like morphology could be generated for different surface layer thicknesses by varying the impact energies.

As was suspected from observations of the experiments involving a layer of bonded particles overlying clay, volatiles were released from the fluid substrate upon impact, as shown by the zone of bubbles in the quarter-space experiments (Fig. 11.5).

IV. IMPACT CRATER MORPHOLOGY AND INTERIOR CHARACTERISTICS OF GANYMEDE

To provide a context for interpreting some of the impact cratering experiment results, we examined Voyager photographs of craters on Ganymede and assessed their general morphology. Four categories were defined:

1. Central-peak craters;
2. Central-pit craters;
3. Flat-floored craters (having neither peaks nor pits);
4. Indistinguishable floor type craters (for cases where small size and/or shadows prohibited the floor type to be defined).

Although subtypes to these basic classes are recognized (Passey and Shoemaker, Chapter 12) such as central pit craters with updomed floors, for the purpose of this study the classes were not subdivided.

Crater counts for Ganymede indicate a pronounced correlation between morphologic type and size. We examined 1082 craters in Galileo Regio (Fig.

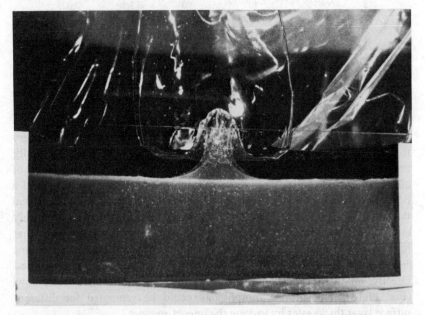

Fig. 11.12. Impact into quarter space multilayer target consisting of an upper stiff layer of silicon oil (η = 0.2 Pa s) underlain by water (η = 10^{-3} Pa s). Note that the material in the central mound is derived mostly from water substrate (shot 800613).

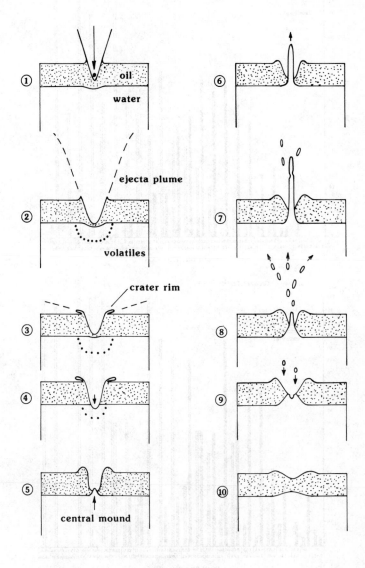

Fig. 11.13. Schematic sequence, drawn from high-speed motion pictures of shot 800629, showing development of central-pit-type crater in oil layer overlying water substrate. Within 200 ms after impact, bubbles exsolved from water. A narrow central plume of water partially broke up as it passed through the base of the oil crater bowl, then receded, with some fragments falling back into the crater bowl.

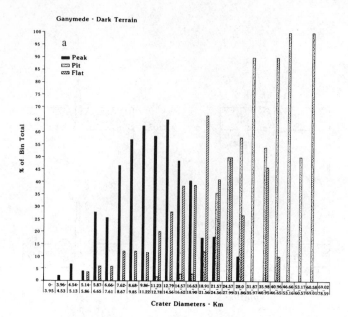

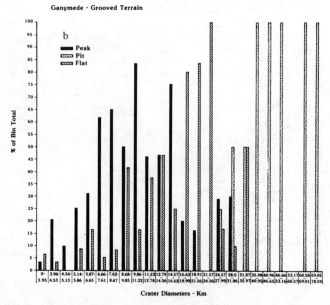

Fig. 11.14. Percentage histograms of crater types (central-peak, flat-floored, central-pit) for the area shown in Fig. 11.1. Not shown for each size range are the undifferentiated craters, i.e., those that could not be classed because of shadow or poor resolution. (a) Old, dark terrain; a total of 879 craters counted in 1.77×10^6 km^2; (b) Young, grooved terrain; a total of 203 craters counted in 1.20×10^6 km^2.

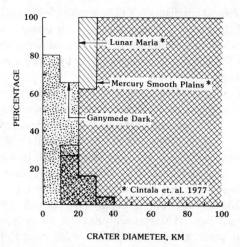

Fig. 11.15. Comparison of size-occurrence of central-peak craters on the Moon and Mercury with those on the dark terrain of Ganymede, showing that central-peak craters are much more prevalent in smaller craters on Ganymede than on the inner planets. (Moon and Mercury data from Cintala et al. 1977.)

11.9) and found the following progression: central peaks predominate in craters < 13 km in diameter, flat floors in craters ~ 16−25 km, and central pits in craters ≳ 32 km (Fig. 11.14). In comparison with the Moon and Mercury (Cintala et al. 1977; Wood and Andersson 1978), central peaks form in significantly smaller craters on Ganymede (Fig. 11.15). We also distinguished between craters in the older heavily cratered terrain and the younger grooved terrain and found similar sequences, although the transitions between crater types occur at smaller diameters in the older terrain. Thus the smallest central-pit craters in the older terrain are ~ 12 km in diameter whereas the smallest pit craters in the young terrain are nearly twice as large, or ~ 25 km in diameter. This relationship may indicate some change in target properties with time that has restricted the formation of central pits in smaller diameter (i.e., lower energy) craters. The sampling here is relatively small (1082 craters in 2.8×10^6 km^2), and this apparent age dependency of pit craters may not occur planetwide. If, however, this correlation proves valid, it may signify differences in the composition of the terrains, or changes with time that can be related to thermal evolution models (e.g., Phillips and Malin 1980).

Other crater counts for Ganymede show a similar progression of morphologic types. Passey and Shoemaker (Chapter 12) classified 1036 craters planetwide and found a sequence comparable to the one just described, with more apparent overlap between crater classes: the smallest craters (diameters ≲ 20 km) have bowl shapes, followed by craters with smooth floors (20–30 km diameter), central peaks 5–35 km) and central pits (16–120 km). Palimpsests, approximately circular features with high albedo and low relief, which are interpreted to be degraded remnants of large craters (Chapter 12, Passey and Shoemaker), occur in the size range of 80–335 km, possibly extending up to 440 km.

The progression from bowl-shaped to flat-floored to central-peak craters has been explained primarily by slow viscous relaxation and by gravitational slumping of the walls of craters formed in ice (Passey and Shoemaker, Chapter 12; Scott 1967). In the case of craters formed early in the history of Ganymede, prompt collapse might have occurred due to the lower effective strength of the satellite's crust at that time. Central-pit craters have been described for Mars where they have been attributed to the release of volatiles from a subsurface ice layer during impact (Wood et al. 1978). Passey and Shoemaker suggest that on Ganymede pits form by the prompt collapse of transient central peaks whose weight becomes too great to be supported by the subjacent material. The predominance of palimpsests in the larger size range has been ascribed to the size-dependence of viscous degradation rates (Johnson and McGetchin 1973; Parmentier et al. 1980).

Our experiments provide additional explanations for morphologic characteristics of Ganymede's craters, as discussed below in Sec. V. The progression from bowl-shaped to flat-floored to central-peak craters to palimpsests has been observed in experimental impacts into homogeneous targets of viscous materials that also possess yield strengths. The observed sequence may reflect changes in the deformational behavior of the surface of Ganymede in response to either varying impact conditions or thermal evolution of the crust. Central-pit formation requires a layered target in our experiments. Palimpsests form through instantaneous relaxation of crater relief in the more fluid target media and for the highest energy impacts, rather than by slow viscous flow.

One of the aims of our experiments was to relate crater morphology to differences in crustal structure at different times in Ganymede's history. Most thermal evolution models for Ganymede (Lewis 1971; Consolmagno and Lewis 1976, 1977, 1978; Reynolds and Cassen 1979; Cassen et al. 1980) involve a differentiated body having a lithosphere composed of Ice I, underlain by slush, a liquid water mantle, and a central zone of dense phases of ice and silicates. The thicknesses of these units and their compositions as a function of time depend upon the mode of planetary accretion, the extent of heating by impacts, and many other factors. According to several models, by $\sim 5 \times 10^8$ yr from the time of accretion, the liquid water phases have solidified (Cassen et al. 1980; Schubert et al. 1981), through a downward thickening lithosphere and an outward expanding solid interior, as shown diagramatically in Fig. 11.16. The rate of liquid mantle loss by freezing depends upon many factors including the original amount of radionuclides

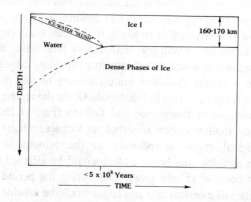

Fig. 11.16. Stylized diagram showing development of Ice-I crust for Ganymede from its time of origin. Depending upon the amount of radioactive elements present and the convective heat flux, during the first half million years the ice crust might have been underlain by liquid water. (Modified from Cassen et al. 1980.)

that provide heat and the presence of salts or other materials that would depress the melting temperature. In any case, it seems likely that early in the history of Ganymede, there would have been impacts that penetrated through the lithosphere into a liquid substrate (somewhat comparable to impacts into crusted magma oceans suggested for the Moon). As the crust thickened with time, only larger impacts would have penetrated to the liquid zone. Photogeological evidence may reflect this progressive thickening and stiffening of the crust of Ganymede. Rimmed furrows in the grooved terrain, for example, probably formed when the lithosphere was between 10 and 50 km thick (Passey and Shoemaker, Chapter 12). Shortly before complete mantle solidification at a depth of 160 to 170 km (Fig. 11.16), a crater at least 200 km in diameter would be required for penetration into the liquid zone. It is conjectural whether observed craters of the size necessary to penetrate such a crust date from the time prior to solidification of the mantle, but the possibility exists, and the morphology may provide some clues.

V. DISCUSSION

Although the experiments described here have been assessed principally in qualitative terms, there are several results that can be applied to the observations of impact craters on Ganymede and Callisto, or that have general implications for the outer planet satellites.

A. Morphologic Sequence

In the laboratory experiments the depth-to-diameter ratio is a function of both the target rheology and the crater dimensions; for a given diameter, with increases in viscosity and/or strength, or decreases in energy, the relative crater depth increases. Assuming that this relationship scales at least qualitatively, impacts of equal energy into slush or thinly crusted liquid water would generate a larger, shallower crater, relative to one formed in more viscous and/or stronger (i.e., rocky) materials. Given the apparently thin crust early in the histories of Ganymede and Callisto (Fig. 11.16; Cassen et al. 1980), the oldest shallow craters observed on Voyager images may reflect as much the original crater morphology as the subsequent slow viscous degradation. In addition, the larger craters would be affected more than the smaller craters because at any given time during the period of thickening crust, only craters of a certain size and larger would be capable of penetrating through the crust.

The smallest observable craters on Ganymede closely resemble lunar bowl-shaped and central-peak craters, whereas the larger craters have central pits and greatly reduced depth: diameter ratios, features not characteristic of the drier terrestrial planets. Furthermore, crater counts for the older dark terrain and younger grooved terrain of Ganymede indicate that morphologic transitions occur at smaller diameters in the dark terrain (Fig. 11.14). These observations may reflect a thickening of the crust on Ganymede during the interval between the formation of the older and younger terrains, so that impacts that penetrated to a weaker or more fluid zone in the early period could no longer reach that level at the later time. The diameter-related transitions may also be caused by changes in the target behavior at higher applied "lithostatic" stresses. In our experiments, craters formed over a range of crater diameters (impact energies) in nearly identical targets showed marked changes in morphology (Fig. 11.4*a,b*). Low-energy impacts on Ganymede might cause the body's crust to behave as a solid and produce bowl-shaped craters, while contemporaneous, higher energy impact might result in a large area behaving as a fluid, resulting in either a Martian-like ejecta flow crater or a crater with low topographic relief.

B. Central-Pit Craters

The experiments suggest three processes that might be involved in the formation of central-pit craters on icy satellites. Devolatilization occurred immediately upon impact into water (Figs. 11.5 and 11.13) and may have occurred during impact into viscous clay targets overlain by weakly bonded sand (Fig. 11.10). If impact into an icy satellite caused a similar degassing, as suggested by Hodges et al. (1980), and these volatiles coalesced under the updomed floor of a crater bowl, they might exert sufficient pressure to penetrate the initial crater from below, leaving a central depression.

The presence of a strong surface over a weaker substrate in several target configurations appears responsible for central pits. In these experiments the impact process occurred in two stages. First, a crater formed in the upper, more competent layer followed almost immediately by depression of the interface between the layers and then subsequent rise of a relatively narrow central plume. The plume then receded below the base of the initial crater floor, leaving a central depression similar to the model suggested by Passey and Shoemaker (Chapter 12). This process required a critical balance between surface layer properties and thickness and impact energy; if the layer was too strong or too thick, relative to the impact energy, the crater showed no influence of the more fluid substrate. For surface layers that were too weak, fluid, or thin, crater relief relaxed almost immediately. If pit formation on Ganymede required that the transient crater bowl penetrate an interior interface to reach a more fluid substrate, then a thickening of the surface layer with time should be reflected by a greater minimum size of central-pit craters in the younger grooved terrain relative to the older dark terrain, a relationship suggested by our crater counts.

In a few experiments in multilayer targets containing icy or weakly bonded sand layers over mud, or oil over water, parts of the rising central peak became detached, rose directly upward and then fell back into the crater bowl, forming a central secondary pit. Impact through a relatively thin ice crust on Ganymede to a more fluid interior might lead to the rise of a central peak that could launch a fragment of the surface layer with such a trajectory that it landed in the middle of the initial crater bowl. Some of the craters on Ganymede show a concentration of secondary craters, dimpling the area immediately surrounding a central pit (Fig. 11.17); these might have formed along with the pit by fallback of ejecta.

The second and third of these models cannot easily account for certain observations of central-pit craters in Voyager images. If central pits require penetration of a satellite's solid crust to a liquid interior, then we would expect to find very few fresh central-pit craters. However, Passey and Shoemaker (Chapter 12) point out several large bright ray craters (assumed to be the youngest, least degraded type) that have central pits. If pit formation required only the presence of an interior interface where thermal properties changed (rather than a zone of liquid water), this condition could be maintained as long as the satellite contained radiogenic heat that reached the surface through a combination of internal convection and surface layer conduction (Phillips and Malin 1980). Formation of central pits as secondary craters due to the break up of a rising central peak does not adequately explain the lack of offset or noncentrally located pits.

The degassing model, which has also been suggested for central-pit craters

Fig. 11.17. A 91-km diameter central-pit crater in the grooved terrain of Ganymede. Dimples surrounding the central pit may represent secondary impacts caused by fallback of ejecta.

on Mars, is more consistent with the observed craters on Ganymede. Craters of sufficient size, formed in a homogeneous ice target, will eventually collapse by slow viscous flow; part of this relaxation process involves updoming of the crater floor (Scott 1967). Volatiles released by the impact might then tend to flow to the center of this domed crater floor. If the concentration of volatiles were great enough, the bubble might break through the crater floor leaving behind a pit. The absence of central pits in small craters might result from low-energy impacts incapable of releasing sufficient volatiles to force their way through an icy surface layer.

C. Palimpsests

Crater palimpsests have been described as being remnants of craters that have lost nearly all their relief by some sort of degradation process. The experiments indicate that palimpsests might form by immediate relaxation of the largest diameter craters rather than by slow viscous creep (Fig. 11.18). This relaxation of the largest diameter craters could occur promptly due to the lower effective viscosity of the target when subjected to the thermal and mechanical stresses associated with a high-energy impact. This explanation can also account for the observations pointed out by Passey and Shoemaker (Chapter 12) of highly degraded palimpsests whose secondary craters are well preserved.

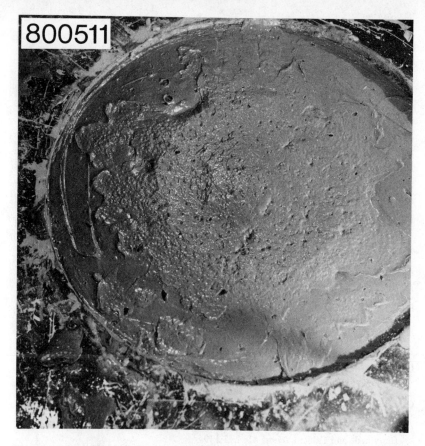

Fig. 11.18. High-energy impact (~ 400 joule, shot 800511) into viscous target. Note that, although there is little topographic expression preserved, the location of the transient cavity is marked by a rough surface texture resulting from outgassing and collapse of the central mound.

D. Crater Statistics

It has been previously suggested (Young et al. 1974; Boyce 1979) that crater size-frequency distributions could be affected by target properties. We find that the crater diameter in our laboratory experiments increases with decreasing viscosity and yield strengths of the targets, all other factors held constant. This relationship holds in experiments involving both homogeneous viscous targets and multilayered targets involving relatively competent material overlying viscous targets. In the latter case (Fig. 11.7), the crater diameter is greater for thinner crustal layers. Although this effect has not yet been assessed quantitatively, one of its consequences would be for surfaces to appear older because of the shift in the size-frequency distribution toward larger *apparent* sizes. This effect must be taken into account when comparing crater size-frequency distributions for Ganymede and Callisto with those for the inner silicate planets.

Acknowledgments. We wish to acknowledge with many thanks the "gun crew" at the NASA-Ames Vertical Ballistic Gun Facility for their tireless cooperation in carrying out the experiments discussed herein. We also thank R. Reynolds and P. Cassen for several discussions of the possible properties and thermal histories of the icy satellites, and E. Shoemaker and M. Cintala for their reviews of an earlier version of the manuscript. This work was supported by the Office of Solar System Exploration, National Aeronautics and Space Administration; J.E.G. was supported by the United Kingdom National Environment Research Council.

August 30, 1980

REFERENCES

Austin, M.G., Thomsen, J.M., and Ruhl, S.F. (1980) Impact cratering dynamics and cratering flow fields on (water) icy planetary surfaces. IAU Coll. 57. *The Satellites of Jupiter* (abstract 6-13).

Bird, R.B., Stewart, W.E., and Lightfoot, E.N. (1960). *Transport Phenomena.* Wiley and Sons, New York.

Boyce, J.M. (1979). Diameter enlargement effects on crater populations resulting from impact into wet or icy targets (abstract). *Reports of Planetary Geology Program. NASA TM* 80339, 119–121.

Carr, M.H., Crumpler, L.S., Cutts, J.A., Greeley, R., Guest, J.E., and Masursky, H. (1977). Martian impact craters and emplacement of ejecta by surface flow. *J. Geophys. Res.* 82, 4055–4065.

Carr, M.H., and the Viking Orbiter Imaging Team. (1976). Preliminary results from the Viking Orbiter imaging experiment. *Science* 193, 766–776.

Cassen, P., Peale, S.J., and Reynolds, R.T. (1980). On the comparative evolution of Ganymede and Callisto. *Icarus* 41, 232-239.

Chabai, A.J. (1977). Influence of gravitational fields and atmospheric pressures on scaling of explosion craters. In *Impact and Explosion Cratering* (D.J. Roddy, R.O. Pepin, and R.B. Merrill, Eds.) pp. 1191-1214. Pergamon Press. New York.

Charters, A.C., and Curtis, J.S. (1962). High velocity guns for free flight ranges. *Publication TM* 62–207, General Motors Corp. Defense Res. Lab., Santa Barbara, Calif.

Cintala, M.J. (1977). Martian fresh crater morphology and morphometry-a pre-Viking view. In *Impact and Explosion Cratering* (D.J. Roddy, R.O. Pepin, and R.B. Merrill, Eds.) pp. 575-591. Pergamon Press, New York.

Cintala, M.J., Head, J.W., and Parmentier, E.M. (1980). Impact heating of H_2O ice targets: Applications to outer planet surfaces. *Reports of Planetary Geology Program. NASA TM* 81776, 347–349 (abstract).

Cintala, M.J., Parmentier, E.M., and Head, J.W. (1979). Characteristics of the cratering process on icy bodies: Implications for outer planet satellites. *Lunar Science* X, 207-209 (abstract).

Cintala, M.J., Wood, C.A., Head, J.W., and Mutch, T.A. (1977). Interplanetary comparisons of fresh crater morphology: Preliminary results. *Lunar Science* VIII, 181-183 (abstract).

Consolmagno, G.J., and Lewis, J.S. (1976). Structural and thermal models of icy Galilean satellites. In *Jupiter* (T. Gehrels, Ed.), pp. 1035–1051. Univ. Arizona Press, Tucson.

Consolmagno, G.J., and Lewis, J.S. (1977). Preliminary thermal history models of icy satellites. In *Planetary Satellites* (J.A. Burns, Ed.), pp. 492–500. Univ. Arizona Press, Tucson.

Consolmagno, G.J., and Lewis, J.S. (1978). The evolution of icy satellite interiors and surfaces. *Icarus* 34, 280–293.

Croft, S.K., Kieffer, S.W., and Ahrens, T.J. (1979). Low velocity impact craters in ice and ice saturated sand with implications for Martian crater count ages. *J. Geophys. Res.* 84, 8023–8032.

Curtis, J.S. (1964). An analysis of the interior ballistics of the constant base pressure gun, *Publication TR* 64-27, General Motors Corp. Defense Res. Labs., Santa Barbara, Calif.

Fink, J.H., Greeley, R., and Gault, D.E. (1980). Fluidized impact craters in Bingham materials and the distribution of water on Mars (abstract). In *3rd Coll. on Planetary Water,* Niagara Falls, New York.

Gaffney, E.S. (1980). Craters in ice and floating ice. IAU Coll. 57. *The Satellites of Jupiter* (abstract 7–5).

Gault, D.E. (1970). Saturation and equilibrium conditions for impact cratering on the lunar surface: Criteria and implications. *Radio Science* 5, 273–291.

Gault, D.E. (1973). Displaced mass, depth, diameter, and effects of oblique trajectories for impact craters formed in dense crystalline rocks. *The Moon* 6, 32–44.

Gault, D.E., and Greeley, R. (1978). Exploratory experiments of impact craters formed in viscous-liquid targets: Analogs for Martian rampart craters. *Icarus* 34, 486–495.

Gault, D.E., and Heitowit, E.D. (1963). The partition of energy for hypervelocity impact craters formed in rock. *Proc. 6th Hypervelocity Impact Symp.* 2, 419–456.

Gault, D.E., Heitowit, E.D., and Moore, H.J. (1963). Some observations of hypervelocity impact with porous media. *NASA TMX*–54, 009.

Gault, D.E., and Moore, H.J. (1964). Scaling relationships from microscale to megascale impact craters. Presented at *7th Hypervelocity Impact Symp.* Tampa, Florida.

Gault, D.E., Quaide, W.L., and Oberbeck, V.R. (1968). Impact cratering mechanics and structures. In *Shock Metamorphism of Natural Materials* (B.M. French, and N.M. Short, Eds.), pp. 87–99. Mono, Baltimore.

Gault, D.E., and Wedekind, J.A. (1977). Experimental hypervelocity impact into quartz sand, II: Effects of gravitational acceleration. In *Impact and Explosion Cratering* (D.J. Roddy, R.O. Pepin, and R.B. Merrill, Eds.), pp. 1231–1244. Pergamon Press, New York.

Gault, D.E., and Wedekind, J.A. (1978). Experimental impact "craters" formed in water: Gravity scaling realized. *EOS* 59, 1121 (abstract).

Greeley, R., Fink, J.H., Gault, D.E., Snyder, D.W., Guest, J.E., and Schultz, P.H. (1980). Impact cratering in viscous targets: Experimental results. *Proc. Lunar Planet. Sci. Conf.* 11. 2075-2097.

Green, J. (1965). Hookes and Spurrs in Selenology. In *Geological Problems in Lunar Research* (F. Whipple, Ed.), *Annals New York Acad. Sci.* 123, 373–402.

Head, J.W. (1976). The significance of substrate characteristics in determining morphology and morphometry of lunar craters. *Proc. Lunar Sci. Conf.* 7, 2913–2929.

Head, J.W., and Roth, R. (1976). Mars pedestal crater escarpments: Evidence for ejecta-related emplacement. In *Symposium on Planetary Cratering Mechanics,* pp. 50–52 (abstract). The Lunar Science Institute, Houston.

Hodges, C.A. Shew, N.B., and Clow, G. (1980). Distribution of central pit craters on Mars. *Lunar Science* XI, 450-452 (abstract).

Holsapple, K.A., and Schmidt, R.M. (1979). A material strength model for apparent crater volume. *Lunar Planet. Sci. Conf.* 10, 2757-2777.

Horner, V., and Greeley, R. (1981). Rampart craters on Ganymede: Implications for the origin of Martian rampart craters. *Lunar Science* XIII (abstract). In press.

Johnson, A.M. (1970). *Physical Processes in Geology.* Freeman, San Francisco.

Johnson, T.V., and McGetchin, T.R. (1973). Topography on satellite surfaces and the shape of asteroids. *Icarus* 18, 612-620.

Killian, B.G., and Germain, L.S. (1977). Scaling of cratering experiments: An analytical and heuristic approach to the phenomenology. In *Impact and Explosion Cratering* (D.J. Roddy, R.O. Pepin, and R.B. Merrill, Eds.), pp. 1165-1190. Pergamon Press, New York.

Lewis, J.S. (1971). Satellites of the outer planets: Thermal models. *Science* 172, 1127-1128.

McKinnon, W.B. (1978). An investigation into the role of plastic failure in crater modification. *Proc. Lunar Planet. Sci. Conf.* 9, 3965-3973.

Melosh, H.J. (1980). Cratering mechanics: Observational, experimental, and theoretical. *Ann. Rev. Earth Planet. Sci.* 8, 65-93.

Murray, J.B. (1980). Oscillating peak model of basin and crater formation. *Moon and Planets* 22, 269-291.

Mutch, T.A. (1972). *Geology of the Moon*. Princeton Univ. Press, Princeton, N.J.

Oberbeck, V.R. (1970). Lunar dimple craters. *Modern Geology* I. 161-171.

Oberbeck, V.R. (1971). Laboratory simulation of impact cratering with high explosives. *J. Geophys. Res.* 76, 5732-5749.

Oberbeck, V.R. (1973). Simultaneous impact and lunar craters. *The Moon*, 6, 83-92.

Oberbeck, V.R. (1975). The role of ballistic erosion and sedimentation in lunar stratigraphy. *Rev. Geophys. Space Phys.* 13, 337-362.

Oberbeck, V.R. (1977). Application of high explosive cratering data to planetary problems. In *Impact and Explosion Cratering* (D.J. Roddy, R.O. Pepin, and R.B. Merrill, Eds.), pp. 45-65. Pergamon Press, New York.

Oberbeck, V.R., and Quaide, W.L. (1968). Genetic implications of lunar regolith thickness variations. *Icarus* 9, 446-465.

Parmentier, E.M., Allison, M.L., Cintala, M.J., and Head, J.W. (1980). Viscous degradation of impact craters on icy satellite surface. *Lunar Science* XI 857-859 (abstract).

Parmentier, E.M., and Head, J.W. (1979). Internal processes affecting surfaces of low-density satellites: Ganymede and Callisto. *J. Geophys. Res.* 84, 6263-6276.

Phillips, R.J., and Malin, M.C. (1980). Ganymede: A relationship between thermal history and crater statistics. *Science* 210, 185-188.

Quaide, W.L., Gault, D.E., and Schmidt, R.A. (1965). Gravitative effects on lunar impact structures. In *Geological Problems in Lunar Research* (F. Whipple, Ed.), *Annals New York Acad. Sci.* 123, 563-572.

Quaide, W.L., and Oberbeck, V.R. (1968). Thickness determinations of the lunar surface layer from lunar impact craters. *J. Geophys. Res.* 73, 5247-5270.

Reynolds, R.T., and Cassen, P.M. (1979). On the internal structure of the major satellites of the outer planets. *Geophys. Res. Letters* 6, 121-124.

Roddy, D.J., Pepin, R.O., and Merrill, R.B. (eds.) (1977). *Impact and Explosion Cratering, Planetary and Terrestrial Implications*. Pergamon Press, New York.

Schmidt, R.M., and Holsapple, K.A. (1980). Theory and experiments on centrifuge cratering. *J. Geophys. Res.* 85, 235-252.

Schubert, G., Stevenson, D.J., and Ellsworth, K. (1981). Internal structures of the Galilean satellites. Submitted to *Icarus*.

Schultz, P.H., and Gault, D.E. (1979). Atmospheric effects on Martian ejecta emplacement. *J. Geophys. Res.* 84, 7669-7687.

Scott, R.F. (1967). Viscous flow of craters. *Icarus* 7, 129-148.

Shoemaker, E.M., and Wolfe, R.F. (1980). Comets and the Galilean satellites. *Bull. Amer. Astron. Soc.* 12, 712 (abstract).

Short, N.M. (1975). *Planetary Geology*. Prentice Hall, New York.

White, J.W. (1971). Examination of cratering formulas and scaling models. *J. Geophys. Res.* 76, 8599-8603.

Wood, C.A., and Andersson, L. (1978). New morphometric data for fresh lunar craters. *Proc. Lunar Planet. Sci. Conf.* 9, 3669-3689.

Wood, C.A., Head, J.W., and Cintala, M.J. (1978). Interior morphology of fresh Martian craters: The effects of target characteristics. *Proc. Lunar Planet. Sci. Conf.* 9, 3691-3701.

Young, R.A., Brennan, W.J., and Nichols, D.J. (1974). Problems in the interpretation of lunar mare stratigraphy and relative ages indicated by ejecta from small impact craters. *Proc. Lunar Sci. Conf.* 5, 159-170.

12. CRATERS AND BASINS ON GANYMEDE AND CALLISTO: MORPHOLOGICAL INDICATORS OF CRUSTAL EVOLUTION

QUINN R. PASSEY
California Institute of Technology

and

EUGENE M. SHOEMAKER
U.S. Geological Survey

Craterform and related features on Ganymede and Callisto include bowl-shaped craters, craters with nearly flat floors, craters with central peaks, craters with central pits, basins, crater palimpsests and penepalimpsests, and giant multiring systems of ridges and furrows. The large majority of all craters > 20 km diameter have a central pit. The pits are interpreted as formed by prompt collapse of transient central peaks. Most craters, in all size ranges, are highly flattened as a consequence of topographic relaxation by slow viscous or plastic flow. During early heavy bombardment, the lithospheres of Ganymede and Callisto were sufficiently thin, owing to the combined effect of impact heating and heat flow from the deep interior, to prevent retention of any recognizable craters. The loss of these craters occurred as a consequence of relatively rapid viscous relaxation of individual craters, smothering of groups of highly flattened craters beneath ejecta deposits, and of regional effects of multiring structures. The oldest retained craters are relatively small (< 10 km) and occur in the polar region of Callisto and near the antapex of motion of each satellite. As the lithosphere of each satellite cooled and thickened, crater retentivity spread as a wave from the polar regions and the antapex toward the apex; at any given location, progressively larger craters were retained with the passage of time. The asymmetric pattern of the thickening and stiffening of the lithosphere, inferred from the pattern of retention of the craters, probably reflects asymmetric development of an insulating regolith and the regional effects of large impact structures.

One of the more startling discoveries of the Voyager missions is the host of craters and related features on Ganymede and Callisto. As the densities of these two largest satellites of Jupiter strongly suggest that they are made up of about one-quarter to one-half H_2O, it was widely supposed that their crusts are composed largely of Ice I. Detailed spectrophotometry, moreover, has confirmed that H_2O ice is an important constituent of the surfaces of these two bodies (Clark 1980; see Chapter 7 by Sill and Clark). The extent to which these icy crusts might preserve observable topographic relief and a decipherable geologic record was largely a matter of conjecture prior to the Voyager missions. The remarkable revelation from the Voyager television pictures is that the cratering record on both Ganymede and Callisto appears to extend back to the period of late heavy bombardment, about four billion years before the present.

A second fact immediately evident from the Voyager pictures is that the forms of the ancient craters have been greatly modified, as might be expected for major topographic features formed in ice. The oldest craters are invariably extremely flattened (Shoemaker and Passey 1979). Relaxation of topographic relief on Ganymede and Callisto by viscous flow or creep, first suggested by Johnson and McGetchin (1973), is basically confirmed by the Voyager evidence. Broad features of the topography have flattened or have disappeared, while small features such as secondary craters and the hummocks of crater rim deposits are preserved, as predicted by viscous relaxation theory.

Although Ganymede and Callisto are similar in size and bulk density, the surfaces of these two bodies are significantly different. Callisto's surface consists essentially of a single type of heavily cratered terrain that evidently dates from a period of heavy bombardment. Ganymede's surface, on the other hand, consists of two very distinct types of terrain, (1) an ancient heavily cratered terrain, and (2) a younger grooved terrain (Smith et al. 1979a,b). Slightly less than half the surface of Ganymede is ancient cratered terrain and more than half is grooved terrain. The ancient cratered terrain of Ganymede is similar to the cratered terrain of Callisto except that the mean crater density on Ganymede is only about half as great.

Craters and related features on Ganymede are remarkably similar to those on Callisto, but the morphology of most craters on Ganymede and Callisto differ in important respects from similar-sized craters on the Moon, Mars, and Mercury. These differences appear to be due, in part, to the contrast in physical properties between the icy satellites and the rocky surfaces of the Moon and terrestrial planets that affect the initial form of the craters, as well as to the slow relaxation of topographic relief on the icy satellites. Despite these differences, it is clear that the craters on Ganymede and Callisto are closely related in origin to the vast majority of craters on the Moon, Mars and Mercury. Ray craters on Ganymede and Callisto, in particular, resemble ray craters on the Moon, both in crater form and in the pattern of the rays. Relationships of rim deposits and secondary craters observed around large relatively fresh craters and basins on Ganymede leave little room for doubt that almost all of the observed craters are either of primary or of secondary impact origin. The collision of comet nuclei has been suggested as primarily responsible for the formation of the ray craters on Ganymede and Callisto (Shoemaker and Wolfe, Chapter 10).

A number of first-order conclusions about Ganymede and Callisto may be obtained directly from a careful study of the forms and relations of their numerous craters. For example, many 10-km craters have extremely subdued relief. The flattening or relaxation of these craters implies very steep thermal gradients or much higher surface temperatures at an early period in the history of both satellites. Many large craters formed initially on the ancient cratered terrain of Ganymede have flattened until the craters are no longer recognizable, but their former presence is indicated by distinctive high albedo patches termed palimpsests. On the other hand, there are also large craters that extend in size up to basins (diameter > 150 km), which were formed relatively early in the cratering history and which have not disappeared. The preservation of the basins indicates that there was a large decrease in thermal gradient and corresponding increase in lithosphere thickness after the 10-km craters had flattened. In this chapter we will first survey the morphologic characteristics of craters and palimpsests on Ganymede and Callisto; then we will explore what might be ascertained from the observations of the craters about the crustal properties of these two icy bodies and, especially, about the evolution of their crustal properties with time.

I. MORPHOLOGY OF CRATERS AND RELATED FEATURES

For purposes of description we have divided the craters and related features on Ganymede and Callisto into several morphological categories. These include (1) bowl-shaped craters, (2) smooth-floored craters, (3) craters with central peaks, (4) craters with central pits, (5) chain craters on Callisto, (6) the Gilgamesh and Western Equatorial Basins on Ganymede, (7) crater palimpsests and penepalimpsests, (8) multiring structures on Callisto, and (9) the Galileo Regio rimmed furrow system on Ganymede.

Bowl-shaped Craters

The smallest craters discernable on Ganymede are simple bowl-shaped craters, generally with resolved raised rims. The diameters of bowl-shaped craters range from < 5 km to ~ 20 km. Essentially all craters smaller than ~ 5 km in diameter appear to fall into the bowl-shaped category. This conclusion may be partially due, however, to the limiting resolution of the imaging system. Craters and other features on Ganymede < 5 km in diameter

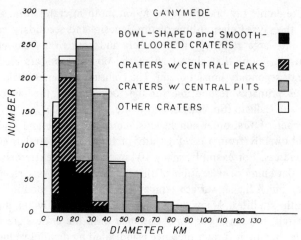

Fig. 12.1. Size-frequency distribution of different morphologic classes of craters on Ganymede. The decrease in the number of craters < 20 km in diameter is due to the limited number of high-resolution images. The category "other craters" represents craters that could not be classified owing to low resolution or obscuring by reseau marks.

are less than ten pixels across in the highest resolution images; details in the crater floors cannot be resolved with this limited number of pixels. Approximately 20% of craters 5 to 10 km in diameter and 30% of craters 10 to 20 km in diameter are bowl-shaped (Fig. 12.1). Simple craters > 20 km in diameter appear to be flat-floored or to have convex floors rather than bowl shapes.

Average depth-to-diameter ratios for fresh appearing bowl-shaped craters that are approximately 10 km in diameter are 1:6 to 1:12. (Depths are measured from the crater rims.) The height of the rim crest above the surrounding surface generally is ≲ 300 m; for very fresh craters the rim height may exceed 500 m. These depths and heights were derived from photoclinometric crater profiles (see Watson 1968; Bonner and Schmall 1973) and from shadow measurements and probably have errors of the order of ± 100 m.

Most of the bowl-shaped craters probably are of primary impact origin, but many craters also occur in swarms around much larger craters and palimpsests. By analogy with crater swarms observed on the Moon and Mercury, the swarm craters are inferred to be secondary craters related to the ejection of material from the much larger primary craters. On Ganymede, the ratio of the diameter of the largest secondary craters to the central primary crater is ~ 1:15, comparable to that observed on the Moon (Shoemaker 1966). Prominent resolved secondary crater swarms are associated with most

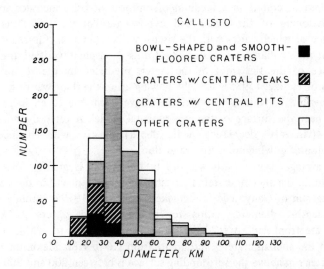

Fig. 12.2. Size-frequency distribution of different morphologic classes of craters on Callisto. The decrease in the number of craters $<$ 30 km in diameter is due to the limited number of high-resolution images. The category "other craters" represents craters that could not be classified owing to image smear or low resolution.

craters $\gtrsim$ 70 km in diameter. The largest secondary craters occur as far as three primary crater diameters beyond the rim of the primary crater.

On Callisto, craters $\lesssim$ 15 km in diameter are difficult to categorize owing to the limited resolution of the Voyager images (Fig. 12.2). Probably many of these craters are bowl-shaped, as on Ganymede.

Smooth-floored Craters

We designate craters with flat or slightly convex floors that do not exhibit either a central peak or central pit as smooth-floored craters. Generally, the smooth-floored craters are larger than the bowl-shaped ones. The transition from bowl shape (or concave floors) to smooth, nearly level floors occurs at crater diameters in the range 15 to 20 km. The diameter for which this transition occurs probably reflects physical properties of the crusts of Ganymede and Callisto at the times the craters were formed. The smooth-floored craters range in diameter from $\sim$ 20 to 40 km. About 20% of Ganymedian craters 20 to 30 km in diameter have smooth floors, compared to only 5% of craters 30 to 40 km in diameter (Fig. 12.1).

Photoclinometric crater profiles show that many smooth-floored craters on Ganymede are distinctly convex or bowed-up in the center (Fig. 12.3). Theoretical studies of the collapse of craters by slow viscous flow have been carried out by Danes (1962, 1965) and Scott (1967). Danes' work indicates that collapse of craters in a medium of uniform viscosity includes both an overall flattening of the crater and a bowing-up of the floor, the time dependence of which varies with the viscosity. Experimental studies by Scott (1967) of the collapse of crater-like forms in asphalt verified the time dependence of the flattening of the craters predicted by the viscous flow equations of Haskell (1935). A slight bowing-up of the floor was observed in these experiments which, however, may have been due to an increase of viscosity near the surface of the experimental model. A vertical gradient of viscosity (viscosity decreasing with depth) clearly will enhance the phenomenon of bowing-up of the crater floor.

The average degree of bowing-up in the center is greater for craters located within the ancient cratered terrain than for craters within the younger grooved terrain of Ganymede (Shoemaker and Passey 1979). Within grooved terrain, depth-to-diameter ratios for smooth-floored craters 20 km in diameter are from 1:6 to 1:20. Depth-to-diameter ratios for similar craters located in the ancient cratered terrain are from 1:6 to 1:80. Maximum height of the crater rim above the surrounding surface is between 200 and 300 m for relatively fresh smooth-floored craters; for degraded craters, the rim heights are substantially less.

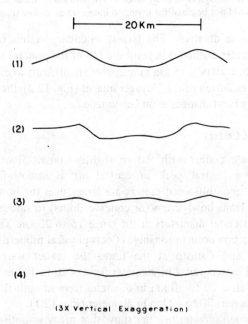

(3X Vertical Exaggeration)

Fig. 12.3. Photoclinometric profiles of four craters on Ganymede with diameters of ~ 20 km. Profile (1) is of a crater located in grooved terrain of the Harpagia Sulci; the other three profiles are for craters located within the ancient cratered terrain of Nicholson Regio.

Smooth-floored craters on Callisto range from ∼ 18 km to 40 km in diameter; they comprise ∼ 30% of craters 20 to 30 km in diameter and ∼ 8% of craters 30 to 40 km in diameter (Fig. 12.2). No photoclinometric profiles have been obtained for craters on Callisto at the time of the writing of this chapter. Measurements of shadow lengths near the terminator, however, reveal that these craters are also anomalously shallow compared to fresh craters of similar size on the Moon. Depth-to-diameter ratios for smooth-floored craters 20 to 40 km in diameter range from ∼ 1:15 to 1:45. Maximum depth below the crater rim for smooth-floored craters on Callisto is ∼ 2 km.

Craters with Central Peaks

On Ganymede, craters displaying a central peak are common in the diameter range of 5 to 35 km. No craters with diameters > 35 km have been found to have a central peak. This observation is in contrast to craters on the Moon, where the majority of fresh craters > 25 km in diameter and essentially all craters > 55 km in diameter have central peaks (Pike 1975). On Mercury, a majority of craters > 40 km in diameter exhibit a central peak (Gault et al. 1975; Cintala et al. 1977). Approximately 70% of Ganymedian craters 5 to 10 km in diameter, 50% of the craters 10 to 20 km in diameter, 7% of craters 20 to 30 km in diameter, and 3% of craters 30 to 40 km in diameter have central peaks (Fig. 12.1). Although the diameter range of craters with central peaks overlaps the diameter range of smooth-floored craters, the smooth-floored craters tend to comprise a larger proportion of craters 20 to 30 km in diameter than do craters with central peaks.

Central peaks range in basal diameter from ∼ 1 km, for small craters, to ∼ 5 km in larger craters. The maximum heights of the peaks are ∼ 700 m, as indicated from shadow measurements. The resolution of these craters is insufficient to show details in the central peaks. Collapsed terrace walls are associated with craters with central peaks. The widths of the resolved terraces vary from ∼ 1 to 4 km. The smallest crater with a central peak that exhibits a resolved terraced wall is ∼ 10 km in diameter.

On Callisto, craters with distinct central peaks have been found within the size range of from 11 to ∼ 40 km in diameter. Approximately 40% of craters 20 to 30 km in diameter and ∼ 15% of craters 30 to 40 km in diameter have central peaks (Fig. 12.2). Depth-to-diameter ratios for craters 30 km in diameter, with central peaks, range from ∼ 1:10 to 1:30.

Craters with Central Pits

Craters that have a central depression or pit are abundant on both the ancient cratered terrain and on the grooved terrain on Ganymede. These craters range in diameter from 16 to > 120 km. Approximately 25% of craters 10 to 20 km in diameter, 60% of craters 20 to 30 km diameter, and ~90% of 30 to 40 km diameter craters have central pits. Essentially all craters > 40 km in diameter have a central pit. From Fig. 12.1, it may be seen that craters with central peaks are replaced by craters with central pits at ~ 20 km crater diameter. It is apparent, from the highest resolution Voyager images of Ganymede, that most if not all craters with central pits also have collapsed terrace walls. The width of the identifiable terrace zones range from ~ 1 to 7 km.

The diameter of the central pit is positively correlated with the crater diameter, as shown in Figs. 12.4 and 12.5 (see also Boyce 1980). From the distribution of the pit diameter versus the crater diameter, the relationship appears to be curvilinear. Using the data for ray craters and craters with high-albedo rims, the best-fit exponential relationship for pit diameter versus crater diameter for Ganymedian craters is

$$d = 1.9 \exp(0.023\, D) \qquad (1)$$

where d is the diameter of the central pit, and D is the crater diameter.

The diameter of the pit relative to the diameter of the crater is also dependent upon the age of the crater. For a given crater diameter, recognizably older craters have central pits with diameters that are larger than the central pits in recognizably younger craters. The smallest pits for a given crater size are generally encountered in ray craters (Fig. 12.6d). Differences in the pit size, as a function of crater age, probably are related to differences in crustal properties at the different times that the craters were formed. All of the craters that have anomalously large pits on Ganymede — pits much larger than predicted by Eq. (1) — are located in the ancient cratered terrain.

Profiles of craters with central pits (Fig. 12.7) reveal that, although the pits are depressions, they generally are surrounded by a raised rim. The depth of the pit below this rim is, in most cases, < 1 km. In large pits, the pit floors are bowed-up in much the same manner as are the crater floors. Among the pits we have measured so far, the floors of the largest pits appear to be very near the level of the crater floors outside of the pits. Measured depth-to-diameter ratios of Ganymedian craters with central pits that are 30 to 50 km in diameter range from 1:9 to 1:30, for craters within grooved terrain. Depth-to-diameter ratios for similar craters within the ancient cratered terrain range from ~ 1:12 to 1:85. The height of the crater rim above the level of the surrounding surface varies from a few hundred m (or less) to ~ 2 km.

Whereas most of Ganymede's craters > 40 km in diameter exhibit a single central pit, one crater, centered at 54° N lat and 192° W long, appears to have a small pit centered within a larger central pit. The diameter of this crater is 115 km; the relatively large rimmed central pit is 50 km in diameter

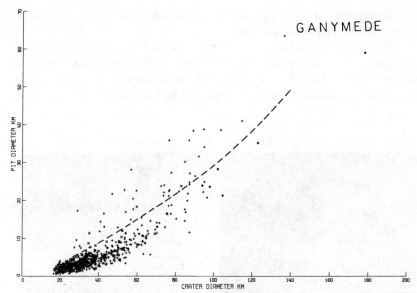

Fig. 12.4. Diameter of central pit versus crater diameter for 640 craters on Ganymede. Craters that fall above the broken line are recognizably older than those below this line. Ray craters generally have the smallest central pit diameter, for a given crater diameter, and fall at the lower boundary of the distribution.

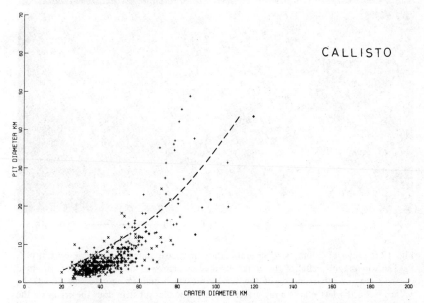

Fig. 12.5. Diameter of central pit versus crater diameter for ∼ 450 craters on Callisto. Craters that fall above the broken line are recognizably older than those below this line; ray craters generally are located at the lower boundary of the distribution.

and the small pit at the center is 4.5 km in diameter. This crater is located in grooved terrain and has bright rays.

Craters with central pits are also common on Callisto. These craters range in diameter from 18 km to $\sim$ 160 km. About 25% of craters 20 to 30 km in diameter and $\sim$ 80% of craters 30 to 40 km in diameter have central pits. Essentially all craters $\gtrsim$ 40 km in diameter have a central pit (Fig. 12.2).

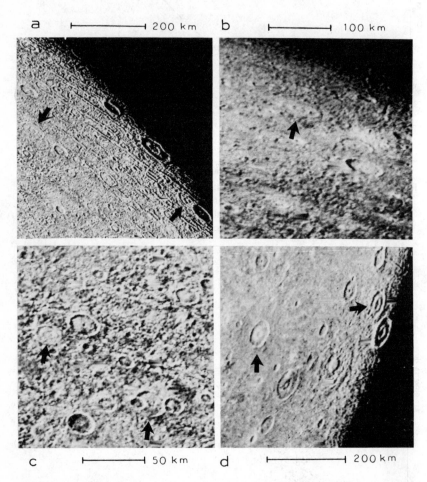

Fig. 12.6. Voyager 2 images of several craters that have anomalously large central pits. The topographic relief of the craters marked by arrows in a, b, and c is more subdued than that of other craters of the same size, which indicates greater age. In (d) the arrow on the left marks a ray crater that has a smaller pit than does the crater on the right of nearly similar size. A type II penepalimpsest, 166 km in diameter, is also visible in (d) to the right and below the center of the frame (FDS 20636.50, 20636.44, 20636.59, and 20631.33).

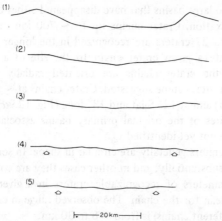

Fig. 12.7. Photoclinometric profiles of five craters with central pits on Ganymede that illustrate various degrees of topographic relaxation. The arrows in profiles (4) and (5) mark the crater rims in these extremely flattened craters. Note that, in the flattened craters, the relative size of the central pit is much larger than in the less relaxed craters (profiles 1, 2, and 3). In profile 1, taken from a crater with a bright rim deposit, the central pit has an unusually high rim. This high rim may be the remnant of a partially collapsed central peak. The vertical scale is the same as the horizontal scale for these profiles.

Crater depths up to 3 km below the crater rims have been calculated from shadow measurements.

As is the case on Ganymede, on Callisto the diameter of the central pit is positively correlated with the crater diameter, and a curvilinear fit to the data is suggested when comparing pit diameter to crater diameter, particularly for the largest craters (Fig. 12.5). On Callisto, however, there is a larger spread in pit diameters for craters > 60 km in diameter, as compared to Ganymede's craters of this size. The smallest pits for a given crater size appear to be associated with the fresh appearing ray craters. The best fit of an exponential function for the pit diameter versus the crater diameter for ray craters and craters with bright rim deposits on Callisto is

$$d = 1.45 \exp(0.028\,D) \tag{2}$$

where d and D are, respectively, the pit diameter and the crater diameter. Most central pits on both Callisto and Ganymede are circular, but some pits have irregular shapes (Fig. 12.8).

Chain Craters on Callisto

Several relatively large chains of craters (catena) have been identified on Callisto (see Table 12.1). These are almost certainly chains of secondary craters related to large basins that have disappeared either by prompt collapse or by slow relaxation. Crater chains up to ~ 700 km in length have been found and up to 27 craters are recognized in the longer crater chains. It is difficult to relate a given crater chain to the site of a former basin, but assuming that the crater chains are oriented radially to the source of impacting ejecta, sites can be suggested. Crater chains at ~ 50° N lat and 350° long (Fig. 12.9*b*) and at 12° S lat and 13° long (Fig. 12.9*d*) may be related to Valhalla. The sites of the original primary basins associated with the other crater chains are not yet identified.

The chain craters generally are circular in shape. In some cases individual craters overlap substantially, and in other cases they are completely separated (Fig. 12.9). Diameters of recognizable craters in a given chain are within ± 15% of the mean for the chain. The observed range in crater diameters, for craters from different chains, is from ~ 8 to 30 km.

Most chain craters ~ 20 km in diameter have central peaks, whereas most chain craters 30 to 35 km in diameter have a slight central depression, possibly a central pit. The morphology of large craters in secondary crater chains is basically the same as for primary craters of the same size. Thus any isolated secondary craters > 10–15 km in diameter are not expected to be morphologically distinguishable from primary craters of the same size.

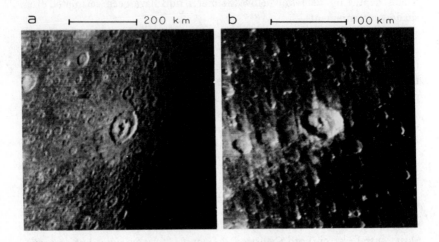

Fig. 12.8. Voyager 1 images of two craters on Callisto with strongly modified central pits. The pit illustrated in (a) appears to be modified either by superposed craters or by unusually extensive and irregular collapse. The form of the central pit in (b) appears to have been influenced by faulting, which extends beyond the confines of the crater rim; considerable enlargement of this pit appears to have occurred along the trend of the fault (FDS 16421.27 and 16426.36).

TABLE 12.1

Prominent Crater Chains on Callisto

No.	Lati-tude[a]	Longi-tude[a]	No. of Craters	Chain Length (km)[b]	Largest Crater Diameter[c] (km)	Smallest Crater Diameter[c] (km)
1	+67	57	14	700	31	24
2	+50	350	6	110	25	18
3	+32	347	27	370	17	15
4	−12	13	9	225	20	14
5	−18	343	10	175	10	8

[a]From U.S. Geological Survey Map I-1239, 1979.
[b]Lengths ± 10%.
[c]Diameters generally ± 2 km.

Gilgamesh and Western Equatorial Basins on Ganymede

One fresh relatively unmodified basin, Gilgamesh, is found on Ganymede, located within grooved terrain (centered at 59° S, 123°W). The floor of the conspicuous central depression is relatively smooth and flat and is ~ 150 km in diameter (Fig. 12.10). Several small peaks are arranged in an arc in the center of the floor; the diameter of the arc is ~ 50 km. Faint radial lineations also appear to be present on the floor (Fig. 12.10c). A poorly defined rim, ~ 175 km in diameter, bounds the central depression. The depression evidently corresponds to the central pit of the impact basin.

A rugged hilly-to-mountainous region surrounds the central depression and extends radially for ~ 500 km. Individual features of the relief have an angular or blocky appearance; the heights of the larger individual blocky massifs are 0.5 to 1.5 km. Several inward facing concentric scarps occur within the outer blocky annulus. The scarps are discontinuous in circumferential extent and have heights from < 1 to ~ 1.5 km. The most prominent scarp at ~ 275 km radius is considered by Shoemaker et al. (Chapter 13) to mark the rim of the Gilgamesh basin.

A great swarm of secondary craters and crater chains surrounds the Gilgamesh basin. These craters extend from a radial distance of ~ 400 km to ~ 1000 km from the center of the basin. The diameters of most secondary craters are < 5 km, but some craters up to ~ 15 km in diameter are present in the swarm.

A second basin, slightly larger than the central depression of Gilgamesh, but greatly flattened by viscous relaxation, occurs on grooved terrain centered at 7°S lat and 115° long (Fig. 12.11). The center of this basin is relatively flat but appears rough, owing to ejecta deposits of craters that postdate this basin. The diameter of this central region is ~ 185 km. A well-defined rim or scarp of height < 600 m surrounds the central region. A rim deposit, composed of low hills and hummocks, surrounds the central basin and extends radially to ~ 265 km; the relief within the rim deposit is generally < 400 m. A swarm of secondary craters surrounds this basin; secondary craters can be found from a radial distance of ~ 150 km to 450 km from the center of the basin. The maximum diameter of the secondary craters is ~ 10 km.

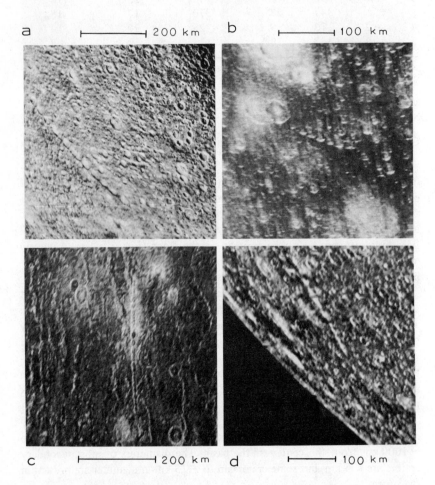

Fig. 12.9. Voyager 1 images of four large crater chains on Callisto. The diameters of the individual craters within a given chain are fairly uniform (FDS 16428.19, 16426.10, 16424.32, and 16424.26).

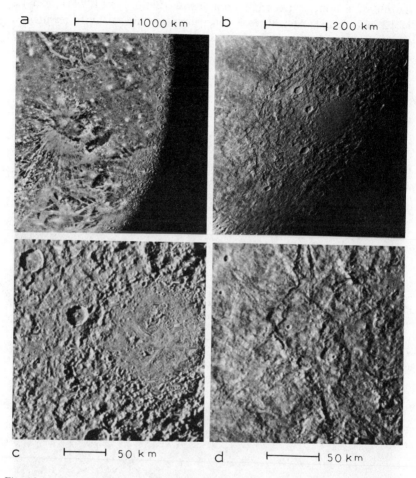

Fig. 12.10. Voyager 2 images of Gilgamesh, the least topographically relaxed basin on Ganymede. The diameter of the relatively smooth central floor is ~ 150 km; the ill-defined rim of the central depression is ~ 175 km in diameter. A high resolution image of the floor (c) reveals radial lineations as well as a roughly arcuate group of central peaks. The central depression is surrounded by a rugged region of a blocky to irregular relief ~ 1000 km in diameter. Inward facing concentric scarps can be seen in this region. In (d) is shown a high resolution image of one of these scarps where its trace apparently cuts across a 20 km diameter crater (just below the center of the frame). Since this crater postdates the formation of the Gilgamesh basin and emplacement of the ejecta rim, the extension of the scarp across this crater is probably a result of renewed movement, possibly due to settling of the ejecta deposit along the fault that produced the scarp. (FDS 20639.04, 20638.14, and 20639.15).

Crater Palimpsests

A crater palimpsest is a roughly circular spot of high albedo that marks the site of a former crater and its rim deposit (Smith et al. 1979b) (see Figs. 12.12 and 12.13). At low resolution palimpsests are relatively featureless. Except for secondary craters and traces of hummocks in the rim deposit, they retain almost none of the original relief of the crater or its rim. In a few instances, a smooth central area can be defined in high resolution pictures of large palimpsests (Fig. 12.12). We infer that this smooth area corresponds

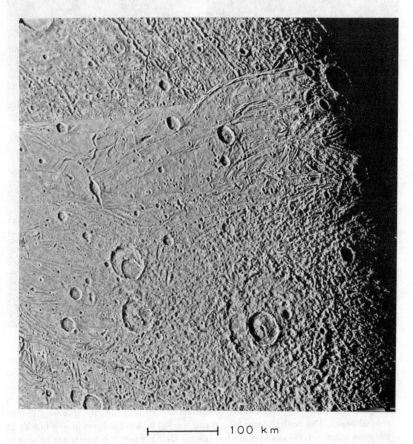

⊢————————⊣ 100 km

Fig. 12.11. Voyager 2 image of the Western Equatorial Basin on Ganymede (lower right part of the frame). This basin is somewhat similar to Gilgamesh but is slightly larger and has much more subdued relief. The diameter of the basin rim is 183 km; the basin is surrounded by a hummocky rim deposit ∼ 530 km in diameter. Prominent secondary craters and crater chains extend beyond the recognizable rim deposit and are superposed on grooved terrain (FDS 20638.39).

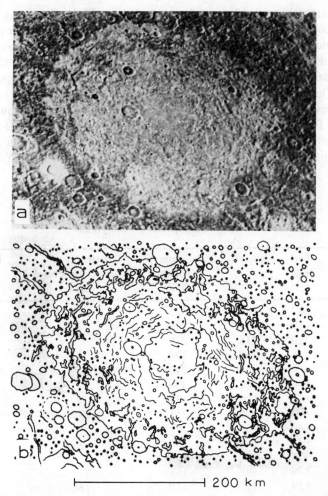

Fig. 12.12. Voyager 2 image and sketch map of a large palimpsest in Galileo Regio on Ganymede. The palimpsest is divisible into three distinct zones, (1) a relatively smooth central region ~100 km in diameter, roughly circular and bounded by a very gentle irregular topographic step (shown with a nearly continuous solid line near center of map), (2) a uniform high-albedo annulus with small semiconcentric ridges or hummocks, and (3) an outer annulus that is mottled in albedo. Vestiges of the prepalimpsest rimmed furrow system extend within the palimpsest. Secondary craters probably are represented among the abundant small craters illustrated in the sketch map; proceeding out from the center, the region of highest density of small craters begins in the annulus of mottled albedo (at ~100 km) and extends beyond the palimpsest. Areas of relatively low albedo are shaded on the sketch map; craters and pits are outlined with solid lines; short irregular solid lines represent subdued ridges; furrows are shown with paired lines; the limit of bright ejecta around two craters is shown with dashed lines (FDS 20638.29).

approximately to the floor of the original crater. Where the smooth area can be recognized, the diameter of the palimpsest is about two to four times the diameter of the central smooth area. Swarms of small secondary craters occur near the outer boundary of large palimpsests, suggesting that the outer boundary corresponds approximately to the limit of continuous ejecta.

Three distinct concentric zones can be recognized from high resolution pictures of one of the largest palimpsests located on Ganymede within Galileo Regio (Fig. 12.12). The central smooth region is essentially featureless and is at nearly the same elevation as the surrounding terrain. A very gentle irregular discontinuous topographic step surrounds this central region. A zone of

TABLE 12.2

Crater Palimpsests on Ganymede

No.	Lati-tude[a]	Longi-tude[a]	Palimpsest Diameter[b] (km)	Diameter of Central Structure[b] (km)	Terrain Type[c]
1	+22	184	335	128	G
2	+19	197	228	–	AC
3	+10	215	370	–	AC
4	+ 6	209	334	–	AC
5	+ 9	203	240	–	AC
6	+ 4	202	306	–	AC
7	+ 8	195	137	–	AC
8	−11	220	211	132	AC
9	+40	160	169	–	AC
10	+38	154	152	–	AC
11	+36	153	196	–	AC
12	+29	146	188	–	AC
13	+19	129	319	186	AC
14	+ 9	141	256	–	AC
15	+11	352	183	–	AC
16	+ 6	353	179	–	AC
17	− 1	356	167	–	AC
18	−17	339	116	58 (34)	AC
19	+10	354	173	–	AC
20	+10	359	102	–	AC
21	+10	358	79	–	AC
22	− 9	15	123	–	AC

[a]From U.S. Geological Survey Map I-1242, 1979.

[b]All measurements ± 10%.

[c]AC = ancient cratered terrain; G = grooved terrain.

TABLE 12.3

Possible Crater Palimpsests on Ganymede

No.	Lati-tude[a]	Longi-tude[a]	Palimpsest Diameter[b] (km)	Diameter of Central Structure[b] (km)	Terrain Type[c]
1	+25	195	235	—	AC
2	−10	211	168	—	AC
3	+37	145	215	—	AC
4	+22	149	156	—	AC
5	− 5	217	436	—	AC
6	+32	9	131	—	AC
7	−11	8	127	—	AC
8	− 3	322	125	70 (36)	AC
9	− 9	342	94	—	AC
10	− 2	342	75	—	AC
11	+28	135	138	—	AC
12	−10	319	155	—	AC

[a]From U.S. Geological Survey Map I-1242, 1979.

[b]All measurements ± 10%.

[c]AC = ancient cratered terrain.

uniform .albedo extends about one central-area diameter beyond this step, and this zone is characterized by a semiconcentric fabric of very low ridges or hummocks with an average radial spacing of ∼ 5 km. Beyond this zone, the palimpsest becomes discontinuous and mottled in albedo. Vestiges of a prepalimpsest rimmed furrow system can be recognized in this outermost zone. Extremely faint lineations along the trends of the rimmed furrows are visible in the central and intermediate zones as well. The outer boundary of the palimpsest is demarcated by the contrast in albedo between the palimpsest and the surrounding ancient cratered terrain. The transition in albedo generally occurs in < 5 km.

A total of 22 readily recognized palimpsests have been identified on Ganymede (Table 12.2). Their outer diameters range from 80 to 335 km. In addition, there are twelve palimpsests that are identified with less certainty, with diameters up to 440 km (Table 12.3). All but one (Fig. 12.13d) of the palimpsests listed are located on ancient cratered terrain.

Palimpsests on Callisto are less obvious than those on Ganymede; the albedo contrast with the surrounding terrain is less than for palimpsests on Ganymede and the outer boundaries are more diffuse and irregular. Table 12.4 gives coordinates and diameters for 15 Callisto palimpsests and possible

TABLE 12.4

Crater Palimpsests and Possible Palimpsests on Callisto

No.	Lat[a]	Long[a]	Diameter[b] (km)	Remarks
1	+42	2	156	
2	+46	348	121	
3	− 5	5	206	
4	− 5	348	169	
5	− 7	337	94	
6	−12	38	93	
7	− 2	25	286	
8	+59	346	87	
9	+11	57	571	Valhalla
10	+80	10	102	
11	+30	39	230	Asgard
12	− 6	247	417	
13	−11	234	272	
14	−19	249	118	
15	−41	224	94	

[a]From U.S. Geological Survey Map I-1239, 1979.

[b]All measurements ± 10%.

palimpsests, with outer diameters from ∼ 80 to 570 km. The resolution of the images of Callisto is insufficient to define morphological zones within these palimpsests. The two largest palimpsests occupy the central regions of two great multiring structures, Valhalla and Asgard (see Figs. 12.16, 12.17).

Penepalimpsests

In addition to palimpsests, there are a number of features on Ganymede that are transitional in form between craters and palimpsests. These features have diverse topographic and albedo characteristics, but all have vestiges of crater rims or other topographic features of the rim deposits; evidently they represent ancient craters that have nearly disappeared by viscous relaxation or creep. Here we will group these features together under the category of penepalimpsests (almost palimpsests). In contrast with the palimpsests, most penepalimpsests occur on grooved terrain rather than on the ancient cratered terrain. For the purpose of description they can be broadly grouped into two categories; type I consists essentially of extremely relaxed craters, and type II comprises more complex structures with a low central dome surrounded by annuli of complicated but very subdued relief.

Type I Penepalimpsests. A variety of features are grouped in this category. One feature (type I-a, Table 12.5) in the ancient cratered terrain of Galileo Regio (Fig. 12.14*a*) closely resembles the palimpsests of this region except that distinct vestiges of the original crater rim are preserved. Surrounding the rim is a roughly circular area of high albedo, about twice the diameter of the crater rim, which is similar to the intermediate and outer zones of the palimpsests.

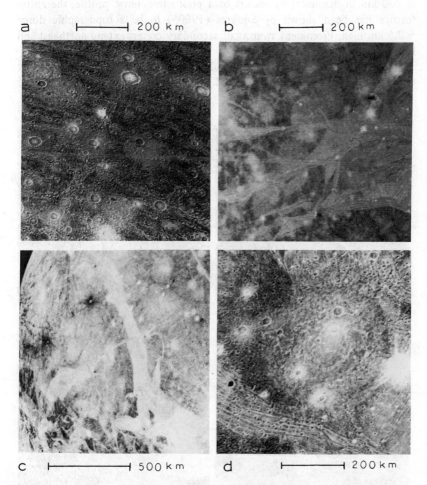

Fig. 12.13. Voyager 1 and 2 images of several palimpsests. (a) shows six palimpsests with diameters < 200 km located in Galileo Regio. Younger grooved terrain units have cut across the palimpsests in (b). The region of Marius Regio in (c) is about half occupied by palimpsests. With the exception of the palimpsest in (d), all are found within the ancient cratered terrain, and appear to be older than the formation of the grooved terrain (FDS 20636.59, 16402.22, 20631.17, and 20635.37).

Three features (type I-b, Table 12.5), one in ancient cratered terrain and two in grooved terrain, are characterized by a wreath of low semiconcentric ridges surrounding a central smooth region. An outer zone, 1.5 to 2 times the diameter of the wreath, is relatively smooth but also contains very subdued semiconcentric ridges. Where it is formed on the grooved terrain there is no conspicuous contrast in albedo between the penepalimpsest and the surrounding surface. In the example illustrated in Fig. 12.14b, the outer zone is 260 km in diameter; by means of a photoclinometric profile, the entire feature has been shown by Squyres (1980b) to be a topographic dome ~ 2.5 km high. Prominent swarms of secondary craters extend northeast and southwest from this penepalimpsest.

One penepalimpsest (type I-c, Table 12.5) is a highly flattened basin that occurs within grooved terrain (Fig. 12.14c). Its relief appears comparable to

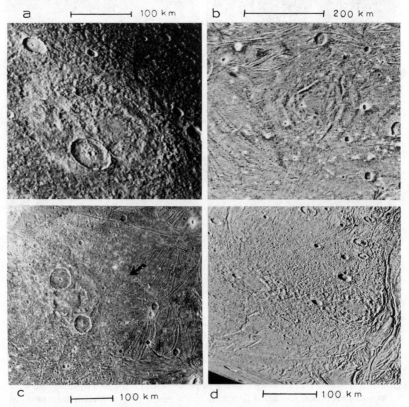

Fig. 12.14. Voyager 1 and 2 images of four Type I penepalimpsests; see text for description. The initial crater relief of all of these features has nearly disappeared, presumably due to viscous flow or creep within Ganymede's crust (FDS 20638.31, 16405.30, 20639.53, and 20640.31).

TABLE 12.5

Penepalimpsests on Ganymede

No.	Lati- tude[a]	Longi- tude[a]	Penepalimpsest Diameter[b] (km)	Diam. Initial Crater (km)	Diam. Central Structure (km)	Penepalimpsest Type[c]	Terrain Type[d]
1	+22	119	266	144	–	I-a	AC
2	−16	120	139	–	–	I-b	G
3	+36	328	260	–	–	I-b	G
4	− 3	323	125	70	–	I-b	AC
5	−36	137	300	138	71	I-c	G
6	−69	265	183	64	–	I-d	G
7	−70	280	171	60	–	I-d	G
8	−28	169	–	103	52 (25)[e]	II	G
9	−14	132	–	127	66 (32)[e]	II	G
10	+ 4	173	–	142	81 (30)[e]	II	AC
11	+36	2	–	149	85 (40)[e]	II	AC
12	−12	102	–	166	90 (38)[e]	II	G

[a]From U.S. Geological Survey Map I-1242, 1979.

[b]All measurements ± 10% due to uncertainty in penepalimpsest boundary.

[c]Refer to text for discussion of penepalimpsest type.

[d]AC = ancient cratered terrain; G = grooved terrain.

[e]Diameter of high albedo central dome.

the type I-a penepalimpsest that lies in the ancient cratered terrain, but because the type I-c penepalimpsest lies in a region of intrinsically higher albedo, there is no evident contrast in albedo between the area immediately surrounding the collapsed basin and the neighboring terrain. A central circular ridge 71 km in diameter is surrounded by an outer very subdued raised rim 138 km in diameter. Local sharp relief is found along the rim. Grooves and ridges in the grooved terrain are obliterated to a distance of ∼ 150 km from the center of the penepalimpsest. We interpret the slightly elevated rim as the original rim of the basin and the inner circular ridge as the rim of a huge flattened central pit. The rim deposit of this basin extends to the limit of obliteration of grooves in the grooved terrain. Abundant secondary craters with maximum diameter ∼ 7 km extend from a radial distance of ∼ 150 km (just beyond the rim deposit) to ∼ 200 km.

Two penepalimpsests on grooved terrain (type I-d, Table 12.5) are characterized by a large central smooth area encompassed by a broad annulus of subdued short semiconcentric ridges and irregular hummocks (Fig. 12.14d). In each case the width of the annulus is roughly the same as the diameter of the central smooth region. The smooth region evidently corresponds to the initial crater floor and the annulus to the original crater rim deposit. Apparently because these penepalimpsests occur in the relatively high albedo grooved terrain, there is no contrast in albedo between the rim deposit and the grooved terrain. Each of these penepalimpsests is superposed on and is younger than some grooved terrain units. One of the penepalimpsests, however, is also older than one unit of grooved terrain that transects the rim deposit (Fig. 12.14d). This latter penepalimpsest, therefore, was formed during the period of the grooved terrain formation.

Type II Penepalimpsests. Five unusual craterform features are included here under the category of penepalimpsest, although a separate designation might be appropriate. They are large highly relaxed craters, approaching basins in size, characterized by unusually large central pits (Fig. 12.15). In the center of each central pit is a circular area of high albedo that appears to have the form of a broad topographic dome, about half the diameter of the central pit. It has been suggested that the high albedo dome may be an icy diapir (Malin 1980) or the result of the freezing of a central lake produced at the time the crater was formed (Croft 1980). Surrounding the dome, but within the pit walls, is an annulus of low albedo marked with roughly radial light streaks. The enclosing central pit walls have high albedo. Considerable relief is preserved in the central parts of these penepalimpsests, but the outer crater rim is extremely subdued.

Two of the type II penepalimpsests occur on ancient cratered terrain (Fig. 12.15c) and three on grooved terrain (Fig. 12.15a, 12.15d). No distinct areas of high albedo surrounding the relaxed craters have been recognized either on grooved terrain or on the ancient cratered terrain.

Multiring Structures on Callisto

Eight systems of multiple concentric ridges have been found on Callisto (Table 12.6); they fall into two size categories, (1) diameter of the outer ridge ring > 500 km, and (2) diameter of outer ridge ring < 200 km. None have been found with outer diameters between 200 and 500 km.

The largest multiring structure, Valhalla, has concentric ridges out to ~ 2000 km from the center (Fig. 12.16c). A central bright palimpsest occupies a region nearly 600 km in diameter. In the center of the palimpsest is a circular, relatively smooth area ~ 350 km in diameter. The superposed crater density on the palimpsest is roughly one-fourth of the crater density of the surrounding surface of Callisto. The innermost concentric ridge occurs at a radius of ~ 200 km; hence the palimpsest overlaps the inner ridges over a radial distance of ~ 100 km.

In detail, the individual ridges in this multiring system are irregular in plan and are discontinuous in circumferential extent. Maximum length of the individual ridge arcs is ~ 700 km; most ridges are between 200 and 500 km in length. Distance between the ridges varies from ~ 20 to 30 km for the inner ridges, to ~ 50 to 100 km for the outer ridges.

On the average, the ridges are ~ 15 km wide, and they appear to be flat-topped (Fig. 12.16d). In a few places, however, a central groove or furrow is resolved within the ridge (McKinnon and Melosh 1981). It is difficult to determine the heights of the ridges of the Valhalla multiring system, owing to the proximity of Valhalla to the subsolar point at the time the pictures were obtained. Comparison with ridges of the Asgard multiring

TABLE 12.6

Multiring Structures on Callisto

No.	Latitude[a]	Longitude[a]	Outer Ring Diameter[b] (km)	Inner Ring Diameter[b] (km)	Ring Spacing		Name
					Inner (km)	Outer (km)	
1	+11	57	4000	400	30	50-100	Valhalla
2	+30	139	1640	163	20-30	50-60	Asgard
3	−53	36	920	130	15	30	Unnamed
4	+45	138	500	?	?	?	Unnamed
5	− 9	222	163	80	7-9	−	Alfr
6	− 3	215	123	25	10	−	Loni
7	+42	213	180	96	5	−	Grimr
8	−41	262	71	35	?	?	Unnamed

[a]From U.S. Geological Survey Map I-1239, 1979.
[b]All measurements ± 10%.

system, the heights of which can be determined by shadows, suggests that the ridge heights in the Valhalla system are < 1 km above the surrounding surfaces.

The second largest multiring system on Callisto, Asgard, has concentric ridges out to 800 km from the center of the system (Fig. 12.17). The center of this system is occupied by a palimpsest $\sim$ 230 km in diameter. The

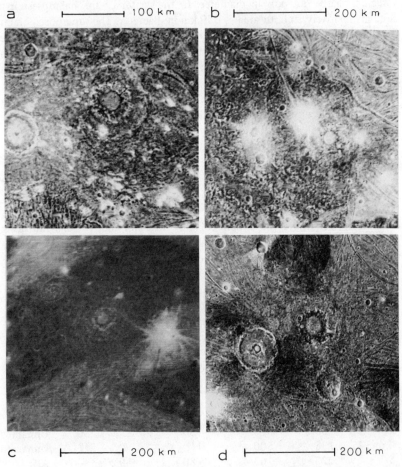

Fig. 12.15. Voyager 1 and 2 images of several type II penepalimpsests. All have a circular central area of high albedo surrounded by an annulus of low albedo with radial light streaks. Two concentric ridges surround this central area. The inner ridge corresponds to the rim of an unusually large central pit, and the outer ridge corresponds to the original crater rim. In some cases, as shown in (a) and (c), the outer rim is bright; however, as shown in (b) and (d), the ridge that corresponds to the initial crater rim is not always bright (FDS 20637.41, 20635.49, 16402.12, and 20637.23).

innermost concentric ridges are located ∼ 80 km from the center. The spacing between the inner ridges is 20 to 30 km, and between the outer ridges generally 50 to 60 km. The heights of most ridges are between 500 and 1000 m above the interridge surfaces.

A third relatively large multiring structure is centered at ∼ 53° S and 36° W (Fig. 12.18d). The outermost ridges are at ∼ 450 km from the center, and the innermost ridges recognizable in the available low resolution images occur at a radius of ∼ 70 km. Coverage of this feature by Voyager pictures is

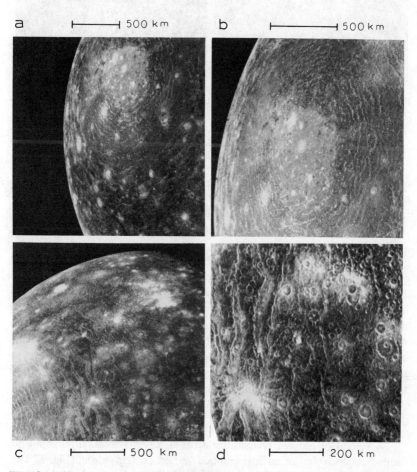

Fig. 12.16. Voyager 1 images of Valhalla, the largest multiring structure on Callisto. The center of the structure is occupied by a high albedo palimpsest 600 km in diameter (b). Ridges occur up to ∼ 2000 km from the center of the feature (c). The individual ridges in the ring system appear to be flat-topped, in (d). (Note that the illumination is from the left and not from the right.) Some images reveal troughs in the tops of the ridges (FDS 16418.42, 16422.11, 16418.58, and 16424.46).

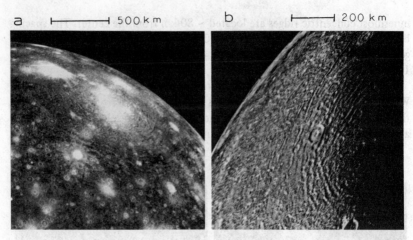

Fig. 12.17. Voyager 1 and 2 images of Asgard, the second largest multiring system on Callisto. This structure is very similar to Valhalla but is only ~1500 km in diameter. The center is occupied by a palimpsest and the ring ridges appear flat-topped. As is the case with Valhalla, the spacing of the ridges is closest near the center of the ring system (FDS 20606.21 and 16428.09).

limited, and details within this system are not resolved. Albedo characteristics of the center of the ring structure are obscured by rays from the very large ray crater Adlinda to the southeast. Another large multiring structure with diameter ~ 500 km overlaps with Asgard and is centered ~ 45° N lat and 138° W long.

Four small multiring structures with outer diameters < 200 km are Alfr and Loni (Fig. 12.18a), Grimr (Fig. 12.18b), and one unnamed structure (Fig. 12.18c). The number of individual circumferential ridges varies from three to more than six. The spacing between ridges is < 10 km. The heights of the ridges cannot be accurately obtained because of the location of these structures with respect to the terminator at the times they were imaged. One of the small multiring structures (Fig. 12.18c) may be analogous to the type II penepalimpsests of Ganymede. The resolution of this feature on Callisto is too low, however, for close comparison with penepalimpsests on Ganymede.

Galileo Regio Rimmed Furrow System

A system of rimmed furrows (Smith et al. 1979b) occurs throughout the Galileo Regio region of ancient cratered terrain on Ganymede, (Fig. 12.19). This system extends over an area > 10^7 km², an area greater than the United States, and can be found in many polygons of ancient cratered terrain that are separated from the Galileo Regio by grooved terrain. The system is > 2000 km across in the direction normal to the trend of the furrows; the area covered by the preserved remnants of this system must be only a fraction of the area of the original system. The rimmed furrow system is so large that it was at first difficult to determine whether the average trends of the furrows follow small circles or great circles; they do follow small circles on the Galileo Regio and on other major segments of ancient cratered terrain south and west. The center of the concentric system of small circles is at ~ 20° S lat and 165° W long, where about half of a faint, poorly defined palimpsest is preserved on the ancient cratered terrain. The southern half of the palimpsest

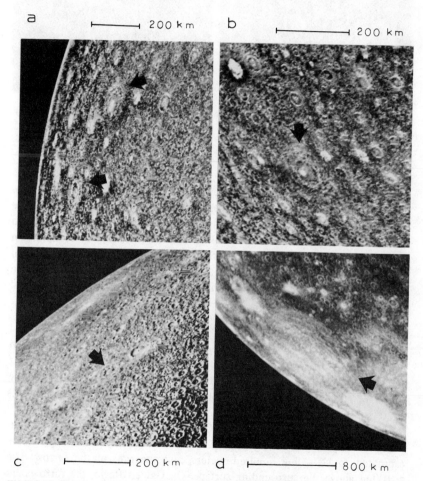

Fig. 12.18. Voyager 1 and 2 images covering five multiring structures on Callisto. Four of these multiring structures are <200 km in diameter (a,b,c), and the spacing of their ring sets is ~5 to 10 km. Alfr (top central) and Loni are shown in (a), Grimr in (b), and an unnamed structure in (c). An unnamed large multiring system with a diameter of ~900 km is shown in (d) (FDS 20617.21, 20616.53, 20619.36, and 16418.14).

a ├────┤ 500 km   b   ├────┤ 100 km

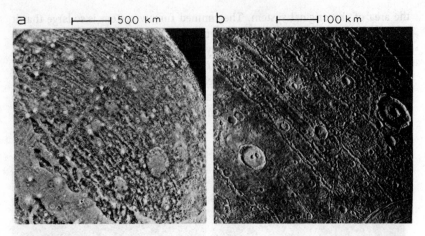

Fig. 12.19. Voyager 2 images of the Galileo Regio rimmed furrow system. The furrows are older than most features on Galileo Regio and they have a curvilinear trend (a); they are ~10 km wide and spaced ~50 km apart (b) (FDS 20631.11, and 20638.35).

is replaced by grooved and reticulate terrain. This palimpsest, ~ 500 km in diameter, is surprisingly small compared to the size of the rimmed furrow system; the original crater may have been smaller than the central depression of Gilgamesh. Since the furrows are older than the vast majority of craters on the ancient cratered terrain, it is perhaps not so surprising that so small an original crater gave rise to the ring system. The lithosphere of Ganymede probably was very thin at the time the furrows were formed (McKinnon and Melosh 1980).

Segments of the rimmed furrow system found on polygons of ancient cratered terrain between latitudes 10° S and 42° N and longitudes 180° and 225° appear to be displaced left laterally and possibly slightly rotated with respect to the part of the system in Galileo Regio. This displacement complicates the problem of defining the exact center of the system, since the center of curvature of the furrows on Galileo Regio (Smith et al. 1979b) is ~ 20° southwest of the center of the palimpsest.

Individual rimmed furrows on Galileo Regio are ~ 10 km wide and hundreds of kilometers long (Fig. 12.19b). The depths of the furrows are estimated to be a few hundred meters, and the rims of the furrows rise ~ 100 m above the surrounding surface. On Galileo Regio, the furrows are relatively uniformly spaced about every 50 km, but they are more closely spaced where well preserved near the center of the system northwest of the central palimpsest. Where age relations are determinable, the furrows almost invariably predate recognizable craters and palimpsests. Elsewhere, a few craters appear to be cut by the furrows.

Ray Craters

Both bright ray and dark ray craters occur on Ganymede. Forty-three dark ray craters have been found; the diameters of these craters range from < 4 km to 60 km (Conca 1981). Roughly two-thirds of the identified dark ray craters are located on grooved terrain, but this large fraction may be due in part to the greater ease in identifying dark ejecta and rays on the relatively high albedo grooved terrain.

Craters on Ganymede with bright rim deposits or rays range in diameter from a few kilometers to 155 km; over 110 such craters > 25 km in diameter have been identified (Fig. 12.20). The mean frequency of craters > 25 km in diameter with bright rim deposits or rays on grooved terrain is 3.3 craters per 10^6 km^2; for similar craters on ancient cratered terrain, the frequency is 2.3 craters per 10^6 km^2. We interpret the one-third lower frequency of craters with bright rim deposits or rays on ancient cratered

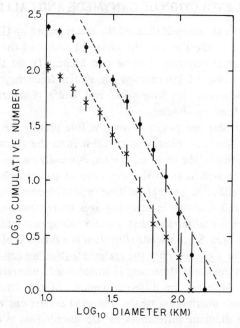

Fig. 12.20. Cumulative size-frequency distribution of bright ray craters on Ganymede. Frequencies of bright ray craters observed in the high resolution Voyager 1 pictures are indicated by x's; the combined frequencies derived from both the Voyager 1 and Voyager 2 coverage are shown with solid circles. Error bars are ± one standard deviation. The dashed lines have a slope of − 2.2 and fit the observations within one standard deviation down to 25 km diameter. The break in the slope at ∼ 25 km ($\log_{10} \approx 1.4$) is interpreted to be the result of incomplete counts at relatively small crater diameters.

terrain to be the result of differential retention times for rays on these two surfaces. Bright rim deposits and rays on the ancient cratered terrain are apparently darkened 40% faster than those of the grooved terrain. The darkening is probably due to the gradual addition of meteoritic debris to the surface, to the concentration of this debris as a lag deposit where net ablation of ice is taking place and where rays are thin, to the gardening of the surface by small impact craters that mix the ray material with the underlying darker regolith.

On Callisto, no craters with dark rays have been found. Craters with bright rim deposits or rays are more abundant than on Ganymede. The maximum diameter of craters with bright rims and rays on Callisto is ~ 120 km; the mean frequency of craters > 25 km in diameter is ~ 4.2 craters per 10^6 km^2. The craters with bright rim deposits or rays on both Ganymede and Callisto generally exhibit the least topographic relaxation.

CRUSTAL EVOLUTION OF GANYMEDE AND CALLISTO

We turn now to an interpretation of the crater record on Ganymede and Callisto and to the implications of the observed forms of the craters with respect to the thermal histories of these two bodies. To do this it will be necessary to make use of information on spatial frequency (density) of craters and of deductions by Shoemaker and Wolfe (Chapter 10) on the cratering rates for these two bodies.

It is crucial to bear two points in mind in interpreting the crater record on these giant icy satellites. First, it is manifest from the varying degrees of topographic relaxation of the observed craters, especially on the most densely cratered terrains, as well as from the existence of palimpsests, that craters have disappeared, either by prompt collapse or slow viscous flow or creep on both Ganymede and Callisto. In assigning ages to terrains on the basis of crater density, ages for all terrains that contain palimpsests must be regarded as *crater retention* ages. Since viscous relaxation is a function of crater size, it is also clear that for a given surface the crater retention age can vary with the size of the craters counted. Moreover, if surfaces are preserved that reflect different thermal histories for different regions of the lithosphere, it is evident that the size distribution of the preserved craters can vary. We shall see that there are dramatic differences in size distribution as a function of both position on the satellite and age of the crater population.

Second, because the orbital velocity of each Galilean satellite is ~ 45 to 60% as high as the average encounter velocity of extraJovian bodies with the orbit of the satellite, there is a very strong gradient in cratering rate from the apex of orbital motion to the antapex (Shoemaker and Wolfe, Chapter 10). At the present time, the ratio of the cratering rate at the apex to the rate at the antapex from impact of comet nuclei ranges from ~ 9.6 on Callisto to

~ 14.9 on Ganymede. These ratios depend, to some extent, on the populations of impacting bodies, but certainly they have always been > 3 on all the Galilean satellites. Hence it is essential to take account of the position on the satellite in interpreting both absolute and relative ages from crater densities. It turns out that regions with comparatively low crater densities near the antapices of Ganymede and Callisto have some of the highest crater retention ages (based on the model of Shoemaker and Wolfe, Chapter 10) for the surfaces of each satellite.

Evolution of Ancient Cratered Terrain

One of the earliest recognizable features preserved on the leading hemisphere of Ganymede is the Galileo Regio rimmed furrow system. As shown by McKinnon and Melosh (1980), the characteristics of the rimmed furrow system provide clues about the thickness of the lithosphere at the time the furrows were formed. The furrows are interpreted by them as graben; if the dips of normal faults bounding these graben are close to 60°, as is typical for many normal faults, they intersect at an average depth of ~ 10 km. This may be taken as an estimate of the thickness of the Ganymedian lithosphere at the time the furrows were formed. The dip of the faults might be steeper and the lithosphere thicker, but from the theory of multiring structures of Melosh and McKinnon (1978) the lithosphere of the Galileo Regio was not thicker than the average spacing between the furrows, ~ 50 km. Hence, the low latitude lithosphere in the leading hemisphere of Ganymede was probably between 10 and 50 km thick at the time the furrows formed, perhaps closer to the 10 km rather than to the 50 km limit.

It is of interest to inquire as to how this lithosphere came into being. What happened prior to the impact that produced the multiring system? One conceivable scenario is that Ganymede was resurfaced by a global event, perhaps a satellite-wide extrusion of water associated with final melting or differentiation of a hydrous mantle a short time before the rimmed furrows formed. The lithosphere could then be envisioned simply as an icy layer refrozen since that event. The crater record on the ancient cratered terrain, however, suggests that this did not occur. On the basis of the cratering time scale and model of Shoemaker and Wolfe (Chapter 10), the crater retention ages of parts of the crust in the trailing hemisphere of Ganymede are substantially greater than on Galileo Regio in the leading hemisphere.

Detailed unpublished studies of the crater densities by J.B. Plescia indicate that, although the mean density of ≥ 10 km craters on the ancient cratered terrain varies slightly with distance from the apex of orbital motion (Table 12.7), the variation in density is not nearly as great as predicted by Shoemaker and Wolfe for a surface of a single age. If only craters > 30 km are considered, there is a larger difference in crater density between areas near

TABLE 12.7

Estimated 10 km Crater Retention Ages for the
Ancient Cratered Terrain on Ganymede

Distance from Mean Apex of Orbital Motion (deg)	Observed Crater Density[a] (per 10^6 km^2)	Derived Satellite-wide Mean Crater Density[b] (per 10^6 km^2)	Calculated Crater Retention Age of Surface[c] (Gyr)
20 – 40	290	165	3.79
40 – 60	258	165	3.79
60 – 80	271	209	3.83
80 – 100	245	245	3.86
100 – 120	270	385	3.94
120 – 140	188	430	3.95

[a]Crater densities for each range of distance from the apex have been obtained by averaging the results of detailed crater counts by J.B. Plescia.

[b]The satellite-wide mean density is obtained from $\overline{F}_{10} = F_{10}(\beta)/[1 + \cos\beta(\delta - 1)/(\delta + 1)]$, where $F_{10}(\beta)$ is the observed integral crater density to 10 km diameter at distance β from the mean apex of orbital motion, $\overline{F}_{10}$ is the corresponding satellite-wide mean density to 10 km diameter (equivalent to the crater density at $\beta = 90°$), and $\delta = 15.0$ is the ratio of the cratering rate at the apex to the cratering rate at the antapex for Ganymede (Shoemaker and Wolfe, Chapter 10).

[c]Crater retention ages greater than 5×10^8 yr are calculated from $\overline{F}_{10}(t) = [R_0 \exp\lambda$ $(t - 3.3\_\text{Gyr})/\lambda + (1.15 \times 10^{-8} \text{ yr}^{-1}) + 5.70]$ craters $\geqslant 10$ km diameter per 10^6 km^2, where $\overline{F}_{10}(t)$ is the mean crater density to 10 km crater diameter on a surface formed at time t, $\lambda = \ln 2/t_{1/2}$, $t_{1/2} = 10^8$ yr, and the exponential component of the cratering rate at 3.3 Gyr, R_0, is 2.63×10^{-8} yr^{-1} (Shoemaker and Wolfe, Chapter 10).

the apex and areas near the antapex, but the difference is still much less than predicted (see also discussion in Chapter 9 by Woronow and Strom). One possible way to explain a more uniform distribution of craters than predicted is to suppose that the rotation of Ganymede was not always synchronous with its orbital motion during part of the decipherable history of cratering. If several stable positions were possible when the rotation was locked, differences in cratering rate over the surface might have been averaged out. Alternatively, it might be supposed that the lithosphere of Ganymede was sufficiently decoupled from the rocky core so that even though the core may always have been tidally locked, the lithosphere was free to slip relative to the core. The size-frequency distribution of craters on the ancient cratered terrain, however, provides direct evidence that the lithosphere was, in fact, approximately fixed with respect to the mean apex during essentially all of the recorded cratering history. On the average, the crater-size distribution is steeper in the trailing hemisphere than in the leading hemisphere; the ratio of small craters to large craters is higher in the trailing hemisphere than in the leading, as would be expected if the trailing hemisphere were older and the crust were everywhere stiffening with time.

The observed distribution of crater densities on the ancient cratered terrain of Ganymede indicates that the crater retention ages of the trailing hemisphere are higher than on Galileo Regio, in the leading hemisphere. The range in crater retention ages can be estimated from the cratering time scale presented by Shoemaker and Wolfe (Chapter 10), which indicates that mean ages increase systematically with increasing distance from the apex of motion (Table 12.7). The range of model ages at 10-km crater diameter is $(16 \pm 3) \times 10^7$ yr. As neither the apex nor the antapex of Ganymede was imaged at high enough resolution for studies of crater density, the full range of crater retention ages is not known but probably is somewhat greater. The rimmed furrow system on Galileo Regio evidently is somewhat younger than the 10-km crater retention age of the surface it cuts, ~ 3.8 Gyr. The mean 10-km crater retention age of Nicholson Regio and nearby polygons of cratered terrain in the trailing hemisphere is ~ 3.9 Gyr. The calculated difference in age depends chiefly on the adopted rate of decay of cratering rate during heavy bombardment (modeled by Shoemaker and Wolfe after the decay rate obtained for the Moon) and is relatively insensitive to errors in the absolute calibration of the cratering time scale.

A similar picture emerges from the study of the crater distribution on Callisto, except that the crater retention ages over the surface of Callisto are systematically higher than on Ganymede. The distribution of large impact structures on Callisto is strikingly nonuniform. The four largest multiring structures (Valhalla, Asgard, and two unnamed structures) are all located in the leading hemisphere; the center of Valhalla is within 40° of the apex, and the center of Asgard is within 55°. Craters > 60 km in diameter are 2.2 times as abundant in areas counted near the apex as near the antapex (Table 12.8); for craters > 25 km in diameter as well as those > 10 km in diameter, the density of craters near the apex is about the same as at the antapex. Hence the size distribution of craters near the apex has a lower slope than the size distribution near the antapex.

If the orientation of the crust of Callisto remained fixed relative to the mean apex of orbital motion and if the crust were everywhere the same age and all craters produced were still preserved, then the ratio of crater density in the region sampled near the apex to the density near the antapex should be ~ 9.6, in contrast to the observed ratios of 2.2 at 60 km, and ~ 1 at 25 km and 10 km crater diameters. We interpret the observations as showing that the crust of Callisto was, in fact, fixed relative to the apex during the period of recorded cratering history, and that the crust has a spatially varying set of crater retention ages (Table 12.8). With the exception of the polar region and

TABLE 12.8

Estimated 25 km and 60 km Crater Retention Ages for Various Locations on the Surface of Callisto

Distance from Mean Apex of Orbital Motion (deg)	Observed Density of 25-km Craters (per 10^6 km²)	Calculated 25-km Crater Retention Age of Surface[a] (Gyr)	Observed Density of 60-km Craters (per 10^6 km²)	Calculated 60-km Crater Retention Age of Surface[a] (Gyr)
40	97	4.07	10.7	4.63
65	98	4.09	9.0	4.03
78	96	4.11	9.8	4.07
81[b]	124	4.16	12.1	4.10
91[c]	107	4.16	5.9	4.02
121[d]	75	4.17	4.3	3.96
123[d]	78	4.19	2.1	4.05
140	112	4.31	4.4	4.11
143	106	4.31	3.9	4.11
167	103	4.39	4.8	4.22
38 Valhalla palimpsest	49	3.96	3.6	3.85
54 Asgard	70	4.04	10.0	4.03

[a] Values of mean crater density used in calculating the crater retention ages are obtained from $\bar{F}_i = F_i(\beta)/[1 + \cos \beta(\delta - 1)/(\delta + 1)]$ where $F_i(\beta)$ is the observed integral crater density to diameter i at distance β from the apex, $\bar{F}_i$ is the corresponding satellite-wide mean density to diameter i (equivalent to the crater density at $\beta = 90°$), and $\delta = 9.58$ is the ratio of the cratering rate at the apex to the cratering rate at the antapex for Callisto. To obtain the equivalent satellite-wide mean crater density at 10 km, F_{10}, from integral crater densities at 25 and 60 km, the size-frequency distribution of craters produced over any given interval of time is assumed to be given by $F_{10} = \bar{F}_i (10/i)^\gamma$ where γ, the size index, is -2.2 (Shoemaker and Wolfe, Chapter 10). Crater retention ages greater than 5×10^8 yr are calculated from $\bar{F}_{10}(t) = 0.52 \, [R_o \exp \lambda (t - 3.3 \, \text{Gyr})/\lambda + (1.15 \times 10^{-8}) \, t \text{ yr}^{-1} + 5.70]$ craters $\geqslant 10$ km diameter per 10^6 km², where $F_{10}(t)$ is the satellite-wide mean crater density to 10 km crater diameter on a surface formed at time t, $\lambda = \ln 2/t_{1/2}$, $t_{1/2} = 10^8$ yr, and the coefficients of the exponential component of the cratering rate at 3.3 Gyr, R_o, is $2.63 \times 10^{-8} \text{ yr}^{-1}$ (Shoemaker and Wolfe, Chapter 10).

[b] Center of area counted is at 78° N lat.

[c] Center of area counted is at 80° N lat.

[d] Area counted probably includes the site of a former palimpsest.

local areas that we suspect are former palimpsests, the crater retention ages at 60 km crater diameter tend to become greater as the antapex is approached. At 25 km crater diameter this trend is even stronger. All of the 25 km crater retention ages are higher than the 60 km crater retention ages, and the difference is greatest on the oldest surfaces, near the antapex. The difference in crater retention age between the apex and the antapex is $\sim 2 \times 10^8$ yr at 60 km crater diameter and 3×10^8 yr at 25 km crater diameter. Unfortunately the resolution of the Voyager images is inadequate to obtain an accurate regional distribution of crater densities to 10 km diameter.

The observed distribution of craters and calculated crater retention ages on the heavily cratered terrain of Ganymede and Callisto follows a pattern that would be expected if craters have disappeared by viscous flow or creep or by prompt collapse and disruption of the lithosphere, and if the retention of craters is, itself, partly but not entirely a function of the cratering history. As the oldest crater retention ages are observed near the antapex, on both Ganymede and Callisto, this indicates that the lithosphere became thick and stiff enough to retain craters earlier near the antapex than near the apex. The implication of this relationship is that the local bombardment history has influenced the cooling of the lithosphere. The primary effect of bombardment on the lithospheric temperature of each satellite probably arose from the production of an insulating regolith from the pulverization and gardening of the surface by small craters (see Chapter 13). For an equilibrium population of craters, a globally uniform mean regolith thickness might be expected. The relatively uniform latitudinal distribution of highly relaxed craters, however, indicates that the regolith must have been thermally annealed at its base (Chapter 13). If a time of the order of $\geq 10^5$ yr was required for the annealing to occur, then the mean thickness of the effective insulating layer would have been proportional to the local cratering rate during heavy bombardment. Hence the top of the lithosphere would have been cooler near the antapex than at the apex of motion. As the cratering rate declined during heavy bombardment, the region of the antapex, where the cratering rate at any given time was much lower than in the leading hemisphere, became cool and stiff enough to support ≥ 10 km craters much earlier than near the apex. At any given location, the lithosphere first stiffened sufficiently to support small craters and then became capable of retaining larger craters.

A second factor that leads to younger crater retention ages in the leading hemisphere is the higher rate of production of multiring structures. Observed crater densities are relatively low throughout a fairly large fraction of the areas of Valhalla and Asgard on Callisto. Production of multiring structures on the thin early lithosphere may have lead to similar regional suppression of the density of recognizable preexisting craters. The distribution of this suppression effect over the satellite should have followed the spatial variation of cratering rate, leading to a more uniform final distribution of retained crater density and a gradient of crater retention ages increasing toward the antapex.

There is a third factor that enhances the abundance of small craters at the antapex, but we judge this effect to be subordinate to the effect of stiffening of the crust with time. Distant secondary craters associated with large craters and basins excavated on the leading hemisphere will be produced on the trailing hemisphere. The occurrence of secondary crater chains shows that many secondaries 10 km in diameter and some up to 30 km diameter can be formed, at least during the late stage of heavy bombardment. It is less clear that equally large secondaries were produced at an earlier stage when the lithosphere was very thin. As many more large craters and basins are formed in the leading than in the trailing hemisphere, the number of distant secondaries produced in the trailing hemisphere will be disproportionately large in comparison with the number of both small and large primaries formed there. This probably leads to significant overestimation of crater retention ages at 10 km diameter over the entire surface and especially near the antapex, but the effect is probably small at 25 km crater diameter, and essentially vanishes for larger craters.

Detailed examination of the local variation of crater density on Ganymede and Callisto provides direct evidence on the manner in which early-formed craters disappeared. On the recognizable palimpsests, the crater density (at crater sizes greater than the largest associated secondary craters) generally is less than on adjacent ancient cratered terrain. Typically, nearly all earlier craters are obliterated in the area covered by each palimpsest, even though 70 to 90% of the area of the palimpsest evidently corresponds to an ejecta blanket. The reason for this efficient obliteration is that nearly all of the prepalimpsest craters were either small enough to begin with or had flattened by viscous flow or creep to sufficiently low relief that the remaining relief was smothered by a comparatively thin ejecta deposit. On Ganymede, the palimpsests are fairly abundant. A large area of ancient cratered terrain in Marius Regio is about one-half occupied by palimpsests (Fig. 12.13c). The mean area occupied by recognizable palimpsests on the ancient cratered terrain of Ganymede is roughly estimated at 25%. Although less easily mapped, palimpsests cover a roughly comparable fraction of the surface of Callisto.

Between the recognizable palimpsests on Ganymede and Callisto, there are also local areas where the crater densities are lower than the regional mean by an amount greater than would be expected by chance. Some of these areas on Callisto are recognizable by their low crater abundances listed in Table 12.8, and similar areas of local low crater abundance have been found by Plescia on Ganymede (see Chapter 13 by Shoemaker et al.). In some cases the occurrence of small palimpsests or large craters in the areas counted is responsible for the low crater abundance, but in other cases there is no obvious cause. Since the albedo contrast between a palimpsest and the surrounding surface evidently fades with time, it seems likely that many areas of anomalously low crater abundance (perhaps 10 to 20% of the ancient cratered terrains) correspond to former palimpsests. If one pushes the cratering history back far enough, of course, all parts of the surface probably were occupied at one time or another by palimpsests.

Our picture of the early evolution of the lithospheres of Ganymede and Callisto can be summarized as follows. During early heavy bombardment, the heating of the lithosphere by the combination of impact and generation of an inulating regolith and heat flow from the deep interior was sufficient to prevent retention of any recognizable craters. As the bombardment rate and heat production by radioactivity waned, the upper part of the lithosphere of each satellite cooled to the point where craters large enough to be recognized in the Voyager images were retained. The earliest retained craters on each satellite are relatively small ($<$ 10 km diameter) and occur either in the polar regions or near the antapex. Crater retention then spread in a wave toward the apex. At any given distance from the apex, progressively larger craters were retained with the passage of time. In detail, the loss of craters was a stochastic process. Individual craters first flattened by flow or creep, then groups of craters were smothered beneath ejecta deposits. Large early craters that penetrated the shallow asthenosphere probably flattened almost immediately after they were formed. At early times, when the lithosphere was very thin ($\lesssim$ 10 km thick), extensive regions probably were resurfaced by the formation of multiring structures.

We have not found any evidence that flooding of the surface by water was an important part of the early evolution of the lithospheres of Ganymede or Callisto nor any evidence that a major fraction of the mantle of either satellite was liquid at any one time during its decipherable history. Owing to combined heating by impact and radioactivity, the shallow thermal gradients on both satellites probably were steep enough to lead to convection of solid ice within their deeper interiors very early in their histories (see Chapter 4 by Cassen et al.). In all likelihood Ganymede and Callisto differentiated during or shortly after accretion; no more than a small fraction of water may have been present at a given moment during this differentiation (Schubert et al. 1981). It should be noted that thermal history calculations for Ganymede and Callisto that ignore heating by impact, both during accretion and possibly during intense heavy bombardment by projectiles derived from sources external to the Jovian system after accretion, probably are unrealistic.

Record of Ray Craters

As on the Moon, the ray craters on Ganymede and Callisto are the youngest features recognized on each satellite. The rays are superimposed on craters that lack rays and on all the recognized types of terrain. The rays must

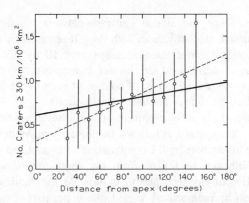

12.21. Density of ray craters $\geqslant 30$ km diameter as a function of distance from the apex ($0°$ N, $90°$ W) on Ganymede. The densities for 84 craters are averaged over $20°$ intervals. The increase in density with distance from the apex indicates that the ratio of the rate of ray removal to rate of production is higher in the leading hemisphere than in the trailing hemisphere. The broken line is a linear least-squares fit to the points plotted, weighted according to the number of points in each distance bin. The fit is No. craters $\geqslant 30$ km / 10^6 km^2 = 0.3063 + 0.0056 $\times$ (distance in degrees), with a correlation coefficient (r^2) of 0.58. The heavy solid line marks the limiting slope assuming that the ratio of ray crater production rate at the apex to the production rate at the antapex, as estimated by Shoemaker and Wolfe, is 9.6, and that ray erasure is due to gardening by projectiles in circular orbits (see text).

fade with time and disappear, evidently as a consequence of processes that influence the average albedo of material exposed locally at the surface. Presumably nearly all primary craters large enough to be resolved on the Voyager pictures had associated rays when they were first formed. These rays are bright for $\sim$ 99% of the newly formed craters on Ganymede and essentially 100% of the craters on Callisto.

On Ganymede, the density of bright ray craters > 30 km in diameter ranges from ~ 0.3 per 10^6 km^2 in the leading hemisphere to ~ 1.1 per 10^6 km^2 in the trailing hemisphere (Fig. 12.21). The observed densities correspond to ray retention ages of $< 5 \times 10^8$ yr in the leading hemisphere, on the basis of the bombardment model and the cratering time scale of Shoemaker and Wolfe (Chapter 10); for the trailing hemisphere, the mean ray retention age is ~ 2 Gyr. On Callisto, the density distribution of ray craters is less clearly related to distance from the apex. If anything, there is a trend of declining density with increasing distance from the apex, but considerable scatter is found in this trend (Fig. 12.22). The mean density of ray craters in the leading hemisphere of Callisto is higher than that observed anywhere on Ganymede. Ray retention ages on Callisto range from 1.1 Gyr in the leading hemisphere to about 3.7 Gyr near the antapex. Although the crater rays on Callisto are retained for a longer period of time than rays on

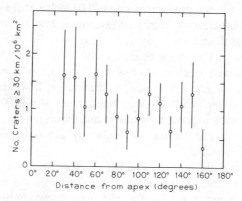

Fig. 12.22. Density of ray craters ⩾ 30 km in diameter as a function of distance from the apex on Callisto. The densities for 79 craters are averaged over 20° intervals. In contrast to the relation found on Ganymede (Fig. 12.21) there is an irregular trend to lower crater densities toward the antapex.

Ganymede, they are generally much fainter. Both the albedo of the rays on Callisto and the contrast in albedo between the rays and the adjacent surface is less than for rays on Ganymede.

The densities and regional variation of density of ray craters on the satellites reflect a balance between the rate of production of ray craters and the rate at which rays disappear. If the rate of ray erasure were everywhere constant, then the ray retention ages would be spatially uniform, and density would reflect the crater production rate. In this case ray craters would be most abundant at the apex of both satellites, and Ganymede would have more ray craters than Callisto, contrary to what is observed.

A range of ray retention ages increasing systematically with distance from the apex would be expected on Ganymede and Callisto, if the fading and disappearance of rays were due entirely to the local production of small impact craters that result in gardening of the surface and contamination of the regolith with dark debris. If the small particles that produce the gardening and contamination of the surfaces had a distribution of orbits like those of the larger bodies that produce the ray craters, and if the ratio of the flux of small particles to the flux of the larger bodies did not change with time, then the rate at which rays fade would be proportional to the rate at which they are produced. In this case the density of ray craters would be spatially uniform on each satellite, and it would be the same on both Ganymede and Callisto. These conditions are closer but still contrary to observation.

If small particles that result in gardening and contamination tend to be on less eccentric orbits than the orbits of the ray-crater-producing bodies, the difference in the impact rates between the apex and antapex will be greater for the small particles than for the larger bodies. In this case the ratio of ray removal rate to the ray production rate will be higher at the apex than at the antapex, and an increasing density of ray craters toward the antapex would be predicted, as found on Ganymede. Moreover, the rate of ray removal would be higher on Ganymede than on Callisto, which would lead to a higher ray crater density on Callisto, as observed. If gardening by small craters were the only cause of ray fading and disappearance, however, a gradient of increasing ray crater density toward the antapex should also be found on Callisto. Since such a gradient is not found, gardening by small craters and contamination of the surface by impacting debris cannot be the only processes of ray erasure.

Assuming that the impacting bodies are all on heliocentric orbits, there is a limit to the gradient in ray crater density between the apex and antapex that can be accounted for by the difference in orbits between the small particles and the larger ray-crater-producing bodies. The maximum possible gradient (ray crater density increasing toward the antapex) would result if all the small impacting particles were on orbits nearly identical with that of Jupiter. The Poynting-Robertson effect will tend to circularize the orbits of small Jupiter-crossing particles, but they cannot be driven to coincide with Jupiter's orbit. The ratio U of encounter velocity with Jupiter's sphere of influence to the orbital velocity of Jupiter is given, in the limit of a circular orbit, by

$$U = [2(1 - \cos i) + 4/9 \, e_0^2]^{1/2} \qquad (3)$$

where i is the inclination of small-particle orbits to the orbital plane of Jupiter, and $e_0 = 0.0464$ is the average eccentricity of Jupiter (see Appendix A of Chapter 10 by Shoemaker and Wolfe). Most of the impacting small particles probably are derived from disintegration of short period comets, and they will tend to inherit the orbital inclination of the respective parent comets. Hence the small particle swarm should have a similar inclination to the short period comets, $\sim 10°.6$; the limiting value of U, then, from Eq (3). is 0.18_7.[a] From the equations presented by Shoemaker and Wolfe at $U = 0.18_7$, the ratio δ_s of the small particle cratering rate at the apex to the rate at the antapex is 26.7 on Ganymede. The ratio σ between ray crater density at the apex and at the antapex is given by $\sigma = \delta\varrho/\delta_s$, where $\delta\varrho$ is the ratio of ray crater production rate between the apex and antapex. Adopting $\delta\varrho = 14.9$, as estimated by Shoemaker and Wolfe, we find $\sigma = 0.56$. This corresponds to a gradient of increasing ray crater density toward the antapex somewhat less steep than the least-squares fit to the observations for Gany-

---

[a]Subscripted figures are used in this chapter to indicate a digit that lies at or just beyond the threshold of significance.

mede (Fig. 12.21), but, within the counting errors, it is consistent with the observations. Using $\sigma = 0.36$, which is the best fit to the observations when the two points with the greatest uncertainty are neglected, and using the limiting value $\delta_s = 26.7$, the maximum value of $\delta\varrho$ suggested by the smaller-crater-gardening-model of ray removal is 9.6. This low value of $\delta\varrho$ would correspond to a distribution of orbits for the large impacting bodies dominated by long period comets rather than a distribution dominated by short period comets as postulated by Shoemaker and Wolfe.

For the small-crater-gardening-model of ray removal, the ratio C_r of mean ray crater density on Ganymede to the mean ray crater density on Callisto is given by $C_r = C_p/\epsilon$, where C_p is the ratio of production of ray craters on Ganymede to the production on Callisto, and ϵ is the ratio of production of small craters on Ganymede to the production on Callisto. From the equations given by Shoemaker and Wolfe it can be shown that ϵ is given by

$$\epsilon = \left(\frac{1 + S_g^2/U^2}{1 + S_c^2/U^2}\right) \left(\frac{S_g^2 + S_{ge}^2 + U^2}{S_c^2 + S_{ce}^2 + U^2}\right)^{\frac{\lambda}{3.4}} \tag{4}$$

where $S_g^2 = 1.389$ is the square of the dimensionless escape velocity from Jupiter at the orbit of Ganymede (ratio of the escape velocity from Jupiter at the orbit of Ganymede to the orbital velocity of Jupiter), $S_{ge}^2 = 0.0440$ is the square of the dimensionless escape velocity from Ganymede, $S_c^2 = 0.7904$ is the square of the dimensionless escape velocity from Jupiter at the orbit of Callisto, and $S_{ce}^2 = 0.0349$ is the square of the dimensionless escape velocity from Callisto and $-\lambda$ is the size index for small craters, provisionally taken to be 2.9 (Chapter 13). Introducing $U = 0.18_7$, the lower limiting value for small particles, we find the maximum value of ϵ is 2.75. If $2.3/1.2_3 = 1.8_7$, found for large crater production by Shoemaker and Wolfe, is adopted for C_p, then the lower limiting value of C_r is found to be $1.8_7/2.75 = 0.68$. The observed value of C_r for craters on ancient cratered terrain down to 25 km diameter is 3.3×1.0^{-6} km$^{-2}/4.8 \times 10^{-6}$ km$^{-2} = 0.69$, in close agreement with the ratio predicted by our model.

If gardening by small craters is the primary process of ray removal on Ganymede, another process or set of processes must be dominant on the trailing hemisphere of Callisto. Ablation of the surface, leading to residual concentration of dark silicates, is one possibility. Development of a lag deposit of dark material might occur through slow sublimation of ice or from differential sputtering of ice and silicates under bombardment of the trailing hemisphere of Callisto by the corotating plasma in the Jovian magnetosphere. Direct evidence of the influence of sublimation due to insolation is found on Ganymede, where bright rays are seen to be generally much fainter at low latitudes than at high latitudes. The slow transfer of H_2O from low latitude

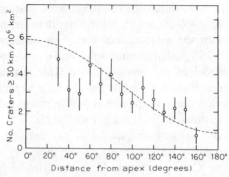

Fig. 12.23. Density of craters $\geqslant 30$ km in diameter with bright rim deposits as a function of distance from the apex on Callisto. The densities for 219 craters are averaged over $20°$ intervals. The decrease in crater density toward the antapex is the trend predicted if the retention time of bright rims is relatively uniform as is shown by the broken line (based on the model of Shoemaker and Wolfe), suggesting that bright rim deposits are not primarily darkened by meteoroid impact. An additional process whose rate is independent of distance from the apex is required to explain the darkening of the rim deposits.

to high has been treated from a theoretical standpoint by Purves and Pilcher (1980) and may account for the polar shrouds found on Ganymede (see Chapter 13). What remains to be accounted for, however, is the absence of similar shrouds on Callisto. Differential sputtering, which has been invoked by Conca (1981) to explain an observed strong concentration of dark ray craters on the trailing hemisphere of Ganymede, may also be a promising mechanism to explain the loss of bright rays on the trailing hemisphere of Callisto. In particular, sputtering may be partly responsible for the apparent low surface density of fine particles on the trailing hemisphere of Callisto indicated by polarization observations (Dollfus et al. 1980). The role of both sublimation and sputtering in ray removal may be greater on Callisto than on Ganymede because of the lower rates of cratering on Callisto.

The distribution of craters with bright rim deposits on Ganymede appears to be independent of distance from the apex, but the observations of density are scattered. The retention ages of bright rims on Ganymede range from ~ 0.8 Gyr in the leading hemisphere to ~ 3.5 Gyr in the trailing hemisphere; the corresponding crater densities range from ~ 2.1 to 3.0 craters > 30 km in diameter per 10^6 km^2 in the leading hemisphere, to ~ 1.6 to 2.6 craters per 10^6 km^2 in the trailing hemisphere. On Callisto, the density of craters with bright rim deposits decreases systematically with increasing distance from the apex (Fig. 12.23). The retention ages for bright rims range from 3.2 Gyr in the leading hemisphere to 3.7 Gyr in the trailing hemisphere. Thus the spatial variation of density of craters with bright rims on Callisto follows approximately, but not precisely, the distribution that would be predicted if the retention time for bright rims were everywhere the same. It is clear that the processes that remove or darken rays work somewhat differently with respect to the albedo of the continuous ejecta deposits. The rim deposits remain noticeably bright long after rays have disappeared, and on Callisto the darkening of the rims is not proportional to the small-crater-gardening rate or to the influx of dark contaminating debris. The very slow ablation of ice, due to insolation, especially at low latitudes, may be an important cause of darkening of the crater rim deposits.

Thermal History of the Lithosphere of Ganymede

The observed topographic relaxation of a crater can, in principle, be made to yield information bearing on the subsurface thermal regime; required, however, is knowledge of the time scale, the original crater form, and the material properties. It should be possible to infer the variation of thermal gradient with time by analysis of the relief of many craters of different ages and different sizes. A different but related approach to determination of thermal history based on crater size statistics has been taken by Phillips and Malin (1980). There are several difficulties in the rigorous application of these concepts to Ganymede and Callisto. First, the relevant physical properties of ice at the very low temperatures of the surfaces of Ganymede and Callisto are unmeasured. The mechanisms of flow or creep of ice at these temperatures have been inferred on theoretical grounds (Goodman et al. 1981) but are unverified. Moreover, there is direct evidence that, at least at the surface, the ice is impure; the composition, abundance, and effect of these impurities on the flow of ice is unknown. Second, while it is possible to make estimates of the ages of surfaces by crater counting, it is difficult to date directly any but the largest individual craters. Hence we are reduced to looking at the distributions of crater shapes on dated surfaces and arriving at the answer by statistical arguments. Finally, the temperature and correlated physical properties vary not only as a function of depth at a given site but also as a function of time. This last circumstance merely raises practical problems of computation rather than any fundamental obstacle. Here we shall attempt to leap over these rather formidable hurdles with some very rough approximations. We do so in the spirit of an exploratory foray to see if the broad outlines of lithospheric thermal history might be ascertained.

We start with the analysis of Danes (1965) of the relaxation of a crater in a Newtonian viscous fluid whose viscosity is uniform in both space and time. At relatively high temperatures ice behaves nearly like an ideal plastic solid rather than a viscous fluid, but for our present exploratory purposes the approximation of a Newtonian viscous fluid may be satisfactory; it has the great advantage of mathematical tractability. The "viscosities" we will derive must be regarded as effective viscosities. Furthermore, they must be regarded as very crude estimates of the average effective viscosity from the surface to depths that are not well defined; also they must be regarded as averaged over time. The most serious defect in our application here of Danes' solution for crater relaxation to craters on Ganymede and Callisto probably lies in the failure to account for very strong vertical gradients in effective viscosity that must exist at any time during the evolution of the lithosphere.

Danes' solution for crater relaxation can be reduced to a series of theoretical profiles at various e-folding times. Profiles obtained by photoclinometry can then be compared with the computed profiles to determine the approximate e-folding degree of relaxation r. If we can set appropriate bounds to the time interval Δt, over which the observed relaxation occurred, the characteristic e-folding time τ is given by

$$\tau = \frac{\Delta t}{r} . \tag{5}$$

The effective viscosity η_{eff} can then be estimated (Danes 1965) from

$$\eta_{\mathrm{eff}} = \frac{1}{1.605} \rho \, \tau \, g(D/2) \tag{6}$$

where ρ is the density of the lithosphere, g is the acceleration of gravity of the satellite, and D is the crater diameter (all units are cgs). For purposes of comparing the results obtained from craters of different sizes we will adopt the mean depth h to which the estimate of η_{eff} applies, as $h = D/2$. In a separate paper we will show that our use of this assumption here gives results similar to the case in which viscosity decreases exponentially with depth, a viscosity profile that corresponds to nearly uniform heat flow as a function of depth.

Next, we will assume that the Arrhenius relation for the temperature dependence of viscosity is applicable to the material of the lithospheres of Ganymede and Callisto (Shaw 1972)

$$\eta = \eta_0 \exp \left[T^* \left(\frac{1}{T} - \frac{1}{T_0} \right) \right] \tag{7}$$

where T is the temperature of the medium, T_0 is the reference temperature for which the viscosity is η_0, and T^* is an "activation" temperature. Two values of T^* are used here; both values are based on an effective viscosity of 10^{14} poise at 270 K as a reference. Assuming that the effective viscosity of ice at 130 K is 10^{25} poise, T^* is 6.28×10^3 K; for a viscosity of 10^{26} poise at 130 K, the corresponding value for T^* is 6.85×10^3 K. Various formulae that have been used to estimate the effective viscosity of ice yield values close to 10^{25} poise at 130 K (e.g., Reynolds and Cassen 1979). The competence of Ganymede's lithosphere to support large craters at a late stage of thermal evolution, however, suggests that 10^{26} poise is a better estimate for effective viscosity at the surface of the lithosphere. The values derived for the activation temperature are fairly typical. For comparison, the viscosity of the upper 100 km of the Earth's crust has been estimated to be in excess of 10^{25} poise (Gordon 1967; McKenzie 1967). Therefore the viscous behavior of ice at very low temperatures ($\sim$ 120 K) is very rocklike.

The estimated thermal gradient $\Delta T/h$ will be taken as

$$\Delta T/h = (T - T_s)/h \qquad (8)$$

where T_s is a representative surface temperature and T is the calculated temperature of the lithosphere at depth h. The annealing temperature at the base of the regolith is estimated by Shoemaker et al. (Chapter 13) to be 130 K. Here we will adopt 130 K as the global temperature at the top of the lithosphere.

The most difficult step in applying Eqs. (5) through (8) is to obtain an appropriate estimate of Δt. From evidence presented earlier, we adopt the model that the lithosphere at any given place was relatively thin at the time the oldest preserved craters were formed and that the thermal gradient decreased monotonically with time. Stochastic changes in the thermal regime associated with the individual larger craters will be ignored for this first analysis. As the thermal gradient was decreasing rapidly during the earliest part of the crater record, Δt will simply be taken as the difference in age between craters that reveal, by their degree of topographic relaxation, a substantial change in the thermal gradient. The success of this procedure depends on the fact that the characteristic relaxation time depends linearly on η_{eff} and that η_{eff} depends exponentially on T. Hence with a steadily declining mean T, most of the observed relaxation occurs early in the history of a given crater. If sufficient determinations of $\Delta T/h$, distributed over time, are obtained for a given region, more precise estimates of Δt can be found, utilizing the first derived thermal history; an improved thermal history is then obtained by iteration. This refinement, however, is not warranted for the very crude estimates presented here.

To illustrate the method, we have taken observations from photoclinometric profiles of craters on the ancient cratered terrain of Nicholson Regio and the adjacent grooved terrain of Harpagia Sulci, a region in the trailing hemisphere of Ganymede that includes areas with some of the highest observed crater retention ages on the satellite. Crater retention ages of the oldest (most flattened) craters at 10 km, 20 km, and 60 km on Nicholson Regio and the age of the most flattened 20 km diameter craters on Harpagia Sulci are from unpublished crater densities by J.B. Plescia. Ages for the oldest ray craters on Nicholson Regio are taken from ray crater densities shown in Fig. 12.21. These various estimates of the ages of the observed craters together with the observed values of crater relaxation r (from the comparison of photoclinometric profiles with the theoretical profiles of Danes) and the derived values of τ, η_{eff}, and $\Delta T/h$ are listed in Table 12.9.

Since the thermal gradient was declining fairly rapidly with time, especially near the beginning of the preserved crater record, the time at which the calculated viscosities and thermal gradients most nearly apply probably is fairly close to the time of origin of the respective craters. For example, the thermal gradient on Nicholson Regio is estimated at 5.5 to 6.7 K km^{-1} at a time close to 3.94 Gyr; the thermal gradient declined to ~ 0.17 to 0.40 K km^{-1} by ~ 2.0 Gyr. As given in Table 12.9 there is one anomaly in an otherwise simple picture of monotonic decrease of thermal gradient with time. The calculated gradient at the time of retention of the oldest 60-km crater on Nicholson Regio ($t_o = 3.76$ Gyr) appears too low. This anomaly almost certainly is an artifact of the assumptions used; our estimate of Δt for collapse of this crater may be several times too long. The values of η_{eff} and $\Delta T/h$ must be considered very rough estimates. Although the absolute values may be considerably in error, the sequence or historical trend of η_{eff} and $\Delta T/h$ is significant.

The thermal gradient in Nicholson Regio apparently decreased by an order of magnitude in the time interval from ~ 4.0 Gyr to 2.0 Gyr. The early decay in the derived thermal gradients rather closely follows the estimated decay in cratering rate, dropping by about half every 10^8 yr during late heavy bombardment. Following heavy bombardment, the decay of the thermal gradient apparently slowed considerably.

A quantitative assessment of impact heating shows that the derived drop in thermal gradient between 3.94 and 3.86 Gyr or at later times cannot be due simply to a decrease in heating by impact at those times. The cratering rate at 3.94 Gyr is ~ 400 craters of $\geqslant 10$ km diameter per 10^6 km^2 per 10^8 yr. The area A covered by these craters is

$$A = \frac{\gamma \pi N}{4} \int\limits_{D_{max}}^{D = 10\ km} D^{(\gamma + 1)}\ dD \qquad (9)$$

where the size index $\gamma = -2.2$ (Shoemaker and Wolfe, Chapter 10), $N = 6.3 \times 10^4$ km$^{-2.2}$ per 10^6 km^2, and D_{max}, the diameter of the largest crater formed, is 150 km. Integration of Eq. (9) yields an area of craters formed in 10^8 yr of 1.4×10^5 km^2, or only 14% of the surface. Integration

TABLE 12.9

Effective Lithospheric Viscosities and Thermal Gradients of Ganymede

Crater	Age (Gyr)	Δt (Gyr)	r	τ (Gyr)	Effective Viscosity[a] (poise)	Thermal Gradient[b]	
						(1) (deg km^{-1})	(2) (deg km^{-1})
Most flattened 10-km crater in Nicholson Regio	3.94	0.08	50	0.0016	2.0×10^{21}	5.5	6.7
Most flattened 20-km crater in Nicholson Regio[c]	3.86	0.10	50	0.0020	5.0×10^{21}	2.4	3.0
Most flattened 60-km crater in Nicholson Regio	3.76	0.15	50	0.0030	2.3×10^{22}	0.6	0.8
Most flattened 20-km crater in Harpagia Sulci	3.61	1.6	6.5	0.25	6.2×10^{23}	0.8	1.4
48-km with bright rim deposit in Harpagia Sulci[e]	(2.0)	0.4	1.0	0.4	2.4×10^{24}	0.17	0.40
123-km crater with rays on Nicholson Regio	1.6	1.6	(6.5)	0.25	3.8×10^{24}	0.04	0.14
Gilgamesh Basin[f] (175 km diameter)	3.5	3.5	(10)	0.35	7.7×10^{24}	0.23	0.30
			(40)	0.09	1.9×10^{24}	0.28	0.35

[a]Viscosity at depth equal to radius of crater.
[b]Gradient (1) is based on assumption that the viscosity of ice at 130 K is 10^{25} poise; gradient (2) is based on assumption that the viscosity of ice at 130 K is 10^{26} poise.
[c]From profile 4, Fig. 12.3.
[d]From profile 1, Fig. 12.3.
[e]From profile 1, Fig. 12.7.
[f]Thermal gradients calculated for Gilgamesh are based on a temperature at the top of the lithosphere of 110 K at 60° latitude rather than 130 K used in calculating the other gradients.

to 1 km crater diameter, a size at which heating of the lithosphere would be very ineffective, would merely double the area covered. The cumulative energy E, delivered to the surface by projectiles forming craters ≥ 10 km in diameter per 10^6 km^2 per 10^8 yr, is

$$E = \frac{\gamma N}{(K_n)^s} \int\limits_{D_{max}}^{D = 10 \text{ km}} D^{(\gamma + s - 1)} \, dD \qquad (10)$$

where K_n is a scaling constant relating the diameter of a crater to the kinetic energy of the projectile W, and s is the inverse scaling exponent

$$D = K_n \, W^{\frac{1}{s}}. \qquad (11)$$

Adopting $K_n = 0.125$ km (kt TNT)$^{-1/3.4}$ and $s = 3.4$ (Shoemaker and Wolfe, Chapter 10), we obtain $E = 2.3 \times 10^{23}$ J, and the power per unit area is $70 \, \mu$W m^{-2}. This may be compared to the drop in power per unit area implied by the drop in thermal gradient between 3.94 and 3.86 Gyr. The mean thermal conductivity coefficient of the lithospheric ice (temperatures from 120 to 190 K) is estimated from measurements summarized by Hobbs (1974) at about 4 W m^{-1} K^{-1}. Hence the change in heat flow corresponding to a change in thermal gradient of 3.7 K km^{-1} is 4 W K^{-1} m^{-1} $\times$ 3.7 $\times$ 10^{-3} km^{-1} which is equal to 15 $\times$ 10^{-3} W m^{-2}. The total rate of delivery of impact energy to the surface of Ganymede at 3.94 Gyr is 200 times less than the derived change in heat flow between 3.94 and 3.86 Gyr. Even if as much as 50% of the impact energy were retained as heat in the lithosphere, heating associated with craters forming at 3.94 Gyr could not account for more than a few parts per thousand of the total heat flow at that time.

It is clear that the thermal gradients on Ganymede, from the time of the earliest crater retention ages, must be controlled by heat flowing from the deep (sublithosphere) interior of the satellite. If the correlation of crater retention ages with proximity to the antapex is related directly to the gradient in cratering rate, the distribution of crater retention ages probably reflects an indirect effect, such as regional variation in the temperature at the top of the lithosphere due to a gradient in thickness of insulating regolith.

A remote possibility remains that the asymmetric cooling and thickening of the lithosphere after 4 Gyr may partly reflect asymmetric impact heating of Ganymede at a time very close to its accretion. The trailing hemisphere of Ganymede probably never was heated as much as the leading hemisphere during early stages of heavy bombardment. If the early flux of projectiles arriving from outside the Jovian system were sufficiently intense, the gradient in crater retention ages on the ancient cratered terrain might reflect an early pattern of impact heating of the entire body.

A history similar to that obtained for Ganymede is applicable to Callisto. The distribution of crater retention ages, increasing toward the antapex, may partly be a reflection of asymmetric regolith development and also partly the result of asymmetric production of multiring structures on the thin early crust. Crater retention ages (at 25 km crater diameter) extend back to 4.3 Gyr (based on a uniform half-life of the cratering rate of 10^8 yr). The greater overall crater retention age of the lithosphere of Callisto, as compared with Ganymede, probably is attributable to two causes, (1) a lower abundance of radioactive heat sources per unit surface area (cf. Cassen et al. 1980 and Chapter 4), and (2) less total heating by impact during accretion and early heavy bombardment.

A comment is in order on the absolute values of the derived thermal gradients in Table 12.9. If we assume that the lithosphere-asthenosphere boundary occurs at η_{eff} of $\sim 10^{18}$ poise, comparable to the boundary on Earth, the temperature of the boundary, from Eq. (7) is ~ 190 K. The thermal gradient of 6.7 K km^{-1} at ~ 3.9 Gyr suggests a lithosphere thickness of $(190-120) / 6.7$ km $\simeq 10$ km. This is close to the lower limiting thickness estimated for the lithosphere of Galileo Regio at a probably comparable stage of thermal evolution. Hence we consider the derived thermal gradient on Nicholson Regio for 3.94 Gyr to be a realistic estimate. The derived gradient for times < 3 Gyr probably are too low, however. The heat flow implied by these gradients is much lower than the equilibrium heat flow expected from plausible abundances of radioactive elements in the rocky core of Ganymede (cf. Cassen et al. 1980 and Chapter 4). The discrepancy is probably due mainly to our failure to take account of the changing influence of the regolith as both the cratering rate and the heat flow declined. As the heat flow decreased, the top of the lithosphere probably cooled significantly (Chapter 13).

The survival of the central depression of the Gilgamesh Basin is also a problem of considerable interest. From crater densities given in Chapter 13 the model age of the basin is ~ 3.5 Gyr. The effective viscosity, averaged over time, required to preserve the depression at its present degree of topographic relaxation is ~ 2 to 8×10^{24} poise, at a depth comparable to the diameter of the depression. This high viscosity (which implies a lithosphere thickness $\geqslant 300$ km) would not be expected at low latitudes. The survival of Gilgamesh apparently is due to its location at 60° latitude. The derived thermal gradient for this latitude (using 110 K as the mean temperature at the top of the lithosphere) is roughly consistent with the derived late stage thermal gradients at the low latitudes of Nicholson Regio. However, if we use the present global heat flow on Ganymede suggested by Cassen et al. for equilibrium with a current level of radioactivity from a chondritic abundance of radioactive elements in the rocky core, the present thickness obtained for the lithosphere below Gilgamesh would be only 80 km.

Origin of Central Pits

The global abundance of craters with central pits on Ganymede and Callisto exceeds, by far, the abundance of similar craters on any of the terrestrial planets. Craters with central pits identified on Mars (Wood et al. 1978; Hodges et al. 1980) and craters with central peak-rings on Mercury and the Moon probably are analogous to the craters with central pits on Ganymede and Callisto. The origin of the central pits and peak-rings has been a subject of debate. We note that craters with central pits on Ganymede and Callisto have terraced walls. On the Moon, there is nearly a one-to-one correspondence between the presence of a central peak and the presence of terraced walls, for fresh craters up to 100 km diameter. The terraces are very probably produced by prompt slumping or collapse along the crater walls. Thus the development of a central peak appears to be strongly correlated with and probably is a consequence of prompt collapse of the crater walls. Our observation that craters with central peaks make up most craters 5 to 20 km in diameter, whereas craters with central pits make up a majority of > 20 km diameter craters on Ganymede and Callisto, suggests that, in > 20 km craters central pits have replaced the central peaks. In other words the circumstantial evidence is very suggestive that central pits have formed by collapse of an initial central peak. High resolution images of Ganymede reveal a number of central peaks which appear to have a small pit at the summit or a pit with an anomalously high rim. We interpret these features as examples of partially collapsed central peaks.

Taking a central peak to be a simple conical solid with a basal radius of ~ 1/7 the crater radius, and a maximum peak height of the order of 1 km (in agreement with the observations of peaks in 20 to 30 km diameter craters), the stress due to the load of the peak at the level of the crater floor is

$$\text{Stress} = \frac{1/3 \, \pi \rho R^2 H g}{\pi R^2} = \frac{1}{3} \, \rho H g \qquad (12)$$

where R is the basal radius of the peak, H is the height, g is gravity, and ρ is the density of the material in the peak. Substituting a value of 1 km for H and the density of Ice I of $0.9 \, \text{g cm}^{-3}$ (the peak may actually consist of metastable higher pressure phases produced by shock) in Eq. (12) yields a stress of 4.3 bar. This may be compared with the ultimate strength of ice. In the temperature range of $0°C$ to $-10°C$ the strength of Ice I varies from ~ 14 to 50 bar (Voitkovskii 1962). Even if the central peak consisted of warm ice, one might expect much larger peaks than those observed to have been preserved. Therefore, for our model of collapse to be physically valid, the strength of the material within and beneath a central peak must be many times less than the strength of pure warm ice.

A possible clue to the implied very low strengths of the materials beneath the central peaks is provided by an experiment reported by Shoemaker et al. (1963). In this experiment, fragments were recovered from an iron projectile fired at hypervelocity into sandstone. As shown by metallography, the temperature of the fragments remained below the melting point of the iron, but certain surfaces of the fragments were coated with a quenched iron melt. The melt was produced by frictional heating of the shear surfaces along which the projectile was pulled apart. Similar frictional heating along shear surfaces in relatively strongly shocked ice in the central peaks of craters on Ganymede and Callisto may well have produced abundant thin veins and veinlets of water. Lubrication of the shear surfaces by water would greatly reduce the bulk strength of the material of the peak and deep-seated material underlying the peak.

We suggest that all central pits in the craters on Ganymede and Callisto are formed by prompt collapse of a transient central peak. A central pit in the Prairie Flat crater, produced by detonation of 500 tons of TNT, evidently was formed in this manner (Roddy 1976; Roddy et al. 1977; D.J. Roddy, personal communication, 1980). When a transient peak is formed whose basal load is much greater than the instantaneous strength of the strongly shocked (and partially melted) subjacent ice, the collapse of the peak is comparable to a case of very low velocity impact. A crater with a raised rim is formed, roughly similar to craters produced by true impact. We do not suggest, however, that any of the material of the peak was necessarily lofted into space above the surface of the satellite or that the peak went through more than one up-down oscillation (cf. Murray 1980).

The occurrence of craters with anomalously large central pits is clearly related to their age. In all cases they probably formed when the lithosphere was relatively thin. A good example is the 60-km highly flattened crater on Nicholson Regio (Table 12.9), which has a 29-km diameter central pit. If the base of the lithosphere occurs at an η_{eff} of $\sim 10^{18}$ poise and a corresponding temperature of ~ 190 K, then the depth to the base of the lithosphere indicated by the thermal gradient given in Table 12.9 is ~ 70 km. We consider the thermal gradient derived for this crater to be too low by more than a factor of 2, however, and the correct thickness of the lithosphere probably was closer to 35 km. The diameter of the crater, in other words, is nearly twice the probable depth to the asthenosphere at the time this crater formed. Assuming the transient cavity produced had a depth of $1/5 D = 12$ km, the uplift of the central peak probably formed a dome in the underlying asthenosphere of the order of 12 km. Hence, the depth to the asthenosphere beneath the transient peak was $(35 - 12)$ km $= 23$ km, somewhat less than the anomalous diameter of the central pit. As the effective viscosity of material at a depth comparable to the diameter of the pit may influence the flow as the pit is formed, it appears reasonable that the proximity of the asthenosphere influenced the growth of this anomalously large central pit.

Acknowledgments. We thank W.B. McKinnon for spirited discourses concerning the description and interpretation of features on Ganymede and Callisto. We are especially indebted to J. Plescia for sharing with us in advance of publication his detailed studies of crater densities on Ganymede. We also thank R. Greeley for a helpful review of this chapter. This work has been partially supported by the Voyager Project and Planetary Geology Program of the National Aeronautics and Space Administration.

Note added in proof: After submission of the manuscript for this chapter, detailed calculations were completed for the relaxation of craters where the viscosity is an exponential function of depth (as predicted for a nearly linear thermal gradient) and the viscosity gradient decays as an exponential function of time. In applying these calculations to the relaxation of craters on Ganymede, account has also been taken of the effect of a regolith of very low thermal conductivity on the temperature at the top of the lithosphere. The existence of this regolith, with a mean thermal conductivity comparable to or somewhat higher than the lunar regolith and a mean thickness > 10 m, is demonstrated by a high density of 1 km diameter craters observed in the highest resolution images (see Chapter 13 by Shoemaker et al.) and the strong degree of relaxation of some craters on the grooved terrain in the south polar region. In the equatorial regions, the mean temperature difference from top to base of the mature regolith appears to have been of the order of 20 K at the time of high heat flow at ~ 3.9 Gyr. When the effects of the regolith and a vertical gradient in lithospheric viscosity are combined, the solution for the time history of the thermal gradient is similar to the rough history presented in Table 12.9 (second column of estimated thermal gradients). Of particular importance is the rapid buildup of regolith on fresh craters in the leading hemisphere. The temperature increase in the subjacent lithosphere due to the buildup of this insulating layer at a time of high heat flow appears adequate to account for the higher rate of relaxation and loss of craters at the apex as compared with the antapex.

February 10, 1981

REFERENCES

Bonner, W.J. and Schmall, R.A. (1973). A photometric technique for determining planetary slopes from orbital photographs. *USGS Prof. Paper* 812-A.

Boyce, J.M. (1980). Basin peak-ring spacing on Ganymede and Callisto: Implications for the origin of central peaks and peak rings. *NASA TM* 81776, 339–342.

Cassen, P., Peale, S.J., and Reynolds, R.T. (1980). On the comparative evolution of Ganymede and Callisto. *Icarus* 41, 232–239.

Cintala, M.J., Wood, C.A., Head, J.A., and Mutch, T.A. (1977). Interplanetary comparisons of fresh crater morphology: Preliminary results. *Proc. Lunar Science Conf.* 7, 181–183.

Clark, R.N. (1980). Spectroscopic Studies of Water and Water/Regolith Mixtures on Planetary Surfaces at Low Temperature. Ph.D. Dissertation, Univ. Hawaii.

Conca, J. (1981). Dark ray craters on Ganymede. *Lunar Planet. Sci.* 12, 172-174 (abstract).

Croft, S.K. (1980). On the origin of pit craters. IAU Coll. 57. *The Satellites of Jupiter* (abstract 6–16).

Danes, Z.F. (1962). Isostatic compensation of lunar craters. *Res. Inst. Puget Sound* RIR–GP–62–1.

Danes, Z.F. (1965). Rebound processes in large craters. *U.S. Geol. Survey Astrogeol. Stud. Ann. Prog. Rept. July 1, 1964 to July 1, 1965* Part A: Lunar and Planetary Investigations, Nov. 1965, 81–100.

Dollfus, A., Mandeville, J.C., and Geake, J.E. (1980). Regolith and cratering on Callisto. IAU Coll. 57. *The Satellites of Jupiter* (abstract 6–6).

Gault, D.D., Guest, J.E., Murray, J.B., Dzurisin, D., and Malin, M.C. (1975). Some comparisons of impact craters on Mercury and the Moon. *J. Geophys. Res.* 80, 2444–2460.

Goodman, D.J., Frost, H.J., and Ashby, M.F. (1981). The plasticity of polycrystalline ice. In preparation.

Gordon, R.B. (1967). Thermally activated processes in the Earth: Creep and seismic attenuation. *Geophys. J.* 14, 33–43.

Haskell, N.A. (1935). The motion of a viscous fluid under a surface load. *Physics* 6, 265–269.

Hobbs, P.V. (1974). *Ice Physics.* Clarendon Press, Oxford.

Hodges, C.A., Shew, N.B., and Clow, G. (1980). Distribution of central pit craters on Mars. *Lunar Sci.* 11, 450–452 (abstract).

Johnson, T.V., and McGetchin, T.R. (1973). Topography on satellite surfaces and the shape of asteroids. *Icarus* 18, 612–620.

Malin, M.C. (1980). Fables in Ganymede tectonics from morphologic studies, IAU Coll. 57. *The Satellites of Jupiter* (abstract 6–3).

McKenzie, D.P. (1967). The viscosity of the mantle. *Geophys. J.* 14, 297–305.

McKinnon, W.B., and Melosh, H.J., (1980). Evolution of planetary lithospheres: Evidence from multiringed basins on Ganymede and Callisto. *Icarus* 44, 454-471.

Melosh, H.J., and McKinnon, W.B. (1978). The mechanics of ringed basin formation: *Geophys. Res. Letters* 5, 985–988.

Murray, J.B. (1980). Oscillating peak model of basin and crater formation. *Moon and Planets* 22, 269–291.

Phillips, R.J., and Malin, M.C. (1980). Ganymede: A relationship between thermal history and crater statistics. *Science* 210, 185–187.

Pike, R.J. (1975). Size-morphology relations of lunar craters: Discussion. *Modern Geology* 5, 169–173.

Purves, N., and Pilcher, C.B. (1980). Thermal migration of water on the Galilean satellites. *Icarus* 43, 51–55.

Reynolds, R.T., and Cassen, P.M. (1979). On the internal structure of the major satellites of the outer planets. *Geophys. Res. Letters* 6, 121–124.

Roddy, D.J. (1976). High-explosive cratering analogs for bowl-shaped, central uplift, and multiring impact craters. *Proc. Lunar Sci. Conf.* 7, 3027–3065.

Roddy, D.J., Ullrich, G.W., Sauer, F.M., and Jones, G.H.S. (1977). Cratering motions and structural deformation in the rim of the Prairie Flat multiring explosion crater. *Proc. Lunar Sci. Conf.* 8, 3389–3407.

Schubert, G., Stevenson, D.J., and Ellsworth, K. (1981). Internal structures of the Galilean satellites. Submitted to *Icarus.*

Scott, R.F. (1967). Viscous flow of craters. *Icarus* 7, 139–148.

Shaw, H.R. (1972). Viscosities of magmatic silicate liquids: An empirical method of prediction. *Amer. J. Sci.* 272, 870–893.

Shoemaker, E.M. (1966). Preliminary analysis of the fine structure of the lunar surface in Mare Cognitum. In *The Nature of the Lunar Surface; Proceedings of the 1965 IAU-NASA Symposium* (W.N. Hess, D.H. Menzel, and J.A. O'Keefe, Eds.), p. 23–78. Johns Hopkins Press, Baltimore.

Shoemaker, E.M., Gault, D.E., Moore, H.J., and Lugn, R.V. (1963). Hypervelocity impact of steel into Coconino Sandstone. *Amer. J. Sci.* 261, 668–682.

Shoemaker, E.M., and Passey, Q.R. (1979). Tectonic history of Ganymede. *EOS* 60, 869 (abstract).

Smith, B.A., and the Voyager Imaging Team. (1979a). The Jupiter system through the eyes of Voyager 1. *Science* 204, 951–972.

Smith, B.A., and the Voyager Imaging Team. (1979b). The Galilean satellites and Jupiter: Voyager 2 imaging science results. *Science* 206, 927–950.

Squyres, S.W. (1980a). Surface temperatures and retention of H_2O frost on Ganymede and Callisto. *Icarus* 44, 502–510.

Squyres, S.W. (1980b). Water Vulcanism on Ganymede? IAU Coll. 57. *The Satellites of Jupiter* (abstract 6–21).

Voitkovskii, K.F. (1962). The mechanical properties of ice. Translated from Russian by The American Meteorological Society and The Arctic Inst. of North America. Air Force Cambridge Research Laboratories, Bedford, Mass.

Watson, K. (1968). Photoclinometry from spacecraft images, *USGS Prof.* Paper 599–B.

Wood, C.A., Head, J.W., and Cintala, M.J. (1978). Interior morphology of fresh Martian craters: The effects of target characteristics. *Proc. Lunar Planet. Sci. Conf.* 9, 3691–3709.

13. THE GEOLOGY OF GANYMEDE

E.M. SHOEMAKER, B.K. LUCCHITTA, D.E. WILHELMS
U.S. Geological Survey

J.B. PLESCIA
Jet Propulsion Laboratory

and

S.W. SQUYRES
Cornell University

*Like the Moon and the terrestrial planets, the crust of Ganymede pre-
serves a geologic record of impact cratering, tectonic deformation, and
extensive eruption of material from its interior. About 40% of the
surface consists of dark cratered terrain dating from a period of heavy
bombardment early in the history of the satellites. Old craters on this
terrain with diameters > 10 km are extremely flattened as a result of
viscous or plastic flow. Large craters have disappeared, leaving bright
smooth palimpsests. The ancient cratered terrain has been extensively
replaced by grooved terrain which evidently formed by progressive*

foundering of the old crust and flooding of the surface by water or brine. Lanes or sulci of the grooved terrain may have formed over rising currents in a satellite-wide deep convection system. A complex pattern of structural cells in the grooved terrain may have formed over small convection cells in the shallow asthenosphere whose pattern was determined by stoping of cold dense slabs of the lithosphere. Ray craters are the youngest geologic structures on Ganymede. The fading of the rays with time appears to result primarily from impact gardening of the surface. Early development of an insulating regolith permitted craters 10 km in diameter in the equatorial region to flatten completely, and larger craters near the poles, which postdate the grooved terrain, to relax. Most large impact craters produced during and soon after the formation of grooved terrain are preserved, though extremely flattened. The lesser degree of topographic relaxation of these craters, compared with older craters in the ancient cratered terrain, shows that the lithosphere was cooling and thickening during the period of grooved terrain formation. A deep stiff shell or lithosphere of ice II may have separated the deep region of convection from a shallower region of convection during this period. Transformation of ice V to ice II at this time could have thrown the upper lithosphere into tension, permitting foundering of old crust.

Among all the satellites of Jupiter, Ganymede's surface and geologic record most nearly resemble those of the Moon and some of the terrestrial planets. Somewhat less than half the area of Ganymede is occupied by a dark relatively heavily cratered terrain (Smith et al. 1979a) that calls to mind the cratered highlands of the Moon, Mercury, and Mars. The principal surface features on this terrain can be shown to date from relatively early solar system history. However, the crater shapes, density, size distribution, and especially the topographic relief of the ancient cratered terrain on Ganymede, differ in important respects from the ancient cratered highlands of bodies in the inner solar system. Most of the remainder of Ganymede's surface is occupied by diverse geologic units, almost all with higher albedo than the ancient cratered terrain; these units have been broadly grouped as grooved terrain (Smith et al., 1979a). A significant part of the grooved terrain actually consists of smooth plains with few or no grooves, but all these various units appear closely related and younger than the ancient cratered terrain. A broad range of crater densities is found on the grooved terrain, generally lower than on the older cratered terrain and roughly comparable to crater densities observed on some plains units of Mercury and Mars and some maria of the Moon.

The diversity of terrains on Ganymede reflects the diversity of processes that have molded its surface. These processes have occurred in a decipherable historical sequence that invites analysis by geologic mapping. We report here the broad outlines of the geologic history of Ganymede, obtained from a first attempt to map the geology at a global scale and to interpret the characteristics of the observed geologic units.

I. CONSTRUCTION OF A GEOLOGIC MAP

Like most planets and satellites so far photographed, Ganymede lends itself to geologic mapping based on the principles of stratigraphy, and has probably yielded more information in a shorter time by this method than have the other bodies. The dark cratered terrain is present in roughly polygonal regions, divided and separated by bands of the lighter grooved terrain. Although these terms derive from the dominant structural features found in the two units, they are lithologic rather than structural. On the preliminary geologic map (see Plates 2 and 3 in the color section), the cratered terrain is mapped as an undivided unit, but it is undoubtedly stratigraphically and structurally complex. Superposed on the cratered terrain is a unit of crater palimpsests; it is the only other mapped unit older than the grooved terrain.

A unit transitional between the cratered and grooved terrains is also mapped: reticulate terrain. Some of the cratered terrain has a reticulate structural pattern that seems related to the grooved terrain. The reticulate pattern occurs at the edges of or in the midst of sets of the grooved terrain, and individual lineaments pass between the two units. Albedo of the reticulate terrain is transitional in places, and some of the unit is as light as the grooved terrain. The mapped reticulate terrain includes both dark and light varieties. Presumably, it is a structural rather than a material unit.

The grooved terrain has elicited the greatest interest because of the complex way in which the surface of Ganymede has been restructured (Smith et al. 1979a,b) and the clues it may provide about internal processes in Ganymede. It is probably both lithologically and structurally distinct from the ancient cratered terrain. On the geologic map, structural units are shown within the grooved terrain; each line is the boundary of a structural cell that contains a unified structural pattern. Most cells contain subparallel sets of grooves, ridge-bordered grooves, or ridges. The sets merge with or are truncated sharply by adjacent cell boundaries. Some sets curve markedly; rotational and translational movements may have occurred within or between cells. Some cells are topographically smooth. Most are light in color, but a few of the smoother ones are as dark as typical ancient cratered terrain.

The grooved terrain has evidently formed at the expense of the ancient cratered terrain. Whether the morphologic varieties of some of the cells reflect material differences as well as structural differences is not clear; the albedo differences between grooved and cratered terrain probably reflect lithologic differences of the underlying material. For the purposes of general geologic mapping, whether the cells are structural or material units does not matter; in either case, the transection of the ancient cratered by the grooved terrain means that the grooved terrain is younger. Subdivision of the grooved terrain by morphology, albedo and color will be a goal of future large-scale mapping efforts.

Two classes of crater materials superposed on other units are mapped: crater material and ray crater material. The former includes deposits around craters that lack visible rays whose albedo is generally similar to the average albedo of nearby terrain. Ray crater material is associated with craters with both bright and dark rays.

Principles of mapping the crater materials are readily adapted from lunar and planetary mapping practice. Deposits are mapped, not the crater depression (which is commonly very shallow on Ganymede). Generally, deposits are mapped as far as they obscure other units. A mapping cutoff of 100 km for the diameter of the deposit has been employed. This convention means that deposits of craters with smaller rimcrest diameters are mapped where deposits are relatively well preserved, that is, mostly deposits of young craters. Dense concentrations of secondary-impact craters, quite like those of the Moon, are included in the mapped deposit. The lunar experience suggests that the bright- and dark-rayed craters are younger than the craters with rim deposits of intermediate albedo, which have had more time to come to equilibrium with the space environment. The craters of a given class, especially the intermediate-albedo craters, probably are not of the same age all over Ganymede. Most mapped craters are superposed on the grooved terrain, but a few intermediate-albedo craters are cut by parts of it. Some craters superposed on grooved terrain are overlain by the deposits or secondary craters of another crater or basin. Basins and craters are not distinguished on the map; for example, the deposits of the giant, fresh, ringed basin Gilgamesh are mapped by the same convention as 100 km wide deposits. As on other planets, basins and craters are members of a size-distribution continuum of craterform features.

II. ANCIENT CRATERED TERRAIN

This dark, heavily cratered terrain is distributed across Ganymede in a series of discrete areas of polygonal, rounded, and irregular outline. As viewed on the map (Plates 2 and 3), these areas are fragments of ancient crust enclosed in a matrix of younger grooved terrain. The largest such remnant, Galileo Regio, is > 3000 km long and forms a conspicuous dark patch in the northern part of the leading hemisphere, recognizable by Earth-based telescopes. The second largest discrete remnant of cratered terrain is in the southern part of the trailing hemisphere and is < 2000 km long and ~ 1000 km wide. This area has been designated part of Nicholson Regio. Curiously, the centers of these two largest remnants are approximately antipodal. Other fragments or polygons of cratered terrain are of widely varying size, extending down to wedges or slices of recognizable cratered terrain less than a few tens of km across. Large remnants tend to be rounded in outline, whereas small remnants are distinctly polygonal and generally elongate.

Crater Characteristics

Besides its distinctly low albedo, the cratered terrain is characterized by relatively abundant craters ranging in diameter from the limit of image resolution to ~ 50 km. There are relatively few craters > 50 km. Typical cumulative densities for craters > 10 km in diameter are 200 to 300 craters per 10^6 km² (Table 13.1), although local densities as high as 400 per 10^6 km² have been found. These densities are an order of magnitude higher than the average crater density of the lunar maria, but lower than typical crater densities of the lunar highlands and the most heavily cratered parts of Mercury and Mars. On the basis of crater density, the cratered terrain is clearly the oldest geologic unit preserved on Ganymede.

The most striking characteristic of most craters on the ancient cratered terrain is their very low topographic relief (Figs. 13.1 and 13.2). As determined by photoclinometry (see Chapter 12), a small fraction of the craters, especially those surrounded by bright rim deposits or rays, have well-defined raised rims and depth-to-diameter ratios comparable to the ratios found for craters on the Moon. Most craters, however, are much shallower than craters of equivalent size on the Moon. A complete spectrum of crater flattening or topographic relaxation is observed, from fresh craters that exhibit negligible relaxation to craters with relief so subdued that they are barely detectable in high-resolution images. This spectrum of flattening is found for craters of all sizes > 8 to 10 km in diameter. Because flattening extends to the threshold of crater detectability, we can safely infer that craters 10 km in diameter and larger have also disappeared as a result of topographic relaxation.

The forms of the flattened craters, as discussed by Passey and Shoemaker (Chapter 12), are precisely those expected to result from plastic or viscous flow in a medium in which effective viscosity decreases with depth. Because the thermal conductivity of ice I is only weakly dependent on temperature (Hobbs 1974), steady heat flow from the interior of Ganymede through an ice lithosphere would lead to a nearly linear geotherm at a given time in the history of Ganymede. Under these conditions, the effective viscosity at any given time would decrease approximately as an exponential function of depth, down to a depth at which heat would be transported by convection rather than conduction (Chapter 4). With such a gradient of effective viscosity, topographic features of long wavelength would relax much more rapidly than features of short wavelength. The floors of craters would flatten and then bow up in the center as the flanks of the crater rims subside. Rim crests would remain relatively sharp, however, during much of the history of relaxation. In highly relaxed craters, a relatively narrow moat would form between the crater rim crest and the broadly arched crater floor. This theoretically predicted form for a highly relaxed crater is widely observed among the flattened craters of the ancient cratered terrain. Thus the observed forms of the craters confirm not only their degradation by flow of the icy lithosphere,

TABLE 13.1

Crater Densities on the Cratered Terrain

PICNO[a]	Center of Counting Area Lat.	Long.	β[b] (deg)	10 km[c]	20 km[c]	30 km[c]	40 km[c]	50 km[c]	Area of Zone Used for Counting (km²)
096911+000	28°N	312°	131	270 ± 44	60 ± 21	18 ± 11	(9 ± 8)	(5 ± 6)	139000
094111+000	10°S	320°	129	140 ± 31	45 ± 17	24 ± 13	15 ± 10	(11 ± 9)	151000
093911+000	15°S	328°	121	155 ± 30	25 ± 12	(7 ± 6)	(3 ± 4)	(2 ± 3)	179000
092311+000	3°S	343°	107	215 ± 40	30 ± 15	8 ± 8	(7 ± 7)	(4 ± 5)	135000
037012−001	15°N	194°	104	275 ± 44	66 ± 21	28 ± 14	(13 ± 10)	(7 ± 7)	142000
092311+000	10°S	350°	100	195 ± 32	68 ± 19	16 ± 9	9 ± 7	4 ± 4	187000
038212−001	10°N	180°	90	200 ± 39	52 ± 20	14 ± 10	8 ± 8	(4 ± 5)	135000
038212−001	0°	177°	87	290 ± 43	28 ± 7	(7 ± 7)	(4 ± 5)	(2 ± 4)	155000
045812−001	31°N	175°	86	270 ± 72	40 ± 28	(15 ± 17)	(7 ± 12)	(4 ± 9)	52000
039512−001	15°S	168°	78	300 ± 33	80 ± 17	31 ± 11	15 ± 7	4 ± 4	280000
047012−001	5°S	170°	80	270 ± 43	85 ± 24	31 ± 14	20 ± 12	(13 ± 9)	145000
044912−001	40°N	155°	71	300 ± 56	55 ± 24	25 ± 16	(7 ± 9)	(4 ± 6)	95000
046112−001	18°N	158°	69	230 ± 53	66 ± 28	27 ± 18	15 ± 14	10 ± 11	82000
046112−001	25°N	155°	67	440 ± 81	200 ± 55	100 ± 39	(68 ± 32)	(50 ± 27)	67000
045212−001	25°N	148°	61	270 ± 63	40 ± 24	(11 ± 13)	(5 ± 9)	(3 ± 6)	68000
046412−001	10°N	150°	61	320 ± 48	94 ± 26	38 ± 16	(22 ± 13)	(14 ± 10)	138000
945212−001	35°N	138°	57	330 ± 68	98 ± 37	55 ± 28	29 ± 20	17 ± 15	71000

TABLE 13.1 (continued)

0464J2–001	15°N	56	260 ± 43	96 ± 26	58 ± 20	(32 ± 15)	(22 ± 13)	137000
0455J2–001	20°N	53	170 ± 31	45 ± 16	21 ± 11	(13 ± 9)	(8 ± 7)	174000
0455J2–001	25°N	50	225 ± 35	62 ± 19	30 ± 13	(15 ± 9)	(10 ± 7)	180000
0464J2–001	12°N	46	400 ± 72	130 ± 41	56 ± 27	23 ± 17	(13 ± 13)	77000
0467J2–001	8°N	43	220 ± 51	70 ± 29	30 ± 19	(15 ± 13)	(10 ± 11)	84000
0594J2–001	17°N	40	270 ± 56	95 ± 33	52 ± 24	(32 ± 19)	(23 ± 16)	87000
0467J2–001	12°N	39	350 ± 71	40 ± 24	16 ± 15	(6 ± 9)	(3 ± 7)	68000
0548J2–001	9°N	31	300 ± 79	75 ± 40	30 ± 25	(18 ± 19)	(11 ± 15)	48000

[a] Picture number of frame used for counting.

[b] Angular distance from apex of motion (90° long and equator).

[c] Number of craters $\geq$ the quoted diameter, normalized to 10^6 km^2. Crater counts for all areas studied were extended to the limits of resolution of the Voyager images used (generally 2 to 3 km). Cumulative crater densities at the quoted diameter are interpolated between the densities at diameters of observed craters which lie closest to the quoted diameter. For example, if the nearest observed diameters to 10 km in an area counted are at 8 km and 12 km, the crater density quoted for 10 km is obtained by interpolation between the observed cumulative densities at 8 and 12 km. Numbers indicated in parentheses were derived from a power function fitted to the observed cumulative crater density distribution. The densities shown in parentheses are used for comparative purposes where the observed cumulative frequency at the quoted diameter drops below one or, in a few cases, where the quoted diameter is close to the resolution limit of the Voyager image studied.

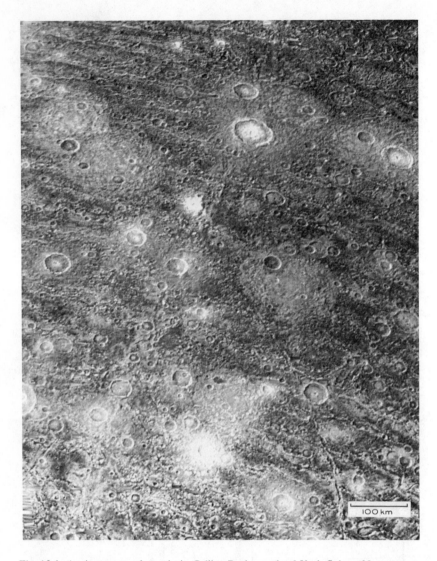

Fig. 13.1. Ancient cratered terrain in Galileo Regio north of Uruk Sulcus. Most craters are extremely flattened. Circular patterns of intermediate albedo (right of center and upper left) are palimpsests, which mark the sites of craters that have disappeared. Note the discontinuous trough that trends from lower right toward center of frame. North is approximately toward the top of the frame in this figure and all following Voyager images. (Voyager 2 image 20636.59 centered at 30°N, 152°W; PICNO 0452J2-001.)

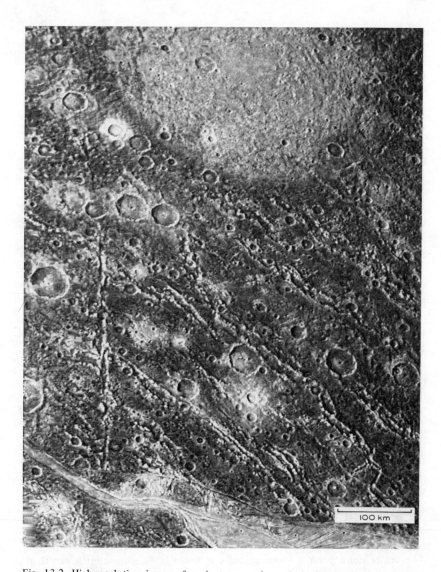

Fig. 13.2. High-resolution image of ancient cratered terrain in SE Galileo Regio. Most craters are highly flattened; a large palimpsest is at top of frame. Density of craters >10 km diameter is distinctly lower on the large palimpsest than on the adjacent cratered terrain. Rimmed furrows that trend diagonally across the frame are part of a great arcuate system in Galileo Regio. Note the discontinuities and kinks in the rimmed furrows. The north-trending trough on left side of frame appears to cut the rimmed furrow system. (Voyager 2 image 20638.33 centered at 10°N, 132°W; PICNO 0546J2-001.)

but also the presence of a gradient of decreasing effective viscosity with depth at the time of crater relaxation.

Crater Palimpsests

In addition to craters, palimpsests are common features of the cratered terrain (Smith et al. 1979*b*). The term palimpsest was first used for parchment from which previous lettering had been scraped or erased, leaving a faint trace under a later inscription. It is applied here to places on Ganymede's surface from which craters have been erased by subsequent crustal relaxation or other processes. The palimpsests, bright patches on the dark cratered terrain (Figs. 13.1 and 13.2), have albedos comparable to that of the grooved terrain. The largest are mapped separately (Plates 2 and 3). They were first clearly recognized from high-resolution Voyager 2 images of Galileo Regio, where they occupy roughly a quarter of the area of cratered terrain. Most palimpsests are roughly circular in outline, but some are oval or irregular (Fig. 13.3). They range from ~ 50 km to nearly 400 km in diameter. Detailed study of the highest resolution pictures of large palimpsests shows that the bright patches represent ejecta deposits of craters that have completely disappeared either by prompt collapse or slow viscous relaxation. Part of the bright region at the center of each palimpsest must correspond to the former crater; in a few cases a distinct smooth area in the center can be recognized (see Passey and Shoemaker, Chapter 12). A number of extremely relaxed but still recognizable craters, termed penepalimpsests by Passey and Shoemaker, further demonstrate the close relation of palimpsests and craters (Fig. 13.4).

Once the characteristics of palimpsests had been established from Voyager 2 pictures, they were readily recognized on Voyager 1 pictures of cratered terrain. On the average, the palimpsests constitute ~ 25% of the cratered terrain, where it has not been so severely dismembered by lanes of grooved terrain as to prevent the tracing of broad features. Locally, on Marius Regio, the fractional area occupied by palimpsests approaches 50% (Fig. 13.5).

The densities of superposed craters suggest that the palimpsests are broadly contemporaneous with the rest of the cratered terrain (Table 13.2). The crater density on a given palimpsest tends to be somewhat lower than that found on the adjacent cratered terrain, and in many cases they are distinctly younger than surrounding surfaces (see Fig. 13.2). The crater density on other palimpsests, however, is not distinguishable from that on surrounding terrain.

The persistence of albedo contrast between the palimpsests and the rest of the cratered terrain, despite the fact that the palimpsests are among the older features of Ganymede's crust, indicates that they probably are underlain by material compositionally distinct from the rest of the shallow substrate of the cratered terrain. The simplest explanation of their moderate

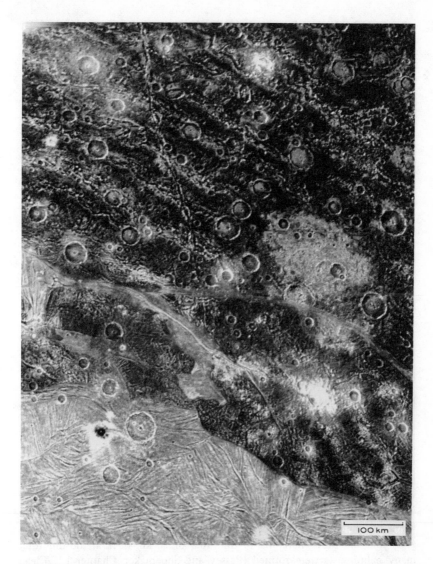

Fig. 13.3. Southern Galileo Regio and part of Uruk Sulcus. Irregular palimpsest to right of center is cut by a narrow sulcus of grooved terrain. Note that neither the palimpsest nor a furrow extending from top to center of frame are offset. Near center of frame, a lobate smooth unit of the grooved terrain overlaps the ancient cratered terrain. (Voyager 2 image 20637.11 centered at, 10°N, 145°W; PICNO 0464J2-001.)

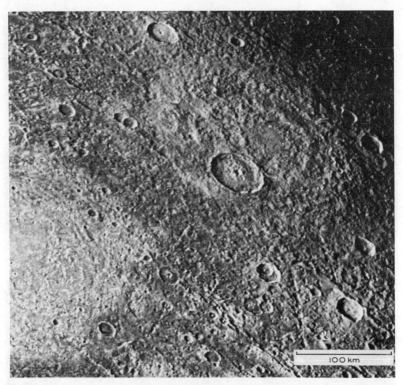

Fig. 13.4. Penepalimpsest in eastern Galileo Regio. Note that vestiges of original crater rim are present in penepalimpsest (to right and above center), whereas in the large palimpsest at lower left the original crater has completely disappeared. (Voyager 2 image 20638.31 centered at 20°N, 125°W; PICNO 0544J2-001.)

albedo is that the palimpsests have been formed by relatively clean ice excavated from substantial depth. Because the craters from which these ejecta deposits came are completely relaxed, the lithosphere evidently was thin when the palimpsests were formed. Modeling of the relaxation process suggests that the lithosphere was of the order of 10 km thick at the time that many palimpsests were formed (Passey and Shoemaker, Chapter 12). Clean ice in the uppermost crater-rim deposits may have been derived from an underlying convecting ice asthenosphere from which silicates had been largely removed by various processes of differentiation. Dark rocky material may have accumulated in the overlying lithosphere from prior bombardment by silicate-bearing planetesimals.

It is interesting that most craters < 50 km diameter on the cratered terrain exhibit no bright rim deposits, although several have relatively bright floors. Most palimpsests apparently were formed around craters somewhat

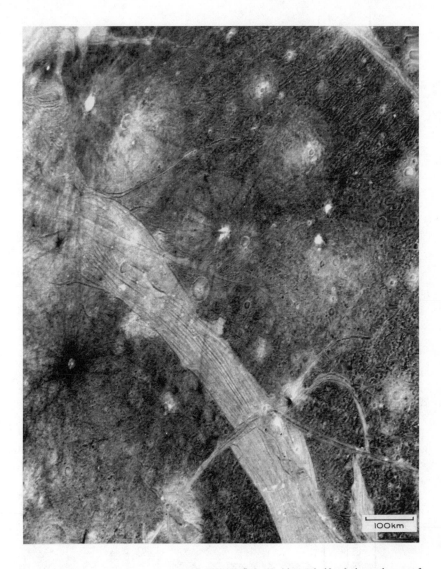

Fig. 13.5. Western Marius Regio and Tiamat Sulcus. About half of the polygon of ancient cratered terrain in upper part of frame is occupied by palimpsests superposed on the NE-trending rimmed furrow system. The abrupt change in width of Tiamat Sulcus across the transecting lane of grooved terrain suggests that differential lateral growth of the sulcus occurred on opposite sides of the transecting lane. (See text for various models of growth.) Note bright ray system near center of frame passing radially outward into dark rays and dark ray system to left of Tiamat Sulcus. (Voyager 2 image 20635.45 centered at 0°, 210°W; PICNO 0378J2-001.)

TABLE 13.2

Crater Densities on the Palimpsests

PICNO[a]	Center of Counting Area		β[b] (deg)	10 km[c]	20 km[c]	30 km[c]	40 km[c]	50 km[c]	Area Used for Counting (km²)
	Lat.	Long.							
0449J2−001	40°N	160°	75	290 ± 106	80 ± 56	55 ± 46	45 ± 42	(27 ± 32)	26000
0449J2−001	39°N	155°	71	180 ± 94	95 ± 68	51 ± 50	(27 ± 36)	(20 ± 31)	20000
0449J2−001	36°N	152°	68	430 ± 134	120 ± 71	(48 ± 45)	(27 ± 34)	(18 ± 27)	24000
0461J2−001	21°N	150°	62	560 ± 203	120 ± 94	(110 ± 90)	(75 ± 74)	(55 ± 64)	14000
0464J2−001	9°N	125°	36	310 ± 115	120 ± 71	86 ± 61	50 ± 46	(31 ± 36)	23000
0542J2−001	18°N	130°	43	80 ± 27	20 ± 14	(6 ± 7)	(3 ± 5)	(2 ± 4)	109000
0485J2−001	19°S	145°	57	190 ± 139	(60 ± 78)	(23 ± 48)	(12 ± 35)	(7 ± 27)	10000
0370I2−001	19°N	197°	106	290 ± 51	35 ± 18	(11 ± 10)	(5 ± 7)	(2 ± 5)	41000

[a]Picture number of frame used for counting.

[b]Angular distance from apex of motion (90° long and equator).

[c]Number of craters $\geq$ the quoted diameter normalized to 10^6 km². See Table 13.1 for explanation of the numbers listed.

> 50 km across, which suggests that the depth of the layer contaminated with dark material is on the order of the depth of the initial excavation cavity for 50 km diameter craters—again about 10 km. Craters superposed on the outer parts of the palimpsests commonly are dark floored, as might be expected from their penetration into a contaminated lithosphere beneath the clean ejecta of the palimpsests. But where excavated near the center of palimpsests, superposed craters generally are relatively bright floored, with rim deposits comparable in albedo to the palimpsests (Figs. 13.1, 13.2, and 13.3).

Ancient Furrows and Troughs on the Cratered Terrain

The cratered terrain is locally cut by linear or sublinear structures some hundreds of km long. The most conspicuous of these belong to a great regional system of rimmed furrows that extends across Galileo Regio (Plates 1 and 2), but is not confined to it; other disconnected parts of the system can be recognized on nearby polygons of cratered terrain, including Marius Regio. Individual furrows are somewhat irregular in form, and commonly offset from one another in a crudely developed en echelon system. The system of furrows on Galileo Regio, however, is broadly arcuate with a center of curvature near the anti-Jovian point and somewhat south of the equator. The disconnected parts of the system in Marius Regio do not appear concentric about the same center as the arcs on Galileo; this misalignment suggests that moderate counterclockwise rotation or left-lateral translation of Marius polygons has occurred relative to those of Galileo. A faint, diffusely bounded palimpsest at least several hundred km in diameter (not mapped) lies at the approximate center of the composite system at ∼ 15° lat, 165° long (Fig. 13.6).

In its broad features, the rimmed furrow system of Galileo Regio, together with the central palimpsest, is similar to the great Valhalla and Asgard multiring systems of Callisto (see Chapter 9). In detailed form, however, the rimmed furrows are markedly different from the ridges and scarps of the Valhalla and Asgard rings. Individual furrows on Galileo are typically 5 to 10 km wide and 50 km to several hundred km long (Fig. 13.2). Their depths are a few hundred m to as much as 1/2 km below the surrounding surface of the cratered terrain. Generally, edges of the furrows are marked by a raised rim or lip that rises ∼ 100 m above the adjacent terrain. Spacing between the furrows is generally between 25 and 100 km. Their pattern and relief strongly suggest that they are a regional system of grabens, as inferred by McKinnon and Melosh (1980). The raised rims probably formed during relaxation of the long-wavelength components of the initial relief; as the graben floors rose, the bounding escarpments probably were rotated and carried upward.

The rimmed furrows clearly predate most large craters and palimpsests on Galileo Regio. Hence they are among the oldest features of this ancient

Fig. 13.6. Faint giant palimpsest at center of Galileo Regio rimmed furrow system. Palimpsest extends nearly the width of the large polygon of cratered terrain in upper half of frame. Note segments of crater spread apart by cross-cutting furrow near the northern edge of giant palimpsest. Southern part of giant palimpsest has been replaced by grooved terrain and reticulate terrain (below center). A penepalimpsest near lower left corner has a dome in the floor of its central pit. (Voyager 2 image 20636.02 centered at 17°S, 165°W; PICNO 0395J2-001.)

segment of crust. On the other hand, determination of the precise age relation of the furrows to most small craters is difficult. In all likelihood the furrows postdate 10 km diameter craters that exhibit a greater degree of topographic relaxation.

Other linear or slightly arcuate structures on Galileo Regio intersect the rimmed furrows at rather large angles (Figs. 13.1, 13.2, and 13.3). Most are discontinuous sets of aligned troughs as much as several hundred km long. They appear to cut the rimmed furrows and thus are probably younger, but they are distinctly older than most craters, and hence are also ancient features of the cratered terrain. Their origin at this writing remains enigmatic.

III. GROOVED TERRAIN

Ganymede's grooved terrain, which occupies ~ 60% of the surface, is the most distinctive feature of the satellite. No close analog of the grooved terrain has been found on any body in the solar system; the conditions under which the grooved terrain formed are probably unique to Ganymede. Several types of surfaces or morphologic units recognized by the Voyager imaging team were designated grooved terrain, smooth terrain, and reticulate terrain (Smith et al. 1979b). These units are spatially associated and appear closely related in origin. Grooved, smooth, and reticulate units are so intimately mixed that their separation on the small-scale geologic map of Plates 2 and 3 is impractical, although some broad areas of reticulate terrain can be mapped separately. Over most of the geologic map, all three morphologic units are grouped as grooved terrain. The grooved terrain occupies most of the polar regions of Ganymede and laces through the intermediate- and low-latitude regions, forming a complex network of broad and narrow lanes called sulci, which dissect the ancient cratered terrain.

Grooved Units

The grooved units are characterized by subparallel grooves with intervening ridges, commonly ≥ 100 km long, spaced ~ 3 to 10 km apart. Topographic relief is generally 300 to 400 m and reaches a maximum of ~ 700 m (Squyres 1980a,1981). Some broad grooves have terraced walls or contain smaller subsidiary grooves. Slopes are gentle; they average ~ 5° and reach a maximum of 20°. Typical photoclinometric profiles across grooved terrain are shown in Fig. 13.7. Individual grooves and ridges terminate by merging, gradual diminution of relief, or abrupt truncation.

The grooves and ridges are assembled into structural cells or sets with straight or curvilinear boundaries. Some cells are only a few tens of km wide and hundreds of km long; others are short and as wide as ~ 100 km. Some are nearly smooth in the center and bounded by deep grooves. The trend of the

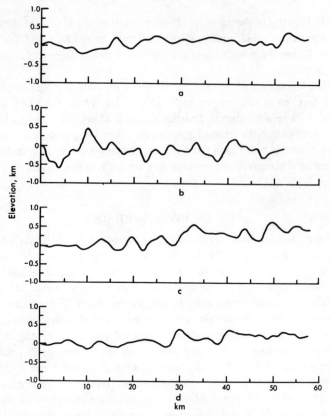

Fig. 13.7. Photoclinometric profiles of grooved units near south pole. The image from which these profiles were derived is shown in Fig. 13.15.

grooves and ridges in some cells is roughly parallel with the long axis of the cell, and in others it is roughly orthogonal to the long axis. A few sets of grooves and ridges are fan-shaped.

The structural cells generally have sharp boundaries that truncate the pattern of grooves and ridges in adjacent cells; the system of truncation is complex (Fig. 13.8). In some cases, an age sequence can be deciphered from the truncation relations, but age relations are often ambiguous. Rarely, two sets of grooves and ridges are superposed, producing reticulate groove-and-ridge patterns (Fig. 13.9).

Some grooves are solitary and deep, forming short and stubby, or long and extended, gashes (Fig. 13.10). They may be straight, curvilinear, or irregular. These solitary grooves occur in both grooved and cratered terrain and locally transect these terrain boundaries. Some single grooves have vaguely circular or semicircular plans that may reflect the imprint of a preexisting crater (Fig. 13.11).

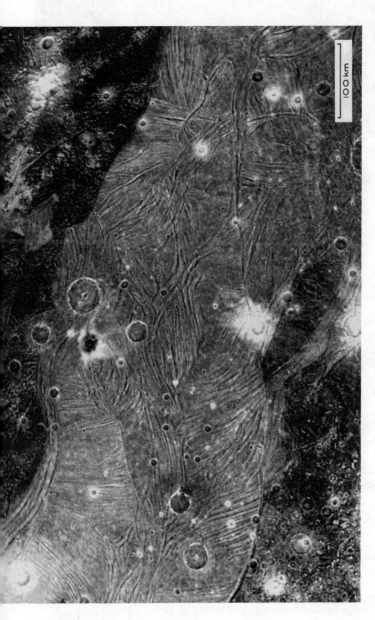

Fig. 13.8. Eastern Uruk Sulcus. Grooved terrain of sulcus contains a complex system of structural cells. Within each cell, the trend of grooves and ridges is approximately parallel. Many cells are bounded by deeper than average grooves that transect the grooves and ridges within the cells. Some cell-bounding grooves are through-going structures that border many cells. Craters on the grooved terrain are much shallower than craters on the Moon but are not as flattened as many craters on the ancient cratered terrain. Note the dark-floored ray crater on grooved terrain (above center). (Voyager image 20637.20 centered at 2°N, 148°W; PICNO 047312-001.)

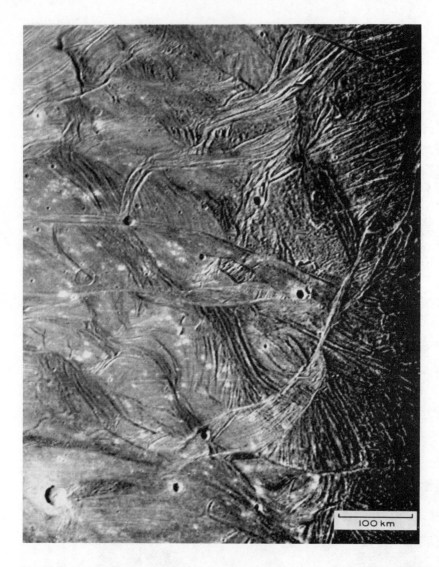

Fig. 13.9. Nun Sulci grooved terrain. Scattered through the grooved terrain are small polygons of intensely grooved, dark, cratered terrain. These remnants of the ancient cratered terrain are so structurally deformed that few craters > 10 km in diameter are recognizable. Much of the grooved terrain consists of smooth units. A highly flattened crater formed on a smooth unit (left of center) is cut by a groove. Note the superposition of two intersecting sets of grooves and ridges (below center) and the pronounced ridge or welt (right of center) that transects older grooves and ridges. (Voyager 1 image 16405.42 centered at 42°N, 305°W; PICNO 0979J1+000.)

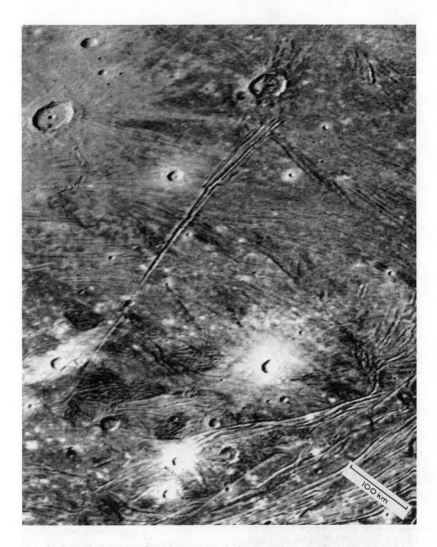

Fig. 13.10. Region in Nun Sulci where long single groove branches into a set of multiple grooves. The long groove cuts older grooved terrain as well as dark cratered terrain; it is one of the youngest structural features in this region. Late-formed single grooves such as this one may represent the initial developmental stage of new structural cells or sulci. Arrested development of this sulcus may indicate that it was formed near the end of the period in which local grooved terrain was formed. An apparent offset in the groove system north of its branching point is an artifact in the television image. (Voyager 1 image 16405.48 centered at 61°N, 335°W; PICNO 0985J1+000.)

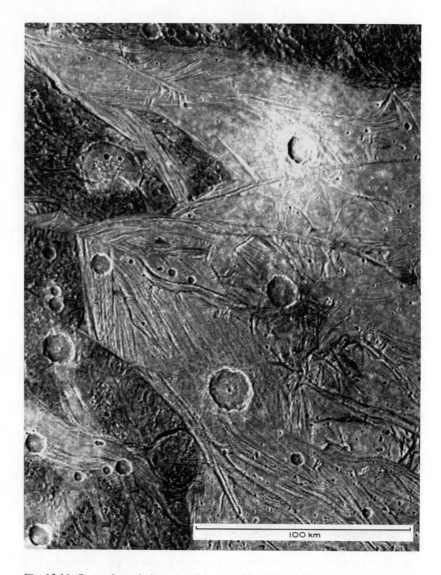

Fig. 13.11. Grooved terrain in equatorial region SE of Galileo Regio. Smooth units occur as small patches that merge with grooved units, or as thin swaths parallel to adjacent grooves. (Voyager 2 image 20638.37 centered at 5°S, 130°W; PICNO 0550J2-001.)

Smooth Units

Smooth plains occur in a few large discrete areas, where they resemble lunar plains (Fig. 13.3), but are more common as smaller patches, either as a facies of a grooved unit, where grooves become shallow and indistinct (Figs. 13.9 and 13.11), or as thin swaths that are parallel to adjacent grooves (Figs. 13.9 and 13.11). Some smooth swaths have sharp, straight edges and are bounded by grooves; others have irregular or indistinct boundaries. Smooth patches adjacent to cratered terrain commonly have irregular or lobate boundaries (Fig. 13.3). The smooth units exhibit very little relief, even close to the terminator; under high Sun the albedo of the smooth units generally matches that of the grooved units, but in a few instances it is as low as that of the cratered terrain. In many places, the distribution and pattern of the smooth units show that they are superposed on and younger than adjacent grooved units. The patterns of the smooth units strongly suggest that they have been emplaced as flows of very fluid material on the surface.

Reticulate Terrain and Grooves on the Ancient Cratered Terrain

In many places, cratered terrain is transitional to grooved terrain. Solitary grooves commonly cross into cratered terrain; some extend several hundred km into polygons of cratered terrain. Elsewhere the cratered terrain has a pervasive grain of subparallel grooves (Fig. 13.12). Grooves occur throughout some polygons of cratered terrain, and are concentrated along the boundaries of other polygons. Reticulate patterns are formed where two pervasive grains of grooves are superposed (Fig. 13.13). Less regular superposition results in hummocky terrain. Discrete, conspicuous examples of reticulate terrain are shown as a separate unit on the geologic map (Plates 2 and 3). These occur near the large, faint palimpsest at the approximate center of curvature of the rimmed furrow system of Galileo Regio.

Large polygons of cratered terrain tend to be least cut by grooves, although some have local strongly disturbed areas. Smaller polygons of cratered terrain are more dissected by grooves; isolated, very small remnants of cratered terrain are invariably heavily grooved. Where small polygons lie close together, the grains of grooves within them commonly trend in different directions. Locally, the trends crudely conform to boundary grooves, but elsewhere groove trends within the cratered terrain and within adjacent grooved terrain are quite different.

Contacts

The boundary between grooved and cratered terrain in most places is sharp, marked by a single groove or groove set. Where a single groove defines the boundary, it is commonly deeper and wider than adjacent grooves. In many places, parallel grooves within a groove set are abruptly truncated by

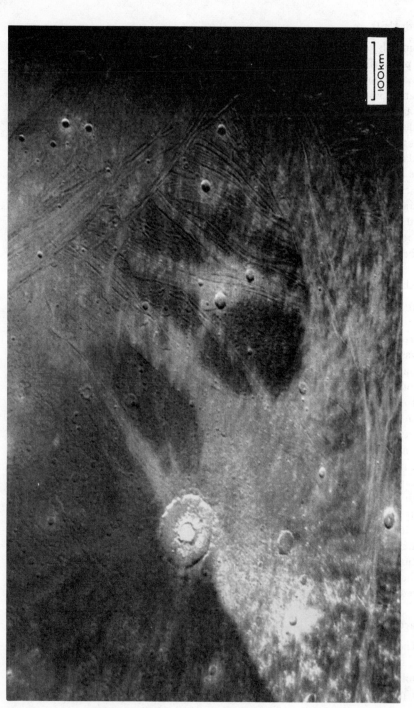

Fig. 13.12. Intensely grooved polygons of cratered terrain near eastern tip of Nicholson Regio. Albedo of rim deposit of large ray crater (left of center) on boundary of Nicholson Regio differs according to the terrain from which the rim materials were excavated; a dark rim is formed where the crater was excavated in dark cratered terrain and a light rim where the crater was excavated in light grooved terrain. (Voyager 1 image 16405.04 centered at 18°S, 305°W; PICNO 0941J1+000.)

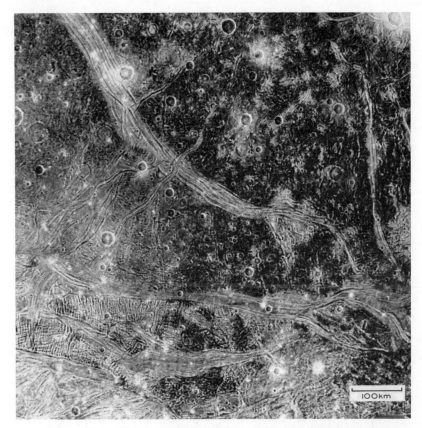

Fig. 13.13. Region of complex geologic relations in vicinity of giant faint palimpsest near center of Galileo Regio rimmed furrow system. Eastern part of giant palimpsest is on left side. In lower left, polygons of reticulate terrain occur between lanes of grooved terrain. North-trending rimmed furrows (center and right) are apparently part of the Galileo Regio concentric rimmed furrow system. Note the lanes of grooved terrain cutting the ancient cratered terrain that taper to points. (Voyager 2 image 20637.29 centered at 18°S, 155°W; PICNO 0482J2-001.)

the boundary groove. Transitional contacts between cratered and grooved terrain are generally found where the adjacent cratered terrain is intensely grooved; some also occur where smooth units directly adjacent to cratered terrain have diffuse margins.

Populations of Craters on the Grooved Terrain

Cumulative crater densities on grooved units, down to 10 km diameter craters, are generally in the range of 25 to 100 craters per 10^6 km^2 at low latitudes (Table 13.3). Note that many of the areas counted include small,

TABLE 13.3

Crater Densities on the Grooved Terrain: Grooved Units

PICNO[a]	Center of Counting Area Lat.	Center of Counting Area Long.	β[b] (deg)	5 km[c]	10 km[c]	20 km[c]	30 km[c]	40 km[c]	50 km[c]	Area Used in Counting (km²)
0949J1+000	5°N	318°	132	340 ± 31	49 ± 12	(6 ± 4)	(2 ± 2)	*	*	352000
0967J1+000	32°N	335°	111	115 ± 23	28 ± 11	8 ± 6	(4 ± 5)	(2 ± 3)	*	216000
0923J1+000	5°N	343°	107	62 ± 21	16 ± 11	(2 ± 4)	*	*	*	138000
0370J2-001	23°N	195°	104	(1400 ± 109)	70 ± 24	(6 ± 7)	(1 ± 3)	*	*	119000
0674J2-001	80°S	280°	100	580 ± 95	140 ± 47	66 ± 32	(33 ± 23)	(21 ± 18)	(15 ± 15)	65000
0965J1+000	35°N	350°	98	260 ± 45	60 ± 21	16 ± 11	8 ± 8	(4 ± 5)	(2 ± 4)	130000
0678J2-001	80°S	230°	98	390 ± 77	150 ± 48	96 ± 38	50 ± 28	(34 ± 23)	(25 ± 20)	65000
0370J2-001	30°N	185°	94	(500 ± 66)	80 ± 26	11 ± 10	(3 ± 5)	*	*	117000
0670J2-001	77°S	160°	86	950 ± 121	220 ± 58	75 ± 34	45 ± 26	20 ± 18	(14 ± 15)	65000
0458J2-001	18°N	173°	83	195 ± 34	54 ± 18	(15 ± 9)	(7 ± 6)	(4 ± 5)	(3 ± 4)	170664
0470J2-001	7°N	168°	78	80 ± 36	28 ± 21	(6 ± 10)	*	*	*	63000
0470J2-001	1°S	157°	67	110 ± 47	26 ± 23	(7 ± 12)	(3 ± 8)	*	*	49000
0497J2-001	27°S	160°	72	410 ± 96	70 ± 40	(10 ± 15)	(3 ± 8)	(1 ± 5)	*	45000
0473J2-001	0°N	150°	60	(210 ± 29)	66 ± 17	34 ± 12	13 ± 7	(9 ± 6)	6 ± 5	242000
0578J2-001	40°S	125°	51	490 ± 65	100 ± 29	17 ± 12	(10 ± 9)	(5 ± 7)	(3 ± 5)	116000

[a] Picture number of framed used for counting.
[b] Angular distance from apex of motion (90° long and equator).
[c] Number of craters ≥ the quoted diameter normalized to 10⁶ km². See Table 13.1 for explanation of the numbers listed.
* Indicates extrapolation of < 1 crater/10⁶ km².

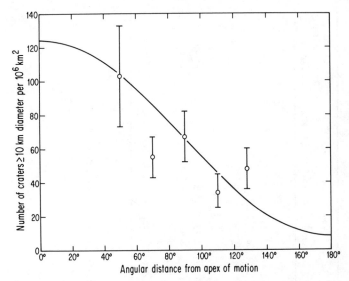

Fig. 13.14. Crater density at low latitude areas in grooved terrain as a function of distance from Ganymede's apex of orbital motion. Crater densities are averaged for 20° intervals of distance from apex. Error bars are one standard deviation. See Table 13.3 for location and other data on the areas counted. Curves show theoretical variation of crater density for surfaces of a single age.

smooth patches and swaths. In the south polar region, densities are as high as ~ 200 craters per 10^6 km^2 and are comparable to crater densities observed on the least heavily cratered regions of ancient cratered terrain. Elsewhere, crater densities of some grooved units are much lower than the average.

Not all the observed variation in crater density reflects differences in age. As shown by Shoemaker and Wolfe (Chapter 10), a gradient in cratering rate occurs between the apex and antapex of orbital motion of the satellite, which may be expected to result in large differences in crater density on surfaces of the same age. To a considerable degree, the variation of crater density on the grooved terrain appears to reflect the theoretically predicted gradient (Fig. 13.14). However, departures from the predicted curve for a surface of uniform age are statistically significant, and indicate a genuine dispersion in age. The grooved terrain in the south polar region, in particular, is clearly older than low-latitude grooved terrain.

Swaths and patches of smooth units in the grooved terrain are mostly too small to obtain statistically stable estimates of crater density. On the whole, crater densities on smooth areas large enough for crater counts tend to be somewhat lower than densities on the grooved units (Table 13.4). The large scatter in crater densities reported in Table 13.4 may be due chiefly to statistical dispersion of the small numbers of craters in the areas counted.

Most craters observed on the grooved terrain exhibit considerable topographic relaxation (Fig. 13.8), but very few have flattened to the threshold of

TABLE 13.4

Crater Densities on the Grooved Terrain: Smooth Units

PICNO[a]	Center of Counting Area		β[b] (deg)	5 km[c]	10 km[c]	20 km[c]	30 km[c]	40 km[c]	50 km[c]	Area Used for Counting (km²)
	Lat.	Long.								
047912–001	21°S	115°	32	450 ± 142	(110 ± 67)	(30 ± 37)	(13 ± 24)	(8 ± 19)	*	22000
057812–001	48°S	120°	55	290 ± 155	(44 ± 59)	(6 ± 22)	(2 ± 13)	*	*	12000
057812–001	45°S	125°	55	110 ± 99	(11 ± 31)	(1 ± 9)	*	*	*	11000
056412–001	28°S	126°	44	270 ± 166	(17 ± 42)	(1 ± 11)	*	*	*	10000
066612–001	64°S	175°	88	1200 ± 343	375 ± 192	(75 ± 86)	(33 ± 57)	(19 ± 43)	(12 ± 34)	10000
066612–001	68°S	189°	93	1200 ± 266	160 ± 97	(26 ± 39)	(8 ± 22)	(4 ± 15)	(2 ± 11)	17000
066012–001	72°S	200°	96	210 ± 68	34 ± 27	(7 ± 12)	(2 ± 7)	(1 ± 5)	*	45000
065812–001	75°S	155°	84	120 ± 96	(19 ± 38)	(3 ± 15)	(1 ± 9)	*	*	13000
0533J2–001	68°S	165°	84	200 ± 68	88 ± 45	40 ± 31	(21 ± 22)	(15 ± 19)	(11 ± 16)	43000

[a]Picture number of frame used for counting.

[b]Angular distance from apex of motion (90° long and equator).

[c]Number of craters $\geq$ the quoted diameter normalized to 10^6 km². See Table 13.1 for explanation of the numbers listed.

*Indicates extrapolations $<$ 1 crater/10^6 km².

detection. Hence, the grooved terrain shows little evidence that craters have been completely lost by flow of the icy lithosphere. In this respect, the grooved terrain is different from the cratered terrain; essentially, everywhere on the grooved terrain the crater population reflects the number of craters actually produced.

On the average, craters on the grooved terrain have greater depth-to-diameter ratios than those on the ancient cratered terrain. The preservation of deeper craters is partly due to the younger overall age of craters on the grooved terrain. But for plausible cratering time scales the greater preserved relief is primarily due to the greater thickness and stiffness of the lithosphere at the time the grooved terrain formed. In particular, the state of preservation of early craters on the grooved terrain > 100 km in diameter indicates that the thermal gradient had decreased and the lithosphere thickened several fold.

The albedo of craters on the grooved terrain may be an important clue to the origin of the terrain. Most craters that have come to photometric equilibrium (i.e., those without rays or bright rims) have floors with about the same albedo as nearby parts of the grooved terrain (Figs. 13.8 and 13.11). This similarity suggests that the material in which the craters were excavated, down to the original crater depth, is similar in composition to the surface material of the grooved terrain. Most craters ⩽ 50 km in diameter on the dark cratered terrain have dark floors (Figs. 13.2 and 13.3). These relations suggest that the materials of the grooved and cratered terrain are compositionally distinct at least to depths of ~ 10 km (the probable maximum transient cavity depths during excavation of 50 km craters). Craters that straddle the boundary of grooved and cratered terrain provide further evidence indicating a contrast in composition. In some cases, either the floor or the ejecta of the part of a crater excavated in one terrain is photometrically distinct from the material excavated from the other terrain (Fig. 13.12).

Age Relations of Grooved Terrain Units

The cratered terrain is transected by all units of grooved terrain, and therefore is everywhere older. Groove sets cut smooth units in some places, and elsewhere are cut or overlapped by smooth units, so no consistent age relation exists between grooved and smooth units. Overlap or truncation of groove sets by smooth swaths and patches seems somewhat more common, however, and crater densities indicate that smooth units are generally younger than grooved units. Solitary grooves commonly cut across all units and boundaries, and appear to be among the youngest tectonic features on Ganymede.

The grooved terrain formed over a geologically significant time span, as shown by the varying crater densities at roughly equal distances from the apex of motion. Distinct episodes of emplacement of grooves and smooth regions are also indicated by intersection of the groove sets with one another and with smooth swaths, and by the superposition of smooth units on grooved units. In a few cases, a crater or the rim deposit of a large crater superposed on the grooved terrain is also cut by grooves (Figs. 13.9 and 13.15). In general, however, the formation of grooved units at any one place appears to have occurred in a relatively short time, because relatively few craters superposed on grooved units are also cut or intersected by later grooves.

Structure of the Grooved Terrain

Because groove sets and craters are truncated by other grooves, specifically by those grooves that lie along boundaries of structural cells or on the contact between grooved terrain and cratered terrain, we can safely infer that the margins of some grooves are faults. The precise nature of displacement along the groove margins, however, is not as easy to establish. Many of the broader grooves, particularly boundary grooves, tend to be flat-floored. They resemble certain very long, narrow terrestrial grabens and some straight lunar rilles. A close search for craters spread apart by cross-cutting grooves has yielded only one unequivocal example (Fig. 13.6). This suggests that the faults bounding most grooves have steep dips. In terms of length-to-width ratio, the Ganymedean grooves resemble narrow grabens of the southern Colorado Plateau (Fig. 13.16), but some individual grooves on Ganymede can be traced for distances of the order of 1000 km, much longer than terrestrial grabens of comparable width. Faults of this great length on the Earth generally have dominantly strike-slip displacement, but the irregular, arcuate to sinuous trace of most Ganymedean grooves precludes large components of strike slip.

The generally branching, irregular, and intersecting ground plan of grooves and groove sets broadly resembles the patterns of normal faults in regions of extensional tectonics on the Earth. At sharp groove-bordered contacts between cratered and grooved terrain, there is almost always a topographic step down on the side of the grooved terrain. Individual grooves cutting the cratered terrain are almost surely formed over down-dropped blocks. Lanes of grooved terrain that taper to a point within a polygon of cratered terrain (Fig. 13.13) can scarcely be formed by transcurrent displacement. Because the average surface level on these lanes is below the surface of the cratered terrain, extensional tectonics is indicated. The simplest interpretation of most ridges on the grooved terrain is that they are raised blocks or horsts remaining between down-dropped blocks underlying the grooves. Where ridges are widely separated, most down-dropped blocks appear to have been covered by a flood of the material that formed the smooth swaths.

In rare instances, the grooved-terrain ridges are distinctly raised above the level of nearby cratered terrain (Fig. 13.17), or converge to form a distinct raised welt on the grooved terrain (Fig. 13.9). These localities provide the

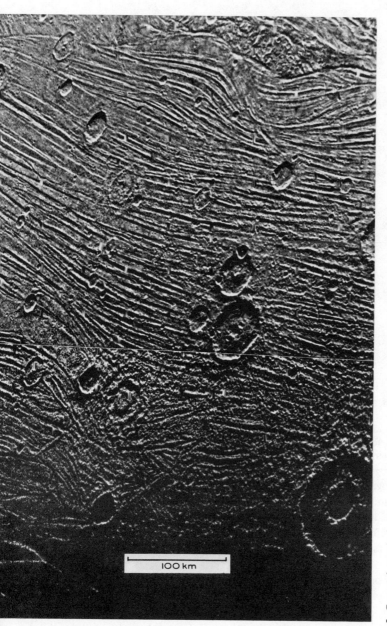

100 km

Fig. 13.15. Grooved terrain in the south polar region of Ganymede. Profiles shown in Fig. 13.7 are from grooved and smooth units near left side. This region has the highest crater density found on the grooved terrain. Note the great diversity in degree of crater flattening. An extremely flattened crater (left of center) appears cut by ridges and grooves. Near the terminator, a population of close-spaced craters 1 to 2 km in diameter is resolved. Most of these small craters must be very shallow (interior slopes <6°) because most are detected only with difficulty at distances >6° from the terminator. The spatial density of these craters indicates that the regolith produced by impact cratering probably averages ~40 m thick on the polar grooved terrain (see text). (Mosaic of Voyager 2 images 20640.41 and 20640.43 centered at 82°S, 170°W; PICNO 0674J2-001 and 0676J2-001, respectively.)

Fig. 13.16. Mesa Butte Graben, Arizona. This long narrow graben (extending from center to lower left) in the southern Colorado Plateau is similar in form but about an order of magnitude smaller than typical grooves on Ganymede. Other grabens nearby, some partly filled with basalt flows, trend at angles to the Mesa Butte Graben. (U.S. Geological Survey high-altitude photograph.)

best recognized evidence for possible local compression on Ganymede. On the other hand, these particular ridges may have been formed by extrusions, or in one case by tilting of fault blocks. No other strong evidence for compressional deformation has been found. Even the most ancient craters on Ganymede retain their circular plan, except where cut and dismembered by grooves.

Although the structure of the grooved terrain appears to have developed largely by vertical displacement, some strike-slip displacement may also have occurred, particularly on faults that bound the structural cells. If significant extension has occurred within the cells, then the cell boundaries probably are transform faults at the places where the boundaries truncate groove and ridge

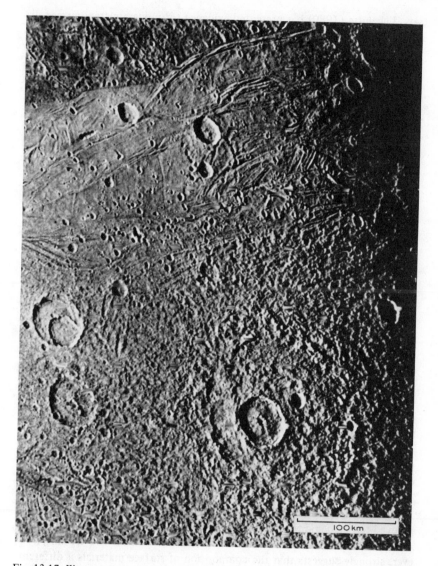

Fig. 13.17. Western equatorial basin. This highly flattened 200 km diameter basin (lower right) is one of the largest craterform features preserved on Ganymede. Beyond the hummocky rim deposit there is a swarm of secondary craters on the grooved terrain. Note the ridges in grooved terrain close to northern boundary (near top) that rise above the adjacent cratered terrain. (Voyager 2 image 20638.39 centered at 5°S, 115°W; PICNO 0552J2-001.)

sets. Lucchitta (1980) has presented evidence for moderate transcurrent displacement at several places. A few localities where evidence suggests transcurrent displacement are shown in Figs. 13.18 and 13.19. None of the evidence for strike-slip displacement, however, is completely convincing. Perhaps the strongest is the apparent offset of the system of ancient rimmed furrows between Galileo Regio and Marius Regio (Plates 2 and 3). A left-lateral offset of a few hundred km may have occurred across Uruk Sulcus, a broad lane of grooved terrain separating these two regions. If so, the net displacement may represent the sum of many small displacements along individual faults in the sulcus.

Origin of the Grooved Terrain

As the grooved terrain occupies more than half the area of Ganymede, it must have formed primarily by replacement of the preexisting cratered terrain. Extensive replacement may have occurred either by transformation of preexisting crust, burial of old crust by new material extruded on the surface, or a combination of these processes.

If Ganymede accreted homogeneously, some expansion of volume and increase of surface area could have accompanied differentiation of the satellite to form an icy mantle and rocky core. However, the maximum increase in surface area for complete differentiation is only ~ 7% (Squyres 1980*b*), and only a small fraction of this total possible expansion may have occurred during the formation of grooved terrain. A possible upper bound on expansion of the satellite at the time of grooved-terrain formation is suggested by the preservation of the huge, cap-like remnant of cratered terrain that is Galileo Regio. McKinnon and Spencer (1981) find that the increase in surface area of the satellite may not have exceeded ~ 2%. Otherwise, the increase in radius of curvature of the surface might have caused tensional hoop stresses in the Galileo Regio cap, and breakup and spreading of the outer parts of the cap.

If the grooved terrain were simply a structural unit, formed by complex faulting of the preexisting crustal materials, replacement of cratered by grooved terrain could readily take place without any net change in the surface area of the satellite. The difference in albedo between the two terrains, however, strongly suggests that the composition of surface materials is different. Probably the surface material in the grooved terrain contains more ice or is less contaminated with dark, rocky constituents than that in the cratered terrain. Moreover, the difference in albedo of mature crater floors in the two terrains suggests that the compositional difference extends to depths of $\gtrsim 10$ km. If these inferences are correct, we are left with only two possibilities: either the composition of the crust underlying the grooved terrain was transformed by some process of regional metamorphism, or the preexisting crust was displaced. If the preexisting crust was displaced, it must have gone down.

Fig. 13.18. Central Uruk Sulcus. Polygons of dark cratered terrain along border of sulcus (near top and below center) appear to have been displaced where cut by narrow lanes of grooved terrain. Left lateral displacement is suggested by geometric relations near top of frame and right lateral displacement by relations below center. (Voyager 2 image 20637.17 centered at 0°, 160°W; PICNO 0470J-001.)

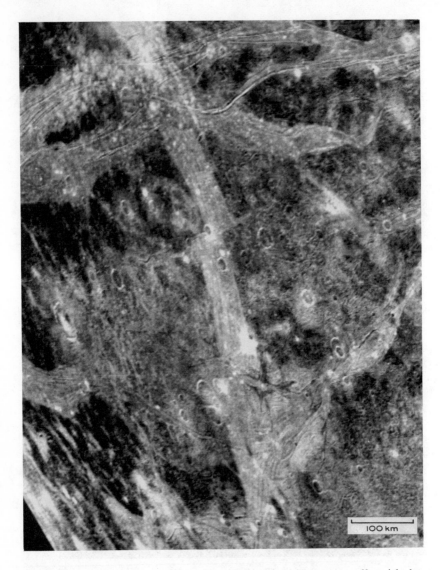

Fig. 13.19. Dardanus Sulcus. Near center of frame, the sulcus appears offset right laterally along a cross-cutting lane of grooved terrain. Alternatively, the Dardanus Sulcus could have developed originally with its present configuration. If surface transport of crustal material has occurred within the sulcus, the cross-cutting lane may
have served as a transform. (Voyager 1 image 16404.44 centered at 12°S, 15°W;
PICNO 0921J1+000.)

It is possible, of course, that replacement of the ancient cratered terrain occurred by a combination of these processes.

The close association of smooth units with grooved units throughout the grooved terrain and the detailed pattern of the smooth units are clues to the replacement mechanism. In many places, the smooth units have lobate boundaries at which they are in contact both with grooved units and cratered terrain. Where the boundaries are lobate, little or no relief is detectable at the contact. These relations strongly suggest that the smooth units were emplaced as flows, and the absence of relief indicates a fluid of very low viscosity. From spectrophotometric observations (Chapter 7) water ice is known to be the primary surface constitutent on Ganymede, and from the bulk density of the satellite and general cosmochemical considerations (Consolmagno and Lewis 1976), water ices are evidently major constituents of the interior. Hence the smooth units can be inferred to have formed by extrusion of water or possibly fluidized systems of water or ice and some highly volatile substance such as methane. If the flows were actually liquid water, then they were probably not contaminated with much suspended rocky debris. Extrusion of uncontaminated water would readily account for the compositional contrast of the smooth units and the ancient cratered terrain.

Widespread and apparently repeated extrusion of water on the surface of Ganymede indicates very specific conditions of density in the lithosphere and asthenosphere at the time of grooved-terrain formation. The mean column density of the combined lithosphere and asthenosphere above the water reservoir cannot have been less than that of the extruded fluid, or the fluid would not have reached the surface. As pure ice I is $\sim$ 7% less dense than pure water over the pressure range of 2 kbar (a likely approximate pressure of the water reservoir), the ice must have been mixed with other constituents sufficient to bring its mean column density up to at least the column density of the extruded fluid. If these constituents were chiefly silicates of the types found in carbonaceous meteorites, $\sim$ 6% by volume of rocky material would be indicated, assuming the reservoir did not lie below the depth of the ice I phase boundary with ice III or ice II. Less contamination with rocky debris would have been required if the water reservoir were at depths corresponding to one of the high-pressure phases of ice. On the other hand, the water may well have carried abundant salts in solution, which would have led to extruded fluid densities substantially higher than the density of pure water.

The smooth units are restricted to the grooved terrain and are not found as isolated patches within polygons of cratered terrain. Because some polygons are extensively cut by grooves, pathways for rising water were presumably present beneath the cratered terrain. The absence of isolated smooth patches therefore suggests that the polygons of cratered terrain had sufficient freeboard to prevent extrusion of water. But the relief between cratered and grooved terrain is extremely small—on the average not more than a few hundred m. These relations suggest that the mean density of the lithosphere and asthenosphere above the water reservoir was adjusted so that the outer ice shell was just barely buoyant. Presumably this fine adjustment was a natural evolutionary result of the early differentiation of the satellite. Initial differentiation of Ganymede probably occurred during late stages of accretion (Schubert et al. 1981). As the outer icy shell formed, its density would have evolved so as to remain just buoyant against liquid regions in the interior. Because most of the shell was actively convecting (Reynolds and Cassen 1979), any parts of the shell denser than the liquid regions would have separated and sunk through these regions. Less dense parts would have remained in the shell. Later extensive reworking of the shell by bombardment and convection could have led to relatively uniform mean column densities over the satellite.

After early differentiation but before formation of the grooved terrain, the density of the lithosphere probably increased slightly due to contamination of the upper part by late falling planetesimals or cometary debris, and to thermal contraction as the lithosphere thickened and cooled. Hence, the early lithosphere probably became unstable relative to the underlying convective asthenosphere. If the density contrast became too great, sections of the lithosphere may have sunk in spasms of convective overturn. However, no record of this sinking has been found on the cratered terrain. A small density contrast could have been sustained against convective overturn by the strength of the lithosphere, so long as it remained everywhere under compression. In this way, the stage may have been set for an episode of slight planetary expansion and the formation of grooved terrain.

A combination of extensional tectonics and extrusion of water appears to offer a satisfactory mechanism for replacement of ancient crust to depths of 10 km or more. Tensional fracturing of the lithosphere probably was required to permit extrusion of water at the surface, in the first place. If the lithosphere was sufficiently dense, individual blocks may have separated and begun to subside into the underlying icy asthenosphere, particularly during major extrusive events. Buoyancy forces would have limited the downward displacement of the subsiding blocks unless the depressions at the surface were filled in. If sufficient extruded water was available to fill the depressions as the blocks sank, and if the crust was sufficiently dense, entire blocks may have been stoped and subsequently absorbed into the asthenosphere. If no water was extruded, or if only a small amount of water was available to fill the depressions, grooves would have remained over the subsided blocks. The initial grooves may have been filled by later extrusions, which would have permitted further subsidence. Repetition of this process could have led to gradual filling-in of the surface of the grooved terrain by relatively clean ice crystallized from repeated extrusions of water. Because clean ice must have a lower density than the subsided old crust, the mean column density of the lithosphere would have progressively decreased, ultimately limiting the foundering of old crust. When this limit was approached, extruded water could have spilled out across broad areas of the grooved terrain and flooded low-lying margins of the cratered terrain.

Analysis of the relaxation of craters (see Chapter 12) suggests that the thickness of the lithosphere at the beginning of grooved-terrain formation was ~ 35 km. (The lithosphere-asthenosphere boundary is taken to be the top of the freely convecting region, assumed to be at an effective viscosity of $\sim 10^{18}$ poise.) The vertical dimension of most subsiding blocks probably corresponded to the full thickness of the early lithosphere. Because the grooves are only 3 to 10 km wide, the individual subsiding blocks evidently were deep thin slabs, typically ~ 10 times as deep as they were wide. If the grooves are in fact formed over grabens, and the intervening ridges correspond to horsts, then the bounding faults must have been steep, generally $> 85°$. Indeed, if faults were initially vertical, subsidence could have taken place with almost no net extension of the crust. Moreover, if a single block were completely stoped, the hole opened up would have permitted collapse of adjacent blocks toward the hole, as long as the hole remained filled by fluid. Such a collapse would permit tilting of the blocks. Rotation and shearing of a stack of thin crustal slabs, somewhat like the shearing of a deck of cards, might have produced some of the topographic forms observed in the grooved terrain. In this case, some ridges would correspond to raised edges of rotated initial surfaces of the slabs, and the grooves to lowered edges.

The broad pattern of the grooved terrain and the dimensions of individual structural cells indicate that the processes producing the grooved terrain operated at two characteristic scales. One scale corresponds to the length of the major lanes or sulci of grooved terrain, typically 2000 to 3000 km. The second scale is the average dimension of the structural cells, ~ 100 km. The former scale and the global distribution of grooved terrain suggest that the lanes of grooved terrain reflect a global driving mechanism, such as a satellite-wide system of deep convection within the icy mantle of Ganymede. The pattern of the structural cells might reflect much more restricted convection in a shallow layer immediately underlying the lithosphere. The thickness of the shallow convection layer may have been of the same order as the mean lateral dimensions of the cells, ~ 100 km. Because melt evidently was transported to the surface along the grooved terrain lanes, it is reasonable to suppose that the lanes formed over warm rising currents in the deep global convection system and that the melt (water or brine) may have been derived by partial melting in the deep convection layer. The complex pattern of structural cells suggests that shallow convection may have been largely decoupled from the deep convection system. Each structural cell may have been formed by a corresponding shallow convection cell. If this hypothesis is correct, the shallow convection may have been driven or the pattern of convection partly determined by the sinking of stoped blocks of cold lithosphere.

These conjectures provide a basis to explain some distinctive features in the grooved-terrain lanes. A remarkable change in width of Tiamat Sulcus occurs where a grooved system transects it (Fig. 13.5). The geometric relations strongly suggest that Tiamat Sulcus has grown laterally, that the transecting groove system served as a transform boundary, and that differential growth occurred on opposite sides of the transform. Three ways in which the sulcus could have grown, while conserving surface area, are:

1. Grooved terrain could have formed by progressive vertical faulting accompanied by flooding, which simply spread laterally from a central graben (faults were formed in succession laterally, but material did not move laterally);

2. Material actually spread laterally from the midline of the sulcus, as in sea-floor spreading on the Earth. If surface area was conserved, spreading would have required generation of new lithosphere at the midline and loss of lithosphere at the boundaries (by subduction or stoping);

3. Stoping of lithosphere occurred at the midline, and the lithosphere collapsed under tension and was stretched and transported from the margin toward the midline. The region of most active faulting was at the contact with the cratered terrain, and the sulcus grew by progressive foundering of old crust at its margin.

The possible dynamics of alternative (1) are obscure; it is not evident why a sulcus should grow laterally by faulting alone, if crustal area is conserved. If subduction or stoping occurred at the margins of the sulcus, as in alternative (2), it is unclear why sets of grooves near the margin are generally parallel with the margin. Generally, such parallelism does not occur at subduction boundaries on the Earth. Alternative (3) seems consistent with the detailed structure of the sulcus, and the two-layer convection model yields a simple dynamical scheme for this alternative.

We suggest that Tiamat Sulcus formed over the axis of a rising current in the deep convecting layer. Water from the lower layer was erupted to the surface along an initial groove. Stoping of the initial midline block localized a convection cell in the upper convecting layer. A sinking current formed under the midline of the growing sulcus and rising currents formed at the margins. The sulcus grew laterally by progressive faulting of the cratered terrain at the margins and by transport of blocks toward the midline. This process may have stopped and started again several times; grooved-terrain sets in the southern part of Tiamat Sulcus are transected by a younger set that cuts across the sulcus at a gentle angle. Consumption of crust by stoping at the midline of the transform groove set that cuts Tiamat Sulcus appears to explain the origin of an adjacent D-shaped segment of cratered terrain west of the sulcus (Fig. 13.5), and the development, by simple extension, of an arc-shaped groove set bounding the D.

A common tendency of structural cells to be smooth in the center and bounded by deep grooves (Figs. 13.11, 13.13, and 13.18) appears consistent with the dynamics suggested for Tiamat Sulcus. The most active faults evidently were those over rising margins of the underlying shallow convection cells; the margins appear to have been the loci of greatest extension, or at least of greatest preserved vertical displacement of the surface within most cells. Inactive fault-line scarps near the center of each cell tended to be buried beneath extruded material, assumed to have been erupted chiefly near the midline of the cell. In rare cases, the entire grooved-terrain lane appears to have been flooded at the end stage of sulcus growth, leaving a single groove along the midline (Fig. 13.3), to mark the position where the crust was stoped. Other deep grooves appear to follow simple transform boundaries, but the reason for continuous depressions along such boundaries is not yet clear.

IV. EARLY POST GROOVED TERRAIN CRATERS AND BASINS

The number of large craters and occurrence of a few basins superposed on the grooved terrain (Plates 2 and 3) indicate that the grooved terrain was probably formed during the waning stages of heavy bombardment. Preservation of the largest craters on surfaces younger than the grooved terrain also indicates that the post grooved terrain lithosphere was much stiffer than the lithosphere in ancient cratered terrain time. The post grooved terrain lithosphere was capable of supporting substantial topographic relief, from the end of heavy bombardment to the present. Some of the earliest large craters formed on the grooved terrain, however, have nearly disappeared, leaving only vestiges of the crater rims. Following Passey and Shoemaker (Chapter 12), we will call these extremely relaxed craters penepalimpsests.

Post Grooved Terrain Penepalimpsests

Simple relaxation of craters by viscous or plastic flow of the lithosphere appears to have produced most penepalimpsests on the grooved terrain. The craters are extremely flattened, but each is surrounded by a well-defined continuous rim deposit that covers groove and ridge topography on the underlying grooved terrain. In most cases, swarms of secondary craters that extend beyond the rim deposit are resolved in the Voyager images. In detail, the penepalimpsests resemble other large craters except that the long wavelength components of crater-form topography are missing. The deposits surrounding these flattened craters are shown by the same mapping convention on Plates 2 and 3 as the deposits around less flattened craters.

The most flattened post grooved terrain crater recognized on Ganymede straddles the border of Uruk Sulcus near $180°$ long (Fig. 13.20). This feature, together with its sharply delimited rim deposit, is listed by Passey and Shoe-

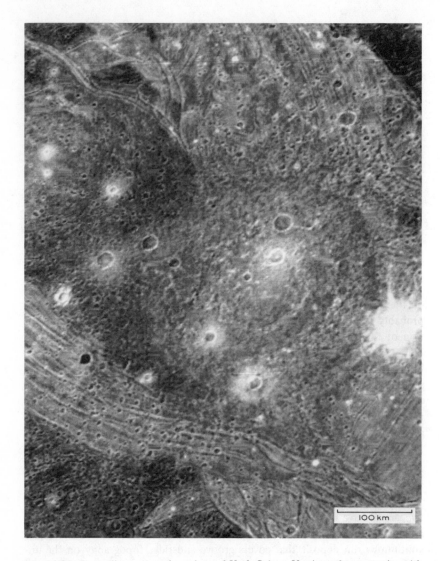

Fig. 13.20. Penepalimpsest on boundary of Uruk Sulcus. Vestiges of a crater rim with radius of curvature ∼ 65 km are preserved near center of penepalimpsest. Note dense swarm of well-preserved secondary craters extending beyond the rim deposit of this extremely flattened crater. (Voyager 2 image 20635.37 centered at $22°N$, $185°W$; PICNO 0370J2-001.)

maker (Chapter 12) as the only palimpsest on grooved terrain. Its precise classification is somewhat ambiguous because partial vestiges of a crater rim with a radius of curvature of ~ 65 km are preserved. On the geologic map it is shown as post grooved terrain crater material. Curiously, the density of craters superimposed on this feature is greater than the crater density on the adjacent grooved terrain, at crater diameters larger than those of associated secondaries, even though the rim deposit and secondary crater swarm of the penepalimpsest are clearly superposed on the grooved terrain. The anomalously high crater density evidently is a statistical fluke, an example of the hazards of determining relative ages from small crater counts. Yet, an age for the penepalimpsest very close to that of the adjacent grooved terrain is strongly suggested by these observations, and is consistent with the extremely relaxed state of the central crater.

Vestigal crater rims at other post grooved terrain penepalimpsests are somewhat better preserved; diameters of these features range from 60 to 166 km. Some of these craters probably are younger than the extremely flattened feature on Uruk Sulcus, but at least one, Hathor, was formed during grooved terrain formation; its rim deposit is cut by younger grooves (Fig. 13.21). More complete preservation of the rim topography of Hathor may be related to the modest size of the crater (64 km in diameter) or its placement at lat −70°, where the crust may have been comparatively cool and stiff when the polar grooved terrain formed.

Most of the largest penepalimpsests (with crater rim diameters between 103 and 166 km) have very large central pits whose diameters range from 52 to 90 km. These pits can easily be mistaken, on casual inspection of the Voyager images, for individual craters. In a number of cases (see e.g. Fig. 13.6), the floors of the central pits have higher than average albedo and appear anomalously domed (type II penepalimpsests of Passey and Shoemaker, Chapter 12). The domes may be underlain by relatively clean masses of ice that either have risen as diapirs (Malin 1980) or have frozen from lakes of water (Croft 1980). Such lakes may have formed by shock melting of the crust, or been erupted into the central pits at the same time the smooth units were formed in the grooved terrain. If the domes are diapirs, ice of relatively low density may have crystallized from intrusions of water beneath the central pits during grooved terrain formation.

Perhaps the most unusual penepalimpsest within grooved terrain occurs in the region of the Nun Sulci at 35° lat, 338° long (Fig. 13.22). Squyres (1980c) has described this feature as a broad dome ~ 260 km in diameter and 2 to 2.5 km high. A wreath of low, semi-concentric ridges surrounding a central smooth region on the dome is considered by Passey and Shoemaker (Chapter 12) to represent remnants of an original crater rim deposit. An extensive swarm of secondary craters around the feature confirms that it originally was an impact crater.

Fig. 13.21. Two penepalimpsests on grooved terrain in south polar region. Rim deposit of Hathor, the penepalimpsest to right of center, is cut by younger grooves and ridges. Near lower right corner of frame, secondary craters of Hathor are superimposed on older units of grooved terrain but are absent from younger units. (Voyager 2 image 20640.31 centered at 70°S, 275°W; PICNO 0664J2-001.)

100 km

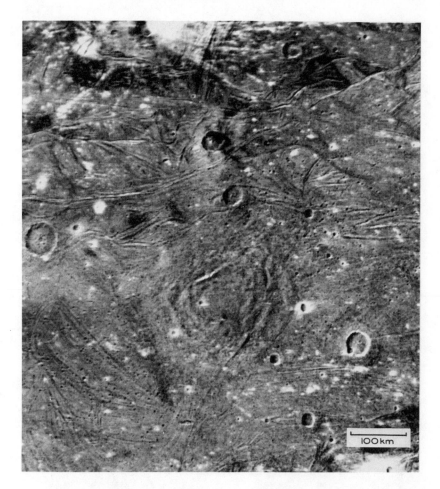

Fig. 13.22. Domed penepalimpsest in region of Nun Sulci. A swarm of secondary craters on the surrounding grooved terrain attests to the original presence of an impact crater near the center of the dome. The dome may have been formed by laccolithic intrusion of water into the lithosphere during the period of grooved terrain formation. (Voyager 1 image 16403.42 centered at 35°N, 328°W; PICNO 0859J1+000.)

Squyres (1980c) suggested that the dome was formed by the combined processes of "ice volcanism" (eruption and freezing of water) and isostatic uplift due to crater excavation in a thin ice crust overlying a liquid water mantle. However, thermal gradients derived from analysis of topographic relaxation of craters (Chapter 12) apparently preclude a shallow mantle of liquid water during or after grooved terrain formation. On the other hand, the isostatic rise of clean ice from an underlying convecting ice asthenosphere could produce the isostatic uplift postulated by Squyres. In this case, we might expect that domes thus formed would be very common. Intrusion of water as a laccolith within the grooved-terrain lithosphere, perhaps at the boundary between new "clean" ice and old foundered crust, would also account for the dome. An intrusion of the required volume could have been rare. The density of the intrusion before freezing presumably was very close to, and perhaps just exceeded, the density of the silicate contaminated asthenosphere. A laccolith would have formed if the intrusion density were greater than that of the local grooved-terrain lithosphere and if the total hydrostatic head of the fluid were insufficient to produce a surface eruption. If the mean column density of the lithosphere decreased during evolution of the grooved terrain, as we suggest, then a few shallow-to-deep intrusions of water may well have occurred near the end of smooth unit formation. A surface dome would have developed, during expansion of the laccolith, as the water froze to ice I.

As shown by the topographic relaxation of large craters, the icy lithosphere of Ganymede cannot have supported a broad dome with 2 to 2.5 km relief from near the time of grooved terrain formation, unless the dome is buoyed up by low-density material in the subsurface. The present dome must be almost in isostatic equilibrium, which could be the case if a large impact had occurred on a thin, relatively dense lithosphere overlying a less dense asthenosphere. Isostatic adjustment would lift the crater floor above the surrounding level. Isostatic equilibrium could also be attained if a very thick laccolith were present. Assuming that the density of the initial lithosphere of the grooved terrain at the Nun Sulci was slightly $< 1.0 \, \text{g cm}^{-3}$ (sufficiently low to have prevented eruption of pure water at the surface), and that the density of the laccolith corresponds to that of pure ice I, $0.93 \, \text{g cm}^{-3}$ at the present prevailing temperatures (see Fig. 13.22), the thickness in km of the laccolith under the center of the dome must be about

$$t = \left(\frac{1.0}{1.0 - 0.93} \right) 2.5 \cong 35 \qquad (1)$$

so that 2.5 km of relief would be in isostatic equilibrium. This thickness is of the same order as the probable thickness of the lithosphere when grooved terrain was first formed.

Gilgamesh Basin

The largest fresh impact feature observed on Ganymede is Gilgamesh, a depression ~ 150 km in diameter (Fig. 13.23), at $-50°$ lat, $128°$ long. Incomplete, ragged, roughly concentric escarpments surround it out to $\geqslant 400$ km from its center. On the Moon, such large ringed structures are called basins

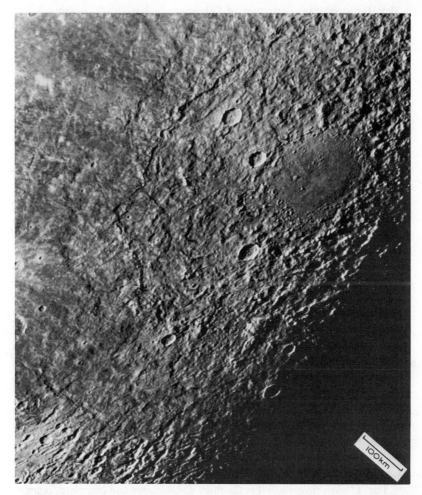

Fig. 13.23. Gilgamesh basin. Note the smooth floor of central basin, blocky material with 1.5 to 2 km relief that surrounds the basin, irregular inward-facing scarp in left central part of frame, and numerous secondary craters in left corner. (Voyager 2 image 20638.14 centered at 61°S, 130°W; PICNO 0527J2-001.)

(Hartmann and Kuiper 1962) or ringed basins. Similar structures are observed on Mars (Wilhelms 1973) and Mercury (Wood and Head 1976; McCauley 1977; Strom 1979). Because its relief (1-1.5 km) is minor, Gilgamesh probably would not have been called a basin if it had been discovered before its lunar analogs. The term is used here because of the lunar precedent and because Gilgamesh has features in common with most lunar basins: depressed central region; rugged ejecta deposit that decreases in relief outward; secondary impact craters, and concentric structure. The most noticeable difference is that the concentric structure of Gilgamesh is less clearly organized into rings than that of most lunar and Mercurian basins. In this respect it resembles most Martian basins, whose ring structures are also hard to discern. However, the lack of well-organized ring structures on Mars may be largely due to erosion and deposition which do not appear to have been important on Ganymede.

The central depression of Gilgamesh has a radius of ~ 75 km. It is very smooth compared to the rest of the structure, with only small-scale, slightly hummocky relief. The depression is surrounded by a blocky, mountainous region that extends ~ 250 to 300 km from the center of the basin. This massif is only crudely concentric and exhibits the most rugged terrain observed on the outer three Galilean satellites. At its approximate outer edge lies the most conspicuous and coherent ring, here called the 275-km ring, a highly irregular inward-facing scarp. Other vaguely concentric arcuate features with more subdued topography lie outside this ring, out to radii of 400-500 km.

Beyond radii 400-500 km are numerous secondary craters. The craters form subradial chains, but the radial structure does not appear as regular as that associated with most lunar basins. Subcircular crater components in these chains generally overlap successively downrange (Fig. 13.24). Their ejecta form herringbone and subradial ridges that probably are coalesced ejecta of intersecting craters (Morrison and Oberbeck 1975). Continuous ejecta bury some secondary craters close to the basin, but in exposed chains the freshest crater is generally uprange. All these features are also observed around lunar basins (Wilhelms 1976).

The Gilgamesh secondaries are conspicuous at distances of ~ 500 km from the basin center, and extend to ⩾ 1000 km. On the Moon, secondaries typically become conspicuous at about twice the radius of the topographic basin rim (such as Montes Cordillera of the Orientale Basin), and extend to about four basin radii from the center (Wilhelms 1976; Wilhelms et al. 1978).

The 275-km ring, since it is the most conspicuous and complete, is analogous to topographic rims of lunar basins. A 275-km topographic basin rim would correlate well with secondaries at 500 km, because similar lunar and Ganymedean gravity should result in similar scaling. It is not known whether the 275-km ring is the boundary of excavation; this question, in fact, has not yet been settled for lunar basins (e.g. McCauley 1977; Hodges and Wilhelms 1978). The boundary of the inner depression, 75 km in radius, probably cannot be the excavation boundary because it is only slightly larger than many craters on Ganymede with far smaller ejecta blankets and secondary fields. The ragged topography between the 75-km and 275-km rings lacks any rimlike forms that might be the excavation boundary. We therefore tentatively favor the 275-km ring as the boundary of excavation for Gilgamesh.

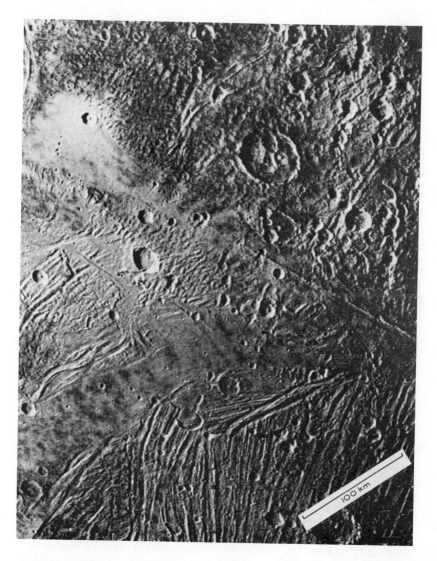

Fig. 13.24. Secondary craters distant from Gilgamesh in south polar region. Secondary craters form irregular chains in upper left and along left side of frame. Note swaths of smooth units in the grooved terrain and a crater (upper left) with a distinct scarp bounding its rim deposit. (Voyager 2 image 20640.25 centered at 75°S, 150°W; PICNO 0658J2-001.)

Gilgamesh is strikingly different from large impact structures on Callisto, such as Valhalla. These have a highly organized concentric structure of incomplete rings, and only moderate topographic relief (Smith et al. 1979a; Passey and Shoemaker Chapter 12). The relatively high relief preserved in Gilgamesh and its gross similarity to lunar basins suggests that the crust of Ganymede at the time of its formation was much thicker and more rigid than the crust of Callisto when Valhalla was formed.

Superposition relations of the extensive secondary craters show that Gilgamesh is clearly younger than the grooved terrain between 80°S lat and 40°S lat. No grooves cut the central depression or rim deposits of Gilgamesh. The age difference between Gilgamesh and the grooved terrain can be evaluated from the density of superposed craters in areas < 400 km from the center, which are essentially free of secondary craters. Craters were counted on the floor of the basin and in two partial annular rings, one extending from 75 to ~ 240 km, the other from 240 to 400 km. The two annuli have comparable crater densities, a cumulative number of ~ 70 craters per 10^6 km^2 for craters > 10 km in diameter (Table 13.5). This density is actually higher than that found in many areas of grooved terrain; it is very close to the average crater density observed on grooved terrain at ~ 65° from apex of motion of the satellite (Fig. 13.14), which is comparable to the distance to the center of Gilgamesh. Within the statistical errors, the observed crater densities are consistent with the known stratigraphic superposition of the Gilgamesh rim on the grooved terrain; however, they suggest that the basin was formed very shortly after the grooved terrain in this region. The crater density on the floor of Gilgamesh is only about half that observed on the rim. The apparent age of the floor seems too low for it to have been formed by flooding by water at the same time that smooth units were emplaced on the grooved terrain. Therefore we attribute the low crater density to complete topographic relaxation of early craters formed in the shock-heated floor material.

Western Equatorial Basin

A highly flattened basin is superposed on the grooved terrain, centered at $-7°$ lat, 115° long (Fig. 13.17). This feature, the western equatorial basin, has a fairly well-defined rim ~ 200 km in diameter that encloses a very shallow central depression roughened by the ejecta of a large superposed crater. The basin is larger than the central depression of Gilgamesh, but much smaller than the ring at 275 km radius that probably corresponds to the initial basin rim. The western equatorial basin has a conspicuous, hummocky rim deposit that extends outward from the rim crest for ~ 265 km. There are abundant secondary craters at radial distances of 150 to 450 km from the center of the basin. The largest secondaries, 10 km in diameter, are about 2/3 the diameter of the largest secondary craters of Gilgamesh. Maximum relief of the rim crest of the western equatorial basin is ~ 600 m. An unpublished photoclinometric profile of the basin by Q.R. Passey shows the floor of the

TABLE 13.5

Crater Densities on the Gilgamesh and Western Equatorial Basins

	Center of Counting Area		β^b (deg)	$5\ km^b$	$10\ km^b$	$20\ km^b$	$30\ km$	Area Used for Counting (km^2)
	Lat.	Long.						
Gilgamesh Basin								
Central depression	58°S	123°	64	340 ± 143	(39 ± 48)	(5 ± 18)	(2 ± 11)	17000
Inner annulus	58°S	130°	66	380 ± 45	79 ± 21	13 ± 10	(5 ± 5)	186000
Outer annulus	54°S	133°	65	420 ± 58	66 ± 23	(4 ± 6)	(2 ± 4)	124000
Western Equatorial Basin	7°S	115°	26	490 ± 110	170 ± 65	48 ± 34	35 ± 29	40000

[a] Angular distance from apex of motion (90° long and equator).

[b] Number of craters ≥ the quoted diameter normalized to 10^6 km^2. See Table 13.1 for explanation of the numbers listed.

central depression to be convex, with a radius of curvature close to that of the satellite. Local relief on the rim is distinctly less than that observed in the inner massif of Gilgamesh.

The crater density in the central part of the western equatorial basin is 170 ± 65 per 10^6km^2, for craters > 10 km in diameter. At $26°$ from the apex of orbital motion, this density exceeds the projected average density for equatorial grooved terrain, but falls within the expected standard deviation for the count. Like Gilgamesh, the age of the basin probably is very close to the age of the grooved terrain on which it was formed. When the crater counts are corrected for distance from the apex, and the stratigraphic relations to the grooved terrain are considered, it appears likely that the age of the western equatorial basin is very close to that of Gilgamesh. If this is true, greater topographic relaxation of the western equatorial basin indicates that the lithosphere was warmer and thinner near the equator than at the high latitude $(60°)$ of Gilgamesh when these two basins were formed.

V. RAY CRATERS

Ray craters are among the landforms of greatest relief, and the rays are prominent markings on Ganymede's surface. The presence of ray craters increases the rough similarity of this remote icy body to the Moon. But unlike the Moon, Ganymede also has dark rays as conspicuous albedo features, although there are much fewer dark than bright ray craters. Both bright and dark rays are superposed on nearly all other geologic units and structures, and no other units have been found that overlap or transect them; therefore the ray craters and their ejecta are the youngest features on Ganymede. Consistent with their younger age, ray craters of a given size exhibit less topographic relaxation than any other craters of the same size, not only because less time has elapsed since they were formed, but probably also because of the thickening and cooling of the lithosphere since the other craters were formed (Chapter 12).

Bright Ray Craters

Observed bright ray craters on Ganymede range in diameter from a few km, the limit of resolution of Voyager images, up to 155 km. The largest ray crater, Osiris, is excavated in a polygon of ancient cratered terrain centered at $-39°$ lat, $161°$ long (Fig. 13.25). It is surrounded by one of the most bright and prominent ray systems and was probably formed fairly recently. Unlike the population of old craters on undisturbed ancient cratered terrain, where there is an anomalous deficiency of large ray craters relative to small ones, the cumulative size-frequency distribution of ray craters approximately follows a simple power function from the smallest ray craters to the largest; the population index (exponent) is ~ -2.2 (see Chapter 12). Within the statistical

Fig. 13.25. Osiris, largest ray crater on Ganymede (top). Albedo of Osiris rays formed on grooved terrain is much higher than that of rays on ancient cratered terrain. This relationship suggests that the rays are composed primarily of material excavated from local secondary craters, and that the compositions of the grooved and ancient cratered terrains are distinct. (Voyager 2 image 20637.59 centered at 45°S, 150°W; PICNO 0512J2-001.)

errors, the index for ray craters is consistent with the population index for craters of all types formed on the grooved terrain.

The proportion of ray craters to other craters on the grooved terrain varies systematically with the distance from the satellite's apex of orbital motion. Near the apex ray craters constitute < 10% of post grooved terrain craters, whereas near the antapex about half of all post grooved terrain craters have rays. A similar variation is found in the distribution of the ray craters on the ancient cratered terrain; the density of ray craters increases almost linearly as a function of distance from the apex (see Chapter 10). At any given distance from the apex, the density on grooved terrain is ∼ 40% higher than on cratered terrain; a difference in ray retention time on the two terrains is indicated. Because the cratering rate from recent impacting bodies (mostly comet nuclei) is shown to be highest at the apex and lowest at the antapex (Chapter 10), and because this predicted spatial variation is apparently reflected by the observed variation in total crater density on low latitude grooved terrain, the reverse variation in density of ray craters must be due to countervailing processes that preferentially erase bright rays near the apex. Passey and Shoemaker (Chapter 12) have shown that the distribution of ray craters on Ganymede can be accounted for if the principal processes of ray erasure are impact and gardening of the surface by small meteoroids in nearly circular heliocentric orbits. The influx of particles on such orbits would be very strongly concentrated at the apex of Ganymede's motion.

Regardless of the mechanisms of ray removal, the observed distribution imples that ray craters in the leading hemisphere are, on the average, much younger than those in the trailing hemisphere. Furthermore, at a given distance from the apex the average age of ray craters on cratered terrain is ∼ 40% less than on grooved terrain. Osiris, which is in the leading hemisphere as well as on a polygon of dark ancient cratered terrain, probably was formed by impact of a comet nucleus ∼ 10 km in diameter within the last several 10^8 yr (see Chapter 10 for estimates of cratering rates).

Voyager images that cover an entire hemisphere of the satellite near full phase clearly show that rays and ray craters tend to be brightest at latitudes ≳ 45° (Fig. 13.26). The high-latitude regions correspond to the apparent polar shrouds or polar caps, actually simply regions in which the average albedo of all types of terrain is highest. In low-resolution pictures the rays also appear more extensive in the polar shrouds, which gives the impression that ray craters are more abundant in the high latitude regions. However, they are found on closer inspection to be about equally abundant at high and low latitudes, at a given distance from the apex. The difference in ray brightness at high and low latitudes is evidently partly related to processes that formed the shrouds or, alternately, lowered the albedo of one region between the shrouds; these processes probably included selective ablation of ice from low latitudes (Purves and Pilcher 1980).

Fig. 13.26. Distant view of Ganymede showing polar shrouds and latitudinal dependence of ray brightness. Large dark region in northern hemisphere is Galileo Regio. Largest bright patch in southern hemisphere is the rim deposit and ray system of Osiris. (Voyager 2 image 20638.35 centered at 10°N, 140°W; PICNO 0548J2-001.)

When examined in detail, the brightness of rays within individual ray systems is found to be related to the substrate on which the rays are formed. Within a few hundred km of the central crater, rays tend to be brightest where formed on grooved terrain and darkest where formed on ancient cratered terrain. A fairly abrupt change in the albedo of the ray system commonly occurs at the substrate contact (Figs. 13.25 and 13.27). This relationship is understandable if the rays are composed dominantly of substrate material excavated from secondary craters (Shoemaker 1962). Gardening of thin rays would also tend to make the rays fade and disappear fastest where superposed on a dark substrate.

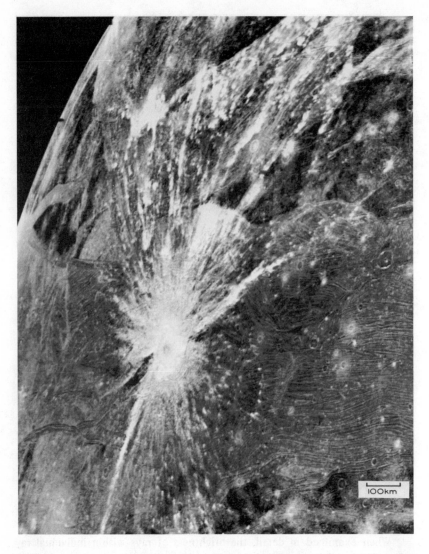

Fig. 13.27. Bright ray crater Tros. Note changes in brightness of the rays where they cross from grooved terrain into dark, ancient cratered terrain. (Voyager 1 image 16404.22 centered at 20°N, 20°W; PICNO 0899J1+000.)

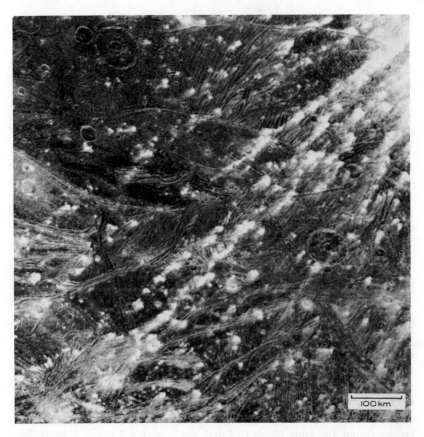

Fig. 13.28. Distal part of Osiris' ray system. Rays at this distance consist predominantly of chains of bright patches. Here the ray albedo is independent of the terrain on which the rays are superimposed. The bright patches may have been formed by deposition of diffuse swarms of strongly shocked (and perhaps partially melted) ejecta. (Voyager 2 image 20630.05 centered at 45°S, 170°W; PICNO 0517J2-001.)

At distances > 500 km, on the other hand, ray brightness appears independent of the substrate. In contrast to the general linearity of long rays found on the Moon and Mercury, the distal parts of very long rays on Ganymede mainly are chains of irregular bright patches (Figs. 13.27 and 13.28). The pattern of the chains of bright patches resembles that of secondary craters, but it is not clear whether resolvable secondary craters occur in the patches. Ejected icy material thrown these great distances was probably partially melted, and the patches formed by deposition of diffuse swarms of slush or possibly refrozen droplets, which landed at velocities of ~ 1 km s^{-1}.

In addition to ray craters, there are many craters that lack rays but have moderately bright rim deposits. These craters are evidently at an intermediate stage in the slow return of surface materials to equilibrium with the space environment. On the average, the bright rim craters are 2 to 3 times as abundant as the bright ray craters. Unlike the ray craters, there is no strong density gradient of bright rim craters from the apex to the antapex. As reported by Passey and Shoemaker (Chapter 12), the distribution is somewhat irregular, and their density is higher than average near the apex.

Perhaps the most remarkable features among the bright ray craters on Ganymede are abrupt transitions from material of very high albedo to material of very low albedo. The darkest surfaces found on the satellite are in rays and in the rims and floors of ray craters. A bright ray crater may have a dark floor (Fig. 13.8), a bright ray system may pass radially outward into a dark ray system (Fig. 13.5), a bright or intermediate albedo crater may be surrounded by dark rays (Fig. 13.5), or a dark crater may be surrounded entirely by dark rays. The distribution of the combined bright and dark ray crater systems appears unrelated to the substrate. They are found in both cratered and grooved terrain.

Dark Ray Craters

About 40 craters with dark rays were identified on Ganymede by Conca (1981). They range in diameter from the limit of resolution in the images examined ($\sim$ 3.5 km) to $\sim$ 60 km. There is no consistent relation between the ray albedo and the albedo of the crater rim or floor. In one case, bright rays and dark rays extend in opposite directions from the same bright rimmed crater. Diversity of ray composition, or of physical characteristics of material ejected from individual craters, is indicated by the observations, but it is not clear whether the diversity originates from the target or the impacting projectile. About the same number of dark ray craters >5 km in diameter are found on the grooved terrain as on ancient cratered terrain. Most are in the dark, low-latitude band, but a few occur in the polar shroud region. There is a fairly strong gradient of increasing numbers of dark ray craters from the region of the apex of orbital motion toward the antapex. In this respect, the distribution of dark ray craters resembles that of bright ray craters.

Conca (1981) suggested that dark rays may have formed by ablation of the surface that produced a residual surface concentration of trace amounts of dark material in the ray deposits. He attributed the ablation to sputtering by high-energy ions in the Jovian magnetosphere. The apparent latitudinal dependence of the frequency of dark rays, however, suggests that ablation by sublimation of ice caused by insolation may be more important than sputtering. The gradient in number of dark ray craters from apex toward antapex is probably related to a gradient in the rate of impact gardening that erases rays. Hartmann (1980) suggested that the dark ray crater material was derived from a restricted range between 3 and 10 km in depth in the crust of Ganymede, but the actual size distribution of dark ray craters shows that this cannot be the case. The upper limit of size for dark ray craters probably reflects the very low frequency with which ray craters are formed. Conca attributed the dark material in the rays to contamination by the impacting projectile. The amount of such contamination is probably much less than the fraction of rocky material in the crust of Ganymede, or at least than that in the crust beneath the ancient cratered terrain. However, grain size or other physical characteristics of the dark contaminant may be important factors in the evolution of dark rays.

VI. THE GANYMEDEAN REGOLITH

The ray craters on Ganymede and their young age indicate that, as on the Moon, certain processes have modified the albedo of the surface with time and erased the rays of older craters. Not only were rays erased, but over a longer time span the albedo of crater rim deposits and crater floors generally evolved toward the average albedo of the surrounding terrain. But in many relatively large craters on the ancient cratered terrain this photometric maturation is incomplete. Both bright and dark rays, rims, and crater floors evidently have faded or returned to photometric equilibrium; therefore contamination of the surface with dark infalling debris cannot be the only process that affected the albedo. Complete blanketing of the surface with extrinsic materials is excluded, because the albedo contrasts between grooved and cratered terrain and between palimpsests and cratered terrain would not have been preserved.

At least four processes probably have contributed to the evolution of the physical state and composition of Ganymede's surface:

1. Gardening of the surface by impact of meteoroids and secondary debris from large craters;
2. Contamination of the surface by infalling meteoritic debris;
3. Transfer or loss of volatile constituents by sublimation at low latitudes with possible redeposition at mid-latitudes;
4. Sputtering by collision of high-energy ions in the Jovian magnetosphere.

We think impact gardening, which formed a regolith of pulverized and mixed surface materials, was the dominant surface process, although the other mechanisms probably had a role in regolith development.

Crater flattening in the polar regions of Ganymede by itself strongly suggests a surface layer with extremely low thermal conductivity. The mean surface temperature drops from $\sim$ 120 K at the equator to $\sim$ 70 K at 80° lat (Squyres 1980d). Assuming that the effective viscosity of ice follows a temperature dependence given by the Arrhenius formula (Chapter 12), this difference of $\sim$ 50 K would correspond to $\sim$ 10 orders of magnitude difference in effective viscosity at the surface. If at any given time the heat flow and thermal gradient through the lithosphere were about the same near the pole and at the equator, and if there were no insulating surface layer, then the polar lithosphere would have been much colder, and thicker, and stiffer than the equatorial lithosphere.

Calculations of crater relaxation by Q.R. Passey (personal communication, 1981), in which the effective viscosity is assumed to decrease exponentially from the surface downward (i.e. to correspond with a nearly uniform thermal gradient), show that essentially no flattening of craters would have occurred in the polar regions during the time required to flatten the most relaxed craters observed in equatorial grooved terrain. Indeed, if there were not an insulating regolith, craters of all sizes at $> 70°$ lat would be stable against viscous relaxation for times much greater than the age of the solar system (in a lithosphere with the maximum thermal gradient estimated by Passey and Shoemaker, Chapter 12). Yet, some of the most flattened craters observed on the grooved terrain are found at $> 80°$ lat (Fig. 13.15). The observed flattening of craters near the pole indicates that the temperature and effective viscosity profiles in the polar lithosphere were not much different, soon after grooved terrain formation, from the corresponding profiles in the equatorial lithosphere. This situation can be readily accounted for if a regolith a few m to tens of m thick and similar in near surface thermal properties to the lunar regolith developed soon after the grooved terrain was formed.

Thermal properties of Ganymede's surface (Hansen 1973; Morrison and Cruikshank 1973) show that it has an extremely low thermal inertia and must consist predominantly of loosely stacked, fine particulate material. Polarimetric properties suggest the presence of both fine ice particles and rock fragments (Dollfus 1975). From the visual phase function of Ganymede, the mean volume density of the particulate material at the surface is estimated by Pang (1981) to be 0.37, very close to the volume density of the regolith at the lunar surface. As found from infrared radiometry during eclipses, the thermal inertia $(K\rho c)^{1/2}$ of most surface material of Ganymede is similar to that of the uppermost lunar regolith. For plausible values of density ρ and heat capacity c, of a pulverized mixture of ice and rock, the mean conductivity K of the uppermost Ganymedean regolith is $\sim 4~\mu W~cm^{-1}K^{-1}$ (Hansen 1973). This may be compared with conductivities in the range 5 to 15 $\mu W~cm^{-1}K^{-1}$ (at mean temperature) for the uppermost lunar regolith obtained by the lunar heat flow experiments conducted at Apollo landing sites 15 and 17 (Langseth et al. 1973; Keihm and Langseth 1973). Hansen (1973) and Morrison and Cruikshank (1973) have noted that the near-surface material on Ganymede cannot be homogeneous. In this respect also, the Ganymedean regolith is like the lunar regolith. Hansen interpreted his radiometric results in terms of a two layer model: a homogeneous surface layer ~ 2 mm thick with extremely low thermal inertia, overlying a thick layer with a high thermal inertia like that of rock or ice. Morrison and Cruikshank made a similar interpretation. Almost certainly, a more realistic representation of the data is given by a regolith like that of the Moon, which contains sparse isolated fragments of rock with high thermal inertia (Vitkus et al. 1968) embedded in a pulverized matrix with a thermal inertia orders of magnitude lower than that of solid rock.

We can expect the mean conductivity of the Ganymedean regolith to increase with depth, due to a decrease in pore space and increase in mean density, with a profile somewhat similar to that of the lunar regolith (Langseth et al. 1973; Keihm and Langseth 1973). Thermal observations of the dawn terminator of Callisto by the infrared interferometer spectrometer (IRIS) experiment on Voyager 2 (Hanel et al. 1979) show that the thermal inertia beneath the extremely insulating material at the surface is about twice the average thermal inertia for comparable depths ($\sim$ 2 to 15 cm) in the lunar regolith. At a depth of 10 cm, the mean conductivity of the lunar regolith is $\sim$ 10 times greater than at the surface; most of the change occurs at $\sim$ 2 cm depth, below which the conductivity rises slowly with depth. The average conductivity at depths between 1 and 2 m is $\sim$ 100 μW cm^{-1}K^{-1} (Langseth et al. 1976). The average conductivity of the lunar regolith probably remains close to this value down to the bedrock contact, at mean depths of 2 to 10 m in the lunar maria. Owing chiefly to the very low conductivity of the upper 2 cm and to the relatively high heat flow at present, the temperatures at 0.5 to 1 m depth in the lunar regolith are several tens of degrees higher than the time-averaged temperature of the surface (Keihm and Langseth 1973). Thus a regolith only 1 m thick, with thermal properties similar to the lunar regolith, would have had a profound effect on the temperature and effective viscosity in the lithosphere of Ganymede, during a period when the heat flow may have been comparable to the present heat flow on the Moon.

Thickness of the Regolith in the South Polar Region

The highest resolution images of Ganymede were obtained as Voyager 2 passed over the south polar region. Craters as small as $\sim$ 1 km in diameter can be detected in these images. Close to the terminator, the surface of the grooved terrain appears corroded and minutely pitted, due to thousands of craters with diameters near the threshold of resolution (Fig. 13.15). A count of these small craters, in a smooth swath $\sim$ 6° from the terminator, gives spatial density 7.9 $\pm$ 0.5 $\times$ 10^4 craters per 10^6 km^2. Most of the craters were detected only as dark spots in the image, and their diameters cannot be accurately measured. Because the surface resolution in the image studied is 1.1 km (line pair spacing at 10% modulation), the threshold of detection must be close to 1 km, which is adopted as the lower limiting diameter of the craters counted.

The observed densities of craters $\geqslant$ 10 km in diameter in three areas of the south polar grooved terrain counted are 140 $\pm$ 47, 150 $\pm$ 48 and 220 $\pm$ 58 per 10^6 km^2; the mean density for the region, at 10 km diameter is 170 $\pm$ 30 per 10^6 km^2. A power function fitted to the densities observed at crater diameters of 1 km and 10 km yields an exponent or population index of $-2.67 \pm .09$, significantly steeper than the index of -2.2 observed for ray craters and for all craters > 10 km diameter on the grooved terrain. The index of the crater size distribution clearly steepens as the 1 km diameter is approached. In this respect, the crater size-frequency distribution resembles the distribution on the lunar maria, which steepens abruptly near 1 km diameter.

The crater density at 1 km diameter on Ganymede's south polar grooved terrain is ~ 7 times higher than the density at Tranquillity Base on the Moon and > 10 times higher than the density at the Apollo landing sites on mare lavas in eastern Oceanus Procellarum or near Hadley Rille. In fact, the density for grooved terrain of 7.9×10^4 per 10^6 km, at 1 km diameter, coincides precisely with the steady-state size-frequency distribution for lunar craters determined by Trask (1966). We can suppose that the steady-state size-frequency distribution of craters on Ganymedean grooved terrain is close to that on the Moon, and that the population of craters at 1 km diameter on the polar grooved terrain therefore is in steady state. Also, because the mean distribution index between 1 km and 10 km crater diameters is much steeper than the steady state index of -2.0, the frequency distribution of craters larger than the steady state limit probably intersects the steady-state distribution near 1 km crater diameter.

If the population of craters on the polar grooved terrain is in steady state at 1 km diameter, these craters should exhibit a complete spectrum of shapes from deep bowls to nearly flat. Average interior slopes should range from $\sim 20°$ for fresh craters to nearly $0°$ for craters almost filled in. At $\sim 3°$ from the terminator, the density of small craters with interior shadows is estimated to be $\sim 50\%$ of the total density of craters observed on the smooth swath at $6°$; relief of the grooves precludes an accurate estimate of total crater density in areas this close to the terminator. At $6°$ from the terminator, only 10% of the small craters counted on the smooth unit have shadows, and at $12°$ from the terminator the density of shadowed craters is reduced to $\sim 3\%$ of the total crater density counted at $6°$. At $12°$ there is a marked reduction in the density of resolved craters, which indicates that the interior slopes of most craters near 1 km in diameter are too low for detection. The frequency distribution of interior slopes indicated by these observations is consistent with the morphology distribution of craters with diameters at and below the steady-state limit on the Moon (Shoemaker 1966). The rough observed slope distribution is also consistent with that predicted theoretically from the erosion and refilling of craters by the cumulative production of still smaller craters (Soderblom 1970; Soderblom and Lebofsky 1972).

Excavation and refilling of craters at sizes below the steady-state limit produces a regolith of impact-generated debris whose thickness-frequency distribution can be roughly estimated from a simple model (Shoemaker et al.

1970; Shoemaker and Morris 1970). The base of the regolith is assumed to be formed by the interconnected floors of the deepest craters that have been completely filled in. To calculate the thickness distribution, it is necessary to estimate the size-frequency distribution of all craters produced, in the size range from the smallest crater whose floor now forms part of the base of the regolith to the crater size at the steady-state limit. In applying this model to predict regolith thicknesses on the lunar maria, Shoemaker and Morris (1970) obtained the small crater production distribution by downward extrapolation of the crater-size distribution observed above the steady-state limit.

Assuming that the small crater production function is of the form

$$'F = \chi_c^\lambda \qquad (2)$$

where F is cumulative frequency of craters $\geq c$, χ the coefficient of the crater size-frequency distribution, c the crater diameter, and λ the population index, the minimum thickness of the regolith h_{min}, corresponding to the original depth of the smallest crater whose floor is now at the regolith base, is given approximately (Shoemaker and Morris 1970) by

$$h_{min} = q \left[c_s^{\lambda+2} + \frac{4(\lambda+2) A_m}{\lambda\pi\chi} \right]^{1/\lambda+2} \qquad (3)$$

where q is the initial depth to diameter ratio of small craters, c_s the crater diameter at limit of steady-state distribution, and A_m the dimensionless ratio between the integrated area of all craters whose base forms part of the regolith to the reference area used for χ. For the lunar maria, the following values of the constants of Eq. (3) yield predicted median regolith thicknesses very close to the observed thicknesses: $q = 0.2$; $\lambda = -2.93$ (average observed crater size distribution at the Surveyor landing sites in the size range from the steady-state limit up to 1 km [Morris and Shoemaker 1970]); and $A_m = 2$. If we adopt these values for the craters < 1 km produced on Ganymede and take, for the polar grooved terrain: $c_s = 1$ km; $\chi = 10^{13.69}/m^\lambda \ 10^6$ km$^2 = 10^{1.69}/m^{\lambda+2}$ (χ is obtained by extrapolation to 1 m crater diameter from the observed frequency at 1 km crater diameter with $\lambda = -2.93$), then from Eq. (3) $h_{min} = 15$ m. The maximum thickness of the regolith h_s is given by $qc_s = 200$ m.

The frequency distribution of regolith thickness is given (Shoemaker and Morris 1970) by

$$H = \frac{100\lambda\pi\chi(1/q)^{\lambda+2}}{4(\lambda+2)A_m} (h^{\lambda+2} - h_s^{\lambda+2}), \quad h_{min} \leq h \leq h_s \qquad (4)$$

where H is cumulative percentage of the surface underlain by a regolith of thickness $\geqslant h$. The median thickness h_{med} is obtained from Eq. (4) at $H = 50\%$. Substitution of the adopted values of the cratering constants, for the polar grooved terrain, gives $h_{med} = 29$ m. The arithmetic mean thickness of the regolith $\bar{h}$, for the distribution represented by Eq. (4), is given (Shoemaker et al. 1970) by

$$\bar{h} = \frac{\pi\lambda\chi q}{4(\lambda+3)\,A_m} \left\{ \left[c_s^{\lambda+2} + \frac{4(\lambda+2)\,A_m}{\pi\lambda\chi} \right]^{\frac{\lambda+3}{\lambda+2}} - c_s^{\lambda+3} \right\} \tag{5}$$

which yields, with the constants adopted for the polar grooved terrain, $\bar{h} = 44$ m.

Of course, the derived median and mean thicknesses of the regolith are model dependent, and particularly depend on the population index λ adopted for the small crater production function. If we use $\lambda = -2.67$ (the observed mean population index between 1 km and 10 km crater diameters on the polar grooved terrain), rather than $\lambda = -2.93$ (observed for craters < 1 km on the Moon), the regolith thicknesses for the polar grooved terrain are $h_{min} = 7$ m, $h_{med} = 17$ m, $\bar{h} = 32$ m, and $h_s = 200$ m. These sample calculations show that, if the index of the small crater production function on Ganymede is at all close to the index observed near 1 km crater diameter on Ganymede or to the index for small craters produced on the Moon, the predicted mean thickness of the regolith formed by impact cratering on the polar grooved terrain is on the order of several tens of m. Similar results can be obtained simply by scaling from the regolith thicknesses observed on the lunar maria according to the ratio of crater density at 1 km crater diameter on the polar grooved terrain to the corresponding crater densities on the lunar maria.

Variations in Regolith Thickness with Latitude and with Distance From Apex of Orbital Motion

For a constant value of λ for the small crater production function, the regolith thicknesses are proportional to the crater density at any diameter above the steady-state limit. Hence the predicted mean thickness of the regolith on low-latitude grooved terrain near $90°$ from the apex, where the 10 km crater density is $\sim 1/3$ the density on the polar grooved terrain (Fig. 13.14), is on the order of 10 m. Near the apex of orbital motion, the regolith on the grooved terrain is predicted to be ~ 20 m thick, on the average; near the antapex the predicted mean thickness is only ~ 1 m. Thus there is a large predicted global variation in thickness of the regolith produced by impact gardening on the grooved terrain. Moreover, if the flux of small impacting bodies is even more concentrated at the apex and depleted at the antapex than is the flux of large bodies, as suggested by the distribution of ray craters (Chapter 12), the global variation in regolith thickness may be more extreme than was predicted by assuming λ constant over the globe. The projectiles responsible for gardening to depths of ≤ 1 m are generally < 10 cm in diameter. Bodies of this size on heliocentric orbits are subject to Poynting-Robertson drag, which tends to make their orbits circular and increase the flux gradient from antapex to apex of the satellite's motion (see Chapters 10 and 12).

Thermal Migration of Water

The polar shrouds show the effects of a latitude-dependent process, presumably governed by insolation. Purves and Pilcher (1980) calculated the migration rate of water on Ganymede for a surface of essentially pure water ice, with no other loss or transport mechanisms for water. In this calculation, the surface ablation rate was ~ 7 m per Gyr at the equator. Net ablation decreases to zero at 17° lat, and ice accumulates at latitudes between 17° and ~ 55°. A peak rate of accumulation of ~ 3 m per Gyr was found at ~ 33° lat. The maximum calculated rates of ablation and accumulation are significant when compared to the predicted impact gardening rates. If the age of the equatorial grooved terrain is ~ 3.1 Gyr to 3.6 Gry, as estimated from crater densities and the cratering time scale of Shoemaker and Wolfe (Chapter 10), the net calculated ablation at the equator would exceed the mean predicted thickness of the regolith formed by impact gardening, for all longitudes except possibly those near the apex of orbital motion. At latitudes near 33°, ice accumulated by thermal migration would constitute ~ 1/10 to ~ 2/3 of the regolith.

The maximum ablation rates of Purves and Pilcher appear to exceed the rate permitted by the lifetime of rays in the equatorial region, up to an estimated ~ 1.1 Gyr on the trailing hemisphere (Chapter 12). Probably their ablation rates are too high at least partly because the effects of surface roughness have been neglected. For a surface with volume density 0.37 (Pang 1981) and with the deep "holes" between surface grains required to explain the observed photometric function, a major fraction of the H_2O molecules released thermally at the top of the regolith will not escape on long trajectories, but will be trapped by collision with nearby grains.

Squyres (1980d) suggested that rocky debris in the regolith could affect the mobility of the ice component, in regions of intense ablation. Where ice ablates from the surface in the equatorial region, a lag deposit of the rocky constituents tends to build up. An ice-depleted, insulating, semipermeable layer formed at the surface would inhibit further ablation. This hypothesis is strengthened by the relatively uniform albedo of the grooved and cratered terrains at latitudes < 35°, which suggests that water has not migrated in the manner calculated by Purves and Pilcher.

An apparent difficulty with the lag deposit hypothesis is that the average abundance of ice at the surface, estimated by Clark (1980) for the leading hemisphere of Ganymede, is ~ 90 wt-%. Laboratory studies by Clark (Chapter 7), on the other hand, show that a frost layer only a few tens of μm thick is sufficient to produce the strong H_2O absorption features he observed on Ganymede. Such a layer might be produced in low latitudes, as water migrates in a series of discrete ballistic hops in the region of net ablation.

Turnover of the regolith by impact gardening tends to erase the surface layer of rocky material postulated by Squyres. During the gardening, most regolith material is exposed at the surface at least once. At latitudes where net ablation occurs, a fraction of the ice would have been lost during this exposure. Thus, depletion of ice must be distributed through the regolith rather than confined to a surface layer. The probability of exposure, however, is a function of depth in the regolith; on the average, material near the surface has been exposed more frequently so there should be a gradient of increasing ice abundance with depth in the regolith.

On the polar shrouds on Ganymede there has probably been almost no migration of water, as calculated by Purves and Pilcher (1980). Above latitudes of $\sim 55°$, exposed ice is stable almost indefinitely. Yet the margins of the shrouds extend to $\sim 40°$ lat on grooved terrain and $\sim 50°$ lat on the ancient cratered terrain (Squyres 1980*d*). Near the margins, there may have been a slight accumulation of ice derived from slow ablation of ice at lower latitudes. If this accumulation is much less than the thickness of the regolith in which it is mixed by gardening, it probably is not readily distinguished photometrically from the region of no water migration.

The principal effect of water migration appears to have been ablation in the equatorial and midlatitude regions, which left a residual enrichment of dark rocky material in the entire regolith in these regions. If this hypothesis is correct, we should speak of the dark equatorial band rather than polar caps or shrouds. Apparently the observed shrouds are merely a result of photometric contrast with the ablated low-latitude band. Albedo differences of geologic units in the low-latitude band evidently are due to differences in abundance of rocky constituents in the subregolith ice, which governed the initial rock/ice ratio of material freshly injected into the regolith by impact gardening. The failure of the simple model of Purves and Pilcher to predict the location of the boundaries between the ablated low-latitude band and the polar shrouds may be primarily due to neglect of other loss or transport mechanisms for water, such as sputtering and photodissociation of the sublimed water molecules followed by escape of both the hydrogen and oxygen.

Effects of Sputtering

Lanzerotti et al. (1978) and Haff et al. (1979) have suggested that sputtering or ion erosion may be significant on the surfaces of the icy Galilean satellites. Their principal experimental basis is a surprisingly high ice sputtering rate by protons and helium ions with energies in the 10 keV-MeV range (Brown et al. 1978; Brown et al. 1980). The maximum rate of sputtering per incident proton was found by Brown et al. (1980) to occur at ~ 100 keV. Based on the energetic proton fluxes measured from the Pioneer 10 flyby, the erosion rate of surfaces on Ganymede composed of 50% ice was calculated by Lanzerotti et al. (1978) to be in the range $\sim 3 \times 10^7$ to $\sim 10^{10}$ H_2O cm^{-2} s^{-1}. This corresponds to surface ablation rates of ~ 30 cm to ~ 100 m Gyr^{-1}. The retention age for rays is estimated to be ~ 1 Gyr in the trailing hemisphere of Ganymede, and, because this estimate probably should be regarded as a minimum age (cf. Chapter 12), it is clear that the upper limit of the erosion rate estimated by Lanzerotti et al. (1978) is precluded by the survival of rays. By analogy with lunar rays, the initial ray deposits on Ganymede are unlikely to be thicker than a few m.

Haff et al. (1979) and Haff (1980) suggested that rocky impurities in the surface material of the icy Galilean satellites might lead to residual concentration of the impurities during sputtering and to armoring of the surface (in the absence of impact gardening). Thus the actual rate of ablation, as in the case of ice sublimation, would be reduced by a lag deposit of rocky debris. In the presence of gardening, on the other hand, the residual enrichment of rocky material due to sputtering must be distributed through the regolith. A very large enrichment, or a corresponding loss of ice, in the regolith on the grooved terrain would be implied by the highest sputtering rate estimated by Lanzerotti et al. (1978), as the calculated ablation over 3.7 Gyr exceeds the thickness of even the polar regolith by a factor of ~ 10. Their lowest rate of ablation, integrated over 3.7 Gyr, is roughly equal to the calculated thickness of the regolith produced by gardening at the antapex.

Several other effects may reduce the rates of surface erosion below those calculated by Lanzerotti et al. (1978). As in the case of ablation due to insolation, the effects of surface roughness must be accounted for. Carey and McDonnell (1976) found that 65-70% of the sputtered atoms might be recaptured on a very rough surface. It has been suggested that a thin O_2 atmosphere on Ganymede might suppress the sputtering rate (Yung and McElroy 1977) or the satellite's net loss due to sputtering (Haff et al. 1979). An observation of a stellar occultation in the ultraviolet by Voyager 1, however, suggests an upper bound for the surface atmospheric pressure of 10^{-8} mbar and a limiting surface density of 6×10^8 cm^{-3} for any atmospheric constituents (Broadfoot et al. 1979). The effect on sputtering or transport of sputtered material of such a very tenuous atmosphere would appear to be minor. P.K. Haff has also suggested (personal communication, 1981) that only a small proportion of the sputtered H_2O may have escaped the satellite; the net erosion would then be much less than indicated by sputtering experiments.

If the sputtering rate of ice on Ganymede is close to the lower bound of

Lanzerotti et al. (1978), then the regolith near the antapex should be greatly enriched in any rocky constituents initially present in the ice or in contaminants added by impact. Near the apex, the enrichment due to sputtering would be less by a factor of $\gtrsim 20$. Hence, a marked difference in albedo or spectral reflectance might be expected. In fact, Pilcher et al. (1972) reported a difference in depth of water-frost absorption bands between the leading and trailing hemispheres of Ganymede, and suggested that the leading hemisphere had a 20% greater frost cover than the trailing hemisphere. This difference is correlated with an observed 15% variation in visual albedo with rotational phase (Millis and Thompson 1975; Morrison and Burns 1976; see also Harris 1961; Johnson 1971). The maximum in the rotational phase curve of albedo occurs at $60°$ rather than at the apex $(90°)$; the minimum is rather broad and poorly defined but appears to include the antapex. From the available Voyager coverage of Ganymede, the displacement of the maximum from $90°$ may be due to the distribution of large polygons of dark ancient cratered terrain, but the broad variation of albedo with phase is not obviously related to the distribution of observed geologic units. Thus, the telescopically observed global variation in regolith composition may reflect residual concentration of rocky debris by sputtering, although sublimation due to insolation could produce the same effect.

Europa exhibits a similar curve of albedo as a function of rotational phase, but with almost twice the amplitude and with the maximum and minimum located almost exactly at the apex and antapex (see Morrison and Burns 1976). Because the flux of high-energy particles in the Jovian magnetosphere is higher at Europa than at Ganymede, we can attribute the observed phase curve to the combined effects of sputtering and gardening.

The amount of concentration of rocky constituents in the regolith of Ganymede may be roughly compatible with the minimum ablation rate of Lanzerotti et al. (1978). From high-precision infrared spectra, Clark (1980) estimated the average abundance of ice in the leading hemisphere of Ganymede at 90 wt-% (significantly higher than the upper estimate of Pilcher et al. 1972). If the trailing hemisphere has $\sim 20\%$ less ice than the leading hemisphere, a more restricted region near the antapex with ~ 1 m of turnover of the regolith since grooved-terrain formation might well have $< 50\%$ ice from the surface down to ~ 1 m depth. For an initial abundance of $\sim 6\%$ rocky constituents (our suggested value, from the hydrostatics of the eruptions that formed the smooth units of the grooved terrain), concentrations to $\sim 50\%$ rocky debris would require $\sim 90\%$ ablation of the regolith at the antapex.

Haff et al. (1979) suggested that surface ablation of the icy satellites by sputtering might explain some peculiar radar reflection characteristics of these bodies (Goldstein and Green 1980; Ostro et al. 1980; Chapter 8). Deep ablation might produce a rugged surface with pinnacles capped by resistant rocky fragments. If ablation due to sputtering is only on the order of 1 m over the last 3 Gyr, however, gardening would have removed such asperities, except possibly in the region of the antapex. Alternatively, the large radar cross section of Ganymede may be due to scattering by coarse rocky material embedded in a regolith that is mostly ice, as first suggested by Goldstein and Morris (1975), or to coarse ice fragments in the regolith (cf. Goldstein and Green 1980; Chapter 8).

A strong argument for ablation due to sputtering on the Galilean satellites can be made from the polarimetric observations of Callisto by Dollfus (1975). These observations indicate that the leading hemisphere of Callisto is largely coated with fine silicate particles, much like the surface of the Moon, whereas the trailing hemisphere's surface is largely devoid of fine particles and must be mainly coarser rocky material (Dollfus 1975; Dollfus et al. 1980). Fine silicate particles must have been deposited by infall of small meteoroids derived from comets (such as those associated with meteor streams and ordinary sporadic meteors) and must also have been produced by impact of these meteoroids on the exposed rocky material. It is therefore difficult to understand how fine silicate particles or dust could be absent from the regolith surface in the trailing hemisphere of Callisto unless some process has removed them. In the region where turnover rate of the regolith is very slow, near the antapex, silicate dust may have been selectively lost by sputtering by ~ 1 MeV protons and also possibly by fast heavy ions in the Jovian magnetosphere (Haff 1981). Similar loss may also have taken place on Ganymede, but light reflected from the polar shrouds and from interstitial ice elsewhere obscures the polarimetric evidence. The polarimetric function for Ganymede appears best represented by a combination of fine ice grains and rock rather than ice grains and silicate dust (Dollfus 1975).

Thermal Annealing of the Regolith

The survival of craters on Ganymede, particularly those with diameters larger than a few tens of km in the equatorial region, indicates that the Ganymedean regolith cannot everywhere resemble the lunar regolith in thermal conductivity to depths of tens of m. If the conductivity of most of the regolith near the south pole is $100\,\mu W\,cm^{-1}K^{-1}$, and if the surface heat flow at ~ 3.7 Gyr was of the order of 1 $\mu W\,cm^{-2}$ (somewhat less than the present measured heat flows of 1.6 to 2.1 $\mu W\,cm^{-2}$ for the Moon (see Langseth et al. (1976), then the thermal gradient through the bulk of the regolith at that time was $\sim 1\,K\,m^{-1}$. A heat flow of $\sim 1\,\mu W\,m^{-2}$ at ~ 3.7 Gyr is consistent with the history of the lithospheric thermal gradients deduced by Passey and Shoemaker (Chapter 12) from studies of topographic relaxation of craters. Given the estimated mean thickness of the regolith near the south pole, ~ 40 m, and assuming that this thickness was almost fully developed by ~ 3.7 Gyr, the mean increase in temperature from the near surface to the base of the regolith was ~ 40 K. To this increase we should add the abrupt increase in mean temperature from the actual surface to a depth of ~ 10 cm. Scaling from a temperature increase of 40 K in the first 10 cm of depth in the lunar regolith, and from the ratio of assumed mean heat flow on Ganymede at 3.7 Gyr to present mean heat flow on the Moon, we have an increase of 20 K. Thus, the total estimated temperature increase from the surface to the base of the polar regolith at 3.7 Gyr is ~ 60 K. This is approximately the difference in mean surface temperature between 87° lat and the equator.

At mean surface temperature ~ 60 K, the temperature at the base of the regolith at 87° S lat is estimated to have been ~120 K when the surface heat flow was 1 μW m^{-2}. Applying the same calculations for the equator at 90° from the apex, where the mean regolith thickness is estimated to be 10 m, we find that the temperature at the regolith base was ~ 150 K. Based on the very strong dependence of effective viscosity on temperature, this ~ 30 K temperature difference at the top of the lithosphere would have erased all craters larger than a few tens of km in diameter in the equatorial region, in the time required for moderate relaxation of craters of the same size near the pole. But the craters on grooved terrain near the equatorial region actually show less topographic relaxation than those on the polar-grooved terrain. Even though the polar-grooved terrain is somewhat older and may have formed at a time of slightly higher heat flow, it is clear that there could not have been a 30 K temperature difference at the top of the lithosphere shortly after grooved terrain formed. Therefore the lower part of the regolith cannot have had thermal properties similar to the lunar regolith in both the polar and equatorial regions.

Morrison and Cruikshank (1973) and Hanel et al. (1979) suggested that the regolith might be thermally annealed. Analysis of the transport of water vapor as a function of temperature suggests that, above a certain threshold temperature, water was mobilized in the regolith. From Purves and Pilcher (1980) and Squyres (1980d), the threshold for mobility appears close to 130 K. Where the temperature at the base of the regolith exceeded this threshold, ice would have tended to sublime near the base and be redeposited at higher, cooler levels. This transport probably filled in pore space at higher levels, up to the threshold isotherm. More important, sublimation probably occurred preferentially at grain contacts, where stress from the load of overlying material was highest; redeposition probably occurred nearby on less stressed grain surfaces. Thus, in the region of vapor transport the regolith tended to become cemented, and cementation would have dramatically increased the conductivity. If the heat flow were constant with time, this process of thermal annealing would have led to a uniform temperature at the top of the annealed part of the regolith in all regions where the regolith was deep enough for annealing.

From the cratering time scale of Shoemaker and Wolfe (Chapter 10) much of the grooved terrain probably formed during the waning stages of heavy bombardment (Chapter 12). Most craters and a significant fraction of the regolith are inferred to have been produced in an interval of a few 10^8 yr after grooved terrain formation. If the lithospheric thermal gradients and surface heat flow decreased substantially after most of the regolith formed, as deduced by Passey and Shoemaker (Chapter 12), then the thermal gradients in the regolith also decreased substantially after the early annealing. The initially high thermal gradients may have caused annealing at much higher levels in the equatorial regolith than would be produced at present temperatures. These levels, or fossil isotherms, would reflect the period of highest heat flow after the thicknesses of the regolith had approached their present values.

Because heat flow has decreased with time, the temperatures at the base of the regolith, or at the fossil annealing isotherm, have approached the temperatures at the surface. As the entire satellite has cooled down, the lithosphere became much cooler and stiffer in the polar regions than at the equator. For this reason, the high-latitude Gilgamesh basin is better preserved than the nearly contemporaneous western equatorial basin.

VI. GEOLOGIC HISTORY OF GANYMEDE

Superposition and intersection of geologic units and structures provide a sequence of major events in Ganymede's history, but to understand their relation to other events in the Jovian system and elsewhere, and to evaluate the processes that have taken place, it is necessary to know the approximate absolute ages of the geologic units on Ganymede. The only method now available for estimating the age of geologic units is from crater-density determinations. Crater counting by itself provides crater ages for various geologic units, which are useful for determining relative ages. To convert crater ages to estimates of absolute age, we must know the cratering history. At this stage in our exploration of the outer solar system, this conversion is difficult, but Shoemaker and Wolfe (Chapter 10) have made a first attempt to assemble the available evidence and construct cratering time scales for the Galilean satellites. Here, we will adopt their proposed time scale for Ganymede as a basis for discussion of its history.

Large uncertainties exist in the cratering time scale for Ganymede; we will use the nominal ages derived from observed crater densities without repeated discussion of the uncertainties. The statistical errors in the crater densities are given in Tables 13.1 to 13.5. Estimated absolute ages obtained from the time scale will be quoted without errors and should be regarded only as model ages for which errors are not defined, probably lower limiting ages. As discussed by Shoemaker and Wolfe (Chapter 10), it is unlikely that the true ages are less than the model ages derived from the adopted cratering time scale.

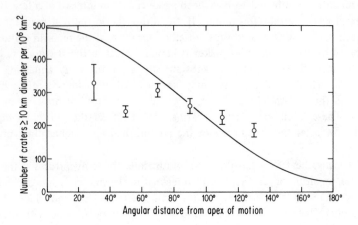

Fig. 13.29. Crater density on ancient cratered terrain as a function of distance from apex of orbital motion. Crater densities are averaged for 20° intervals of distance from apex. Error bars are one standard deviation. See Table 13.1 for location and other data on areas counted. Curves show theoretical variation of crater density for surfaces of a single age.

History of the Ancient Cratered Terrain

Large differences in crater density are observed on the ancient cratered terrain (Table 13.1; Fig. 13.29). Figure 13.29 shows that the lowest crater densities are found in the trailing hemisphere and the highest in the leading hemisphere, but the observed distribution of densities does not follow the theoretical curve for a surface of uniform age, of density variation with distance from the apex of orbital motion. As discussed by Passey and Shoemaker (Chapter 12), the average apparent age of the surface is lowest in the leading hemisphere and highest in the trailing hemisphere. In dealing with ages derived from crater density on the ancient cratered terrain, note that probably many craters ⩾ 10 km in diameter have been lost by topographic relaxation. Thus, these derived ages represent crater retention ages on the ancient cratered terrain, not the age of the crust on which the craters were formed. The overall distribution of observed crater densities apparently represents a crater population nearly in equilibrium between crater production by impact and crater loss by impact and topographic relaxation (see Chapter 12).

An equilibrium distribution of crater densities would not be expected to be uniform if large craters have played a major role in the elimination of small craters. Wherever a large crater and its ejecta deposit locally eliminated smaller craters, a finite time was required to build up the small crater density again. An equilibrium distribution of small crater densities therefore will show a spread that reflects the spread in times since the last large crater was formed at each place. A detailed inspection of the individual areas counted to determine crater density on the ancient cratered terrain reveals that many areas where 10 km craters are low in density have recognizable large craters. In other areas of low density, however, no large craters are recognizable. We infer that large craters now completely erased by viscous relaxation were present in these areas. In many areas of low crater density there are distinctive inflections in the crater size-frequency distribution. These inflections indicate that small craters have been preferentially lost, presumably by ejecta blanketing of the surface from large craters that subsequently disappeared. The areas listed in Table 13.1 were selected because they do not contain recognizable palimpsests. We infer nevertheless that palimpsests once were present in all these areas, and that with time the albedo of the former palimpsests decreased and they disappeared, probably as a result of moderately heavy surface contamination by dark material from many small impacting bodies during the heavy bombardment in which the ancient cratered terrain was formed.

The average crater density on the ancient cratered terrain is far below typical crater densities of the lunar highlands and also below the density on Callisto (Chapter 12). Such a low equilibrium density may be attributable to the major role of viscous relaxation in crater loss. The lower equilibrium density on the ancient cratered terrain of Ganymede, compared to Callisto, indicates that Ganymede's lithosphere was thinner at times of equivalent bombardment rates. Because the cratering rate probably was higher at any given time on Ganymede than on Callisto (Chapter 12), a significant delay in cooling and thickening of the lithosphere on Ganymede is implied by the comparatively low crater density of the ancient cratered terrain. This difference in thermal history between Ganymede and Callisto may be due partly to Ganymede's greater size and mass and partly to its greater collisional heating during late stages of accretion (Chapter 12).

If viscous relaxation had a large role in determining the equilibrium distribution of crater densities on the ancient cratered terrain, we might expect the highest mean crater density in the leading hemisphere, where the crater production was highest. The mean density at 10 km crater diameter is higher in the leading hemisphere than in the trailing hemisphere, but only by $\sim 30\%$. This difference is $\sim 1/4$ of the predicted difference in cratering rate (see Chapter 10, and Fig. 13.29). The small difference between hemispheres could have resulted from a change or changes in their position relative to the orbital motion of the satellite, as a result of energetic impacts. A more likely explanation is that the local impact rate indirectly affected the rate of viscous relaxation of craters. For plausible values of the heat flow during ancient cratered terrain time, an insulating regolith probably was required for viscous loss of craters 10 km in diameter. The relaxation rate of a freshly formed 10 km crater therefore may have depended directly on the buildup rate of an insulating layer. The mean rate of regolith buildup probably was about twice as great, on the average, in the leading hemisphere as in the trailing hemisphere, because the cratering rate was higher. A thicker regolith, together with the higher rate of obliteration of small craters by large craters in the leading hemisphere, appears to explain the small difference between the two hemispheres in crater populations on the ancient cratered terrain.

Assuming that the present leading and trailing hemispheres have remained fixed with respect to the orbital motion of the satellite, the mean crater retention age of the ancient cratered terrain varies systematically with distance from the apex of motion. In the range $20°$ to $40°$ from the apex, the mean crater retention age is 3.8_0[a] Gyr, whereas at $120°$ to $140°$ from the apex the mean crater retention age rises to 3.9_8Gyr (Chapter 10). The rimmed furrow system in Galileo Regio, which can be traced to within $40°$ of the apex, evidently is ∼ 3.8 Gyr old. Because the furrows, which average ∼ 10 km in width, have not disappeared by viscous relaxation, they cannot be older than the oldest 10 km craters at $40°$ from the apex. The furrows are clearly older than all recognized palimpsests on Galileo Regio, however, and apparently are older than most craters > 20 km in diameter.

The regolith at any given place on the ancient cratered terrain probably was formed and destroyed many times. Here we restrict the term regolith to the particulate surface layer of low thermal conductivity, in the same sense as the term was initially applied by Shoemaker et al. (1969) and Shoemaker and Morris (1970) to the similar surficial layer on the Moon formed primarily by surface gardening by many small craters. Ejecta deposits surrounding large craters must have repeatedly buried previously formed regoliths on the ancient cratered terrain. These deposits probably were very different from the regolith in their physical properties. Because they were excavated from subregolith ice and were shock heated as well, large ejecta deposits formed during ancient cratered terrain time probably were rapidly annealed. Old buried regoliths would also have been annealed. Just as the crater population apparently was in equilibrium, the thickness distribution of regolith at the surface of the ancient cratered terrain probably was in equilibrium. Because a finite time was required to build up a new insulating regolith after each major cratering event, moreover, the polar lithosphere probably was cooler and thicker on the average than the equatorial lithosphere in ancient cratered terrain time.

---

[a]Subscripted figures are used in this chapter to indicate a digit that lies at or just beyond the threshold of significance.

History of Grooved Terrain

Unlike the ancient cratered terrain, nearly all craters > 1 km in diameter produced on the grooved terrain appear to be preserved. A few craters may have been lost in the south polar region, where the oldest grooved terrain is found and where the presence of penepalimpsests shows that some of the largest preserved craters were nearly lost. Consistent with the crater preservation, the global distribution of crater densitives reflects the global variation in cratering rate. However, there is considerable scatter around a theoretical crater density curve for a surface of uniform age, which indicates that there are real variations in the age of the grooved terrain. Crater densities in the range $40°$ to $60°$ from the apex indicate an average age ~ 3.6 Gyr; at $60°$ to $90°$ from the apex the typical age is ~ 3.1 Gyr (except near the poles), at $90°$ to $120°$ it is ~ 3.3 Gyr, and at $120°$ to $150°$ it is ~ 3.8 Gyr. In the south polar region, the estimated mean age of the grooved terrain is close to 3.8 Gyr. Thus, although substantial statistical uncertainty exists for crater counts of individual areas, a signficant age range in grooved terrain is indicated. The estimated mean age is ~ 3.4 Gyr in the leading hemisphere (exclusive of the polar regions) and ~ 3.6 Gyr in the trailing hemisphere. Grooved terrain apparently formed first in the polar regions, where it occupies most of the surface, then in the trailing hemisphere, and finally in the leading hemisphere. The earliest grooved terrain was formed during waning heavy bombardment, but grooved terrain formation probably extended beyond the heavy bombardment period.

Better preservation of the relief of craters on the grooved terrain, relative to that on the ancient cratered terrain, shows that the grooved terrain formed during a period of cooling and thickening of the lithosphere. Hence, it is interesting that the grooved terrain formed first in the polar regions, where cooling of the lithosphere was probably the most advanced at any given time. Moreover, cooling of the lithosphere in the trailing hemisphere may have preceded cooling in the leading hemisphere, due to the slower rate of regolith buildup on freshly cratered surfaces in the trailing hemisphere. These relations suggest that regional cooling of the lithosphere, or perhaps of deeper layers in the satellite, may have triggered the formation of grooved terrain.

The possible causal connection between cooling and formation of the grooved terrain is an incentive to search for ways in which cooling might lead to tensional tectonics in the lithosphere and possibly to a two-layer convection regime in grooved-terrain time. It has been proposed that a substantial global expansion resulting from internal differentiation of Ganymede may have caused grooved terrain to form (Squyres 1980b). Although the outer portion of Ganymede probably underwent very early melting and differentiation due to accretional heating, differentiation of the deep interior may have taken considerably longer (Schubert et al. 1981). Extension of the crust might have resulted from renewed differentiation of ice and rocky constituents in a deep undifferentiated shell in the satellite (D.J. Stevenson, personal communication 1981). However, it is difficult to understand why an internal episode of differentiation would correlate with cooling of the outer shell of the satellite. Because the phase transitions from ice VII and VIII to lower pressure polymorphs are strongly exothermic, internal differentiation would more likely be reflected by increased heat flow with consequent thinning and softening of the lithosphere. Therefore, we propose an alternative hypothesis.

We assume that Ganymede has been at least partly differentiated and that the outer part is mainly ice. The structure of this outer part will depend on the position of the geotherm on the phase diagram for ice (Hobbs 1974). Second, we assume that, at any given temperature, ice II is much stiffer than ice I. Unpublished experimental results by K.A. Echelmeyer and B. Kamb (personal communication 1981) suggest that this might be true, but further studies are required to verify this assumption. Our third assumption is that at sufficiently high temperature the effective viscosity of ice II becomes low enough that convection can occur through the ice II layer on Ganymede. If ice II remains too rigid for convection to occur at any temperature, then the geotherm must pass through the ice III stability field or lie on the ice III-ice II phase boundary (see Figs. 13.30 and 13.32) until heat flow from the deep interior drops to a level at which the heat transport can be accommodated by conduction through ice II. The latter possibility only slightly modifies the following scenario for the thermal evolution of Ganymede.

Stage I, Ancient Cratered Terrain Time ($\sim$ 4.0 to 3.8 Gyr)

As estimated from the viscous relaxation of craters (Chapter 12), the shallow lithospheric thermal gradient declined from $\sim$ 5 to $\sim$ 6 K km^{-1} early in this stage to $\sim$ 1.5 K km^{-1} at its end. The lithosphere thickened from $\sim$ 10 km at the beginning of this stage to $\sim$ 35 km at the end. (The base of the lithosphere is taken to be the level at which the effective viscosity drops to $\sim$ 10^{18} poise, Chapter 12.) In the equatorial region, 10 km diameter craters were retained from the beginning of stage I, and $\sim$ 60 km diameter craters were retained from its end.

Large, through-going convection cells that circulated from the top of the rocky core of the satellite to a boundary layer near the base of the lithosphere are inferred to have been active in stage I. The asthenosphere consisted of warm layers of ice I, ice II, ice V, and ice VI plus mixed-phase transition layers (Fig. 13.30). To allow convection, the geotherm must have been essentially adiabatic from the top of the core to the boundary layer near the base of the lithosphere. Steep gradients in the geotherm, one positive and one negative, must have occurred in the mixed phase regions (Fig. 13.30). Temperatures of the convecting asthenosphere were constrained to remain high by the presence of ice II, because stiffening of cool ice II would have impeded

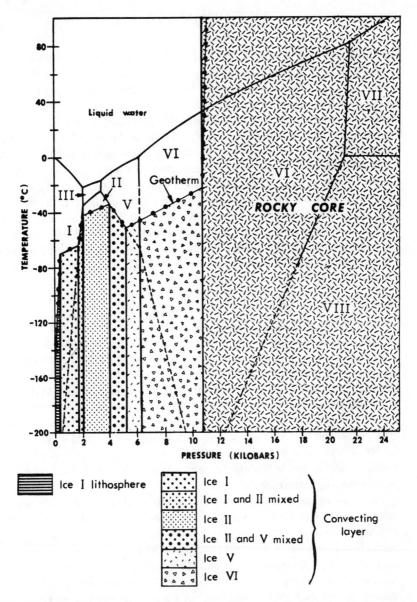

Fig. 13.30. Model of geotherm and structure of Ganymede in ancient cratered terrain time (∼ 3.9 Gyr). One convective layer and one lithospheric layer of ice are postulated. Approximate pressure at top of rocky core, and the steep thermal gradient shown in the core, are based on the assumption that the satellite is fully differentiated. Above the core is a convecting ice asthenosphere which is layered as shown in the diagram. The geotherm in the asthenosphere is assumed essentially adiabatic. In the thin ice I lithosphere, the thermal gradient is governed by conductive transport of heat from the interior. Phase relations shown for the stable polymorphs of ice are from Hobbs (1974); solid lines indicate measured phase boundaries and dashed lines are phase boundaries estimated by extrapolation.

the convective heat transport. The temperature at the top of the ice II layer would have been just high enough to permit ice II to convect. From the base of the ice I layer to the lithospheric boundary layer, the thermal gradient probably was closely adiabatic. Ice I in this region probably was much softer than at the threshold of effective viscosity that permitted convection.

In our calculation of the pressure at the top of the core (Fig. 13.30), the satellite is assumed to have been fully differentiated into an ice mantle and rocky core; the core is assumed to consist chiefly of phyllosilicates with bulk density 2.7 g cm^{-3}. The geotherm in the upper part of the rocky core is assumed to have followed a conductive heat transport gradient. If the satellite were only partially differentiated, a layer of mixed rock and ice would have lain between the ice mantle and the rocky core (Schubert et al. 1981). In this case, the base of the ice asthenosphere would have been shallower, and the geotherm in the layer of mixed ice and rock probably would have been nearly adiabatic.

Transition Stage ($\sim$ 3.8 Gyr)

The time of transition varies with position on the satellite. When a critical thermal gradient of $\sim$ 1.5 K km^{-1} in the shallow lithosphere was reached, heat flow from the deep interior was accommodated by conduction through ice II, and a double lithosphere structure became stable (Fig. 13.31). The upper lithosphere, in ice I, extended to a depth of $\sim$ 35 km. At the end of the transition stage, the ice I convection layer extended from the base of the ice I lithosphere to $\sim$ 130 km in depth. The thermal gradient was adiabatic in this convection layer. A lower lithosphere was formed in the top of the ice II layer that extended from $\sim$ 130 km to $\sim$ 150 km in depth. The geotherm in the ice II lithosphere approximately followed the ice I-ice II phase boundary, but was slightly flatter.

Because of negative latent heat of transformation of ice I to ice II, collapse of ice I to ice II proceeded spontaneously. The rate of transformation was limited only by the rate at which the heat was carried away by conduction at the top of the ice II lithosphere, by convection in the overlying ice I convection layer, and by conduction through the upper lithosphere. Because initial temperatures probably were well above the limiting flow temperatures in the ice I convecting layer, the excess heat would have been convected rapidly to the ice I lithosphere.

Just prior to thermal collapse to the double-lithosphere configuration, the heat flow from the deep interior of the satellite must have dropped to $\sim$ 1 X 10^{-6} joule cm^{-2}s^{-1}, which corresponds to a thermal gradient of $\sim$ 1.5 K km^{-1} near the top of the ice I lithosphere. A slight delay in the onset of collapse due to asymmetry of the heat sources in the rocky core could have caused a large difference in crater densities on the subsequently formed grooved terrain, because the collapse evidently occurred during the waning of

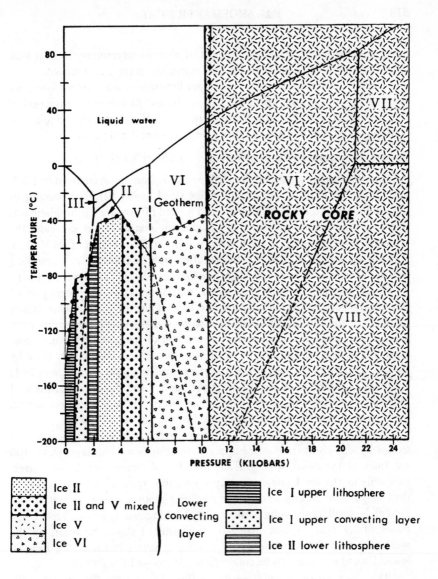

Fig. 13.31. Model of geotherm and structure of Ganymede in grooved terrain time (~ 3.5 to 3.7 Gyr). Two separate convecting layers of ice and two lithospheric layers of ice are postulated. Existence of the lower lithosphere is postulated on the basis of preliminary laboratory observations of K.A. Echelmeyer and B. Kamb (unpublished) on the stiffness of ice II. The geotherm is assumed to be adiabatic in the convecting layers and to follow conductive heat transport gradients in the two lithospheric layers. Temperatures in the ice I convecting layer have dropped since ancient cratered terrain time due to stiffening of the upper part of the ice II layer and establishment there of a conductive heat transport regime. Thickness of the ice I lithosphere has increased significantly. Slow conversion of ice V to ice II as the interior of the satellite cools results in net expansion that throws the ice I lithospheric shell into tension. (See Fig. 13.30 for explanation of rocky core and phase relationships shown for ice.)

heavy bombardment. Thus, the effects of a minor asymmetry in heat flow from the core would have been strongly magnified in the crater record.

During the transition stage, the upper lithosphere was thrown into compression by transformation of ice I to ice II, and to thermal contraction of the cooling phases. But the lithosphere was sufficiently thin so that net contraction probably was taken up by very minor regional strain.

Stage II, Formation of Grooved Terrain ($\sim$ 3.8 to $\sim$ 3.1 Gyr)

The thermal gradient in the shallow lithosphere decayed from $\sim$ 1.5 K km^{-1} to $\sim$ 0.5 K km^{-1} during this period (see Chapter 12). When cooling of the ice I layer and ice II lithosphere had leveled off, the ice I lithosphere was thrown into tension. As the thermal gradient continued to decrease, the principal volume change in the interior was the transformation of ice V to ice II in the upper part of the lower convecting layer. Expansion accompanying this transformation was only slightly offset by thermal contraction in the ice II lithosphere. The beginning of tension marked the beginning of this stage whose end occurred when the base of the ice I lithosphere finally dropped to the top of the ice II lithosphere and the upper convecting layer disappeared. Until it disappeared, the position of the base of the ice I convecting layer (i.e., the top of the ice II lithosphere) remained nearly fixed by the low (adiabatic) thermal gradient in the convecting layer. Hence, very little ice I was transformed to ice II during the cooling and thickening of the ice I lithosphere. At the end of the stage, when the two lithospheres were joined, the base of the ice II lithosphere extended to a depth of $\sim$ 200 km.

Evolution of structural cells in the grooved terrain apparently was controlled by ice I convection in the upper convecting layer. This layer was $\sim$ 100 km thick at the onset of stage II, and the lateral dimensions of the convection cells in the ice I convecting layer probably were of the same order as their height. Net extension of the crust during grooved terrain formation was extremely small, amounting perhaps to $\sim$ 0.1% increase in surface area. Large plates of ice I lithosphere may have been shifted about, however, by the motion in many cells in the upper convecting layer, which would have allowed limited extension to be concentrated in grooved terrain sulci.

The reservoirs supplying the eruptive fluids that formed the smooth units in grooved terrain time probably were located at depths where the geotherm was closest to the liquidus. As shown in Fig. 13.31, this would have been in the ice II convecting layer, but may have extended into the ice II-ice V mixed layer. At these depths, Fig. 13.31 suggests a temperature difference of $\gtrsim$ 20 K between the geotherm and the pure water liquidus. If this model is correct, dissolved salts or other constituents must have been present to lower the freezing point of the aqueous liquid below the geotherm.

Grooved terrain formation must have ceased either when the upper convecting layer disappeared or when it became too thin for convection to displace the thickening ice I lithosphere above. Because the surface temperature was lower at the poles, stage II probably came to an end sooner there than at the equator, which may partly account for the higher crater density and greater age of polar grooved terrain. If the rocky core was a little cooler at the poles, stage II would also have begun first in the polar regions. The ice I lithosphere may have been thickest at the poles at the onset of grooved terrain formation. This may have enhanced the density contrast between the lithosphere and upper convecting layer and more efficiently driven subsidence of lithosphere slivers, thus accounting for more extensive development of grooved terrain in the polar regions.

Stage III ($\sim$ 3.1 Gyr to Present)

The shallow lithospheric thermal gradient continued to decrease from ~ 0.5 K km^{-1} to ~ 0.3 K km^{-1}, until the combined ice I-ice II lithosphere thickened to ~ 300 km in low latitude regions at the present time (Fig. 13.32). In the last 3 Gyr, net volume change of the satellite was small because conversion of ice I to ice II nearly offset the volume change due to continued conversion of ice V to ice II. Net cooling of the entire lithosphere may have thrown the upper part of the lithosphere into compression. The present asthenosphere probably consists mainly of ice VI with possibly a thin layer of ice V and a mixed layer of ice II and ice V near the top.

Note that this scenario for grooved terrain formation is dependent on the assumption that ice II is much stiffer than ice I. This assumption is required for two-level convection, since soft ice II would permit through-going convection across phase boundaries. If such deep convection occurred at the time of grooved terrain formation, the geotherm may have lain much farther below the solidus than is indicated in Fig. 13.31. In this situation, it would be much less likely that any liquid was present to flood the surface. Determination of the rheology of ice II is thus critical to the evaluation of competing models of grooved terrain formation.

Post Grooved Terrain History

The beginning of post grooved terrain events depends on position on the satellite. In the south polar region, the high density of post grooved terrain craters evidently reflects early high bombardment rates; the density appears too high to be explained by a steady cratering rate over geologic time. Consistent with the high cratering rate, more craters cut by later grooves are recognized on the polar grooved terrain than on grooved terrain elsewhere.

The post grooved terrain Gilgamesh and western equatorial basins record the impact of two of the largest bodies that struck Ganymede near the end of heavy bombardment. Probably not by coincidence, both basins occur in the leading hemisphere, where the flux of impacting bodies was highest. Their preservation demonstrates that by the time they were formed the lithosphere

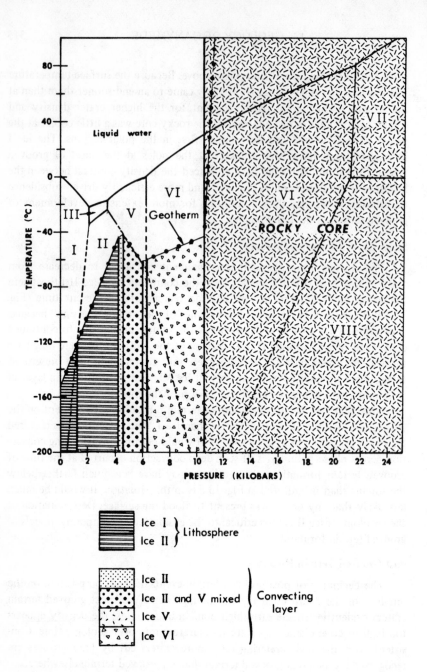

Fig. 13.32. Model of geotherm and structure of Ganymede at present time; one convective and one lithospheric layer are again postulated. The combined ice I and ice II lithosphere extends to pressure of ~4 kbar (~300 km in depth). Estimated mean conductive thermal gradient in the lithosphere is ~0.3 to 0.5 K km^{-1}. The geotherm in convecting ice asthenosphere is assumed adiabatic. (See Fig. 13.30 for explanation of rocky core and phase relations shown for ice.)

was much thicker and stiffer than in ancient cratered terrain time, when craters as small as 10 km in diameter were flattened beyond recognition by viscous relaxation. Continued thickening and stiffening of the lithosphere through most of the rest of geologic time probably was required to support the relief of the Gilgamesh basin as we see it today (see Chapter 12). Based on the density of smaller craters superposed on the rim of Gilgamesh, the age of the basin is estimated at $\sim$ 3.5 Gyr. However, the secondary craters of Gilgamesh are superposed on grooved terrain that appears from crater density estimates to be somewhat younger. The true ages of Gilgamesh basin and the nearly contemporaneous western equatorial basin probably lie in the range 3.2 Gyr to 3.5 Gyr, based on the cratering time scale of Shoemaker and Wolfe. If this time scale is approximately correct, the interval from the time of a thin lithosphere over most of the globe, when craters as small as 10 km were flattened, to the time when the lithosphere was capable of supporting basins, was $\sim$ 0.3 to 0.6 Gyr. Regardless of the absolute duration of this interval, it demonstrably coincides with the episode of grooved terrain formation over the satellite. The conclusion seems inescapable that the grooved terrain formed during a period of dramatic stiffening of the lithosphere.

The late history of the surface of Ganymede chiefly involves bombardment at a relatively slow rate over the last 3 Gyr. The ray craters are the youngest features produced by this bombardment. On the basis of global variation in ray crater density, the oldest ray craters in the trailing hemisphere are $\sim$ 1.1 Gyr and the oldest ray craters near the apex are $\sim$ 0.5 Gyr (Chapter 12).

Acknowledgments. We wish to thank P.K. Haff, B. Kamb, Q.R. Passey, and D.J. Stevenson for sharing insights and results of their work in advance of publication. P.D. Spudis and L.A. Soderblom made helpful suggestions in technical reviews of this chapter, and D.B. Wier and M. Strobell provided much assistance in editorial review. We are also indebted to C.S. Shoemaker, without whose assistance the manuscript could not have been completed in time to be included in this book. Finally, we thank M.S. Matthews and D. Morrison for their patience and for accepting the manuscript, which was submitted at an unconscionable late date.

October 1981

REFERENCES

Broadfoot, A.L. and the Voyager Ultraviolet Spectrometer team. (1979). Extreme ultraviolet observations from Voyager 1 encounter with Jupiter. *Science* 204, 979-982.

Brown, W.L., Augustyniak, W.M., Brody, E., Cooper, B., Lanzerotti, L.J., Ramirez, A., Evatt, R., and Johnson, R.E. (1980). Energy dependence of the erosion of H_2O ice films by H and He ions. *Nuclear Instr. Methods* 170, 321-325.

Brown, W.L., Lanzerotti, L.J., Poate, J.M., and Augustyniak, W.M. (1978). "Sputtering" of ice by MeV light ions. *Phys. Rev. Letters* 40, 1027-1030.

Carey, W.C., and McDonnell, J.A.M. (1976). Lunar surface sputter erosion: A Monte Carlo approach to microcrater erosion and sputter redeposition. *Proc. Lunar Sci. Conf.* VII, 913-926.

Clark, R.N. (1980). Ganymede, Europa, Callisto, and Saturn's rings: Compositional analysis from reflectance spectroscopy. *Icarus* 44, 388-409.

Conca, J.L. (1981). Dark-ray craters on Ganymede. Proc. Lunar Planet Sci. Conf. XII. In press.

Consolmagno, G.J., and Lewis, J.S. (1976). Structural and thermal models of icy Galilean satellites. In *Jupiter,* (T. Gehrels, Ed.), pp. 1035-1051, Univ. Arizona Press, Tucson.

Croft, S.K. (1980). On the origin of pit craters. IAU Coll. 57. *The Satellites of Jupiter* (abstract 6-16).

Dollfus, A. (1975). Optical polarimetry of the Galilean satellites of Jupiter. *Icarus* 25, 416-431.

Dollfus, A., Mandeville, J.C., and Geake, J.E. (1980). Regolith and cratering on Callisto. IAU Coll. 57. *The Satellites of Jupiter* (abstract 6-6).

Goldstein, R.M., and Green, R.R. (1980). Ganymede: Radar surface characteristics. *Science* 207, 179-180.

Goldstein, R.M., and Morris, G.A. (1975). Ganymede: Observations by radar. *Science* 188, 1211-1212.

Haff, P.K. (1980). A model for the formation of thin films in dirty ice targets by sputtering: Application to the satellites of Jupiter. *Proc. Symp. Thin Film Interfaces and Interactions,* (J.E.E. Boglin and J.M. Poate, Eds.), vol. 80-2, pp. 21-28, The Electrochemical Soc., Princeton.

Haff, P.K. (1981). Erosion of surfaces by fast heavy ions. *W.K. Kellogg Radiation Lab. Preprint,* Calif. Inst. of Tech., Pasadena.

Haff, P.K., Watson, C.C., and Tombrello, T.A. (1979). Ion erosion on the Galilean satellites of Jupiter. *Proc. Lunar Planet. Sci. Conf.* X, 1685-1699.

Hanel, R., and the Voyager Infrared Spectrometer team (1979). Infrared observations of the Jovian system from Voyager 2. *Science* 206, 952-956.

Hansen, O.L. (1973). Ten-micron eclipse observations of Io, Europa, and Ganymede. *Icarus* 18, 237-246.

Harris, D.L. (1961). Photometry and colorimetry of planets and satellites. In *Planets and Satellites,* (G.P. Kuiper and B.M. Middlehurst, Eds.), pp. 272-342. Univ. Chicago Press, Chicago.

Hartmann, W.K. (1980). Surface evolution of two-component stone/ice bodies in the Jupiter region. *Icarus* 44, 441-453.

Hartmann, W.K., and Kuiper, G.P. (1962). Concentric structures surrounding lunar basins. *Commun. Lunar Planet. Lab.* 1, 51-66.

Hobbs, P. (1974). *Ice Physics.* Clarendon Press, Oxford, England.

Hodges, C.A., and Wilhelms, D.E. (1978). Formation of lunar basin rings. *Icarus* 34, 294-323.

Johnson, T.V. (1971). Galilean satellites: Narrowband photometry 0.30 to 1.10 microns. *Icarus* 14, 94-111.

Keihm, S.J., and Langseth, M.G. (1973). Surface brightness temperatures at the Apollo 17 heat flow site: Thermal conductivity of the upper 15 cm of regolith. *Proc. Lunar Sci. Conf.* IV, 2503-2513.

Langseth, M.G., Keihm, S.J., and Chute, Jr., L.L. (1973). Heat flow experiment. *Apollo 17 Preliminary Sci. Rept.* 9, 1-23. NASA SP-330, Washington.

Langseth, M.G., Keihm, S.J., and Peters, K. (1976). Revised lunar heat-flow values. *Proc. Lunar Sci. Conf.* VII, 3134-3171.

Lanzerotti, L.J., Brown, W.L., Poate, J.M., and Augustyniak, W.M. (1978). On the contribution of water products from Galilean satellites to the Jovian magnetosphere. *Geophys. Res. Letters* 5, 155-157.

Lucchitta, B.K. (1980). Grooved terrain on Ganymede. *Icarus* 44, 481-501.

Malin, M.C. (1980). Fables in Ganymede tectonics from morphologic studies. IAU Coll. 57. *The Satellites of Jupiter* (abstract 6-3).

McCauley, J.F. (1977). Orientale and Caloris. *Phys. Earth Planet. Interiors* 15, 220-250.

McKinnon, W.B., and Melosh, H.J. (1980). Evolution of planetary lithospheres: Evidence from multiringed basins on Ganymede and Callisto. *Icarus* 44, 454-471.

McKinnon, W.B., and Spencer, J. (1981). Deformation of Galileo Regio. *Lunar Planet. Sci. Conf.* XII, Part 2, 695-696 (abstract).

Morrison, D., and Burns, J.A. (1976). The Jovian satellites. In *Jupiter,* (T. Gehrels, Ed.), pp. 991-1034. Univ. Arizona Press, Tucson.

Morrison, D., and Cruikshank, D.P. (1973). Thermal properties of the Galilean satellites. *Icarus* 18, 224-236.

Morrison, R.H., and Oberbeck, V.R. (1975). Geomorphology of crater and basin deposits–emplacement of the Frau Mauro Formation. *Proc. Lunar Sci. Conf.* VI, 2503-2530.

Ostro, S.J., Campbell, D.B., Pettengill, G.H., and Irwin, I.S. (1980). Radar observations of the icy Galilean satellites. *Icarus* 44, 431-440.

Pang, K.D. (1981). Microstructure and particulate properties of the surfaces of Io and Ganymede: Comparison with the solar system bodies. *Proc. Lunar Planet. Sci. Conf.* XII. In press.

Pilcher, C.B., Ridgway, S.T., and McCord, T.B. (1972). Galilean satellites: Identification of water frost. *Science* 178, 1087-1089.

Purves, N.G., and Pilcher, C.B. (1980). Thermal migration of water on the Galilean satellites. *Icarus* 43, 51-55.

Reynolds, R.T., and Cassen, P.M. (1979). On the internal structure of the major satellites of the outer planets. *Geophys. Res. Letters* 6, 121-124.

Schubert, G., Stevenson, D.J., and Ellsworth, K. (1981). Internal structures of the Galilean satellites. Submitted to *Icarus.*

Shoemaker, E.M. (1962). Interpretation of lunar craters. In *Physics and Astronomy of the Moon,* (Z. Kopal, Ed.), pp. 283-359. Academic Press, New York.

Shoemaker, E.M. (1966). Preliminary analysis of the fine structure of the lunar surface in Mare Cognitum. In *The Nature of the Lunar Surface,* (W.B. Hess, D.H. Menzel, and J.A. O'Keefe, Eds.), pp. 23-78. Johns Hopkins Press, Baltimore.

Shoemaker, E.M., Batson, R.M., Holt, H.E., Morris, E.C., Rennilson, J.J., and Whitaker, E.A. (1969). Observations of the lunar regolith and the earth from the television camera on Surveyor 7. *J. Geophys. Res.* 74, 6081-6119.

Shoemaker, E.M., Hait, M.H., Swann, G.A., Schleicher, D.L., Schaber, G.G., Sutton, R.L., Dahlem, D.H., Goddard, E.N., and Waters, A.C. (1970). Origin of the lunar regolith at Tranquillity Base. *Proc. Apollo 11 Lunar Sci. Conf.,* pp. 2399-2412.

Shoemaker, E.M., and Morris, E.C. (1970). Surveyor final reports. Geology: Physics of fragmental debris. *Icarus* 12, 188-212.

Smith, B.A., and the Voyager Imaging Team. (1979a). The Jupiter system through the eyes of Voyager 1. *Science* 204, 951-972.

Smith, B.A., and the Voyager Imaging Team. (1979b). The Galilean satellites and Jupiter: Voyager 2 Imaging Science results, *Science* 206, 927-950.

Soderblom, L.A. (1970). A model for small-impact erosion applied to the lunar surface. *J. Geophys. Res.* 75, 2655-2661.

Soderblom, L.A., and Lebofsky, L.A. (1972). Technique for rapid determination of relative ages of lunar areas from orbital photography. *J. Geophys. Res.* 77, 279-296.

Squyres, S.W. (1980a). The topography of Ganymede's grooved terrain. IAU Coll. 57. *The Satellites of Jupiter* (abstract 6-20).

Squyres, S.W. (1980b). Volume changes in Ganymede and Callisto and the origin of grooved terrain. *Geophys. Res. Letters* 7, 593-596.

Squyres, S.W. (1980c). Topographic domes on Ganymede: Ice volcanism or isostatic upwarping? *Icarus* 44, 472-480.

Squyres, S.W. (1980d). Surface temperatures and retention of H_2O frost on Ganymede and Callisto. *Icarus* 44, 502-510.

Squyres, S.W. (1981). The morphology and evolution of Ganymede and Callisto. Ph.D. Dissertation (Cornell University, Ithaca, New York).

Strom, R.G. (1979). Mercury: A post-Mariner 10 assessment. *Space Sci. Rev.* 24, 3-70.

Trask, N.J. (1966). Size and spatial distribution of craters estimated from Ranger photographs. *Ranger VIII and IX, Part 2, Experimenters Analyses and Interpretations. Jet Propulsion Lab. Tech. Rep.* 32-800, 52-263.

Vitkus, G., Garipay, R.R., Hagemeyer, W.A., Lucas, J.W., Jones, B.P., and Saari, J.M. (1968). Lunar surface temperatures and thermal characteristics. *Surveyor VII, A Preliminary Report*, pp. 163-180. NASA SP-173, Washington.

Wilhelms, D.E. (1973). Comparison of martian and lunar multiringed circular basins. *J. Geophys. Res.* 78, 4084-4095.

Wilhelms, D.E. (1976). Secondary impact craters of lunar basins. *Proc. Lunar Sci. Conf.* VII, 2883-2901.

Wilhelms, D.E., Oberbeck, V.R., and Aggarwal, H.R. (1978). Size frequency distributions of primary and secondary lunar impact craters. *Proc. Lunar Sci. Conf.* IX, 3735-3762.

Wood, C.A., and Head, J.W. (1976). Comparison of impact basins on Mercury, Mars, and the Moon. *Proc. Lunar Sci. Conf.* VII, 3629-3651.

Yung, Y.L., and McElroy, M.B. (1977). Stability of an oxygen atmosphere on Ganymede. *Icarus* 30, 97-103.

14. THE GEOLOGY OF EUROPA

BAERBEL K. LUCCHITTA AND LAURENCE A. SODERBLOM
U.S. Geological Survey

The surface of Europa consists of two basic units, plains and mottled terrain. The plains units are generally bright, of uniform color, and smooth. The mottled terrain units have a dense population of spots and patches of dark material ranging in size from a few tens of km down to the resolution limit of a few km. The brown mottled terrain along the terminator displays a hummocky topography of pits and hollows a few km wide. Both plains and mottled terrain units are transected by numerous varied bands, usually brown or dark gray, that appear to have formed as dark material in-filling fractures, grabens, or gaps in the crust. The light central stripes found in some dark bands may result from dike-like filling of fracture by fluids that expanded on freezing. Numerous narrow ridges visible near the terminator may have similarily formed as dikes. In one area the ridges radiate from a fresh crater, suggesting formation on impact fractures. A large brown spot appears to be a crater scar analogous to crater palimpsests or collapsed craters found on Ganymede. A few small (10 to 30 km diameter) impact craters superposed on Europa's surface are confidently identified in the highest resolution images. Plains materials are the oldest geologic units; other units, including mottled terrains and spots and stripes, formed by their disruption or replacement. The gray predates the brown mottled terrain. Ridges appear to be the youngest features on Europa and locally postdate even young, fresh impact craters. The dark stripes and dark spots in mottled terrain units suggest that debris-laden materials were transported, perhaps as muddy slurries, through Europa's icy crust

[521]

to the surface. Such movement could have occurred if the outer ice shell were no thicker than a few tens of km. A thin ice shell suggests that most of Europa's water has remained in the interior, possibly tied up in hydrated silicates. Alternatively, evidence of rafting and rotation of sections of the crust suggests that the ice shell is thick enough in places to have been decoupled from the subjacent silicate sphere.

Europa, second closest to Jupiter of the four large Galilean satellites, is about the size of Earth's Moon (diameter 3126 ± 10 km). Europa orbits Jupiter with a period of 3.55 days at a radius $\sim$ 600,000 km. Beginning about 1960, a number of telescopic observations suggested that Europa's surface is composed dominantly of water ice (Kuiper 1961; Moroz 1966; Gromova et al. 1970; Kieffer 1970; Pilcher et al. 1972; Fink et al. 1973; Morrison and Cruikshank 1974; Johnson and Pilcher 1977; Pollack et al. 1978). Such a composition would explain its high reflectivity of $\sim$ 70% (0.5 to 1 μm). An absorption feature in the ultraviolet and $\sim$ 30% global variation in albedo (Morrison and Morrison 1977; Nelson and Hapke 1978) both suggested that varying amounts of impurities are mixed with the water ice. Whereas the Moon and Io have densities $\sim$ 3.5 g cm^{-3}, Europa's density is $\sim$ 3.0 g cm^{-3} (Morrison and Cruikshank 1974). The lower density and the evidence for water ice at its surface led to several models (Lewis 1971; Consolmagno 1975; Consolmagno and Lewis 1976,1977, and 1978; Fanale et al. 1977) in which Europa was composed mostly of silicates with densities as high as 3.5 g cm^{-3} (approximately the bulk density of Io and the Moon) and as much as 20% water. Further, because of the radiogenic heat supplied by the silicates, these models proposed that Europa had differentiated into a silicate sphere surrounded by a shell of water ice perhaps 100 km deep. Based on their estimate of the rate of heat conduction out of the interior, Fanale et al. (1977) proposed that Europa's water ice shell would be liquid below a depth $\sim$ 40 km. Subsequently, Cassen et al. (1979, 1980) proposed that solid-state convection in a frozen ice shell could bring heat from the silicate body to the surface rapidly enough to keep the ice shell frozen. A more recent model (Ransford et al. 1980) suggested that most of Europa's water may be tied up in its interior as hydrated silicates such as serpentine, chlorite, and brucite, and only a relatively small amount may have outgassed to produce a thin ice shell of perhaps only a few km thick. As we shall discuss, geologic evidence on Europa's surface allows us to distinguish between these alternatives.

Visual telescopic observations of Europa show it as bland on a global scale. The equatorial belt bounded by about $\pm$ 45° lat was known to be darker than the polar regions (Murray 1975); spectrophotometric measurements of the full disk showed the leading hemisphere somewhat brighter than the trailing hemisphere (Johnson and Pilcher 1977; Morrison and Morrison 1977; Nelson and Hapke 1978). These Earth-based observations of subtle albedo variations in the equatorial region and the presence of brighter polar regions were confirmed in March 1979 by Voyager 1's global images, which range in resolution from 30 to 100 km/line pair (lp). The best Voyager 1 images show dark stripes in the form of great and small circles, crisscrossing the entire globe (Smith et al. 1979*a*). They are 10 to 20 km wide, and stand out against the bright regions but are difficult to see in the darker terrains. Voyager 1 images also showed Europa to be divided into patches of darker and brighter terrain that differ in albedo by ∼ 20%. Four months later Voyager 2 passed within 204,000 km of Europa, providing multispectral coverage at resolutions as high as 4.5 km/lp for ∼ 20% of the body. The combined best coverage of Europa from the two Voyagers (Fig. 14.1) encounters was used to generate the global map shown in Fig. 14.2.

During the Voyager 2 encounter with Europa, the subspacecraft point was near the terminator (∼130° long), enabling acquisition of detailed topographic information in that area. Because of the rapid rotation of Europa, the terminator moved about 15° to the west during the ∼ 6 hr of high-resolution imaging. This shift allowed examination of features as they moved through the terminator region and permitted distinction of topographic forms from albedo variations. Observations near the terminator indicated that relief is extremely low on Europa's surface, varying by no more than a few hundred meters.

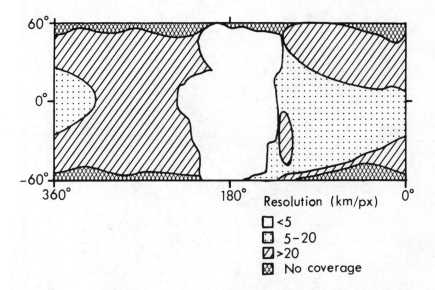

Fig. 14.1. Resolution (km/pixel) of Voyager 1 and 2 images of Europa: □ < 5; ⊞ 5-20; ▨ > 20; ▩ no coverage.

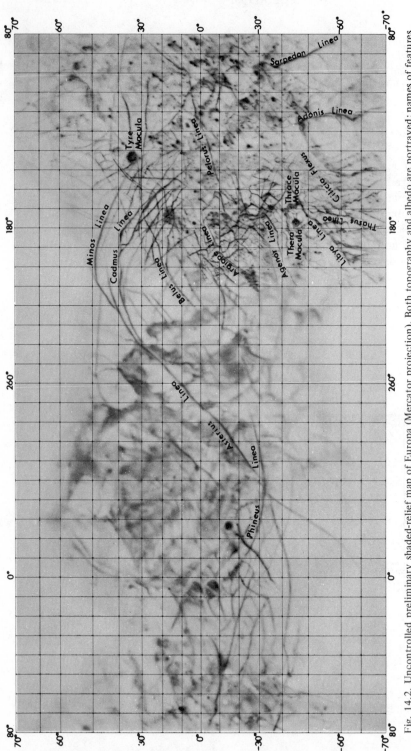

Fig. 14.2. Uncontrolled preliminary shaded-relief map of Europa (Mercator projection). Both topography and albedo are portrayed; names of features discussed in the text are given. Ten degrees at the equator represent ~270 km on the ground. Voyager 2 high-resolution coverage between ~130° and ~210° long; elsewhere resolution of images covers a wide range. The airbrush base was prepared by J.L. Inge, U.S. Geological Survey at Flagstaff, Arizona.

I. MAP UNITS AND LINEATIONS

Enhanced high-resolution multispectral images acquired by the Voyager 2 Imaging System (Plate 4 in the color section) were used to prepare a geologic map covering the area between 130° and 210° long (Plate 5). A detailed definition of mapped units and linear features is given in Appendices A and B; here we discuss geologic interpretations of the data. The geologic map of Europa is unusual in that mapped units appear to result mostly from structural breakup or replacement of the preexisting crust; generally the surfaces of terrestrial planets are dominated by superposition of stratigraphic units. The surface of Europa is divided into two major classes of units, mottled terrains and plains (Smith et al. 1979b). These units exhibit numerous spots, irregular patches, and complex lineations, all of highly variable albedo and color, although most are darker than the major units. In addition it is transected by numerous lineations marked by albedo contrasts. Long, narrow ridges, both curved and straight, can be seen near the terminator, and a few rimmed circular depressions are interpreted as impact craters. The geologic units have been subdivided on the basis of color and albedo; a brief summary of their colorimetric properties is included.

Mottled terrain materials

These materials have been subdivided into brown and gray units (Appendix A); their colorimetric properties are compared in Fig. 14.3. The brown mottled terrain occurs as a light-brown region with small variation in overall albedo but numerous darker small brown spots and a uniformly hummocky topography with a roughness wavelength of a few km (Figs. 14.4 and 14.5). Lineations tend to break up and disappear where they enter this unit (Plate 4 and Figs. 14.5 and 14.6). The narrow ridges visible near the terminator are also scarce in this terrain. The gray mottled terrain is coarser in texture; resolved components range in size from a few km to several tens of km. This gray unit is made up of isolated dark patches with uniform albedo against a background of gray material and is transected by numerous intact lineations (Plate 4).

Whereas the brown mottled terrain is roughly textured and hummocky, the gray mottled terrain appears to consist of plains materials with a high density of isolated dark patches. Unfortunately, unlike the brown terrain unit, the gray unit does not occur at the terminator in the images available, so its detailed topography is unknown. The similarity of the spectral signatures of the gray mottled terrain and the plains, as discussed below, suggests that they are closely related.

As Fig. 14.3 shows, the normalized spectral reflectance of the gray mottled terrain in the mapped area closely resembles the normalized spectra of the dark and bright plains and bright craters. All show a uniformly rising

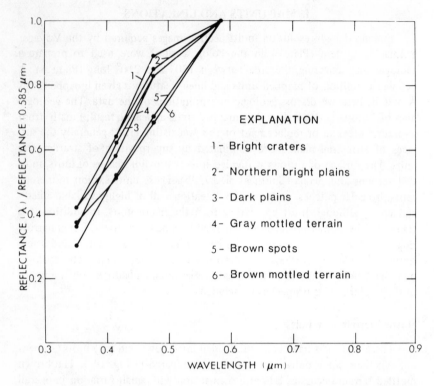

Fig. 14.3. Normalized reflection spectra of color units on Europa. Voyager 2 multi-spectral images (see Plate 4 in the color section) were used to extract radiometric information in four spectral bandpasses with effective wavelengths (black dots) of 0.351 μm, 0.413 μm, 0.479 μm, and 0.585 μm, with spectral widths of ～60 μm. The radiometric values were normalized to 0.585 μm to remove photometric effects. Each spectrum represents three to nine measurements.

Fig. 14.4. Index picture of that part of Europa covered by high-resolution images. Area shown is identical to that covered in Plate 4 in the color section and extends from ～130° to ～210° long and from +60° lat across the south pole to −80° lat on the opposite hemisphere. This picture, like Plate 4, is composed of the violet, blue, and orange spectral bands of Voyager 2 images. The various irregular surface markings visible in this image and on Plate 4 are subdivided into terrain units, as shown on the geologic map in Plate 5 and described below in Appendix A. Also shown here and on Plate 4 are linear markings sketched in Figs. 14.7, 14.8, 14.9, and 14.11, and described in Appendix B. Delineated areas give the approximate location of Figs. 14.5, 14.6, and 14.10. Also indicated are a possible crater palimpsest (A) (Tyre Macula, Fig. 14.2) and small fresh craters (Fig. 14.12); (B), bowl-shaped crater (Fig. 14.12a); (C), crater with central peaks and bright ejecta blankets (Fig. 14.12b); (D), multiring crater (Fig. 14.12c); and (E), dark-halo crater (Fig. 14.12d).

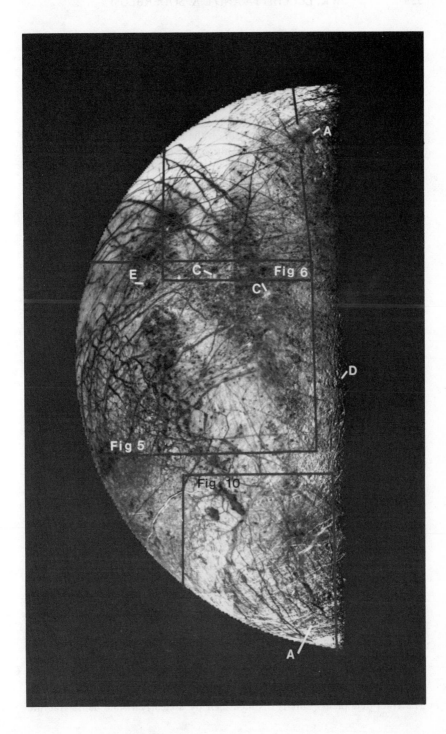

reflection curve from the ultraviolet through the midvisible and a roll-over between ~ 0.5 and 0.6 μm. In contrast, the brown mottled terrain and dark-brown patches display steeply rising curves throughout the Voyager imaging spectral range. This configuration explains the redder appearance of the brown terrain unit relative to the gray. According to Johnson et al. (1981), who assembled a global multispectral map from Voyager far-encounter images, the mottled terrain on the leading hemisphere (centered at (90° long) is relatively bright in the ultraviolet (normal albedo ~ 35%) compared with the mottled terrain on the trailing hemisphere, which is relatively dark in this wavelength. Both the brown and gray mottled terrain units apparently belong to the mottled terrain classified as UV-bright by Johnson et al. (1981) and represent a subtle subdivision of that class.

The difference between the leading and trailing hemispheres might result from processes related to the impact of ions trapped on Jupiter's magnetosphere (Lanzerotti et al. 1978). Sputtering would be more active on the trailing hemisphere because plasma trapped in the Jovian magnetosphere rotates faster than Europa revolves in its orbit; the ions impact the trailing hemisphere with a velocity of ~ 100 km s^{-1}. Sulfur, injected into the magnetosphere by Io, is a possible contaminant for Europa's trailing hemisphere (Eviatar and Siscoe 1981; Johnson et al. 1981).

Plains units

The plains units are classified by albedo and the nature of individual superposed lineations. Mapped units are bright plains, dark plains, fractured plains, and undivided plains (Plates 4 and 5); Table 14.2). Figure 14.3 shows that the normalized spectral reflectances of the plains units are similar. As was known from Earth-based observations (Murray 1975), the north and south polar plains are brighter than the equatorial. The difference may be due to darkening in the equatorial region by ions (trapped in the magnetosphere) that have bombarded and contaminated the surface. These ions may have selectively removed ice by sputtering and left a lag deposit of silicates (Lanzerotti et al. 1978; Haff and Watson 1979). Alternatively, a lag in the equatorial area may have been caused by evaporation of ice (Squyres 1980), and brightening in the polar area may have been due to poleward migration and deposition of ice, as proposed for Ganymede by Purves and Pilcher (1980). The orange color of stripes in the north polar area (Plate 4) may also be caused by the brightening by frost deposits of stripes that appear dull brown elsewhere. In such a case, the frost would have to be extremely thin and uniform. The dark plains unit, north of the equatorial region, could have acquired its relatively low albedo from the darkening processes mentioned above or from dense fractures near and below the limit of resolution. The fractured plains unit is distinguished by a clearly visible dense fracture

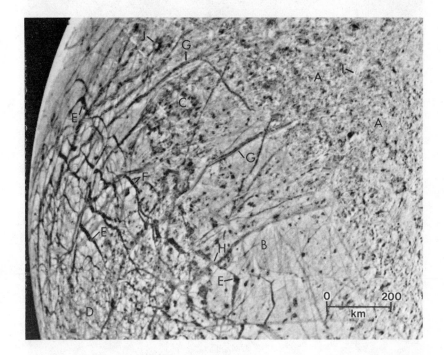

Fig. 14.5. Mottled terrain and plains units. Brown mottled terrain unit (A) has sharp
boundary with plains unit (B). Gray mottled terrain unit at (C) and fractured plains
unit, characterized by short curvilinear fractures and pentagons and hexagons at (D).
Dark, sharply delimited, locally wedge-shaped bands (E) are crossed by thin bright
stripes (F). Triple bands (G) disappear near boundary with brown mottled terrain. In
places, triple bands are diverted where they cross dark bands (H). A bright-ray crater
is at (I), a dark-halo crater at (J). Lineations in center and top part of image are
straight; those at left and bottom are curved. See Fig. 14.4 for location of image
(1255J2-001F, clear filter).

system that has produced pentagons, hexagons (Plates 4 and 5; Fig. 14.5),
and short curvilinear fractures that terminate where they butt against each
other (Pieri 1980).

Linear and curved markings

The two major types of terrain units, mottled terrain and plains, that
make up the surface of Europa are transected by varied linear and curved
markings. Detailed descriptions are given in Appendix B. The lineations vary
in size, color, and pattern. Some are global, as was first recognized on Voyager 1 images (Smith et al. 1979a); Voyager 2 showed others restricted to
local areas; all appear controlled by fractures (Pieri 1980; Finnerty et al.
1980).

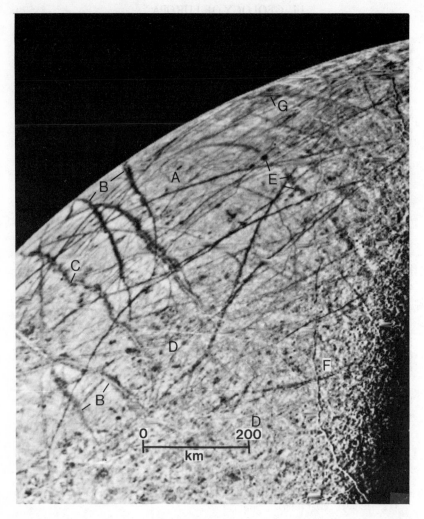

Fig. 14.6. Northern plains unit. Plains unit (A) crossed by numerous straight lineations and triple bands (B) consisting of central light stripes flanked by dark bands or spots. Some meander slightly (C). Triple bands disappear near boundary with mottled terrain unit (D) or continue as aligned spots (E). A ridge is at (F); the orange band Minos Linea (Fig. 14.2) is barely visible at (G). See Fig. 14.4 for location of image (1364J2-001F, clear filter).

The lineations are straight to curved streaks, stripes, or bands, and range from thin sharp lines to broad faint swaths. Some are uniform in albedo; others are composed of dark bands with bright central stripes, here called triple bands. Pieri (1980) analyzed the lineations and concluded that they can be subdivided into sets possibly reflecting separate origins; we also recognize lineation sets perhaps of different origins. Our classification is generally similar to that of Pieri (1980) but is expanded to include some colorimetric properties. We also attempt to determine the relative ages of sets based on their transection relations, because we feel that establishing the chronology of the lineation sets is essential to understanding the stress fields involved in their formation.

Dark wedge-shaped bands. Straight and curved, locally wedge-shaped bands of this set (Pieri 1980) are the most conspicuous markings in the region photographed at high resolution by Voyager 2 (Plate 4; Figs. 14.5 and 14.7). The normalized reflection spectrum of these bands is similar to that of the brown mottled terrain. These coarse dark bands occur in a belt trending southeast across the south equatorial part of the mapped area. The sharply delimited edges of the dark bands cut across most other stripes and bands, suggesting that the dark bands are relatively young. The wedge shape of several of the bands and the apparent displacement of crossing lineations suggest that rotation and expansion of the crust were involved in forming the bands. This possibility was explored by Schenk and Seyfert (1980), who concluded that counterclockwise rotation of a few degrees by crustal plates lying south of the dark bands could account for opening of fractures and infilling by darker, contaminated water or ice from below. They place the pole of rotation at $\sim -38°$ lat, $174°$ long in an area north of the irregular dark patch Thrace Macula (Fig. 14.2). They further suggest that some form of incipient plate tectonics may be responsible for this crustal deformation. We concur with their conclusion that slight rotation of crustal plates, opening gaps in the crust, may have taken place. But the observation that most of the central light stripes of the triple bands transect the dark, wedge-shaped bands (Fig. 14.5) suggests that the light central stripes formed later than the outer dark zones of the triple bands. In a few places the central stripes appear diverted and have a jog where they cross the wedge-shaped bands; evidently the central stripes follow older lines of weakness.

The dark wedge-shaped bands are generally curvilinear in the western part of the belt and straight in the northeastern part (Fig. 14.5). The curvilinear bands are in the area characterized by pentagons and hexagons, which Pieri (1980) interpreted as caused by an isotropic stress field. The straight bands are in the region characterized by tetrahedrons, which Pieri suggested might be caused by oriented stress fields. The observation that young, dark, wedge-shaped bands are both curved and straight suggests that they are fractures that opened along old lines of weakness; the observation that they have uniform albedo and continuous color suggests that they resulted from a single episode. Probably they formed from a late single stress field but fractured along old lines of weakness established in response to earlier variable stress fields (Lucchitta et al. 1978).

Pieri (1980) interpreted the densely fractured plains area and adjacent terrains of curved bands, pentagons and hexagons as due to isotropic stress conditions. Helfenstein and Parmentier (1980) suggested that these stresses

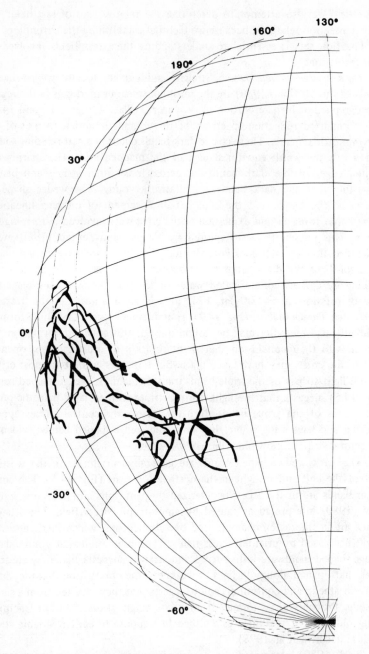

Fig. 14.7. Sketch map of dark, wedge-shaped bands in mapped area of Europa. For description see Appendix B.

and resulting fracture patterns possibly originating from tidal pushing and pulling centered on the anti-Jovian point at $0°$ lat, $180°$ long. Finnerty et al. (1980) pointed out that tidal stresses are < 10 bar, probably insufficient to rupture the surface. They and Ransford et al. (1980) suggested that tensional stresses over a region of upwelling in the silicate interior, having been influenced by Europa's tidal lock with Jupiter, caused the fractures in the anti-Jovian region.

Triple bands. The second most conspicuous lineation type on Voyager 2 images is triple bands, consisting of pairs of dark bands flanking a central, narrow bright stripe (Figs. 14.5 and 14.6). The distribution of these bands is shown in Fig. 14.8. We include in this group the orange stripe, Minos Linea, identified as a triple band on clear-filter images, and the white stripe in the south, Agenor Linea (Fig. 14.2). The bands include the global lineations following great circles observed on Voyager 1 images (Smith et al. 1979*a*; Fig. 14.2).

The triple bands locally merge with single brown streaks, suggesting that many faint lineations may be triple bands below the limit of resolution. Parts of the triple bands cut through other lineations and parts are cut by them, suggesting that individual triple bands formed over extended time periods. Where they intersect the dark wedge-shaped bands, as discussed above, the dark outer bands are locally offset (Fig. 14.5 at H) but the central light stripes clearly cut across and thus are younger than the wedge-shaped bands in several places (Fig. 14.5 at F); the time of formation of triple bands evidently spans at least the period of opening and filling of the wedge-shaped fractures. That the stress fields forming these global triple lineations existed for an extended period of time is further suggested by an observation, elaborated below, that several of these bands appear to emanate from dark circular spots seen on global images and interpreted as craters. If these stress fields were long lived, sporadic impacts could have released them locally and thus could have controlled the location of fractures in some areas.

For the most part, the triple bands disappear where they enter the brown mottled terrain; a few can be traced by sparse aligned spots (Fig. 14.6). Their disappearance suggests that either (1) they are older than the brown mottled terrain, and disintegrated during its formation, or (2) the physical properties of the brown mottled terrain are sufficiently different from the plains to prevent the extension of fractures through that region.

The central stripes of the triple bands appear, in many places, to be ridges (Malin 1980). One central stripe can be traced into the terminator area, where it clearly continues as a ridge. This observation suggests that some of the ridges described below, which are seen only in the terminator area, are the central stripes of triple bands whose albedo contrast cannot be seen near the terminator, and conversely that ridges are probably present in areas of high sun angles where their relief cannot be recognized.

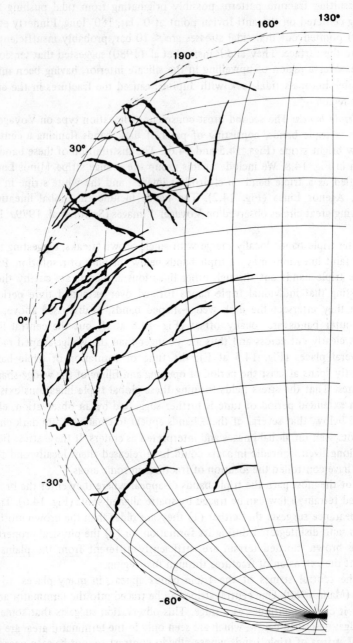

Fig. 14.8. Sketch map of triple bands in mapped area of Europa. Note northwest trends north of equator and southwest trends south of equator. For description see Appendix B.

The origin of the triple bands has been discussed by Finnerty et al. (1980), who proposed that they are due to global expansion, perhaps oriented along tidally induced lines of weakness. The resulting cracks were filled by aqueous fluids, contaminated by silicate xenoliths, that rose from the hydrous silicate mantle and formed breccia dikes of dark material. Clean water was later forced out of these cracks because of the increased volume of the liquid upon freezing, thus forming the bright central dike-like ridges. This mechanism is attractive because it accounts for most of the features observed in the triple bands. Another possibility is that dark flanking stripes of triple bands may not be composed of replacement materials but simply be highly cracked areas along the global fractures, which reflect light less efficiently. The central stripes then would represent dikes of unfractured ice that filled central cracks. This latter hypothesis is less likely, as the dark-brown stripes of triple bands have a different spectral signature than do the surrounding plains, and therefore appear composed of a different material.

The triple bands trend northwest north of the equator and southwest south of the equator. The split along the equator and the global extent of the fractures suggest that global stresses were responsible for their formation. The problem was investigated by Helfenstein and Parmentier (1980), who concluded that the intersection of calculated stress trajectories and observed fracture patterns precludes tidal despinning as a mechanism for their formation; they proposed instead that the global lineations may have formed as conjugate shear fractures due to stresses from cyclical tidal deformation resulting from orbital eccentricity. Ransford et al. (1980) and Finnerty et al. (1980) suggested that the fractures were caused primarily by global expansion, but they concurred that the trends may be controlled by preexisting tidal stress patterns. Our observations that these fractures follow stress patterns that apparently existed for long periods of time also support some form of tidal control.

Gray bands. This lineation set consists of gray, curvilinear bands in the southern part of the mapped area (Plates 4 and 5; Figs. 14.9 and 14.10). The gray bands are relatively old; they are transected by other lineations and ridges. Their origin appears local rather than global, and could be related to mechanisms producing circular structures on planetary bodies, such as basin-scale impacts, volcano tectonic structures, or structures centered on former poles of tidal deformation. Even though lack of detailed information precludes a firm interpretation of their origin, the gray bands do appear to represent a well-defined structural set.

Materials of dark spots and patches

Brown to dark gray materials occur in small spots and large patches (Plates 4 and 5) as well as in the various bands described above. Materials of this color and albedo also occur as a halo surrounding a small (15-km diame-

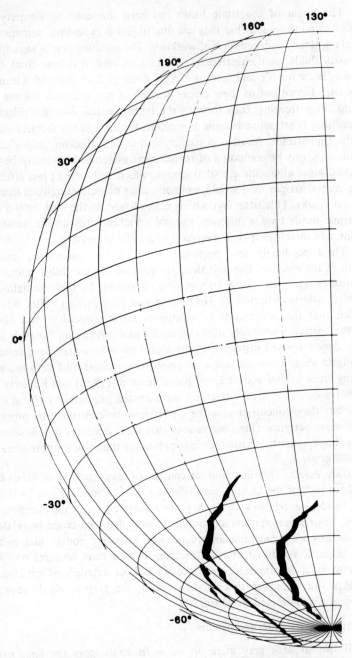

Fig. 14.9. Sketch map of gray bands in mapped area of Europa. Bands appear roughly concentric to a point near $-65°$ lat, $110°$ long. For description see Appendix B.

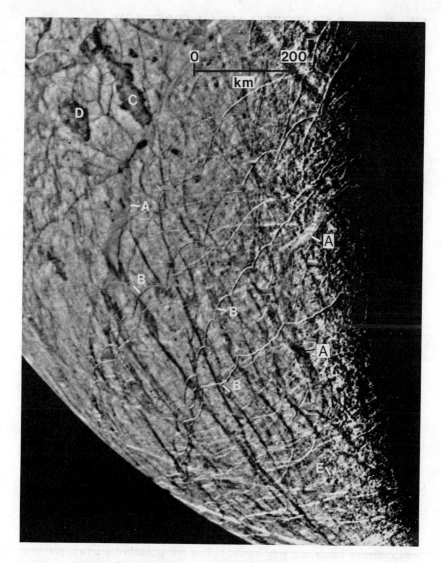

Fig. 14.10. Southern plains unit. Wide, curvilinear gray bands (A) are also seen on Plate 4 in the color section. Conspicuous cycloidal ridges (B) are superposed on gray bands. Brown patch Thrace Macula is at (C), and Thera Macula at (D) (see also Fig. 14.2). A bowl-shaped crater is at (E). See Fig. 14.4 for location of image (1364J2-001F, clear filter).

ter) crater. The spectral diagrams (Fig. 14.3) clearly show that these materials, like the brown mottled terrain, are relatively low in albedo toward the shorter wavelengths and rise approximately linearly to the orange wavelengths. Orange stripes in the north (Minos Linea in Fig. 14.2 and Plate 4) and south (flanking stripes on Agenor Linea in Fig. 14.2 and Plate 4) may well be composed of similar material brightened by some unknown process, for instance deposition of a thin frost. This hypothesis is supported by the similarity of form and orientation of stripes at high latitude to those in the equatorial regions.

Most brown material is probably of internal origin, as it is associated with numerous structural lineations. Brown material is also found as aligned spots in the brown mottled terrain along the extension of structural bands (Fig. 14.6). Concentrations of brown materials are especially prevalent where structures such as triple bands intersect, and brown material appears to have been excavated from the subsurface and deposited on the rim and in rays of a small crater (Plates 4 and 5 and Fig. 14.4).

We agree with Finnerty et al. (1980) that the brown material flanking central light stripes in the triple bands is probably composed of dike-like intrusions of water contaminated with silicate materials. The brown patches and spots may likewise be of intrusional origin and may represent replacement of the crust by contaminated frozen water. Alternatively, the brown material may be of extrusional origin and represent a surface deposit of effusive water; this is less likely because of the difficulty of extruding heavy muddy water onto the lighter ice crust.

Ridges

Although numerous ridges can be identified clearly only along the terminator and south polar region where illumination is low and relief is enhanced, such ridges are probably ubiquitous on Europa's surface (Plate 4; Figs. 14.10 and 14.11). The ridges are the youngest of the tectonic systems in the mapped area, for they cut all other lineations and stripes; the youth of at least some ridges is further supported by the apparent control of their radial pattern by a fresh-looking, relatively young impact crater at the terminator (Plate 4; Figs. 14.4, 14.11, 14.12c). The ridges have straight segments in the northern and central parts of the map area and cycloid patterns in the south (Malin 1980). These regions of straight and curvilinear patterns are similar to those of other lineations on Europa, and suggest that the ridges, like dark wedge-shaped bands and gray bands, follow preestablished fracture patterns.

Like the triple bands, the ridges in the northern part of the map area trend northwest and in the south, southwest. In addition to indicating that some ridges may be central stripes of triple bands, these trends suggest that the two sets are closely related. However the plot of ridges on a Mercator-projection map by Pieri (1980) shows that the ridges form small circles centered near the anti-Jovian point, whereas the triple bands form global lineations on a great circle. The difference in pattern suggests that many ridges are not related to triple bands but form an entirely separate set of structures on Europa's surface. However, the circularity of the ridges may be an artifact of the illumination, as the subsolar point is also roughly centered on the anti-Jovian point and shadows radial to this point tend to be enhanced.

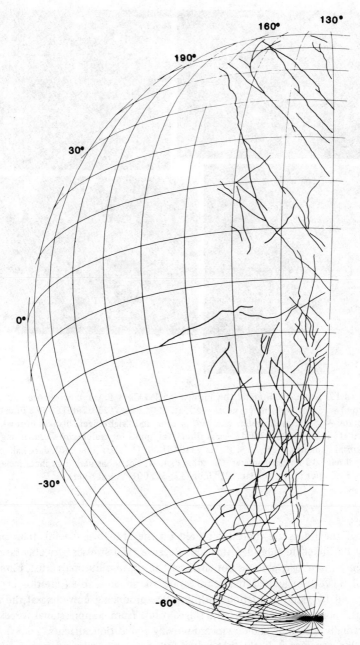

Fig. 14.11. Sketch map of ridges in mapped area of Europa. Note radial pattern sur-rounding crater near −25° lat, 135° long. Also note northwest trends north of equator and southwest trends in south. For description see Appendix B.

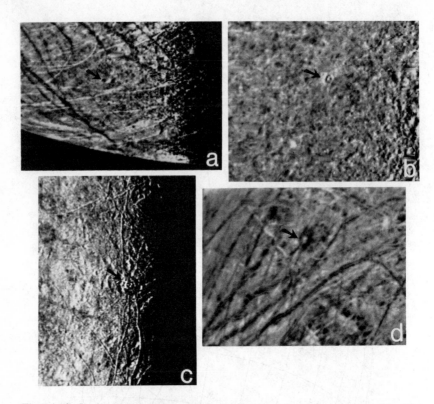

Fig. 14.12. Small, fresh craters on Europa labeled Class 1: (a), bowl-shaped crater (arrow) ~15 km in diameter, near −85° lat, 200° long (1234J2-001F, blue filter); (b), crater ~15 km in diameter with central peak and bright ejecta blanket (arrow), −6° lat, 152° long (1368J2-001F, clear filter); (c), multiring crater (arrow), inner ring ~30 km in diameter, −25° lat, 135° long (1368J2-001F, clear filter); (d), dark-halo crater (arrow) ~15 km in diameter. On color Plate 4, ejecta has dark gray inner facies and brown outer facies, +0° lat, 181° long (1255J2-001F, clear filter).

If the ridges are indeed centered on the anti-Jovian point, their origin may be linked to tidal deformation. Because tidal stresses probably amount to only a few bar, perhaps not sufficient to rupture Europa's crust, Finnerty et al. (1980) suggested that convection of an anhydrous interior became oriented by tidal interaction with Jupiter, producing upwelling at the anti-Jovian point. In their model the ridges result from compressional forces due to radially spreading and downward-moving convection currents.

We consider it likely that ridges formed by a similar mechanism as the central bright stripes of triple bands, that is, by intrusion of clean ice into fractures. This hypothesis is supported by the observation that some central stripes merge with ridges and that some ridges and triple bands, which contain central stripes, have similar trends. If ridges are indeed intrusions into fractures, they, contrary to the Finnerty et al. (1980) model, would probably be related to tensional stresses.

Craters

The impact features that we tentatively identify on Europa fall into two general classes:

1. Craters a few to a few tens of km in diameter that display rims, central peaks, and ejecta blankets (termed class 1),
2. Large, flat, brown, circular spots 100 km or larger in diameter (termed class 2).

We are fairly confident in the identification of five craters of the first class (Plate 5). They include sharp-rimmed, bowl-shaped craters (Figs 14.4 and 14.12a); craters with bright rims, diffuse bright ejecta blankets, and bright central peaks (Fig. 14.12b); a multiringed crater with several concentric rims and troughs (Fig.14.12c); and a bright crater with a dark halo (Fig. 14.12d). Three small bright spots with ray-like patterns are also interpreted as craters whose rims are near the limit of resolution. The multiringed crater is near the terminator and appears to control the radial pattern of ridges in its vicinity (Plate 4; Figs. 14.11 and 14.12c). Control of ridge pattern by impact was suggested by Fagin et al. (1978) for certain ridges in the lunar maria that apparently formed on radial fractures. Evidently the ridges near the multiringed crater on Europa are as young as the crater and are similarly controlled by it.

The best example of the second class of features appears as a flat, dark, circular patch > 100 km in diameter in the northern part of the mapped area (Plates 4 and 5; Fig. 14.4). This feature, Tyre Macula, is probably some form of crater or basin scar similar to those described as crater palimpsests on Ganymede by Smith et al. (1979a) and Shoemaker et al. (Chapter 13). Tyre Macula consists of vague multiple concentric rings with no detectable surface relief; it differs from most crater palimpsests on Ganymede which have one rim circle with a surrounding flat bright area, and resembles only a few very old palimpsests in Galileo Regio.

In global images (Figs. 14.2 and 14.13), Tyre Macula appears as a dark circular spot approximately centered on some of the global-scale dark bands. At high resolution, these bands become tangent to Tyre Macula. Similarly, other dark, circular patches with nearly radial lineation patterns can be seen in the global map and images (Plate 4; Figs 14.2 and 14.13), and may well be

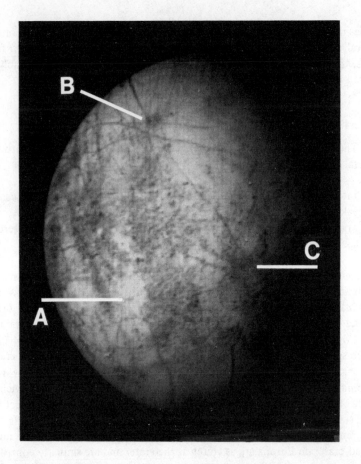

Fig. 14.13. Best global image of Europa's leading hemisphere. Area covered by high-resolution images is left third of illuminated region. Multiringed crater that diverts ridges is the bright spot at (A) (also shown in Fig. 14.12c). This crater is located on the terminator of high-resolution images. The brown spot Tyre Macula (Fig. 14.2) with tangential lineations is at (B), and represents craters of Class 2. Another brown spot is apparently at the center of radial lineations at (C). (Average of violet, green, and clear filter images 1549, 1559, 1557J2-001.)

crater related; many dark global stripes and bands appear to originate at these suspected crater scars. Impact craters occurring sporadically across Europa's surface may have generated weak points in the crust. Centered on these weak points, structural patterns such as ridges or triple bands formed as stresses were released by impact.

As already mentioned, we are reasonably certain that the smaller crater features, 10- to 30-km in diameter, are of impact origin. We are less certain that

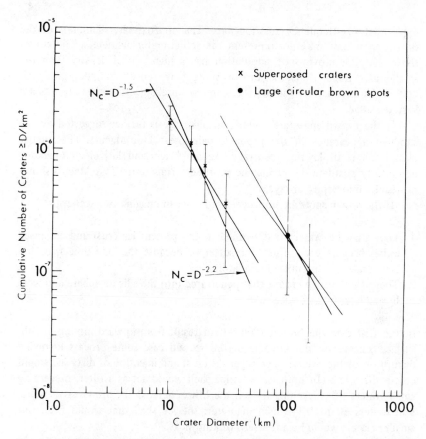

Fig. 14.14. Size-frequency distribution of probable and possible impact features on Europa. Cumulative numbers of craters and brown spots per unit area larger than a given diameter are plotted. The superposed craters are Class 1 craters confidently identified on Voyager 2 high-resolution images. The plot of the brown spots (Class 2 craters) includes the large crater palimpsest Tyre Macula (+30° lat, 145° long) in the Voyager 2 high-resolution area and a second, probably similar feature at −15° lat, 335° long. The effective area of global coverage for the brown spots is assumed to be 10^7 km². Also plotted are two power functions representing crater production functions. The first ($Nc \sim D^{-1.5}$, Nc, cumulative number of craters and D, diameter) is representative of typical production functions for the Moon and Mars. The second ($Nc \sim D^{-2.2}$) is a model production function for the Galilean satellites from Shoemaker and Wolfe (see Chapter 10).

the large circular brown patches, in which we include the probable crater palimpsest Tyre Macula, were formed by impact. Regardless of their origin, the brown patches appear to be separate from class 1 craters, as shown in Fig. 14.4. In this figure, diagonals drawn through the data are power functions representing model size-frequency distributions. The approximate diameters of the brown spots used may not represent the actual crater diameters. The curves show that the brown-spot population has a higher areal density than the small-crater population; the brown spots may reflect an older group of craters. Unfortunately, the statistics are so unreliable that these results are very tentative.

If the brown spots are craters, an older age is further suggested by the subdued appearance of the probable palimpsest Tyre Macula. The smaller class 1 craters display rims, central peaks, and bright and dark ejecta blankets, and clearly postdate the plains and mottled terrain units. Tyre Macula shows no visible topography or rays.

If the brown spots are impact related, we can imagine two scenarios:

1. They may be craters that formed in the present ice crust and collapsed either because early crust was softer or because the long time span and large crater size permitted relaxation.
2. They may be large craters that penetrated into the silicate sphere and were buried later by a new ice shell.

In the first case the brown stain would result from upward mixing of sub-surface silicate by the impact; in the second case some process involving disruption of the overlying younger ice crust and injection of dirty ice would create the strain. In both cases an ice shell would act as a filter, preventing small craters from reaching the silicate lithosphere. If the brown spots represent craters 50 to 100 km in diameter, the ice shell that would filter out smaller craters would be a few tens of km thick.

II. GEOLOGIC PROCESSES

A major characteristic of the surface geology of the terrestrial planets is largely absent on Europa; namely, evidence for the development of the crust by successive deposition of layers on the surface. The stratigraphic succession of units such as volcanic flows, crater ejecta, and eolian deposits that dominates the upper crusts of the terrestrial planets displays common characteristics in the nature of boundaries between these surface units. Such characteristics include burial and truncation of older units and structures by younger, superposed units; topographic forms commonly control the boundaries. Instead, on Europa we see an assemblage that resembles magmatic intrusions or dike-filling in tectonic fractures (Finnerty et al. 1980). Stripes, patches, and ridges transect one another, usually displaying little deviation caused by pre-existing structures or topography. The material units on Europa appear to continue at depth and are not simple surface layers in an onion skin type of geology. In other words, if a few km of Europa's surface were stripped away, the global appearance would remain generally unchanged. The only clear stratigraphic units are a few ejecta blankets (Figs. 14.12*b* and *d*). The brown spots or patches might be local extrusions of material onto the surface; most are too small to make definite judgments possible. However, the close association of brown material with stripes and bands suggests that many patches also represent local replacement of the crust by material from underneath. A few large patches show crude polygonal patterns suggestive of structured control. Our terrestrial analog for Europa's general geology is igneous intrusion from deep-seated plutons.

The other dominant process in our model of Europa's geology is that of tectonic disruption of units already in place. Examples are the fractured plains and the gray mottled terrain units, which may be plains modified by coarse fractures with an especially high concentration of patches. The origin of the brown mottled terrain unit is more controversial; we also suggest formation by tectonic disruption of older plains units resulting in small-scale hummocky topography. The breakup may have been accompanied by addition of brown material, presumably by intrusion from below, developing many small spots and giving the mottled appearance. Other evidence that disruption was involved in the origin of the brown mottled terrain unit is that only faintly aligned remnants of triple bands and other lineations can be traced from the adjoining plains into this unit. An alternative explanation is that the brown mottled terrain unit has different physical properties than the plains units, so that lineations simply did not form there. However, if this were the case an abrupt termination of lineations at the boundary would be expected.

III. GEOLOGIC HISTORY

On the basis of transection relations among material units and structural features, we conclude that Europa's geologic history can be divided into four general stages, some of which are related to the emplacement of major units including plains material, brown material, and ridge material. The stages of emplacement of these materials overlap to some degree.

1. The earliest stage was the development of the bright plains unit; all subsequent units appear to have developed through disruption, displacement, replacement, or in a few cases superposition of this unit.
2. The next major stage was the development of the dark plains, the fractured plains, and possibly the gray mottled terrain units. All appear to have resulted from modification of the plains by the development of gray albedo features. The gray hues could result from intense fracturing, intrusion of gray material, or alteration or aging of other material, for

example old brown material. Such aging could come from sputtering or gardening that alters optical properties over time.

3. The emplacement of brown material in spots, patches, and bands is the next major event and appears to overlap the development of gray units. The brown materials intruded preexisting plains, cracks, and fractures; they are the major component of the triple bands and wedge-shaped dark bands. Some brown material may have erupted locally onto the surface, but the evidence is ambiguous. Brown material was also excavated from the substrate and deposited on the surface during apparent impact events. Near or following the end of emplacement of brown material, the region of the brown mottled terrain was disrupted, producing complex topography in which the vestigial imprints of triple bands can barely be seen. Brown material was possibly added to the mottled terrain units at the time of their disruption. Thus, the brown mottled terrain is a young unit, even though the materials involved are probably a mixture of old plains and younger brown material.

4. The last episode includes emplacement of the ridges which, as suggested by Finnerty et al. (1980), for the central bright stripe of the triple bands, were probably also intruded as bright dike-like structures.

Like the development of units distinguished by color and albedo, the formation of tectonic patterns can be divided into an approximate time sequence. (1) The earliest recognized deformation produced the curvilinear and concentric gray bands (Fig. 14.9). (2) Triple bands (Fig. 14.8) appear to have formed throughout much of Europa's history, overlapping formation of other sets. Some apparently developed on lines of weakness released by sporadic impacts. The formation of these global lineation patterns may reflect a period of expansion of Europa due to dehydration of the core and hydration of the mantle (Ransford et al. 1980); such expansion would have had to continue for most of the history recorded in Europa's present crust. (3) Relatively late crustal expansion combined with rotation of plates to cause the opening of fractures south of the anti-Jovian area forming the dark, wedge-shaped bands (Fig. 14.7). (4) The central stripes of some triple bands, as well as narrow ridges (Fig. 14.11), were emplaced in the last stage.

Throughout this tectonic history we see young tectonic features following zones of weakness of preexisting tectonic sets. For example, the fairly early crust of Europa developed fractures that define pentagons, hexagons, and curved traces. This pattern dominates only the southern half of the map area (Lucchitta et al. 1981) whereas the northern half is dominated by tetrahedrons and straight fracture traces (Pieri 1980). Later the dark wedge-shaped bands adopted variously the old curvilinear or straight fractures. Also, bright central stripes formed in the center of broad dark bands that apparently had developed previously.

IV. CRUSTAL MODELS

One of the central questions of Europa's surface evolution concerns the structure of the upper 100 km of the crust. Two general ideas have been proposed, which differ in the degree to which water is thought to have been released from the interior. In the first, the thick-ice model, tectonic and intrusive activity takes place in a thick brittle ice shell decoupled from the silicate sphere by warmer convecting ice or water (cf. Fanale et al. 1977; Cassen et al. 1980). In the second, the thin-ice model, tectonic activity takes place in a hydrated brittle lithosphere 200 to 300 km thick overlain by an ice crust ~ 25 km thick (Finnerty et al. 1980). In the following discussion we attempt to establish which hypothesis is borne out by the surface features.

In thick-ice models, an ice shell as thick as 100 km overlies a relatively dry silicate sphere (Lewis 1971; Consolmagno 1975; Consolmagno and Lewis 1976,1977; Fanale et al. 1977). Heat is either supplied radiogenically from the silicate interior or generated directly in the ice shell by a forced tidal eccentricity with Ganymede (Cassen et al. 1980; see also Chapter 4), much as has been proposed for Io (Peale et al. 1979). Water may be present at the base of the ice (cf. Fanale et al. 1977), although Cassen et al. (1979, 1980) estimated that solid-state convection could transfer heat to the surface fast enough to freeze any underlying water. Whether warm ice or water is present at depth is probably a moot point; either water or warm ice would provide a zone of weakness, a slip interface like the Earth's low-velocity zone, over which Europa's ice crust could glide and become decoupled from the silicate interior.

From observations of Europa's surface character we can place limits on the thickness of the ice shell. If sections of the shell have slid around and rotated, producing for example the wedge-shaped dark markings, the shell cannot be thinner than a few km. On the other hand, the shell probably cannot be thicker than a few tens of km, based on the observation of abundant brown material at the surface, suggesting transport of subsurface silicate debris, maybe as slurries, to the icy surface. If the ice layer were as thick as 100 km, migration of clear water or denser mud-laden water through a thick shell of much lighter ice to the surface seems difficult. The discontinuity of lineations across the brown mottled terrain also suggests that the ice is only a few tens of km thick; we have suggested that the loss of lineations in the brown mottled terrain resulted from physical disruption of the ice shell. Perhaps the mottled terrain lodged on a relatively high underlying bedrock area. If such regional bedrock highs exist, the relief between them and regional lows cannot be more than a few km (Johnson and McGetchin 1973; Smith et al. 1979b), again suggesting a likely thickness of the ice shell of at most a few tens of km. Finally, if the probable crater palimpsest Tyre Macula and other brown spots seen on a global scale represent impacts that have penetrated the ice shell where smaller craters have not, their sizes suggest an ice shell a few tens of km thick.

In the thin-ice model, Ransford et al. (1980) and Finnerty et al. (1980) envisioned that most of the water ice remains in Europa's mantle tied up in a cold, rigid, hydrated zone in the upper 200 to 300 km, and that a shell of brittle ice ~ 25 km thick is firmly anchored to this silicate lithosphere. In this model, planetary expansion due to formation of hydrated minerals, perhaps modified by internal convective processes in the silicate interior, was thought mostly responsible for the tectonic development of the crust. Most of the evidence we find supports this model. For instance, evidence presented above that the ice shell cannot be thicker than a few tens of km is consistent with their thickness estimate and the idea that perhaps 80% of the water has remained in hydrated minerals in the silicate interior. Although evidence indicates to us that the ice shell was in some areas and at some time decoupled from the silicate subsurface by warm ice (Cassen et al. 1980) or even water (Fanale et al. 1977), permitting relaxation of craters and local slippage and rotation of ice rafts, we agree with Finnerty et al. (1980) that the global lineations probably represent structures developed mainly in the upper lithosphere. If they were fractures developed only in the ice, the degree of expansion of the ice shell suggested by the widths of the global bands appears unreasonable. More likely the bands are the surface reflection of grabens filled with dark material or fractures within the silicate lithosphere filled by breccia dikes, as suggested by Finnerty et al. (1980), implying far less expansion than does formation of opened fractures entirely within the ice shell.

The most controversial question in our discussion of Europa's crustal history is probably the time scale for the development of the geologic features observed. The consensus is that the present surface cannot date back to the period of heavy bombardment $\gtrsim 4$ Gyr ago, as most of the surfaces of Ganymede and Callisto apparently do. Had the features we see on the icy surface today existed during the bombardment, the subtle network of lineations could not have survived. The underlying silicate crust also probably formed after cessation of the intense bombardment. If the ice shell is only a few tens of km thick, a dense population of large impact craters and basins on the silicate surface would have affected the patterns of the subsequent tectonic sets. Fractures would have followed or been deviated around numerous old impact structures. Evidently development of both the ice and silicate crustal layers continued well past the termination of the intense bombardment ~ 4 Gyr ago.

The difficult question is whether the surface we now see is a few Gyr old or substantially less. Shoemaker and Wolfe (Chapter 12) have estimated that the flux of impacting bodies producing craters $>$ 10 km diameter probably has been several times higher for Europa than for the Moon over recent time. The frequency of class 1 craters (Fig. 14.14) superposed on Europa's icy crust is several times lower than the frequency of similar sized craters formed on typical lunar maria. From these two considerations Shoemaker and Wolfe conclude that the superposed class 1 craters seen on Europa accumulated during the last 30 to 200 $\times$ 10^6 yr. The implication is that either (1) the surface we observe is only $\sim 10^8$ yr old, or (2) the surface is much older (10^9 yr or more) but the craters are continually disappearing, vanishing by viscous collapse over a time span of $\sim 10^8$ yr.

In the first case, in which the surface was actually recreated $\sim 10^8$ yr ago, the ice shell could now be brittle and frozen; rotation and slippage of parts of the shell could have occurred over a short period while the shell was being reformed, and the class 1 craters observed could represent the entire population accumulated since the present surface was created. Such a circumstance is certainly conceivable. Yoder (1979) has suggested that the tidal lock among the Galilean satellites may be as young as a few hundred million yr; recreation of the surface could be tied to that event. The second case, plastic deformation of craters continuing to the present, is possible if the thermal gradient in the ice is high (a few deg/km as suggested by Fanale, 1977) and Shoemaker and Wolfe (Chapter 12). Crystalline solids begin to creep at temperatures substantially below the melting temperature, and as the average temperature at Europa's surface is ~ 93 K (Gaffney and Matson 1980), ice warmer than half of the melting temperature could exist at a depth of only ~ 10 km; ice would creep below this depth.

Distinction between these alternatives is extremely difficult with the resolution of the Voyager images. If the shell is brittle and the class 1 craters represent the net accumulation, we would expect a normal population of smaller craters, following typical production functions seen elsewhere. If the shell contains a soft layer and craters relax viscously, we would expect to see a disproportionally large population of small craters about 1 km in diameter. Further, if class 1 craters are continually disappearing through such viscous relaxation, we might expect blemishes and scars of collapsed craters disrupting the complex network of lineations; the Voyager resolution is inadequate to make this test.

V. SUMMARY

From analysis of the Voyager 2 near-encounter images of Europa, we arrive at the following conclusions:

1. Unlike the surfaces of the terrestrial planets and other Galilean satellites, Europa lacks evidence of a horizontally stratified crust. Rather, Europa's geology appears characterized by disruption of the crust and intrusions into an icy shell, possibly by breccia of ice and rock or by icy slurries.

2. The surface of Europa consists of two general types of terrain units:

plains and mottled terrain. The plains units are oldest; all other terrain units appear to have developed through their disruption and replacement. Of the mottled terrain units, the gray unit is older than the brown; stripes and bands cross the gray unit but are disrupted in the brown. Dark-gray and brown spots and patches, and dark material in stripes and bands, were emplaced throughout much of the history of Europa's present crust. Emplacement of gray materials may have preceded that of brown materials, or brown materials may have become gray with age. Light stripes running along the center of many dark bands may be late intrusions of clean ice.

3. Numerous straight and curved lineations, streaks, stripes, and bands cross Europa's surface on a global and local scale. Most lineations appear related to fractures in Europa's crust. The lineation patterns are subdivided into sets with apparently different structural origins. Straight fractures tend to dominate the northern and east-central regions of the area covered by Voyager 2 images, curvilinear fractures the west-central and southern regions. This difference in pattern appears fairly old, and associated fractures appear rejuvenated by later structural activity. Gray curvilinear bands are centered on a point near 110° long, −65° lat, and are probably also old. Sharply delimited dark bands look young and appear to be fractures opened by translational and rotational movement of slabs of crust. Ridges may form small circles centered on the anti-Jovian point, although lighting conditions may artificially enhance this pattern. The ridges are generally the youngest of all lineations, because they cross most others. Global bands consisting mostly of dark bands flanking central light stripes form great circles and apparently have developed during most of the existence of Europa's present crust. Europa's tidal deformation seems to influence many structural patterns.

4. Five fresh craters in the 10 to 30-km diameter range are visible; one appears to divert ridges that radiate from it. A large circular brown spot of little relief could be a crater palimpsest similar to palimpsests on Ganymede; global bands appear tangential to this spot. On low-resolution global images, other brown spots also appear to control radiating dark bands, suggesting that impact-induced fracturing has controlled the location of some structural features.

5. The dark spots, stripes, and bands that appear to have replaced sections of Europa's crust suggest that material was transported to the surface from the subjacent silicate lithosphere. Depth to the silicate subcrust is probably no more than a few tens of km because dense ice and mud slurries would have difficulty reaching the surface through a thicker crust of lighter ice. On the other hand, the evidence for rafting and rotation of the crust suggests that the ice shell must have been thick enough to permit local decoupling at some depth of ice slabs near the surface from the silicate subcrust. Collapse of large craters also indicates that viscous relaxation of the crust may have taken place locally or at some time in the past.

6. The apparent low density of craters superposed on Europa's surface suggests that the surface is $\sim 10^8$ yr old. This young age may be due to resurfacing or regeneration of the crust within the last 10^8 yr, or to the collapse of topographic features by cold flow and by annealing of the ice within this time period. The resolution of Voyager 2 images is inadequate to make a choice between these hypotheses.

Acknowledgments. This work was supported by the NASA Voyager Project and by the NASA Planetary Geology Program. We wish to thank D. Pieri and A. Finnerty for careful reviews.

APPENDIX A.
Description of Map Units

Brown mottled terrain occupies large region in north-central part of map area (Plates 4 and 5) and locally extends into terminator area, where it appears hummocky (elements typically 5-20 km in diameter). It is finely mottled by numerous small brown or gray spots, with irregular or elongate depressions in places and a local network of fine, light-gray stripes. Boundary is sharp in most places (Fig.14.5). Most triple bands in plains units disappear near boundary, but some can be traced into mottled terrain where they continue as vague color streaks or aligned brown spots (Fig. 14.6).

Gray mottled terrain occurs mainly in west-central part of map area (Plates 4 and 5) and is characterized by regions of gray material surrounding irregular dark patches 10-50 km wide. Boundary with adjacent plains is ill defined, locally transitional to dark plains and fractured plains. This unit is traversed in places by triple bands whose dark stripes locally merge with dark patches.

Plains occupy central and southeastern parts of map area (Plates 4 and 5) and have smooth, brightly reflecting surfaces transected by variety of stripes and bands. In east-central part, superposed stripes are long and straight, crossing one another at varied angles; in west-central and southern parts, stripes are curved and tend to terminate against one another (Fig. 14.5). This unit is transitional with other plains units.

Bright plains occupy northern and southern parts of map area (Plates 4 and 5) and are distinctly brighter than plains units elsewhere. Superposed stripes range in size from broad bands to faint streaks at limit of resolution. Lineations are generally straight, crossing, and more orange than lineations in equatorial area. Boundary with dark plains is sharp.

Dark plains adjoin northern bright plains unit to southwest and surround northern part of gray mottled terrain unit (Plates 4 and 5). Albedo is slightly lower than other plains units but higher than gray mottled terrain. Superposed stripes are similar to those on bright plains, but grayer.

Fractured plains occur in southwestern part of map area (Plates 4 and 5) and are marked by intricate network of fine gray streaks, generally curved and short, terminating against one another (Fig. 14.5). Individual streaks are conspicuous, except where so densely spaced that they blend at limit of resolution. There are abundant brown spots and patches in places. This unit is transitional to gray mottled terrain unit.

Brown spots and bands are irregular patches and linear bands that contrast in albedo with background plains or mottled terrain units by ~ 20% and range in hue from orange through brown to dark gray (Plates 4 and 5); orange hue is more common in northern and southern plains. They include materials of sharp-edged, wedge-shaped bands trending southeast across west-central part of map area (Fig. 14.5) and of brown stripes and triple bands, whose parallel dark bands and aligned spots flank light-gray center stripes (Figs. 14.5 and 14.6). Spots and patches occur throughout mottled terrains and plains but with higher concentration in mottled terrain units. Examples include two irregular patches 100-200 km wide (Thrace and Thera Maculae, Fig. 14.2). Some spots align to form lineations, others occur along extension of triple bands into brown mottled terrain unit (Fig. 14.6) or form patches at intersections of triple bands in the gray mottled terrain unit. Dark-gray and brown material occurs on rim of a dark-halo crater (Fig. 14.12*d*).

Light bands occur throughout the map area as thin, bright streaks and central bright stripes of triple bands (Plates 4 and 5). They are most easily identified in gray mottled terrain unit. Only two bands wide enough to map are in the southwest, including Agenor Linea (Fig. 14.2).

Gray bands occur in the southeastern part of map area as smooth gray swaths (Plates 4 and 5) transected by ridges (Fig. 14.10). Boundaries are fairly sharp.

Craters include seven craters with ejecta blankets or halos, with sizes ranging from 15 to 30 km in diameter. Central peaks occur in three craters, of which two have bright ejecta blankets (Figs. 14.4 and 14.12*b*) and the third, near the terminator, has multiple rings (Figs. 14.4 and 14.12*c*); one crater is bowl-shaped (Figs. 14.4 and 14.12*a*). Additionally, three ray-like patches with ≤ 10 km diameter were mapped as craters.

Dark-halo crater (15 km diameter) has a bright interior, brown, ray-like ejecta blanket and dark-gray patch on rim (Plates 4 and 5 and Figs. 14.4 and 14.12*d*).

Crater palimpsest is a circular flat structure (110 km diameter) with multiple concentric rings (Plates 4 and 5 and Fig. 14.4).

APPENDIX B.
Description Of Linear And Curved Markings

Dark, wedge-shaped bands (Fig. 14.7) are sharply delimited, straight or curved, dark-brown, subparallel bands trending southeast from ~ 0° lat, 200°

long to ~ −40° lat, 160° long across west-central part of map area (Fig. 14.5). Bands are as much as 25 km wide, segments are connected in places by northerly trending cross links. Brown material of dark-brown bands and dark stripes of triple bands merge at most intersections, but central light-gray stripes of triple bands are superposed (Fig. 14.5).

Triple bands (Fig. 14.8) are pairs of orange, dark-brown, or gray bands (each 2 to 7 km wide) flanking central light-gray, narrow (a few km wide) stripes (Plate 4 and Figs. 14.5 and 14.6); central stripes are locally flanked by dark spots rather than continuous stripes. These bands may fade into single brown streaks; many are straight and subparallel, some meander slightly, others intersect or bifurcate (Fig. 14.6). They occur mostly in the ± 50° lat belt and are conspicuous where crossing plains and gray mottled terrain units, but terminate near the boundary with brown mottled terrain unit or continue as aligned brown spots (Fig. 14.6). They trend northwest north of the equator and southwest south of the equator (Fig. 14.8). They are both superposed on and cut by other linear markings; central light stripes are superposed on dark-brown bands and some central stripes merge with ridges visible near terminator. Triple bands include Agenor Linea in the south and Minos Linea in the north (Fig. 14.2). Agenor Linea has a conspicuous wide bright stripe in center, flanked intermittently by orange bands (Plate 4); albedo contrast on clear images is high. Minos Linea has poorly developed central stripe and conspicuous wide orange bands (Plate 4); albedo contrast on clear images is low (Fig. 14.6).

Gray bands (Fig. 14.9) are several swaths of interconnected curvilinear bands apparently centered at a point east of southern part of terminator near −65° lat, 110° long, (Plate 4 and Fig. 14.10). They form slightly raised plateaus at terminator and are transected by numerous conspicuous ridges.

Ridges (Fig. 14.11) are observed mostly in terminator area, and vary in size from thin ridges composed of aligned hummocks to prominent ridges casting shadows (Fig. 14.10), as much as 200 m high (D. W. G. Arthur, U.S. Geological Survey, 1981, personal communication) and several km wide. Some form doublets, others fork at acute angles. Ridges are mostly straight and trend northwest in the north and are cycloid and trend southwest and west in the far south (Malin 1980) (Plate 4). In the equatorial area they trend dominantly north-south. Hue is orange in the north, gray elsewhere. Away from terminator, ridges locally merge with faint gray streaks. They are conspicuous in plains units, less numerous in brown mottled terrain unit. Ridges are generally superposed on other structures, but diverted by the crater on terminator in the equatorial area to vaguely radial and tangential trends (Figs. 14.4, 14.11, and 14.12c).

June 1981

REFERENCES

Cassen, P., Peale, S. J., and Reynolds, R. T. (1980). Tidal dissipation in Europa: A correction. *Geophys. Res. Letters.* 7, 963-970.

Cassen, P., Reynolds, R. T, and Peale, S. J. (1979). Is there liquid water on Europa? *Geophys. Res. Letters.* 6, 731-734.

Consolmagno, G.J. (1975). *Thermal History Models of Icy Satellites.* Master's Thesis, Massachusetts Institute of Technology, Cambridge.

Consolmagno, G. J.,and Lewis, J. S. (1976). Structural and thermal models of icy Galilean satellites. In *Jupiter* (T. Gehrels, ed.), Univ. Arizona Press, Tucson, pp. 1035-1051.

Consolmagno, G. J., and Lewis, J. S. (1977). Preliminary thermal history models of icy satellites. In *Planetary Satellites* (J. A. Burns, ed.), Univ. Arizona Press, Tucson, pp. 492-500.

Consolmagno, G.J., and Lewis, J.S. (1978). The evolution of icy satellite interiors and surfaces. *Icarus* 34, 280-293.

Eviatar, A., and Siscoe, G. (1981). Effects of Io ejecta on Europa. *Icarus.* In press.

Fagin, S.W., Worall, D.M., and Muehlberger, W.R. (1978). Lunar mare-ridge orientations: Implications for lunar tectonic models. *Proc. Lunar Planet. Sci. Conf.* 9, 3473-3479.

Fanale, F. P., Johnson, T. V., and Matson, D. L. (1977). Io's surface and the histories of the Galilean satellites. In *Planetary Satellites,* (J. A. Burns, ed.), Univ. Arizona Press, Tucson, pp. 379-405.

Finnerty, A. A., Ransford, G. A., Pieri, D. C., and Collerson, K. D. (1980). Is Europa's surface cracking due to thermal evolution? *Nature* 289, 24-27.

Fink, U., Dekkers, N. H., and Larson, H. P. (1973). Infrared spectra of the Galilean satellites of Jupiter. *Astrophys. J.* 179, L154-L155.

Gaffney, E.S., and Matson, D.L. (1980). Water ice polymorphs and their significance on planetary surfaces. *Icarus* 44, 511-519.

Gromova, L., Moroz, V. I., and Cruikshank, D. P. (1970). Spectrum of Ganymede in the region 1-1.7 microns. *Astron. Circ.* 569, 6-8.

Haff, P.K., and Watson, C.C. (1979). Ion erosion on the Galilean satellites of Jupiter. *Proc. Lunar Planet. Sci. Conf.* 10, 1685-1699.

Helfenstein, P., and Parmentier, E. M. (1980). Fractures on Europa: Possible response of an ice crust to tidal deformation. *Proc. Lunar Planet. Sci. Conf.* 11, 1987-1998.

Johnson, T. V., and McGetchin, T. R. (1973). Topography of of satellite surfaces and the shape of asteroids. *Icarus* 18, 612-620.

Johnson, T. V., and Pilcher, C.B. (1977). Satellite spectrophotometry and surface compositions. In *Planetary Satellites* (J.A. Burns, ed.), Univ. of Arizona Press, Tucson, pp. 232-268.

Johnson, T. V., Soderblom, L. A., Mosher, J. A., Danielson, G. E., Cook, A. F., and Kuperman, P. (1981). Global multispectral mosaics of the icy Galilean satellites. Submitted to *J. Geophys. Res.*

Kieffer, H. (1970). Spectral reflectance of CO_2-H_2O frosts. *J. Geophys. Res.* 75, 501-509.

Kuiper, G. P. (1961). *Planets and Satellites.* Univ. Chicago Press, Chicago, Vol. 3.

Lanzerotti, L. J., Brown, W. L., Poate, J. M., and Augustyniak, W. M. (1978). On the contribution of water products from Galilean satellites to the Jovian Magnetosphere. *Geophys. Res. Letters* 5, 155-157.

Lewis, J. S. (1971). Satellites of the outer planets: Their physical and chemical nature. *Icarus* 15, 175-185.

Lucchitta, B. K., Soderblom, L. A., and Ferguson, H. M. (1981). Structures on Europa. *Proc. Lunar Planet. Sci. Conf.* 12. In press.

Malin, M. C. (1980). Morphology of lineaments on Europa. IAU Coll. 57. *The Satellites of Jupiter* (abstract 7-2).

Moroz, V. I. (1966). Infrared spectrophotometry of the moon and the Galilean satellites of Jupiter. *Sov. Astron. A.J.* 9 999-1006.

Morrison, D., and Cruikshank, D. P. (1974). Physical properties of the natural satellites. *Space Sci. Rev.* 15, 641-739.

Morrison, D., and Morrison, N. D. (1977). Photometry of the Galilean satellites. In *Planetary Satellites* (J. A. Burns, ed.), Univ. Arizona Press, Tucson, pp. 363-378.

Murray, J. B. (1975). New observations of surface markings on Jupiter's satellites. *Icarus* 25, 397-404.

Nelson, R. M., and Hapke, B. W. (1978). Spectral reflectivities of the Galilean satellites and Titan, 0.32 to 0.86 micrometers. *Icarus* 36, 304-329.

Peale, S. J., Cassen, P., and Reynolds, R. T. (1979). Melting of Io by tidal dissipation. *Science* 203, 892-894.

Pieri, D. C. (1980). Lineament and polygon patterns on Europa. *Nature* 289, 17-21.

Pilcher, C. B., Ridgway, S. T., and McCord, T. B. (1972). Galilean satellites: Identification of water frost. *Science* 178, 1087-1089.

Pollack, J. B., Witteborn, F. C., Erickson, E. F., Strecker, D. W., Baldwin, B. J., and Bunch, T. E. (1978). Near-infrared spectra of the Galilean satellites: Observations and compositional implications. *Icarus* 36, 271-303.

Purves, N. G., and Pilcher, C. B. (1980). Thermal migration of water on the Galilean satellites. *Icarus* 43, 51-55.

Ransford, G. A., Finnerty, A. A., and Collerson, K. D. (1980). Europa's petrological thermal history. *Nature* 289, 21-24.

Schenk, P. M., and Seyfert, C. K. (1980). Fault offsets and proposed plate motions for Europa. *EOS* 61, 286 (abstract).

Smith, B. A., and the Voyager Imaging Team (1979*a*). The Jupiter system through the eyes of Voyager 1. *Science* 204, 951-972.

Smith, B. A., and the Voyager Imaging Team (1979*b*). The Galilean satellites and Jupiter: Voyager 2 imaging science results. *Science* 206, 927-950.

Squyres, S. W. (1980). Surface temperatures and retention of H_2O frost on Ganymede and Callisto. IAU Coll. 57. *The Satellites of Jupiter* (abstract 6-4).

Yoder, C. F. (1979). How tidal heating in Io drives the Galilean orbital resonance locks. *Nature* 279, 767-770.

15. THE GEOLOGY OF IO

GERALD G. SCHABER
U.S. Geological Survey

Io's surface displays a wide variety of colors resulting from the spectral reflectance of various allotropes of sulfur, as well as volcanic vent craters, fissures, and other morphologic forms attributed to volcanic processes that continue to shape its surface. Complete burial of craters attributed to impact origin is associated with Io's volcanism over billions of years. A preliminary geologic map of 34.8% of the surface of Io has been compiled using best-resolution (0.5 to 5 km/line pair) Voyager 1 images. Nine volcanic units are identified, including materials of mountains, plains, flows, cones, and crater vents, in addition to six types of structural features. Photogeologic evidence indicates a dominantly silicate composition for the mountain material, which supports heights of ≥ 9 km. Sulfur flows of diverse viscosity and sulfur-silicate ($< 15\%$ silicate) mixtures are thought to compose the extensive plains. Pit-crater and shield-crater vent-wall scarps reach heights of > 2 km, and layered-plains boundary scarps have estimated heights of 150 to 1700 m indicating a material with considerable strength. A cumulative size-frequency distribution plot for 170 volcanic craters with diameters > 14 km is similar to the curves for impact craters on other bodies in the solar system, attesting to a similar nonrandom distribution of crater diameters and a surplus of small craters. Within the mapped area, Io's equatorial zone ($0°$ to $-30°$ lat, $240°$ to $30°$ long) has ~ 2 times as many vents per unit area as the south polar zone ($-60°$ to $-90°$ lat, $240°$ to $30°$ long). A total of 151 lineaments and grabens is recognized; their four dominant azimuthal trends form two nearly orthogonal sets spaced $110°$ apart (N $85°$E, N $25°$W and N $45°$E, N $55°$W). The mapped area discussed in this chapter *lies within the longitudinal zone, 250° to 323°, of least abundant SO₂ frost, where other sulfurous components dominate the upper surface layers.*

Io is one of the remarkable bodies in the solar system. Its surface is characterized by a wide variety of colors (yellow, orange, red, brown, black, and white), apparently resulting from the presence of various allotropes of sulfur (see Plate 7 in color section), as well as volcanic vent craters, fissures, and other morphologic forms attributed to volcanic processes that still dominate the shaping of its surface. Nine active volcanic plumes (Morabito et al. 1979; Smith et al. 1979*b*) and associated thermal anomalies (Pearl et al. 1979; Chapter 19 by Pearl and Sinton) observed on Io during the Voyager encounters have confirmed the presence of active volcanism (Peale et al. 1979; Chapter 16 by Strom and Schneider); spectral-reflectance observations with Voyager instruments have verified the dominance of sulfur and related compounds on the surface (Soderblom et al. 1980; Chapter 20 by Fanale et al.). Complete burial of craters attributed to impact origin is associated with Io's prolonged volcanism over billions of years. Such craters dominate the surface of every other planetary body in the solar system between Mercury and the satellites of Saturn, with the exception of Europa which is very sparsely cratered (Johnson et al. 1979; Chapter 9 by Woronow and Strom; Chapter 14 by Lucchitta and Soderblom).

Prior to Voyager, speculations about unique geologic conditions on Io were mainly based on observations of unusual spectral reflectance signatures and anomalous thermal behavior, and on detection of clouds of neutral atoms and ions, including NaI, KI, and SII, apparently emanating from Io at its orbital radius from Jupiter. Early speculations concerning surface materials including sulfur and related compounds (Kuiper 1973; Wamsteker et al. 1974; Nash and Fanale 1977), and interpretations of the anomalous thermal behavior of Io summarized by Morrison (1977), are discussed in Chapters 7, 19, and 20 in this book.

Following Voyager encounters, a controversy arose over the presence or absence of conventional silicate volcanic materials (e.g., basaltic lava flows, pyroclastic cinders) because of the lack of evidence based on spectral reflectance signatures for such silicates (Carr et al. 1979; Hapke 1979; Masursky et al. 1979; Sagan 1979; Smith et al. 1979*a*; Schaber 1980*b*; Soderblom et al. 1980; Chapter 7 by Sill and Clark). In this chapter, I address from a photogeolic approach the question of sulfur versus silicate origin of observed landforms and surface deposits. The evidence from spectral observations for a global mantling of sulfur and related volatile compounds is compelling, but the nature of remote spectral reflectance measurements requires only the upper mm of the surface to be covered with sulfur to mask the much weaker absorption bands of silicate materials (see Plate 7). Preliminary geologic analysis of Io's landforms and surficial deposits indicates that sulfur-silicate mixtures could dominate many observed features,

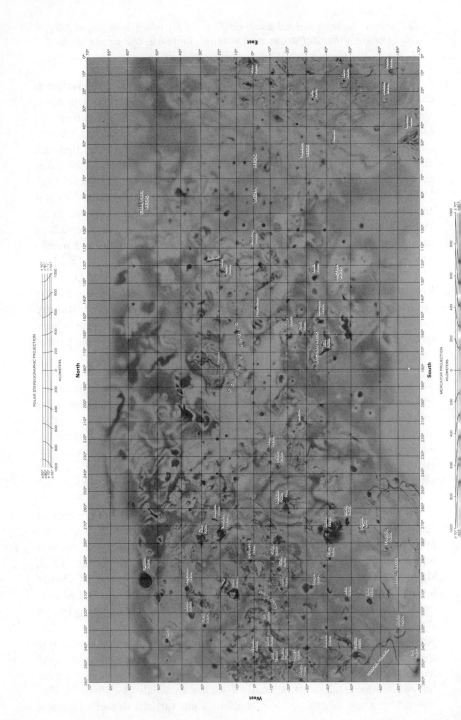

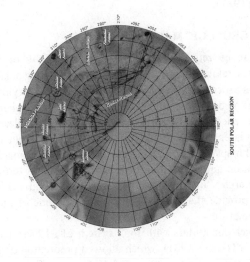

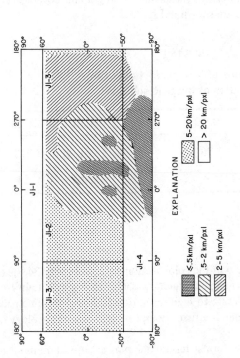

Fig. 15.1. Preliminary pictorial map of Io (U.S. Geological Survey 1979) showing outline of area mapped geologically (Plate 6) and names of Ionian surface features approved by the Working Group for Planetary System Nomenclature of the International Astronomical Union (1980). Lower map is Mercator projection; map at upper right is polar stereographic projection; index map at upper left indicates distribution of Voyager picture resolution. Note that index map is shown in different longitudinal format than is Mercator map to indicate location of Io quadrangle maps (1:5,000,000 scale) being prepared for publication (Batson et al. 1980).

especially landforms that support considerable relief or high-angle slopes, as discussed by Clow and Carr (1981).

I. GEOLOGIC MAPPING

The picture-taking sequence required during a spacecraft flyby of a planet or satellite necessitates taking images with constantly changing solar phase angles and ranges. This procedure results in images with greatly varying geometries, surface albedos, and resolution elements. A task undertaken immediately after Voyager encounters with Io and the other Galilean satellites was to render the images in more usable forms such as uncontrolled, preliminary airbrush pictorial maps (Batson et al. 1980). These first-step maps are hand drawn into Mercator and polar stereographic projections, using reasonably accurate ephemerides of the satellites in conjunction with spacecraft tracking data to provide the scale, latitude, and longitude of the subspacecraft point for each picture.

Shortly after the Voyager encounters with Io, a system of 319 surface control points was established (Davies 1981), which allowed production of a controlled airbrush pictorial map accurate within 2° lat and long. This map (Fig. 15.1) was a base for the geologic mapping of Io described in this chapter. The Voyager pictures used in constructing the base and in mapping the geology are listed in Table 15.1; available photographic coverage and resolution of Voyager pictures are shown in Fig. 15.1.

TABLE 15.1

**Total Number of Voyager Pictures of Io
as a Function of Image Resolution[a]**

	< 0.5[b]	0.5-2	2-5	5-20	> 20
Voyager 1	33	96	75	61	278
Voyager 2	–	–	–	200	54

[a]After Batson et al. (1980).
[b]Column headings are in km/pixel.

The geologic map in Plate 6 in the color section represents only 34.8% of the surface area of Io. The mapped region was chosen by availability of Voyager 1 image data at moderate to high resolution (0.5 to 5 km/line pair) (Table 15.1, Fig. 15.1). The map is a first attempt to delineate the geology of part of Io.

Geologic mapping and less formal terrain mapping are useful methods to interpret the general spatial distribution of surface units and tectonic structures on a planet or satellite following spacecraft reconnaissance. However, mapping of Io must be done without two techniques particularly valuable in mapping the Moon, Mercury, and Mars, on which the number of impact craters superposed on a particular terrain was used to ascertain the general stratigraphy and/or age relations with respect to other terrains. On Io, continuous volcanic deposition has buried almost all impact craters (see Chapter 19 by Johnson and Soderblom). Stereographic viewing overlapping pairs of pictures is also used in conventional planetary mapping, but on Io the imaging sequence (selected for maximum coverage during the flyby) resulted in little overlap of stereographic pictures. Thus, apparent stratigraphic relations of surface units were established during initial geologic mapping mostly by monoscopic photogeologic techniques; surface color and albedo, and tectonic features such as faults, lineaments, and erosional scarps, also helped to establish local stratigraphic histories.

Nine geologic units, in addition to vents, scarps, and other tectonic features (Fig. 15.2), were identified during initial geologic mapping (see Plate 6). Size-frequency and areal distributions of volcanic vents, and the azimuthal trends of tectonic features, will be discussed. The geologic units are divided into three groups, all illustrated in Voyager pictures: mountain, plains, and vent materials. The percentages of the geologic map occupied by each unit and by surfaces obscured by active plumes are given in Table 15.2.

TABLE 15.2

Relative Areal Abundance of Geologic Units on Io

Geologic Unit	% Mapped Area[a]
Intervent plains	39.6
Pit-crater flow	20.2
Shield-crater flow	9.7
Layered plains	9.3
Walls and floors of pit craters, shield craters, and fissures	4.0
Mountain material	1.9
Fissure flow	1.2
Eroded layered plains	0.7
Crater cone	0.1
(Obscured by vent plume materials)	13.3
TOTAL	100.0

[a]Mapped area is 34.8% of Io's total surface area. (See footnote (a) in Table 15.3.)

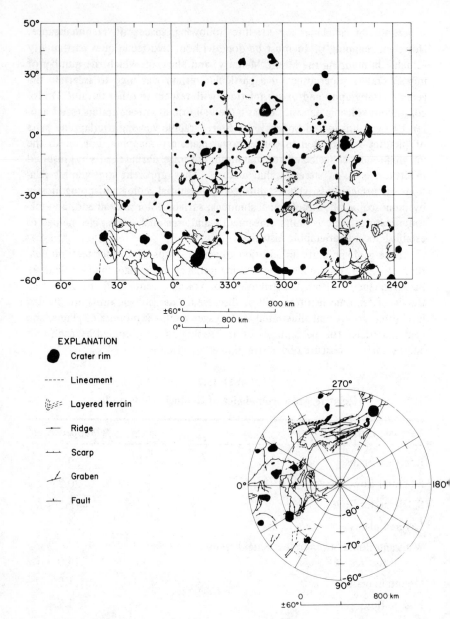

Fig. 15.2. Preliminary geologic map of Io showing only vent wall and floor material (black) and tectonic features, for ease in location (see Plate 6).

The isolated high massifs of mountain material are interpreted as formed from the oldest exposed geologic unit because of their tectonic disruption, morphology, and stratigraphic relations with surrounding surfaces. Mountain material appears equally well represented in the equatorial and south polar regions (Plate 6).

Plains materials units are represented by intervent plains, layered plains, and eroded layered plains. Layered plains with high, steep boundary scarps are present throughout the mapped region of Io, but are larger and thus best observed near the south pole. Eroded layered plains are restricted to the south polar region.

Vent materials are subdivided into five units: vent floor and wall, pit-crater flow, shield-crater flow, fissure flow, and crater cone. Shield craters and their deposits are present only in the equatorial and midlatitude regions of active volcanism (Strom et al. 1979). The youngest deposits on Io represented by thin fumarolic haloes surrounding vents have not been mapped as separate geologic materials; they are addressed by Strom and Schneider (Chapter 16) in their discussions of active and recently active vents (Plate 7).

Lineaments are widespread, but distinct grabens with clearly defined central troughs are recognized only south of $-20°$ lat, possibly because the lower sun angle results in greater contrast in the Voyager pictures.

The surface area of Io discussed here is mainly reddish orange and lies within the longitudinal zone suggested by Nelson et al. (1980) to contain the least amount of SO_2 frost (white), according to spectral reflectance observations (Plate 7). The surface layers may be dominated by sulfur and other sulfurous components at least to 1 mm thickness.

II. ELEVATION ESTIMATES

Heights of faults, scarps, and volcanic crater vent walls have been estimated by Arthur (1981) using an essentially photometric method based on the recorded digital brightness values (DN) in the original data. The format of the printouts or dumps of the data tapes duplicates the geometry of the pictures, with corresponding lines and columns of DN. Interpolated DN values along a line can be sequenced as a photometric profile, and the profile of a shadow is a depressed segment with sloping rather than vertical terminal sections due to instrumental blurring.

In his work on Viking pictures of Mars, Arthur (1980) found further degradation due to the atmosphere and ambiguities caused by asperities, near the lower end of Martian shadows. However, these uncertainties can usually be resolved and the shadow terminals identified with the steepest parts of the curves. The shadow length is then estimated as the distance between these terminal points, and the relative height computed from the navigation support data in the same way as lunar relative heights are determined from shadows; the principal steps are scaling, compensation for foreshortening (phase), and application of the solar elevation above the local planetary horizon.

This simple technique is useless for fault scarps and other low-relief features whose shadow lengths are very short (a few or even $<$ 1 pixel). In these cases, the terminal curves on the photometric profile overlap, and the lowest value never drops to the DN value that should hold for the shadow. A simple integration technique solves this problem. In the simplest case, the shadow band separates two surfaces equally bright in the picture, as when a scarp crosses a level plain. It is assumed that all radiation in the profile is conserved and merely displaced. The profile is a level curve with a dip representing the narrow shadow; the level of the dip is taken by assumption from the interiors of nearby large shadows.

Heights of selected mountain massifs are estimated with Voyager images showing the massifs in profile on the terminator, almost on the limb, simply by trigonometry and Voyager image-data records, after the location of the true and invisible limbs, the terminator, and the base of the camera-facing slope of the massif are established. This radial slope is assumed the same as the tangential slopes in profile measured directly against the dark field on the limb profile of the massif. No shadows are involved in this estimate, unlike that of the scarp heights discussed above.

III. GEOLOGIC DESCRIPTION

Mountain Material

The mountain unit is thought to be the oldest recognized unit, characterized by $\geq 9 \pm 1$ km relief and by a tectonically disrupted, topographically rugged surface (Figs. 15.3 to 15.7; Plate 6). It is probably primarily composed of silicate volcanic material, because it supports significant relief (Carr et al. 1979; Masursky et al. 1979; Schaber 1980b; Clow and Carr 1980) (Figs. 15.6, 15.7). The unit is commonly associated with patches of layered plains material, and occurs in several locations rimming large, shallow, circular volcanic depressions within the intervent plains (Fig. 15.7). Pit craters with steep walls are common on the mountain material unit, indicating that the mountains themselves are vent sources (Figs. 15.5, 15.7, and 15.8). The mountain unit is evenly distributed in lat and long, contrary to earlier suggestions (Smith et al. 1979b), and makes up 1.9% of the mapped area (Schaber 1980b).

The most geologically informative exposures of mountain material are those centered around the large unnamed circular vent in the intervent plains at $-6°$ lat, 267° long (Fig. 15.7). Here, three large massifs seem associated by their position with the 172-km-diameter vent as rim deposits or as individual volcanic constructs of high yield-strength material.

Lobate scarps associated with an exposure of mountain material just north of Creidne Patera resemble large landslides or terrestrial and Martian lava flows (Figs. 15.4, 15.8). Similar lobate fronts have not been recognized elsewhere. The apparent thicknesses (not estimated) of these flow-like features indicate a high yield-strength material, and their surfaces are fretted or grooved, suggestive of some form of erosion or innate flow structure. The boundary of the grooved surface texture southeast of the lobate scarps is abrupt, occurring along a distinct lineament, possibly a fault or fissure-vent source for the flow material (Fig. 15.8).

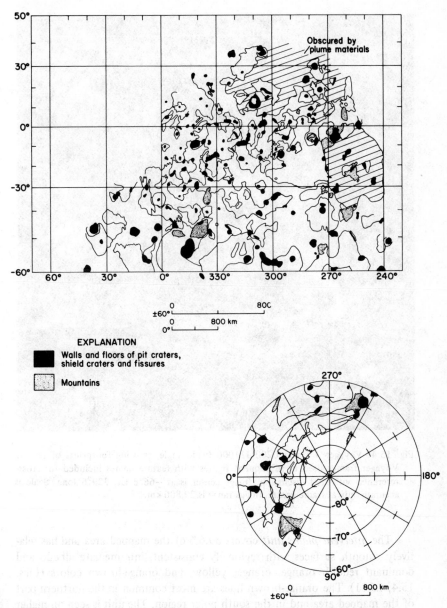

Fig. 15.3. Preliminary geologic map of Io showing only vent wall and floor material (black) and mountain material (patterned) for ease in location (see Plate 6).

Plains Materials

Layered planar surfaces are clearly dominant on Io (Table 15.2). They are divided into three major units: intervent plains, layered plains, and eroded layered plains (Plate 6, Figs. 15.4, 15.7, 15.9, 15.10, and 15.11).

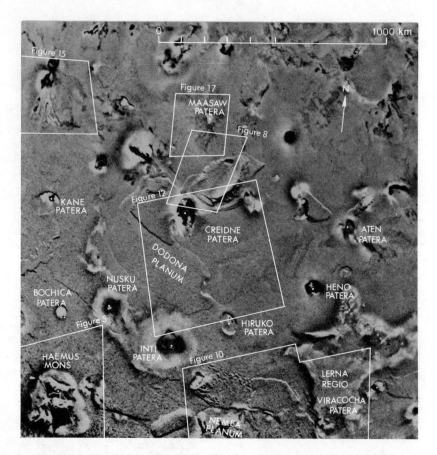

Fig. 15.4. Voyager 1 frame 0200 J1+000 (wide angle) showing footprints of several
Voyager frames used in subsequent figures with feature names included for cross-
reference with text discussion. Image center is at −66°8 lat, 329°6 long. Scale is
accurate only at frame center. Image range is 29,800 km.

The *intervent plains unit* covers 39.6% of the mapped area and has relatively smooth surfaces with regionally consistent, intermediate albedo and
dominant reddish orange, orange, yellow, and orange-brown colors (Figs.
15.4, 15.11). The orange-brown hues are most common in the northern part
of the mapped area and in the south polar region. The unit is seen on higher
resolution Voyager 1 images to contain abundant low-relief scarps (estimated
< 10 to 100 m heights) ranging from straight to sinuous. This unit provides the
major component of the spectral reflectance and photometric observations
described for Io from both Earth-based and Voyager data (Soderblom et al.
1980; Chapter 7 by Sill and Clark, Chapter 20 by Fanale et al.; Plate 7).

The unit is thought to be composed of stratified materials that originated
as frozen fallout from volcanic plumes, probably imbedded with local fuma-

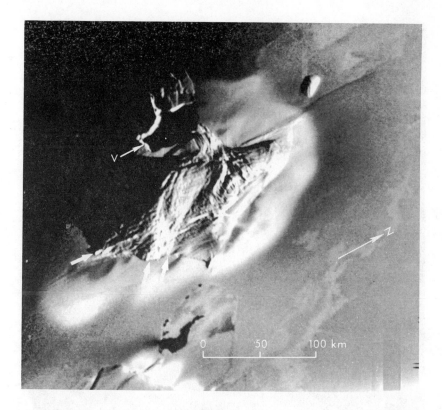

Fig. 15.5. Section of Voyager 1 frame 0157 J1+000 showing details of Haemus Mons (see Fig. 15.4). Arrows are aligned along lineament trends; large pit crater is indicated at V. Note high-albedo halo of suspected plume and fumarole materials surrounding base of mountain. Scale is accurate only at frame center. Image range is 56,000 km.

role materials and flow deposits from pit craters, shield craters, and fissure vents. Spectral reflectance observations indicate that the unit is probably composed of a combination of sulfur and its compounds (including SO_2 frost), although estimated extremely low-average surface density and high porosity suggest that altered physical forms of these materials dominate the uppermost layers of an Ionian regolith (Nelson and Hapke 1978; Nash and Nelson 1979; Nelson et al. 1980; Soderblom et al. 1980; Pang et al. 1981; Matson and Nash 1981; Morrison and Telesco 1980; Chapter 20 by Fanale et al.).

Clow and Carr (1980) assessed the yield strength of pure sulfur under surface conditions on Io and showed that pure sulfur cannot hold up the steep walls of pit craters formed in the intervent plains; they suggested a significant amount of high yield-strength materials such as silicates.

The *layered plains unit* has an extensive, smooth, flat-surface, boundary scarps ranging in height from 150 to 1700 m, and abundant grabens either

Fig. 15.6. Section of Voyager 1 frame 0129 J1+000 showing two mountains on the limb
of Io outside the geologically mapped area at −28° lat, 228° long. *LP* indicates
layered plains unit. Each mountain has estimated height of 9 ± 1 km (see Sec. II).
Nearer mountain is ∼150 km long. Image range is 74,700 km.

parallel or transverse to the scarps (Figs. 15.10 to 15.12). The unit covers
9.3% of the mapped area and, though found at various lat and long, is best
developed near the south pole where individual exposures cover areas as large
as 2 × 10⁵ km² (McCauley et al. 1979), and are most readily observed due
to low sun angles (Figs. 15.10, 15.11).

The layered plains unit is thought to have the same origin and composition as the intervent plains unit, differing only in the presence on the layered
plains unit of boundary scarps of greatly varied height. Some scarps appear to
represent at least temporary erosional limits controlled by normal faults and
grabens; scalloped or fretted contacts such as those in Figs. 15.10 and 15.12,
may indicate scarp erosion that has progressed beyond controlling fault structures (McCauley et al. 1979).

Fig. 15.7. Enlarged section of Voyager 1 frame 0125 J1+000 showing exposures of mountain unit *M* surrounding a shallow 172-km-diameter vent depression at $-6°$ lat, $267°$ long. *LP* indicates layered plains unit, *V*, possible vents. Note topographic irregularity and presence of tectonic lineaments (wider arrows). Image range is 76,600 km.

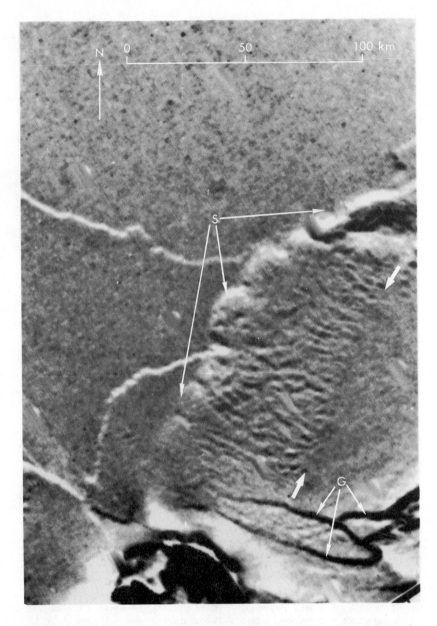

Fig. 15.8. Enlarged section of Voyager 1 frame 0075 J1+000 (see Fig. 15.4) showing detailed lobate, flow-like scarps *S* at base of mountain unit. Note sharply defined boundary (arrows) of fretted or channeled surface on flowlike feature, indicating possible fissure vent or fault. Deep grabens *G* occur at south end of mountain material (see Fig. 15.12). Note layered plains scarps west of *S*. Image center is at −47° lat, 342° long. Image range is 108,700 km.

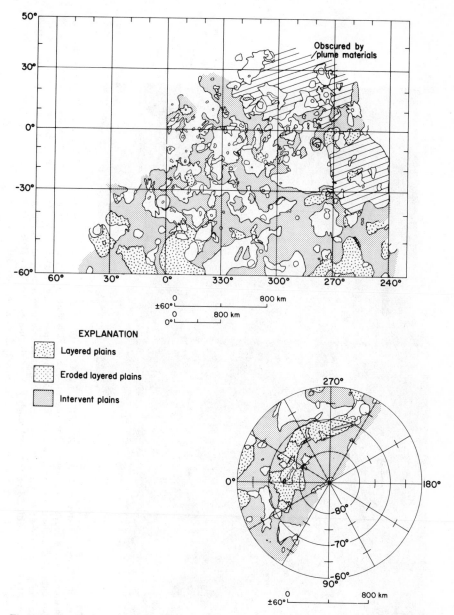

Fig. 15.9. Preliminary geologic map of Io showing only plains units (patterned) for ease in location (see Plate 6).

McCauley et al. (1979) discussed an erosional process whereby Io's SO_2 rich plains materials are stripped back by a sapping of SO_2 similar to the sapping of CO_2 ice or water ice envisioned in the formation of the channels and box-canyon networks on Mars. Reservoirs of liquid SO_2 have also been

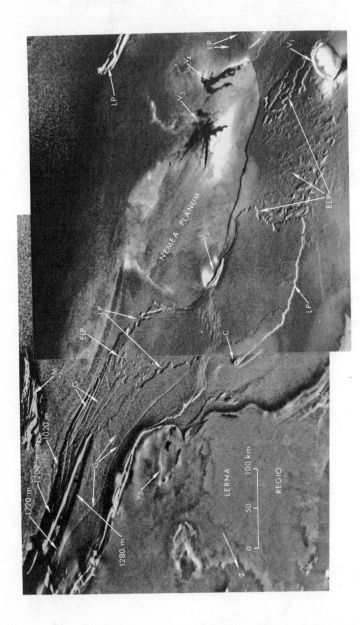

Fig. 15.10. Photomosaic of Voyager 1 frames 0141 J1+000 (left) and 0147 J1+000 (right) showing layered plains unit *LP*, eroded layered plains unit *ELP*, and mountain unit *M* in the south polar region of Io (see Figs. 15.1, 15.4, **Plate 6**). Scarp heights (in m) of units were estimated (±15%) by Arthur (see text). Note abundant grabens *G* on layered plains unit. Mountain unit to east (top) is at −76° lat, 240° long. Eroded layered plains unit is best developed east of pit crater vent V_1 and is the infrared hot spot described by Pearl and Sinton in Chapter 19. Vent V_2 on layered plains scarp may represent SO_2 ice fog; additional vents (V_3 through V_5) are also indicated. Note scalloped erosion of some boundary scarps of layered plains. A volcanic cone 3 to 4 km in diameter is on layered plains scarp at *C*. Scale is accurate only in center of mosaic. Image ranges are 65,900 km (0141 J1) and 62,100 km (0147 J1). Sun-angle is ~4°.

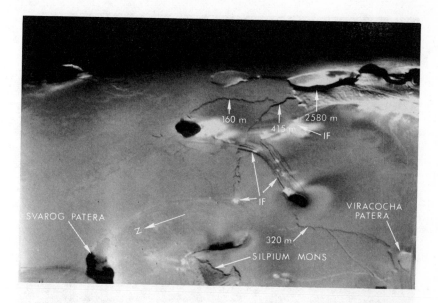

Fig. 15.11. Section of Voyager 1 frame 0113 J1+000 showing complex of layered plains, mountain, and intervent plains units centered at −56°.7 lat, 239°.7 long (image center). Svarog Patera is area of Voyager infrared hot spot (Chapter 19 by Pearl and Sinton). Possible SO$_2$ ice fog *IF* areas are indicated along fractures and layered plains scarps (see Fig. 15.1; Plate 6).

proposed to explain what might be ice clouds issuing from fractures and scarps in Io's crust (Soderblom 1980) (Fig. 15.11). This hypothesis assumes that liquid SO$_2$ would have easiest access to the surface along scarps of fault origin; as it reaches the surface, perhaps forced out by artesian pressures, the confining pressure drops below a critical point and the liquid explodes into an ice fog. The strongest evidence for SO$_2$ sapping or surface devolatilization of the layered plains is seen in the eroded layered plains unit.

Semicircular outcrops of layered plains such as those north of Nemea Planum (Fig. 15.10) are common on Io. They may result from control of boundary scarps by arcuate faults and grabens, or from outward erosion (SO$_2$ sapping) from a central point such as a center of deposition or a region of high heat flow.

The *eroded layered plains unit* is thought to be isolated remnants of volatile-related collapse of the layered plains unit. The unit occupies only 0.7% of the mapped area and appears restricted to the south polar region, but this distribution may well be an artifact of low sun angles in this region. The erosional process that destroyed these plains apparently operates within upper layers by the removal of volatile materials. Recent volcanic activity suggested by bright fumarole deposits of pit craters and fissures along boundary scarps of the layered plains unit (shown in Figs. 15.10, 15.11) lends strength

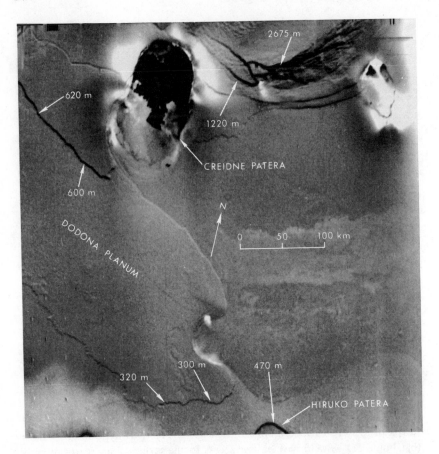

Fig. 15.12. Voyager 1 frame 0145 J1+000 showing heights of layered plains scarps and mountain material fault scarps (1220 m and 2675 m) near Creidne Patera (−53° lat, 346° long). (See Figs. 15.1, 15.4; Plate 6.) Frame center is location of infrared hot spot in Voyager data (see Chapter 19 by Pearl and Sinton and Fig. 15.11). Sun angles range from 26° (south edge) to 40° (north edge). Image range is 62,900 km.

to the hypothesis (McCauley et al. 1979) that SO_2 is the dominant erosion mechanism. However, erosion that forms the eroded layered plains not only operates along established boundary scarps but disrupts the entire surface of a layered plains. Hot gases and fumarole materials from the pit craters V_1 and V_3 shown in Fig. 15.10, may have been important in devolatilizing and eroding layered plains into the eroded layered plains unit. Vents V_1 and V_3 and the eroded layered plains unit between them shown in Fig. 15.10 are identified as a hot spot indicating active volcanism (Chapter 19 by Pearl and Sinton; see Fig. 19.13). There is clear photogeologic evidence that devolatilization is a major process in forming eroded layered plains; it may provide our best clue to the relative roles of sulfur and silicates in the extensive Ionian plains units.

Vent Materials

Five separate units associated with volcanic vents have been mapped (Plate 6; Figs. 15.3, 15.13). Younger deposits characterized by diffuse aureoles or haloes that surround vents are not mapped; they are ephemeral and too complex at this mapping scale (Chapter 16 by Strom and Schneider).

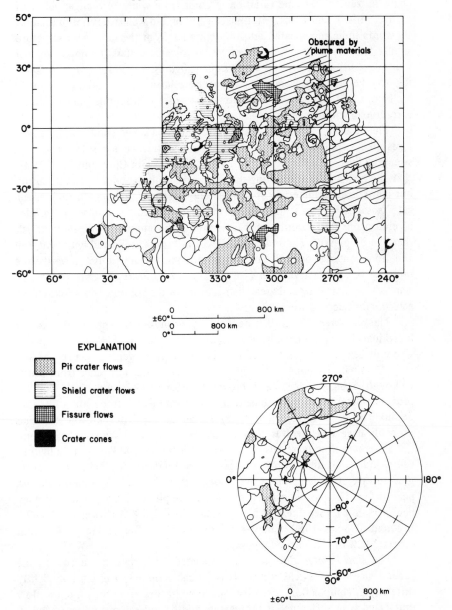

Fig. 15.13. Preliminary geologic map of Io showing only flow units of vent materials (patterned) for ease in location (see Plate 6).

The *vent wall and floor unit* includes the walls and floors of pit craters, shield craters, and fissures (Plate 6; Fig. 15.3). None of these craters is suspected to be of impact origin. The materials of this unit, thought to be the same as those of the intervent plains, layered plains, and eroded layered plains units, are strong enough to support steep vent walls > 2 km high. The unit is characterized by extreme variation in color from white to orange, yellowish orange, brown, and black. Multiple-stage floor and wall structures in many of the vents indicate repeated eruption episodes from the same vent structure; similar features are common in terrestrial volcanic craters. Examples of such multiple vents are illustrated in subsequent figures. More of the vents responsible for the plumes observed by Voyager are linear than circular (Chapter 16 by Strom and Schneider), indicating that fissure eruptions may dominate early Ionian volcanic activity.

The absence of wall slumps of pit craters and shield craters under conditions like explosive venting provides strong evidence for substantial wall strength (Figs. 15.4, 15.5, 15.11, 15.12). I agree with Clow and Carr (1981) that the plains units supporting the steep vent walls may not be simple sulfur, but contain some materials of considerably higher yield-strength such as silicate glass. The devolatilization of the layered plains unit into the eroded layered plains unit described above does limit the amount of silicate materials that can be present. Thermal anomalies or hot spots associated with vent wall and floor material do not have sufficiently high temperature to verify the presence of silicate lava lakes (see Chapter 19 by Pearl and Sinton). Discussion of size-frequency and areal distributions of the volcanic vents follows the geologic unit descriptions.

The *pit-crater flow unit* is deposited near pit-crater vents; thicknesses of individual flows are unresolvable at Voyager resolution (Figs. 15.13, 15.14). The unit generally extends from one side of pit-crater vents as massive coalescing flows with extreme color and albedo variations; colors range from white, yellowish orange, orange, and brownish red to brown and black, and are probably determined by surface deposits of various allotropes of sulfur (Soderblom et al. 1980; Chapter 20 by Fanale et al.; Plate 7). Flow materials probably have a high sulfur content, but some silicate lavas may be present.

Pit-crater flow materials can be traced as far as 700 km from individual vents, indicating high eruption rates or very fluid eruptive material or both; the eruptive style is certainly less violent than that of the fumarole haloes, near-surface flows of volcanic gases, and active plume ring deposits (not mapped in Plate 6). (See Cook et al. 1979; Smith et al. 1979a; Strom et al. 1979; Lee and Thomas 1980; Chapter 18 by Kieffer.) The pit-crater flow unit covers 20.2% of the mapped region.

The *shield-crater flow unit* is associated with rims and occurs in the general vicinity of shield craters. It is probably similar in origin but slightly different in composition, viscosity, and flow history from the pit-crater flow unit. The main difference is the narrow widths and sinuous paths of shield-crater flows (exemplified at Ra and Maasaw Paterae) indicating migration

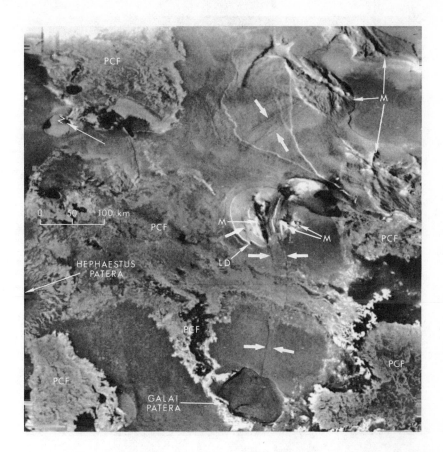

Fig. 15.14. Voyager 1 frame 0115 J1+000 showing numerous pit-crater flows *PCF* and well-developed faults (arrows) associated with mountain unit *M*. Evidence of lateral displacement along fault at *LD*. Mountain unit near top right corner of figure shown in more detail in Fig. 15.7 (see Plates 6 and 7; Fig. 15.13). Note pit-crater flow originating from north side of Galai Patera. Image center is at $-3°1$ lat, $278°0$ long. Image range is 82,500 km.

down moderately steep slopes (magnitude unknown) from the shield summits and possibly lower rates of eruption (Figs. 15.15 to 15.17). Individual flow lobes can be traced as far as 300 km; color and albedo vary along their length. Colors are similar to those of the pit-crater flow unit and may be related to quenching of various temperature-dependent sulfur allotropes along the flow path (Sagan 1979; Soderblom 1980; Chapter 7 by Sill and Clark).

Shield constructs with associated flow materials are limited to the zone between $+30°$ and $-45°$ lat in the mapped region (Plate 6; Fig. 15.13). Their concentration may coincide with the dominance of active volcanic plumes in the equatorial zone (Chapter 16 by Strom and Schneider). The shield-crater flow unit covers 9.7% of the mapped area.

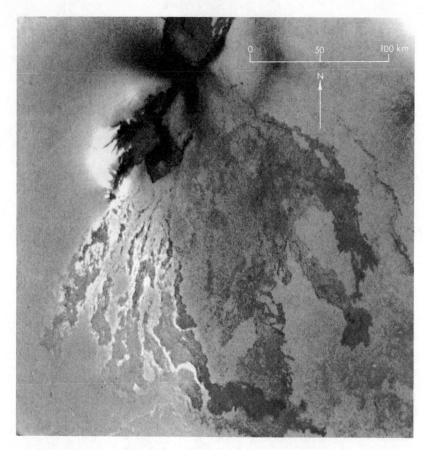

Fig. 15.15. Voyager 1 frame 0184 J1+000 showing details of unnamed shield-crater vent and flow unit at $-34°5$ lat, $2°9$ long. Area of lighter flows (east side) is 10,400 km^2; area of darker flows (west side) is 17,100 km^2. Image range is 39,100 km (see Figs. 15.1, 15.4; Plate 6).

Composition of the shield flows is not yet known, but sulfur and sulfur compounds are evident in the spectral reflectivity data from their surfaces. The narrow, sinuous shield-crater flows (Fig. 15.16) resemble silicate lava flows from well-known terrestrial shields such as Mount Etna in Sicily (Figs. 15.17, 15.18). The Pleistocene and Holocene lava field from Mount St. Helens (Washington State) and the Holocene lavas on Mount Etna are shown to scale in Fig. 15.17. Lava flows on Mount Etna dating from 1900 to 1974 are shown in Fig. 15.18, along with summit caldera, parasitic flank vents, and lava flows dating from Pleistocene time to 1899.

Length-to-width ratios for the well-defined, 200-km-long flows on Ra Patera (Fig. 15.16) range from 20:1 to 50:1 over unknown surface slopes. Twentieth-century Hawaiiite lava flows on Mount Etna (Fig. 15.18) have similar length-to-width ratios (20:1 to 40:1) for flows 7 to 10 km long ex-

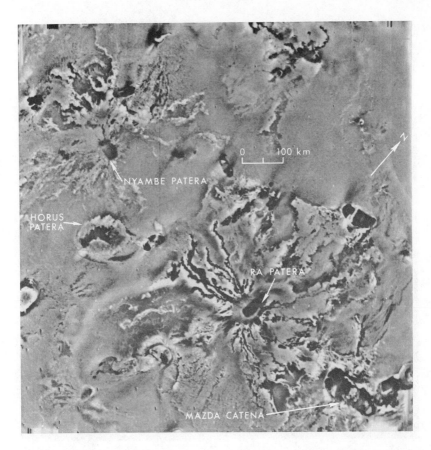

Fig. 15.16. Voyager 1 frame 0043 J1+000 showing good examples of shield volcanic constructs with summit shield craters (Ra and Nyambe Paterae). Note narrow, sinuous flowlike materials emanating from Ra Patera. Image center is at −9°4 lat, 324°1 long. Image range is 129,900 km (see Plates 6 and 7; Fig. 15.1).

truded onto slopes of ~ 8°. However, no compositional similarities are proposed because of difficulties in relating flow morphologies, surface slopes, and assumed yield strengths to composition (see Moore et al. 1978).

Two unusual discoid shield constructs are located on Io at − 13° lat, 350° long (Figs. 15.1, 15.19; Plate 6). Inachus Tholus and Apis Tholus are reminiscent of the much larger Olympus Mons on Mars; all three are characterized by circular outlines and the presence of summit calderas and abrupt basal scarps (heights on Io not estimated). No lava flow lobes are observed on the slopes of these shields, though the south side of Apis Tholus appears overlapped by younger shield flow deposits from the west (Fig. 15.19). Similarly, younger shield-crater flows from the north have been diverted around the basal scarp of Apis Tholus. As in the Olympus Mons basal scarp, those associated with Apis and Inachus Tholi are of undetermined origin.

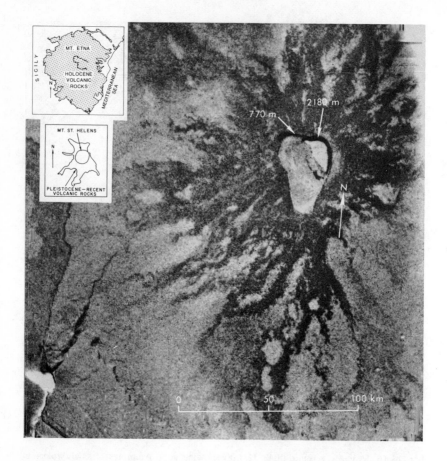

Fig. 15.17. Voyager 1 frame 0199 J1+000 showing shield-crater vent and radial, low-albedo flows of Maasaw Patera. Depths of shield-crater wall scarps are shown (in m). Deepest caldera pit (2180 m) is probably the youngest part of summit crater complex. Insets show (to scale) the extent of the Holocene volcanic flows of Mount Etna in Sicily (Consiglio Nazionale delle Ricerche 1979) and the Pleistocene and Holocene flows of Mount St. Helens in Washington State. Maasaw Patera is one of the smallest of the mapped vents. Flows from Maasaw Patera are of unknown composition but are thought to be either sulfur or sulfur-rich, highly fluid silicate lavas. Image center is at $-52°6$ lat, $330°0$ long. Image range is 30,200 km.

The *fissure-flow unit* is associated with elongate fissure vents that may have fault control (Plate 6; Fig. 15.13). This type of vent and deposit is rare on Io (1.2% of mapped area) compared with the abundant nearly circular pit-crater and shield-crater vents. All four exposures of the fissure-flow unit shown in Fig. 15.13 originate from northwest-southeast fissures (see Sec. V).

The best documented fissure-flow deposit is just west of Mazda Catena at $-4°$ lat, $320°$ long (Fig. 15.20). Its source is a raised, dikelike feature 92 km long from which thick flows emanate in both directions normal to the strike

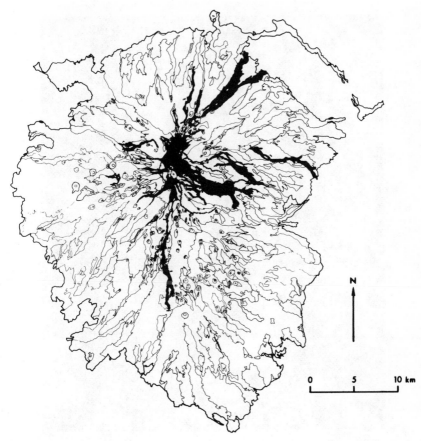

Fig. 15.18. Generalized geologic map of Mount Etna in Sicily showing distribution of 20th-century lava flows to 1974 (black) and volcanic deposits dating back to early Holocene time (uncolored). (Map after Consiglio Nazionale delle Ricerche 1979.) Compare narrow widths of flows with flows from Maasaw Patera shown in Fig. 15.17. Parasitic pyroclastic cones and their summit craters are also shown on the flanks of Etna. Rim of summit caldera is indicated with heavy line and sawtooth pattern.

of the fissure. Thicknesses of the flows are substantial but cannot be measured because of the high sun angle (Arthur, personal communication 1981). The flows are grooved along the flow direction and have high, steep, lobate scarps. Just north of these, other flow materials with similar surface morphology and thickness, thought to be of similar composition, emanate along a 100-km-wide front (Fig. 15.20). A circular pit-crater vent has formed along this fissure. The thickness and surface morphology of these two flow sequences appear unique; the fluid material must have had considerably higher viscosity and yield-strength than most pit and shield-crater flows on Io. The high lobate scarps forming the flow edges resemble those on Mars border-

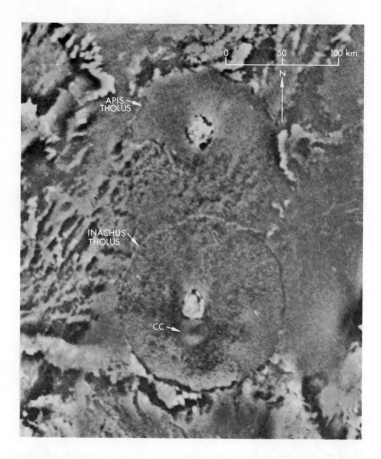

Fig. 15.19. Enlarged section of Voyager 1 frame 0071 J1+000 showing Apis Tholus and
Inachus Tholus, unique disk-shaped shield volcanoes at −17° lat, 350° long. Note
bright summit craters and diffuse crater cone *CC* on south side of summit crater of
Inachus Tholus. Basal scarps of the Tholi resemble, at a smaller scale, that of Olympus
Mons on Mars. Note diversion of flow by basal scarp on north side of Apis Tholus
and overlap of Apis by shield-crater flow from west. Image range is 111,600 km.

ing flows of inferred high SiO_2 content (Schaber 1980a). The flow surface
and its surroundings north of the fissure dike shown in Fig. 15.20 are orange;
south of the fissure dike the flow surface is whitish orange (Plate 7). These
colors are probably caused by surface coatings of sulfur and SO_2 and may not
be the color of the actual flow materials.

The *crater cone unit* forms raised rims around pit and shield craters.
Cones appear similar in form, but not in scale, to common terrestrial pyroclastic cones of silicate ash and cinders (Plate 6; Fig. 15.13). This unit forms
only 0.1% of the mapped area and is best observed associated with the pit-
crater vent Amaterasu Patera at +36° lat, 306° long (Fig. 15.21). This crater

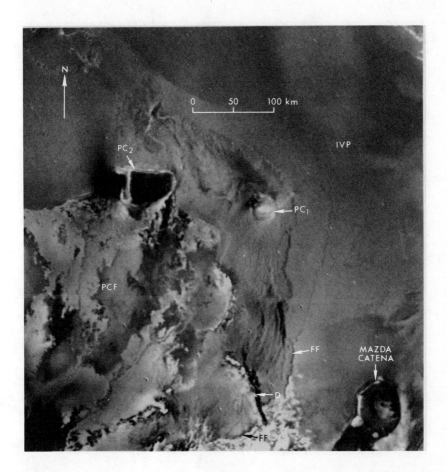

Fig. 15.20. Enlarged section of Voyager 1 frame 093 J1+000 showing thick 100-km-wide fissure flow *FF* emanating from raised dikelike feature *D* at $-4°$ lat, $320°$ long (see Figs. 15.1, 15.13; Plate 6). Thickness of flow is not measurable because of high sun angle of $75°$. Note grooving on flow surface along flow direction. Younger flow with similar surface texture emanates from circular pit crater at PC_1. Intervent plains surface *IVP* lies east of fissure-flow unit. Pit-crater flows *PCF* with unresolvable relief extend from south side of pit crater PC_2. Image range is 96,700 km.

has a distinctive terraced wall that may show strandlines from former lakes of ash, sulfur, or silicate lava. Another, smaller crater cone is seen atop Inachus Tholus (Fig. 15.19); a second ring-shaped feature is isolated on the intervent plains unit at $-45°$ lat, $330°$ long (Fig. 15.22).

IV. SIZE–FREQUENCY AND AREAL DISTRIBUTIONS OF VENTS

The mapped area of Io contains 170 vents (Plate 6; Fig. 15.3) > 14 km in diameter. A cumulative size-frequency distribution plot of these vents assuming a perfectly circular outline for assignment of crater diameters was pre-

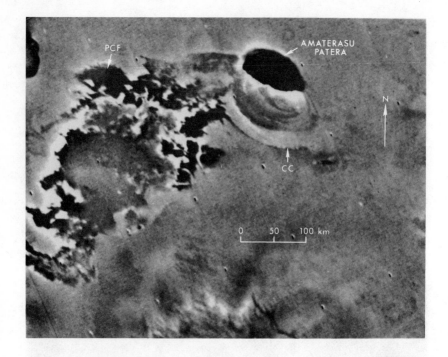

Fig. 15.21. Enlarged section of Voyager 1 frame 0015 J1+000 showing crater-cone unit *CC* that forms south rim of Amaterasu Patera. Pit-crater flow unit *PCF* emanates from west side of patera. Note highly varied albedo of flow unit and absence of well-defined flow units. South wall of crater shows evidence of stepped terraces, possibly indicating high lava- or sulfur-lake levels or successive eruptions of crater-cone ash rings. Amaterasu Patera is at +36° lat, 306° long. Image range is 148,900 km (see Figs. 15.1, 15.13; Plates 6 and 7).

pared after calculating their area (vent wall and floor unit) (Fig. 15.23). Craters within craters were considered as single vents. The curve in Fig. 15.23 is remarkably similar to conventional size-frequency distribution plots of impact craters; like these plots, the Ionian curve has a nonrandom distribution of crater diameters and a surplus of small craters.

A total of 587 terrestrial craters (1 to 12 km in diameter) of explosive volcanic origin was shown by Simpson (1966,1967) to have a percentage size-distribution curve very similar to those observed for lunar and Martian craters. This overall similarity of crater size-distribution curves led Simpson (1966) and other early workers to incorrectly assume volcanic origin for most observed craters on the Moon and Mars.

The resemblance of size-frequency and percentage-frequency statistics of impact craters to those of volcanic craters emphasizes the need to integrate studies of crater origin with crater-count research on solid solar system bodies (Greeley and Gault 1979). Frequencies of both impact and volcanic craters

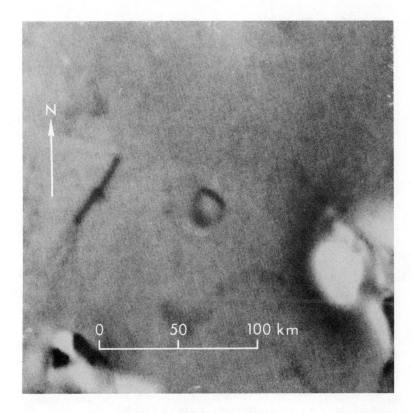

Fig. 15.22. Enlarged section of Voyager 1 frame 0196 J1+000 showing a 20-km-diameter discoid-ring crater cone lying on the intervent plains at −45° lat, 330° long. This feature is extremely rare within the mapped region of Io. Image range is 31,900 km.

have an excess of small craters with a predictable falloff to larger crater sizes. For impact craters, this phenomenon is a function of the collisional evolution of planetesimals, asteroids, and cometary debris; impact velocity; and gravitational acceleration. The size of a volcanic explosion crater is a complex function of magma temperature and gas saturation, overburden strength and density, and gravitational acceleration (Smith et al. 1979a,b; Chapter 18 by Kieffer).

The cumulative size-frequency curve (Fig. 15.23) for volcanic vents in the mapped area of Io has slope −3.7 between vent diameters of 182 to 70 km, slope −1.8 between diameters 70 to 35 km, and slope only −0.33 for diameters 35 to 15 km. The inflection of the curve below diameters of 35 km is similar to those observed on size-frequency plots for impact craters attributed to resolution limitations or resurfacing, and a similar explanation may apply to the Ionian curve, in which mantling and obscuration of smaller vents by active plumes and recent fumarolic deposits may contribute to this effect.

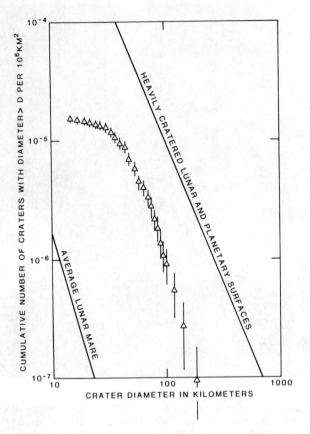

Fig. 15.23. Log-log cumulative size-frequency plot of Io's volcanic vents compared with conventional impact-crater cumulative size-frequency curves for average lunar mare and heavily cratered surfaces on the Moon and other planetary bodies. Total volcanic vents counted within mapped area is 170. One-σ error bars are shown.

Equal-area bins were established within the map area enclosed by 0° to −90° lat and 270° to 0° long to assess the spatial distribution of volcanic vents within latitudinal and longitudinal bands. The area selected for this analysis represents only 35.9% of the geologic map and 12.5% of the surface area of Io, and thus may not be representative of the entire satellite. Table 15.3 includes a summary of these statistics as well as latitudinal statistics on mean vent diameters and standard deviations of vent diameters.

It can be seen from Table 15.3 that the equatorial zone has 1.1 times the number of vents per unit area as the midlatitude zone and 1.8 times more than the south polar zone (see footnote (a), Table 15.3). A previously unrecognized increase in the number of vents between 330° and 0° long appears greatly influenced by an equatorial concentration of vents within this longitudinal band. The significance of longitudinal variation in volcanic vents is

TABLE 15.3
Ionian Vent Statistics[a]

	Area (10^4 km^2)	Number of Vents	Vent Concentration (1 vent per unit area) (10^4 km^2)	Mean Vent Diameter	Std. Dev. of Vent Diameters
Total Area[b]					
0° to −90° lat 270° to 0° long	5.2	93	5.6	54	27
Latitude Bands					
Equatorial zone 0° to −30° lat 270° to 0° long	2.6	51.5	5.0	48	25
Midlatitude zone −30° to −60° lat 270° to 0° long	1.9	34.5	5.6	52	25
South polar zone −60° to −90° lat 270° to 0° long	0.65	7	9.2	65	19
Longitude Bands					
0° to −90° lat 270° to 300° long	1.73	27	6.4	—	—
0° to −90° lat 300° to 330° long	1.73	25	6.9	—	—
0° to −90° lat 330° to 0° long	1.73	41	4.2	—	—

NOTES TO TABLE 15.3

[a]Vent statistics presented here differ significantly from those published in Schaber (1980*b*, Table II) because of an error in earlier establishment of equal-area bins shown in Schaber (1980*b*, Fig. 23).

[b]Area used for vent distribution statistics contains highest resolution Voyager pictures and represents 12.5% of total surface area of Io and 35.9% of geologic map area (Plate 6).

limited by the small range of longitude assessed in this study. Mean vent diameter appears to increase steadily from the equatorial band (48 km) to the south polar band (65 km), but standard deviations overlap, indicating little or no statistical significance.

The slightly increased concentration of vents within the equatorial band may be related to volcanic activity observed during the Voyager encounters (Chapter 16 by Strom and Schneider; Chapter 19 by Pearl and Sinton). Overall, volcanic vents are evenly distributed over the surface of Io (Masursky et al. 1979); within the specific area described here, they are much more randomly distributed than are volcanic craters on the Earth, Moon, and Mars, which have a dominantly structural control. The poor structural control of vents on Io must reflect anomalous thermal conditions of the near surface associated with a high degree of interior melting (Chapter 4 by Cassen et al.; Chapter 18 by Kieffer).

As noted earlier, the shield craters and their deposits appear concentrated within ±45° of the equator. The highest concentration of mapped shield deposits includes the region surrounding the sub-Jupiter point, defined at 0° lat, 0° long (Plate 6; Fig. 15.13). One could argue that gravitational forces from Jupiter might generate a unique type of volcanism (perhaps siliceous shield construction) that is less concentrated in other areas in the vicinity of the sub-Jupiter point. However, no active volcanism was observed in this vicinity. The volcanic plumes and hot spots observed during the encounters were restricted to ± 45° of the equator, but were distributed widely in longitude (Chapter 16 by Strom and Schneider; Chapter 19 by Pearl and Sinton). The suspected concentration of shield volcanism in the region of the sub-Jupiter point may result from photogeologic bias introduced by the restricted moderate to high resolution photographic coverage of Io's surface (Fig. 15.1). Detailed geologic analysis of the entire satellite will be necessary to clarify this issue.

The photogeologic evidence as well as plume and hot-spot analyses appear to support a model for at least latitudinal periodicity of Ionian volcanism. Periodicity has been discussed by Consolmagno (1981) and Pearl and Sinton (Chapter 19). The heat flux (between 1 and 2 W m^{-2}) for Io is higher than predicted by the simple tidal dissipation model of Peale et al. (1979). (See Matson et al. 1981; Sinton 1981; Morrison and Telesco 1980.) In Consolmagno's model, volcanism occurs ~ 10% of the time, preferentially near the poles (expected from the tidal heating model), but also in or near the equatorial regions, as currently observed. His model predicts a crust 15 km thick and an upper limit to the volcanic periods of 16,000 yr.

A proper analysis of any model proposed for periodicity in Ionian volcanism is not yet possible because of the many unknown parameters regarding the interior of the body (Schubert et al. 1980; Chapter 4 by Cassen et al.). However, Earth-based observational evidence for continued thermal emission gathered over the past decade and Voyager observations of an equatorial concentration of volcanism are certain. Yet the photogeologic evidence for recent volcanism in the south polar region, based primarily on the presence of brightly haloed volcanic vents such as Nusku and Inti Paterae (Figs. 15.1 and 15.4) is convincing.

V. SCARPS, LINEAMENTS, AND GRABENS

Extremely high resurfacing rates of 10^{-3} to 10^{-1} cm yr^{-1} have been proposed to account for the complete burial of impact craters formed on the surface of Io (Johnson et al. 1979; see also Chapter 17 by Johnson and Soderblom). Similar rates have been calculated from studies of mass flow of materials in suspension from active plumes (Lee and Thomas 1980). Erosional and tectonic scarps, lineaments, and grabens are relatively abundant at all latitudes and longitudes on Io, despite the youthful surface (Fig. 15.2). These features are less well defined in the equatorial regions than in the south polar region because of less favorable solar illumination (compare Figs. 15.10 and 15.16). Despite the difficulties associated with high (75°) sun angles, a system of grabens connecting several volcanic vents and exposures of mountain material is recognizable at $-8°$ lat, 284° long (Fig. 15.15). Normal faulting appears to be dominant in this region, but evidence for some lateral displacement is noted on Fig. 15.14.

South of $-46°$ lat, tectonic features are mainly grabens clearly defined at low sun angles (Figs. 15.2, 15.10, 15.11). These tensional structures are concentrated on the layered plains unit and appear to control at least some of the layered plains boundary scarps, as discussed earlier. The grabens appear to be simple tensional features with only rare en echelon complexity; individual structures are even in width and usually linear or very broadly curved (Figs. 15.10, 15.11). Intersections of grabens are uncommon.

Scarps on Io with heights of 200 m and 600 m should be buried by younger volcanic deposits after 10,000 yr and 30,000 yr respectively, assuming an estimated average resurfacing rate of 10^{-2} cm yr^{-1} (Chapter 17 by Johnson and Soderblom). The absence of partly buried scarps and grabens in many areas indicates that erosional and tectonic processes may regionally dominate over volcanism, especially in the south polar area where no active plumes were observed (Chapter 17 by Strom and Schneider), and where pit-crater, shield-crater, and fissure-flow units are less abundant than in the equatorial region (Plate 6).

Both grabens and lineaments mostly trend northwest and northeast (Fig. 15.2), thus exhibiting a common planetary grid pattern that probably results from tidal flexing and characterizes almost all silicate and icy bodies observed so far in the solar system. The tectonic azimuth rosette shown in Fig. 15.24 indicates the presence of two nearly orthogonal (110° separation in azimuth) sets of lineaments and grabens; the approximate east-west trend (N 85° E) is not observed on the Moon and Mercury but is described on Mars and Venus (Harp 1974; Cordell and Strom 1977; Masursky et al. 1980).

The mountain material is the most disrupted by tectonic stresses (Figs. 15.5, 15.7); almost every exposure of this unit has at least one nearly orthogonal set of fractures that control the irregular topography on this material. Slump fractures common to many mountain unit exposures have resulted in large-scale terracing along the steepest slopes (Fig. 15.12).

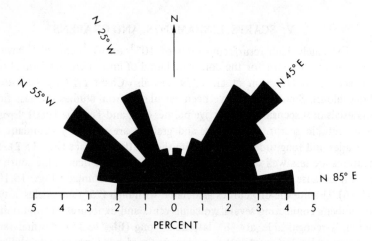

Fig. 15.24. Tectonic azimuth rosette compiled from lineaments and grabens mapped on Io (see Fig. 15.2). Two nearly orthogonal structural trends occur with 110° separation (N 25° W to N 85° E and N 55° W to N 45° E). Length of histograms is given as percent of total number of lineaments and grabens measured. Total number of data points is 151.

VI. SULFUR VERSUS SILICATES

The diverse surface morphologies on Io are described in this chapter as an attempt to provide photogeologic evidence for the presence of silicates, and to assess their possible role in the development of Io's surface. We know from bulk density (3.53 g cm^{-3}) that Io is probably dominantly composed of fer-romagnesian silicates and that it has a substantial iron or iron sulfide core. The evidence from spectral reflectance data for sulfur, sulfur compounds (including SO_2), and alkali sulfides on Io's surface is stronger than any similar evidence for silicates, at least in the uppermost layers. These spectra also severely limit the amount of H_2O that can be present on Io, thus further impeding proof of the existence of conventional silicate volcanism (Pollack et al. 1978; Fanale et al. 1977; Smith et al. 1979a).

The photogeologic evidence for at least a mixture of silicates and sulfur compounds to km-scale depths is strongest for the mountain unit, based principally on the heights attained by exposed mountain material. On Io, only the steepest slopes of some exposures seem to show large-scale shear failure (Figs. 15.5, 15.12).

The strongest photogeologic evidence against a substantial amount of silicate material composing the extensive layered planar deposits is the way the eroded layered plains unit is created from the layered plains unit (Fig. 15.10). The dissemination of the layered plains unit by what appears to be a devolatilization process provides little evidence that these planar units are composed of any significant amount of silicates. The absence of any substantial residuum following the devolatilization process implies a maximum of ~ 15% nonvolatile materials. It is possible, however, that this amount of silicates could provide the additional wall strength necessary to sustain the steep scarps observed in the pit-crater and layered-plains scarps.

The scarcity of observable slumping indicative of shear-failure along such scarps is puzzling, even considering the reduced gravity (18% that of Earth), and assuming 10 to 15% silicate materials. Calderas situated in dominantly silicate materials on Earth (e.g., Kilauea in Hawaii) and Mars (e.g., Olympus Mons, Arsia Mons) almost always have multiple, concentric fractures and slump terraces from repeated volcanism. Tectonic activity on Io is indicated by abundant faults; tidal disruption of the entire satellite has been proposed by Peale et al. (1979) and Cassen et al. (Chapter 4).

Photogeologic documentation of diverse volcanic landforms on Io does not provide the hard evidence required to verify the existence of silicate materials; compositional implications are uncertain because of the scarcity of similar terrestrial volcanic features composed dominantly of sulfur. Only a few small sulfur flows (as much as 1.4 km long, 3 m wide, and a few meters thick) are known associated with fumaroles in Japan, the Galapagos Islands, and Hawaii (Watanabe 1940; Skinner 1970; Colony and Nordlie 1973; Carr et al. 1979).

It has been proposed (Sagan 1979) that the unique color-viscosity-temperature relations and insulation properties of sulfur (Tuller 1954) should provide characteristic flow morphologies usable to differentiate sulfur from silicate dominated flows on Io. Below 433 K, sulfur has a viscosity of ~ 1 poise, but viscosity rapidly increases above 433 K as a result of polymerization, and reaches a maximum of 932 poise at ~ 463 K (Tuller 1954). In general, sulfur flows have lower viscosity than typical terrestrial lava flows of silicate composition because of the low liquidus temperature for sulfur (393 K) compared to silicate lavas (~ 1473 K).

Schonfeld and Williams (1980), noting apparently highly irregular margins on some Ionian lava flows, proposed that sulfur flows on Io could remelt underlying solid sulfur deposits, creating a characteristic nonsilicate flow morphology. Terrestrial lava flows of silicate composition can sometimes melt portions of the country rock, but the effect on the overall flow morphology is very small. Schonfeld and Williams (1980) suggested that remelting of underlying surfaces may be more significant on Io; melting of solid sulfur by molten sulfur is possible because the heat of fusion of sulfur is 16 times less than that of typical silicate lavas.

No remelting effect has been documented during the photogeologic mapping described in this chapter, but resolution available in Voyager pictures may be insufficient to detect remelting of older deposits by younger ones. Lava flows of unknown composition from a pit crater and two shield craters on Io have been shown to be similar to recent Hawaiiite flows on Mount Etna (Figs. 15.15 to 15.18). Irregular margins are also typical of most terrestrial and Martian silicate flows, and are determined primarily by rates of extrusion, slope gradients, and viscosity of the lava.

Walker (1973) reported that the average effusion rate of lava, not simply its viscosity, is the principal factor in determining the length of terrestrial silicate lava flows. His correlations of effusion rate of flow length for Mount Etna lavas are more convincing than his data for Hawaiian and other volcanoes. Recently, Malin (1980) investigated the correlation between flow length of Hawaiian lavas and rates of effusion and flow volume, showing little support for a direct relation between flow length and effusion rate in Hawaii, but revealing the expected relation between flow length and total volume of material extruded. Malin suggested that factors such as cross-sectional area and volume may be as important as effusion rate; cross-sectional area is dependent on complex factors such as slope, viscosity, and cooling rate.

The equivalent length-to-width ratios described earlier for the flows of Ra Patera (Fig. 15.16) and Mount Etna (Fig. 15.18) may also reflect proportional cross-sectional areas and thus imply similar fluid behaviors. The thickness, and thus the cross-sectional areas, of the Ra Patera flows cannot be determined. Silicate flow materials are most likely present on Io in the shield-crater flow unit concentrated in the equatorial region (Plate 6; Figs. 15.13, 15.16).

Flow deposits on Io, mapped as pit-crater flows, shield-crater flows or fissure flows, cover a wide area. The largest deposits cover about the same total area ($\sim 10^4$ to $\sim 10^6$ km^2) as major flood-lava and ignimbrite deposits on the Earth, Moon, and Mars (Schaber 1973; Fielder and Wilson 1975; Williams and McBirney 1979; Greeley and Spudis 1981; Scott and Tanaka 1981). Isolated deposits associated with an unnamed crater (Fig. 15.15), Maasaw Patera (Fig. 15.17), and Amaterasu Patera (Fig. 15.21), cover somewhat smaller areas of respectively 2.7×10^4 km^2, 1.3×10^4 km^2 and 6.7×10^4 km^2, similar to those of the Jurassic Karroo flood lavas in South Africa (5×10^4 km^2); the Pleistocene Lake Toba in Sumatra (2.5×10^4 km^2); and New Zealand ignimbrites (Williams and McBirney 1979). However, the Ionian examples are considerably less extensive than the Deccan flood basalts in India (Upper Cretaceous to Oligocene—5×10^5 km^2), Parana or Cerra Geral in Brazil (Cretaceous—10^6 km^2), Columbia River basalts (Miocene—1.3×10^5 km^2), or the youngest fissure and shield lava deposits on Mars (1 to 5×10^5 km^2) (Fielder and Wilson 1975; Williams and McBirney 1979; Scott and Tanaka 1980).

The volume of material erupted from these Ionian vents cannot be quantitatively assessed because of our inability to measure flow thicknesses, even from Voyager images of highest resolution. Volumes of terrestrial and Martian flood basalt deposits range from $\sim 10^3$ to $\sim 10^5$ km^3, and those of terrestrial and proposed Martian (Scott and Tanaka 1981) ignimbrite deposits are as large as $\sim 10^4$ km^3 and $\sim 10^6$ km^3, respectively.

A few examples of apparently thick fissure flows on Io, as in Fig. 15.20, may indicate the existence of lava-flow materials of different composition and rheological properties from those discussed above. Thick flows may, for example, represent the morphologic characteristics of sulfur extruded at temperatures between 433 K and 463 K (where viscosity is greatest), or alternatively may imply a high-silica flow only mantled by sulfurous deposits.

Topography and size data on 655 terrestrial and 13 Martian volcanic edifices were tabulated by Pike (1978), listed by general type of construct and chemical composition of erupted material. The best terrestrial analog to the large rimless craters on Io appears to be a calc-alkalic ash-flow plains caldera typified by mean diameter 21 km (diameter range 7 to 72 km), mean crater depth 450 m, mean edifice height 560 m, and circularity index 0.48 ± 0.13 (Pike 1978). Examples of such volcano-tectonic depressions can be found at Yellowstone in Wyoming; Aso in northeast Kyushu, Japan; and Rotorua in New Zealand. The depressions are characterized by voluminous sheets of ignimbrite of rhyolitic or dacitic composition and typically have little lava (Pike 1978).

The inability to distinguish silicates beneath a thin mantle of sublimate deposits with spectral reflectance data, and the absence of suitable terrestrial sulfur-flow analogs, indicate that detailed theoretical and laboratory studies (including simulations) of sulfur-flow behavior must be undertaken before many of these compositional questions can be answered. The existence of silicate volcanic materials on the surface of Io and their possible role in development of observed landforms are subjects open to continuing debate. Project Galileo to Jupiter later in this decade should provide additional evidence pertaining to the sulfur-silicate debate.

VII. CONCLUSIONS

The Voyager 1 encounter with Io has revealed a unique body whose surface morphology, color, and composition are dominated by volcanism related to tidal heating possibly in combination with other mechanisms. Preliminary mapping of the volcanic vents, geologic units, and tectonic features has provided insight into resultant landforms.

Voyager 1 images obtained during closest approach to Io show landforms indicating a wide range of compositions, possibly including silicate volcanic materials in addition to sulfur and related fumarole materials. The mountain material clearly represents a type of volcanism not currently active, as indicated by this unit's isolated exposure and eroded, tectonically disrupted surface. The heights of the mountain material strongly indicate a much greater density and strength than that of pure sulfur. This material is probably dominated by silicate tuff, ash, or ignimbrites, and may represent silicate volcanism with possibly significant amounts of sulfur. The source of the mountain material is probably shield volcanism on the silicate crust where it penetrates the sulfur-rich upper layers of Io. Thick-fissure vent flows and anomalous, thick, lobate scarps associated with the mountain unit exposures indicate fluid materials with uniquely high viscosity that might be attributed either to sulfur erupted between 433 K and 463 K, or to silicate materials of extremely high silicate content.

We cannot rule out the probability that the layered plains deposits and vent-flow materials that make up ~ 95% of the mapped surface area of Io are composed of some mixture of sulfur and silicate ash or lava, as implied by the scarp heights supported by these materials. The devolatilization of the layered plains into eroded layered plains, however, implies a low silicate content for the plains material. The crater cones greatly resemble terrestrial pyroclastic cones, but could conceivably be formed during low-energy fire-fountaining of mostly sulfur and SO_2. More violent eruptive episodes, such as those observed during the Voyager encounters result in immense plumes that spread deposits over very large areas. The best terrestrial analog (based on topography and size, not necessarily on origin) for the flat, rimless pit craters on Io may be calcalkalic ash-flow plains calderas (sometimes called volcano-tectonic depressions).

There exists a slight excess in vents per unit area in the equatorial zone and a slight deficiency of vents in the south polar region. This observed increase in the number of vents > 14 km in diameter within the equatorial zone may reflect the global concentration of active plumes and hot spots near the equator.

The restriction of shield craters and their associated deposits to the equatorial zone (± 45° lat), and the apparent concentration of such deposits in the vicinity of the subJupiter point, may imply that volcanism here is different, perhaps more siliceous, than that anywhere else in the mapped area. The presence of massive disseminated layered plains in the southern latitudes may imply that erosional processes here progress at a higher rate than resurfacing by volcanic materials. However, recent violent volcanic activity in the southern latitudes is indicated by large pit craters with well-defined haloes of fumarole or plume deposits, e.g., Nuska and Inti Paterae (Fig. 15.4).

The similarity of cumulative size-frequency distribution curves for Io's vents to those reported for other, impact cratered surfaces reflects the nonrandomness of the Ionian volcanic crater sizes and the excess of small vent sizes. A similar excess of small vents has been recognized for terrestrial explosive volcanic craters; the diameters of such craters on both bodies are thought to reflect a bias toward lowest energy eruption levels. Both are a complex function of magma temperature and gas saturation, strength and density of overlying layers, and the acceleration of gravity.

Acknowledgments: Funding for this research was furnished to the U.S. Geological Survey by the National Aeronautics and Space Administration (Solar System Exploration Division: Solar System Science Branch; Planetary Geology Program). The author recognizes D.W.G. Arthur for his calculation of the heights of surface features on Io and A.L. Dial and R.C. Kozak for their aid in establishing the size and spatial distribution of volcanic vents and geologic units. This chapter is essentially a revision of a paper on Io that appeared in *Icarus* (Schaber 1980b). Although it contains additional material, this chapter includes some text and a number of illustrations taken essentially intact from that earlier work. The editor of this book, D. Morrison, and I, are grateful to Pergamon Press (Washington, D.C.), copyright holder of the earlier paper, and to the University of Arizona Press (Tucson, Arizona), for permitting reprinting of materials.

May 26, 1981

REFERENCES

Arthur, D.W.G. (1980). Precise Mars relative altitudes: In *Reports of the Planet. Geol. Program: 1929-1980, NASA TM 81776*, p. 358 (abstract).

Arthur, D.W.G. (1981). Vertical dimensions of the Galilean satellites: In *Reports of the Planet. Geol. Program: 1980, NASA TM 82385*, pp. 12-13 (abstract).

Batson, R.M., Bridges, P.M., Inge, J.L., Isbell, C., Masursky, H., Strobell, M.E., Tyner, R.L. (1980). Mapping the Galilean satellites of Jupiter with Voyager data. *Photogrammetric Engineering and Remote Sensing* 46, 1303–1312.

Carr, M.H., Masursky, H., Strom, G.H., and Terrile, R.S. (1979). Volcanic features of Io. *Nature* 280, 729–733.

Clow, G.D., and Carr, M.H. (1980). Stability of sulfur slopes on Io. *Icarus* 44, 268–279.

Colony, W.E., and Nordlie, B.E. (1973). Liquid sulfur at Volcan Azufre, Galapagos Islands. *Economic Geol.* 68, 371–380.

Consiglio Nazionale delle Ricerche (1979). Carta geologica del Monte Etna, 1:50,000. *Litografia Artistica Cartografica*, Firenze, Italy.

Consolmagno, G.J. (1981). An Io thermal model with intermittent volcanism. *Lunar and Planet. Sci.* XII, 175–177. (abstract).

Cook, A.F. II, Shoemaker, E.M., and Smith, B.A. (1979). Dynamics of volcanic plumes on Io. *Nature* 280, 743–746.

Cordell, B.M., and Strom, R.G. (1977). Global tectonics of Mercury and the Moon. *Phys. Earth Planet. Letters* 15, 146–155.

Davies, M.E. (1981). Coordinates of features on the Galilean satellites. *J. Geophys. Res.* In press.

Fanale, F.P., Johnson, T.V., and Matson, D.L. (1977). Io's surface and the histories of the Galilean satellites. In *Planetary Satellites* (J.A. Burns, Ed.), pp. 379–405. Univ. Arizona Press, Tucson.

Fielder, G., and Wilson, L. (1975). Volcanoes of the Earth, Moon, and Mars. St. Martins Press, New York.

Greeley, R. and Gault, D.E. (1979). Endogenic craters on basaltic lava flows: Size-frequency distributions. *Proc. Lunar Planet. Sci. Conf.* 10, 2919-2933.

Greeley, R., and Spudis, P.D. (1981). Volcanism on Mars. *Rev. Geophys. Space Phys.* 19, 13–41.

Hapke, B. (1979). Io's surface and environs; A magmatic-volatile model. *Geophys. Res. Letters* 6, 799–802.

Harp, E.L. (1974). Fracture systems of Mars. *Proc. First Intern. Conf. New Basement Tectonics*. Utah Geological Assoc., Salt Lake City, pp. 389–408.

International Astronomical Union (1980). Working group for planetary system nomenclature. In *International Astronomical Union Transactions* 17B, 285–304.

Johnson, T.V., Cook, A.F., Sagan, C., and Soderblom, L.A. (1979). Volcanic resurfacing rates and implications for volatiles on Io. *Nature* 280, 746–750.

Kuiper, G.P. (1973). Comments on the Galilean satellites. *Comm. Lunar Planet. Lab.* 10, 28–34.

Lee, S., and Thomas, P. (1980). Near surface flow of volcanic gases on Io. *Icarus* 44, 280–290.

Malin, M.C. (1980). Lengths of Hawaiian lava flows. *Geology* 8, 306–308.

Masursky, H., Eliason, E., Ford, P.G., McGill, G.E., Pettengill, G.H., Schaber, G.G., and Schubert, G. (1980). Pioneer-Venus radar results: Geology from radar images and altimetry. *J. Geophys. Res.* 83, 8232–8260.

Masursky, H., Schaber, G.G., Soderblom, L.A., and Strom, R.G. (1979). Preliminary geological mapping of Io. *Nature* 280, 725–729.

Matson, D.L., and Nash, D.B. (1981). Io's atmosphere: Pressure control by subsurface regolith coldtrapping. *Lunar Planet. Sci.* XII, 664–666 (abstract).

Matson, D.L., Ransford, G.A., and Johnson, T.V. (1981). Heat flow from Io (J1). *J. Geophys. Res.* In press.

McCauley, J.F., Smith, B.A., and Soderblom, L.A. (1979). Erosional scarps on Io. *Nature* 280, 736–738.

Moore, H.J., Arthur, D.W.G., and Schaber, G.G. (1978). Yield strengths of flows on the Earth, Moon and Mars. *Proc. Lunar Planet. Sci. Conf.* 9, 3351–3378.

Morrison, D. (1977). Radiometry of satellites and rings of Saturn. In *Planetary Satellites,* (Burns, J.A., Ed.), pp. 269–301. Univ. Arizona Press, Tucson.

Morrison, D., and Telesco, C.M. (1980). Io: Observational constraints on internal energy and thermodynamics of the surface. *Icarus* 44, 226–233.

Nash, D.B., and Fanale, F.P. (1977). Io's surface composition based on reflectance spectra of sulfur/salt mixtures and proton-irradiation experiments. *Icarus* 31, 40–70.

Nash, D.B., and Nelson, B.W. (1979). Spectral evidence for sublimates and adsorbates on Io. *Nature* 280, 763–766.

Nelson, R.M., and Hapke, B.W. (1978). Spectral reflectance of the Galilean satellites and Titan, 0.32 to 0.86 micrometers. *Icarus* 36, 304–329.

Nelson, R.M., Lane, A.L., Matson, D.L., Fanale, F.P., Wash, D.B., Johnson, T.V. (1980). Io: Longitudinal distribution of SO_2 frost. *Science* 210, 784–786.

Pang, K.D., Lumme, K., and Bowell, E. (1981). Interpretation of phase curves of Io and Ganymede: Nature of surface particles. *Lunar and Planet. Sci.* XII, 799–801 (abstract).

Peale, S.J., Cassen, P., and Reynolds, R.T. (1979). Melting of Io by tidal dissipation. *Science* 203, 892–894.

Pearl, J., Hanel, R., Kunde, V., Maguire, W., Fox, K., Gupta, S., Ponnamperuma, C., and Raulin, F. (1979). Identifications and gaseous SO_2 and new upper limits for other gases on Io. *Nature* 280, 755–758.

Pike, R.J. (1978). Volcanoes on the inner planets: Some preliminary comparisons of gross topography. *Proc. Lunar Planet. Sci. Conf.* 9, 3239-3273.

Pollack, J.B., Witteborn, F.C., Erickson, E.F., Strecker, D.W., Baldwin, B.J., and Bunch, T.E. (1978). Near-infrared spectra of the Galilean satellites: Observations and compositional implications. *Icarus* 36, 271–303.

Sagan, C. (1979). Sulfur flows on Io. *Nature* 280, 750–753.

Schaber, G.G. (1973). Lava flows in Mare Imbrium: Geologic evaluation from Apollo orbital photography. *Proc. Lunar Sci. Conf.* 4, 73–92.

Schaber, G.G. (1980a). Radar, visual and thermal characteristics of Mars: Rough planar surfaces. *Icarus* 41, 159–184.

Schaber, G.G. (1980b). The surface of Io: Geologic units, morphology, and tectonics. *Icarus* 43, 302–333.

Schonfeld, E. and Williams, R. (1980). Morphological effects of unusual "lavas" on Io and Europa. *Lunar Planet. Sci.* XII, 993–994 (abstract).

Schubert, G., Stevenson, D.J., and Ellsworth, K. (1981). Internal structures of the Galilean satellites. *Icarus.* In press.

Scott, D.H. and Tanaka, K.L. (1980). Mars Tharsis region: Volcanotectonic events in the stratigraphic record. *Proc. Lunar Planet. Sci. Conf.* 11, 2403–2421.

Scott, D.H. and Tanaka, K.L. (1981). Ignimbrites on Mars. *J. Geophys. Res.* In press.

Simpson, J.F. (1966). Additional evidence for the volcanic origin of lunar and martian craters. *Earth Planet. Sci. Letters* 1, 132–134.

Simpson, J.F. (1967). The frequency distribution of volcanic crater diameters. *Bull. Volcanologique* 30, 335–336.

Sinton, W.M. (1981). The thermal emission spectrum of Io and a determination of the heat flux from its hot spots. *J. Geophys. Res.* In press.

Skinner, B.J. (1970). A sulfur lava flow on Mauna Loa. *Pacific Sci.* 24, 144–145.

Smith, B.A., Shoemaker, E.M., Kieffer, S.W., and Cook, A.F., II (1979*a*). The role of SO_2 in volcanism on Io. *Nature* 280, 738–743.

Smith, B.A. and the Voyager Imaging Team (1979*b*). The Jupiter system through the eyes of Voyager 1. *Science* 204, 951–971.

Soderblom, L.A. (1980). The Galilean moons of of Jupiter. *Sci. Amer.* 242, 88–100.

Soderblom, L.A., Johnson, T.V., Morrison, D., Danielson, G.E., Smith, B.A., Veverka, J., Cook, A.F., II, Sagan, C., Kupferman, P., Pieri, D., Mosher, J., Avis, C., Gradi, E.J., and Claney, T. (1980). Spectrophometry of Io: Preliminary Voyager 1 results. *Geophys. Res. Letters* 7, 963–966.

Strom, R.G., Terrile, R.J., Masursky, H., and Hansen, C. (1979). Volcanic eruption plumes on Io. *Nature* 280, 733–736.

Tuller, W.N., Ed. (1954). *The Sulfur Data Book.* McGraw-Hill, New York.

U.S. Geological Survey (1979). Preliminary pictorial map of Io. *U.S. Geol. Survey Misc. Inves. Map I–1240.*

Walker, G.L. (1973). Lengths of lava flows. *Roy. Soc. London Philos. Trans.* A274, 107–118.

Wamsteker, W., Froes, R.L., and Fountain, J.A. (1974). On the surface composition of Io. *Icarus* 23, 417–424.

Watanabe, T. (1940). Eruptions of molten sulfur from the Siretoko-Iosan Volcano, Hokkaido, Japan. *Japanese J. Geol. and Geog.* 17, 289–310.

Williams, H., and McBirney, A.R. (1979). *Volcanology.* Freeman, Cooper and Co., San Francisco.

16. VOLCANIC ERUPTION PLUMES ON IO

ROBERT G. STROM

and

NICHOLAS M. SCHNEIDER
University of Arizona

Nine eruption plumes were observed over a period of six and one half days during the Voyager 1 encounter with Io. During the Voyager 2 encounter, four months later, eight of these eruptions were still active. The largest plume viewed by Voyager 1 became inactive sometime between the two encounters. A major eruption occurred at Surt sometime between the two encounters and deposited an ejecta blanket comparable in size to those associated with the largest plumes. The plumes range in height from ~ 60 to > 300 km with corresponding ejection velocities of about 0.5 to 1.0 km s⁻¹. Where topographic information exists (Plumes 1 and 8), plume sources are located on level plains rather than topographic highs and consist of either fissures or calderas. With the exception of Plume 1 (Pele), the brightness distribution monotonically decreases from the core to the top of the plume. The shape and brightness distribution together with the pattern of the surface deposit of at least Plume 3 (Prometheus) can be simulated by a ballistic model in which the ejection velocity is constant (0.5 km s⁻¹) and ejection angles vary from vertical down to 55° from vertical. The brightness distribution of Plume 1 is probably better explained by a shock front near the top of the plume. Numerous surface deposits similar to those associated with active plumes probably mark the sites of recent eruptions. The distribution of active and recent eruptions apears to be concentrated in the equatorial regions indicating that the volcanic activity is more frequent and intense there than in the polar regions. This suggests that the depositional rate is greater and the surface age younger in the equatorial regions, possibly accounting for the darker polar zones. The geologic setting of certain plume sources and the very large reservoirs of volatiles required for the active eruptions suggests that sulfur volcanism rather than silicate volcanism is the most likely driving mechanism for the eruption plumes.

The detection of active volcanism on Io by the Voyager spacecraft established Io as the most volcanically active planet-sized object so far detected in the solar system. Voyager 1 and 2 images have revealed nine active eruptions evidenced by large volcanic plumes viewed at the limb regions of the satellite. In general the plumes appear larger in the ultraviolet and in forward-scattered rather than back-scattered illumination, suggesting the presence of a shell of very fine-grained particles at high altitudes. The dimensions of one plume (Loki) changed significantly during a relatively short time period, indicating that short-term fluctuations in the intensity of the eruptions may be common. It is also possible that smaller eruptions ($<$ 50 km high) were active during the two encounters but were not observed because of a combination of poor resolution and limited limb coverage.

Of the nine active eruptions viewed by Voyager 1, eight were still active during the second encounter four months later. The largest plume (Plume 1, Pele) was not observed during the second encounter although its source area was viewed near the limb. Apparently this eruption ceased sometime during the four-month interval between the two encounters. Also sometime between the two encounters a tenth major eruption apparently deposited a circular blanket of ejecta surrounding the dark caldera, Surt. The source region of this eruption was viewed by Voyager 2 at an arc distance of about 18° from the limb, but no plume was observed. Therefore, if it was still active during the second encounter it must have been less than $\sim$ 100 km high. The areal extent of the associated ejecta blanket is comparable to those of Plumes 1 and 2 (300 and 165 km high, respectively), suggesting, but certainly not proving, that the eruption which produced the deposit may have begun and terminated, or at least decreased in intensity, sometime during the four-month interval between encounters. Fig. 16.1 is a montage of the nine plumes viewed during the first encounter. Plume 9 is so close to Plume 2 that it is not readily visible in this figure (it is, however, evident in the brightness contours shown in Fig. 16.7a).

Table 16.1 lists a variety of parameters associated with a selection of photographs of the volcanic plumes, including the measured plume heights and widths. The list is not complete and includes only the better photographs from which the most accurate measurements could be made. Only a few of the best Voyager 2 Io "movie" frames of Plumes 5 and 6 are listed, although over 70 were taken. The corrected heights and widths take into account the location of the source with respect to the limb. Uncertainties in the measurements result from irregularities at the top and sides of the plumes, the exact position of the base of the plume with respect to the limb, and the diffuse nature of some plumes; they also apply to the corrected values.

TABLE 16.1

Voyager 1 and 2 Plume Observations

Plume No.	Site Name	Coordinates Lat.	Long.	S/C	Frame FDS No.	Time (1979) Day	Hour	Min	Filter[a]	Res. km/Pix	Arc Dist[b] from Limb (°)	Measured Height(km)	Corrected[c] Height (km)	Measured Width(km)	Corrected[c] Width(km)
1	Pele	-19.4	256.8	1	16368.28	63	19	19.7	Cl	7.5	-13	250±16	298	1162±106	1242
1	Pele	-19.4	256.8	1	16368.50	63	19	37.3	UV	7.3	-10	284±16	312	1115±103	1162
2	Loki	19.0	305.3	1	16375.34	64	1	0.6	Cl	4.6	- 1	165±10	166	300±25	300
2	Loki	19.0	305.3	1	16375.44	64	1	8.6	UV	4.6	+ 1	210±16	211	560±50	560
2	Loki	19.0	305.3	1	16377.52	64	2	50.9	Cl	4.1	+18	79±11	172	—	—
2	Loki	19.0	305.3	2	20513.48	186	1	56.7	UV	44.0	+16	308±62	382	352±62	—
2	Loki	19.0	305.3	2	20513.51	186	1	59.1	Cl	44.0	+16	264±62	338	352±62	—
2	Loki	19.0	305.3	2	20592.13	188	16	40.7	UV	28.2	+15	127±40	182	424±40	—
2	Loki	19.0	305.3	2	20592.23	188	16	48.7	Cl	28.2	+15	169±40	224	479±40	—
2	Loki	19.0	305.3	2	20621.30	189	16	6.3	UV	12.4	+ 1	211±17	211	508±17	508
2	Loki	19.0	305.3	2	20621.33	189	16	8.7	Cl	12.4	+ 1	148±17	148	445±17	445
3	Prometheus	- 2.9	153.0	1	16377.48	64	2	47.8	Cl	4.2	+ 5	70±6	77	258±6	272
4	Volund	21.5	177.0	1	16382.17	64	6	22.9	V	3.5	-10	70±6	98	94±8	—
4	Volund	21.5	177.0	1	16382.23	64	6	27.7	UV	3.5	-10	58±11	96	155±12	—
5	Amirani	27.2	118.7	1	16372.36	63	22	38.1	Cl	5.5	+15	50±13	114	110±13	184
5	Amirani	27.2	118.7	1	16372.50	63	22	49.3	UV	5.4	+14	81±17	137	275±18	387
5	Amirani	27.2	118.7	1	16375.28	64	0	55.9	Cl	4.6	-10	37±17	65	—	—
5	Amirani	27.2	118.7	2	20621.30	189	16	6.3	UV	12.4	- 6	62±17	72	273±17	—
5	Amirani	27.2	118.7	2	20621.33	189	16	8.7	Cl	12.4	- 6	43±17	53	61±17	—
5(F)[d]	Amirani	27.2	118.7	2	20666.12	191	3	51.9	Cl	10.8	-10	86±15	114	150±15	200
5(F)	Amirani	27.2	118.7	2	20667.00	191	4	30.3	Cl	10.8	- 9	75±15	97	205±15	—
6	Maui	18.9	122.4	1	16375.28	64	0	55.9	Cl	4.6	- 7	55±6	68	300±14	343
6	Maui	18.9	122.4	1	16372.50	63	22	49.3	UV	5.4	+18	55±15	148	—	—
6	Maui	18.9	122.4	1	16372.36	63	22	38.1	Cl	5.5	+19	33±13	e	88±18	—
6	Maui	18.9	122.4	2	20621.30	189	16	6.3	UV	12.4	- 3	54±17	57	285±17	—
6	Maui	18.9	122.4	2	20621.33	189	16	8.7	Cl	12.4	- 3	74±17	76	185±17	—
6(F)	Maui	18.9	122.4	2	20666.12	191	3	51.9	Cl	10.8	-14	32±15	88	204±15	—
6(F)	Maui	18.9	122.4	2	20667.00	191	4	30.3	Cl	10.8	-12	65±15	105	281±15	—

TABLE 16.1 (continued)

Voyager 1 and 2 Plume Observations

Plume No.	Site Name	Coordinates		S/C	Frame FDS No.	Time (1979)			Filter[a]	Res. km/Pix	Arc Dist[b] from Limb (°)	Measured Height(km)	Corrected[c] Height (km)	Measured Width(km)	Corrected[c] Width(km)
		Lat.	Long.			Day	Hour	Min							
7	Marduk	-27.9	209.7	1	16389.21	64	12	2.2	B(wa)	11.3	- 5	83±16	90	200±16	—
7	Marduk	-27.9	209.7	2	20608.01	189	5	19.1	UV	17.9	- 6	45±25	55	230±25	—
7	Marduk	-27.9	209.7	2	20608.05	189	5	22.3	Cl	17.9	- 6	54±25	64	160±25	—
8	Masubi	-45.2	52.7	1	16390.28	64	12	55.8	Cl	1.1	- 3	58±3	61	60?	—
8	Masubi	-45.2	52.7	2	20641.52	190	8	23.9	Cl	12.8	+ 7	51±18	64	153±18	177
9	Loki	16.9	300.6	1	16375.34	64	1	0.6	Cl	4.6	+ 4	[f]12±6	16	—	—
9	Loki	16.9	300.6	1	16375.44	64	1	8.6	UV	4.6	+ 5	[f]13±6	20	—	—
9	Loki	16.9	300.6	2	20621.33	189	16	8.7	Cl	12.4	+ 6	[f]25±12	35	—	—

[a]Cl = Clear, UV = Ultraviolet, V = Violet, B = Blue, wa = wide angle camera.

[b]Error on all values is ±0.5; + = in front of limb, - = beyond limb.

[c]Takes into account the position of source with respect to limb.

[d]F = viewed in forward scattering; all others backscattering.

[e]This part of Plume 6 may be a relatively low-angle ejection arm viewed obliquely (see text for explanation).

[f]Brightest part only (see text for explanation).

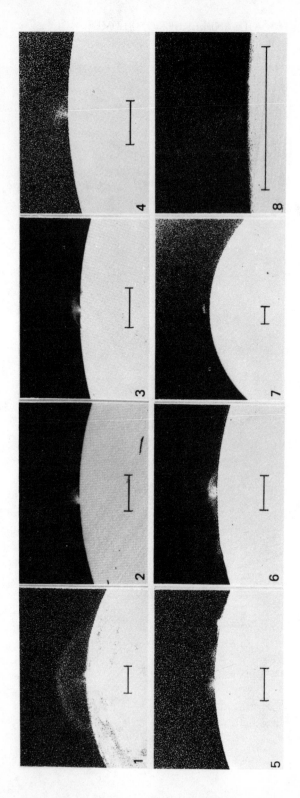

Fig. 16.1. Montage of eight eruption plumes viewed by Voyager 1. The scale bar represents 200 km. Plume 1 was taken through the ultraviolet filter, Plume 4 through the violet filter and Plume 7 through the wide-angle camera's blue filter; all others are clear-filter pictures. Plume 9 is too close to Plume 2 to be visible in this picture.

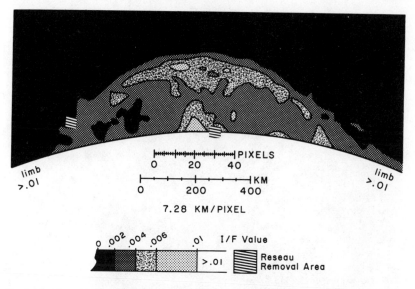

Fig. 16.2. Ultraviolet smoothed brightness contours of Plume 1 (Pele). This is the only plume which shows a bright top. The lateral extension beyond the main part of the plume to the right (north) corresponds to the extended diffuse region seen in Fig. 16.3. In this and all other brightness contour plots $I/F = \pi \times$ measured radiance/solar irradiance (GDS 16368.50, ultraviolet filter).

I. DESCRIPTION OF PLUMES AND THEIR SOURCE AREAS

Plume 1.

The largest eruption viewed by Voyager was that associated with Pele. Its height and width were ~ 300 and 1200 km, respectively. The ultraviolet brightness contours of the plume (Fig. 16.2) show an umbrella shape with a bright core and top. This is the only plume that has a bright top with the possible exception of Plume 8. No matter through which filter the plumes were viewed, the brightness of all others monotonically decreases from the core to the top. Furthermore, unlike most of the other plumes, Plume 1 has about the same dimensions when viewed through both the clear and the ultraviolet filters. In a simple ballistic model, the ejection velocity required to reach a height of 300 km is ~ 1.0 km s $^{-1}$. Another characteristic of this plume is that on the south side (left in Fig. 16.2) the lateral boundary between the surface and plume is sharp and well defined, but on the north side it is ill defined and appears to continue northward at a height of ~ 30 km.

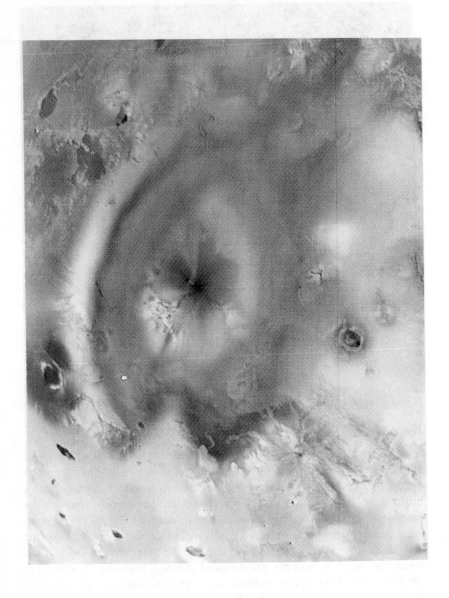

Fig. 16.3. Heart-shaped deposit laid down by Plume 1 (Pele). The north-south diameter of the darkest ring is ∼ 1000 km, which corresponds closely to the diameter of Plume 1. The very diffuse region to the north and east of this feature corresponds to the lateral extension of Plume 1 seen in Fig. 16.2. North is at the top. (FDS 16377.52, clear filter.)

The source region (Pele) of this plume forms the most dramatic surface marking on Io. It consists of an elliptical heart-shaped albedo marking formed of light and dark diffuse rings surrounding a black radial core (Fig. 16.3). The southern and western margins of the "heart" are relatively sharp and well defined, while the eastern and northern margins are very diffuse. The diameter of the plume coincides very closely with the outermost dark ring, which may represent a more concentrated deposit of pyroclastic material. The diffuse northern and eastern boundary appears to be the result of material deposited beyond about one plume radius from the outer ring. This outer diffuse zone coincides with the lateral extension of the plume beyond the main umbrella shown in Fig. 16.2 and is probably material deposited by this part of the plume, implying an asymmetry in the ejection conditions with a significant horizontal velocity component directed to the north and east of the main umbrella. The multiple rings associated with this eruption may record stages in its evolution similar to growth rings in trees. As pointed out earlier, the outermost dark ring coincides closely with the observed plume diameter, and the inner rings may represent early stages or fluctuations in the intensity of the eruption. A comparison between Voyager 1 and 2 images of the surface markings shows that the dark inner core did not change significantly between the two encounters, although the ring geometry did. The core may be primarily a surface deposit that has retained its form since its initial deposition. Possibly the inner dark core records an initial low intensity eruption that subsequently grew and deposited larger and larger rings until it ceased sometime between the two encounters.

The primary source of Plume 1 appears to be a black elongated depression ~ 24 km long and 8 km wide (see Fig. 16.4) surrounded for $\sim 180°$ with a dark, diffuse pyroclastic deposit some 200 km in radial extent. This black depression occurs at the eastern end of an east-west trending trough ~ 60 km long that forms the northern margin of a fractured plateau. To the west about 25 km are two black spots situated on opposite sides of the trough. These black spots seem to be the source of the dark pyroclastic deposit to the west separated from the larger eastern deposit by a light region (see Fig. 16.4). It is not possible to determine if all three sources were active and contributing to Plume 1, or just the larger one.

Although the eruption site of Plume 1 is associated with a fractured plateau ~ 200 km across, it occurs in a flat area at the margin of the plateau and is tectonically controlled by an east-west trending graben. Other eruption sites with dark pyroclastic deposits are also found either at the margins of the plateau or on the floor of the central graben. All appear to be tectonically controlled. The plateau itself does not appear to be a volcanic construct similar to those found on the Earth and Mars; rather it seems to be a fractured uplift with the location of eruptive vents controlled by the fractures. No constructs are associated with the eruption sites, indicating that

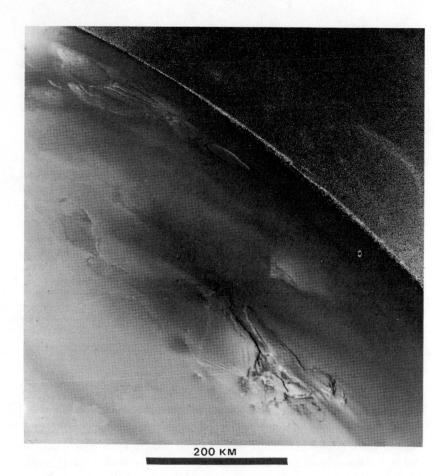

200 KM

Fig. 16.4. High resolution picture of the source region of Plume 1 (Pele). In this picture
the sky has been separately processed to show the filamentary structure near the top
of the plume. The primary source of Plume 1 is the dark elliptical caldera $\sim 24 \times 8$
km situated at the base of the fractured plateau. (FDS 16391.30, clear filter.)

there has not been any long-term repetitive activity at these locations.
Furthermore, the oblique view of Plume 1 (Fig. 16.4) shows no column of
dark material rising from the source as seen in similar views of plumes at
Loki, Prometheus, and Maui. This indicates that the material being ejected is
optically thin and of low particle density. Possibly this type of activity
represents the final phase of an eruption.

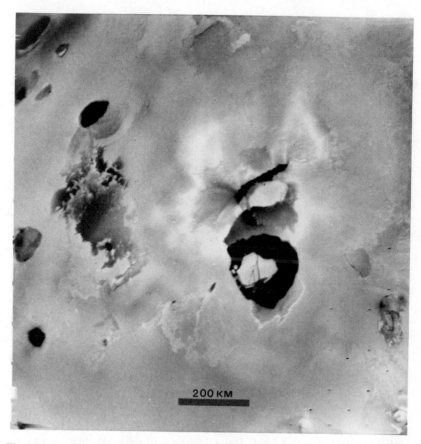

Fig. 16.5. Nearly vertical view of the source area of Plumes 2 and 9 (Loki). A dark curtain of ejecta (Plume 2) is issuing from the western edge of the black strip Loki. This curtain of material is seen obliquely in Fig. 16.6. Plume 9 is the small dark diffuse spot located at the extreme eastern end of Loki. North is at the top. (FDS 16389.42)

Plumes 2 and 9.

A black strip termed Loki is the site of at least two major eruptions which form Plumes 2 and 9. Loki is ~ 180 km long and 20-50 km wide with an irregular margin (Fig. 16.5). It is located ~ 150 km north of a black annulus (Loki Patera) ~ 200 km in diameter, which is a warm spot ~ 180 K above the background temperature of ~ 130 K (Pearl et al. 1979; Pearl and Sinton, Chapter 17). Plume 2 originates from a 50 km long fissure that marks the

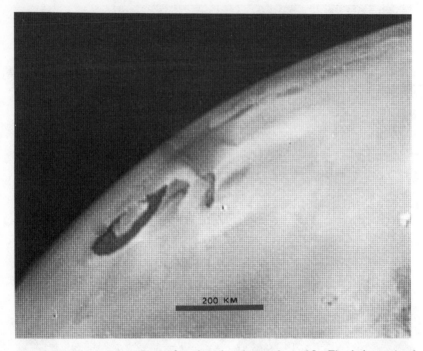

Fig. 16.6. Oblique view of Plume 2 projected against surface of Io. The dark curtain of ejecta corresponds to the dark area at the western end of Loki seen vertically in Fig. 16.5. The portion of the ejecta closest to the limb is optically thinner than that near the vent region. (FDS 16377.52)

extreme western end of Loki. There, a curtain of optically thick pyroclastic material is being ejected in a westerly direction along a looping path suggestive of a ballistic trajectory (Figs. 16.5 and 16.6). Preliminary estimates of the range (~ 100 km) and height (~ 50 km) of this curtain suggest an ejection velocity of ~ 470 m s^{-1} and an ejection angle of $\sim 63°$ from the horizontal. Image enhancement of the picture shown in Fig. 16.6 indicates that the plume consists of two components: an optically thick portion mentioned above and an optically thin component that extends ~ 80 km above the limb. The latter component probably represents a low-density cloud ejected at a higher velocity (~ 750 m s^{-1}). Figures 16.7a and b are brightness contours of Plume 2 derived from pictures taken through the clear and ultraviolet filters. The characteristics of Plume 2 are quite different from those of Plume 1. Plume 2 is more diffuse and lacks a well-defined umbrella shape. Furthermore, like the other plumes, the brightness distribution monotonically decreases from a maximum in the core to a minimum at the

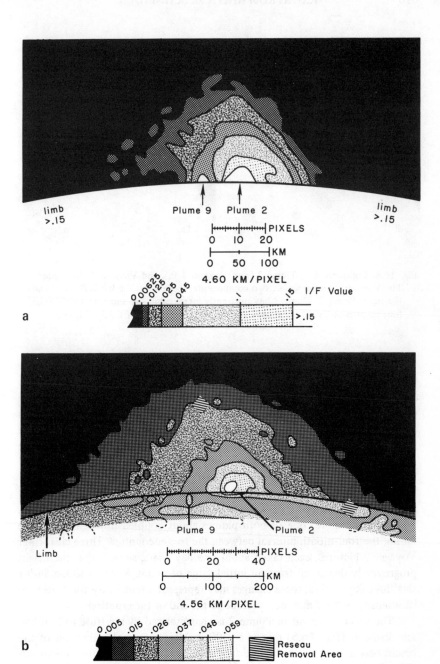

Fig. 16.7. (a) Clear filter smoothed brightness contours of Plumes 2 and 9. (FDS 16375.34) (b) Ultraviolet filter smoothed brightness contours of Plumes 2 and 9. (FDS 16375.44)

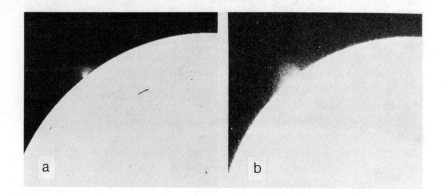

Fig. 16.8. Comparison of Plume 2 from Voyager 1 (a), and Voyager 2 (b) encounters. The Voyager 2 frame shows two components; the one on the left is Plume 9 and on the right, Plume 2. Plume 9 has apparently increased in size since the first encounter four months earlier. (Left image FDS 16375.34; right image FDS 20621.33.)

top. Viewed through the ultraviolet filter the plume is very diffuse but appears higher by a factor of ~ 1.3 and wider by a factor of ~ 2. The preferential scattering in the ultraviolet may be due to very fine particles above the main part of the plume (Collins 1981). It was in this plume that the Voyager 1 IRIS experiment detected gaseous SO_2 (Pearl et al. 1979). On the first encounter Plume 2 was considerably larger than Plume 9 and dominated the overall structure and brightness distribution (see Figs. 16.7a and b). By the second encounter, however, Plume 9 had grown considerably so that at least the bright central core was comparable in size to that of Plume 2 (Fig. 16.8). Furthermore, one or both plumes about doubled in height sometime during the four-month interval between the two encounters. Three clear-filter Voyager 2 pictures, each taken about one day apart, suggest that the height progressively decreased from an initial value of ~ 330 km to 148 km during this three-day period; these changes may represent a temporary fluctuation in its intensity or they may be a prelude to the end of the eruption.

The surface deposit of Plume 2 is characterized by a diffuse halo of 850 km diameter (Fig. 16.5) that is ~ 300 km larger than the diameter of the hemispherical part of the plume in the ultraviolet. The brightness contours (Fig. 16.7), however, show lateral extensions that may be particles with a horizontal velocity component that are being deposited at distances of a few 100 km beyond the main plume. Through the clear filter, the plume width is

~ 240 km where it intersects the limb (see Fig. 16.7). Seen vertically, this width corresponds to a diameter midway between the outer boundary of the dark curtain of ejecta and the surrounding bright zone. Although more irregular, this bright zone is similar to the bright annulus surrounding Prometheus (Plume 3) and also occurs at a similar position with respect to the lateral extent of the plume. Perhaps these white zones are frozen SO_2 deposited around the periphery of the plumes when the temperature fell below its freezing point. This deposit is not seen on the dark annulus to the south because at ~ 290 K the surface is well above the melting point of SO_2.

Plume 9 is located at the eastern end of Loki, which shows a small dark diffuse splotch somewhat similar to the dark curtain of ejecta associated with Plume 2 (Fig. 16.5). This feature corresponds to the small bright spot shown in the brightness contours of Fig. 16.7a. Similar light and dark splotches occur all along the northern boundary of Loki; possibly low intensity eruptions are taking place along this entire border.

Significant changes in the deposition pattern of the pyroclastics occurred during the four-month interval between the two encounters (Plate 8 in the color section). In particular the light and dark pattern associated with Plume 9 changed and enlarged considerably. This enlargement is probably the result of the growth of Plume 9 mentioned earlier. Also changes are evident in the albedo markings associated with Plume 2. Furthermore, the albedo of the northern portion of Loki Patera has increased, probably due to the accumulation of pyroclastics during the four-month period between encounters.

Because of the lack of topographic information associated with Loki, it is not possible to determine its structural characteristics. The black material may be an extrusive occupying a long irregular depression or caldera, or it may consist of flows extruded from a fissure forming its western and northern boundaries. The very diffuse nature of Plume 2 may be related, at least in part, to the vent geometry. Plume 2 originates at a 50 km long irregular fissure, while the very symmetrical, umbrella-shaped Plume 1 originates at an elliptical depression.

Plume 3.

Prometheus is one of the most distinctive surface markings on Io. It is characterized by a bright annulus ~ 330 km in diameter surrounding a darker region 230 km in diameter with irregular black markings (Fig. 16.9). Oblique pictures (Fig. 16.10) show dark ejection arms issuing from the center of the ring. Several of these arms impinge on the bright annulus.

Plume 3 was photographed at moderate resolution (8 km/line pair) only through the clear filter. It has a more or less symmetrical umbrella shape and is ~ 77 km high and ~ 270 km wide. Unlike Plume 1, which has a similar overall shape, the brightness distribution of Plume 3 shown in Fig. 16.11

Fig. 16.9. Voyager 1 image showing the source areas of Plumes 3, 4, 5, 6, and 7 and
several diffuse deposits similar to those associated with active plumes (arrows). (FDS
16368.28, clear filter.)

monotonically decreases from the core to the top. The width of the plume is
only 40 km larger than the diameter of the inner ring of Prometheus, and
therefore most of the bright annulus lies outside the limits of deposition of
the plume as seen through the clear filter. However, plumes usually appear
larger in the ultraviolet, and it is possible that such a picture would show a
width comparable to the outer diameter of the bright annulus. As suggested
for the bright area associated with Plume 2, this bright annulus may be a
frozen SO_2 deposit.

The vent area from which Plume 3 issues is not visible because of a
combination of poor resolution and obscuration by the dark ejected material.
However, the overall symmetry of the plume and the umbrella shape suggests

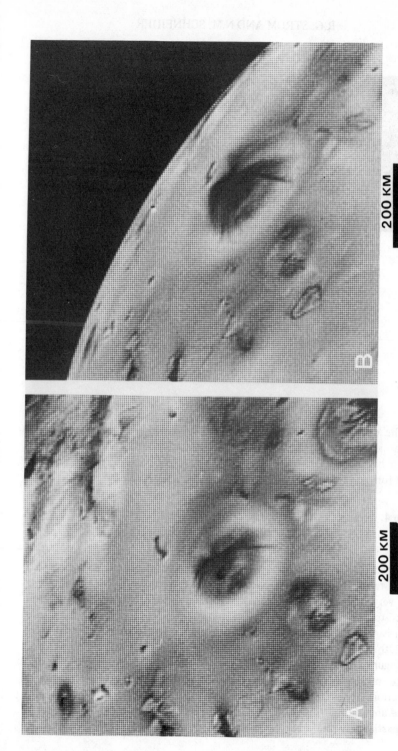

200 KM

200 KM

Fig. 16.10. Voyager 1 oblique views of Plume 3. The left-hand image (A) was taken 2.3 hours before the right-hand image (B). (Left image FDS 16372.36; right image FDS 16375.28)

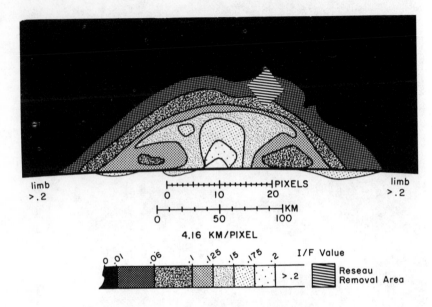

Fig. 16.11. Clear filter smoothed brightness contours of Plume 3. (FDS 16377.48)

that the vent may be similar to that associated with Plume 1 — a roughly circular area rather than a long fissure. Computer simulations of Plume 3 using a ballistic model also suggest that a disk or point source is a possible model for the vent (see Sec. II).

Plume 4.

Only two pictures of Plume 4 (Volund) were obtained by Voyager 1, one through the violet filter and the other through the ultraviolet. Both have a resolution of ∼ 7 km. Plume 4 was also apparently imaged by Voyager 2 on a picture taken for optical navigation purposes (Strom et al. 1981). This plume has a very irregular shape somewhat like a mushroom with an elongated top slanted to one side (Fig. 16.12). The maximum height is ∼ 120 km and the width is ∼ 95 km. This is one of the few plumes whose height is greater than its width.

Figure 16.13 shows the two highest resolution pictures (11 km) of the source region of Plume 4, taken ∼ 2.3 hr apart. They are orthographic projections centered on Volund and show a marked change in the appearance of the area during this 2.3-hour interval. The entire area surrounding Volund and particularly the area to the north appears blurred and softened in the

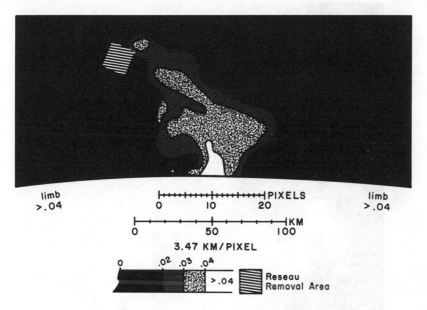

Fig. 16.12. Violet filter smoothed brightness contours of Plume 4. (FDS 16382.17)

later image as if it were partially obscured by a cloud of material. In all earlier views of this region well-defined light and dark patches and lineations can be seen. The limb picture of Plume 4 (Fig. 16.1) was taken ~ 5 hr after the later image (Fig. 16.13). In these images the plume is much too small to account for these very widespread changes. A possible source area for an additional plume occurs in the complexly marked region north of Volund where most of the obscuration is centered; however, this area was very near the limb on the images showing Plume 4, and no plume was observed there. There are three possible explanations for the marked changes in the appearance of this region. First, the difference in phase angle between the two frames is ~ 12°, and it is possible that the apparent change is a phase effect due to different photometric properties of the material in this region. A second possibility is that there was a brief (<7 hr) intensification of the Volund eruption large enough to obscure surface detail over an area comparable to that of Plume 1. Alternatively, a new large eruption, possibly centered in the complexly marked region north of Volund, may have started and then terminated within a period of ~ 7 hr. If either of the latter two possibilities is correct, then very short-term eruptions or intense fluctuations are probably common on Io. A fluctuation in the intensity of Plume 2 occurred during the second encounter, and possibly a similar phenomena occurred at Surt.

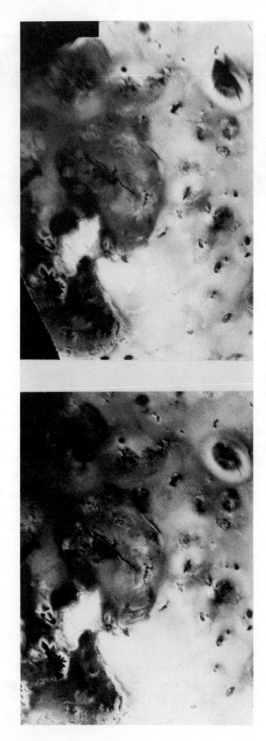

1000 KM

Fig. 16.13. Voyager 1 orthographic projections of the source area of Plume 4 (Volund). The left-hand image was taken 2.3 hr before the right-hand image. Most of the surface detail west and north of Volund in the left view has been obscured in the right view. (Left image FDS 16372.36; right image FDS 16375.28)

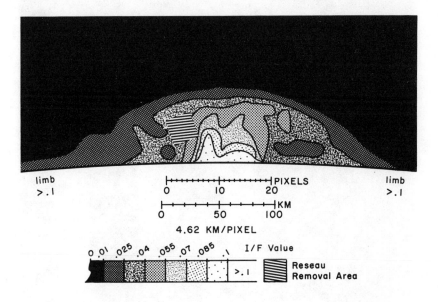

Fig. 16.14. Clear filter smoothed brightness contours of Plume 6. The bright central core of this plume has two components that correspond to the two dark eruption jets seen in the oblique view in Fig. 16.15. (FDS 16375.28)

Plumes 5 and 6.

Plumes 5 and 6 are associated with two irregular black spots (Amirani and Maui) that are surrounded by light and dark diffuse halos (Fig. 16.9). These eruption sites are separated by ~ 300 km, but connected by an irregular black line. Fig. 16.14 shows the brightness contours of Plume 6 (Maui). Although the overall structure of this plume is somewhat similar to Plume 3, there is a notable difference. In particular Plume 6 has a broad central core ~ 50 km wide showing two distinct peaks, corresponding to two dark optically thick jets of material seen on an oblique image (Fig. 16.15). Furthermore, the combined width of these jets is nearly the same as the bright core of the plume. A more nearly vertical view of Maui shown in Fig. 16.9 also displays a dark streak directed to the east, which is probably one or both of these eruptive jets. Apparently Plume 6 consists of at least two major eruptions at either end of a ~ 50 km long portion of an irregular black spot forming Maui. The resolution is not sufficient to determine whether these eruptions are similar to Plumes 2 and 9 and associated with a fissure active along most of its 50-km length, or whether they represent two discrete

Fig. 16.15. Voyager 1 oblique view of Plumes 5 (Amirani) and 6 (Maui). Plume 6 consists of two dark streaks that coincide with the two bright components that make up its central core shown in Fig. 16.14. (FDS 16372.36)

sources associated with calderas, but the fact that they are preferentially directed to the east suggests that they may be fissure eruptions similar to those associated with Loki.

Plumes 7 and 8.

Plume 7 was imaged only twice during Voyager 1 encounter. Both frames were taken near closest approach with the wide-angle camera through the blue filter. It is a mushroom-shaped cloud ~ 90 km high and 200 km wide (Fig. 16.1). The source region, termed Marduk, consists of three dark markings connected by an irregular black line (Fig. 16.16). The surface configuration of the northernmost markings appears to have changed during a 2.5-hr period and therefore this feature is the most likely source.

Only two images of Plume 8 were taken by Voyager 1. Both are clear-filter narrow-angle pictures taken at closest approach with a resolution of about two km. At this time Plume 8 (Masubi) was near the terminator on the dark limb and the images of the plume are extremely faint. Its height is ~ 60 km, but its shape and width are very uncertain due to the faintness of the images. The highest resolution picture of the Masubi source area is shown

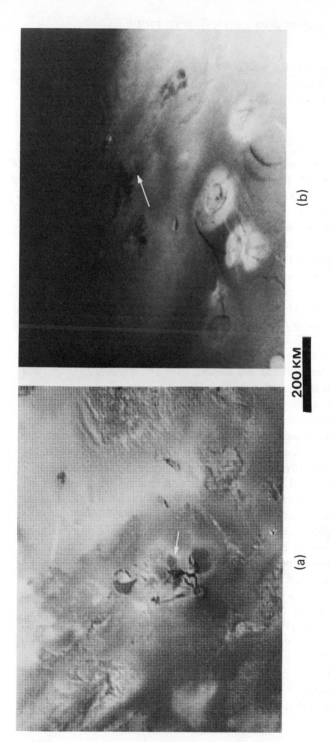

(a)   (b)

200 KM

Fig. 16.16. Voyager 1 highest resolution images of the source areas of Plumes 7 (Marduk) and 8 (Masubi). Marduk (a) consists of three diffuse areas connected by a black band. The northernmost component (arrow) is thought to be the vent area because the dark region to the right appears to have changed in a later frame. Masubi (b) is indicated by the arrow and consists of a dark diffuse area near the end of a black flow. (Left image FDS 16375.28; right image FDS 16392.37)

in Fig. 16.16*b*. It is a dark irregular V-shaped feature possibly consisting of flows. A short lateral extension is surrounded by a dark and light diffuse halo; the distal end of this extension is probably the vent area.

Surt (Site 10).

Sometime during the four-month interval between the Voyager 1 and 2 encounters a large eruption apparently began, centered on a black spot at $+ 45°.2$ latitude and $337°.7$ longitude (Surt). A comparison between Voyager 1 and 2 images of this region is shown in Plate 8 in the color section. In the Voyager 1 image the area is characterized by circular to irregular light patches surrounded by dark regions. On the Voyager 2 image the same region has changed dramatically. It now consists of a circular diffuse area ~ 1000 km in diameter centered on a black spot ~ 50 km in diameter. Most of the light patches have been obscured by dark material, or their pattern has been significantly altered. The caldera Manua Patera (M in Plate 8), clearly visible in the Voyager 1 image, has almost totally disappeared in the Voyager 2 image. Numerous other changes can be seen. This eruption was apparently detected on 11 June 1979 by Sinton (1980*a*) who observed a 5 μm radiation outburst and decay near the longitude of Surt. Unfortunately Voyager 2 did not view Surt closer to the limb than an arc distance of $\sim 18°$. Because no plume was observed, an upper limit on the plume height of ~ 100 km can be set, if the eruption was still active during the second encounter. However, the areal extent of the obscuration is comparable to dimensions of deposits associated with Plumes 1 and 2, which are 300 and 160 km high, respectively, suggesting that the eruption may have begun and terminated, or at least decreased in intensity, sometime between the two encounters. Probably this decrease and/ or termination occurred during the one-month interval between Sinton's Earth-based observation and the Voyager 2 encounter. This observation indicates that the onset and intensity fluctuations of volcanic eruptions on Io must be very frequent, i.e., on a time scale of months of less. Furthermore, major changes in surface markings also must occur on similar time scales, and it is probable that the albedo and color patterns viewed by the Voyagers will be very different in a few years. The highest resolution picture of Surt is an oblique view obtained by Voyager 1 at a resolution of $\sim$3 km (Fig. 16.18). A 50-km diameter caldera (S in Fig. 16.17) corresponds to the similar-sized black spot seen on the Voyager 2 image in Plate 8. This caldera is apparently the source of the 1000-km diameter deposit that surrounds it on the Voyager 2 image. During the first encounter the floor and rim of this caldera contained a few small dark spots, but, in general, it had an albedo similar to its surroundings. Apparently the Surt eruption was accompanied by the extrusion of black material that covered the caldera floor.

Fig. 16.17. Voyager 1 oblique view of Surt (S). Surt is a 50-km diameter caldera, the site of the large eruption which took place between the Voyager 1 and 2 encounters. The crater Manua Patera (arrow) and many of the albedo patterns were obscured by the deposit laid down by this eruption (see Plate 8 in the color section). (FDS 16388.58)

II. PLUME DYNAMICS

The motions of the particles in the plumes are directly linked to the eruption mechanism and conditions at the vent. An understanding of the dynamics and dominant physical processes should give significant insights into the eruptive process and the nature of the source.

Cook et al. (1979) introduced the ballistic and shock front models as possible explanations for the shape and structure of the plumes, favoring the shock model. This section summarizes the analysis of plume dynamics in the context of these two models, but incorporates the corrections of Strom et al.

(1981). In this analysis, Voyager plume images are compared with computer simulations and theoretical predictions of the models to decide which one best fits the eruptions on Io.

Before the models and simulations are discussed, it should be emphasized that the geometry and brightness structure of Io's eruption plumes vary considerably. Apparently the vent geometry significantly affects the shape and brightness distribution; other parameters such as ejection velocity, angle and direction, and particle size distribution probably also affect plume characteristics. One simple dynamical model probably cannot explain all plumes. In this section only rather simple preliminary models are compared to the observed characteristics of Plumes 1 and 3.

A number of simplifying assumptions have been made for the preliminary modeling done to date. Both models assume a constant flow of gas and entrained particles from the vent. As the material leaves the immediate vicinity of the vent, the motions of the gas and particles become decoupled. Most of the plumes are $\lesssim 100$ km high and $\lesssim 300$ km wide, and therefore the change in gravity with height and curvature of the surface are negligible and not considered. However, the 300 km height and 1200 km width of Plume 1 result in a decrease in gravity of 25% at the top of the plume, and a surface curvature of $38°$ at its base. Therefore, any conclusion concerning the dynamics of Plume 1 must be viewed with caution until a more rigorous analysis is completed. Furthermore, ejection velocity is held constant in both models.

In the ballistic model, particles follow parabolic trajectories dictated by the initial conditions at the vent, so that

$$z = r \cot i - \frac{g r^2}{2v_0^2} \csc^2 i \qquad (1)$$

where z is the height, r is the radial distance from the vent, i is the ejection angle from the vertical, g is the surface gravity, and v_0 is the initial ejection velocity. The limiting envelope for a set of such trajectories is

$$z_M = \frac{v_0^2}{2g} - \frac{g r_M^2}{2v_0^2} . \qquad (2)$$

If the limiting ejection angle of the set (i_0) is closer to the vertical than $45°$ (which gives maximum range), the envelope is described by Eq. (2) at the top and by Eq. (1) at the sides, with $i = i_0$.

If $r_M = 0$ in Eq. (2), the ejection velocities can easily be derived from the plume heights listed in Table 16.1. With the exception of Plume 1 and a few Voyager 2 Plume 2 measurements, all the velocities fall below 1.0 km s^{-1}. A few Plume 2 measurements give velocities from 0.75 to 0.90 km s^{-1}, but the vast majority of the plumes have velocities in the range 0.45 to 0.65 km s^{-1}.

In the shock front model, the gas and particles initially follow the same parabolic trajectories as in the ballistic model, but the driving gas density is assumed to be significantly greater, such that a parcel of rising gas will collide with a falling parcel. In equilibrium, this collision takes place at a dome-shaped shock front. Cook et al. (1979) and Strom et al. (1981) show the equation for this surface to be

$$z_M = \frac{4v_0^2}{9g} - \frac{g r_M^2}{2v_0^2} .$$ (3)

If the shock front is not supported by gas flow at all points, the envelope may drop off more rapidly at the sides. A two-dimensional analogy is the shock front created by water running up a tilted plate. The thickness of the water layer on the plate corresponds to gas density in the plume.

A dense layer of gas will be formed above the shock front. A molecule in this layer will be subjected to the upward force of a pressure gradient, but a simple energy calculation shows that gravity confines the molecules near the shock front. Assume that roughly all the bulk velocity lost in crossing the front is converted to thermal velocity, and consider a trajectory near the top of the plume. Such a molecule will attain its greatest height if its thermal velocity is upwards, i.e., very close to its original trajectory. Thus, in this best case a molecule on the average will just reach the ballistic envelope given by Eq. (2). Since most molecules will have random velocities in less favorable directions, they will be confined to a region between the two envelopes. Comparisons of Eqs. (2) and (3) show that the distance between the two is only 1/8 the height of the shock envelope. The net result is a dense outer shell of gas, which flows down the outside of the shock front. This simple analysis ignores such problems as phase changes across the shock front, gas-particle energy exchange, and detailed pressure balancing over the shock front. These will undoubtedly affect the shape and density structure of the shocked layer, but will not qualitatively affect the plume structure. If the density of the shocked layer is low, the particles will pass through with unchanged trajectories, and the shock front will not be observable. Further discussion will be limited to the other extreme, where the dense layer will trap the particles and carry them down the outside of the front.

The ballistic and shock models arise from significantly different driving gas densities but lead to plumes that are similar in many respects. The envelopes described in Eqs. (2) and (3) differ only by a small constant. This constant is significant only if one ignores the thickness of the shocked layer and rigidly accepts the constant velocity assumption. Also, a downward bend at the sides of the envelopes can be interpreted either as an unsupported shock front of a ballistic plume with a limiting ejection angle less than $45°$ from the vertical. A comparison of envelope shapes, therefore, cannot be used to decide between the two models. In this section we will show that both models predict ring-shaped deposits.

Fortunately, the two models predict significantly different density structures. Despite the simplicity of the ballistic model, it is difficult to derive the structure analytically. This problem is overcome with the use of Monte Carlo computer simulations. A large number of trajectories were chosen randomly according to a variety of parameters, and represented by a set of points spaced equally in time.

Plume 3 was chosen for comparison with the simulations because (1) its envelope is relatively symmetrical, (2) it was imaged at relatively high resolution, and (3) its brightness structure is relatively symmetrical and well shown by the brightness contours. (Figure 16.11 shows the brightness contours of Plume 3.) A simulation of the profile of Plume 3 must reproduce four general characteristics:

1. A relatively symmetrical parabolic envelope;
2. A wide ($\sim$ 20 km) bright core;
3. A monotonic decrease in brightness from the core to the top of the plume;
4. Darker areas on each side of the central core surrounded by brighter areas.

Furthermore, Plume 3 exhibits a surface deposit characterized by a bright white ring near the outer margins of the plume (see Fig. 16.9). A model simulation should also be capable of reproducing this ring, which may be a concentration of SO_2 frost deposited by the plume. Other plumes are also associated with light and dark diffuse rings that probably represent concentrations of pyroclastics and/or SO_2 frost.

A number of ballistic simulations were attempted with various geometries of the vent area, limits on ejection angles, angle-frequency distributions, and velocity-frequency distributions. Clearly it is not possible to match the structure of the plume exactly with the simple models used in these initial simulations; asymmetries and irregularities in the shape of the plume envelope will require more detailed models than have been attempted to date. Only qualitative comparisons can be made between the brightness structure in the Voyager image and the density in the simulations. Nevertheless, at least two simple ballistic models give remarkably good fits to the observed plume characteristics. In all simulations the ejection velocity was held constant at 0.5 km s^{-1}, which is the approximate velocity derived by applying the ballistic equation to Plume 3. The simulations that best fit the brightness contours shown in Fig. 16.11 are ones in which material is ejected isotropically into a cone from $0°$ down to $55°$ from vertical. Figures 16.18 and

16.19 show two simulations with different source geometries. In Fig. 16.18 the source area is a set of points distributed uniformly over the surface of a 15-km radius disk. In Fig. 16.19 the vent is a point source. Considering the simplicity of the models, both simulations duplicate rather closely the parabolic shape, the monotonic decrease in brightness from the core to the top of the plume, and the darker regions on either side of the core. However, in the point source model (Fig. 16.19a) the width of the bright core is much narrower than shown in the brightness contours. In both the point and disk source simulations, the source is precisely at the limb. However, in the picture from which the brightness contours were derived, the source was five degrees in front of the limb and therefore the apparent basal width of the core is, in fact, its width seven km above the source. When the base of the plume is shifted seven km upward on the point source simulation (white line in Fig. 16.19a), the width of the bright core is more similar to that in the brightness contour plot (Fig. 16.11). A similar shift on the disk source simulation does not significantly change the core width. Therefore, either source model will fit the brightness distribution of Plume 3, but a point source is probably the more geologically realistic model. The restriction that velocity is independent of ejection angle prevents the simulation from matching exactly the shape of the envelope of Plume 3. The simulation would better match the plume by introducing a ∼ 5 % variation in velocity as a function of ejection angle.

Both source model simulations predict an annular concentration of particles at maximum range as shown in Figs. 16.18b and 16.19b. The annular concentration occurs at maximum range with isotropic ejection because dr/di goes to zero at this point, and therefore ejecta are concentrated in a thinner ring at maximum range. The annulus is wider in the disk source simulation because discrete sources are distributed uniformly over a rather broad area. This wider annulus is more like the width of the bright ring surrounding Prometheus. However, a wider annulus could be produced in the point source model by a small ejection velocity dispersion with standard deviation ∼ 5 % of ejection velocity, or by fluctuations in the intensity of the eruption, or both.

The simulations indicate that rather simple ballistic models are capable of reproducing the general characteristics observed in Plume 3. Furthermore, these models place rather tight constraints on the eruption conditions. For example, there can be very little variation in ejection velocity because the addition of a thermal speed ≳ 10 % of the bulk flow speed will result in a plume shape poorly matching that observed. Also, the range of ejection angles must exceed 45° (maximum range) from vertical but be considerably less than 90° from the vertical. (In the simulations this range was from 0° to 55°.) If the angles are less than ∼ 40° from the vertical, the brightness distribution does not match that observed in Plume 3, and the deposit is not concentrated in a ring. The angles used here are highly dependent on the assumption of

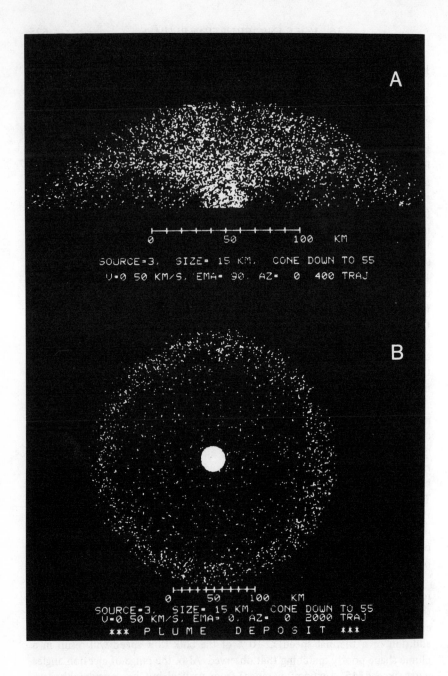

Fig. 16.18. Computer simulation of a ballistic model for Plume 3, where the ejection velocity is 0.5 km s^{-1} and the ejection angle varies from vertical 0° down to 55°. In this model the vent is a 15-km radius disk with multiple point sources. The profile (A) is very similar to the Plume 3 shape and brightness distribution shown in Fig. 16.11. The surface concentration of ejecta is shown in the bottom simulation (B).

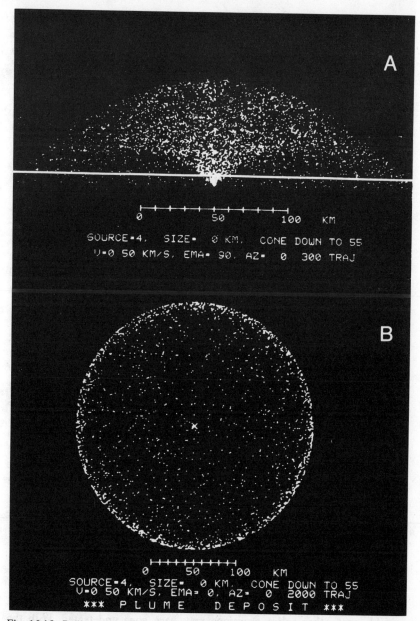

Fig. 16.19. Ballistic simulation with the same parameters as used in Fig. 16.18 except that here the vent is a point source. The brightness distribution (A) is similar to that of Plume 3 shown in Fig. 16.11, but the core region is too narrow. However, since the source area of Plume 3 was 5° in front of the limb, the brightness contours are actually seven km above the vent (white line). When this effect is taken into account the width of the core corresponds more closely to that shown in Fig. 16.11.

constant ejection velocity. If the velocity is allowed to be a slowly varying monotonic function of angle, there will be an angle that gives maximum range. As long as the limiting ejection angle from vertical is greater than this angle, we can match the roughly parabolic envelope, the slight brightening of the outside edges surrounding the darker lateral areas, and the ring deposits. Finally, if the angle-frequency distribution departs significantly from isotropic, neither the observed plumes nor their deposits can be reproduced. For example, a distribution peaked toward the vertical will produce a tall narrow plume with a deposit concentrated around the vent area.

Plume 1 also has a symmetrical umbrella shape and a ring-shaped deposit similar to that of Plume 3. However, Voyager images show a bright outer envelope comparable in brightness to its core. This is apparent in the ultraviolet brightness contour (Fig. 16.2) and two other clear-filter images. To date no ballistic simulation has produced the bright outer envelope together with the ringed surface deposit. However, other factors not included in the initial simulations may be important. For example, with a plume of this height, the change in gravity with altitude and the longer flight times ($\sim$ 20 min) may be important. Until these have been more thoroughly explored, we tentatively propose a shock front model based on the bright outer envelope. As mentioned above, the equations of motion have not yet been solved for the shock model. However, the density will increase across the shock front, and the parabolic shell of denser material will appear as a bright outer envelope in projection. The ring deposits clearly can be reproduced by material flowing down the shock front. Therefore, the bright outer envelope of Plume 1, together with the filamentary structure of the upper regions as pointed out by Cook et al. (1979) and shown in Fig. 16.4, suggest that the structure of Plume 1 may be the result of a shock front. The possibility of a shock front in Plume 1 has some interesting implications. Apparently the density of the driving gas in this plume is significantly greater than in Plume 3, especially considering the much greater size of Plume 1. This suggestion is consistent with the much greater ejection velocity and this could imply significant differences in the eruption conditions or in the nature of the materials being erupted.

III. DISTRIBUTION OF PLUMES AND RECENT PYROCLASTIC DEPOSITS

Although Voyagers 1 and 2 imaged nine eruption plumes on Io, it is possible that other active eruptions may have gone undetected. With the possible exception of Plume 9, all of the observed plumes are > 50 km high. Probably all plumes greater than this height have been detected by the combined coverage of Voyagers 1 and 2, but plumes less than $\sim$ 20 km high may have gone undetected because the limb coverage on which such small plumes would have been detected occurs only on Voyager 1 images and is very limited. The presence of small dark pyroclastic deposits strongly suggests that recent low-intensity eruptions have occurred on Io. From the diameter of these deposits one can estimate that the eruptions producing them may have been no higher than $\sim$ 10–20 km (Strom et al. 1981). Most of these deposits visible on the high resolution coverage occur at longitudes corresponding to gaps in the limb coverage; it is conceivable that some of these vents may have been active during the Voyager encounters.

The surface deposits associated with active eruptions are generally characterized by (1) a very dark or black central area often consisting of diffuse radial to subradial streaks, (2) one or more diffuse bright or dark rings surrounding the central core, and (3) a broad diffuse region surrounding the rings. Numerous areas, other than plume sources, display similar characteristics and are almost certainly the sites of recent eruptions. Several of these areas are shown in Fig. 16.9. With the possible exception of Plume 1, all active eruptions display a bright circular to irregular ring surrounding the central core. This bright deposit appears to be missing from the apparently inactive source areas. Perhaps, if the bright deposit is frozen SO_2, its lifetime is relatively short and it has sublimed away (Chapter 7 by Sill and Clark).

In addition to these broad diffuse regions, there are numerous small ($<$ 40 km diameter) dark diffuse spots and streaks that are seen only on the higher resolution pictures (better than 4 km). These relatively small spots have the same characteristics as the cores of active eruptions but lack the broader and lighter surrounding diffuse regions (Fig. 16.20). Most of these spots occur on the rims of calderas and probably represent recent (perhaps in some cases active), relatively small-scale eruptions. It is uncertain whether the eruption mechanism is the same for these small deposits as it is for the larger deposits associated with active plumes. Almost all of the small dark deposits occur on the rims of large calderas, and, where resolution allows, the deposits appear to be almost equally distributed on the floor and the exterior of the caldera. In Fig. 16.20 two such deposits are shown at the rim of an 80-km diameter caldera.

Figure 16.21 is a preliminary map showing the locations and approximate diameters of active plumes, pyroclastic deposits similar to those associated with active plumes, and the small dark diffuse spots discussed above. The active plumes are more or less randomly distributed with respect to longitude, but are concentrated within about ± 45° of the equator (Strom et al. 1979). Either the intensity or frequency of eruptions, or both, is greater in the equatorial regions than in the polar implying that, on the average, the deposition rate is also greater in the equatorial regions. Therefore, the equatorial regions in general are probably younger than the polar and this, at least in part may account for the darker regions at the poles, which may be subjected to radiation darkening over a longer period of time.

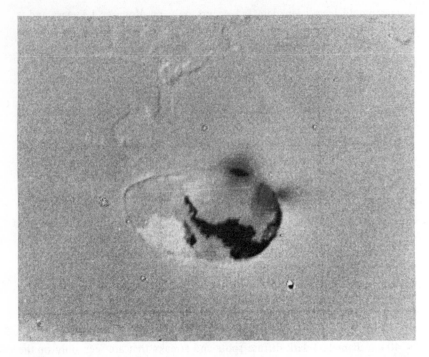

Fig. 16.20. Voyager 1 image of two dark diffuse halos on the rim of an 80-km diameter caldera. These diffuse markings are probably pyroclastic deposits from relatively small eruptions. (FDS 16391.40)

The pyroclastic deposits similar to those associated with active plumes also appear to be concentrated in the equatorial region but randomly distributed with respect to longitude. This apparent distribution may be partly the result of a combination of poor coverage and resolution over large parts of the north polar regions. The concentration of small dark diffuse spots between ~ 150° and 360° longitude is the result of observational bias, since this area was imaged at the highest resolutions. Within this area the dark halos appear to be more abundant in the equatorial regions, which is consistent with the finding of Schaber (1981) that calderas are about six times more abundant in the equatorial than in the south polar region. These results also suggest that volcanic activity is more common in the equatorial regions.

IV. SUMMARY AND CONCLUSIONS

The volcanic eruption plumes on Io range in height from ~ 60 to 300 km, indicating ejection velocities of ~ 0.5–1.0 km s^{-1}. However, fluctuations in the intensity of eruptions can occur on relatively short time scales (hours to

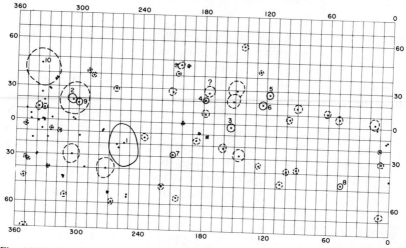

Fig. 16.21. Map showing the distribution and diameter of active plumes (solid circles), diffuse deposits similar to those associated with active plumes (dashed circles), and small dark diffuse spots having diameters $\lesssim 30$ km (black dots). See text for interpretation.

months). Some of the plumes show a faint lateral extension beyond the main plume, which indicates a horizontal velocity component. Where ultraviolet photographs are available, the plumes show two components: an inner core, which is bright at longer wavelengths, and an outer component, which is bright in the ultraviolet. Photometric studies of Plume 2 by Collins (1981) indicate that the outer ultraviolet component consists of extremely fine particles (0.01–0.1 μm radius), while the inner component consists of larger particles (1–1000 μm radius). The shapes of the plumes appear to be related to vent geometry; highly symmetrical umbrella-shaped plumes seem to originate from circular to elliptical vents, or calderas, while very diffuse plumes appear to originate from long fissures. Only images of Pele and Masubi (Plumes 1 and 8) are of sufficient resolution and proximity to the terminator to derive topographic information. The source of Plume 1 is a black elliptical caldera ~ 24 km long and 8 km wide situated on a flat plain at the base of a fractured uplift. Plume 8 is associated with a black flow which occurs on a flat plain with no visible topographic relief. If these sources are typical, eruption plumes do not appear to be situated on topographic highs.

The shape and brightness distribution of at least Plume 3 can be explained by two relatively simple ballistic models in which the ejection angles vary from vertical $0°$ down to $55°$ with a constant ejection velocity of ~ 0.5 km s^{-1}. The vent geometry in these models can be either a point source or a disk containing multiple point sources. Both models result in annular concentrations of ejecta at maximum range similar to the ring deposits associated with most plumes. Plume 1 has a bright outer envelope which does not occur in other plumes. To date, no ballistic simulation has been able to duplicate this feature; it is possible that the bright envelope is the result of a concentration of particles at a shock front.

Numerous surface deposits similar to those associated with active plumes probably represent relatively recent eruptions comparable in size to the observed plumes, and in addition small black diffuse halos probably represent low-intensity eruption sites, some of which may have been active during the encounters. All observed plumes occur within $45°$ of the equator; it is possible that the sites of relatively recent eruptions are also concentrated in, but not restricted to, the equatorial regions. If the frequency of volcanic activity is greater in the equatorial than in the polar regions, the equatorial regions are younger than the polar regions; this may, at least in part, account for the darker polar areas. Exposure to severe radiation, particularly the high-energy proton fluxes which occur at Io, will turn sulfur compounds to a dark reddish-brown (Chapter 7 by Sill and Clark).

The volcanic eruptions responsible for the large plumes on Io differ significantly from those on Earth. Although volcanoes on Earth can remain active for many years or even centuries, violent activity is periodic and short-lived (a few hours or days). Terrestrial steam-blast eruptions have produced ejection velocities of ~ 0.5–0.6 km s^{-1} (Gorshkov 1959; Fudali and Melson 1970), similar to velocities associated with most plumes on Io. However, these eruptions are of very short duration (a few minutes). On Io at least eight eruptions have continued at a high level of activity with ejection velocities of ~ 0.5–1.0 km s^{-1} for at least four months and probably much longer, suggesting that large reservoirs of volatiles (e.g., S_2, SO_2) are available to continuously supply gases to the vent areas. Sulfur dioxide has been observed in absorption against a hot local background at Loki by Pearl et al. (1979). In the absence of a suitable driving volatile for silicate volcanism, it is more plausible to consider S_2 and/or SO_2 volcanism. In this connection it should be noted that the two sources for which there is topographic information (Pele and Masubi) are situated on level plains rather than topographic highs.

Acknowledgments. We sincerely thank R.J. Terrile, A.F. Cook, and C. Hansen for their valuable contributions to this chapter. Discussions with S.W. Kieffer, G.G. Schaber, and members of the Voyager Imaging Science Team concerning various aspects of the eruption plumes are gratefully acknowledged. Most of the processing of the pictures in this chapter was done by the Image Processing Laboratory of the Jet Propulsion Laboratory.

October 1980

REFERENCES

Collins, S.A. (1981). Spatial color variations in the volcanic plume at Loki, on Io. *J. Geophys. Res.* In press.

Cook, A.F., Shoemaker, E.M., and Smith, B.A. (1979). Dynamics of volcanic plumes on Io. *Nature* 280, 743–746.

Fudali, R.F., and Melson, W.G. (1970). Ejecta velocities, magma chamber pressure and kinetic energy associated with the 1968 eruption of Arenal Volcano. *Bull. Volc.* 35, 383–401.

Gorshkov, G.S. (1959). Gigantic eruption of the Volcano Bezymianny. *Bull. Volc.* 20, 77–112.

Pearl, J. and the Voyager Infrared Science Team. (1979). Identification of gaseous SO_2 and the new upper limits for other gases on Io. *Nature* 180, 755–758.

Schaber, G.G. (1980). The surface of Io: Geologic units, morphology, and tectonics. *Icarus* 43, 302-333.

Sinton, W.M. (1980*a*). Io: Are vapor explosions responsible for the $5-\mu m$ outbursts? *Icarus* 43, 56–64.

Sinton, W.M. (1980*b*). Io's 5 micron variability. *Astrophys. J.* 235, L49-L51.

Smith, B.A., Shoemaker, E.M. Kieffer, S.W., and Cook, A.F. (1979). The role of SO_2 in volcanism on Io. *Nature* 280, 738-743.

Strom, R.G., Terrile, R.J., Masursky, H., and Hansen, C. (1979). Volcanic eruption plumes on Io. *Nature* 280, 733-736.

Strom, R.G., Schneider, N.M., Terrile, R.J., Cook, A.F., and Hansen, C. (1981). Volcanic eruptions on Io. *J. Geophys. Res.* In press.

Witteborn, F.C., Bergman, J.D., and Lester, D.F. (1980). Spectrophotometry of the Io 5 micron brightening phenomenon. IAU Coll. 57. The Satellites of Jupiter (abstract 4–4).

17. VOLCANIC ERUPTIONS ON IO: IMPLICATIONS FOR SURFACE EVOLUTION AND MASS LOSS

T. V. JOHNSON
Jet Propulsion Laboratory

and

L. A. SODERBLOM
U.S. Geological Survey

Active volcanism on Io results in a continual resurfacing of the satellite. Analysis of required burial rates to erase impact craters, the mass production in the observed plumes, and the energy requirements for the volcanic activity suggest resurfacing rates of 10^{-3} to 10 cm yr^{-1} in recent geologic time. If this rate is typical of the last 4.5 Gyr, then extensive recycling of the upper crust and mantle must have occurred. The currently estimated loss rate of S, O, and Na from Io into the magnetosphere corresponds to only a small fraction of the resurfacing rate and should not have resulted in either extensive erosion or total depletion of any of the escaping species.

The discovery of active volcanism on Io was one of the major surprises of the Voyager Jupiter encounters. Since Io's size and mass are essentially the same as the Moon's, most pre-Voyager models of its thermal history resembled current ideas about the Moon's evolution, that is, early activity and volcanism driven by accretional heat and short-lived radionuclides, and more gradual heating by long-lived radionuclides. Little or no large-scale volcanic activity later than $\sim$ 3 Gyr ago is recorded on the lunar surface. The suggestion by Peale, Cassen, and Reynolds just prior to the Voyager 1 encounter that tidal heating may have played a dominant role in Io's history revised this view drastically (Peale et al. 1979), but even this hypothesis did not immediately predict the extensive nature of the current activity.

Voyager 1's high-resolution images of Io provided clear evidence for a geologically very young surface, totally devoid of identifiable impact craters even in the highest resolution images. Many structures of clearly volcanic origin were also identified even before the recognition of active eruptions in progress during Voyager's flyby (see Morabito et al. 1979; Smith et al. 1979b). Studies of the plume characteristics and arguments based on estimates of cratering flux suggested that Io is being resurfaced at rates of $\sim$ 0.1 cm yr^{-1} or possibly even faster (Johnson et al. 1979). In this chapter we review the various lines of evidence for resurfacing rates and discuss some of the consequences of these rates for loss of material from Io and the evolution of its surface.

I. GENERAL CHARACTERISTICS

There are at least two basic types of volcanic activity that are clearly responsible for covering up the surface of Io at the current time: (1) active eruptive plumes, and (2) volcanic flows. The active plumes contain particulates that are deposited over a wide area surrounding the vents, ranging from a few hundred kilometers up to a thousand or more (Strom et al. 1980; Chapter 16 by Schneider and Strom). Analysis of the light scattered from the plumes suggests a range of particle sizes that affect the visible appearance of the plumes, from $\sim$ 1 μm particles in the denser parts of the plumes to $\leqslant$ 0.01 μm in the surrounding envelope of at least Plume 2, Loki (Johnson et al. 1979; Collins 1981). Larger particles are also undoubtedly present but do not scatter as much light. The plumes are apparently driven by sulfur compounds, such as SO_2 and possibly sulfur gas, and should contain condensed particles of these substances as well as material entrained in the plume from the vent area (Smith et al. 1979a; Chapter 18 by Kieffer). The large eruptive plumes are confined generally to $\pm$ 45° latitude, although there are fresh-appearing deposits at higher latitudes (Strom et al. 1980; Chapter 16 by Schneider and Strom). Also, there is evidence of lower-level, geyser-like activity that may occur more generally (Carr et al. 1979; McCauley et al. 1979; Cook et al. 1980).

There are widespread volcanic flows on Io's surface as well, mostly associated with large caldera structures. These are reasonably uniformly distributed over the surface (see Masursky et al. 1979; Carr et al. 1979; Chapter 15 by Schaber). The composition of these flows has been a subject of debate, with both sulfur and sulfur-contaminated basaltic flows being suggested to explain these features (Smith et al. 1979a; Sagan 1979; Masursky et al. 1979; Carr et al. 1979).

II. RESURFACING RATES: ERUPTIVE PLUMES

In order to draw conclusions about the average rates of resurfacing on Io based on Voyager observations, we obviously have to make the fundamental assumption that the level of activity seen and the state of the surface are typical of what has been going on over geologic time. Although there is evidence for some variability in magnetospheric phenomena on time scales of months to years, which may be associated with Io's volcanic activity, there are several reasons for believing that the Io observed by Voyagers 1 and 2 was not in an atypical state by orders of magnitude. First, in the extreme case, it is obviously unlikely that we should have just happened to observe Io during the only volcanic episode in its history. If such an episode had just started, older, cratered surfaces should still be present. Second, all the proposed energy sources – tidal, radiogenic heating, and magnetospheric (e.g., Gold 1979) – for the observed high level of volcanic activity should have operated with basically the same efficiency over geologic time, although they all might be expected to produce some time variations in the surface expressions of underlying energetics. Third, groundbased infrared observations of Io suggest that volcanic activity comparable to that observed by Voyager has been going on over the last decade (Witteborn et al. 1979; Sinton 1981; Matson et al. 1981; Morrison and Telesco 1980). Finally, the range of geologic features observed, in different stages of evolution and with complex superposition relationships, suggest that the current activity has been common at least during recent geologic time (the last few hundred thousand or a million years). We thus make the assumption in the following discussion that the level of volcanic activity and surface deposition observed by Voyager can be used at least to an order of magnitude to extrapolate over Io's entire history. Of course, one value of the exercise is to ascertain if this assumption leads to a *reductio ad absurdum,* forcing us to change our view of the continuity of Io's volcanic activity.

A calculation that gives a minimum resurfacing rate for the current epoch is the estimate of the rate at which material is being deposited on the surface by the plumes observed by Voyager. The major uncertainty in this approach is in determining from the images the mass of material in the visible plumes. Johnson et al. (1979) made a preliminary attempt to estimate the particulate mass from the assumption of small (~ 1 μm) particles and a fraction, f, of a cylindrical volume filled with material of optical depth unity. Figure 17.1 shows a summary of these results for the eight active plumes well observed by Voyager 1. These results suggested rates of at least several 10^{-4} cm yr^{-1}, and the authors pointed out several factors that might lead to even higher rates, including contributions from other particle sizes, small vents below Voyager's resolution, gas condensation, and surface flows.

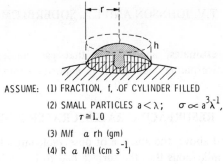

ASSUME: (1) FRACTION, f, OF CYLINDER FILLED

(2) SMALL PARTICLES $a < \lambda$; $\sigma \propto a^3\lambda^{-1}$, $\tau \cong 1.0$

(3) M/f a rh (gm)

(4) R a M/t (cm s^{-1})

PLUME#	M/f x 10^9 gm	R/f x 10^{-12} cm s^{-1}
1	>63.0	>66.0
2	>4.7	>8.3
3	>4.0	>8.3
4	>1.6	>2.8
5	>3.6	>7.1
6	>4.5	>8.8
7	>4.9	>7.8
8	>2.2	>4.6
TOTAL	>89	>113

$f > 10\% \longrightarrow R > 3.5 \times 10^{-4}$ cm yr^{-1}

Fig. 17.1. The geometry of an idealized plume illustrated together with the assumptions made for estimating the mass of material in each plume. Table 17.1 then gives the mass and resurfacing rate for each plume parameterized by the geometric filling factor f. (From Johnson et al. 1979.)

Subsequent analysis of spectrophotometric data for the bright ultraviolet envelope around the Loki plume indicated that very small particles in the 0.01 μm to 0.1 μm range may account for even more mass in this plume than the 1-μm particles (Collins 1981). This analysis yields a total mass of these small particulates of 10^{12} to 10^{14} g, some 10^4 to 10^5 times the mass of large particles given in Fig. 17.1 (for $f = 0.10$). Using the same ballistic flight time of $\sim 10^3$s for these particles leads to a mass rate of 10^9 to 10^{11}g s^{-1} or $\sim 10^{-2}$ to 1 cm yr^{-1} of deposition by small particles from this one plume alone. As pointed out in Johnson et al. (1979), this is a lower limit estimate since it ignores the potentially greater mass (> 1 μm) particles. Since Loki appears to have the most extensive surrounding envelope of bright ultraviolet scattering of any of the plumes, it is conceivable that the bulk of material being supplied to Io's surface during the Voyager encounters was in this plume.

An independent estimate of flow rates from Plume 1 (Pele) has been made by Lee and Thomas (1980) based on analysis of adjacent albedo markings. They derive mass rates of 10^7 to 10^9g s^{-1}, consistent with the

Johnson et al. estimates, and suggest that gas condensation from an intermittently active plume could result in deposition rates of 10^{-1} to 10^{-2} cm yr^{-1}.

III. RESURFACING RATES: CRATER BURIAL

As mentioned above, the absence of observable impact craters on Io was one of the first indications that the surface had to be geologically young. In order to estimate a resurfacing rate from these observations, it is necessary to use an assumed crater production flux to calculate the number of craters of a given size that should have been seen in Voyager images. The entire surface of Io was imaged at low resolution only, and the highest resolution images covered only a small fraction of Io's surface. When this result is combined with the fact that small craters are far more numerous than large craters, a size range can be defined that will have maximum sensitivity for placing a limit on the age of the surface. Figure 17.2 shows as a function of crater diameters: (a) the areal coverage (in km^2) of Voyager 1 images on which craters of diameter D or larger could have been detected; (b) the expected number of craters of size D or greater that would have been seen on a 10^6 yr old surface assuming a lunar cratering flux; and (c) the resurfacing rate

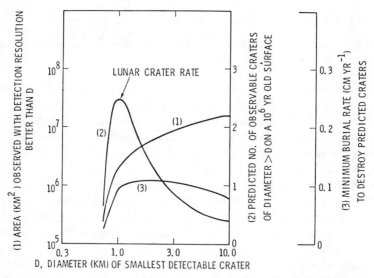

Fig. 17.2. Estimate of resurfacing rate to destroy impact craters. Shown as a function of crater diameter are: (a) the areal coverage, in km^2, of Io obtained by Voyager 1 on which a crater of diameter D or larger could be detected; (b) the number of craters larger than D that would have been seen in the Voyager 1 picture collection if Io's surface were 10^6 yr old, assuming a lunar cratering size-frequency spectrum and rate; and (c) the minimum rate of burial (or erosion) required to remove the craters so that none are now visible. (From Johnson et al. 1979.)

TABLE 17.1

Relative Cratering Rates on the Galilean
Satellites for $v_\infty = 8$ km s^{-1} [a]

Satellite	Flux Concentration by Jupiter Gravity	Mean Impact Velocity (km s^{-1})	Crater Production Rate Relative to Callisto
Callisto	3.3	14	1.0
Ganymede	5.0	18	2.0
Europa	7.3	21	2.8
Io	11.1	26	5.3

[a]From Smith et al. 1979.

required to insure that all visible craters on a 10^6 yr old surface would be buried. The burial rate assumes burial to twice the crater depth (where depth $\simeq 0.2D$).

The required resurfacing rate from this analysis is $\sim 10^{-1}$ cm yr^{-1} for a lunar cratering rate. One of the major uncertainties in this estimate is our lack of knowledge of the actual cratering rate in the Jovian system. Estimates based on the current population of asteroids and comets suggest rates at Jupiter within an order of magnitude of those in the inner solar system (Shoemaker 1977, 1980; Wetherill 1977; Chapter 10 by Shoemaker and Wolfe). In addition, the effects of Jupiter's gravity on the rates at different satellites must be taken into account. Table 17.1 shows the way in which crater production rates are expected to vary among the satellites due to this effect; using this table, the cratering record on the other satellites can be used to place limits on Io's cratering history. In particular, several impact craters have been observed on Europa (Chapter 14 by Lucchitta and Soderblom). If Europa's surface is less than 4.5 Gyr old, these craters suggest that cratering rates of Europa must have been at least 10^{-1} times the lunar rate, leading to a rate at Io of several times 10^{-1} the lunar *or more* due to gravitational focusing (see Table 17.1). The cratering argument thus leads to estimates that are consistent with the lower limits estimated from current plume deposition rates and suggest even higher rates, despite considerable uncertainty in the crater flux estimates.

A remaining difficulty with the crater argument is the question of how long craters are retained on the surface. In an extreme model with a liquid sulfur ocean overlain by a thin crust of solid sulfur and solid and liquid SO_2 layers, topographic features will be rapidly degraded by creep deformation in the weak crustal materials (Smith et al. 1979a; Johnson and McGetchin

1973). Under these conditions, impact craters might be deformed beyond recognition in very short times, geologically. There are reasons for believing that such extreme conditions do not prevail over the entire surface, at any rate. Io has significant topography, with heights of several to ten kilometers; it is, in fact, the only Galilean satellite to show topography above ~ 1 km. There are also deep calderas, with estimated depths of several kilometers. It is possible that all these structures are very young indeed, or that they are continually renewed in some manner. However, the number and variety of the structures combined with calculations of expected thermal profiles for a liquid sulfur model (Clow and Carr 1980) suggest that the crust is stronger in many places than a uniform liquid "ocean" crust model would predict. Smith et al. (1979a) in fact suggest that underlying silicate topography may be close to the surface or protrude through in places. It is more difficult to evaluate the possible effects of intermediate cases of crustal viscosity. For 4 to 5 km craters to be lost in 10^6 yr, viscosities of $\lesssim 10^{22}$ poise are required. Even if deformation in a weak crust has significantly altered the preservation of craters for these lengths of time, the direct estimates of eruptive deposition given above are still valid.

IV. RESURFACING RATES: ENERGETICS

Other potentially useful limits on resurfacing are the amount of energy required to drive the eruptions and the total heat flow from Io. Io's heat flow has been estimated from a number of different types of infrared observations: historical observations of eclipse cooling and brightness temperature as a function of wavelength (Matson et al. 1981); 2-10 μm infrared emission while in eclipse (Sinton 1981; Morrison and telesco 1980); and Voyager infrared spectrometer observations of hot spots on Io's surface (Hanel et al. 1979; Pearl 1980). All these estimates agree and suggest very high heat flow values, ~ 2 W m^{-2} (48 μcal cm^{-2} s^{-1}) of $\sim 10^{14}$ W total radiated power. Observations of short-period transient events at 5 μm also support a high flow, although the power in each event is low compared with the total heat flow (Witteborn et al. 1979; Sinton 1981; Matson et al. 1981). For comparison, thin-shell tidal heating models give only ~ 0.5 Wm^{-2} (Peale et al. 1979; Reynolds et al. 1980; Chapter 4 by Cassen et al.). Very high dissipation rates are required to provide the observed levels of heat flow, implying at least a partially molten interior (see Matson et al. 1981; Chapter 19 by Pearl and Sinton).

We can compare this energy output with the energies implied by the estimates of resurfacing rates. First, the power required to supply kinetic energy to the materials in the plumes is

$$P = \frac{1}{2} \dot{m} v^2 . \qquad (1)$$

For $v \simeq 1$ km s^{-1}, a mass flux of 10^9 g s^{-1} (equivalent to $\sim 10^{-2}$ cm yr^{-1} resurfacing), requires a power of 5×10^{11} W, or less than one percent of the observed output. Thus, mass fluxes $> 10^{11}$ or 10^{12} g s^{-1} (or resurfacing > 2 to 20 cm yr^{-1}) are probably ruled out as long as the mass is all brought out in eruptive plumes with high velocities.

Since surface flows may be as important or more important than the eruptive plumes for resurfacing, we can also estimate the energy involved in transport of liquids to the surface, following Reynolds et al. (1980),

$$h = \frac{dl}{dt}\rho \,(H_f + C_p \,\Delta T) \tag{2}$$

where h is heat flux in ergs cm^{-2} s^{-1}, dl/dt is the resurfacing rate in cm s^{-1}, ρ is density (g cm^{-3}) and H_f and C_p are the heat of fusion and heat capacity of the mobilized material. ΔT is the temperature difference between the melting point and the surface temperature. Reynolds et al. point out that this is a minimum energy requirement since it neglects conducted heat. Assuming a minimum resurfacing rate of 10^{-1} cm yr^{-1} and silicate volcanism, they derive $h > 200$ ergs cm^{-2} s^{-1} or 0.2 W m^{-2}. For sulfur, $h > 16$ ergs cm^{-2} s^{-1} or 0.016 W m^{-2}. Neither of these estimates is a problem if the actual heat flow is ~ 2 W m^{-2}. Another way of looking at this calculation is that resurfacing rates of 1 to 10 cm yr^{-1} could supply the observed heat flux and greater rates would probably violate the observed infrared fluxes depending on what the mobilized fluid is.

V. RESURFACING RATES: SUMMARY

From the previous three sections, we find that resurfacing rates $\lesssim 10^{-3}$ cm yr^{-1} are unlikely if the current level of eruptive activity is reasonably typical. Likewise, rates $\gtrsim 10$ cm yr^{-1} are probably ruled out by energetics, if the infrared estimates of average heat flow are correct. Cratering arguments suggest rates intermediate between these extremes, on the order of 10^{-1} cm yr^{-1}. In the following sections we examine the implications of this range of possible rates for mass loss from Io.

VI. IMPLICATIONS OF MASS LOSS

Since the discovery of sodium emission from Io and the identification of its source as a cloud of escaping atoms, we have known that significant amounts of material were being lost from Io and being injected into the magnetosphere (Brown 1974; Trafton et al. 1974; Matson et al. 1974). The species known to be escaping from Io in either neutral or ionized form as of 1980 are: Na, K, S, O and possibly SO_2. The required supply rates of these species can be estimated in a number of ways. The brightness and shape of the neutral sodium cloud can be modeled assuming ionization rates based on electron impact and escape velocity profiles. The results of these models show that rates of $\sim 10^8$ atoms cm^{-2} s^{-1} ($\sim 4 \times 10^{25}$ atoms s^{-1}) are needed to supply the cloud (Brown and Yung 1976; Smythe and McElroy 1978). In periods of high plasma density, the rate may be increased to $\sim 10^9$ atoms cm^{-2} s^{-1} to offset higher ionization rates, resulting in a cloud of approximately constant characteristics (Goldberg et al. 1980; Carlson, personal communication, 1980). Little is known about the neutral K surrounding Io except that it is present in approximately solar proportions compared to Na, suggesting a proportionally smaller supply rate.

The required rates of heavy ion injection into the magnetosphere, particularly for S and O, have been estimated both from considerations of the energetics of the ultraviolet emissions from the plasma torus, and from the characteristics of the plasma temperature and rotation; these arguments are connected to some extent (see Chapters 22 and 23). Briefly, if each newly created heavy ion supplies full corotational energy to the plasma torus, then a few time 10^{10} ions cm^2 s^{-1} (or $\sim 10^{28}$ s^{-1}) are required to supply the radiated ultraviolet energy of $\sim 3 \times 10^{12}$ W (Broadfoot et al. 1979); Shemansky 1980). Arguments based on analysis of plasma flow data suggest that deviation from corotational flow may be due to mass loading by a somewhat higher injection rate, $\sim 10^{30}$ amu s^{-1}, or several times 10^{28} atoms s^{-1} of amu = 20 to 30 (Hill 1980). Several other aspects of magnetospheric energy balance also seem to require high injection rates (Eviatar and Siscoe 1980). A limit on the amount of ionization occurring near Io comes from ultraviolet observations that show no excess emission from O or S when Io was in the slit of the Voyager UVS experiment (Shemansky 1980; see also Chapter 22 by Pilcher and Strobel). This places a limit of several times 10^{27} ions s^{-1} being produced near Io by this process.

A final possible process for loss of material from Io is escape of fine particulates or dust from the eruptive plumes. Although direct ballistic ejection of material is unlikely due to the energetics of the eruptions (see Smith et al. 1979a; Chapter 18 by Kieffer), escape of small grains charged by interaction with the Jovian plasma may be possible (Johnson et al. 1980). It is hard to estimate the magnitude of this supply directly (although clearly it has to be some small fraction of the total mass rate supplied to the plumes), but an indirect estimate comes from calculations of the effect of these dust particles in producing the Jupiter ring by impacts with ring "parent bodies" (Morfill et al. 1980; Chapter 2 by Jewitt). Morfill et al. estimate that only $\sim$ 10 g s^{-1} of 10^{-5} cm radius dust is required to achieve a fit to the ring data. Even if significantly more mass were escaping in even smaller particles by this mechanism, a very steep population index would be required to make this loss competitive with the 10^6 g s^{-1} loss of S and O from Io.

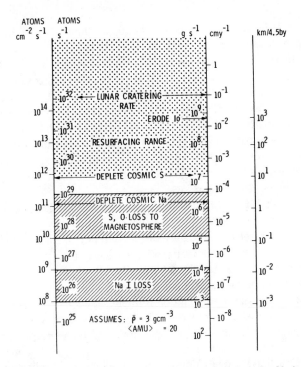

Fig. 17.3. A scale of rates for supply and loss of material from Io. Various rates and critical values are discussed in the text.

In Fig. 17.3 we attempt to relate these various estimates to proposed resurfacing rates and to the physical and chemical results of continual supply or loss of material over geologic time. The left-hand scale gives either loss or supply rates in both atoms s^{-1} and atoms cm^{-2} s^{-1} from Io's surface. To translate these results to mass erosion or deposition rates, we need to assume an average atomic mass and density. Slightly different assumptions are needed for different comparisons; for instance, calculations of the mass in a thin layer of volcanic ejecta or condensation should use a density close to unity while the estimate of the number of atoms lost, if a layer of Io 10 km deep were stripped off, should use something closer to the mean density of Io ($\sim$ 3 g cm^{-3}). In the figure we have chosen 3 g cm^{-3} and used a mean atomic weight of 20 amu for convenience — it thus applies within a factor of two for the three most prominent elements, S (32), O (16) and Na (23). The right-hand scale gives first g^2 s^{-1} by directly applying Avogadro's number. The second scale on the right translates these mass flux numbers to cm yr^{-1} and km $(4.5 \text{ Gyr})^{-1}$ of deposition or erosion using 3 g cm^{-3}.

We can now use the scales in Fig. 17.3 to locate conveniently the rates of resurfacing and magnetospheric loss discussed in the previous sections. First, we plot the range and resurfacing rates, from the lower limit for large particle plumes near 10^{-3} cm yr^{-1} to the energy, upper limit of ~ 10 cm yr^{-1}. The magnetospheric loss rate estimates for S and O are shown as a shaded zone across the figure. Sodium loss rates are shown as a separate zone. Also plotted are the rates required to deplete Io of its cosmic complement of S and Na in geologic time. This is not an absolute limit, of course, but if required loss rates were far in excess of these levels we would have to conclude either that those rates had not applied over geologic time or that Io is significantly enhanced in these volatile elements over so-called cosmic or solar values. The loss rate which would "erode" all the atoms in Io is also shown. Again, this limit would alert us to an inconsistent set of assumptions if any proposed loss rate approached or exceeded it.

The data in Fig. 17.3 allow us to draw several conclusions, despite the large uncertainties in some of the quantities discussed above.

(1) Even the lowest estimates of the amount of material involved in resurfacing imply a supply of material far in excess of the largest estimates of loss to the magnetosphere. Thus, virtually all of the material supplied volcanically to the surface must remain there.

(2) Given (1), the resurfacing rates imply a great amount of recycling of materials in Io's crust and/or mantle. The lowest rates suggest reworking of at least the upper ~ 100 km over geologic time, and the more probable higher rates imply either complete reworking of Io's entire mass or many, many cycles of a shallower zone (more likely). The obvious implications of this degree of recycling for outgassing of Io and loss of the more volatile compounds have been discussed (Johnson et al. 1979; Smith et al. 1979a; Kumar 1979).

(3) The magnetospheric loss rates imply the loss over geologic time of the equivalent of only a few hundred meters to perhaps a few kilometers of material from Io's surface. These are not particularly interesting erosion rates on a planet as active as Io, where resurfacing rates are orders of magnitude higher. However, the material lost does not reflect the bulk composition of the satellite and consists of relatively volatile species. Thus the magnetospheric loss rates may play an important role for surface and atmospheric processes (see Chapter 20 by Fanale et al.). There appears to be no problem supplying the required amounts of S, Na, and O from a chondritic (or cosmic composition) Io, however, as long as the resurfacing allows the supply of new material from depth or, alternatively, if the continual recycling of a large fraction of Io's bulk has concentrated these materials in the upper layers of the crust and mantle. It should be noted, however, that the higher magnetospheric loss estimates come within an order of magnitude of the rates required to deplete a cosmic complement of S over geologic time. Any hypothesis requiring a greatly increased loss compared to this must also treat the logical possibilities of intermittent activity or sulfur enrichment mentioned above.

VII. CONCLUSIONS

A number of independent lines of evidence lead to the conclusions that Io is being resurfaced at a rate $> 10^{-3}$ cm yr^{-1} but probably not $\gtrsim 10$ cm yr^{-1}. This implies extensive recycling of Io's crust and mantle. Magnetospheric loss does not represent a significant fraction of the material involved in Io's eruptive activity, and continual loss of material at essentially the same rate is not in conflict with physical or chemical limits. Further groundbased observations and analysis of Voyager data should define all of these estimates considerably, and Galileo orbiter observations in the late 1980s should give us a much better picture of all the processes involved in the evolution of Io's surface and atmosphere.

Acknowledgments. Discussions with many of our colleagues during the IAU Colloquium 57, *The Satellites of Jupiter,* are gratefully acknowledged. In addition, several preprints sent to us prior to publication were of value in writing this chapter. We particularly thank S. Lee, P. Thomas, G. Siscoe, S. Kumar, and D. Hunten. A portion of this work presents results of research carried out under a contract at the Jet Propulsion Laboratory, California Institute of Technology, and under the Voyager Project and the Planetary Geology Program, of the National Aeronautics and Space Administration.

REFERENCES

Broadfoot, A.L. and the Voyager Ultraviolet Spectrometer Team. (1979). Extreme ultraviolet observations from Voyager 1 encounter with Jupiter. *Science* 204, 979–982.

Brown, R.S. (1974). Optical line emission from Io. In *Exploration of the Planetary System* (A. Woszczyk and C. Iwaniszewski, Eds.), pp. 527–531. D. Reidel, Dordrecht, Holland.

Brown, R.A. and Yung, Y.L. (1976). Io, its atmosphere and optical emissions. In *Jupiter* (T. Gehrels, Ed.), pp. 1102–1145. Univ. Arizona Press, Tucson.

Carr, M.H., Masursky, H., Strom, R.G. and Terrile, R.J. (1979). Volcanic features of Io. *Nature* 280, 729–733.

Clow, G. and Carr, M.H. (1980). Stability of sulfur slopes on Io. *Icarus* 44, 268–279.

Collins, S.A. (1981). Spatial color variations in the volcanic plume at Loki, on Io. Submitted to *J. Geophys. Res.*

Cook, A.F., Smith, B.A., Danielson, G.E., Johnson, T.V. and Synott, S.P. (1980). Volcanic origin of the eruptive plumes on Io. *Science.* In press.

Eviatar, A. and Siscoe, G.L. (1980) Limit on rotational energy available to excite Jovian aurora. *Geophys. Res. Letters* 7, 1085–1088.

Gold, T. (1979). Electrical origin on the outbursts of Io. *Science* 206, 1071–1073.

Goldberg, B.A., Mekler, Y., Carlson, R.W., Johnson, T.V. and Matson, D.L. (1980). Io's sodium emission cloud and the Voyager 1 encounter. *Icarus.* In press.

Hanel, R. and the Voyager 1 Infrared Spectrometer Team (1979). Infrared observations of the Jovian system from Voyager 1. *Science* 204, 972–976.

Hill, T.W. (1980). Corotation lag in Jupiter's magnetosphere: Comparison of observations and theory. *Science* 207, 301–302.

Johnson, T.V., Cook II, A.F., Sagan, C. and Soderblom, L.A. (1979). Volcanic resurfacing rates and implications for volatiles on Io. *Nature* 280, 746–750.

Johnson, T.V. and McGetchin, T.R. (1973). Topography on satellite surfaces and the shape of asteroids. *Icarus* 18, 612–620.

Johnson, T.V., Morfill, G., Grün, E. (1980). Dust in Jupiter's magnetosphere: An Io source? *Geophys. Res. Letters* 7, 350–308.

Kumar, S. (1979). The stability of an SO_2 atmosphere on Io. *Nature* 280, 758–760.

Lee, S. and Thomas, P. (1980). Near surface flow of volcanic gases on Io. Submitted to *Icarus.*

Masursky, H., Schaber, G.G., Soderblom, L.A. and Strom, R.G. (1979). Preliminary geological mapping of Io. *Nature* 280, 725–729.

Matson, D.L., Johnson, T.V. and Fanale, F.P. (1974). Sodium D-line emission from Io: Sputtering and resonant scattering hypothesis. *Astrophys. J.* 192, L43–L46.

Matson, D.L., Ransford, G.A. and Johnson, T.V. (1981). Heat flow from Io (JI). *J. Geophys. Res.* 86, 1664–1672.

McCauley, J.F., Smith, B.A. and Soderblom, L.A. (1979). Erosional scarps on Io. *Nature* 280, 736–738.

Morabito, L.A., Synnott, S.P., Kupferman, P.N. and Collins, S.A. (1979). Discovery of current active extraterrestrial volcanism. *Science* 204, 972.

Morfill, G.E., Grün, E. and Johnson, T.V. (1980). Dust in Jupiter's magnetosphere: Origin of the ring. *Planet. Space Sci.* 28, 1101-1110.

Morrison, D. and Telesco, C.M. (1980). Io: Observational constraints on internal energy and thermophysics of the surface. *Icarus* 44, 226–233.

Peale, S., Cassen, P. and Reynolds, R. (1979). Melting of Io by tidal dissipation. *Science* 203, 892–894.

Pearl, J.C. (1980). The thermal state of Io on March 5, 1979. IAU Coll. 57. *The Satellites of Jupiter* (abstract 4–1).

Reynolds, R.T., Peale, S.J. and Cassen, P. (1980) Io: Energy constraints and plume volcanism. Submitted to *Icarus.*

Sagan, C. (1979). Sulfur flows on Io. *Nature* 280, 750–753.

Shemansky, D.E. (1980). Radioactive cooling efficiencies and predicted spectra of species of the Io plasma torus. *Astrophys. J.* 236, 1043–1054.

Shoemaker, E.M. (1977). Astronomically observable crater forming projectiles. In *On Impact and Erosion Cratering* (D.J. Roddy, R.O. Pepin and R.B. Merrill, Eds.), pp. 617–628. Pergamon, New York.

Sinton, W.M. (1981). Thermal emission spectrum of Io and a determination of the heat flux from its hot spots. *J. Geophys. Res.* 86, 3122–3128.

Smith, B.A., Shoemaker, E.M., Kieffer, S.E. and Cook II, A.F. (1979a). The role of SO_2 in volcanism on Io. *Nature* 280, 738–793.

Smith, B.A. and the Voyager Imaging Team (1979b). The Jupiter system through the eyes of Voyager 1. *Science* 204, 951–972.

Smythe, W.H. and McElroy, M.B. (1978). Io's sodium cloud: Comparison of models and two-dimensional images. *Astrophys. J.* 226, 336–346.

Strom, R.G., Schneider, N.M., Terrile, R.J., Cook II, A.F. and Hansen, C.J. (1980). Volcanic eruptions on Io. Submitted to *J. Geophys. Res.*

Trafton, L., Parkinson, T. and Stacy, W. Jr. (1974). The spatial extent of sodium emission around Io. *Astrophys. J.* 190, L85–L89.

Wetherill, F.W. (1977). The nature of the present interplanetary crater forming projectiles. In *On Impact and Erosion Cratering* (D.J. Roddy, R.O. Pepin and R.B. Merrill, Eds.), pp. 613–615. Pergamon, New York.

Witteborn, F.C., Bregman, J.D. and Pollack, J.P. (1979). Io: An intense brightening near 5 micrometers. *Science* 203, 643–646.

18. DYNAMICS AND THERMODYNAMICS OF VOLCANIC ERUPTIONS: IMPLICATIONS FOR THE PLUMES ON IO

SUSAN WERNER KIEFFER

U.S. Geological Survey

The plumes on Io are assumed volcanic in origin. Observable characteristics, such as temperature, shape and velocity, depend in a complex way on the subsurface features of the volcanic systems. A generalized model is developed in which a volcanic system is considered to be conceptually separable into five regions: supply region, reservoir, conduit, crater and plume. The fluid dynamic and thermodynamic behavior of fluids moving through these regions is examined. Within each region, fluid flow is governed by equations of conservation of mass, momentum, and energy in conjunction with an approximating equation-of-state. Four fluids relevant to volcanism on Earth and Io are considered: H_2O, CO_2, SO_2, and S. Based on plausible Iothermal gradients, both SO_2 and S can erupt from supply regions that give a much broader range of qualitative thermodynamic paths than is common in terrestrial eruptions: low-entropy volcanism, involving boiling of the reservoir fluid, is similar to terrestrial volcanism, whereas high-entropy volcanism, involving condensation, has not been common on Earth. The ascent of sulfur from a reservoir on Io to the surface is nearly isothermal, so inferred temperatures of surface emplacement are close to temperatures in the reservoirs. Orange and yellow sulfur originate in relatively cool (~ 400 K) shallow reservoirs presumed to be from layers of recycled pyroclastic debris. Black sulfur originates in hotter, presumably deeper, reservoirs and may indicate the temperature of the silicate substrate assumed to underlie the thiosphere. Plumes can form only if the vapor pressure of the volatile phase is sufficient to allow vesiculation; below 700 K sulfur probably cannot form a plume. Both boiling and condensation of SO_2 and S may occur in an Ionian eruption; only for the very highest temperature reservoirs, or for plumes laden with pyroclastic fragments, is the ascent to ambient atmospheric conditions unaffected by phase changes. The regions within the volcanic systems in which the phase changes take place are estimated for a variety of initial reservoir conditions; characteristic velocities (reservoir, sonic, limiting) may vary by three orders of magnitude with the phase composition. For most compositions, the velocities and corresponding pressures that develop in a narrow conduit are sufficient to cause formation of a surface crater. If

*the erupting fluid traverses a crater deeper than ~ 1 km before entering
the Ionian atmosphere, a balanced plume, in which initial plume pressure is close to ambient, will form; Plume 3 (Prometheus) may be an
example. If the erupting fluid emanates directly from a conduit, fissure,
or shallow crater, an overpressured (underexpanded) plume will form.
It will have a complex structure due to internal expansion, compression, and shock waves; Plume 2 (Loki) may be an example.*

I. VOLCANIC SYSTEMS ON IO

The association of high-velocity plumes on Io with surface features that
strongly resemble terrestrial, lunar, and Martian volcanic landforms has led to
the consensus that the plumes are manifestations of active volcanism (see
Gold 1979 for theory of nonvolcanic origin and Cook et al. 1980 for a reply).
However, the Ionian volcanoes present some new and perplexing aspects.
On one hand, if the plumes emanate from magmatic sources of silicate composition (proposed by Carr et al. 1979; Hapke 1979; Masursky et al. 1979), a
magmatic chemistry unknown elsewhere in the solar system is indicated by
the ubiquitous presence of sulfur dioxide in a magmatic system apparently
devoid of water and carbon dioxide (Pearl et al. 1979). Furthermore, no
magmatic silicates are indicated either in spectral characteristics of the planet
or thermal data, since the highest reported temperatures are ~ 600 K (Witteborn et al. 1979), an unusually low temperature if the surface manifestations
of volcanism are directly associated with silicate magmatism.

On the other hand, if a magmatic system is postulated with a chemistry
that satisfies the spectral and thermal data (e.g., the sulfur-sulfur dioxide
model of Smith et al. 1979 and the sulfur models of Consolmagno 1979,
Hapke 1979, Reynolds et al. 1980a,b, Sinton 1980, and Chapter 16 by Strom
and Schneider), problems arise in explaining the remarkable similarity of
heights, shapes, and slopes of the Ionian landforms to those associated with
silicate magmatism on other planets (Carr et al. 1979, Masursky et al. 1979,
Schaber 1980, Chapter 15 by Schaber). All proposed sulfur or sulfur-sulfur
dioxide systems are based either directly or indirectly on a sulfur magma
source. Observed surface temperatures of ~ 600 K at hot spots (see Chapter
19 by Pearl and Sinton) are several hundred degrees higher than the near-
surface sulfur liquidus temperature 393 K, whereas highly superheated magmas on Earth are rare. Available experimental data and theories based on
observations of terrestrial volcanism and specifically on silicate magmatism,
do not provide an adequate base for predicting the behavior of plumes that
have unusual composition or in which multiple phase changes from vapor
to liquid to solid can occur, or of plumes that erupt into an environment of
different pressure, temperature, and gravity than that of Earth. Terrestrial
phenomena provide only a narrow base for visualizing the behavior of volcanoes erupting under Ionian conditions. For example, if Old Faithful geyser

(which erupts to 30 m height) were to erupt under Ionian gravity, it would rise to $\sim$ 180 m; if it erupted into an Ionian atmosphere of 10^{-12} bar pressure, it would rise to more than 38 km and would probably be interpreted by earthlings as a significant volcanic plume if captured in images like those by the Voyager 1 spacecraft.

The concepts developed in this chapter allow calculation of maximum and minimum flow velocities for various proposed volcanic regimes. Some aspects discussed are:

1. The effect of system geometry on velocity distribution;
2. Heat and mass transfer among the gas, solid, and liquid phases present in the expanding fluid;
3. Phase changes and their relation to the fluid flow.

Ideally the observed plume characteristics (velocity, velocity dispersion, ejection angle) could be used to determine source characteristics (pressure, temperature, composition), through the models, but at present the observed properties are too few, and even the simplest model of volcanism is too complex, to allow such a treatment. Three situations prevent uniform treatment of the currently popular models of Ionian volcanism: (1) the hypotheses themselves have been presented in different levels of detail; (2) data on proposed volcanic fluids are uneven in abundance and quality; and (3) properties of the fluids themselves vary in complexity. These volcanic models are speculations, in places beyond justification by current data on volcanic processes, on volcanic fluids, or on Io, but perhaps the models will provide a basis for further experiments and observational programs.

A. Constraints on Models of Ionian Volcanism

A model for volcanic processes on Io must satisfy the following observations and generally accepted inferences (for conciseness, references are generally to other chapters in this book rather than original citations).

(1) The Io bulk density 3.5 g cm^{-3}, with the likely implication that the planet is composed dominantly of ferromagnesian silicates, possibly with a small differentiated (Fe-S-O) core.

(2) A molten interior (inferred from tidal dissipation models and electrical heating models), with a thin, solid, outer shell balancing conductive and convective heat flow from the interior with radiative heat losses to space.

(3) An atmosphere containing SO_2, possibly but not definitely in equilibrium with SO_2 as surface frost (see Chapter 20 by Fanale et al. and Chapter 21 by Kumar and Hunten) and depleted in H, C, N, and their compounds. If the atmosphere were SO_2 in equilibrium with surface SO_2 frost at 130 K, the surface atmospheric pressure would be on the order of 10^{-7} bar as was measured by the Voyager IRIS experiment near Loki. At 110 K, the pressure would be $\sim 10^{-9}$ bar and at 90 K, $\sim 10^{-14}$ bar, although disk-wide pressures

may be substantially lower than pressures near the plumes (see Chapter 20, and Matson and Nash 1981).

(4) Surface topography and geomorphic forms resembling silicate volcanic landforms on other planets, including craters, calderas, and lava flows (Chapter 15 by Schaber).

(5) Planetary resurfacing rates on the order of 10^{-3} to $10 \, \mathrm{cm \, yr^{-1}}$ (Chapter 17 by Johnson and Soderblom).

(6) Sudden onsets of sources of energy (corresponding to a 600 K blackbody emission occupying the area of a circle 52 km in diameter) (Witteborn et al. 1979), probably due to volcanic processes (Chapter 19 by Pearl and Sinton).

(7) Material of low albedo (resembling lava lakes) occupying many calderas, having in at least one case a temperature $> 150 \, \mathrm{K}$ above the ambient temperature of the surrounding surface; initial temperature $\sim 600 \, \mathrm{K}$ for this material, if it is sulfur, to account for its black color; a number of hot spots with temperatures $> 500 \, \mathrm{K}$ for which there is telescopic evidence; average heat flow $\sim 1.5 \, \mathrm{W \, m^{-2}}$ (Chapter 7 by Sill and Clark; Chapter 19 by Pearl and Sinton).

(8) Eruption durations of as long as 4 months; indications of temporal variations in eruptive velocity (Chapter 16 by Strom and Schneider).

(9) Plume heights extending to 300 km, implying ejection velocities as great as $1.05 \, \mathrm{km \, s^{-1}}$; ejecta distributions requiring slightly higher velocities (Chapter 16 by Strom and Schneider).

(10) Plume shapes ranging from symmetrical umbrellas (Plumes 1 and 3 from Pele and Prometheus) to diffuse clouds (Plume 2, Loki); plume properties that are complex as a function of the wavelength at which they are observed (Collins 1980; Chapter 16 by Strom and Schneider).

(11) Plume-ejection angles of $0°$ to $55°$ from vertical; ejecta-velocity distribution apparently isotropic for Plume 3, Prometheus (Chapter 16 by Strom and Schneider); ejecta distributions confined to relatively narrow annuli surrounding the vents.

B. The Model of a Volcanic System

In order to model the effects on volcanic processes of source environment, composition, and surface boundary conditions, a volcanic system is here conceptually separated into 5 regions (illustrated schematically in Fig. 18.1):

(1) *The supply region,* the region of host rocks that supplies the erupting fluid, which may be only a minor component of the total material in the supply region;

(2) *The reservoir,* the region in which the erupting fluid is concentrated prior to eruption and from which eruptions occur directly by vesiculation or vaporization. The fluid is assumed to be at rest or at low velocities in this

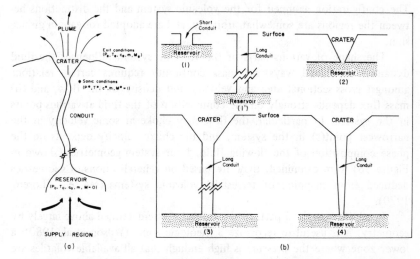

Fig. 18.1. Geometry of volcanic systems, terminology, and notation used in this chapter. (a) Schematic drawing of a volcanic system showing five regions discussed in text. Thermodynamic properties of pressure P, temperature T, sound speed c, mass of ratio of solids to vapor m, and Mach number M are shown. Properties in the reservoir bear the subscript o; those at the exit of the conduit, the superscript $*$; and those at the surface level of the crater, the subscript e. (b) Schematic illustrations of the subsurface geometries examined for fluid-flow calculations. (1) An idealized short, straight (or slightly convergent) frictionless conduit from reservoir to surface. (1′) A straight (or slightly convergent) conduit long enough that frictional effects are important. (2) A divergent crater or conduit from reservoir to surface; frictional effects are assumed unimportant. This shape is plausible for shallow phreatic eruptive maars. (3) A long, straight (or slightly divergent or convergent) conduit from reservoir to crater. This geometry is the simplest that might represent most large terrestrial volcanic systems: deep-seated maars, volcanoes, and diatremes. (4) A slightly divergent long conduit from reservoir to strongly divergent surface crater. This shape is typical of many published cross sections of deep-seated volcanic vents that have a surface crater or maar overlying a long conduit.

region compared to the final eruption velocities, because the amount of vesiculation is assumed to be small in the reservoir compared to the total vesiculation possible;

(3) *The conduit,* the channel through which the erupting material accelerates to a limiting speed determined by the narrowest point in the conduit; it is assumed to be straight, slightly converging, or slightly diverging, as suggested by numerous studies of terrestrial volcanic conduits (see e.g. Lorenz 1970);

(4) *The crater,* a strongly divergent part of the system at the top of the conduit through which the fluid emerges at the surface; and

(5) *The plume,* the region occupied by the erupted material in the planetary atmosphere.

The configuration assumed for the volcanic system and the distinctions between the regions are somewhat arbitrary, and are adopted for ease in discussion.

The shape of various parts of the volcanic system influences the fluid dynamics in several ways: (i) mass continuity requires certain relations amongst cross-sectional area and velocity and density of the flow, and (ii) mass flux depends strongly on the composition of the fluid at various points in the system. In particular, the flow may choke at sonic velocity at the narrowest point(s) in the system, and the choke velocity depends on the phase composition of the flowing fluid. Four system geometries (shown in Figure 1b) were examined; they are based on generalizations of geometries deduced from mapping of terrestrial volcanic systems (see, e.g., Lorenz 1970).

In addition to the 5 parts of the volcanic system defined above largely by geometry, it is useful to recognize 3 rheologic zones (Wilson et al. 1980): a lower zone where the pressure is high enough that all available volatiles are completely dissolved in the liquid phase; a middle zone where the pressure is low enough that some exsolution of the volatiles can occur so that the magma is a liquid with entrained gas bubbles; and an upper zone where the pressure is low enough that the magma is fragmented and consists of exsolved gas with entrained liquid or solid pyroclastic materials. Wilson et al. (1980) called the boundary between the lower and middle zones the exsolution surface, and the boundary between the middle and upper zones the fragmentation surface. Although I consider some dynamics of the lower and middle zones, velocities obtained in these zones are generally low compared with velocities obtained above the fragmentation surface (e.g. see Wilson et al. 1980, Fig. 4), and the plume dynamics are most directly related to the high-velocity upper zone.

This discussion is divided into sections dealing with the 5 regions shown in Fig. 18.1. The supply region controls the chemistry and petrology of the available fluid and may be a large, heterogeneous subsurface region. Unfortunately, little is known about the supply regions of Ionian volcanoes; constraints and plausible models are discussed in Sec. II. The thermodynamic conditions of the fluid in the reservoir constrain the fluid dynamics of the eruption by determining the initial energy available for the flow. The initial thermodynamic state and the thermodynamic path followed during ascent determine the stable phases of the expanding fluid. These phases and their dependence on reservoir conditions and thermodynamic paths are discussed in Sec. III. Detailed equations of state used to model the thermodynamic properties of the erupting fluid are given in Appendix A. A comparison of equilibrium and nonequilibrium thermodynamics during fluid flow is given in Appendices B and C. The shape of the volcanic conduit influences the eruption dynamics. It is assumed in this chapter that at depth the conduit is approximately one-dimensional. A model for flow in such a conduit is given in Sec. IV. As the flow ascends through the conduit, the dynamic pressures can be sufficient to erode the top layers of the regolith, forming a crater. A model for this process and for the effect of the crater on the expansion process is presented in Sec. V. The properties of the plumes formed by erupting fluid depend on conditions in all other regions of the volcano and the atmospheric and gravitational conditions of the planet. Implications of the models for properties of the plumes are discussed in Sec. VI.

II. THE SUPPLY REGION

A. Internal Structure of Io

The composition and physical state of the outer layer(s) of Io are controversial; proposed materials range from mixed hot and cold silicates with only trace of S or SO_2 to mixed hot and cold S and SO_2 with trace silicates.

On the one hand, Carr et al. (1979) suggested that the near-surface zone of the planet consists of interbedded sulfur and silicate lavas (Fig. 18.2a). They proposed that sulfur is relatively mobile in the upper few km of this zone and is molten below ~ 4 km, and that silicates may be molten at greater depths. Volcanism then could recycle both sulfur and silicates, perhaps on different time scales. Sublimates of either sulfur or sulfur dioxide vapors could account for the diffuse light markings on the surface of the planet. Carr et al. suggested that the volcanism is produced by degassing of near surface silicate lavas charged with S or SO_2.

On the other hand, Smith et al. (1979) proposed that the near-surface zone consists of continually recycled, interbedded sulfur and sulfur dioxide pyroclastics and interbedded sulfur lavas, not including and nearly everywhere obscuring older silicate rocks (Fig. 18.2b). To account for the fact that a S-SO_2 system could not be in chemical equilibrium with molten ferromagnesian silicates (see Haughton et al. 1974), they proposed a buffer zone of differentiated iron-poor silicates that isolate the active sulfur system from the ferromagnesian silicates making up the bulk of the planet. As shown in Fig. 18.2b, mechanically and chemically distinct zones on Io can be conveniently hypothesized that are analogous to those commonly defined on Earth: a mechanically rigid lithosphere overlying a mechanically weak asthenosphere, and a crust, mantle, and core, three zones of different chemical (rather than mechanical) properties. Depending on the Ionian thermal gradients (Iotherms), at the base of the Ionian lithosphere there may be a thin zone of solid ferromagnesian silicates of mantle composition, rather than of crustal composition, like the zone of rocks of mantle composition at the base of the terrestrial lithosphere. Smith et al. (1979) suggested three zones of pyroclasts in the Ionian crust: a surface zone in which sulfur and sulfur dioxide are both solid with the SO_2 forming a permafrost within the sulfur;

(a) S – Silicate Model

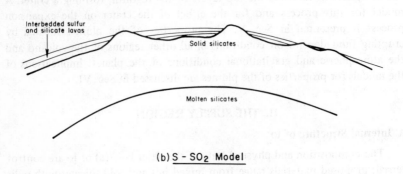

(b) S – SO₂ Model

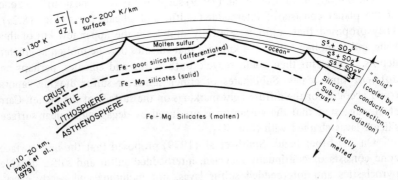

Fig. 18.2. Schematic cross-sectional models of the structure of Io. (a) This model, emphasizing near-surface silicates, is drawn from several sources, chiefly Carr et al. (1979). (b) The S-SO₂ model of Smith et al. (1979); some terminology from this chapter is added on the left. Total crustal thickness is assumed ∼ 35 km (Peale et al. 1979; Cassen et al., Chapter 4). Although the outer crust is referred to as solid, in contrast to the tidally melted interior, only the silicate component is solid; both S and SO₂ may occur in solid, liquid, or vapor phases.

an aquifer, in which the SO_2 is a liquid phase within a solid matrix of S; and possibly a thin zone in which the SO_2 is a vapor within a solid matrix of S. At greater depths the sulfur matrix melts, forming pools and oceans of molten sulfur on the basal zone of differentiated silicates; Sagan (1979) suggested the name "thiosphere" for this region.

In general, the thermal regime of the planet places broad constraints on the volcanic processes by determining the extent of solid, liquid, and vapor phases. The thicknesses of the zones in the model of Smith et al. (1979) depend on the chosen Iotherms and heat fluxes. Within the molten interior of

Io, the rates of heat generation and transfer depend on tidal dissipation of energy (as proposed for the Earth by Shaw 1969,1970 and Gruntfest and Shaw 1974; for the Moon by Wones and Shaw 1975, and for Io by Peale et al. 1979), on induced electrical heating (Drobyshevski 1979; Ness et al. 1979; Yanagisawa 1980), and convection. However, Smith et al. assumed that within Io's crust the rate at which temperature increases with depth is determined solely by conductive heat transfer. (Clearly, in the vicinity of the volcanoes and anywhere that aquifer fluids or magma flow, convective transfer may dominate.) In a conductive regime, the rate at which temperature increases with depth depends on the thermal conductivity coefficient of the crustal layers and on the local heat flux. A typical conductivity of the crust is assumed to be $\sim 3 - 5 \times 10^{-3}$ W cm^{-1} K^{-1}; if the heat flow is partly by convection, the effective value may be substantially higher. The local heat flux is very difficult to estimate. Smith et al. (1979) considered heat fluxes ranging from 70 to 200 erg cm^{-2} s^{-1}, with a preferred value of 100 mW m^{-2}, of which they assumed that 1/4 escaped through volcanism, leaving 75 mW m^{-2} to be transferred through the crust by normal conduction and non-volcanic convection. From these values, they concluded that plausible Io-thermal gradients range from 70 to 200 K km^{-1}. These Iotherms and, for comparison, terrestrial geotherms, are shown superimposed on the pressure-temperature phase diagrams of H_2O, CO_2, SO_2 and S in Fig. 18.3. These graphs illustrate the effect of the reduced Ionian surface pressure and cooler crustal temperatures on the stable, low-pressure phases that might be expected in volcanism: e.g., ice instead of liquid water, and solid instead of liquid SO_2 and CO_2.

The Iotherms used by Smith et al. (1979) suggest that the base of the SO_2 permafrost zone is at depths between 500 and 1000 m, that the base of the SO_2 aquifer is 500 to 1000 m deeper, and that the top of the molten sulfur zone is $\sim$ 1500 m below the surface. Recalculation of the crustal heat flow by Cassen et al. (Chapter 4) and reexamination of groundbased measurements of Io's thermal emission (Matson et al. 1980) suggests that some modification of these thicknesses may be necessary. The volcanic products carry more heat than initially estimated, maybe nearly the total heat flow of the planet. If this is the case, then the Iotherms of Smith et al. (1979) may be regarded as high upper limits on the gradient in nonvolcanic areas; i.e. the crust may be much colder at depth than initially thought (see further discussion in O'Reilly and Davies 1980). Lower thermal gradients would increase the thicknesses of the permafrost, aquifer, and vapor zones in the Smith et al. model.

At the crux of the disagreement on crustal structure is the lack of agreement on possible geochemical, petrologic, and dynamic relations between silicate and sulfur phases on Io. Even with substantial experimental work on the silicate-sulfur phase relations, their applicability to a planet whose ther-

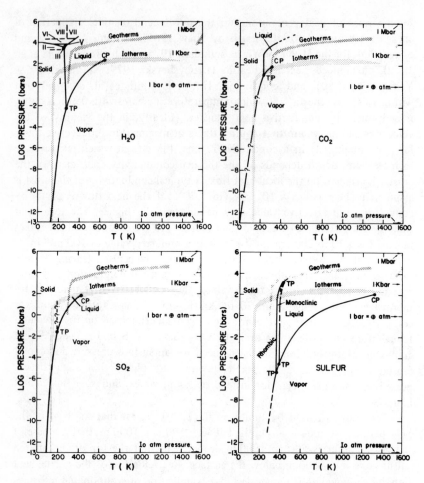

Fig. 18.3. Pressure-temperature phase diagrams showing superposed terrestrial and Ionian thermal gradients (geotherms and Iotherms). Data for the phase diagrams are from these sources: H_2O (Fletcher 1970); CO_2 (ASHRAE 1972); SO_2 (Giauque and Stephenson 1938; Braker and Mossman 1971; Wagman 1979, personal communication); sulfur (Tuller 1954). Shaded area of terrestrial geotherms includes four continental geotherms (Boyd and Nixon 1973; MacGregor and Basu 1974; Bird 1975; Mercier and Carter 1974), as discussed in Kieffer and Delany (1979). The small piece of geotherm in the upper right corner of each diagram is a midocean ridge geotherm based on an assumed adiabatic gradient of 0.5 K km^{-1}. The Iotherms are from Smith et al. (1979).

mal history, extent of outgassing, and internal dynamics are unknown is controversial. The experimental work of Haughton et al. (1974) and observations of Moore and Calk (1971) on submarine basalts demonstrate that mafic silicate magmas can hold $\leqslant 1\%$ sulfur, depending on S_2 and O_2 fugacities and on ferrous iron content; if more sulfur than can be dissolved in the melt is present, a sulfide melt coexists with the silicate melt. Such sulfide melts, rich in dissolved metals, are denser than the silicate melts, and tend to sink to form discrete sulfide-rich layers, as are found at the base of many terrestrial layered intrusions. If the early planetwide melting of Io postulated by Peale et al. (1979) did occur, these sulfide melts may have separated out to form a core; scenarios for this are discussed by Consolmagno and Lewis (1980), Lewis (1981), and Consolmagno (1981). Thus, if silicate melts are present, any sulfur remaining near the surface would have to be from outgassing of the $< 1\%$ in equilibrium with the underlying silicates, unless some kinetic, non-equilibrium mechanism (such as that postulated by Hapke 1979) brings the core-sulfur or sulfides dynamically through the silicate layers.

On Earth, although degassing of silicate magmas can produce sulfur-rich globules within basalt (Skinner and Peck 1969; Moore and Calk 1971) or fumerolic deposits (viz. those at Kilauea in Hawaii), or even occasional sulfur lava flows (Watanabe 1940), the volume ratios of sulfur to silicate are always extremely small because of the small amount of sulfur available from a silicate magma. Thus, even though Carr et al. (1979) believed that repeated extrusion of sulfur-saturated silicate lavas would result in continued enrichment of sulfur in the upper part of the crust, they stated that the dominant rocks in the crust must be of silicate composition. However, Smith et al. (1979) argued that fresh injections of silicate magma into such a crust would lead to melting of the sulfur present, causing sulfur volcanism and segregation of sulfur from silicates. Also, because of buoyancy differences between the sulfur and the silicates, this melting of sulfur would prevent the rise of silicates necessary for silicate volcanism. Smith et al. (1979) therefore preferred a model that keeps the silicates at the base of the crust as the hidden heat source for volcanism that recycles on the S and SO_2.

B. Geochemistry of the Volcanic Fluids

The models proposed to date for Ionian volcanism differ most in the assumed composition of the volatile phase, and implicitly in other ways such as initial temperature and depth of the reservoir, relation of the gas phase to the heat source, state of the components in the volatile phase prior to their vaporization, and proposed geometry of subsurface stratigraphic relations. There are three obvious source regions for magma or volatiles to drive Ionian eruptions (shown schematically in Fig. 18.2). The source regions are listed below from shallowest to deepest.

(1) *Crustal zones of recycled pyroclastics containing SO₂ as the volatile phase.* On the basis of the initial identification of SO_2 on Io's surface and in its atmosphere, Smith et al. (1979) proposed that SO_2 is the driving volatile phase in the plumes. The Smith et al. (1979) model that proposes a shallow crustal source for the volcanic fluid considers the volcanism as phreatic,[a] with the volatile phase SO_2 driven by heat from a deeper sulfur magma that is in turn heated at depth by silicates. A volcanic episode might be initiated as fresh sulfur was intruded into fissures in the crust, rising to the surface to form sulfur volcanoes and lava flows. Initially, the sulfur intrusions would be expected to freeze along the vent walls, sealing the fissure against inflow of liquid SO_2 in the aquifer. If the sulfur withdrew from the conduit (e.g., due to foundering and subsequent bobbing up of the crust or migration of the sulfur magma elsewhere on Io), SO_2 would flow into the vent and initiate a phreatic eruption and a SO_2 plume. This suggested mechanism is analogous to that of the 1924 eruption of Kilauea in Hawaii (Finch 1947) and to that postulated by Sheridan et al. (1981) and Sheridan and Wohletz (1981) for the Pelean eruption of Vesuvius; other scenarios for phreatic eruptions are possible.

(2) *Crustal zones containing sulfur as the volatile phase.* The ubiquitous presence of elemental sulfur on the surface of Io has been deduced from colorimetric data (Hapke 1979; Sagan 1979; Soderblom et al. 1980; Chapter 7 by Sill and Clark). Therefore, several volcanic models with sulfur as the driving volatile phase in the plumes have been proposed. Consolmagno discussed the problem of retaining any volatiles other than sulfur on Io and suggested that rising hot silicate magma vaporizes sulfur to form S_2 plumes. Hapke (1979, personal communication 1980) postulated that sulfur is generated at the base of the silicate crust at the silicate liquidus temperature (~ 1500 K) and that the plumes consist of S vapor from this source. Sinton (1980) suggested catastrophic vapor explosions of submerged pools of liquid sulfur in order to explain the 600 K source observed by Witteborn et al. (1979). Reynolds et al. (1980a,b) proposed that silicate magma intrudes into an overlying sulfur layer and vaporizes the sulfur (in contrast to the behavior proposed by Smith et al., in which the silicates warm the sulfur which in turn mobilizes as a liquid phase and intrudes the overlying pyroclastics to cause vaporization of the SO_2).

---

[a] In terrestrial geologic literature, phreatic refers to water in the zone of saturation by groundwater, and a phreatic eruption is one caused by the conversion of groundwater to steam by heat from a magmatic source. In this chapter phreatic refers to the hypothesized SO_2 liquid groundwater in an aquifer on Io. Magma is used with the prefix sulfur or silicate to indicate the molten phases of these substances. Fluid will be used to indicate a substance capable of flowing, either a liquid or gas.

These sulfur models differ from the S-SO$_2$ model of Smith et al. both in the composition of the volatile phase(s) and in the source or reservoir temperature. In contrast to the minimum reservoir temperature of 393 K of Smith et al., the above sulfur models require reservoir temperatures of 600 to 1500 K. As discussed below and pointed out by Reynolds et al. (1980a,b), at the lower temperature the required ejection velocities of 1.0 km s^{-1} are only marginally met by any vapor, but at higher temperatures either S or SO$_2$ can produce the required ejection velocities.

(3) *Crust and mantle zones containing silicates, with a dissolved volatile phase.* The similarity of morphologic forms on Io, including heights, slopes and shapes of craters, to geomorphic features associated with silicate volcanism on Earth led Masursky et al. (1979) and Carr et al. (1979) to propose that both sulfur and silicate volcanism occur on Io. These workers suggested that dark deposits interpreted as ash are formed either by erosion of the volcanic vents by the gas jets that form the plumes or by rapid degassing of near-surface silicate lavas charged with S or SO$_2$.

Density and buoyancy problems are inherent in all models of Ionian volcanism because of the different densities of the crustal components (density of solid silicates $\sim$ 2.59 g cm^{-3}; density of solid sulfur 2.0 g cm^{-3}; liquid sulfur 1.8 g cm^{-3}; liquid SO$_2$ 1.5 g cm^{-3}). For example, in the Smith et al. model of crustal structure, a solid sulfur layer overlying the liquid sulfur ocean would be unstable. Their model accounts for the different relative densities of the layers of the crust, proposing that the porosity and low density of SO$_2$ within the solid-sulfur crust provide positive buoyancy to maintain a stable crust over the sulfur ocean. Liquid sulfur intrusions occur only where negative freeboard is present due to lowering of porosity.

The major criticisms of the S-SO$_2$ model have been that: (1) thermodynamic equilibrium (assumed in the calculations) is not possible (Reynolds et al. 1980a,b), and the required eruption velocities cannot be obtained from a low-temperature SO$_2$ system; (2) a S-SO$_2$ crust is not consistent with the observed cliff heights (Clow and Carr 1980); (3) the volcanic landforms are similar to those of silicates (Schaber 1980). The major problems with the sulfur models are that SO$_2$, not S, was observed as a plume constituent, although, as pointed out by S. Peale (personal communication, 1981) vapor or particulate S would not have been observed, and that if SO$_2$ is abundant in the crust, rising sulfur would drive off SO$_2$ which has a much higher vapor pressure than sulfur at a given temperature, and so would become the dominant volatile in the plume dynamics. The major problems with models that propose near-surface silicate melts as sources for the plumes are to account for (1) the buoyancy of silicate melt within a low-density pyroclastic crust (Smith et al. 1979); (2) the absence of spectral and thermal indications of liquidus-temperature silicates on the surface of Io; and (3) the unknown chemistry of the silicate melt.

TABLE 18.1

Summary of Material Properties of H_2O, CO_2, SO_2 and S

	H_2O	CO_2	SO_2	S
Solid				
ρ (g cm^{-3})	0.9	1.6	2	~2
M	18	44	64	32
α (K^{-1})	10^{-4}	(not available)	(not available)	~3 × 10^{-4}
Liquid				
ρ (g cm^{-3})	1	0.9 (273 K)	1.5 (263 K)	1.8 (400 K)
Vapor				
ρ (g cm^{-3})	5.5 × 10^{-4} (1 bar, 373 K)	0.1 (34 bar, 273 K)	0.003 (1 bar, 289 K)	0.004 (1 bar, 718 K)
C_p (J g^{-1} K^{-1})	1.9	0.83 (289 K)	0.62 (298 K)	~1, $T > 1300$ K: complex at lower T (see text)
C_v (J g^{-1} K^{-1})	1.43	0.64 (289 K)	0.48 (298 K)	(see text)
γ	1.30	1.30	1.29	1.29 for S_2 (see text): 1.10 for S_6 (see text)
R (m^2 s^{-2} K^{-1})	462	189	130	130 for S_2 (see text): 43 for S_6 (see text)
α (K^{-1})	4 × 10^{-3} (~ 350 K)	37 × 10^{-3} (~ 350 K)	3.9 × 10^{-3} (~ 350 K)	(not available)

TABLE 18.1 Continued

Summary of Material Properties of H_2O, CO_2, SO_2 and S

	H_2O	CO_2	SO_2	S
Relevant phase transitions				
Triple point(s)	273 K, 0.006 bar (many others)	216 K, 5.1 bar	197 K, 0.01 bar	rh, mon, vap: 368.5 K, 5×10^{-6} bar; rh, liq, vap: 386 K, 1.7×10^{-5} bar; mon, liq, vap: 392 K, 2.4×10^{-5} bar
Critical point	647 K, 221 bar	304 K, 73 bar	430.5 K, 77.8 bar	1313 K, 116 bar
Heat of fusion ($J\,g^{-1}$)	334 (273 K)	~ 586 (140 – 216 K)	116 (198 K)	rh to liq: 66 (383 K)
Heat of vaporization ($J\,g^{-1}$)	57 (373 K)	234 (273 K)	389 (263 K)	rh to liq: 392 (368.5 K)

C. Physical and Thermodynamic Properties of Assumed Magmatic Fluids

The main focus of this chapter is on plume dynamics and therefore on thermodynamics and fluid dynamics at high levels within the volcanic system. A major problem is to simplify the complex pressure-temperature-volume-composition relations of the proposed magmas to relations amenable to fluid dynamics analysis. I assume that the pressure changes and accelerations in the deep parts of the system are small compared to those in the upper part of the system, and that details of exsolution of volatiles from magma upon ascent do not strongly influence the plume dynamics. This assumption is supported by the many terrestrial studies that have shown that high-level magmatic fluids are superheated vapors (usually steam) carrying varying amounts of liquid and solid particulate debris derived either from the deep-seated magma or from the walls surrounding the volcanic system. (See e.g. Smith 1960,1980; Smith and Bailey 1966; Walker et al. 1971; McBirney 1973; Wilson 1976,1980; Wilson et al. 1978,1980; Sparks 1978; Self et al. 1979.) I therefore model the magmatic fluid as a vapor phase (SO_2 or S for Ionian conditions; H_2O or CO_2 for terrestrial conditions), either pure as a reference material or (more realistically) carrying entrained solid or liquid fragments. These entrained fragments are simply called pyroclastics in this chapter, without regard to their possible origin from magma or wall rock. Even with this simplified representation of volcanic fluids, the fluid dynamics analysis is complex and is derived only for the specific case where the vapor or vapor-pyroclastic mixtures are modeled by an equation of state in the algebraic form of a perfect gas law. However, as shown in Appendix A, this simplification allows a reasonable approximation of processes occurring within fluids other than perfect gases.

Physical and thermodynamic properties of the fluids discussed in this chapter are listed in Table 18.1. Most are self-explanatory and are included in common references on physical properties of gases, solids and liquids; but, two of the most important properties required for the calculations, the gas constant R and the isentropic exponent γ, are not readily available for sulfur and require some explanation to justify the values chosen here.

The gas constant R is simply the universal gas constant $\tilde{R}$ (8314 J kg-mol^{-1} K^{-1}) divided by the molecular weight of the molecule M. The molecular weights of SO_2, H_2O, and CO_2 are well defined, except at high temperatures where the effective composition and hence molecular weight of the gases may change with temperature because of molecular dissociation. The molecular weight of sulfur vapor is not well defined because sulfur vapor is a dissociating system at all temperatures considered here (Meyer 1968). The vapor consists of short chains of sulfur atoms ranging in effective structure from S_2 to S_8. At 1 bar pressure, below 700 K, S_8 accounts for > 50% of the vapor; at 900 K, S_2, S_6, S_7, and S_8 are about equally abundant; by 1100 K,

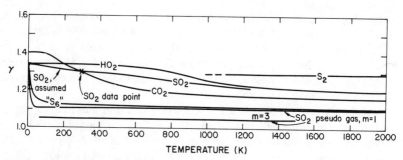

Fig. 18.4. Temperature dependence of the isentropic exponent γ for vapors and two SO_2 pseudogases considered in this study.

S_2 accounts for $> 50\%$ of the vapor (Rau et al. 1973). Increased pressure increases the average molecular weight of the vapor (West 1950). The flow dynamics of sulfur vapor will depend strongly on the kinetics of reaction among the constituent molecules as the vapor changes composition. As limiting cases, two extremes are considered: (1) metastable expansion of S_2 vapor from high temperature (R = 130 J/kg-mol K); and (2) expansion of vapor with an average composition of S_6 (R = 43.3 J/kg-mol K).

From statistical mechanics (e.g. Thompson 1972), the isentropic exponent γ can be related to the number of active degrees of freedom f, by $\gamma = (f + 2)/f$. Because f is a function of temperature, so is γ. The measured variation of γ with temperature for H_2O and CO_2 (taken from Thompson 1972), and the values calculated below for SO_2, SO_2 with a load of entrained pyroclastics and for S are shown in Fig. 18.4.

At low temperatures, where only translational and rotational modes of molecules are active, the value of γ approaches a limiting value γ_{lt} determined by the number of these transitional and rotational modes: $f = 5$, $\gamma = 1.4$ for diatomic S_2 and for the linear dumbbell CO_2; $f = 6$, $\gamma = 1.33$ for nonlinear triatomic molecules, H_2O and SO_2. As the temperature increases γ decreases, because more vibrational degrees of freedom become active. At room temperature the measured value of γ for SO_2 is 1.29 (Braker and Mossman 1971). A constant value of 1.3 is used for γ of SO_2. At high temperatures, γ of S_2 would approach 1.29, corresponding to $f = 7$ (three translation, two rotation, two vibration). At low temperatures, γ of S_6 should approach 1.33, corresponding to $f = 6$ (three translational and three rotational modes). At high temperatures, $f = 18$ (three degrees of freedom per atom in S_6) so that γ approaches $20/18 = 1.1$. Because S_7 and S_8 have more atoms per molecule and hence more degrees of freedom, their γ would approach even lower limits at high temperature (1.09 for S_7, 1.08 for S_8). The exact temperature dependence cannot be modeled without detailed spectroscopic data and, because S_6 is a hypothetical composition for the vapor, detailed calculations are not warranted. A constant value of 1.1 is used for γ of S_6 in this chapter, substantially lower than that of either S_2 or SO_2 (for which $\gamma = 1.3$). Note that complex molecules such as S_2O_3 or polymerized $S_2O\text{-}SO_2$ that might be postulated to be present on Io would be characterized by low γ's according to the above arguments.

As shown in Appendix A, a low value of γ is also produced in vapors by the inclusion of entrained solid fragments; e.g., SO_2 carrying entrained pyroclastics with mass solid-to-vapor ratio of $m = 0.5{:}1$ has an effective $\gamma = 1.15$; if $m = 1{:}1$, $\gamma = 1.1$, and if $m = 3{:}1$, $\gamma = 1.04$ (see Fig. 18.4). γ is 1.1 for S_2 vapor with $m = 1$. Therefore, SO_2 with only a small mass fraction of pyroclastics would, if modeled as a homogeneous pseudogas, dynamically flow like pure sulfur vapor. That is, the calculated pressure, temperature and density distributions would be identical, and the flows would differ only in a secondary effect not considered here, such as flow separation of solid and vapor phases.

II. THE RESERVOIR REGION

A. Ascent of Magmatic Fluids: Adiabatic, Isentropic, and Isenthalpic Approximations

The thermodynamic paths possibly followed by magmatic fluids during ascent from the supply region to the reservoir and during eruption affect the way in which the initial reservoir conditions are stated in the following section. Therefore, consider plausible paths of magma rising from the reservoir to the surface. For further discussion, see Lang (1972), Rumble (1976), Marsh (1978), and Kieffer and Delany (1979).

Conservation of energy for a mass element of a vertically flowing fluid gives

$$d\dot{Q} - d\dot{W} = \dot{m}\left[dh + d\left(\frac{u^2}{2}\right) + g\,dz\right] \tag{1}$$

where $d\dot{Q}$ is incremental rate of heat transfer; $d\dot{W}$ incremental rate of work performed; $\dot{m}$ mass flow rate (equal to $\rho u A$, where ρ is density, u velocity, and A cross-sectional area of the flow); dh is the incremental enthalpy $[d(u + P/\rho)$, where P is the pressure]; $d(u^2/2)$ incremental specific kinetic energy; g acceleration of gravity; and dz incremental elevation change.

Although all energy sources and sinks in this equation are available in a volcanic process, some simplification is necessary for direct application. Generally it is assumed that no outside mechanical work is done on or by the ascending magma, so $d\dot{W} = 0$. (However, situations are conceivable where this is not valid; e.g. if the bounding walls of a magma chamber collapse inward during an eruption $d\dot{W}$ would not be zero; Shaw and Swanson (1970) discuss this process.) It is also generally assumed that heat transfer to or from the system is negligible, i.e. the walls bounding the volcanic system are perfect insulators and/or the ascent or eruption processes occur so rapidly that negligible heat is transferred to or from the system, so that $d\dot{Q} = 0$. The ascent is then adiabatic and, in the absence of work, Eq. (1) reduces to

$$dh + d\left(\frac{u^2}{2}\right) + g\,dz = 0 \ . \tag{2}$$

The second law of thermodynamics requires that

$$\rho T\left(\frac{DS}{Dt}\right) = \Gamma - \bar{\nabla} \cdot \bar{q} \tag{3}$$

where $\bar{q} = -k\bar{\nabla}T$, ($k$ is thermal conductivity, and Γ is a viscous dissipation function [see Thompson 1972]). In this equation T is absolute temperature; D/Dt material derivative, S entropy; $\bar{q}$ heat flux vector; and $\bar{\nabla}$ spatial derivative operator. The equations of fluid dynamics are greatly simplified for inviscid fluids with negligible thermal conduction effects so that

$$\frac{DS}{Dt} = 0 \ . \tag{4}$$

In this case the flow is isentropic. Note that adiabatic flow is isentropic only if the effects of viscosity and thermal conduction are neglected. Kieffer and Delany (1979) have demonstrated that for many volcanologic problems, the effects of viscosity and thermal conduction can be ignored when volume changes due to viscous dissipation and thermal conduction are small compared to overall volume changes caused by pressure changes. In such cases, the flow is quasi-isentropic, Eq. (4) can be used, and further analysis of fluid flow is greatly simplified.

For steady-state flow of an inviscid, nonconducting fluid, Eq. (2) becomes

$$h + \frac{u^2}{2} + gz = \text{const.} \tag{5a}$$

or

$$h_1 + \frac{u_1{}^2}{2} + gz_1 = h_2 + \frac{u_2{}^2}{2} + gz_2 \ . \tag{5b}$$

As pointed out by Lang (1972), Eq. (5b) clearly shows that in the absence of external work an adiabatic process involving vertical movement cannot be isenthalpic. For slow flows over short distances, the kinetic and potential energy changes may be small compared to the enthalpy changes; then and only then is the flow approximately isenthalpic:

$$h_1 \sim h_2, \text{if } u_1 \sim u_2 \text{ and } z_1 \sim z_2 \; . \tag{6}$$

These considerations show that the thermodynamic history of a magma and its volatile phases is complex, and adiabatic, isentropic, or isenthalpic approximations considerably oversimplify this history. Nevertheless, such approximations are necessary to combine thermodynamic and fluid dynamic calculations. I assume that at all stages the magmatic ascent is adiabatic and approximately isentropic to allow the simplest analysis of fluid dynamics. The above discussion and that in Kieffer and Delany (1979) suggest that this assumption is reasonable for the volatile-rich systems postulated for Io. However, because of the large velocities obtainable upon eruption, an assumption of isenthalpic flow is not generally justified.

With these assumptions about the thermodynamic path of the magma, the thermodynamic state of reservoir fluids through the eruption process can be specified. Most of this chapter will be devoted to analyzing the near-surface behavior of the volatile-rich materials, but I also compare briefly the thermodynamics of emplacement of silicate and sulfur magmas as liquid phases, leading to certain speculations about the source regions of the magmas on Io.

For adiabatic, isentropic ascent of a body rising without undergoing phase changes, the temperature change is simply given by

$$\left| \left(\frac{\partial T}{\partial z} \right)_S \right| = \frac{g \, \alpha \, T}{C_P} \tag{7}$$

(see Yoder 1976, p. 102). In this equation, g is the gravitational acceleration; α volumetric thermal expansion; T absolute temperature; and C_P heat capacity at constant pressure. Note that the adiabatic gradient depends on both material (through α and C_P) and planetary conditions (through T and g). For a given material the adiabatic gradient on Io will be 18% that of Earth because Ionian gravitational acceleration (179 cm s^{-2}) is 18% of Earth's. For a typical silicate rock, the parameters in Eq. (7) are: $\alpha = 2 \times 10^{-5}$ deg^{-1}, $T = 1500$ K, $C_P = 11.7 \times 10^6$ ergs g^{-1} deg^{-1}. For sulfur, they are $\alpha = 3 \times 10^{-4}$ deg^{-1} (International Critical Tables 1926-30), $T = 400$ K and $C_P = 10^7$ ergs g^{-1} deg^{-1}. The terrestrial adiabatic gradient for silicate magma at 1500 K is therefore 0.25 K km^{-1}; for sulfur magma at 400 K it is 1.17 K km^{-1}. The corresponding Ionian gradients are 0.045 K km^{-1} for silicate magma, and 0.21 K km^{-1} for sulfur magma.

What do these gradients imply about thermal relations between reservoir or supply region fluids and magma erupted at the surface? Assume first that the reservoir contains completely melted basaltic magma (or its high-pressure equivalent), that the magma ascends through 10 km of the Earth's crust, and that the magma has the same values of α and C_P given above for rock. The total temperature change upon adiabatic ascent, without phase changes or viscous dissipation, would be 2.5 K, i.e., the ascent would be nearly isothermal. The magma would emerge at the Earth's surface as a superheated melt, because the surface liquidus temperature of basalt is about 75 K lower than the liquidus temperature at 3 kbar (Yoder 1976). Evidence for superheated melts, however, is generally not found on the Earth. Although there could be kinetic reasons, the generally accepted explanation is that the material at depth is not initially basalt liquid (or its high-pressure equivalent), but is solid basalt (or its high-pressure equivalent) brought to the temperature of incipient melting. As this initially solid material rises (due to buoyancy), it retains a temperature in excess of the solidus temperature, which decreases toward the surface. This heat is available for melting, and is approximately equivalent to an excess enthalpy of

$$\Delta h \simeq C_P \, \Delta T \, . \tag{8}$$

If, for example, the solidus temperature at 10 km were 75 K higher than the surface solidus temperature, then $\sim 9 \times 10^8$ erg g^{-1} would be available for melting. The typical heat of melting is 5×10^9 erg g^{-1}, so $\sim 18\%$ of the material could melt and it would emerge at the surface as a magmatic mixture of liquid and solid phases, at a temperature between the solidus and liquidus temperatures. Thus, the ascent history of silicates is generally constrained to temperatures between the liquidus and solidus because of the large latent heat of melting, so superheated melts are generally not found on Earth.

The case would be quite different for a body of sulfur undergoing a similar history. Consider the (fictitious) case of the ascent of an initially solid mass of sulfur through 10 km of the Earth's crust. At 3 kbar pressure, sulfur melts at 460 K; at 1 bar, it melts at 370 K (orthorhombic-monoclinic phase change is ignored). The latent heat of melting of monoclinic sulfur is 3.8×10^8 erg g^{-1}, a factor of 13 less than that of silicate rocks. At a terrestrial sulfur adiabatic gradient of 1.17 K km^{-1}, sulfur initially undergoing incipient melting at 469 K would cool adiabatically to 448 K; therefore it would have available for melting an excess enthalpy of $\Delta h = 7.8 \times 10^8$ erg g^{-1}, more than twice that required for complete melting. Thus in spite of greater adiabatic cooling of sulfur than of silicates upon ascent, half the initial energy would be preserved as superheat because the latent heat of melting is small.

Consider now the ascent of two bodies of sulfur from two source regions on Io (Fig. 18.2): (1) initially solid sulfur from the base of the crust, warmed to incipient melting by the sulfur ocean; and (2) initially liquid sulfur from the sulfur ocean. The change of melting temperature of sulfur over the depths postulated for the solid-sulfur crust (a few km) is small, the adiabatic temperature change of a body rising through such a zone would be small, and therefore the temperature of a body of sulfur initially at incipient melting conditions would just follow the melting curve, possibly not gaining enough heat to melt and certainly not becoming appreciably superheated. If the sulfur melted and rose all the way to the surface, the result would be the orange and yellow sulfur deposits on the Ionian surface, as these are the colors of sulfur quenched from $\sim$ 400 K (Sagan 1979).

By a similar argument, it can be shown that initially liquid sulfur in the sulfur ocean would rise to the surface nearly isothermally (under the assumption that α and C_p of liquid sulfur are approximately the same as for solid sulfur). As the black lava lakes on Io are suggested to be sulfur quenched from temperatures between 500 and 600 K, a reservoir or source temperature of 500-600 K is inferred from the above argument.

According to the model of Ionian crustal structure in Fig. 18.2, the black lava lakes arise from the deeper sulfur ocean, i.e. from reservoirs of sulfur between the crust of recycled S and SO_2 pyroclastics and an underlying differentiated silicate substrate. The temperature of such sulfur reservoirs must be nearly isothermal, because in the presence of modest temperature gradients convection will occur and thoroughly mix the sulfur in the reservoir. Consider a simple model of a sulfur reservoir bounded above by the solid sulfur crust at $\sim$ 400 K, and below by a silicate subcrust. Assume that the temperature of the silicate subcrust is 600 K. Vigorous convection will occur in the sulfur reservoir if the Rayleigh number

$$Ra = \frac{\alpha g \Delta T h^3}{\kappa \nu} \tag{9}$$

is $> 10^5$ (Elder 1976). In this equation, ΔT is temperature difference across the layer, h thickness of the layer, κ thermal diffusivity, and ν kinematic viscosity (ratio of viscosity to density). For sulfur, with κ estimated here as 10^{-2} cm^2 s^{-1} and ν as 100 stokes (Sagan 1979), the Ra of a layer 1 m thick would be 2×10^7, i.e. it would freely convect. In fact, all layers thicker than 20 cm would convect vigorously if the 200 K gradient were present.

The temperature in a freely convecting body will be nearly uniform, except for a thin top layer in equilibrium with crustal temperatures; the average temperature will be within a few percent of the basal temperature (Elder 1976). Thus, where samples of the sulfur ocean are locally exposed, e.g., in the black lava lakes, they reveal not only the averaged temperature of the sulfur reservoirs at depth but also the temperature of the underlying substrate, in this case the proposed silicate subcrust.

In summary, consideration of adiabatic ascent of sulfur from two proposed reservoirs in the Ionian crust suggests that the sulfur would experience different thermodynamic histories depending on whether it was initially solid or liquid in the reservoir. Solid sulfur, undergoing incipient melting in the reservoir, would rise nearly isothermally at a temperature near the surface liquidus temperature, and if quenched would form orange or yellow surface deposits. Liquid sulfur would rise nearly isothermally and if initially superheated relative to the shallow melting temperature of 393 K would emerge with the superheat preserved. The black lava lakes may indicate reservoir and silicate subcrust temperatures of ~ 600 K.

B. Thermodynamic Conditions in the Reservoir and Styles of Volcanism

If the fluid ascending from the reservoir expands quasi-isentropically, initial conditions in the reservoir can be specified as a function of entropy, as well as of pressure and temperature. Then the thermodynamic history of the system can be examined on a plot of pressure versus entropy (P-S) or temperature versus entropy (T-S) with entropy as the abscissa, so constant entropy paths are vertical lines. T-S diagrams are used here and given for SO_2 S, H_2O, and CO_2 in Fig. 18.5 through 18.8. These graphs are useful even if entropy is not constant; irreversibility of processes upon ascent due to friction or shock waves will cause the actual paths to deviate from the vertical toward higher entropy, and therefore the ascent paths will not be vertical but will slant toward the lower right or jump discontinuously (across shock waves) to the right.

A comparison of Ionian volcanic reservoir conditions and thermodynamic paths on the SO_2 and S temperature-entropy (T-S) diagrams (Figs. 18.5 and 18.6) with terrestrial conditions for H_2O and CO_2 (Figs. 18.7 and 18.8) demonstrates the possible wide range of behavior shown by Ionian volcanoes. A similar span of qualitative behavior in a terrestrial H_2O system can be obtained only by considering phenomena ranging from shallow geothermal activity (geysers, fumeroles) to geologically rare eruptions of mantle origin (carbonatite eruptions).

Consider first SO_2 as a driving volatile phase for Ionian eruptions. From the P-T diagram presented in Fig. 18.3 and the presumed Iotherms, five reservoir conditions can be defined leading to significantly different eruption dynamics. The differences arise because of the wide range of initial entropies plausible for reservoirs at different postulated pressures and temperatures (Fig. 18.5). The reservoirs are shown schematically in Fig. 18.9. Typical thermodynamic states for the reservoirs might be the following.

Reservoir I. SO_2 liquid in the aquifer at $P \sim 15$ bar, $T < 300$ K, as shown schematically in Fig. 18.9a. The specific case of $\rho > 9$ bar, $T = 350$ K, $S_0 = 2.60$ J g^{-1} K^{-1} is examined.

Reservoir II. SO_2 liquid on the saturation curve, in contact with sulfur at the sulfur liquidus temperature at 1.5 km depth. $P = 40$ bar, $T = 393$ K, as in Fig. 18.9b. $S_0 = 2.90$ J g^{-1} K^{-1}.

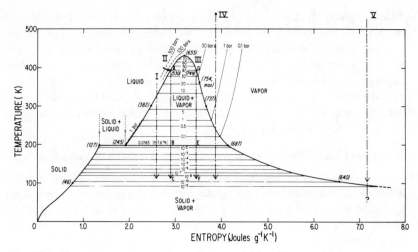

Fig. 18.5. Temperature-entropy phase diagram of SO_2. Heavy solid lines are phase boundaries; heavy dashed or dotted lines are extrapolated beyond data sources; light lines are isobars. I constructed the diagram from liquid-vapor equilibrium data of Braker and Mossman (1971), adjusted so that entropy equals 0.0 for the solid phase at 0 K by using triple-point entropy and enthalpy of fusion from Giauque and Stephenson (1938). The solid sublimation curve was calculated from the heat capacities from Giauque and Stephenson by numerical integration. The vapor curve on the right was calculated from heats of sublimation provided by D.D. Wagman (1979, personal communication). The 1-bar and 30-bar isobars, and all isobars in the liquid + vapor region, were interpolated from vapor-pressure data listed by Braker and Mossman (1971); the 0.1-, 100-, and 300-bar isobars are from Canjai and Manning (1967). Isobars in the solid + vapor region are interpolated from vapor-pressure data provided by Wagman (1979, personal communication). The heavy arrow is the pressure-temperature path of the sulfur liquidus, which intersects the SO_2 liquid-liquid + vapor phase boundary at point A. Isentropes I - V are discussed in the text as examples of low- and high-entropy volcanism. Specific enthalpies are shown in parentheses.

Reservoir III. SO_2 vapor on the saturation curve, in contact with sulfur at the sulfur liquidus temperature at 1.5 km depth. $P = 40$ bar, $T = 393$ K, as in Fig. 18.9b. $S_o = 3.46$ J g^{-1} K^{-1}.

Reservoir IV. SO_2 superheated vapor in contact with superheated sulfur liquid at 1.5 km depth. $P = 40$ bar, $T = 600$ K, as in Fig. 18.9c. $S_o = 3.80$ J g^{-1} K^{-1}.

Reservoir V. SO_2 superheated vapor in contact with or degassing from a hypothetical silicate melt at 1.5 km depth. $P = 40$ bar, $T = 1400$ K, as in Fig. 18.9c. $S_o = 7.13$ J g^{-1} K^{-1}.

Reservoir V'. SO_2 superheated vapor continuously in contact with silicate melt and/or pyroclastics, originating at 1.5 km depth. $P = 40$ bar, $T = 1400$ K, $S_o = 7.13$ J g^{-1} K^{-1}, as in Fig. 18.9c. The mass ratio m of solids-to-vapor is taken as 1.

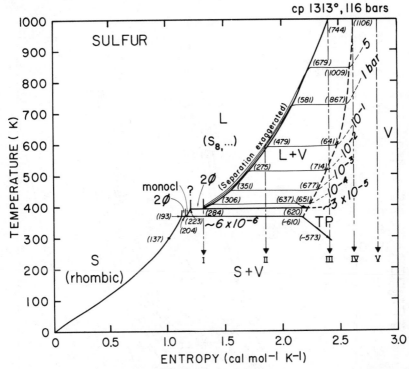

Fig. 18.6. Temperature-entropy phase diagram of sulfur. Data are from the following sources: solid rhombic sulfur and liquid sulfur (JANAF 1965a,b,1966,1967); phase transitions at 368.6 K, 392 K, and 718 K (Rossini et al. 1952); entropy of vaporization of the liquid (West 1950); and critical parameters from Samsonov (1968). Other boundaries are schematic. Heavy solid lines indicate phase boundaries; heavy dashed lines are extrapolated beyond data sources; light lines represent isobars. The isentropic paths *I-V* represent low- and high-entropy thermodynamic paths discussed in the text. Specific enthalpies are shown in parentheses.

The eruption dynamics of material ascending from reservoirs I and II would differ from the dynamics of that from reservoirs III, IV, V, and V' because the progression of phase changes would be different (Fig. 18.5). If material were to ascend with initial entropy less than the critical-point entropy $(3.20 \text{ J g}^{-1} \text{ K}^{-1})$ it would originate either in the liquid (or supercritical) field (as reservoir I) or on the (liquid + vapor) phase boundary (as reservoir II), and it would boil as the pressure decreased upon ascent until the pressure became equal to the triple-point pressure (0.0165 bar). At the triple point the liquid would freeze and, upon further decrease of pressure and temperature, the remaining vapor condense to a solid phase. If the material were to ascend with exactly the critical-point entropy $(3.2 \text{ J g}^{-1} \text{ K}^{-1})$, the liquid to (liquid + vapor) transition would be suppressed because no latent

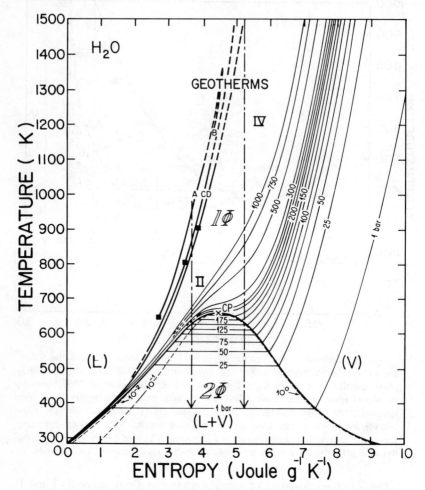

Fig. 18.7. Temperature-entropy phase diagram of H_2O. Heavy solid line indicates the liquid-vapor phase boundary; light lines represent isobars. Four terrestrial geotherms for pure water are shown (the same as those shown schematically in Fig. 18.3). Conditions at the base of the crust are shown on each by a square. Paths II and IV represent adiabatic, isentropic ascent from crustal and mantle conditions, respectively; the labels II and IV are used by analogy with similar phenomenological paths for SO_2 (Fig. 18.5). Note that for crustal volcanism the ascent results in phase changes from supercritical fluid to liquid to liquid + vapor (as shown by path II), whereas for eruptions from the upper mantle phase changes from supercritical fluid to vapor to vapor + liquid are possible (as shown by path IV). (See Kieffer and Delany 1979, for details.)

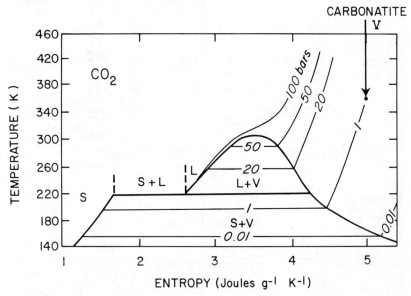

Fig. 18.8. Temperature-entropy phase diagram of CO_2. Heavy solid lines indicate the phase boundaries; heavy dashed lines are estimated; light lines represent isobars. Data from ASHRAE Handbook (1972). The thermodynamic path V represents isentropic ascent from a hypothetical terrestrial carbonatite reservoir at 1000° C, 3 kbar (10 km). (See Kieffer and Delany 1979).

heat difference exists at the critical point. If the initial entropy were between the critical-point entropy and the entropy of the vapor at the triple point (4.1 J g^{-1} K^{-1}), as in the cases of reservoirs III and IV, the material would pass through these equilibrium states successively: supercritical fluid; vapor; vapor + liquid; vapor + liquid + solid at the triple point; vapor + solid; and solid at O K. If the initial entropy were greater than the triple-point entropy of the vapor (> 4.1 J g^{-1} K^{-1}), as in the case of reservoirs V and V', then the liquid field would be bypassed and the only phase change would be vapor-to-solid, if the temperature and pressure decreased sufficiently to allow freezing. It is also possible that for fluids with entropies between those of the critical point and the vapor triple point (reservoir IV), a metastable (vapor-to-solid) transition could occur instead of the (vapor-to-liquid) transition. This metastable transition would be due to the subcooling generally required for nucleation and growth of a condensed phase. (See Appendix C for further discussion.)

Note from Fig. 18.5 that decompression of SO_2 fluids from reservoirs I-IV to ambient atmospheric pressure (10^{-7} to 10^{-12} bar) results in a multiphase solid and vapor mixture at ~ 100 K. All these fluids initially at different reservoir conditions emerge to the same pressure and temperature; only

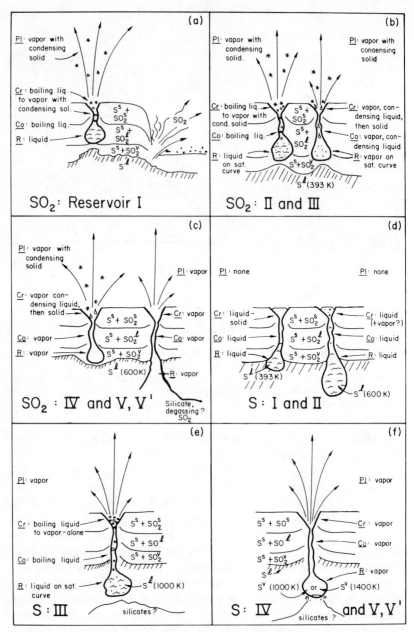

Fig. 18.9. Schematic drawings of models of Ionian volcanic activity. Initial conditions for the fluids in each reservoir are given in the text. In each figure, a plausible position for the reservoir is shown within the Ionian crust proposed by Smith et al. (1979) (see Fig. 18.2). The annotation shows the composition of erupting fluid in the reservoir R, conduit Co, crater Cr and plume Pl. Liquid phase is indicated by the wavy pattern, boiling flow by circles, condensing flow by drops, and condensation of a solid phase by snowflakes. Vapor is unpatterned. Within the liquid sulfur zone, the relatively cool (~ 393 K) top boundary layer is indicated by the diagonal pattern; hotter sulfur at depth is unpatterned.

by measuring the solid to vapor mass ratio (and velocities) at ambient surface conditions could the different initial reservoir conditions be differentiated; e.g., reservoir I or II fluids would have lower velocities and vapor fraction than reservoir III or IV fluids.

Volcanism associated with reservoirs I and II is here called low-entropy volcanism because the initial or stagnation entropy is less than the critical-point entropy; the characteristic phase change is boiling from liquid to vapor. This type of volcanism occurs in the eruption of terrestrial geysers or, perhaps, in deep-seated phreatic eruptions (see Kieffer and Delany 1979). McCauley et al. (1979) concluded that low-entropy SO_2 eruptions from reservoir I could explain the range of observed frost deposits at the base of tectonic scarps on Io; e.g., as shown schematically in Fig. 18.9a. At artesian pressures of 15 to 30 bar, liquid near the base of a scarp would expand to a mixture of solid and vapor moving 350 m s^{-1} at atmospheric pressure and could be deposited as far as 70 km from the scarp. At slightly higher initial values of entropy, as in reservoir II, low-entropy eruptions might cause low-level plumes, as shown schematically in Fig. 18.9. Smith et al. (1979) assumed that the minimum-energy volcanic system on Io would be one driven by sulfur at its liquidus temperature (reservoir II); they estimated the limiting velocities of the ejecta by using a simple model of reversible adiabatic, isentropic flow with chemical and thermodynamic equilibrium maintained. Velocities as high as 433 m s^{-1} could be obtained by expansion of the liquid + vapor mixture to the triple point; velocities as high as 1030 m s^{-1} would be obtained if all the latent heat of freezing from the triple point to 0 K were added to the flowing vapor phase. Although thermodynamic and chemical equilibrium could not be maintained as the vapor phase condenses out at low temperatures in such a system, the decrease in velocity caused by nonequilibrium effects, as well as by internal friction and shock waves, could be offset by entrainment of hot pyroclastics in the fluid or by an elevation in the initial temperature of the reservoir.

Volcanism associated with reservoirs III, IV, and V is called high-entropy volcanism because the initial or stagnation entropy is higher than the critical-point entropy; the characteristic phase change of high-entropy volcanism is condensation from vapor to liquid or vapor to solid. In the model of Smith et al. (1979), high-entropy volcanism was postulated to result from segregation of SO_2 as a vapor phase in the reservoir, shallow intrusion of sulfur magma into the pyroclastic zone, or entrapment of hot sulfur pyroclastics in the SO_2 flow, as shown schematically in Fig. 18.9b and c. In this case, condensation rather than evaporation or boiling of the fluid is the dominant process as the fluid ascends. Velocities of 637 m s^{-1} would be obtained by expansion of the gas + condensing-liquid phase from reservoir III at the sulfur liquidus temperature of 393 K to the triple-point conditions ($P = 0.0165$ bar; $T = 197.6$ K); velocities as high as 1224 m s^{-1} would be obtained by equilibrium expansion to 0 K. Higher velocities would of course be obtained for either low-entropy or high-entropy volcanism if the initial temperature in the reservoir were higher, e.g., in reservoirs IV and V. Such situations could occur if SO_2 were in contact with sulfur superheated above its liquidus temperature, or with hot or molten silicates, or if SO_2 is degassing directly from a silicate magma (Fig. 18.9c). The qualifications on idealized flow velocities discussed above also apply to high-entropy flow.

The temperature-entropy diagram of sulfur (Fig. 18.6) is more complex than that of SO_2 (Fig. 18.5) and cannot now be constructed completely due to lack of information on the entropy of all the polymorphic forms. Nevertheless, many qualitative relations considered above for SO_2 hold for a volcanic system in which sulfur is the driving volatile phase (see schematic diagram, Figs. 18.9d,e,f). Six representative sulfur reservoirs are considered here:

Reservoir (I). Sulfur liquid on the saturation curve at $P = 40$ bar (corresponding to 1.5 km lithostatic overburden) at the liquidus temperature 393 K, as in Fig. 18.9d. $S_0 = 1.3$ J g^{-1} K^{-1}.

Reservoir (II). Sulfur liquid at $P = 40$ bar (corresponding to 1.5 km lithostatic overburden) superheated above the liquidus temperature to 600 K, as in Fig. 18.9d. $S_0 = 1.8$ J g^{-1} K^{-1}.

Reservoir (III). Sulfur liquid on the saturation curve at 1.5 km depth; $P = 40$ bar, $T \sim 1000$ K, $S_0 \cong 2.4$ J g^{-1} K^{-1}, as in Fig. 18.9e.

Reservoir (IV). Sulfur vapor on the saturation curve at 1.5 km depth; $P = 40$ bar, $T \sim 1000$ K, $S_0 \cong 2.65$ J g^{-1} K^{-1}, as in Fig. 18.9f. Assume S_0 as average vapor composition.

Reservoir (V). Sulfur superheated vapor at $P = 40$ bar, $T = 1400$ K. $S_0 = 2.8$ J g^{-1} K^{-1}, as in Fig. 18.9f. Assume S_2 as average vapor composition.

Reservoir (V'). Sulfur superheated vapor carrying entrained sulfur fragments, initially at $P = 40$ bar, $T = 1400$ K, $S_0 = 2.8$ J g^{-1} K^{-1}, as in Fig. 18.9f. Assume S_2 as average vapor composition. The mass ratio m of solid-to-vapor is taken as 1.

In Fig. 18.6, consider first the ascent of sulfur from reservoir I conditions. As the pressure on such a fluid decreased, the fluid would remain approximately on the melting curve, as discussed above. At pressure $\sim 3 \times 10^{-5}$ bar, triple point conditions would be encountered; the liquid would freeze isobarically with the evolution of a small amount of vapor, if the kinetics of reaction permitted. Further pressure decrease would cause condensation of more solid from the vapor.

If the molten sulfur were initially superheated to 600 K at a few tens of bar pressure, as in reservoir II, the fluid would enter the liquid + vapor field when the pressure had decreased to $\sim 10^{-1}$ bar. Vapor formation and expansion could begin if (1) bubbles were able to nucleate and (2) if the pressure of the vapor phase were sufficient to allow the bubbles to grow. Bubble nucleation, though a much studied subject, is not well understood even for simple fluids such as H_2O. Major contributions to the theory of bubble formation and growth in magmas have been given by Verhoogen (1951), Shimozuru et al. (1957), McBirney and Murase (1970), Bennett (1974), and Sparks (1978). To discuss whether or not plumes will develop from a given reservoir fluid, I optimize conditions for plume formation by assuming that nucleation does occur (e.g., that the activation energy is low or there are preexisting nucleation sites). For the fluid to form a plume there must be a sufficiently large increase in volume of the vapor phase, i.e. of the bubbles, to allow fragmentation. Specific mechanisms of bubble growth and magma fragmentation (relative roles of liquid tensile strength, bubble surface tension, viscous resistance or inertia of the liquid) are controversial. However, the consensus is that vapor pressures on the order of bar to tens of bar are required to disrupt magmas with properties comparable to those of sulfur. Ascent of sulfur from reservoir II and probably from any reservoir with an initial temperature less than 700 K therefore would not produce a plume but rather a dense or finely vesicular sulfur flow.

Consider next eruption from reservoirs III and IV, analogous to SO_2 reservoirs II and III. The liquid in sulfur reservoir III would begin boiling upon inception of upward motion; the vapor fraction would increase upon ascent until at $\sim 10^{-2}$ bar all liquid would have boiled off into sulfur vapor. The vapor would then expand as the stable phase to ambient pressure.

The fluids in reservoirs IV, V and V' are initially vapors and would remain so throughout the ascent.

Reynolds et al. (1980a,b) considered examples of high-entropy, vapor-alone sulfur volcanism, such as from reservoir V, as a driving mechanism for plumes. Adiabatic decompression of sulfur plumes of S_2 composition from 1000 K, 1200 K, and 1500 K to final temperatures of 300 K would produce velocities of 0.88, 1.00, and 1.16 km s^{-1}, respectively. On this basis, Reynolds et al. argued that sulfur is energetically more likely to drive the plumes than SO_2. Sulfur at 1000 to 1500 K is more energetic upon decompression than SO_2 at 393 K because the sulfur is appreciably hotter. At low temperatures, phase changes influence the eruption dynamics of both S and SO_2; as pointed out by Smith et al. (1979) special circumstances of thermodynamic equilibrium would be required to obtain the measured velocities in excess of 1 km s^{-1} from an SO_2 reservoir at 393 K. At the same temperatures, sulfur would not produce any plume. On the other hand, as shown by the T-S graph for SO_2 (Fig. 18.5), expansion of SO_2 from temperatures in excess of 1000 K provides the same freedom from constraints of thermodynamic equilibrium as expansion of S from these temperatures, because expansion of the vapor phase alone is stable. Final velocities obtained by expansion of SO_2 vapor into a vacuum from 1000, 1200 and 1500 K are 1.06, 1.16, and 1.30 km s^{-1} respectively.

In summary, except for the highest temperature systems postulated

(> 1000 K), initial reservoir conditions and surface boundary conditions on Io are such that low-temperature phase changes (boiling and condensation) will affect the fluid dynamics of volcanic eruptions. These phase changes affect the flow in two ways: (1) latent heat is absorbed or released as the phase changes occur, and (2) the mass ratio of vapor to liquid or solid phases changes during eruptions. For example, if condensation occurs latent heat release warms the remaining vapor above temperatures it would obtain if expanding alone, but the decrease in vapor quantity from formation of a condensed phase lowers expansion work that can be performed. Since these effects are difficult to treat quantitatively, only simple approximations will be used in this chapter. A discussion of an equilibrium model for flow with phase changes can be found in Appendix B and of nonequilibrium effects in Appendix C.

IV. THE CONDUIT

A. Equations of Flow and Equations of State

I assume that the flow in the volcanic conduit is quasi-one-dimensional, i.e., the flow occurs within a tube that can have changing area but does not have large curvature, and that variations in flow characteristics across the conduit can be adequately represented by an averaged value. I also assume that multiphase flow is homogeneous, so that the flow is governed by the equations for single-phase flow with the use of suitably averaged thermo-dynamic quantities. The flow is then governed by equations of mass conti-nuity, momentum, and energy, in conjunction with an equation of state for the expanding fluid. For quasi-one-dimensional steady flow of an inviscid fluid in a vent of changing area, the equations of motion are (see Thompson 1972):

$$\frac{1}{\rho}\frac{d\rho}{dz} + \frac{1}{u}\frac{du}{dz} + \frac{1}{A}\frac{dA}{dz} = 0 \qquad \text{(continuity)} \qquad (10)$$

$$\rho u \frac{du}{dz} + \frac{dP}{dz} \pm G + \frac{2}{D}f\rho u^2 = 0 \quad \text{(momentum)} \qquad (11)$$

$$\frac{d}{dz}\left(h + \frac{u^2}{2} + \Psi\right) = 0 \qquad\qquad \text{(energy)} \qquad (12)$$

where ρ is density, u velocity, A area, G gravitational acceleration or other body force, P pressure, h enthalpy, Ψ a field potential ($\Psi_{,k} = -G_k$) and f Fanning friction factor ($f = 2\tau/\rho u^2$ where τ is shear stress at the boundary of the fluid).

The effect of gravity is ignored in calculation of fluid flow properties between the reservoir and the planetary surface; the acceleration of gravity is assumed small compared to the accelerations caused by volumetric expansions. The effect of friction is not considered because it causes minor pressure changes compared with those undergone by the fluid as it travels from the reservoir (at tens of bar pressure) into the atmosphere (at pressure between $\sim 10^{-7}$ and 10^{-12} bar). For example, as discussed in Sec. V, if pure SO_2 vapor decompressed from 40 bar through either a short conduit (Fig. 18.1b,1) or a frictionless conduit 1 km long (Fig. 18.1b,1$'$), the pressure at the exit would be 22 bar. However, if the long conduit had a wall friction factor of $f = 0.005$ (for discussion of friction factors, see e.g., McGetchin, 1968,1973; Wilson et al. 1980), frictional choking would reduce the pressure from 22 to 14 bar. Given the many variables in Ionian volcanic processes and uncertainties in defining friction parameters for a vent of unknown geometry traversing a crust of unknown chemistry and physical properties, the likely magnitude of frictional effects does not warrant detailed treatment.

Of Eqs. (10,11,12), the most commonly cited is the energy condition Eq. (12) which, from the Gibbs equation

$$dh = VdP \tag{13}$$

can be rewritten as

$$VdP + d\left(\frac{u^2}{2} + \Psi\right) = 0 \tag{14}$$

Here V is volume. Two equivalent forms of this equation are

$$h + \frac{u^2}{2} + \Psi = \text{const.} \tag{15}$$

(discussed above as Eq. 5), or

$$\int_{P_o}^{P} VdP + \frac{u^2}{2} + \Psi = \text{const.} \tag{16}$$

where P_o is any reference state. These are forms of Bernoulli's equation for a compressible fluid. For an incompressible fluid, V is constant along a streamline, and Eq. (16) reduces to

$$P + \frac{1}{2}\rho u^2 + \rho\Psi = \text{const.} \tag{17}$$

This equation was used by Gorshkov (1959), Fudali and Melson (1971), McBirney (1973), and Melson and Saenz (1973) to calculate volcanic reservoir pressures from ejection velocities; this procedure led to an overestimate of reservoir pressures as demonstrated by Self et al. (1979)

The equation of state required to supplement the conservation equations is difficult to formulate for a fluid undergoing continuous phase changes because there is continuous transfer of mass between phases. I treat the volcanic fluids as if the ratios of solid, liquid, and gas phases were constant, at least over parts of the flow, e.g. in the conduit. For some flow conditions, this assumption may be quite good; during the initial (hot) part of ascent from SO_2 reservoir V or sulfur reservoir V, the fluid might be simply the superheated vapor phase of the volatile plus entrained nonreacting pyroclastics, or during the final (cold) part of the ascent of most fluids, reaction kinetics might be so slow that the composition would be frozen in at a fixed solid-to-vapor ratio. For other parts of the flow in which the composition is changing, I use an average solid (or liquid)-to-vapor mass ratio and, while considering the transfer of specific heat between phases, I neglect the effect of the latent heat transfer. The assumption of fixed average composition is much worse in these cases.

The equation of state of mixtures of fixed composition can be modeled with a law of form of a perfect gas law (Wallis 1969)

$$PV = R_{mix} T \tag{18a}$$

$$PV^{\gamma_{mix}} = \text{const.} \tag{18b}$$

In these equations, P, V and T are as defined above and R_{mix} and γ_{mix} are a gas constant and an isentropic exponent. For a simple perfect gas, the gas constant R_{gas} is given by the difference in isobaric and isochoric specific heats C_P and C_V

$$R_{gas} = C_P - C_V \tag{19}$$

and γ_{gas} is given by their ratio

$$\gamma_{gas} = \frac{C_P}{C_V} \cdot \tag{20}$$

As shown by Wallis (1969) and in Appendix A, gas-particle mixtures can be represented by the same laws (Eqs. 18a,b), with R and γ appropriately modified to account for mass of the entrained particles and heat transfer. The effect of mass addition is to modify the gas constant

$$R_{mix} = \frac{R_{gas}}{1 + m} \tag{21}$$

where m is mass ratio of solids to vapor. If no heat is transferred from particles to gas, e.g., if the particles are large, then

$$\gamma_{mix} = \gamma_{gas} \, . \tag{22a}$$

If heat is transferred so that the gas and particles are always in thermal equilibrium, e.g., if the particles are small, then

$$\gamma_{mix} = \frac{C_{P_{gas}} + m \, C_s}{C_{V_{gas}} + m \, C_s} \tag{22b}$$

where C_s is heat capacity of the solid phase ($C_{Ps} \sim C_{Vs}$). With these modifications of R and γ, the behavior of a gas-particle system will obey all the laws of fluid dynamics developed for perfect gases. For example, an enthalpy increment dh is given by

$$dh = C_p dT \tag{23}$$

and an entropy increment by

$$Tds = dh - VdP \, . \tag{24}$$

From these the equation for isentropic changes of state can be derived

$$PV^\gamma = \text{const.} \tag{25}$$

Alternative relations are

$$\frac{P}{P_0} = \left(\frac{T}{T_0}\right)^{\gamma/(\gamma-1)} \tag{26a}$$

$$\frac{\rho}{\rho_0} = \left(\frac{T}{T_0}\right)^{1/(\gamma-1)} \tag{26b}$$

$$\frac{P}{P_0} = \left(\frac{\rho}{\rho_0}\right)^\gamma \, . \tag{26c}$$

Although these simplifications are necessary for current modeling of volcanic eruption dynamics, note that their use ignores many important phenomena associated with phase changes.

The most important variable governing eruption dynamics is the local sound speed of the erupting fluid. The ratio of flow velocity to local sound speed (the Mach number M) determines the flow dynamics regime, the importance of compressibility, and the presence or absence of shock waves. The sound speed is given by

$$c^2 = (\partial P/\partial \rho)_S \tag{27}$$

and for a fluid obeying a perfect gas law is

$$c = (\gamma RT)^{1/2} = (\gamma PV)^{1/2} \tag{28}$$

where γ and R are specified for the gas or mixture under consideration.

B. Characteristic Velocities

Three velocities characterize the erupting fluid: (1) reservoir sound velocity, given by Eq. (28) with γ, P, V, T taken as reservoir values; (2) sonic velocity; and (3) limiting velocity, discussed below.

Velocities obtainable upon expansion of a gas or pseudogas from a reservoir are simply related to the initial enthalpy and can be given in terms of the reservoir sound speed for a fluid obeying perfect gas laws if phase changes are ignored. The initial enthalpy is

$$h_o = C_P T_o + \text{const.} \tag{29}$$

where the constant is set equal to zero when the reference state is taken as that of the vapor phase at 0 K. If gravitational accelerations are unimportant, the energy equation becomes

$$h_o + \frac{1}{2} u_o{}^2 = h_1 + \frac{1}{2} u_1{}^2 \tag{30}$$

or, using equation (23),

$$C_P T_o + \frac{1}{2} u_o{}^2 = C_P T_1 + \frac{1}{2} u_1{}^2 \ . \tag{31}$$

Equation (31) demonstrates that during isentropic flow the temperature of the gas drops as its speed increases. If a solid phase did not condense from the vapor and the gas cooled as a metastable phase to 0 K, this equation shows that the limiting velocity would be

$$u_{\lim} = \sqrt{2 C_P T_o} \tag{32}$$

because all initial enthalpy would be converted to kinetic energy as the gas cooled to 0 K. It should be emphasized that in the postulated metastable expansion of the vapor to 0 K, the possibility of release of latent heat from condensation of the vapor is not considered, and the enthalpy that should be represented as

$$h = C_P T + \text{const.} \tag{33}$$

(where the constant gives the latent heat terms) is taken relative to a reference state of the vapor phase, not the 0 K solid, by the choice of the constant as zero.

With the relations

$$c^2 = \gamma RT \tag{34}$$

and

$$C_P = \gamma R/(\gamma - 1) \tag{35}$$

the energy equation for steady flow of a perfect gas can be written

$$c^2 + \frac{\gamma - 1}{2} u^2 = c_o^{\,2} \tag{36}$$

where the unscripted variables c and u are sound speed and flow velocity at any point in the flow, and c_o is initial sound speed in the reservoir. Because c_o is usually taken for a reservoir initially at rest ($u_o = 0$), as in Eq. (36), it is frequently called the stagnation sound speed. From this equation the maximum or limiting speed of flow of a pure gas expanding without condensation into a vacuum is

$$u_{lim} = \left(\frac{2}{\gamma - 1}\right)^{1/2} c_o \tag{37}$$

not to be confused with the sonic flow velocity that can be obtained from Eq. (36). At the sonic condition $u = c$ by definition, so

$$u_{sonic} = \left(\frac{2}{\gamma + 1}\right)^{1/2} c_o = u^* \quad . \tag{38}$$

The sonic velocity is a limiting flow velocity at some places in the flow (e.g. at a throat in the conduit) because for normal fluids a transition from subsonic to supersonic flow occurs only under restricted geometric conditions. These conditions can be derived by combining the equations of conservation of continuity and momentum (Eqs. 10 and 11) to give

$$\frac{1}{u}\frac{du}{dz} = \frac{1}{M^2 - 1}\frac{1}{A}\frac{dA}{dz} \tag{39}$$

where M is the Mach number, u/c (see Thompson 1972, p. 278). If the flow passes continuously from subsonic to supersonic (as it will in expansion from rest in a volcanic reservoir to high velocities in the near-vacuum environment of Io's surface if the geometry permits), M must pass through unity. Equation (39) demonstrates that the acceleration can remain finite for $M = 1$ only if $dA/dz = 0$, i.e. the transition through $M = 1$ can occur only at an area minimum or constriction in the volcanic system. In my volcanic geometry (Fig. 1), this minimum is the conduit. Thus, whereas the flow is limited to the sonic or choked velocity in a straight or converging conduit, the sonic velocity is not a universal limiting flow velocity, because the flow can continue to accelerate in diverging parts of the system (viz., supersonic, converging-diverging, or deLaval nozzles). Note that Eq. (39) allows subsonic flow to remain subsonic, or supersonic flow to remain supersonic, upon passing through a constriction (throat). A generalized form of Eq. (39) containing gravitational and frictional terms is given by Wilson et al. (1980).

The above requirement that sonic conditions be obtained at an area minimum has led to the common assumptions that (1) the exit flow from orifices is sonic if reservoir pressures are suitably high, and (2) the sonic flow velocity is a limiting flow velocity. If volcanic flow were through a one-dimensional cylindrical conduit, as shown in Fig. $1b,1'$, the conduit would be in effect an elongated converging nozzle, and it is reasonable to assume that the flow would be sonic at the exit plane. This flow is called choked. However, the flow could expand at increasing velocities into the atmosphere above the vent if the exit pressure exceeded atmospheric pressure. Thus assumption (1) can be true without implying assumption (2).

For perfect gas or pseudogas flow, the choked velocity, pressure, temperature, and density, in terms of reservoir conditions, are:

$$c^{*2} = \frac{2}{\gamma + 1} c_0^2 \tag{40}$$

$$P^* = \left(\frac{2}{\gamma + 1}\right)^{\gamma/(\gamma-1)} P_0 \tag{41}$$

$$T^* = \left(\frac{2}{\gamma + 1}\right) T_0 \tag{42}$$

$$\rho^* = \left(\frac{2}{\gamma + 1}\right)^{1/(\gamma-1)} \rho_0 . \tag{43}$$

The mass flux is controlled at the choked part of the volcanic system, that is, in the conduit, and is given by

$$\dot{m} = \rho^* u^* A^* \tag{44}$$

and the energy flux can be specified if the energy per unit mass e is given, for example, if

$$e = C_p \Delta T^* + L \qquad (45)$$

where ΔT is temperature above the reference temperature at sonic conditions, and L includes the latent heat of relevant phase changes.

Equation (44) shows the strong dependence of the mass flux on the density and sonic velocity of the flow at the choke point (in the conduit). Because these quantities depend on the phase composition of the fluid as it passes the choke point, the composition must be specified as a function of position in the volcanic system. In detail, this calculation is difficult because the effects of gravity, friction, and flow accelerations must be calculated for a specific configuration. However, the flow composition can be estimated from the T-S phase diagrams (Figs. 18.5 and 18.6 for SO_2 and S); the compositions for the model flows in various parts of the volcanic system are shown in Fig. 18.9. In this figure, note that only four reservoir fluids can be treated as vapors (SO_2 reservoir V; S reservoir V) or pseudogases (SO_2 reservoir V'; S reservoir V') throughout the entire ascent. For these reservoirs, sonic and limiting velocities are calculated from perfect gas laws (Eqs. 28, 37, 38). The fluids from SO_2 reservoirs I and II would boil as soon as the pressure began to decrease upon ascent. They are assumed to choke at sonic conditions for boiling flow; the sonic conditions are estimated by comparison with boiling H_2O. The fluid from SO_2 reservoir III begins condensing upon ascent. Because some supersaturation is required before condensation begins (see Appendix C), it is assumed that the vapor from this reservoir reaches sonic velocity in the conduit before appreciable condensation begins, and that condensation to liquid and/or solid phases occurs primarily in the crater. The fluids from SO_2 reservoirs I, II, III and IV all begin condensing liquid or solid phases in the crater or plume. The fluids from SO_2 reservoirs V and V' and S reservoirs V and V' emerge as vapors. It is assumed that S from reservoirs I and II emerges as a liquid and freezes.

With these compositions in mind, the characteristic velocities for the fluids in SO_2 and S reservoirs I-V' are given in Table 18.2 with 3 examples of H_2O fluids for comparison. The reservoir sound speed is for the fluid at rest in the reservoir, the sonic velocity u^* is for the conduit fluid, and the limiting sound speed is for theoretical decompression of reservoir fluid to zero pressure, either as a pure vapor (metastable in some cases) or with complete conversion of initial enthalpy of the reservoir fluid into kinetic energy. These limiting velocities should be considered reference quantities for idealized conditions, not estimates of actual flow velocities. For example, in addition to the effects of nonequilibrium phase kinetics and dissipation discussed above, the small but significant atmospheric pressure around the plumes might cause the plume velocities to fall short of the limiting velocities. From Table 2 in Smith et al. (1979), the difference in velocities obtained upon decompression to 10^{-7} bar rather than 0 bar is nearly 400 m s^{-1} for SO_2 decompressing from reservoir III.

TABLE 18.2

Characteristic Velocities of Model Reservoir Fluids[a]

Characteristic Velocity	Reservoir Fluid (1) c_o	Conduit Fluid (2) u^*	Idealized Limiting Case (3) u_{lim}	
Reservoir				
SO_2-I[b]	~1500	~1-10	(a)	964
			(b)	407
SO_2-II[c]	< 30	< 30	(a)	1030
			(b)	430
SO_2-III[d]	258	240	(a)	1224
			(b)	637
			(c)	666
SO_2-IV[e]	318	296	(a)	1364
			(b)	786
			(c)	821
SO_2-V[f]	486	452	(a)	1800
			(b)	–
			(c)	1255
SO_2-V'[g]	316	309	(a)	–
			(b)	–
			(c)	1465
S-I,II[h]	~2000	–	–	
S-III[i]	~2000	~10	(a)	1224
			(b)	595
S-IV[j]	218	213	(a)	1490
			(b)	974
S-V[k]	486	453	(a)	1637
			(b)	1255
S-V'[ℓ]	317	309	(a)	–
			(b)	1418
H_2O liquid at 50 bar, 500 K[m]	1500	1-10	(a)	–
			(b)	1400
H_2O sat. liquid at 1 bar, 373 K[n]	1-10	< 1-10	(a) >1000	
			(b)	370
H_2O superheated steam, 1400 K, 1 bar[o]	892	845	(a)	–
			(b)	2630

[a]See Fig. 18.9 for summary of phase changes in fluid upon ascent. 1, 2, 3 below apply to columns 1, 2, 3 respectively.

[b](1) The reservoir fluid is liquid; assume $c_0 \sim c_0$ (H_2O). (2) Assume that boiling begins before flow reaches sonic conditions of the liquid phase; sonic conditions then occur at $P \lesssim P_{sat}$ on isentrope, $9 - 5$ bar. u^* estimated by comparison with boiling H_2O; see Kieffer (1977). (3) (a) Assume all initial enthalpy (465 J g^{-1}) converted into kinetic energy; (b) Assume only the enthalpy difference (83 J g^{-1}) between initial and triple point conditions available for kinetic energy.

[c](1) Reservoir fluid is saturated liquid at 40 bar. c_0 estimated by comparison with saturated H_2O; see Kieffer (1977). (2) Assume that the flow is choked by boiling at about 40 bar; u^* estimated by comparison with boiling H_2O; see Kieffer (1977). (3) (a) Assume all initial enthalpy (530 J g^{-1}) converted into kinetic energy; (b) Assume only the enthalpy difference between initial and triple point conditions (90 J g^{-1}) available for kinetic energy.

[d](1) Reservoir fluid is saturated vapor at 40 bar, $\gamma = 1.3$ (Eq. 28). (2) Assume that the flow reaches sonic conditions either as a supersaturated vapor (see Appendix C) or a vapor with such small condensation that the effect of condensate on sonic velocities is negligible (Eq. 38). (3) (a) Assume all initial enthalpy (749 J g^{-1}) converted into kinetic energy; (b) Assume only the enthalpy difference between initial and triple point conditions (200 J g^{-1}) available for kinetic energy; (c) Assume metastable expansion of vapor-alone to 0 K (Eq. 37).

[e](1) Reservoir fluid is superheated vapor, $T_0 = 600$ K, $\gamma = 1.3$ (Eq. 28). (2) Assume that the flow reaches sonic conditions as a pure vapor (Eq. 38). (3) (a) Assume all initial enthalpy (940 J g^{-1}) converted into kinetic energy; (b) Assume only the enthalpy difference between initial and triple point conditions (309 J g^{-1}) available for kinetic energy; (c) Assume metastable expansion of vapor-alone to 0 K.

[f](1) Reservoir fluid is vapor alone, $\gamma = 1.3$ (Eq. 28). (2) Equation (38). (3) (a) Assume all initial enthalpy (1600 J g^{-1}) available for kinetic energy; (b) Fluid does not expand through triple point; (c) Assume metastable expansion of vapor alone to 0 K (Eq. 37).

[g](1) Fluid modeled as pseudogas with heat transfer, $m = 1$, $\gamma = 1.1$, $R = 65$ J kg-mol^{-1} K^{-1} (Eq. 28). (2) Equation (38). (3) (a) No estimate of limiting velocity was made with an enthalpy that included latent heats of transition from the vapor to the 0 K solid; (b) Fluid does not expand through the triple point; (c) Assume metastable expansion of vapor-alone to 0 K (Eq. 37).

[h](1) Reservoir fluid is liquid; c_0 estimated somewhat greater than c_0 of H_2O because of greater density of S. (2) and (3) Assume low-velocity, subsonic oozing, no plume.

[i](1) Reservoir fluid is liquid; see footnote h (1) above. (2) Conduit fluid is liquid plus vapor; u^* estimated to be same magnitude as for boiling water; see Kieffer (1977). (3) (a) Assume all initial enthalpy (750 J g^{-1}) converted into kinetic energy; (b) Assume only the enthalpy decrease between the initial condition and the onset of condensation (177 J g^{-1}) available for kinetic energy.

[j](1) Reservoir and conduit fluids are vapor alone. Assume S_6, $\gamma - 1.1$, $R = 43.3$ J kg-mol^{-1} K^{-1} (Eq. 28). (2) Equation (38). (3) (a) Assume all initial enthalpy (~ 1110 J g^{-1}) available for kinetic energy; (b) Assume metastable expansion of S_6 vapor alone to zero pressure (Eq. 37).

[k](1) Reservoir and conduit fluids are vapor alone. Assume S_2, $\gamma = 1.29$, $R = 130$ J/kg-mol K (Eq. 28). (2) Equation (38). (3) (a) Assume all initial enthalpy (~ 1340 J g^{-1}) available for kinetic energy; (b) Assume expansion of S_2 vapor alone to zero pressure (Eq. 37).

[l](1) Fluid is modeled as S_2 pseudogas with $m = 1$, $C_{Pv} = 0.576$ J g^{-1} K^{-1}; $C_{Vv} = 0.44$ J g^{-1} K^{-1}; $C_s = 1$ J g^{-1}; $\gamma = 1.1$. $R = 65$ J kg-mol^{-1} K^{-1}; c_0 from Eq. (28). (2) Equation (38). (3) (a) No estimate was made of limiting velocity with an enthalpy that included latent heats from the vapor to the 0 K solid; (b) Assume metastable expansion of vapor-alone to 0 K (Eq. 37).

[m](1) Reservoir fluid is liquid. (2) Assume that boiling begins before flow reaches sonic conditions of liquid phase; sonic conditions then occur at $P \lesssim P_{sat}$ on isentrope, at ~ 10

bar (see Fig. 18.7). u^* from Kieffer (1977). (3) (a) A velocity based on inclusion of latent heats of solid phases was not calculated; (b) Assume only the enthalpy difference (991 J g^{-1}) between the initial and triple point conditions available for kinetic energy. [n](1) Reservoir fluid is saturated liquid; see Kieffer (1977). (2) Flow assumed to reach sonic conditions while boiling. (3) (a) Exact enthalpy of fluid relative to 0 K solid not available; this estimate, based on adding heats of vaporization and fusion to values of enthalpy relative to the triple point, is taken from Keenan et al. (1969); (b) Assume only the enthalpy difference (71 J g^{-1}) between initial and triple point conditions available for kinetic energy. On Io, a simple ballistic equation would give plume height $h = v^2/2g = 38$ km. [o](1) Reservoir fluid is vapor. $\gamma = 1.23$ at 1400 K (Fig. 18.4), $R = 462$ J kg-mol^{-1} K^{-1}. (2) Conduit fluid is vapor alone (Eq. 38). (3) (a) No estimate was made of limiting velocity with an enthalpy that included latent heats from the vapor to the 0 K solid; (b) Equation (37).

The characteristic velocities of erupting fluids can vary by several orders of magnitude, depending primarily on phase composition and secondarily on temperature. For SO_2 the reservoir sound velocity is on the order of 1500 m s^{-1} if the fluid is liquid, a few tens of m s^{-1} if the liquid is on the saturation curve, and a few hundred m s^{-1} if it is a vapor or pseudogas. The sonic velocity is a few to a few tens of m s^{-1} if the liquid is boiling and a few hundred m s^{-1} if the fluid is a vapor pseudogas. The limiting flow velocity can be as low as a few hundred m s^{-1} if only part of the latent heat contribution is considered (as for SO_2 reservoirs I and II) or as high as 1500 m s^{-1} for a pseudogas expanding nearly isothermally.

These considerations of phase composition and velocity allow an order of magnitude estimate of mass flux to be made, based on volcanic eruption dynamics theory. In the case of SO_2 vapor erupting from reservoir III, the initial vapor density in the reservoir is ~ 0.1 g m^{-3}; the sonic density is 0.06 g cm^{-3}. The sonic velocity is (from Table 18.2) 240 m s^{-1}. For a conduit radius (at choked conditions) of 5 m, the mass flux is 10^9 g s^{-1}. Ten volcanoes erupting continuously at this flux would resurface the whole planet at a rate of 1 cm yr^{-1}, a figure consistent with the resurfacing rates inferred by Johnson et al. (1979) and Johnson and Soderblom (Chapter 17), and with laterial surface mass flow rates inferred by Lee and Thomas (1980). Variations in the number of plumes, duration of eruptions, vent area, and mass of entrained debris could easily cause the mass fluxes to vary by an order of magnitude, but the model fluxes clearly indicate high resurfacing rates due to volcanic activity. Note that the mass flux is controlled by conduit radius, not crater size.

V. THE CRATER

A. Formation of a Crater

Volcanic vents through which violent eruptions occur are unlikely to be entirely straight walled all the way from the reservoir to the surface; conduits of terrestrial cinder cones or maars generally flare toward the surface as funnel-shaped craters. Enlarged surface craters develop by erosion in response to high velocities and dynamic flow pressures, and in turn affect flow velocities and pressures.

The calculations in Sec. IV suggest that craters develop because fluid velocities and pressures in a straight-walled conduit are large. In such a conduit (which might represent a fissure through which an eruption begins), the fluid would obtain a steady-flow velocity equal to the sonic velocity

$$u_s = \left(\frac{2}{\gamma + 1}\right)^{1/2} c_o .$$ (46)

The sonic pressure in the fluid at this velocity is (e.g., Thompson 1972)

$$P^* = P_o \left(\frac{2}{\gamma + 1}\right)^{\gamma/(\gamma + 1)}.$$ (47)

For example, the sonic pressure would be 25 bar for pure SO_2 vapor originating in a reservoir at 40 bar pressure at 393 K, with $\gamma = 1.3$ (ignoring friction and assuming metastable expansion through a small amount of supersaturation to this pressure; see the T-S diagram of Fig. 18.5 and a discussion in Appendix B). If the SO_2 were carrying a dead load of sulfur or large silicate clasts (so that heat transfer was not important), the above pressure would also apply; if the SO_2 were carrying fine-grained sulfur or silicate ash with heat transfer, the exit pressure would be only slightly lower, 24 bar for $m = 3$, for which $\gamma = 1.04$. Thus, if a conduit were straight walled all the way to the surface of Io, the flow would exit highly overpressured relative to the ambient surface pressure (10^{-7} to 10^{-12} bar) and also, as shown below, overpressured relative to cohesive strengths of the regolith.

Flow that is overpressured relative to ambient pressure will, if unconstrained, expand laterally through decompression or rarefaction waves according to the Prandtl-Meyer expansion (Fig. 18.10). The angle through which unconstrained flow turns depends on: (1) relative pressures of flow and atmosphere; (2) Mach number as the flow exits the vent; and (3) thermodynamic properties of the fluid. For the specific case of choked flow, where the exit $M = 1$, the angle of turning in an expansion to a given final M is called the Prandtl-Meyer angle ν, given by the following expression (Zucrow and Hoffman 1976):

$$\nu = -\left(\frac{\gamma + 1}{\gamma - 1}\right)^{1/2} \tan^{-1}\left[\frac{(M^2 - 1)(\gamma - 1)}{(\gamma + 1)}\right]^{1/2} + \tan^{-1}(M^2 - 1)^{1/2}.$$ (48)

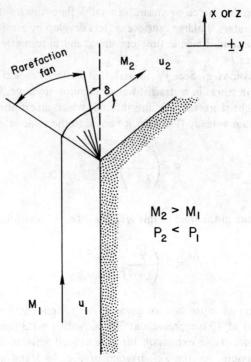

Fig. 18.10. Schematic drawing of flow expansion from Mach number M_1 to M_2 or from velocity u_1 to u_2.

For expansion into a complete vacuum ($M = \infty$), the maximum turning angle is

$$v_{max} = \pm \frac{\pi}{2} \left[\left(\frac{\gamma + 1}{\gamma - 1} \right)^{1/2} - 1 \right] . \tag{49}$$

For expansion of air, with $\gamma = 1.4$, v_{max} is $130°$; for SO_2 with $\gamma = 1.3$, v_{max} is $159°$; and for SO_2 carrying enough pyroclastic ash that $\gamma = 1.1$, v_{max} is $322°$.

The maximum turning angle would not be obtained on Io because: (1) the surface of the planet inhibits turning angles greater than $90°$ from the vertical. Cook et al. (1979) pointed out that unless all gas freezes on contact with the surface, reflected waves are set up that confine the margins of the flow; (2) at high volumetric expansions even in the absence of liquefaction and condensation, the flow becomes so rarified that molecular collisions cannot maintain equilibrium and the expansion process terminates or freezes

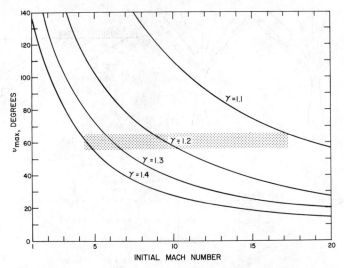

Fig. 18.11. Maximum flow deflection angle ν_{max} for a two-dimensional flow, as a function of the initial Mach number M for fluids with $\gamma = 1.4$, 1.3, 1.2, and 1.1. The stippled band shows the approximate maximum deflection angle of ejecta at Prometheus (Plume 3).

(Thompson 1972, p. 302; see also Sec. VI); (3) the ambient atmospheric pressure on Io ($< 10^{-7}$ bar), though negligible by terrestrial standards, is large enough to constrain the expansion. Variations in ambient pressure due to day-night temperature variations and to local atmospheres created by the eruptions themselves could appreciably affect plume divergences, so temporal variations in plume behavior due to atmospheric variations should be expected.

The maximum possible flow-deflection angle δ, as a function of the initial Mach number (at the exit of the conduit or crater), is shown in Fig. 18.11 for fluids with $\gamma = 1.4$, 1.3, 1.2, and 1.1. The measured deflection angle of Plume 3 is also shown in this figure (see Chapter 16 by Strom and Schneider). From this figure we can infer that plume 3 cannot be erupting from a one-dimensional conduit or fissure for which $M_1 = 1$ into a vacuum, $M_2 = \infty$, because the measured turning angle is too small. Allowance for a finite final pressure (e.g., 10^{-7} bar) instead of a vacuum environment, reduces the limiting angles by only a few to a few tens of degrees. Therefore, we can conclude from this figure that the initial Mach number of the flow as it enters the plume must be > 5, and probably is in the range of 10 to 20 for the more plausible volcanic fluids.

How and where in the volcanic system are such high Mach numbers obtained? $M = 1$ must be obtained in the narrowest part of the volcanic system—here defined as the conduit (Fig. 18.1); expansion to lower pressures and acceleration to higher velocities can occur only if the conduit diverges,

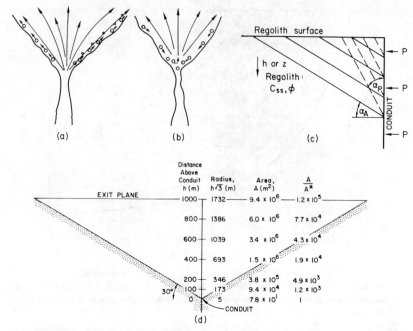

Fig. 18.12. Schematic drawings of erosional processes in a volcanic crater. (a) Erosion by
   direct abrasion and transport of material out of the crater. (b) Erosion by slumping
   into of material not in contact with fluid of sufficient energy to remove it directly.
   the flow. (c) Possible planes of failure of a material obeying a Mohr envelope when a
   pressure P is applied to a vertical wall of the material. The solid lines at an angle of
   α_A with the horizontal are parallel to planes of failure for low pressures; material fails
   by downward motion along these planes. The dashed lines at an angle α_P with the
   vertical are parallel to planes of failure for high pressures; material fails by upward
   motion along such planes; $\alpha_P = \alpha_A$. (d) Drawing of nominal crater geometry assumed
   for an Ionian volcano. The conduit has radius 5 m; the crater has depth 1000 m.

$dA/dz > 0$ where A is cross-sectional area of the conduit and z is the vertical
coordinate. The most plausible configuration is that the conduit flares into a
surface crater so that $dA/dz > 0$. Expansion to supersonic velocities $(M > 1)$
then occurs in the crater.

   Surface craters can form (1) directly, by outward abrasional transport of
wall material in contact with the erupting fluid (Fig. 18.12a), or (2) indirectly, by slumping into the flow of material not in contact with fluid of
sufficiently great energy to directly remove it (Fig. 18.12b). Both processes
can be observed in terrestrial volcanism. Relatively rare large, deep-seated
eruptions have flow pressures sufficiently high that lateral expansion is appreciable. In such eruptions, crater or fissure walls are directly eroded by flow-
abrasion. (This process probably operated at Mount St. Helens during the
initial vent-forming episode of 27 March 1980, and during both the lateral

blast of 18 May and the subsequent Plinian phase of the eruption, to form the large amphitheater in the summit of the mountain.) Small, shallow phreatic eruptions (e.g., those of March-April at Mount St. Helens) have flow pressures comparable to atmospheric pressure and, therefore, give plumes that do not expand to fill the crater created by previous, more violent eruptions. Such plumes contribute to erosion of a summit crater only by entraining debris that slumps from the wall into the conduit. These two models of erosion (direct flow abrasion and indirect slumping and entrainment) roughly correspond to failure of the regolith in compression and tension. If flow pressures are low tensile modes of failure may dominate.

Expansion of a plume in a regolith without shear strength should erode the walls back to angle of repose either directly by flow abrasion or indirectly by slumping and entrainment; even shallower angles would result if the flow were highly overpressured. Therefore, in a regolith of no strength plume divergence angles should be $\geq 60°$ from the vertical. On the other hand, if the regolith had very large strength compared to the pressure in the plume, the plumes would be confined to near-vertical. In this case, the exit Mach number would be unity and the expansion would be through all angles up to the Prandtl-Meyer angle, or through the angle at which expansion becomes frozen due to extreme rarefaction as discussed above. Clearly, plume shape depends closely on crater shape. I propose below a simple model for evolution of crater shape from a narrow conduit into a crater formed by direct abrasion.

Assume that the yield properties of the regolith can be modeled by a simple Mohr envelope (Jaeger 1969). The shear stress that the material can support τ is related to the cohesion C_{ss} (shear stress that the material can sustain at zero normal stress), to the normal stress, σ, and to the internal angle of friction ϕ by

$$|\tau| = C_{ss} - \sigma \tan \phi \qquad (50)$$

For convenience, define

$$\alpha = 45° + 1/2\, \phi . \qquad (51)$$

Consider two ways to estimate ϕ and C_{ss} for the Ionian regolith. First, assume that the S-SO$_2$-silicate pyroclastics common to all models of the upper layers of Io resemble terrestrial permafrosts. The apparent angle of internal friction, ϕ, of many permafrosts is $\sim 30°$, similar to that of sand (once the permafrost has failed the friction between the rock or sand particles determines its cohesive behavior). This would give $\alpha = 60°$. However, permafrost strength C_{ss} varies greatly with confining pressure, temperature, ice content, and clastic size and shape (because strength depends on temperature and amount of ice, as well as on internal friction between sand grains). Thus, many factors, unknown for the surface of Io, affect the strength of the regolith: low surface temperatures favor high strength at the surface, but as temperature increases with depth, the strength decreases. The unconsolidated and porous nature of pyroclastic deposits favors low strength at the surface, but increasing compaction with depth increases the strength. The comparison with permafrosts suggests that a strength of 5 bar might be reasonable. An alternate method of estimating strength may ultimately come from a detailed analysis of the height of cliffs in the volcanic plume regions and a criterion for cliff height stability in terms of strength. For example, Melosh (1977) gives stable cliff heights if

$$\frac{\rho g h}{C_{ss}} \leqslant 5 . \tag{52}$$

For cliff heights currently estimated ~ 1 km (Chapter 15 by Schaber), this criterion gives a strength of 5 bar.

Assume that an Ionian eruption begins when a narrow fracture propagates from depth to the surface. The reservoir fluid will accelerate up this fracture (a proto-conduit) until sonic conditions are reached. As demonstrated in Sec. IV, flow pressures at sonic conditions are on the order of tens of bar, greater than plausible regolith strengths and greater than hydrostatic pressure in the upper layers of the regolith. As shown in Fig. 18.12c, the fluid will exert a horizontal pressure on the conduit walls. Jaeger (1969, p. 83) has shown that if

$$P > 2C_{ss} \tan \alpha + \rho g z \tan^2 \alpha \geqslant \rho g z \tag{53}$$

the regolith will fail by upward slip along planes inclined at α to the vertical (labeled α_P in Fig. 18.12c). A surface crater with shallow internal slopes would be formed (for $\phi \simeq 30°$, the slopes would be 60° from the vertical) as regolith material is ejected outward. The flow would diverge into the crater and pressures strongly decrease with height above the crater floor (as discussed quantitatively in Sec. VI). (If, as proposed for deep-seated terrestrial eruptions by Shoemaker et al. 1962, the pressure were to drop to less than hydrostatic, failure would occur if

$$P < \frac{\rho g z - 2C_{ss} \tan \alpha}{\tan^2 \alpha} \leqslant \rho g z . \tag{54}$$

Material would slip downward along planes inclined at α to the horizontal; this is labeled α_A in Fig. 18.12c.)

The initial angle of the interior slopes of the crater formed by upward ejection of regolith material is coincidentally near the angle of repose of slopes of cohesionless material. Therefore, volcanic debris that fell out of the flow into the crater or broke off the upper slopes of the crater and was too large to be carried outward by the low-pressure flow, would roll down the slopes toward the center of the crater. Secondary enlargement of the crater, after the initial crater-forming event discussed above, would occur primarily by indirect erosion—slumping of material into the crater and transport outward when the material became sufficiently comminuted to be ejected from the crater by the low-pressure flow within the crater.

An analogous process occurred at Mount St. Helens during March and April 1980. The initial eruption of 27 March formed a summit crater. The exact geometry of this crater is not well known because the event was not observed. However, the slope of the interior walls was $< 60°$ from the vertical. Abundant large boulders and rocks were notably absent from the slopes of the mountain outside the crater. The above analysis suggests that, because the crater formed in igneous rocks of high cohesive strength (several hundred bar), the dynamic pressures in the eruption were less than those given by Eq. (53) and that failure occurred along planes at $70°$ (assuming $\phi \sim 50°$) to the horizontal rather than the vertical. Subsequent enlargement of the crater through March and April occurred almost entirely by slumping of debris (rocks and ice) into the center of the crater, where it was ground and comminuted by repeated small eruptions before being ejected from the crater as small particles. This sequence is qualitatively similar to that proposed above for Io: initial formation of a crater by processes dependent upon the relation between regolith material properties and flow properties, and subsequent removal of material only after inward slumping and secondary abrasional grinding processes.

For purposes of illustration in the rest of this chapter, the nominal crater for Ionian volcanoes is taken to have interior walls at $60°$ from the vertical. The geometry of the crater assumed for modeling is shown in Fig. 18.12d. The nominal depth is 1 km, with radius and area specified by knowledge of the slope angle as $60°$ from vertical. The crater is assumed to overlie a conduit of radius 5 m, a typical dimension for a large terrestrial dike.

B. Steady Flow Through the Conduit Into the Crater

The discussion of conduit conditions demonstrated that an erupting fluid would enter a crater at choked or sonic conditions. For a vapor alone or with a small mass fraction of entrained pyroclastics, pressures would be on the order of several bar or tens of bar and velocities $\gtrsim 100$ m s^{-1} at the entrance to the crater for the reservoirs discussed here. As the mass fraction of pyroclastics increases, the velocity would decrease. For boiling fluids, sonic pressures are comparable, but velocities are on the order of tens of m s^{-1}. In this section I examine changes of pressure, temperature, and density as the fluid traverses the crater. The discussion cannot at this time include details of phase changes in the erupting fluid, although they are important.

Within the surface crater, the flow is assumed to be quasi-one-dimensional. It is then governed by the equations for one-dimensional flow with area change (e.g. Thompson 1972)

$$T = (1 + \frac{\gamma - 1}{2} M^2)^{-1} T_o \tag{55}$$

$$P = (1 + \frac{\gamma - 1}{2} M^2)^{(\gamma-1)/\gamma} P_o \tag{56}$$

$$\rho = (1 + \frac{\gamma - 1}{2} M^2)^{(\gamma-1)} \rho_o \tag{57}$$

where M is obtained implicitly from

$$\frac{A}{A^*} = \frac{1}{M} \left[\left(\frac{2}{\gamma + 1}\right)\left(1 + \frac{\gamma - 1}{2} M^2\right) \right]^{\frac{\gamma + 1}{2(\gamma-1)}} \tag{58}$$

and

$$M^* = \left[\frac{(\gamma + 1) M^2}{2 + (\gamma - 1) M^2} \right]^{1/2} . \tag{59}$$

The quantities are tabulated in many textbooks for commonly occurring values of γ (e.g., $\gamma = 1.4$, 1.3, 1.2, 1.1 in Zucrow and Hoffman 1976). Where needed, I have generated tables for the pseudogases with γ very close to unity. Note that the results scale directly with A/A^*; the following discussion is based on $A^* = 78$ m^2 ($r^* = 5$ m), chosen to be plausible for conduit dimensions based on studies of terrestrial dikes.

The velocity of erupting fluid increases, and the pressure and temperature decrease, as the fluid flows through the diverging crater. The expansion is determined, in the absence of shocks, by the ratio of flow area A at any point in the crater to the sonic area A^* of the conduit feeding the crater. As shown in Fig. 18.12d, this ratio is $> 10^3$ at all distances > 100 m above the conduit, e.g., the ratio is 1.2×10^3 at 100 m, 3×10^4 at 500 m, and 10^5 at 1 km. What effect does the areal expansion have on the fluid flow variables?

Consider first pure-vapor flow: e.g., SO_2 with $\gamma = 1.3$. From perfect-gas tables, at 100, 500, and 1000 m above the crater bottom, the pressure ratios are, respectively, 1×10^{-5}, 2×10^{-7}, 4×10^{-8}. The corresponding temperature ratios are $T/T_o = 0.08$, 0.03, and 0.02. If the reservoir pressure were ~ 40 bar, then the vapor would leave a crater 1 km deep at a pressure of 1.6×10^{-6} bar, slightly higher than the maximum proposed Ionian atmospheric pressure (10^{-7} bar), but possibly comparable to pressures generated by temporary overpressuring of the atmosphere by the eruptions. The Mach number for the flow would be 18.5 as it left the crater.

Consider as another example one of the erupting fluids that can be modeled as a pseudogas with $\gamma = 1.1$. At 100, 500 and 1000 m above the crater bottom, the pressure ratios are respectively, 7×10^{-5}, 1×10^{-6} and 4×10^{-7}. The corresponding temperature ratios are 0.42, 0.30 and 0.26, much higher than for the pure-vapor flow. For reservoir pressures on the order of 40 bar, the exit pressure would be 1.6×10^{-5}, an order of magnitude higher than for the case of vapor alone. The Mach number for the flow would be 7.5.

Because I have postulated that the likely eruptive fluids on Io can be modeled as pseudogases with γ between 1 and 1.3, these two examples suggest that unless extremely deep and highly divergent craters are present on Io or the atmospheric density is considerably greater than 10^{-7} bar, the erupting fluids will enter the Ionian atmosphere at pressures ranging from nearly ambient pressure to many orders of magnitude greater.

The range of possible plume temperatures at the surface is also wide. For SO_2 emanating from reservoir III at 400 K, the temperature 100 m into the crater would be 108 K and at 1 km would be 52 K. If the reservoir were at 1400 K (reservoir V), the temperatures at 100 and 1000 m would be 112 and 28 K, respectively. If the fluid were a pseudogas from SO_2-V′, it would be considerably hotter at corresponding heights in the crater; the temperature at 100 m would be 560 K, and at 1000 m, it would be 364 K.

The density of the plume fluids decrease drastically through the crater; typically at 100 m the density has decreased by $\sim 10^{-4}$ and at 1000 m, by 10^{-6}.

The very low temperatures and densities obtained in the plumes suggest that we should ask whether the expansion process might terminate because the gas becomes so rarified that molecular collisions within the plume are not available to maintain equilibrium. Thompson (1972, p. 302) suggests that, to a first approximation, termination of equilibrium expansion might occur when the local mean free path Λ becomes equal to the local diameter of the flow d: $\Lambda \sim d$. Because mean free path is inversely proportional to fluid density, it can be written as

$$\Lambda = \Lambda_0 \, \rho_0 / \rho = \frac{k \, d^* \, \rho_0}{\rho} \qquad (60)$$

where Λ_0 is normalized to the conduit diameter d^* at sonic conditions

$$\Lambda_0 = k \, d^* \quad \cdot \qquad (61)$$

According to Thompson (1972, p. 302), the density ratio at which molecular collisions are not available is approximately

$$\frac{\rho_\infty}{\rho_0} \sim 3k^2 . \qquad (62)$$

Typical molecular mean free paths at $400\,\mathrm{K}$ are $\sim 10^{-6}$ cm and, with a conduit diameter $\sim 10\,\mathrm{m}$, the constant k is $\sim 10^{-9}$. Therefore, molecular collisions do not cease until $\rho_\infty/\rho_0 = 3 \times 10^{-18}$, a value much lower than obtained through eruption into the Ionian atmosphere. Because of the very large dimensions of the system, molecular collisions would still occur with sufficient frequency within the flow to maintain equilibrium, despite the very low temperatures and seemingly low densities of the erupting fluid.

In summary, fluid flow through a conduit would not allow pressures to drop to the ambient Ionian atmospheric pressure. Near the Ionian surface, highly overpressured erupting fluids would erode the regolith into a surface crater. The fluids would accelerate supersonically through such craters as the pressure, temperature, and density decreased. A wide range of surface emergence pressures, temperatures, densities, and velocities are possible depending on crater depth and thermodynamic properties of the fluid. In the next section, I propose plausible plume structures associated with the different surface emergence conditions described above.

VI. THE PLUMES: IMPLICATIONS OF THE MODEL

The detailed structure of plumes on Io depends on the pressure of the erupting fluid relative to atmospheric pressure, and the Mach number of the fluid as it leaves the conduit or crater at the surface. Discussion in the previous sections emphasizes that exit pressure, velocity, and temperature are complex functions of the composition of the volcanic fluids, their initial condition in the reservoir, and the shape of the volcanic system. Even in an imaginary, strictly one-dimensional system, exit conditions are complex functions of the assumed thermodynamic path and nonequilibrium thermodynamic processes, and depend particularly on heat and mass effects of entrained pyroclastics. In a real, multi-dimensional system of irregular geometry, the shape of the system affects the location of choke points and strongly affects the velocities of the plumes as they emerge from the planetary surface. The models discussed for conduit flow and crater flow suggest that volcanic fluids are likely to leave the Ionian surface and enter its atmosphere at pressures either (1) comparable to the ambient atmospheric pressure, or (2) much higher. For convenience, I call the plume structure associated with (1) a balanced plume, and that associated with (2) an overpressured or underexpanded plume. Similar terminology is used in aeronautics literature, the source of most of the following discussion. Note that the following conclusions apply to the nominal conduit-crater geometry discussed in Sec. V.

The models suggest that balanced plumes occur when the fluid expands through a relatively deep (> 1 km) crater, and that overpressured plumes occur if the crater is shallow or negligible. Although factors other than areal divergence (such as hydrostatic pressure changes and friction) can affect pressure decay in the fluid as it travels from reservoir to surface, large changes in the shape of volcanic systems are by far the most influential factor in the determination of plume structure.

Balanced Plumes

Because the pressure of the fluid in balanced plumes is equal to ambient atmospheric pressure, particle and vapor trajectories are not strongly modified by the atmosphere after ejection from the surface. Effects not considered here that may affect the structure of balanced plumes are:

1. Shock waves within the crater that can occur to form balanced plumes;
2. Phase changes within the plume;
3. Boundary-layer development along crater walls and at the boundary with the atmosphere;
4. Heating of the plume by radiation above the surface;
5. Additional plume expansion caused by geometric divergence of the fluid as it erupts into the atmosphere, leading to possible underpressuring of the plume with altitude;
6. Change of atmospheric density with height.

The latter two effects may be somewhat compensatory: as the plume diverges with altitude, the pressure drops, and the atmosphere might tend to compress the plume inward because a pressure gradient from the atmosphere into the plume could be present. However, if the density of the atmosphere decreases rapidly, the plume might remain in a more or less balanced condition as it rises. These effects are not currently resolvable for Io by our spacecraft observations, but they could provide a considerably detailed structure for the plume.

In the discussion so far, the flow has been assumed one-dimensional (Fig. 18.13a); streamlines remain parallel to the conduit, spreading in area as the flow traverses the crater. Such a plume would look like a fire hydrant upon eruption. The assumptions of one-dimensional flow, although adequate for predicting the magnitude of the fluid flow variables, are clearly inadequate for predicting directions of flow as it emerges from craters with divergences as large as those proposed here, i.e. craters whose walls diverge as much as 60° from vertical. The flow would be turned by rarefaction waves emanating from the corners where the conduit and crater intersect, as shown in Fig. 18.10, and would tend to diverge into a cone whose boundaries parallel the crater walls, rather than following the conduit walls.

Detailed analysis of turning of flow by corner rarefaction waves is beyond the scope of this chapter, but a simple modification of the theory of one-dimensional flow used permits an estimate of the flow directions and variables within the plume. Assume that flow streamlines remain straight as the flow diverges from conduit to crater, but that the streamlines diverge

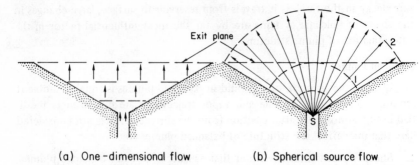

(a) One-dimensional flow   (b) Spherical source flow

Fig. 18.13. Schematic drawings of crater conditions for balanced-plume flow. (a) One-dimensional flow, with streamlines remaining parallel to their initial direction in the conduit. Flow properties are constant over planes (dashed lines). (b) Spherical source flow, with streamlines remaining straight but emanating from a source point S. Flow properties are constant over hemispheres (dashed lines). Hemispheres (1) and (2) intersect the exit plane of the crater at the center and edges, respectively.

from an apparent source point that would be the intercept point if the crater walls were extended into the conduit, as shown in Fig. 18.13b. The flow properties are then constant along spherical surfaces at constant radii from the apparent source of the streamlines (e.g. see Zucrow and Hoffman 1976, p. 220). All properties calculated for a given plane of area A_p in the crater actually apply to a hemispherical surface of equal area A_s. If the plane is at distance R from the source, a simple geometric calculation shows that a hemisphere of equal area will be at radius R' given by

$$R' = \frac{R \sin \alpha}{(2 - 2 \cos \alpha)^{1/2}} . \tag{63}$$

Here α is the slope of the crater walls. For the slope of $60°$ used in this chapter, equivalent properties occur along a hemisphere at distance $R' = 0.87$ R from the plane at distance R.

What is the distribution of velocities over a plane covering the exit surface of the crater? Across the exit plane (Fig. 18.13b) directions of velocity vectors would range from $60°$ at the edges to $0°$ at the center, following radii to the apparent flow source. The angular distribution of velocities would be isotropic, but the velocities across the plane would not be uniform. (Here we ignore boundary-layer effects that would tend to reduce the velocities toward the edges of the flow.) Velocities in the center of the flow would correspond to those on a hemispherical surface generally lower in the crater (surface 1 in Fig. 18.12b), and velocities toward the edges, to hemispheres progressively higher in the crater (e.g., surface 2 in Fig. 18.13b). In supersonic flow, the velocities will increase as the areas of the hemispheres increase, and the veloci-

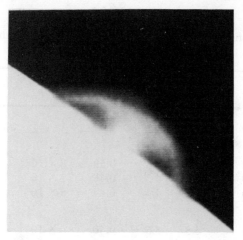

Fig. 18.14. Voyager photograph of Plume 3 (Prometheus), Voyager 1 FDS 1637748.

ties will therefore be greater in the outer regions of the surface plane across the crater. Hemispheres tangent to the surface plane vary in area by a factor of 4 (from radius $R = z$ at the center, where z is the height of the surface above the source point, to $R = z/\cos 60° = 2z$, for the sphere intersecting the edge of the surface plane). At small Mach numbers ($M = 1$ to 3), such an area change would give a large change in velocity across the exit plane (e.g., for $\gamma = 1.3$, if the ratio A/A^* changed from 1 to 4, M would change from 1 to 2.7, and the dimensionless velocity u/c_0 would double from 1 to 2). However, at large Mach numbers obtained in craters more than a few hundred meters deep the velocity changes associated with such an area change are small (e.g., as A/A^* changes from 300 to 1200, M changes from 7 to 9, but u/c_0 changes only by 2% from 2.6 to 2.66). Therefore, the velocity distributions across planes deep within the crater may have large variations, but those across the exit plane of deep craters will appear nearly isotropic in distribution and uniform in magnitude.

Because of the balanced pressure condition between the plume and the atmosphere, the particles and vapor in a balanced plume will follow ballistic trajectories originating at or near ground level. The plume shape is determined by, and reflects, crater shape; Prandtl-Meyer expansion outside the crater is not significant. Equations for the height and range of material following ballistic trajectories on Io are given by Strom and Schneider in Chapter 16, as is detailed calculation for Plume 3 (Prometheus). I propose that the symmetric, umbrella-shaped plumes on Io, Plume 1 (Pele) and Plume 3 (Prometheus), are balanced plumes (Fig. 18.14) that, according to the modeling presented here, emanate from surface craters on the order of 1 km deep and several km in diameter.

TABLE 18.3

**Estimates of Flow Variables in Model Ionian Craters
at Height h above Crater Bottom**

Variable	M (u/c)	M^* (u/c_0)	T/T_0	P/P_0	ρ/ρ_0
Reservoir and height (m)					
SO$_2$-I[a]					
$h =$ 100	4.8	3.8	0.59	8×10^{-5}	1×10^{-4}
500	6.0	4.2	0.48	2×10^{-6}	5×10^{-6}
1000	6.5	4.4	0.44	5×10^{-7}	1×10^{-6}
SO$_2$-II[b]					
$h =$ 100	5.5	3.5	0.40	4×10^{-5}	1×10^{-4}
500	6.5	3.8	0.32	4×10^{-6}	1×10^{-5}
1000	7.0	3.9	0.29	1×10^{-6}	4×10^{-6}
SO$_2$-III[c]					
$h =$ 100	6.0	3.2	0.27	4×10^{-5}	2×10^{-4}
500	8.2	3.4	0.17	1×10^{-6}	6×10^{-6}
1000	9.3	3.5	0.13	2×10^{-7}	1×10^{-6}
SO$_2$-IV[d]					
h $h =$ 100	7.0	3.0	0.17	2×10^{-5}	1×10^{-4}
500	10.0	3.1	0.09	6×10^{-7}	6×10^{-6}
1000	12.0	3.2	0.06	7×10^{-8}	1×10^{-6}
SO$_2$-V[e]					
$h =$ 100	8.9	2.66	0.08	1×10^{-5}	2×10^{-4}
500	14.9	2.73	0.03	2×10^{-7}	8×10^{-6}
1000	18.5	2.74	0.02	4×10^{-8}	2×10^{-6}
SO$_2$-V$'$[f]					
$h =$ 100	5.3	3.5	0.42	7×10^{-5}	2×10^{-4}
500	6.9	3.8	0.30	1×10^{-6}	5×10^{-6}
1000	7.5	3.9	0.26	4×10^{-7}	1×10^{-6}
S:I,II[g]	—	—	—	—	—
S:III[h]					
$h =$ 100	4.7	3.8	0.64	1×10^{-4}	1×10^{-4}
500	5.8	4.3	0.54	3×10^{-6}	5×10^{-6}
1000	6.3	4.5	0.50	5×10^{-7}	1×10^{-6}
S-IV[i]					
$h =$ 100	5.2	3.5	0.42	7×10^{-5}	2×10^{-4}
500	6.9	3.8	0.30	1×10^{-6}	5×10^{-6}
1000	7.5	3.9	0.26	4×10^{-7}	1×10^{-6}

TABLE 18.3 continued

Estimates of Flow Variables in Model Ionian Craters
at Height h above Crater Bottom

Variable	M (u/c)	M^* (u/c_0)	T/T_0	P/P_0	ρ/ρ_0
S-V[j]					
$h = $ 100	8.9	2.66	0.08	1×10^{-5}	2×10^{-4}
500	14.9	2.73	0.03	2×10^{-7}	8×10^{-6}
1000	18.5	2.74	0.02	4×10^{-8}	2×10^{-6}
S:V'[k]					
$h = $ 100	5.3	3.5	0.42	7×10^{-5}	2×10^{-4}
500	6.9	3.8	0.30	1×10^{-6}	5×10^{-6}
1000	7.5	3.9	0.26	4×10^{-7}	1×10^{-6}

[a]Assume $u \sim 0$ until boiling begins. Assume flow choked at 9 bar at entrance to crater. Boiling from 9 bar to 0.0165 bar, condensation to P_f. Vapor fraction x increases from 0 to 0.45 at triple point, then decreases to 0.37 at $P_f = 10^{-7}$ bar. Δh (between 9 bar and 0.0165 bar) $= 380$ J g^{-1}; u (triple point) $= 277$ m s^{-1}. Δh (between triple point and 10^{-7} bar) $= 295$ J g^{-1}; u (10^{-7} bar) $= 500$ m s^{-1}. To obtain estimate of flow variables in crater, assume that, over the whole range, the fluid is a heat-transferring pseudogas with $\bar{x} = 0.33$, $m = 2$ ($\gamma = 1.06$). $T_0 = 350$; $P_0 = 9$ bar (saturation pressure); $\rho_0 \sim (1 + m) \rho_{ov}$, where ρ_{ov} is density of the vapor at the saturation curve at 9 bar, $\rho_0 \sim 0.06$ g cm^{-3}. Triple point conditions are reached below 100 m. If ambient atmsopheric pressure near the volcano is 10^{-7} bar, $P_{atm}/P_0 = 6 \times 10^{-9}$, so this flow is still overpressured as it leaves a crater 1 km deep.

[b]Assume flow choked by boiling at 40 bar at entrance to a crater. Boiling from 40 bar to 0.0165 bar, condensation to P_f. Vapor fraction x increases from 0 to 0.56 at triple point and then decreases to 0.45 at $P_f = 10^{-7}$ bar, Δh (between 40 bar and triple point) $= 86$ J g^{-1}; v (triple point) $= 414$ m s^{-1}. Δh (between triple point and 10^{-7} bar) $= 103$ J g^{-1}; $u_f = 615$ m s^{-1}. To obtain estimate of flow variables in the crater, assume that, over the whole range, the fluid is a heat-transferring pseudogas with $\bar{x} = 0.5$, $m = 1$ ($\gamma = 1.1$). $T_0 = 393$ K; $P_0 = 40$ bar; $\rho_0 \sim (1 + m) \rho_{ov}$, where ρ_{ov} is density of the vapor at saturation conditions at 40 bar, $\rho_0 \sim 0.22$ g cm^{-3}. Triple point conditions are reached below 100 m.

[c]Condensation of vapor to liquid from 40 bar to 0.0165 bar, of vapor to solid from 0.0165 bar to P_f. Vapor fraction x decreases from 1.0 in reservoir to 0.70 on liquid side of triple point, increases to 0.76 as the liquid freezes isobarically at the triple point, and then decreases to 0.58 at $P_f = 10^{-7}$ bar. Δh (between 40 bar and 0.0165 bar) $= 194$ J g^{-1}; u (triple point) $= 630$ m s^{-1}. Δh (between triple point and 10^{-7} bar) $= 138$ J g^{-1}; v (10^{-7} bar) $= 814$ m s^{-1}. To obtain estimate of flow variables in crater, assume that, over the whole range, the fluid is a heat-transferring pseudogas with $\bar{x} = 0.66$, $m = 0.5$, $\gamma = 1.15$. $T_0 = 393$ K, $P_0 = 40$ bar, $\rho_0 \sim (1 + m) \rho_{ov}$, where ρ_{ov} is density of vapor at saturation conditions at 40 bar, $\rho_0 \sim 0.17$ g cm^{-3}. Triple point conditions are reached below 100 m.

[d]Pure vapor flow from 40 bar to 0.4 bar, condensation of liquid from 0.4 to 0.0165 bar (if no supersaturation), condensation of solid to P_f. Vapor fraction x is 1.0 to 0.4 bar, decreases to 0.85 on the liquid side of the triple point, increases to 0.88 with isobaric

crystallization at the triple point, and then decreases to 0.65 at $P_f = 10^{-7}$ bar. Δh (between 40 bar and 0.0165 bar) = 309 J g^{-1}; v (triple point) = 786 m s^{-1}. Δh (between triple point and 10^{-7}) = 362 J g^{-1}; v (10^{-7} bar) = 1158 m s^{-1}. To obtain estimate of fluid flow variables in crater, assume that, over the whole range, the fluid is a heat-transferring pseudogas with $\bar{x}$ = 0.85, m = 0.176, γ = 1.20. T_0 = 600 K, P_0 = 40 bar, $\rho_0 \sim (1 + m) \rho_{OV}$, where ρ_{OV} is density of the vapor at 40 bar, 600 K; $\rho_0 \sim 0.07$ g cm^{-3}. Triple point conditions are reached below 100 m.

[e]Pure vapor flow throughout ascent, γ = 1.3. P_0 = 40 bar, T_0 = 1400 K, $\rho_0 \sim 0.02$ g cm^{-3}. Triple point conditions are reached below 100 m.

[f]Pure vapor flow throughout ascent, γ = 1.1. P_0 = 40 bar, T_0 = 1400 K, $\rho_0 \sim 0.04$ g cm^{-3}.

[g]No plume; see text.

[h]Assume flow choked by boiling at 10 to 40 bar at entrance to crater. Boiling from 40 bar to $\sim 2 \times 10^{-2}$ bar (see Fig. 18.6); vapor-alone from $\sim 2 \times 10^{-2}$ bar to P_f taken as 10^{-7} bar. Vapor fraction ranges from 0 at initial condition to 1.0 at 2×10^{-7} bar. Δh (initial condition to 2×10^{-2}) = 44 J g^{-1}; if converted to kinetic energy, u = 297 m s^{-1}. Δh (between 40 bar and 10^{-7} bar) estimated as 135 J g^{-1}; if converted to kinetic energy, v_f = 520 m s^{-1}. To estimate flow variables in the crater, assume that the initial vapor formed at ~ 600 K is S_6, γ = 1.1, R = 43.3 J kg^{-1} K^{-1}. The boiling fluid would fragment, giving a vapor-droplet mixture that could be modeled as a pseudogas, with γ lowered because of the effective mass of the droplets. The droplets would disappear as the pressure reached 2×10^{-2} bar, and the effective γ would increase back toward 1.1. However, the temperature simultaneously decreases and more S_7 and S_8 constituents form in the vapor. To account for the heavy effective molecular weight throughout the process, I represent the fluid with an average γ = 1.05. P_0 = 10 to 40 bar. T_0 = 1000 K, $\rho_0 \sim 2$ g cm^{-3}.

[i]Pure vapor flow throughout ascent, assume S_6, γ = 1.1. P_0 = 10-40 bar, T_0 = 1000 K, $\rho_0 \sim 1$ g cm^{-3} (extrapolated from West 1950).

[j]Pure vapor flow throughout ascent, assume S_2, γ = 1.29. P_0 = 40 bar, T_0 = 1400 K, $\rho_0 \sim 0.2$ g cm^{-3} (extrapolated from West 1950).

[k]Pure vapor flow throughout ascent, assume S_2,, m = 1, γ = 1.1. P_0 = 40 bar, T_0 = 1400 K, $\rho_0 \sim 0.4$ g cm^{-3}.

Overpressured Plumes

The structure of plumes highly overpressured relative to ambient pressure is complex; the pressure gradient from the plume into the atmosphere causes expansion of the plume after it leaves the planetary surface. In response to the expansion, the atmosphere, in turn, presses in on the flow; the net result is a complex plume structure, illustrated schematically in Fig. 18.15. Because details of such flows on Io cannot be resolved with available observations, only the major points are mentioned here to illustrate the dynamics; see Kieffer (1981,1982) for detailed discussion.

If the erupting fluid is overpressured (underexpanded) when it reaches the surface, it flows through rarefaction waves that emanate from the edges of the conduit or crater (Fig. 18.15). At the boundary between the plume and the ambient atmosphere, the pressure is assumed equal to ambient atmospheric pressure. The magnitude of angular divergence of the flow as it leaves the vent, and the velocity it ultimately attains, are determined by the initial

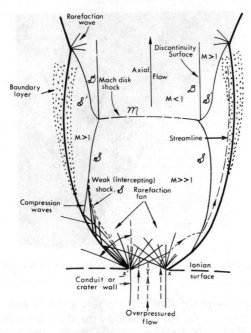

Fig. 18.15. Schematic drawing of structure in overpressured plume (after JANNAF 1975). Upon leaving the Ionian surface (through a plane at X'X), the flow is initially deflected through the Prandtl-Meyer angle ν by rarefaction waves (the rarefaction fans emanating from the corners X and X'). These waves cross the flow and reflect from the flow boundary (shown by a heavy line) as compression waves that coalesce into weak intercepting shocks S and S'. These weak shocks are connected across the flow by the Mach disk shock M. The flow pattern between X'X and the Mach disk shock can repeat downstream. Discontinuity surfaces D can develop, separating axial flow from deflected flow near the boundary layer. The boundary layer of flow with ambient atmosphere (stippled region) increases in width downstream. Supersonic ($M > 1$ or $M \gg 1$) and subsonic ($M < 1$) regions of the flow are shown. Dashed lines are schematic streamlines of particles in the flow.

Mach number of the flow, ratio of flow pressure to atmospheric pressure at the vent, and thermodynamic properties of the fluid. Near the vent, the flow is deflected through the Prandtl-Meyer angle ν by the rarefactions, as discussed in Sec. V.A. As the fluid passes through the rarefaction waves, velocity increases and pressure decreases; the pressure can even drop below ambient atmospheric pressure in a zone near the axis of the plume.

The rarefaction waves cross the flow and reflect off the constant pressure boundary between the flow and the atmosphere. The reflected waves are compression waves that coalesce into weak lateral shocks, called intercepting shocks because they intercept the initial expansion waves, which in turn connect across the flow into a strong shock, called a Mach disk shock. These compression waves and shocks form in response to the pressure gradient between the atmospheric boundary and the underpressured zone near the axis

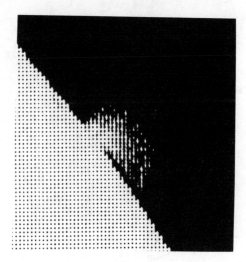

Fig. 18.16. Voyager photograph of Plume 2 (Loki), Voyager 1 FDS 1637534.

of the flow, and cause the flow boundary to curve inward away from the initial Prandtl-Meyer angle. The flow below the Mach disk shock is supersonic ($M \gg 1$) and above it is subsonic ($M < 1$). In the laboratory, the pattern described here (shown in Fig. 18.15) is observed to occur repeatedly with increasing elevation, forming a plume that resembles a string of sausages.

Bold scaling of rocket data (e.g. see JANNAF 1975) suggests that the scale of the structure described above, from the surface to the Mach disk shock, would be several tens of km for the nominal crater with a surface diameter of 3.5 km modeled here. I therefore think that the erupting fluid would flow supersonically at velocities on the order of 1 km s^{-1} up to a height of several tens of km above the Ionian surface where it would be strongly decelerated by the Mach disk shock, and that the fluid would flow laterally from the vent at high velocities through all angles from vertical to nearly horizontal until decelerated by the lateral intercepting shocks. The net result would be a diffuse, ill-structured plume with a broader range of ejection angles than the balanced plume and a strongly nonuniform distribution of velocity magnitudes through any given plane near the surface. Even though the fluid is at higher pressure when it leaves the planetary surface than the fluid in a balanced plume, overpressured plumes may not be as high as the balanced plumes because of the strong shock waves in the plume a few tens of km from the vent. Thus, I suggest that the diffuse, ill-structured plumes observed on Io such as Plume 2 (Loki) are overpressured plumes (Fig. 18.16). The preceding discussion suggests that they emanate from shallow craters or narrow conduits. Perhaps they are associated with immature eruptions that have not had time to form surface craters, or with eruptions through rocks of sufficient strength that large, shallow craters have not been formed by erosion.

In summary, plume structures on Io may reveal more about the shape of volcanic systems from which they emanate than about the nature of the erupting vapor. I have proposed that the symmetric, umbrella-shaped plumes result from fluids approximately at ambient pressure on reaching the surface, and that the diffuse irregular plumes result from fluids strongly overpressured when they reach the surface. The conditions of balance or overpressure are determined mainly by the shape of the system and can be obtained with either SO_2 or S, with or without entrained pyroclastics, originating at temperatures from 400 to 1400 K.

VII. CONCLUSIONS

Conclusions of this chapter are: (1) a given volcanic reservoir can produce a wide variety of surface manifestations, depending on conduit-crater details, and (2) the observational constraints on Ionian plume characteristics allow a wide variety of postulated reservoir conditions. Volcanic plume structures on any planet are the result of a series of complex processes, undergone by a reservoir fluid of unknown and probably complex composition and physical and thermodynamic properties, in systems of irregular shape. Until we understand the material properties of erupting fluids, possible thermodynamic paths during their ascent, constraints of shape of plausible volcanic system geometries, and plume-atmosphere interactions, we should not overinterpret the limited spacecraft data on Ionian plumes.

Acknowledgments. I am grateful to S.J. Peale, N.M. Schneider, H.R. Shaw, and P.R. Toulmin, III, for critical reviews of the original manuscript, to S.J. Peale for kindly sharing data on sulfur. I also thank the editors of this book, fellow authors, and family who have been tolerant of the havoc that the eruptions of Mount St. Helens in 1980 created in the preparation of this manuscript. This chapter is dedicated to T. McGetchin, whose many contributions to volcanology and infectious enthusiasm for exploration of the planets are fittingly symbolized by the volcanoes of Io.

APPENDIX A

A PSEUDOGAS MODEL FOR MULTIPHASE FLUIDS THAT HAVE NO MASS TRANSFER BETWEEN PHASES

Several situations exist in which the volcanic fluid might be modeled either as a perfect gas or as a gas phase carrying a fixed mass load of solid (or liquid) particles (see text and Appendices B and C). Here it is demonstrated that such mixtures can be modeled with an equation of state of the form of a perfect gas law with an appropriately modified gas constant and isentropic exponent. This model has been used in engineering applications for some time, but is new to volcanology, so a complete derivation is given. The derivation follows that of Wallis (1969), who called a fluid that could be so described a "pseudo gas."

Two limiting cases of thermal behavior are examined, (1) no heat is transferred between the gas and the condensed phases, and (2) thermal equilibrium between the phases is complete. The first type of behavior is approached by a fluid carrying relatively large entrained fragments; if heat transfer is mainly by conduction, then in a time of ascent t a typical particle with radius $> (8\kappa t)^{1/2}$ (where κ is thermal diffusivity) will not conduct much of its heat to the gas. (See Carslaw and Jaeger 1959, for thermal-conduction models.) If t is assumed to be on the order of 30 s for volcanic ascent (e.g., through distance 3 km at average velocity 100 m s^{-1}), and κ is the diffusivity for pure conduction of a typical silicate ($\sim 10^{-2}$ cm^2 s^{-1}), particles larger than ~ 1.5 cm will not transmit much of their heat to the gas; the behavior of the gas-particle mixture then could be modeled by case 1 below. For smaller particles, the behavior will approach that of case 2.

Case 1. A Pseudogas with no Heat Transfer between Phases

Assume (after Wallis 1969) that the equation of state of the gas phase is given by

$$P_g V_g = R_g T \tag{64}$$

where P_g is pressure in the gas phase, V_g specific volume of the gas phase, R_g gas constant of the gas, and T absolute temperature. The density of the mixture is given by

$$\frac{1}{\rho_{mix}} = \frac{x}{\rho_s} + \frac{(1-x)}{\rho_g} \tag{65}$$

where ρ_{mix} is density of the mixture, ρ_s density of the solid, ρ_g density of the gas, and x mass fraction of the solid phase. If the particle density is very large compared to the gas density, and m is the mass ratio of the particles to gas,

$$\frac{\rho_{mix}}{\rho_g} = \frac{V_g}{V_{mix}} \cong 1 + m \tag{66}$$

where V_{mix} and V_g are specific volumes of mixture and gas phase, respectively. Because

$$V_g \cong (1 + m) V_{mix} \tag{67}$$

the equation of state of pseudogas, from Eq. (64), becomes

$$P_{mix} V_{mix} = \frac{R_g}{(1 + m)} T. \tag{68}$$

Therefore, the effect of adding solid (ash) or inert liquid particles to vapor is modification of the vapor gas constant to account for increased density of the pseudogas mixture.

If the gas and particles have the same velocities (no phase lag), then because a homentropic (everywhere adiabatic and isentropic) expansion of the vapor follows the law

$$P V_g^{\gamma_g} = \text{const.} \tag{69}$$

the mixture follows the law

$$P V_{mix}^{\gamma_g} (1 + m)^{\gamma_g} = \text{const.} \tag{70}$$

with the substitution of Eq. (67) for V_g. If $(1 + m)$ is constant (i.e. no mass transfer occurs between the phases and no mass is added to the flow), this law can be rewritten

$$P V_{mix}^{\gamma_g} = \text{const.} \tag{71}$$

Therefore, in the absence of heat or mass transfer, the mixture expands homentropically exactly as does the vapor phase alone, except that R is modified and the volume of the mixture is used. The value of γ for the mixture is taken equal to the value for the gas phase $\gamma_{mix} = \gamma_g$.

Case 2. A Pseudogas with Heat Transfer and Thermal Equilibrium between Phases

If entrained ash particles are sufficiently small that thermal conduction or radiation can transfer heat between phases, then a model in which thermal equilibrium is assumed to be maintained between the phases is more appropriate. Again, the flow is assumed homentropic, so entropy s is at all times constant:

$$dS = dS_s + dS_g = 0 \tag{72}$$

where subscripts s and g are respectively the solid (inert liquid) and gas phases. If the heat capacity of the solid is C_{Ps}, and assumed equal to C_{Vs}, and that of the gas is C_{Pg}, the entropy change of the solid particles due to transfer of a small amount of heat dQ to or from the gas is

$$dS_s = \frac{dQ}{T} = C_{Ps}\, m \frac{dT}{T} \; . \tag{73}$$

The corresponding entropy change for the gas is

$$dS_g = \frac{dQ}{T} = C_{Pg} \frac{dT}{T} - R_g \frac{dP}{P} \; . \tag{&74}$$

Addition of these two equations and use of Eq. (72) gives

$$dS_s + dS_g = \frac{dT}{T}\,(C_{Pg} + mC_{Ps}) - R_g \frac{dP}{P} = 0 \; . \tag{75}$$

With integration

$$T^{-1}\, P^{(\gamma - 1)/\gamma} = \text{const.} \tag{76}$$

where

$$\frac{\gamma - 1}{\gamma} = \frac{R_g}{C_{Pg} + m\, C_{Ps}} \; . \tag{77}$$

From the definition of R_g

$$R_g = C_{Pg} - C_{Vg} \tag{78}$$

Eq. (77) can be solved for γ

$$\gamma = \frac{C_{Pg} + m\, C_{Ps}}{C_{Vg} + m\, C_{Ps}} \; . \tag{79}$$

By comparison of this expression with the definition of γ for a perfect gas, specific heats of the pseudogas at constant pressure and volume are

$$\frac{C_{Pg} + m\, C_{Ps}}{1 + m} \quad \text{and} \quad \frac{C_{Vg} + m\, C_{Ps}}{1 + m} \; . \tag{80}$$

Therefore, if heat transfer between the phases is rapid the pseudogas behaves as if it had a gas constant

$$R_{mix} = \frac{R_g}{1 + m} \qquad (81)$$

and an isentropic exponent

$$\gamma_{mix} = \frac{C_{Pg} + m\, C_{Ps}}{C_{Vg} + m\, C_{Ps}} \; . \qquad (82)$$

Pseudogases modeled by either Case 1 or 2 obey all the laws of fluid dynamics derived for perfect gases. In particular, the sound speed of a pseudogas is

$$c_{mix} = (\gamma_{mix} R_{mix} T)^{1/2} \; . \qquad (83)$$

APPENDIX B

THERMODYNAMIC CONSIDERATIONS FOR EQUILIBRIUM EXPANSIONS OF BOILING LIQUIDS AND CONDENSING VAPORS

The models used in the text as approximations of thermodynamic properties of the volcanic fluids assume that the fluid is either a single-phase vapor (perfect gas) or a multiphase mixture of constant phase composition (pseudogas). Here I summarize some effects of phase changes on the dynamics of flowing fluids, assuming that phase changes always occur under thermodynamic equilibrium conditions. In Appendix C, the effects of nonequilibrium thermodynamic conditions are summarized.

Low-entropy (Boiling) Fluids

In isentropic expansion from low-entropy reservoirs, as from SO_2 reservoirs I and II or S reservoir III, the first phase change in the fluid is due to boiling. For a boiling fluid flowing isentropically up a vertical conduit, conservation of energy gives

$$h_1 + \frac{u_1^2}{2} = h_2 + \frac{u_2^2}{2} \qquad (84)$$

where h_1 and h_2 are enthalpies of the liquid-vapor mixture at two different levels in the conduit, and u_1 and u_2 are the velocities of the mixture at those levels. Potential energy changes are assumed to be negligible. The flow is assumed homogeneous with no slip between liquid and vapor phases. If the kinetic energy is small, as in terrestrial geothermal wells where it rarely exceeds 5% of the enthalpy, a reasonable approximation is that the flow is isenthalpic ($h_1 \sim h_2$), as discussed in the text for a perfect gas. However, for Ionian volcanoes, with u_1 taken as 0, u_2 would be given by equation (84) as

$$u_2 \cong \sqrt{2(h_1 - h_2)} \ . \tag{85}$$

Thus, a velocity corresponding to any final mixture enthalpy (i.e. composition) can be specified. For any appreciable vapor fraction (e.g. 10%) the final flow velocity is typically hundreds of m s^{-1}.

The sound speed of a boiling mixture is given by

$$c = \left(\frac{\partial P}{\partial \rho}\right)^{1/2} = \left(\frac{K}{\rho}\right)^{1/2} \tag{86}$$

where K is bulk modulus of the mixture. Because the mixture has the compressibility of the gas phase (low K) and approximately the density of the liquid phase (high ρ), this expression indicates that the sound speed of a boiling mixture can be very low. For example, the sound speed of water with a few percent steam fraction is only a few to a few tens of m s^{-1} (Kieffer 1977); the sound speed of boiling SO_2 should be equally low. Thus, while accelerating from rest in a reservoir to the limiting velocity given by Eq. (84), a two-phase boiling fluid would choke at relatively low flow velocities compared with the choked flow velocities of a perfect gas (a few hundred m s^{-1}). This choking phenomenon is a well-known problem in terrestrial geothermal exploitation because the low velocity at which the flow chokes limits the mass flux (see Eq. 44) and therefore the energy flux obtainable from a well. Compared with perfect gas flow, therefore, a boiling fluid will choke at low flow velocities in the conduit and will accelerate to an appreciable final velocity only upon entering the divergent part of the summit crater.

Condensing Vapors

In flow from high-entropy reservoirs such as SO_2 reservoirs III and IV, the first phase change is condensation. An energy equation analogous to Eq. (84) holds for the flow of condensing vapors, as shown in the following derivation (from Altman and Carter 1956).

Assume that an isentropically expanding vapor enters the two-phase liquid-vapor field at temperature T_1 and pressure P_1, with entropy $S_g^{T_1}$ (an analogous assumption would hold for a vapor-solid transition). For equilibrium expansion, subsequent temperatures and pressures will be given by the vapor-pressure equation. The only unknown is the composition, which can be

determined by the chord rule and the requirement of constant entropy
($\Delta S = 0$). Let X_g be the mole fraction vapor, and S_g and S_ℓ the molar entropies of the vapor and liquid phases, respectively. $\Delta \bar{S} = 0$ implies that at any
temperature T

$$S_g^{T_1} = X_g S_g^T + (1 - X_g) S_\ell^T . \tag{87}$$

Denoting $S_g - S_\ell$ as ΔS_v (molar entropy of vaporization) this equation becomes

$$S_g^{T_1} = S_g^T - (1 - X_g) \Delta S_v^T . \tag{88}$$

Solving for the mole fraction of liquid

$$X_\ell = (1 - X_g) = \frac{S_g^T - S_g^{T_1}}{\Delta S_v^T} = \frac{R \ln(P_1/P) - \int_T^{T_1} C_{Pg} \, \mathrm{d}\ln T}{\Delta S_v^T} . \tag{89}$$

If C_{Pg} is constant from T_1 to T, this simplifies to

$$X_\ell = 1 - X_g = \frac{R \ln(P_1/P) - C_{Pg} (\ln T_1/T)}{\Delta H_v/T} \tag{90}$$

where the equilibrium relation

$$\Delta S_v^T = \Delta H_v/T \tag{91}$$

is used. The enthalpy change per mole of mixture in this temperature interval
is

$$H_1 - H = \int_T^{T_1} C_{Pg} \, \mathrm{d}T + (1 - n_g) \Delta H_v^T \tag{92}$$

where n_g is number of moles of gas. The kinetic energy increase in this region
of condensation is

$$\frac{1}{2} M(u^2 - u_1^2) = H_1 - H \tag{93}$$

where M is molecular weight of the condensing substance.

The maximum possible velocities are obtained for a vapor condensing to
a solid phase at $0 \, \mathrm{K}$, $H = 0$; this equation was used in the Smith et al. (1979)
model to calculate limiting velocities for SO_2 from reservoir conditions III

and IV. These results and others for SO_2 and S reservoir conditions I through V' are shown in Table 18.2. Note that these velocities are upper limits on obtainable velocities, not necessarily obtainable velocities.

The phenomenon of choking again becomes important in condensing flow. The effects of latent heat release on the flowing vapor depend on the Mach number of the flow when condensation begins; if the flow is subsonic, as would be likely if reservoir conditions are near saturation conditions, then the heat addition results in acceleration of the flow to Mach 1, where it chokes. Further heat can be released only after divergent expansion to lower pressures and higher Mach numbers. If the flow is supersonic at the onset of condensation, as it might be for flow from SO_2 reservoirs III and IV, the heat addition causes the pressure to remain higher than if the condensation were not occurring.

Phase changes, condensation or boiling in a flowing vapor alter flow conditions in two ways: (1) latent heat is released or absorbed by the phases, and (2) the number of moles of the vapor phase is increased or decreased. These two effects act in opposition directions and both must be considered to calculate flow properties; e.g., if condensation occurs latent heat release warms the remaining vapor above the temperature it would have if expanding without phase changes, and therefore the velocity is higher; however, the decrease in vapor abundance lowers the expansion work that can be performed. Altman and Carter (1956) concluded that the net effect is usually a slight increase of velocity when condensation occurs.

APPENDIX C

NONEQUILIBRIUM EFFECTS IN BOILING AND CONDENSING VAPORS

The effects of nonequilibrium thermodynamic processes in the flow are discussed here to provide perspective on the limitations of the equilibrium and metastable expansion theories used in the text. Details of nucleation and growth of new phases from the initial fluid phase cannot be considered quantitatively in the fluid-flow calculations because so little data are available, even for the relatively well studied H_2O. Emphasis is placed here on the high-entropy condensation of vapor to liquid and solid phases, because the author considers it most relevant to the high-velocity plumes on Io. As discussed by Kieffer and Delany (1979), high-entropy volcanism of H_2O is rare on Earth but has probably occurred in the eruptions of kimberlites. Discussion of nonequilibrium effects in low-entropy boiling flow can be found in Kieffer (1977). Four questions must be addressed: (1) what the nonequilibrium effects are; (2) over what range of P-T they occur; (3) what their effect is on the flow; (4) how they can be approximated.

Nonequilibrium Effects in Low-entropy (Boiling) Flow

Nonequilibrium effects occur in boiling flow because heat and mass transfer across the liquid-vapor interface occurs at a finite rate. For example, consider the magnitude of nonequilibrium affects on the sound speed of a boiling mixture (assume that flow velocities will be similarly affected because of the direct relation between sound velocities and flow velocities). The sound speed of a fluid is affected by boiling because a vapor phase drastically alters the compressibility of the fluid. The equilibrium sound speed of boiling water is a few m s^{-1}, in contrast to 1500 m s^{-1} characteristic of liquid water and to 440 m s^{-1} characteristic of pure steam (Kieffer 1977). If equilibrium is not obtained, the sound speed can be as much as an order of magnitude lower. Thus, nonequilibrium effects may be expected to have nearly an order of magnitude effect on flow dynamics. The effects will be present at all conditions in which boiling occurs. Kieffer (1977) has discussed approximations for the complete equilibrium and complete disequilibrium limits.

Nonequilibrium Effects in High-entropy (Condensing Flow)

Nonequilibrium effects in condensing flow occur because supersaturation of the vapor is generally required before the condensed phases nucleate. This section shows the range of pressure and temperature conditions over which two-phase flow and kinetic effects may be important, and estimates valid pressure-temperature ranges for the equilibrium and metastable approximations used in the text. The general discussion here applied to SO_2 is from Zel'dovich and Razier (1966).

Because expansions are well described by the specific volume of the system and kinetic effects are generally specified in terms of temperature or temperature differences, a temperature-volume phase diagram is convenient for this analysis (Fig. 18.17, for SO_2). Consider the isentropic ascent of a fluid initially a vapor, such as SO_2 at 26 bar, 400 K at a point A of Fig. 18.17. The initial expansion of the fluid is through the vapor-alone field. The isentrope intersects the vapor curve $(CP - Y)$ relatively steeply, and the two-phase assemblage (liquid + vapor) becomes the equilibrium assemblage at lower temperatures. It can be demonstrated that two-phase assemblage becomes the equilibrium assemblage, as follows. The isentrope is a power law in the form

$$T \sim V^{-(\gamma-1)}. \tag{94}$$

To a good approximation the vapor curve is represented by

$$V_{vap} = B T e^{U/RT} \tag{95}$$

$$T = \frac{U}{R} \left(\ln \frac{V_{vap}}{BT} \right)^{-1} \tag{96}$$

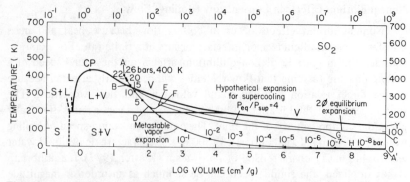

Fig. 18.17. Temperature-volume relations in the SO_2 system. Phase boundaries are shown by heavy lines: S, solid; L, liquid; V, vapor; CP is the critical point. The three curves are, from top to bottom, the two-phase equilibrium expansion adiabat; a schematic equilibrium adiabat for presumed $100°$ supercooling, taking into account the kinetics of the vapor-liquid phase change (the liquid-solid phase change is ignored); and the adiabat of metastably cooled vapor.

where U is an average heat of vaporization (taken as 388 J g^{-1} for SO_2) over the liquid-vapor portion of the saturation curve; R is the gas constant; and B is an approximately constant coefficient. (For SO_2 B varies from $1.0 - 2.06 \times 10^{-5}$ as the volume varies from $7 - 1.3 \times 10^4$ cm^3; 1.5×10^{-5} was used as the average value.) Therefore, because the isentrope has a strong temperature dependence on volume (Eq. 94) and the saturation curve has only a weak dependence (Eq. 96), the two curves intersect at a steep angle (at point B of Fig. 18.17), and the two-phase assemblage will become the equilibrium assemblage.

If supersaturation occurs, the vapor will expand along the metastable vapor isentrope (BA'). This isentrope represents the smallest volumes possible at any given temperature during expansion and therefore is one limiting approximation to the thermodynamic behavior, the metastable limit discussed in the text. On the other hand, if centers of condensation and liquid droplets form, the volume of the vapor phase is greater at a given temperature because latent heat released transfers to and warms the vapor. The equilibrium mixture volume is larger than the metastable vapor volume, in spite of the volume change associated with the transformation. The equilibrium isentrope of the two-phase mixture is given by

$$dS = 0 = \left[C_{V1}(1-x) + C_{V2}x \right] \, dT + RT(1-x)\frac{dV}{V} - \left[U - (C_{V2} - C_{V1})T \right] \, dx \; . \quad (97)$$

Here C_{V1} is specific heat of the vapor at constant volume (taken as 0.48 J/g/C); C_{V2} specific heat of the liquid (taken as 1.33 J g^{-1} C^{-1}); x relative mass fraction of liquid to (vapor + liquid), and V volume of the two-phase mixture. U, C_{V1}, and C_{V2} are assumed constant. The volume of the two-phase mixture in terms of the components is

$$V = V_{\text{liq}} \, x + V_{\text{vap}} \, (1 - x) \tag{98}$$

and for $x \ll 1$

$$V \approx V_{\text{vap}} \, (1 - x) \ . \tag{99}$$

Eq. (95) for the vapor volume gives

$$\frac{V}{1 - x} = BT e^{U/RT} \tag{100}$$

or

$$x = 1 - \frac{V}{BT} \, e^{-U/RT} \quad . \tag{101}$$

The differential dx is therefore

$$dx = \frac{-e^{-U/RT}}{BT} \, dV + dT \left[\frac{-U \, V}{R \, B \, T^3} \, e^{-U/RT} + \frac{V}{B \, T^2} \, e^{-U/RT} \right] \tag{102}$$

Substitution of Eqs. (101) and (102) into Eq. (97) and rearrangement of terms gives

$$\frac{dV}{dT} = \left\{ \left[U - (C_{V2} - C_{V1})T \right] \left[\frac{-UV}{R \, B \, T^3} \, e^{-U/RT} + \frac{V}{BT^2} \, e^{-U/RT} \right] \right.$$
$$\left. - \left[C_{V1} \left(\frac{V}{BT} \, e^{-U/RT} \right) + C_{V2} \left(1 - \frac{V}{BT} \, e^{-U/RT} \right) \right] \right\} \div \tag{103}$$
$$\left\{ \frac{R}{B} \, e^{-U/RT} + \left[U - (C_{V2} - C_{V1})T \right] \frac{e^{-U/RT}}{BT} \right\} \ .$$

This equation is easily integrated numerically on a small calculator from initial conditions V_0 and T_0 on the saturation curve to any condition V and T in the two-phase region. For the isentrope considered here, the initial conditions at the two-phase boundary were taken as $T_0 = 330 \, \text{K}$ and $V_0 = 34.59 \, \text{cm}^3$. The resulting two-phase equilibrium isentrope is BC on Fig. 18.17.

It can be seen from this figure that, at a given temperature, the difference in volume between systems undergoing metastable and equilibrium expansion can be as much as 5 orders of magnitude. This volume difference is largely due to heating of the vapor phase by the release of latent heat. Because volume and velocity changes are related through the conservation equations, there must be corresponding velocity differences between a gas expanding metastably and in equilibrium.

The thermodynamic history of adiabatic decompression of a real substance lies between the two extremes described above because of the kinetics of condensation. After passing the saturation state (point B of Fig. 18.17), a vapor will continue to expand, following the metastable vapor isentrope, until formation of condensation nuclei. Their rate of formation is a very strong function of degree of supersaturation; if little undercooling occurs a few centers will nucleate and grow, and if a large degree of undercooling occurs many centers will nucleate. Ions and dust particles in volcanic ejecta will likely create favorable conditions for the rapid formation of condensation centers. However, Zel'dovich and Razier (1966) point out that the presence of foreign particles is not necessary and that in a pure supersaturated vapor condensation centers will appear naturally as the result of agglomeration of molecules into molecular complexes. After reaching a certain critical size, these complexes become stable and grow into droplets of liquid. Therefore, as the degree of undercooling increases, the number of nuclei of the liquid phase increase rapidly. Figure 18.17 shows a schematic case BD for 100° supercooling; the liquid-solid phase change is ignored in these calculations.

The nucleation of growth centers results in release of latent heat; as a result, the temperature of the mixture increases back toward the saturation curve (along path DE). The rate of nucleation of liquid centers decreases, but the rate of condensation continues to accelerate because of the increase in surface area of drops to which the vapor molecules attach. The formation of new nuclei eventually ceases and further condensation proceeds only by attachment of molecules to the previously formed drops. Thus, as a rule all condensation centers are born at the beginning of the condensation process.

The number of condensation centers depends on the maximum attainable supersaturation and is determined by the interplay of opposing effects: cooling of the vapor by expansion and heating by release of latent heat in condensation. The rate of droplet growth is also controlled by two opposing effects, adhesion and evaporation. In an adiabatic expansion, condensation always exceeds evaporation; the slower the process, the more the condensation and the closer the approach to equilibrium (Zel'dovich and Razier 1966).

Maximum velocities would be obtained only if the fluid regained and followed the two-phase equilibrium curve (FC in Fig. 18.17), a path corresponding to conversion of all mass in the system to the solid phase at 0 K and to release of all available specific and latent heat. Thus, condensation must proceed rapidly while the gas is sufficiently dense that molecular collisions maintain thermodynamic equilibrium, in order to approach the condition of complete condensation. In a real expansion (such as DEG) appreciable departure from thermodynamic equilibrium will take place at very low volumes and temperatures: the spacing between molecules becomes so rarified that the vapor can no longer find liquid or solid nuclei to attach to (GH on Fig. 18.17; the position of GH shown here is schematic; as discussed in the text (Sec. V.B), it depends on the scale of the system). Thus, the expansion freezes and some vapor expands to infinite volume without releasing its latent heat; the substance at 0 K is a disequilibrium mixture of vapor and condensate.

Referring to Fig. 18.17, consider the qualitative effect of thermo-dynamics and kinetics on the flow. Fluids (such as H_2O) commonly expand in a supersaturated metastable vapor state to $P \sim 1/7\, P_{sat}$ during adiabatic expansion; mercury vapor has been known to expand to supersaturation factor 2000 without condensing (Hill 1966, p. 593). As nucleation of conden-sate occurs in a flowing supersaturated vapor, a condensation shock generally forms, a discrete zone in the flow where the latent heat is suddenly released. The effects of heat release depend on the Mach number of the flow. For subsonic flow, $v_1 < c_1$, $v_2 < c_2$, $P_2 < P_1$, $u_2 > u_1$, and condensation pro-duces rarefaction rather than a shock (Landau and Lifshitz 1959, p. 497). For supersonic flow, $v_1 > c_1$, $v_2 > c_2$, $P_2 > P_1$ and $u_2 < u_1$. (In these expres-sions state 1 is before the zone of heat release and state 2 is after the zone.)

Kinetics affect condensation during flow by replacing the equilibrium process of continuous heat addition to the vapor with successive dumps of condensate into the flow, each dump increasing the mass that the flow must carry. Because the condensate particles formed by this process are generally less than 1000 Å in diameter, their thermal energy lost during subsequent adiabatic cooling in the flow is effectively transmitted to the vapor. To a first approximation, until the mass fraction of condensate becomes appreciable, the main effect of condensation on the flow is pumping of latent heat into the diminishing supply of vapor; the secondary effect is the mass loss of vapor.

In summary, in condensing flow nonequilibrium effects arise because supersaturation is required for nucleation of condensed phases. The flow may expand as a metastable vapor to pressures perhaps an order of magnitude below the saturation pressure and temperatures of tens or 100 deg below saturation temperature. In this range of pressures and temperatures the fluid might accurately be modeled as a metastable vapor; in the case of SO_2 shown in Fig. 18.17, the vapor-liquid transition might even be ignored and only a vapor-solid transition modeled for simplicity. When nucleation and growth of the condensed phase does occur, it occurs in a series of dumps of latent heat and mass into the flow rather than by continuous equilibration of phases through the decreasing pressure and temperature conditions. Thus, although the total energy release may be the same overall as given by Eq. (93), the flow dynamics is controlled by the successive energy dumps. Between dumps the mass ratio of condensate to vapor is constant, and in this chapter is modeled by the pseudogas equation of state for heat-transferring particles.

October 1981

REFERENCES

Altman, D., and Carter, J.M. (1956). Expansion processes. In *High Speed Aerodynamics and Jet Propulsion*, vol. 2, *Combustion Processes*, (B. Lewis, R.N. Pease, and H.S. Taylor, Eds.), pp. 26-63. Princeton Univ. Press, Princeton, New Jersey.

ASHRAE (American Society of Heating, Refrigeration, and Air-Conditioning Engineers). (1972). *Handbook of Fundamentals*, New York.

Bennett, F.D. (1974). On volcanic ash formation. *Amer. J. Sci.* 274, 648-661.

Bird, P. (1975). Thermal and mechanical evolution of continental convergence zones: Zagros and Himalayas. *Ph.D. Dissertation*, Mass. Inst. of Tech., Cambridge.

Boyd, F.R., and Nixon, P.H. (1973). Structure of the upper mantle beneath Lesotho. *Carnegie Inst. Washington Yearbook* 72, 431–449.

Braker, W., and Mossman, A.L. (1971). Sulfur dioxide. In *Matheson Gas Data Book* (5th ed.), pp. 513–519. E. Rutherford, New Jersey.

Canjai, L.N., and Manning, F.S. (1967). Thermodynamic properties of sulfur dioxide. In *Thermodynamic Properties and Reduced Correlations for Gases*, pp. 169–174. Gulf Publishing Co., Houston.

Carr, M.H., Masursky, H., Strom, R.G., and Terrile, R.J. (1979). Volcanic features on Io. *Nature* 280, 729–733.

Carslaw, H.S., and Jaeger, J.C. (1959). *Conduction of Heat in Solids* (2nd ed.) Oxford, New York.

Clow, G.D., and Carr, M.H. (1980). Stability of sulfur slopes on Io. *Icarus* 44, 268–279.

Collins, S.A. (1981). Spatial color variations in the volcanic plume at Loki, on Io. *J. Geophys. Res.* 86, 8621–8626.

Consolmagno, G.J. (1979). Sulfur volcanoes on Io. *Science* 205, 397–398.

Consolmagno, G.J. (1981). Io: Thermal models and chemical evolution. *Icarus* 47, 36–45.

Consolmagno, G.J., and Lewis, J.S. (1980). The chemical-thermal evolution of Io. IAU Colloquium 57, *The Satellites of Jupiter* (abstract 7-11).

Cook A.F. II, Shoemaker, E.M., and Smith, B.A. (1979). Dynamics of volcanic plumes on Io. *Nature* 280, 743–746.

Cook, A.F. II, Shoemaker, E.M., Smith, B.A., Danielson, G.E., Johnson, T.V., and Synnott, S.P. (1980). Volcanic origin of the eruptive plumes on Io. *Science* 211, 1419–1422.

Drobyshevski, E.M. (1979). Magnetic field of Jupiter and the volcanism and rotation of the Galilean satellites. *Nature* 282, 811–813.

Elder, J. (1976). *The Bowels of the Earth*. Oxford Univ. Press, London.

Finch, R.H. (1947). The mechanics of the explosive eruption of Kilaauea in 1924. *Pacific Sci.* 1, 237.

Fletcher, N.H. (1970). *The Chemical Physics of Ice*. University Press, Cambridge.

Fudali, R.F., and Melson, W.G. (1971). Ejecta velocities, magma chamber pressure and kinetic energy associated with the 1968 eruption of Arenal Volcano. *Bull. Volcan.* 35, 383–401.

Giauque, W.F., and Stephenson, C.C. (1938). Sulfur dioxide. The heat capacity of solid and liquid. Vapor pressure. Heat of vaporization. The entropy values from thermal and molecular data. *J. Amer. Chem. Soc.* 60, 1389–1394.

Gold, T. (1979). Electrical origin of the outbursts on Io. *Science* 206, 1071–1073.

Gorshkov, G.S. (1959). Gigantic eruption of the volcano Bezymianny. *Bull. Volcan.* 20, 77–109.

Gruntfest, I.J., and Shaw, H.R. (1974). Scale effects in the study of earth tides. *Trans. Soc. of Rheology* 18, 287–297.

Hapke B. (1979). Io's surface and environs: A magmatic-volatile model. *Geophys. Res. Letters* 6, 799–802.

Haughton, D.R., Roeder, P.L., and Skinner, B.J. (1974). Solubility of sulfur in mafic magmas. *Econ. Geology* 69, 451–467.

Hill, P.G. (1966). Condensation of water vapor during supersonic expansion in nozzles. *J. Fluid Mech.* 25(3), 593–620.

Jaeger, J.C. (1969). *Elasticity, Fracture and Flow*. Methuen and Co. Ltd., London.

JANAF (Joint Army, Navy and Air Force Project Principia of the Advanced Research Projects Agency). (1965a). *Thermochemical Tables,* 2nd ed., Nat. Stand. Ref. Data Ser. 37, Nat. Bur. of Stand., Washington, D.C.

JANAF (Joint Army, Navy and Air Force Project Principia of the Advanced Research Projects Agency). (1965b). *Thermochemical Tables* (first addendum), Therm. Res. Lab. Dow Chem. Co., Midland, Michigan.

JANAF (Joint Army, Navy and Air Force Project Principia of the Advanced Research Projects Agency). (1966). *Thermochemical Tables* (second addendum), Therm. Res. Lab. Dow Chem. Co., Midland, Michigan.

JANAF (Joint Army, Navy and Air Force Project Principia of the Advanced Research Projects Agency). (1967). *Thermochemical Tables* (Third addendum), Therm. Res. Lab. Dow Chem. Co., Midland, Michigan.

JANAF (Joint Army, Navy, NASA, and Air Force. *Handbook of Rocket Exhaust Plume Technology*). (1975). Chemical Propulsion Information Agency Publication 263. Johns Hopkins Univ., Laurel, MD.

Johnson, T.V., Cook, A.F. III, Sagan, C., and Soderblom, L.A. (1979). Volcanic resurfacing rates and implications for volatiles on Io. *Nature* 280, 746–750.

Keenan, J.H., Rages, F.G., Gill, P.G., and Moore, J.G. (1969). *Steam Tables.* John Wiley and Sons, New York.

Kieffer, S.W. (1977). Sound speed in liquid-gas mixtures: Water-air and water-steam. *J. Geophys. Res.* 82, 2895–2904.

Kieffer, S.W. (1981). Blast dynamics at Mount St. Helens on 18 May 1980. *Nature* 291, 568–570.

Kieffer, S.W. (1982). Fluid dynamics of the May 18 blast at Mount St. Helens. *U.S. Geol. Survey Prof. Paper* 1950, 379–400.

Kieffer, S.W., and Delany, J.M. (1979). Isentropic decompression of fluids from crustal and mantle pressures. *J. Geophys. Res.* 84, 1611–1620.

Kumar, S. (1979). The stability of an SO_2 atmosphere on Io. *Nature* 280, 758–760.

Landau, L.D., and Lifshitz, E.M. (1959). *Fluid Mechanics.* Pergamon, Oxford.

Lang, A.R. (1972). Pressure and temperature gradients in ascending fluids and magmas. *Nature* 238, 98–100.

Lee, S.W., and Thomas, P.C. (1980). Near-surface flow of volcanic gases on Io. *Icarus* 44, 280–290.

Lewis, J.S. (1981). Io: Geochemistry of sulfur. Submitted to *Icarus.*

Lorenz, V. (1970). An investigation of volcanic decompressions. III. Maars, tuff-rings, tuff-cones, and diatremes. Progress Report on NASA Research Grant NGR-38-003-012 (unpublished).

MacGregor, I.D., and Basu, A.R. (1974). Thermal structure of the lithosphere: A petrologic model. *Science* 185, 1007–1011.

Marsh, B.D. (1978). On the cooling of ascending andesitic magma. *Phil. Trans. Roy. Soc. Lond. A.* 288, 611–625.

Masursky, H., Schaber, G.G., Soderblom, L.A., and Strom, R.G. (1979). Preliminary geological mapping of Io. *Nature* 280, 725–729.

Matson, D.L., and Nash, D.B. (1981). Io's atmosphere: Pressure control by subsurface coldtrapping. *Lunar Planet. Sci.* XII, 664-666 (abstract); also *J. Geophys. Res.* In press.

Matson, D.L., Ransford, G.A., and Johnson, T.V. (1980). Heat flow from Io. IAU Coll. 57. *The Satellites of Jupiter* (abstract 4-2).

McBirney, A.R. (1973). Factors governing the intensity of explosive andesitic eruptions. *Bull. Volcanologique* XXXVII, 443–453.

McBirney, A.R., and Murase, T. (1970). Factors governing the formation of pyroclastic rocks. *Bull. Volcan.* 34, 372–384.

McCauley, J.F., Smith, B.A., and Soderblom, L.A. (1979). Erosional scarps on Io. *Nature* 280, 736–738.

McGetchin, T.R. (1968). The Moses Rock dike: Geology, petrology and mode of emplacement of a kimberlite-bearing breccia-dike, San Juan County, Utah, Ph.D. Dissertation, Calif. Inst. of Tech., Pasadena.

McGetchin, T.R., and Ulrich, W.G. (1973). Xenoliths in maars and diatremes with inferences for the Moon, Mars and Venus. *J. Geophys. Res.* 78, 1833–1853.

Melosh, H.J. (1977). Crater modification by gravity: A mechanical analysis of slumping. In *Impact and Explosion Cratering* (D.J. Roddy and R.B. Merrill, Eds.), pp. 1245–1260. Pergamon, New York.

Melson, W.G., and Saenz, R. (1973). Volume, energy and cyclicity of eruptions of Arenal Volcano, Costa Rica. *Bull. Volcan.* 37, 416–437.

Mercier, J.C., and Carter, N.L. (1974). Pyroxene geotherms. *J. Geophys. Res.* 80, 3349–3362.

Meyer, B. (1968). Elemental sulfur. In *Inorganic Sulfur Chemistry* (G. Nickless, Ed.), pp. 241–258. American Elsevier, New York.

Moore, J.M., and Calk, L. (1971). Sulfide spherules in vesicles of dredged pillow basalt. *Amer. Mineral.* 56, 476–488.

Ness, N.F., Acuna, M.H. Lepping, R.P., Burlaga, L.F., Behannon, K.W., and Neubauer, F.M. (1979). Magnetic field studies at Jupiter by Voyager 1. *Science* 204, 42–47.

O'Reilly, T.C., and Davies, G.F. (1980). Magma transport of heat on Io: A mechanism allowing a thick lithosphere. *Geophys. Res. Letters* 8, 313–316.

Peale, S.J., Cassen, P., and Reynolds, R.T. (1979). Melting of Io by tidal dissipation. *Science* 203, 892–894.

Pearl, J., Hanel, R., Kunde, V., Maguire, W., Fox, K., Gupta, S., Ponnamperuma, C., and Raulin, F. (1979). Identification of gaseous SO_2 and new upper limits for other gases on Io. *Nature* 280, 755–758.

Rau, H., Kutty, T.R.N., Guedes de Carvalho, J.R.F. (1973). Thermodynamics of sulfur vapor. *J. Chem. Thermodynamics* 5, 833–844.

Reynolds, R.T., Peale, S.J., and Cassen, P.M. (1980a). Sulfur vapor and sulfur dioxide models of Io's plumes. IAU Colloquium 57. The Satellites of Jupiter (abstract 3-9).

Reynolds, R.T., Peale, S.J., and Cassen, P.M. (1980b). Io: Energy constraints and plume volcanism. *Icarus* 44, 234–239.

Rossini, F.D., Wagman, D.D., Evans, W. H., Levine, S., and Jaffee, I. (1952). *Selected Values of Chemical Thermodynamic Properties,* Nat. Bur. Stand. Circular 500, Washington, D.C.

Rumble, D. (1976). The adiabatic gradient and adiabatic compressibility. *Annual Report of the Director of the Geophysical Lab, Carnegie Inst. Wash. Yearb.* 75, 651–655.

Sagan, C. (1979). Sulfur flows on Io. *Nature* 280, 750–753.

Samsonov, G.V. (1968). *Handbook of Physiochemical Properties of the Elements* (translated from Russian). Plenum Press, New York.

Schaber, G.G. (1980). The surface of Io: Geologic units, morphology, and tectonics. *Icarus* 43, 302–333.

Self, S., Wilson, L., and Nairn, I.A. (1979). Volcanian eruption mechanisms. *Nature* 277, 440-443.

Shaw, H.R. (1969). Rheology of basalt in the melting range. *J. of Petrology 10* (3), 510–535.

Shaw, H.R. (1970). Earth tides, geobal heat flow, and tectonics. *Science* 168, 1084–1087.

Shaw, H.R., and Swanson, D.A. (1970). Eruption and flow rates of flood basalts. In *Columbia River Basalt Symposium, 2nd, Proceedings* (E.H. Gilmour, and D. Stradling, Eds.), pp. 271–299. Eastern Washington State College Press, Cheney, Wash.

Sheridan, M.F., and Wohletz, K.H. (1981). Hydrovolcanic explosions: The systematics of water-pyroclast equilibration. *Science* 212, 1387–1389.

Shimozuru, D., Nakamuda, O., Seno, H., Noda, H., and Taneda, S. (1957). Mechanism of pumice formation. *Bull. Volcan. Soc. Japan* 2, 17.

Shoemaker, E.M., Roach, C.H., and Byers, F.M., Jr. (1962). Diatremes and uranium deposits in the Hopi Buttes, Arizona. In *Petrologic Studies: A Volume to Honor A.F. Buddington,* pp. 327–355. Geological Soc. of Amer., Boulder, Colorado.

Sinton, W.M. (1980). Io: Are vapor explosions responsible for the 5-μm outbursts? *Icarus* 43, 56–64.

Skinner, B.J., and Peck, D.L. (1969). An immiscible sulfide melt from Hawaii. *Econ. Geol. Monogr.* 4, 310–322.

Smith, B.A., Shoemaker, E.M., Kieffer, S.W., and Cook, A.F. II. (1979). The role of SO_2 in volcanism on Io. *Nature* 280, 738–743.

Smith, R.L. (1960). Ash flows. *Geol. Soc. Amer. Bull.* 71, 795–842.

Smith, R.L. (1980). Ash-flow magmatism. *Geol. Soc. Amer.* Spec. Paper 180, 5–27.

Smith, R.L., and Bailey, R.A. (1966). The Bandelier Tuff: A study of ash-flow eruption cycles from zoned magma chambers. *Bull. Volcanologique* 29, 83–104.

Soderblom, L., Johnson, T., Morrison, D., Danielson, E., Smith, B., Veverka, J., Cook, A., Sagan, C., Kupferman, P., Pieri, D., Mosher, J., Avis, C., Gradie, J., and Clancy, T. (1980). Spectrophotometry of Io: Preliminary Voyager I results. *Geophys. Res. Letters* 7, 963–966.

Sparks, R.S.J. (1978). The dynamics of bubble formation and growth in magmas. A review and analysis. *Jour. Volcan. and Geothermal Res.* 3, 1–37.

Thompson, P.A. (1972). *Compressible Fluid Dynamics.* McGraw-Hill, New York.

Tuller, W.N. (1954). *The Sulphur Data Book.* McGraw-Hill, New York.

Verhoogen, J. (1951). Mechanics of ash formation. *Amer. J. Sci.* 249, 729–739.

Walker, G.P.L., Wilson, L., and Bowell, E.L.G. (1971). Explosive volcanic eruptions. I. The rate of fall of pyroclasts. *Geophys. J. Roy. Astron. Soc.* 22, 377-383.

Wallis, G.B. (1969). *One-dimensional Two-phase Flow.* McGraw-Hill, New York.

Watanabe, R. (1940). Eruptions of molten sulfur from the Siretoko-Iosan Volcano, Hokkaido, Japan. *Japan. J. Geol. Geog.* 17, 289–310.

West, J.R. (1950). Thermodynamic properties of sulfur. *Industrial and Engineering Chemistry* 42, 713–718.

Wilson, L. (1976). Explosive volcanic eruptions. III. Plinian eruption columns. *Geophys. J. Roy. Astron. Soc.* 45, 543–556.

Wilson, L. (1980). Relationships between pressure, volatile content and ejecta velocity in three types of volcanic explosion. *J. Volcanol. Geothermal Res.* 8, 297–313.

Wilson, L., and Head, J.W. III. (1981). Ascent and eruption of basaltic magma on the Earth and Moon. *J. Geophys. Res.* 86, 2971–3001.

Wilson, L., Sparks, R.S.J., and Walker, G.P.L. (1980). Explosive volcanic eruptions. IV. The control of magma properties and conduit geometry on eruption column behavior. *Geophys. J. Roy. Astron. Soc.* 63, 117–148.

Witteborn, F.C., Bregman, J.D., and Pollack, J.B. (1979). Io: An intense brightening near 5 micrometers. *Science* 203, 643–646.

Wones, D.R., and Shaw, H.R. (1975). Tidal dissipation: A possible heat source for mare basalt magmas. *Lunar Science* VI, (abstract).

Yanagisawa, M. (1980). Can electromagnetic induction current heat Io's interior effectively? *Lunar and Planet. Sci. XI,* 1288–1290 (abstract).

Yoder, H.S. (1976). *Generation of Basaltic Magma.* National Academy of Sciences, Washington, D.C.

Zel'dovich, Ya. B., and Razier, Yu. P. (1966). *Physics of Shock Waves and High-Temperature Hydrodynamic Phenomena.* 2 vol. Academic Press, New York.

Zucrow, M.J., and Hoffman, J.D. (1976). *Gas Dynamics.* 12 vol. John Wiley, New York.

19. HOT SPOTS OF IO

J. C. PEARL
Goddard Space Flight Center

and

W. M. SINTON
University of Hawaii

High-temperature regions related to volcanism on the surface of Io have not only been observed by the Voyager 1 IRIS infrared spectrometer, but have been observed for the last decade in groundbased measurements. Detailed comparison of the IRIS observations with the Voyager imaging data reveals two kinds of associations: (1) features at < 400 K, which generally correlate with dark calderas and flows; and (2) a small 650 K area at the source of the Pele plume. A third type of thermal behavior has been observed only in groundbased data; seemingly due to sudden, perhaps explosive, events, this is characterized by 600 K color temperatures. One of these events is apparently responsible for the remarkable changes that occurred between the Voyager flybys in the area containing the Surt caldera. Some recent groundbased observations suggest that there may be molecular or aerosol absorptions associated with these transient events. The very high temperature in the vicinity of Pele indicates the presence of a surface material other than sulfur. The bulk of the active thermal emission arises from roughly twenty major sources. Data by a number of observers encompassing 2.2 to 50 μm have been interpreted to determine the disk-averaged heat flow from the interior of Io with values ranging from 1.5 to 2.0 W m^{-2}, the first figure being the best determined. These heat fluxes are higher than can be accounted for by present tidal dissipation theories.

In the early 1970s infrared detectors were developed with sufficient sensitivity to permit groundbased measurements of the thermal emission from the Galilean satellites. Initial effort centered on verification of the satellite temperatures, which could be predicted from their known albedos and the assumption of equilibrium with the incident sunlight. Of greater interest was the possibility of learning something of the physical nature of the satellites' surfaces by observing their thermal responses to the rapid changes in insolation provided by frequent eclipses by Jupiter. Except for Io, the early studies showed expected disk temperatures; it was also learned that the surfaces of all of the Galilean satellites had remarkably low thermal inertias. (See the review of this early work by Morrison [1977].) One discrepancy in the early measurements concerned the disk temperature of Io observed outside of eclipse. Although the 20-μm brightness temperature of the satellite was found to be 128 K, in approximate agreement with the calculated equilibrium temperature for a slowly rotating body of Io's albedo at its distance from the Sun, the 11-μm brightness temperature was 138 K (Morrison et al. 1972; Hansen 1973). In even earlier work, Gillett et al. (1970) found a still higher temperature of 149 K at 8.4 μm.

An additional problem was encountered in understanding the eclipse results. Whereas Io cooled to 102 K at 20 μm by the end of a total eclipse by Jupiter, it cooled to only 125 K when observed at 11 μm (Hansen 1973; Morrison and Cruikshank 1973; Morrison 1977). These discrepancies led to disagreement of the thermophysical parameters derived from the data at the two wavelengths. Using a two-layer model, the two groups agreed concerning the thermal inertia (chiefly the thermal conductivity) of the uppermost surface layer, but they disagreed as to its thickness above a higher conductivity subsurface material. Sinton (1973) made an attempt to understand both of these discrepancies by supposing an NH_3 atmosphere. However, near-infrared spectra by Fink et al. (1976) failed to reveal the presence of the gas, and this explanation soon became untenable. In the intervening years the discrepancies were all but forgotten.

The discovery of intensely active volcanism on Io by Voyager cameras (Smith et al. 1979*b*; Morabito et al. 1979) and of associated "hot spots" by the Voyager IRIS instrument (Hanel et al. 1979; Pearl et al. 1979) has led to a resolution of these difficulties. In a system composed of a number of hot spots on a cooler background, the hot spots affect the thermal data more at 11 μm than at 20 μm due to the shift of the Planck function to shorter wavelengths with increasing temperature; this effect is most pronounced in eclipse data, since the loss of sunlight leaves the hot spots virtually unaffected while the remainder of the surface cools significantly.

By quantifying these properties, it is possible to estimate the temperatures and sizes of spatially unresolved hot spots. Furthermore, this provides a means of monitoring Io's volcanic activity from the Earth. Indeed,

Matson et al. (1981) have used published data obtained in the early 1970s to model the total heat flow from the hot spots.

I. SIZE AND TEMPERATURE

A. Deconvolution of Size and Temperature Information

To date, all detected hot spots have been smaller than the spatial resolution of the observing instrumentation. It has therefore been necessary to deconvolve contributions from the various hot spots in order to determine the properties of the thermal sources (Hanel et al. 1979; Matson et al. 1981; Sinton 1981).

If it is assumed that all surfaces have unit emissivity and uniform but unequal temperatures, then the observed spectral radiance from n hot spots within the field of view may be represented by the weighted average

$$I_\lambda = \sum_{i=1}^{n} A_i B_\lambda (T_i) + \left[1 - \sum_{i=1}^{n} A_i \right] B_\lambda (T_b) \qquad (1)$$

where A_i is the fractional area at temperature T_i; T_b is the temperature of the passive background, which covers most of the surface; and B_λ is the Planck function. This relation contains $2n + 1$ adjustable constants: the n areas A_i, their n associated temperatures T_i, and the background temperature T_b. In applying this relation it is important to use data over as wide a spectral range as possible, since, according to the displacement law, the signatures of successively hotter sources shift to progressively shorter wavelengths. Corrections to this expression may be required for such factors as surface emissivity, a nonuniform background temperature (caused for example, by variations in albedo within the field of view), or the presence of reflected solar radiation in the near infrared.

The limitations of using Eq. (1) for deriving source temperatures and areas should be kept in mind. The breadth of the Planck function, the overlap of Planck functions for different temperatures, and the presence of noise in the data limit the precision of such fits. Increasing the number of terms beyond two or three may reduce residuals, but the parameters so determined may lack direct physical significance. An illustration of these points is provided by considering the spectrum shown in Fig. 19.1.

A minimum of three blackbodies is required to satisfactorily fit the data. As derived from Eq. (1) by using a least-squares procedure over the spectral range $180 < \nu < 2200$ cm^{-1}, the three temperatures (and fractional weights) for the fit are 114 K (0.899), 175 K (0.100), and 654 K (5.77 × 10^{-4}). The precision of the 654 K estimate can be investigated by establishing the sensitivity of rms residuals to fits made using various fixed values of the high

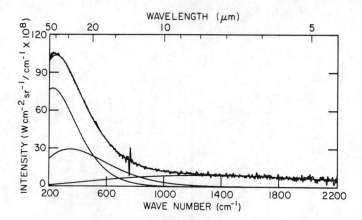

Fig. 19.1. Voyager spectrum of region of Io containing Pele. The ringing centered near 760 cm^{-1} is caused by interference. The smooth curve through the data is a least-squares fit using three weighted blackbody spectra. The lower curves are the three components; from left to right, their temperatures (and fractional areas) are 114 K (0.899), 175 K (0.100), and 654 K (5.77 $\times$ 10^{-4}).

temperature and determining the other four parameters by least squares. The uniqueness of interpretation of the 654 K estimate is addressed by trying to establish whether the high-temperature portion of the spectrum can be fit only by a single isothermal source, or whether it could be fit equally well by a source with a horizontal thermal gradient. To approximate the thermal gradient a pair of sources with temperatures fixed symmetrically about 654 K is used in place of the single isothermal hot source; the remaining five parameters are determined by least squares. After obtaining solutions corresponding to several temperature differentials, the sensitivity of the rms residuals to the choice of differentials is investigated. The results of these exercises are summarized in Table 19.1.

A plot of the fit using a 604 K hot source is found to be visibly poorer than one for the optimum parameters, while a plot using a 704 K source is not. Assuming that this is representative of differences that would result if repeated measurements of the same area were fitted, acceptable fits appear to have rms residuals no more than $\sim$ 1.3% greater than the minimum. The high temperature is therefore relatively well determined as 654$^{+60}_{-45}$ K. Uncertainties in hot spot area and radiated power associated with the uncertainty in temperature are approximately 30% and 2%, respectively.

For cases with temperature differentials across the hot spot, the residuals lie extremely close to the minimum; differentials greater than ±200 K are acceptable. This indeterminacy is reflected as uncertainties in the estimated hot spot area and radiated power of > 92% and > 17%, respectively. We therefore conclude that a mean temperature for this hot spot is rather well determined, but that the interpretation of the mean is highly ambiguous;

TABLE 19.1

Temperatures and Areas Fitted to Hot Component of Voyager Spectrum Including Pele

Fit[a]	Temperature T_1 (K)	Relative Area a_1	Temperature T_2 (K)	Relative Area a_2	Total Relative Area a_{tot}	Total Relative Radiated Power p	Relative rms of Fit
1	604	1.36	—	—	1.36	0.99	1.016
2	654	1	—	—	1	1	1
3	704	0.76	—	—	0.76	1.02	1.010
4	754	0.35	554	0.82	1.17	1.04	1.000
5	854	0.27	454	1.65	1.92	1.17	1.002

[a]Data for fit 2 are from an unconstrained least-squares fit of three blackbodies to the spectrum shown in Fig. 19.1; actual values for the highest temperature component of this fit are: $T = 654$ K, area = 110 km^2; radiated power = 1.16×10^{12} W, and rms residual = 1.39×10^{-8} W cm^{-2} sterad^{-1}/cm^{-1}. The remaining entries are normalized to these values. Fits 1 and 3 are from three-blackbody fits with the high temperature constrained at the indicated values of T_1. Fits 4 and 5 are from four-blackbody fits with the two high temperatures constrained at the indicated values of T_1 and T_2.

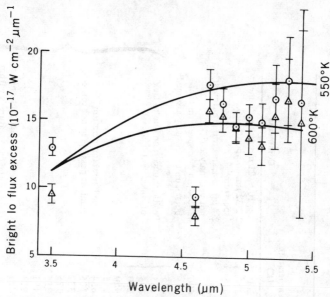

Fig. 19.2. Excess infrared flux from Io during the brightening of 20 Feb. 1978 (from Witteborn et al. 1979). Two approximate fits using weighted blackbody curves are also shown. With the 600 K temperature and unit emissivity, the radiating area is ~ 2 × 10⁻⁴ of the disk of Io. Circles and triangles represent the excess flux relative to spectra taken 5 hr before and 20 hr after the brightening, respectively.

large lateral temperature gradients can be present in the source without producing a discernable signature in the measured portion of the spectrum.

B. Observations and Derived Parameters

A chronology of observations, together with estimates of hot spot sizes and temperatures, is presented in Table 19.2. The first record of a localized thermal enhancement of Io was apparently made in the 2 to 4 μm spectral region by Fink et al. (1978). However, the first suggestion of a possible thermal origin for a similar observation was made by Witteborn et al. (1979). Based on abnormal brightness observed in the spectral range from 4.5 to 5.4 μm (Fig. 19.2), the presence of a region at 600 K with a projected area of 2 × 10⁻⁴ of Io's surface (assuming an emissivity of 0.25) was considered as one possible inference. Subsequently, during a patrol to look for similar events, Sinton (1980a) detected a second enhancement of comparable magnitude using a 5-μm photometer. Since no spectral coverage was obtained during this observation, no direct temperature and size estimates of the source were possible. However, based on a model of the event as a vapor explosion, the source was estimated to be 50 km in diameter with a temperature of ~ 600 K (Sinton 1980b); radical changes in a caldera of this size (Surt), which were consistent with the model, were found in Voyager 2

TABLE 19.2

Observations of Hot Spots on Io

Date of Observation	Iographic Position or Central Meridian[a] (deg)	Feature	Equivalent Radius (Normal Viewing)[b] (km)	Temperature (K)	Comments
1969-1972	—	—	260-580	200-250	8.4, 10.6, 21 μm disk radiometry[c]
1974-1979	—	—	227 / 17	294 / 615	2.2-22 μm composite disk spectrum[d]
1/26/78	23	—	—	—	2-4 μm disk spectrum (outburst); not yet analyzed[e]
2/20/78	68	—	26	600	4.7-5.4 μm disk spectrum (outburst); unit emissivity assumed[f]
3/5/79	(30,210)	Dark region in NW Colchis Regio	9 / 55	385 / 165	Positions uncertain due to lack of concurrent imaging
3/5/79	(27,119)/ (19,122)	Amirani/Maui	13 / 49	395 / 200	
3/5/79	(13,310)	Loki Patera	21 / 121	450 / 245	
3/5/79	(−19,257)	Pele	6 / 80	654 / 175	Large lateral temperature gradient possible in hot source[g]
3/5/79	(−40,272)	Babbar Patera	5.3 / 72	322 / 175	
3/5/79	(−41,288)	Ulgen Patera	14 / 78	355 / 191	
3/5/79	(−48,267)	Svarog Patera	30	221	
3/5/79	(−52,344)	Creidne Patera	20	231	Corrected for the fact that only 80% of the dark feature is within instrument field of view; much of surface covered by lighter material[g]

TABLE 19.2 Continued

Observations of Hot Spots on Io

Date of Observation	Iographic Position or Central Meridian[a] (deg)	Feature	Equivalent Radius (Normal Viewing)[b] (km)	Temperature (K)	Comments
3/5/79	(−80,320)	Flow in Nemea Planum	17	225	Voyager observation
6/11/79	(45,338) 30	Surt	25	600	5 μm disk radiometry (outburst); caldera altered between 8/5/79 and 7/9/79[h]
12/10/79	18	—	—	—	4.6–5.3 μm disk spectrum (outburst); interpretation uncertain (see text)[i]
12/15/79	340	—	27	523–574	2.2, 3.8, 4.8 μm disk photometry of eclipses[j]
12/22/79	340	—	26	493–590	
12/24/79	340	—	34	442–577	
3/20/80	19	—	—	—	4.6–5.3 μm disk spectrum (outburst); low SNR[i]
4/12/80	10	—	438	200	3.4, 4.8, 10.2, 20, 30 μm disk photometry of eclipse[k]
			54	350	
			11	600	

[a]Feature coordinates from Davies and Katayama (1981).
[b]Except for outbursts, results from disk determinations are normalized to the entire surface of Io.
[c]Matson et al. (1981).
[d]Sinton (1981).
[e]Fink et al. (1978).
[f]Witteborn et al. (1979).
[g]Voyager observations.
[h]Sinton (1980a).
[i]Witteborn et al. (1981).
[j]Sinton et al. (1980).
[k]Morrison and Telesco (1980).

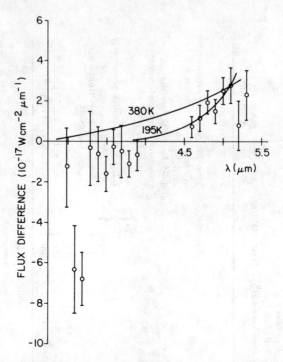

Fig. 19.3 Excess infrared flux from Io during the brightening of 10 Dec. 1979 (modification of a figure from Witteborn et al. 1981). Two weighted blackbody curves are also shown. The 195 K curve fits the data well, but requires that the source cover 88% of the disk of Io. The 380 K curve results if an area of 8×10^{-4} of the disk is assumed, together with the requirement that the curve pass through the datum at 5.1 μm. The resulting absorption between 3 and 5 μm is similar to that observed in quiescent disk spectra.

pictures (see Fig. 18 of Smith et al. 1979c). Another enhancement, from 4.6 to 5.3 μm (Fig. 19.3), was observed by Witteborn et al. (1981). They consider possible thermal explanations of the data to range between two extremes: a widespread region at 180–250 K (we obtain 88% of the disk at 195 K, assuming unit emissivity), or a much smaller and warmer source obscured by material with an absorption band centered near 4 μm. A fifth enhancement was also detected by Witteborn et al. (1981); however the signal to noise ratio in the spectrum was too low to allow derivation of hot-spot size and temperature.

Near-infrared thermal emission has also been detected from Io during eclipses. Sinton et al. (1980) observed three eclipses of Io as it entered Jupiter's shadow, and found that it did not disappear at 4.8, 3.8, or even 2.2 μm. For half an hour after entrance into the shadow, essentially constant flux levels were measured in all three channels. The residual flux cannot be due to solar radiation refracted or scattered through the Jovian atmosphere, since such an effect would decrease rapidly after immersion (Smith et al. 1977). Color temperatures near 500 K and projected source areas of 5×10^{-5} of Io's disk were derived from the three possible combinations of the filter band data (Sinton et al. 1980). A tendency for the derived temperatures to increase as the median wavelength of the pairs of filters decreases suggests that a mixture of source temperatures is actually present.

Since the spectrum of a collection of sources with widely differing temperatures will tend to be dominated by only a few in any narrow spectral interval, a range of source temperatures can be fairly assessed only by considering a thermal spectrum as nearly complete as possible. Sinton (1981) has produced a disk spectrum from 2.2 to 22 μm by combining data from various groundbased measurements. In order to reduce the time span covered by the included measurements, and to provide good spectral resolution, the early photometry at 8.4, 10.6, and 21 μm was not used. Instead, intermediate bandwidth photometry made in 1974 was combined with 16 to 22 μm spectral data obtained in 1979 to determine the spectrum, shown in Fig. 19.4. Three terms in Eq. (1) are adequate to produce a satisfactory fit. The

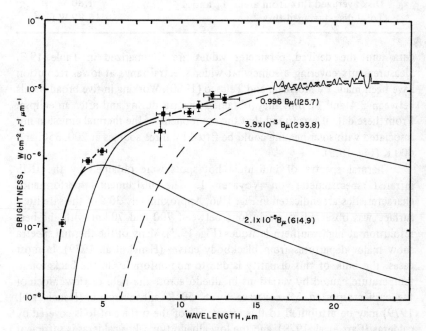

Fig. 19.4. Thermal emission spectrum of Io from combined groundbased observations (from Sinton 1981). The heavy curve through the data is a fit composed of the three weighted blackbody functions drawn below; the fractional weights and temperatures of each curve are as indicated (see also Table 19.3).

TABLE 19.3

**Data and Model Parameters for Composite
Disk Spectrum of Io**

Wavelength (μm)	Spectral Radiances (μW cm^{-2} sterad^{-1} μm^{-1})
2.2	0.120
3.8	0.844
4.8	1.4
11.1	5.9
20	13.5

Model Parameters	
A_1 (fraction of disk)	2.10×10^{-5}
A_2 (fraction of disk)	3.87×10^{-3}
T_1	614.9 K
T_2	293.8 K
T_3	125.7 K
Disk averaged flux from areas A_1 and A_2	1.80 W m^{-2}
Total disk averaged flux	15.89 W m^{-2}

data and the derived parameter values are summarized in Table 19.3. Measurements covering a somewhat wider spectral range at lower resolution have been made by Morrison and Telesco (1980). Working in five broad bands between 3.4 and 30 μm, they obtained spectra during and after an eclipse. From these data it was found that the component of the thermal emission not associated with solar heating could be fitted by three sources at 200, 350, and 600 K (Fig. 19.5).

Thermal spectra of individual hot spots were obtained by the IRIS infrared spectrometer on Voyager 1. The instrument's performance characteristics are indicated in Fig. 19.6. Approximately 30% of the satellite's surface was observed at resolutions between 700 and 70 km with the best resolution at high southern latitudes (Fig. 19.7). Many of the thermal spectra show major departures from blackbody curves (Hanel et al. 1979). In most cases, the bulk of this disparity is due to nonuniformity in the background temperature caused by variations in albedo across the field of view. Much of the earlier reported 5% coverage by $\sim$ 50 K enhancements (Hanel et al. 1979) may be attributed to this effect; 5% of the surface of Io is covered by calderas (Carr et al. 1979), and the low albedo typical of calderas is sufficient to raise their midday temperatures by $\geqslant$ 30 K above the temperature of the brighter, cooler surroundings. In addition to these passive enhancements, however, several major thermal sources have been identified in the data. Two-

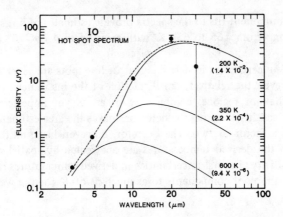

Fig. 19.5. Spectrum of hot spot emission from Io as observed by Morrison and Telesco (1980) during the total phase of the eclipse on 12 Apr 1980. The background emission (87 K) has been removed from the spectrum. The dashed curve is a fit composed of the three weighted blackbody functions drawn below; the temperatures and fractional weights of each are as indicated.

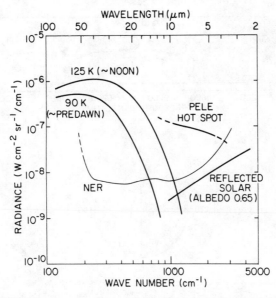

Fig. 19.6. Spectral range and sensitivity of the Voyager 1 infrared spectrometer. The noise level of the instrument is represented by the thin line labeled NER (noise equivalent radiance). The spectral resolution of the spectrometer is 4.3 cm^{-1}. The instrument also includes a single-channel radiometer covering the spectral range from 5,000 to 30,000 cm^{-1}. The approximate range of conditions encountered on Io is schematically represented by the heavy curves.

and three-component fits to the spectra provide estimates of hot spot temperatures ranging from 165 to 654 K, with areas up to 4.7×10^4 km^2 (Table
19.2).

Formal uncertainties in derived areas of hot spots are generally less than
±10%. However, the relatively small fraction of the instrument field of view
filled by the hot regions, combined with an as yet imperfectly mapped
variation in sensitivity across the detector, makes the area estimates uncertain
by perhaps a factor of two. The detector, however, has a uniform spectral
response so the derived temperatures are unaffected by spatial variations in
overall sensitivity. Formal uncertainties in derived temperatures range from 1
to 5 K for the cool components to as much as 60 K for the warmest component.

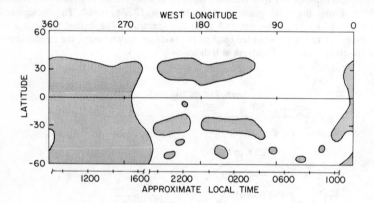

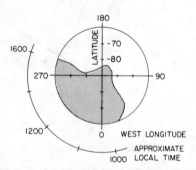

Fig. 19.7. The 30% of Io observed by the Voyager infrared spectrometer during the close
encounter phase. Initial observations (spatial resolution ~700 km) were in the region
near latitude +30°, between longitudes 270° and 360°. Mapping proceeded to the
south, with the best spatial resolution (~70 km) at high southerly latitudes.
Low-latitude dark side observations were made as the spacecraft receded from the
satellite, and were therefore of lower spatial resolution (~600 km).

The source at Loki Patera, previously reported at 290 K (Hanel et al. 1979; Pearl et al. 1979), has been resolved into two components into portions at 245 and 450 K. The hottest observed source is associated with Pele (Fig. 19.1). Three components adequately fit the spectrum, but large temperature differentials across the hot source are not ruled out (Table 19.1). The absence of detectable luminosity, however, appears to limit the amount of material present with temperatures much in excess of 750 K.

II. MORPHOLOGY AND DISTRIBUTION

The availability of simultaneously obtained Voyager termal and imaging data allows the matching of hot sources and specific geologic features (Table 19.2). Identification is made by comparing the hot spot areas A_i derived from Eq. (1) with areas of features on the images. This procedure is somewhat imprecise, however, because of the previously mentioned uncertainties in the derived areas.

The hot spots most frequently appear to correlate with dark, well-defined, caldera-like structures; examples are Loki, Ülgen, Svarog and Creidne Paterae (Figs. 19.8–19.11). The dark elongated depression considered to be the primary source of Plume 1 (Chapter 16 by Strom and Schneider) is likely the hot region detected at Pele in the infrared (Fig. 19.12). The existence of dark features with temperatures near 300 K is a problem with a sulfur crust model. Even if rapidly quenched, pure sulfur maintained at these temperatures will quickly convert to its normal yellow orthorhombic form (Chapter 7 by Sill and Clark). Maintenance of the dark color suggests the presence of impurities, of some process preventing crystallization, or an entirely different composition. Loki, Svarog, and Creidne Paterae show irregular patches within the calderas; a highly stretched version of Fig. 19.11 reveals that half of Creidne Patera is so covered (see Plate 3 of McCauley et al. 1979). This patchiness suggests temperature-related color changes, encroachment of a crust, or deposition of higher albedo material. Even condensation of SO_2 could occur on some of the lower temperature components of several hot spots, since they have mean temperatures below the 198 K condensation temperature.

The specific association of features with the lower temperature components of the hot spots is not always straightforward. While at Loki Patera the warm region probably encompasses the bulk of the dark material in the caldera, and at Pele and at Babbar Patera (Fig. 19.13) the warm regions may be the pyroclastic deposits, at Ülgen Patera no obvious feature of the appropriate size exists.

Flow patterns or pyroclastic deposits are associated with many hot spots, although Svarog Patera has neither. The hot feature in Nemea Planum (Fig. 19.14) consists of a radiating flow pattern with only a small dark center.

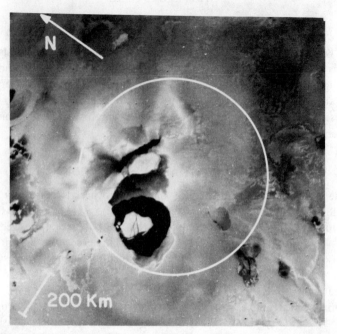

Fig. 19.8. IRIS footprint and the hot feature at Loki Patera. In this and subsequent images, the position of the IRIS footprint may be in error by 10% of its radius. (Voyager frame number 019J1+000.)

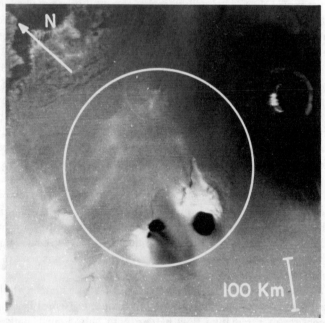

Fig. 19.9. IRIS footprint and the hot feature at Ülgen Patera. (Voyager frame number 109J1+000.)

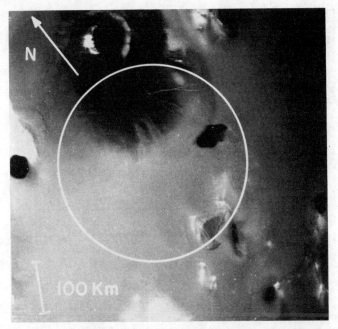

Fig. 19.10. IRIS footprint and the hot feature at Svarog Patera. (Voyager frame number 111J1+000.)

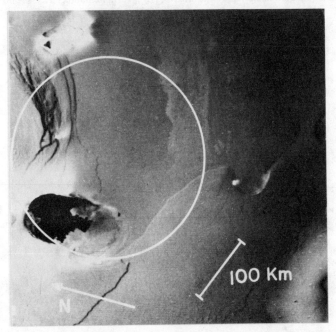

Fig. 19.11. IRIS footprint and the hot feature at Creidne Patera. Note that part of the feature is not within the field of view. A highly stretched version of this frame shows the dark feature to be about half covered with somewhat lighter patches. (Voyager frame number 145J1+000.)

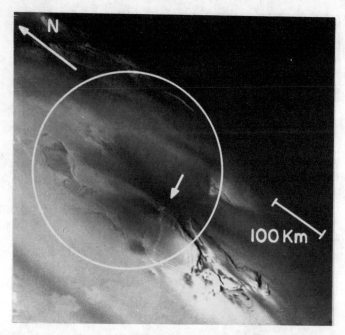

Fig. 19.12. IRIS footprint and the hot feature at Pele. The hottest component is the small dark region indicated by the arrow. (Voyager frame number 127J1+000.)

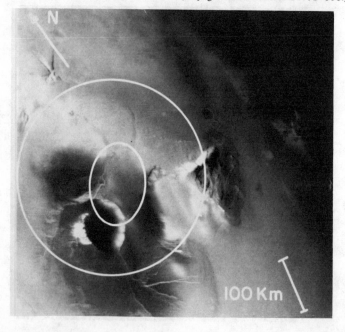

Fig. 19.13. IRIS footprints and the hot feature at Babbar Patera. The high-temperature component appears strongly in the spectrum taken at the higher spatial resolution, and is presumably located within the crater. The position of the smaller footprint may be uncertain by 20% of its semimajor axis. (Voyager frame number 121J1+000.)

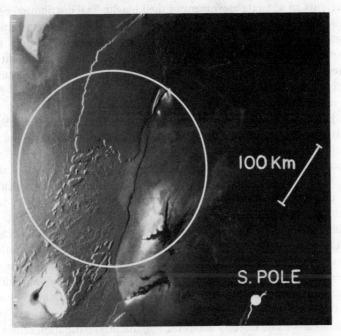

Fig. 19.14. IRIS footprint and the hot feature in Nemea Planum. (Voyager frame number 147J1+000.)

Dark, more or less radial pyroclastic deposits accompany Pele and Babbar Patera. Bright fan-shaped deposits appear to have formed from material issuing from parts of Ulgen and Creidne Paterae. Artesian venting of SO_2 can produce deposits of this type (McCauley et al. 1979). In both cases the paterae themselves are probably too warm for deposition of SO_2 (the area of Ülgen Patera with a mean temperature of 191 K could include the bulk of the dark material within the caldera at 198 K or above, since the temperature determination is relatively insensitive to small thermal differentials across the source).

Babbar Patera and Ülgen Patera both lack the extensive ring deposit of Pele, but the sources of other active plumes identified as hot spots (Loki, Maui, and Amirani) and Surt do have associated pyroclastic rings. These rings and the many volcanic vents that cover Io provide means of estimating the global distribution of hot spots. Strom and Schneider (Chapter 16) have found that the active plumes and the pyroclastic deposits similar to those associated with active plumes are distributed more or less randomly in longitude, but are concentrated between latitudes ±45°. Small dark diffuse spots and streaks having the same characteristics as those of the cores of active eruptions also seem to be concentrated near the equator; their small size prevents an accurate assessment of their longitudinal distribution because of inadequate imaging resolution at some longitudes. The distribution of geologic landforms associated with volcanic vents, including craters, cones, and flows, has been studied by Schaber (Chapter 15) over an area slightly less than one hemisphere; no longitudinal asymmetry is found, but the numbers of vents in the latitudinal bands $0°$ to $-30°$, $-30°$ to $-60°$, and $-60°$ to $-90°$, respectively, are in the ratios $7:5:1$.

It has been pointed out by Witteborn et al. (1981) that the five observed outbursts all occurred in the first quarter of Io's orbit following eclipse. More precisely, the data of Table 19.2 show an absence of outbursts over a minimum range of longitudes between $158°$ and $288°$, i.e., an arc of at least $130°$. Although this appears to be a large gap, a Monte Carlo calculation shows that a void larger than this has better than three chances out of four of occurring if the outbursts are randomly distributed in longitude; the most probable maximum longitudinal arc between two adjacent events in a random selection of five events is about $145°$. Thus the observed distribution of outbursts is not inconsistent with the apparently random longitude distribution of pyroclastic deposits discussed by Strom and Schneider in Chapter 16. It should be pointed out, however, that it is geometrically possible for all five events to have originated at a single source.

Little can be said directly about the longitudinal distribution of hot spots from Voyager observations, because of limited longitudinal coverage; the latitudinal coverage is more extensive. After correcting for differences in longitudinal sampling, and averaging the hot spot counts between $+30°$ and $-30°$ lat, the distribution of Voyager sources can be compared with the vent distribution found by Schaber; the numbers of hot spots in the three $30°$ wide latitudinal bands at progressively higher latitudes are in the ratio $2:4:1$. However, two observational effects make direct comparison between the Voyager and Schaber statistics difficult. First, the Voyager ratios are based on 9 major hot spots, whereas Schaber's analysis included 93 vents. Second, there is a selection effect in the Voyager data due to the flyby trajectory; the higher spatial resolution and the lower temperature of the passive background surface in the polar region (relative to low latitudes) accentuate smaller, cooler thermal anomalies in the data. Nevertheless, at mid and high latitudes, the latitudinal distribution of prominent hot spots is consistent with the vent distribution. At lower latitudes the apparent disparity between relative numbers of hot spots and vents suggests that additional thermal emission in the equatorial region arises from subdued sources. This is supported in that the heat flux from the nine hot spots is $\sim 75\%$ the flux obtained by integrating over the entire area sampled by the Voyager observations.

Normalized to the area sampled, the integrated heat flux gives a mean value consistent with determinations from groundbased observations (Sec. VI). Also, if scaled from the 30% of the surface sampled by Voyager to the

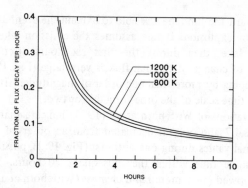

Fig. 19.15. The fractional decay rate of 5-μm flux from a sulfur lake with initial temperatures as indicated (from Sinton 1980b).

total surface of Io, the power radiated by the high-temperature component at Pele agrees well with the power from sources at ∼ 600 K as determined by Morrison and Telesco (1980), and by Sinton (1981). These results indicate that the major active thermal sources in the area mapped by Voyager were detected, and that the sample is representative of the planet as a whole. Of the ∼ 75% of the total flux which originates from the hot spots in Table 19.2, over 95% arises from just six features. Thus, the bulk of Io's active thermal output appears provided by about twenty major sources, rather than by numerous small ones.

Although the Voyager data provide the only high spatial resolution observations of hot spots to date, it is possible to resolve and locate the hot spots on the disk of Io and to determine their individual characteristics by observing occultations of Io by the Moon. On the average these occultations occur once a year. The limit of resolution in this type of experiment is set by diffraction at the lunar limb; at 3.5 μm this amounts to ∼ 50 km at the distance of Io. From any one location on the Earth, the fluxes from strips of Io parallel to the lunar limb may be obtained. At a different terrestrial location, the strips will crisscross the first set, since the lunar limb will be at a different position angle. If the source of emission on Io consists of a few discrete hot spots, the interpretation should be unique with only two position angles of the lunar limb. With large infrared telescopes and available detectors, it is now possible to obtain maps of the hot spots in this way.

III. VARIABILITY

The most spectacular variable thermal features observed to date are the 5-μm outbursts (Fink et al. 1978; Witteborn et al. 1979; Sinton 1980a; Witteborn et al. (1980). Only Sinton (1980a,b) has obtained time resolved measurements of the phenomenon. After correcting the data for rotation of the source toward the limb, he was able to explain the event as the aftermath of a large vapor explosion. If one assumes the emission takes place from a sulfur magma lake, then during the first six hours after the event the calculated rate of change of thermal flux is very large (Fig. 19.15), with the total flux dropping by more than one order of magnitude. After this, cooling continues on a time scale of the order of a day or two.

The observations of Witteborn et al. (1979) indicate that the brightness of Io between 4.7 and 5.4 μm decreased from an observed high of 3 to 5 times the normal values during one outburst (Fig. 19.2), possibly reaching its normal level some time within the subsequent 20 hours. Two additional observations showed much more rapid changes (Witteborn et al. 1981). In the event of 10 Dec. 1979 the 5-μm flux dropped from a maximum of 1.6 times normal (Fig. 19.3) to its background level in two hours; in the event of 20 March 1980 the flux dropped from 1.5 times normal to near normal in 80 min. Since the positions of the responsible features on Io are unknown, it cannot be determined whether the returns to normal brightness were due to actual decay of the outbursts, to rotation of the sites across the limb, or to a combination of both. The earlier enhancement recorded by Fink et al. (1978) did not change during the observation. However, it was absent 24 days later in observations of Io at the same orbital longitude.

During each of the Voyager encounters with Io the spacecraft was close enough for the imaging system to resolve large plumes for several days. Minor fluctuations in Plumes 3, 4 and 5 occurred on a scale of a few hours (Strom et al. 1981). In the four months between the two flybys, the Surt caldera associated with the outburst observed by Sinton altered significantly, the plume associated with Pele (Plume 1) disappeared, and the plume at Loki (Plume 2) added a new component (Plume 9) and increased in size by 45% (see Fig. 18 of Smith et al. 1979c).

The outbursts and the observed changes in plumes together provide evidence for local changes in Io's thermal emission which can be abrupt, and which occur several times per year. The five outbursts observed over a period of about two years suggest that the minimum frequency of these occurrences is about two per year. Based on the asumption that about half of the large calderas on Io are produced explosively, and on an estimation of 10 yr of visibility before burial, Sinton estimates that outbursts must occur at a rate of ~ ten per year (Sinton 1980b).

Morrison and Telesco (1980) have pointed out that their observations apparently occurred when Io was relatively quiescent, since their 3.4- and 4.8-μm data imply a flux equal to about half that observed by Sinton (1980a) during an outburst. However, at longer wavelengths, where the bulk of Io's active thermal emission occurs, they found no evidence for variability; their 10- and 20-μm brightness temperatures were identical to those measured in 1972 by Hansen (1973) and by Morrison and Cruikshank (1973), respectively. Multiwavelength observations of eclipses such as those made by

Morrison and Telesco are invaluable, since these data specify, at each eclipse, the hemispheric average heat flow of Io; after a number of these observations have been made, we may then knowledgeably discuss the possible time variation of the heat flow.

IV. POSSIBLE ABSORPTION FEATURES

In addition to their thermal output, it is possible that hot spots are also characterized by production of various gases and particulate materials. Gases expanding from hot vents will cool adiabatically; initially entrained fine particulates will cool by radiation (particles with $r < 10$ μm will typically cool with time constants on the order of $\lesssim 10$ s). Consequently, high thermal contrast will quickly be established between these materials and the underlying hot surface; if the gases and/or particulates have active infrared bands, strong absorption features may appear in the thermal spectra of the hot spots. These features may then provide information regarding the composition and evolution of such hot regions.

Several hot spots evident in the Voyager infrared data have been identified with areas from which plumes originate (see Table 19.2). One such hot spot (Loki Patera) shows the spectral signature of gaseous, and possibly solid, SO_2 near 1350 cm^{-1} (Pearl et al. 1979). Although the 0.2-cm-atm abundance of gaseous SO_2 derived from the data can be accounted for by the vapor in equilibrium with deposits of solid SO_2 adjacent to the hot spot, the quoted uncertainty of a factor of two leaves open the possibility that a contribution from the plume may exist. Two other lines of evidence also support this possibility. First, if solid SO_2 is present in the plume as a result of condensation (Pearl et al. 1979; Caldwell et al. 1980), then particle nucleation is more likely to occur in a cooling environment, such as an adiabatically expanding gas cloud, than in an atmosphere having a substantial temperature inversion, as should exist in the presence of strong heating by ultraviolet radiation and charged particles (Chapter 21 by Kumar and Hunten). Second, a detailed examination of images of the plume at Loki by Collins (1981) indicates that the total mass of particulates in the plume may be as much as two orders of magnitude greater than the early estimate of Johnson et al. (1979). Assuming that the abundance of gas in the plume is comparable to the mass of particulates (Smith et al. 1979a), the amount of gas may then be comparable to the amount of SO_2 derived from the Voyager data. If the driving gas is SO_2, it may therefore contribute significantly to the observed absorption feature.

It is noteworthy that none of the other Voyager hot spot spectra show the signature of SO_2 (e.g., Fig. 19.1). This can be interpreted in several ways:

1. That the plume at Loki, if driven by SO_2, is unique;
2. That if all plumes are driven by spectroscopically active gases such as SO_2, the amounts of gas are below the detection threshold of the

Voyager infrared instrument (approximately 0.004 cm atm of SO_2 for conditions at Loki and Pele);

3. That the plumes are driven by gases such as S_2, which are inactive in the infrared.

Considering Pele, the 0.004 cm atm detection threshold for SO_2 and the hot spot area derived from Table 19.2 together require that a minimum of 3×10^{10} g of gas be present in the plume to be detectable from Voyager. If 10% of the plume volume is particulates, the data of Johnson et al. (1979) indicate a lower limit of 6×10^9 g for the particulate mass in the plume. Assuming an equal mass of gas, taken to be SO_2, it is therefore possible that interpretation 2 is valid, at least for Pele.

Since the vibrational mode of diatomic sulfur is optically inactive, any S_2 in a plume cannot be detected in the infrared. However, even at temperatures as high as 700 K, the equilibrium composition of saturated sulfur vapor is dominated by S_6, S_7, and S_8, rather than by S_2 (Berkowitz and Marquart 1963). Furthermore, once gas is vented, condensation should occur within the plume. Since polymeric and crystalline sulfur have infrared active bands (MacNeill 1963; Wiewiorowski et al. 1965; Nimon et al. 1967; Meyer and Stroyer-Hansen 1972; Gardner and Rogstad 1973; Steudel and Rebsch 1974), diagnostic sulfur signatures should appear in plume spectra if sulfur is present. A rough estimate of the Voyager detection threshold can be made. Lacking accurate optical constants, an absorption coefficient of 50 cm^{-1} for the 11.8-μm band of rhombic sulfur appears consistent with the data of MacNeill (1963). For the conditions under which Voyager viewed Pele, this requires a total plume content of approximately 10^{13} g of solid S_8 for detection, which is about three orders of magnitude greater than the mass of particulates estimated by Johnson et al. (1979), assuming that condensates occupy 10% of the plume volume. Comparable limits can be derived from the other infrared bands of rhombic sulfur. Thus, while condensation produces infrared active forms of sulfur, they may be undetectable by the Voyager instrument because of the weakness of their spectral features.

Near-infrared absorption features seem to be present in two spectra of outbursts. In the observations of 20 Feb. 1978, Witteborn et al. (1979) find a deep sharp feature near 4.6 μm and a broad, shallow one between 4.7 and 5.3 μm (Fig. 19.2); an additional feature at 1.56 μm is also mentioned. In the observations of 10 Dec. 1979, completely different absorptions appear (Witteborn et al. 1981); a narrow one from 3.1 to 3.2 μm, and possibly a broad one from roughly 3.0 to 4.8 μm (Fig. 19.3). None of these has been identified. Based on the ratio of band depth to indicated noise in the data, the 3.1- and 4.6-μm features should be real; the broad features are more questionable. A 3.0 to 4.8 μm feature was present in the spectrum taken on 10 Dec. 1979 if the requirement for a relatively small source is imposed on the data; without this requirement, a blackbody fit can be made if the spot temperature and area are, respectively, ~ 195 K and 88% of Io's disk. Such a large area seems unreasonable for a thermal source. On the other hand, if one considers the spot to be, for example, 100 km in diameter, a size more compatible with those of other hot spots, then a temperature of 380 K fits the datum at 5.1 μm, while leaving a broad 3.0 to 4.8 μm absorption. Several interpretations of this feature are possible:

1. It is spurious;
2. It is due to the emissivity of hot surface material;
3. It represents absorption by an obscuring cloud of erupted material;
4. It indicates the appearance or enhancement of a disk absorption feature between the time the reference spectrum and the hot spot spectrum were obtained.

The similarity of this feature to an absorption present in quiescent disk spectra (Clark and McCord 1980) suggests that, if interpretation 3 is correct, the material expelled in this event is similar to material already widely distributed on the surface. For case 4 the feature may have nothing to do with the thermal event; the presence of the strong 3.1-μm feature, where the thermal energy from the spot is small, may support this interpretation.

If real, the variety and variability of the spectral features associated with hot spots are potentially very significant. They suggest not only different mechanisms and/or compositions for individual hot spots, but the possibility of studying time-dependent processes in the evolution of these phenomena. Further observations are needed, and sensitive spectral data over the lifetime of a hot spot would be particularly useful.

V. PROCESSES

Events associated with different thermal processes may show different thermal histories. In a random set of observations, such differences should show up in the relative frequency of occurrence of various temperatures. For example, observations of vapor explosions will be skewed toward low temperature measurements, since the rate of cooling is highest when the temperature is highest (Fig. 19.15); observations of stable hot vents on the other hand will cluster around the vent temperature. We take the data of Table 19.2 as an approximately random sampling of the hot spots. (There is bias in the data set, however, since the detection systems are not uniformly sensitive to all temperatures, and since the derived temperatures are, in some sense, mean values to which sources with different temperatures may contribute.) The frequency of occurrence of various temperatures in the data set is shown in Fig. 19.16. There is a suggestion of bimodality, with a majority of the observations clustering below 400 K, and most of the others at ~ 600 K. Together with the fact that the high-temperature results may be sorted into transient and stable events, these data suggest at least three different origins:

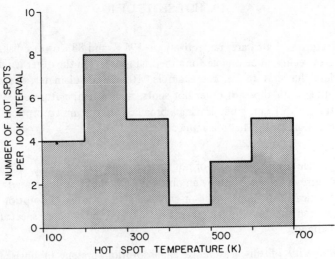

Fig. 19.16. The data of Table 19.2 plotted as the number of hot sources observed per 100 K interval versus source temperature.

1. Relatively stable, low temperature (< 400 K) sources. The low melting point of sulfur (385 K) suggests a simple explanation for the temperature cutoff if gently effusing sulfur flows are common. However, even high-temperature flows such as those of terrestrial silicates, after rapid initial cooling and crust formation, attain low surface temperatures and cooling rates. Thus heat transfer calculations should be performed for both silicate and sulfur flows *in vacuo* before the paucity of sources between 350 and 500 K can be considered significant evidence against the presence of silicate or other high-melting-point flows. Because of crust formation and low thermal conductivity, even large lakes of magmatic liquid will quickly develop low surface temperatures. Sinton (1980*b*) has shown that the sulfur lake formed after a vapor explosion behaves in this way; initial rapid cooling (Fig. 19.15) drops the surface temperature to ~ 350 K in less than a day (due to a computational error, the temperatures in Fig. 3 of Sinton [1980*b*] are uniformly too high by ~ 100 K). Subsequent cooling proceeds at a much lower rate.

2. Transient high-temperature (~ 600 K) sources. Vapor explosions of the type discussed by Sinton (1980*b*) fall into this category, though gaseous and/or particulate ejecta may partially mask their emissions (Witteborn et al. 1980). Other possibilities include rapid refilling of calderas, catastrophic flows following the breaching of crater walls, and hot ash flows.

3. Relatively stable high-temperature (~ 600 K) sources. Sinton et al. (1980) observed three eclipses of Io over a period of nine days and found a nearly constant flux from sources near 560 K. Voyager results indicate that a sizeable hot source (~ 654 K) is associated with Pele. Sinton et al. consider the high-temperature features to be small and numerous, such as

active volcanic vents or cracks in caldera floors caused by convection or crust subduction. The Voyager observation at Pele indicates that large localized sources also exist.

The relatively minor changes in the Pele plume geometry during the Voyager 1 flyby suggest that the high-temperature feature did not reside on a sulfur crust. If it had, the sizeable (~ 110 km^2) area at a temperature nearly twice the melting point of sulfur should have caused massive melting and rapid alteration of the vent geometry. An underlying material with a high melting point, such as silicates or more refractory sulfur compounds (perhaps sulfides or sulfates), therefore seems likely.

In the search for outbursts, numerous near-infrared observations of Io have been made over a wide range of orbital phase angles (Witteborn et al. 1981; D.P. Cruikshank, unpublished results; W.M. Sinton, unpublished results). However, the five recorded events have all occurred when Io was between 18° and 68° orbital longitude. The probability that five random events should fall into such a narrow range is $\sim 10^{-4}$. Sinton (1980) has suggested that some triggering mechnism may be operating. Since the recorded outbursts have all occurred at small orbital longitudes, they may be causally related to eclipses, but it is hard to think of a mechanism by which this is likely. Another possibility is an association with the time variable body tides. Moonquakes are most likely to take place when lunar tidal stresses are at a maximum (Lammlein 1977). Roughly equal contributions to the tidal stresses arise from the variations in Earth-Moon separation, and from the latitudinal and longitudinal optical librations. For Io, the dominant contributions arise from variations in the Jupiter-Io separation due to the forced eccentricity of Io's orbit and the accompanying longitudinal librations. However, the precession of the longitude of perijove is so rapid (Peale et al. 1979) that it prevents a correlation between tidal extrema and the orbital longitudes of the observed outbursts. The existing correlation with orbital longitude remains unexplained.

VI. ESTIMATION OF GLOBAL HEAT FLUX

Groundbased measurements provide the most direct means for estimating the global heat flux of Io. Although these observations lack spatial resolution, they do sense hot spot emission from all areas except those very close to the poles. Since the major hot spots are preferentially distributed between ±45° latitude, the poor viewing of high latitudes is no disadvantage. In addition, telescopic observations already extend back to the early 1970s, so their continuation will allow monitoring of Io's heat flux on a time scale of decades.

Matson et al. (1981) made the first estimate of the heat flux from the hot spots using the published data of Morrison and Cruikshank (1973), Morrison (1977), Gillett et al. (1970), and Hansen (1973). Matson et al. assume that the discrepancies between 8.4, 10.6, and 21 μm measurements both in and out of eclipse are due to the previously unrecognized radiation by the hot spots of Io. Their estimate of 2 ± 1 W m^{-2} for the disk-averaged heat flow is obtained by setting constraints on this quantity from a number of different considerations (Fig. 19.17). In their first consideration, the fluxes for the illuminated disk at 8.4, 10.6, and 21 μm are used with Eq. (1) to derive the temperature and area of the hot spots. The Stefan-Boltzmann relation is then applied to obtain the disk-averaged heat flux from the hot spots. This method is imprecise because of the relatively narrow spread of wavelengths and the uncertainties in the observational data. In a second method they assume that the background surface temperature (the majority of the disk area which is heated solely by the Sun) can be determined from a knowledge of the Russell-Bond albedo. The third method uses the 10.6- and 21-μm fluxes from Io just prior to its emergence from a total eclipse. Because the majority of the surface is continually renewed by fallout from the volcanoes, they assume that it behaves as a homogeneous

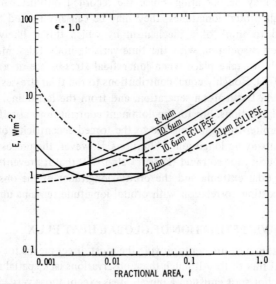

Fig. 19.17. A parametric representation of the global heat flux from Io as a function of the fractional area covered by hot spots (from Matson et al. 1981). Unit emissivity is assumed. The solid curves represent solutions based on eliminating the passive background flux from the measured illuminated disk fluxes. (Since the background temperature is close to the measured 21-μm brightness temperature, a heat flux determination based on the 21-μm datum is subject to large errors.) The dashed curves represent solutions based on eliminating the passive background flux from the measured eclipsed disk fluxes. The boxed region represents the acceptable range of solutions: (2 ± 1) W m^{-2}, with (1.5 ± 1) % of the disk covered by hot spots, rather than standard deviations.

semiinfinite medium rather than the vertically inhomogeneous medium that earlier investigators (Hansen 1973; Morrison and Cruikshank 1973) had assumed to explain their eclipse observations. In addition to the homogeneous surface, Matson et al. assume that it is interspersed with hot spots that supply relatively more flux at 10.6 than at 21 μm. From the last two methods they also obtain 2 ± 1 W m^{-2} for the average heat flux.

Another estimate has been obtained by Sinton (1981). Whereas the Matson et al. estimate uses data from ca. 1970, Sinton uses the composite spectrum shown in Fig. 19.4, which combines data obtained in the time interval 1974–1979. The results from Table 19.3 have been used to compute the heat fluxes from the hot spots. The combined heat flux, weighted by the areas A_1 and A_2, is found to be 1.8 ± 0.6 W m^{-2}.

In April 1980, Morrison and Telesco (1980) observed an eclipse reappearance of Io through 5 filters at wavelengths extending from 3.4 to 30 μm (Fig. 19.5). They measured the background with a 30 μm filter, which includes a much smaller fraction of hot spot radiation than do the 10- and 20-μm filters. They found a background eclipse temperature of 87 K with upper and lower limits of 90 and 85 K, respectively. Their derived model parameters for the hot spots are included in Table 19.2. They have also compared their 10- and 20-μm brightness temperatures with the older 10- and 20-μm results and find agreement. Thus over an 8-yr baseline there is no evidence for variation in the 10–20 μm region. Their 3.5- and 4.8-μm fluxes, however, were significantly less than those found by Sinton et al. (1980), providing evidence of variability of Io's flux at these short wavelengths. However, since most of the hot spot emission occurs at longer wavelengths, the variability observed at short wavelengths does not appreciably affect the overall heat flow. Their value of the heat flow is 1.5 ± 0.3 W m^{-2}, for which they give a firm lower limit of 1.0 W m^{-2}.

Three sets of Voyager data are being analyzed to evaluate the global heat flow. The first involves a disk measurement made $\sim$ 23 hr before Io encounter; the second, a nearly full disk observation made $\sim$ 7 hr later; and the third an integration over the portion of Io observed during the encounter phase (Fig. 19.7). From these measurements the mean heat flux is $\sim$ 2 W m^{-2}, but careful analysis of the passive background contribution must be completed before a precise value is obtained.

VII. DISCUSSION OF THE HEAT FLOW

The four independent determinations of the heat flux (see Table 19.4) agree within their statistical errors and lead to considerable confidence that the mean heat flow from regions on the scale of a hemisphere is between 1 and 2 W m^{-2}; expressed in customary geophysical units, these values are 24 and 48 μCal cm^{-2} s^{-1}. By comparison to known heat fluxes from other solid planets, this value is immense. The mean heat flux from the Moon, due to

TABLE 19.4

Average Heat Flux and Model Parameters of Hot Spot Emission

Hot Spot Temperature (K)	Fractional Coverage	Equivalent Single Source Diameter (km)	Disk Average Heat Flux W m^{-2}	Date	Longitude (deg)	Method
200 – 250	5 × 10^{-3} – 2.5 × 10^{-2}	130 – 250 }	2 ± 1	1969 – 1972	various	broadband disk radiometry; eclipse[a]
294	3.9 × 10^{-3}	227 }	1.8 ± 0.6	1974 – 1979	various	total spectrum[b]
615	2.1 × 10^{-5}	17 }				
	(see text)		~ 2	4,5 Mar 1979	(see text)	Voyager observations
200	1.4 × 10^{-2}	438 }				
350	2.2 × 10^{-4}	54	1.5 ± 0.3	12 Apr 1980	10	eclipse[c]
600	9.4 × 10^{-6}	11 }				

[a]Matson et al. (1981).
[b]Sinton (1981).
[c]Morrison and Telesco (1980).

passive diffusion of heat through the crust, is 0.4 μCal cm^{-2} s^{-1} (Langseth et al. 1976). The flux from the Earth, a relatively active body, is 2 μCal cm^{-2} s^{-1}, of which only 40% is associated with active processes such as volcanoes and sea floor spreading centers (Williams and Von Herzen 1974).

If the disk averaged results for Io apply uniformly to the planet, then the total power coming from the interior is between 4 and 8 $\times$ 10^{13} W. A number of mechanisms has been considered for the source of this power. Radioactive (Peale et al. 1979), electromagnetic (Ness et al. 1979; Colburn 1980), and plasma (Matson et al. 1981) sources are at least an order of magnitude too small. The most significant mechanism discussed to date is tidal dissipation (Peale et al. 1979; see Cassen et al., Chapter 4). Here the resonances between the motions of Io, Europa, and Ganymede produce forced eccentricities in the orbits of the satellites. The resulting time-variable tidal strain leads to internal heating through viscous dissipation. The observed high value of the heat flux can be obtained by adjusting the tidal energy dissipation factor (Q) of Io, but the required dissipation is untenable if the current eccentricity of Io's orbit is an equilibrium value determined by a balance of the effects of dissipation in Jupiter and Io (Yoder 1979). As Cassen et al. (Chapter 4) point out, the satellites would have been pushed farther from Jupiter in 4.6 $\times$ 10^9 yr than their present distances. Hence the solution of one enigma, the old 10 to 20 μm discrepancies, has led to yet another enigma: apparent incompatibility with the present orbital configuration.

One possible way to avoid the problem is to assume that the heat flow from Io is variable, and that we happen to view it at a time when it is abnormally large. The measurements, however, span a decade and do not allow much variation within this period, but if such variation is possible, the large amount of energy stored relative to the rate at which Io is currently losing heat would permit very long periods. Complete elucidation of the heat source remains a significant outstanding problem resulting from the discovery of active volcanism on Io.

Acknowledgments. We thank P. Cassen, D. Matson, L. Soderblom and C. Telesco for helpful comments on the manuscript. We also thank H. Masursky, R. Samuelson, R. Strom, and F. Witteborn for useful discussions.

June 3, 1981

REFERENCES

Berkowitz, J., and Marquart, J.B. (1963). Equilibrium composition of sulfur vapor. *J. Chem. Phys.* 39, 275–285.

Caldwell, J., Cess, R.D., Owen, T., Slobodkin, L.S., Buyakov, I.F., and Triput, N.S. (1980). Further interpretation of Voyager infrared observations by means of laboratory spectra of SO$_2$ frost. IAU Coll. 57. *The Satellites of Jupiter* (abstract 4–16).

Carr, M.H., Masursky, H., Strom, R.G. and Terrile, R.J. (1979). Volcanic features of Io. *Nature* 280, 729–733.

Clark, R.N., and McCord, T.B. (1980). The Galilean satellites: New near-infrared spectral reflectance measurements (0.65–2.5 μm) and a 0.325–5 μm summary. *Icarus* 41, 323–339.

Colburn, D.S. (1980). Electromagnetic heating of Io. *J. Geophys. Res.* 85, 7257–7261.

Collins, S.A. (1981). Spatial color variations in the volcanic plume at Loki, on Io. *J. Geophys. Res.* In press.

Davies, M.E., and Katayama, F.Y. (1981). Coordinates of features on the Galilean satellites. *J. Geophys. Res.* In press.

Fink, U., Larson, H.P., and Gautier, T.N. III (1976). New upper limits for atmospheric constituents on Io. *Icarus* 27, 439–446.

Fink, U., Larson, H.P., Lebofsky, L.A., Feierberg, M., and Smith, H. (1978). The 2–4μ spectrum of Io. *Bull. Amer. Astron. Soc.* 10, 580 (abstract).

Gardner, M., and Rogstad, A. (1973). Infrared and Raman spectra of cycloheptasulfur. *J.C.S. Dalton* 6, 599–601.

Gillett, F.C., Merrill, K.M., and Stein, W.A. (1970). Albedo and thermal emission of Jovian satellites I–IV. *Astrophys. Letters* 6, 247–249.

Hanel, R., and the Voyager IRIS Team (1979). Infrared observations of the Jovian system from Voyager 1. *Science* 204, 972–976.

Hansen, O.L. (1973). Ten-micron eclipse observations of Io, Europa, and Ganymede. *Icarus* 18, 237–246.

Johnson, T.V., Cook, A.F. II, Sagan, C., and Soderblom, L.A. (1979). Volcanic resurfacing rates and implications for volatiles on Io. *Nature* 280, 746–750.

Lammlein, D.R. (1977). Lunar seismicity and tectonics. *Phys. Earth Planet. Int.* 14, 224–273.

Langseth, M.G., Keihm, S.J., and Peters, K. (1976). Revised lunar heat-flow values. *Proc. Lunar Sci. Conf.* 7, 3143–3171.

MacNeill, C. (1963). Infrared transmittance of rhombic sulfur. *J. Opt. Soc. Amer.* 53, 398–399.

Matson, D.L., Ransford, G.A., and Johnson, T.V. (1981). Heat flow from Io (J1). *J. Geophys. Res.* 86, 1664–1672.

McCauley, J.F., Smith, B.A., and Soderblom, L.A. (1979). Erosional scarps on Io. *Nature* 280, 736–738.

Meyer, B., and Stroyer-Hansen, T. (1972). Infrared spectra of S_4. *J. Phys. Chem.* 76, 3968–3969.

Morabito, L.A., Synnott, S.P., Kupferman, P.N., and Collins, S.A. (1979). Discovery of currently active extraterrestrial volcanism. *Science* 204, 972.

Morrison, D. (1977). Radiometry of satellites and of the rings of Saturn. In *Planetary Satellites* (J.A. Burns, ed), pp. 269–301. Univ. Arizona Press, Tucson.

Morrison, D., and Cruikshank, D.P. (1973). Thermal properties of the Galilean satellites. *Icarus* 18, 224–236.

Morrison, D., Cruikshank, D.P., and Murphy, R.E. (1972). Temperatures of Titan and the Galilean satellites at 20 microns. *Astrophys. J.* 173, L143–L146.

Morrison, D., and Telesco, C. (1980). Observational constraints on the internal energy source of Io. *Icarus* 44, 226–233.

Ness, N.F., and the Voyager Magnetometer Team (1979). Magnetic field studies at Jupiter by Voyager 1: Preliminary results. *Science* 204, 982–987.

Nimon, L.A., Neff, V.D., Cantley, R.E., and Buttlar, R.O. (1967). The infrared and Raman spectra of S_6. *J. Mol. Spec.* 22, 105–108.

Peale, S.J., Cassen, P., and Reynolds, R.T. (1979). Melting of Io by tidal dissipation. *Science* 203, 892–894.

Pearl, J., Hanel, R., Kunde, V., Maguire, W., Fox, K., Gupta, S., Ponnamperuma, C., and Raulin, F. (1979). Identification of gaseous SO_2 and new upper limits for other gases on Io. *Nature* 280, 755–758.

Sinton, W.M. (1973). Does Io have an ammonia atmosphere? *Icarus* 20, 284–296.

Sinton, W.M. (1980a). Io's 5-μm variability. *Astrophys. J.* 235, L49–L51.

Sinton, W.M. (1980b). Io: Are vapor explosions responsible for the 5-μm outbursts? *Icarus* 43, 56−64.

Sinton, W.M. (1981). The thermal emission spectrum of Io and a determination of the heat flux from its hot spots. *J. Geophys. Res.* In press.

Sinton, W., Tokunaga, A., Becklin, E., Gatley, I., Lee, T., and Lonsdale, C.J. (1980). Io: Ground-based observation of hot spots. *Science* 210, 1015−1017.

Smith, B.A., Shoemaker, E.M., Kieffer, S.W., and Cook, A.F. II, (1979a). The role of SO_2 in volcanism on Io. *Nature* 280, 738−743.

Smith, B.A., and the Voyager Imaging Team (1979b). The Jupiter system through the eyes of Voyager 1. *Science* 204, 951−971.

Smith, B.A., and the Voyager Imaging Team (1979c). The Galilean satellites and Jupiter: Voyager 2 imaging science results. *Science* 206, 927−950.

Smith, D.W., Greene, T.F., and Shorthill, R.W. (1977). The upper Jovian atmosphere aerosol content determined from a satellite eclipse observation. *Icarus* 30, 697−729.

Steudel, R., and Rebsch, M. (1974). Infrared and raman spectra of cyclo dodecasulphur (S_{12}). *J. Mol. Spec.* 51, 189−193.

Strom, R.G., Schneider, N.M., Terrile, R.J., Cook, A.F. II, and Hansen, C. (1981). Volcanic eruptions on Io. *J. Geophys. Res.* In press.

Wiewiorowski, T.K., Matson, R.F., and Hodges, C.T. (1965). Molten sulfur as a solvent in infrared spectrometry. *Anal. Chem.* 37, 1080.

Williams, D.L., and Von Herzen, R.P. (1974). Heat loss from the earth: New estimate. *Geology* 2, 327−328.

Witteborn, F.C., Bregman, J.D., and Lester, D.F. (1981). Spectrophotometry of the Io 5 micron brightening phenomenon. Submitted to *Icarus*.

Witteborn, F.C., Bregman, J.D., and Pollack, J.B. (1979). Io: An intense brightening near 5 micrometers. *Science* 203, 643−646.

Yoder, C.F. (1979). How tidal heating in Io drives the Galilean orbital resonance locks. *Nature* 279, 767−770.

20. IO'S SURFACE: ITS PHASE COMPOSITION AND INFLUENCE ON IO'S ATMOSPHERE AND JUPITER'S MAGNETOSPHERE

F. P. FANALE, W. B. BANERDT,
L. S. ELSON, T. V. JOHNSON, and R. W. ZUREK

Jet Propulsion Laboratory

Groundbased and spacecraft observations suggest that the surface of Io is largely covered with an anhydrous mixture of sulfur allotropes. Another suggested important component (and one that links the surface with the atmosphere) is SO_2 frost. Also, some Na- (and K-) containing phase must be a significant component of the surface; the sulfide salts Na_2S and K_2S appear to be the candidate phases most harmonious with the simultaneous constraints imposed by Io's optical properties and Earth-based and in situ observations of Io's neutral cloud and the plasma torus. Io SO_2 meteorology is a complex problem. Plasma data argue for an SO_2-dominated atmosphere, and several observations seem nearly compatible with a simple model involving local buffering by SO_2 frost. Such a model, in which Io's atmosphere is considered to be saturated everywhere, is certainly a useful approximation of the actual situation, but leads to the expectation that Io has developed ubiquitous thick SO_2 frost coverage at high latitudes, whereas Io's near-polar regions do not generally show optically thick frost coverage. We suggest a variant of this simple local buffering model having the following characteristics. (1) The global SO_2 gas abundance is primarily controlled by buffering in the brightest coldest regions. (2) The net SO_2 flux across the disk is limited by "regional cold trapping" on high-albedo regions and possibly by the resistance of a tenuous non-SO_2 residual atmosphere. (3) The continuing migration of SO_2 toward cooler regions and those lacking SO_2 sources is opposed by SO_2 destruction and planetary ejection process, including sputtering, thus preventing buildup of thick ubiquitous SO_2 coverage. The outflow of SO_2 from the Loki plume alone may be comparable to the total global SO_2 poleward flux, suggesting that the global SO_2 regime could be affected by the location and activity of major sources as well as frost-atmosphere equilibrium.

This chapter deals with evidence and interpretations pertaining to the surface phase composition of Io and the mechanism(s) by which Io's surface influences its atmosphere. We also briefly discuss the mechanism(s) by which Io's surface and/or atmosphere supplies neutral and ionic species to the region around Io and ultimately to the Jovian magnetosphere. We will discuss the surface phase composition first. Groundbased spectral observations of Io's surface influenced its interpretation earliest and most strongly; observations of Io's atmosphere by Pioneer 10 provided a basis for atmospheric modeling, while both early groundbased observations and Pioneer observations suggested a very strong compositional linkage between Io and the surrounding magnetosphere of Jupiter. With the arrival of Voyager 1 and several nearly simultaneous, independent geophysical insights and laboratory breakthroughs came an explosion of information concerning the properties of Io's surface and the nature of its interaction with its atmosphere and the Jovian magnetosphere, resulting in considerable revision of pre-Voyager working hypotheses on these subjects. Following our discussion of surface phases we consider how they affect the atmosphere and magnetosphere.

We attempt in this chapter to describe the understanding of Io-related phenomena in the post-Voyager period and to point out the many new ideas and questions that are arising as these data are analysed. It should be emphasized that many of the inferences we offer in this chapter with regard to surface composition are based on two kinds of data: Earth-based data of excellent spectral resolution, but only hemispheric spatial resolution; and imaging data with quite good spatial resolution, but very low spectral resolution. Thus, although we try to synthesize both types of data with plasma and other information to constrain the global phase assemblage present, we admit that future observations may well reveal small spatial domains with very different spectral reflectance from any observed so far, and indicate the local presence of phases we claim here to have generally eliminated on the basis of available plasma or spectral data. The next leap in our knowledge of the surfaces of the Galilean satellites, and their atmospheres and interaction with the Jovian magnetosphere, will probably be provided by the Galileo mission, which will provide multiple close encounters with the four Galilean satellites by a fully instrumented orbiter, although in the meantime many advances can be expected from continuing groundbased and Earth-orbital observations of Io.

I. IO'S SURFACE PHASE COMPOSITION

Early attempts to interpret the visible and infrared reflectance spectrum of Io identified elemental S as a likely dominant component of its surface (Fanale et al. 1974; Wamsteker et al. 1974). Later, a more detailed spectroscopic study of mixtures of elemental S with other compounds led to the suggestion that various allotropes of elemental S probably constituted $\geqslant$ 50% of the surface (Nash and Fanale 1977) and that high-temperature allotropes of S might have spectral properties very different from those of the familiar "flowers of sulfur" (S_8) (Nelson and Hapke 1978). When Brown (1974) reported the discovery of the neutral Na cloud around Io, the supply mechanism for the Na was suggested by Matson et al. (1974) to be sputtering of Io's surface by charged particles in the magnetosphere. Fanale et al. (1974) concluded that Io's surface is a mixture of elemental sulfur and a Na-containing phase that has two spectral properties: high albedo, and an absence of deep spectral features (primarily multiatomic vibrational combinations and overtones) in the wavelength region (0.3-2.5 μm), for which reflection data of moderate quality were available at that time. Sulfates or halides seemed the best candidates.

Fanale et al. (1974) suggested that Io's sulfur-rich surface was the product of internal differentiation and degassing of a warm or hot interior. However, if long-lived nuclides were the only heat source, then Io would be unlikely to get any hotter than a hypothetical chondritic Earth Moon, even if solid-state convection were neglected (Fanale et al. 1974, 1977*a*). Specifically, they found that the basalt solidus would probably lie $\geqslant$ 400 km deep for most of Io's history. They therefore suggested that instead of ordinary basaltic volcanoes, Io's surface probably formed by mild thermal degassing processes such as release of H_2S gas, followed by photodissociation, H escape, and plating out of elemental S on the surface. Another process suggested by Fanale et al. was transport of sulfates to the surface by hydrothermal solutions followed by partial reduction to S by impact and/or proton bombardment.

A major observational constraint was supplied when two groups of investigators measured Io's spectral reflectance in the 2.5-5.0 μm region (Pollack et al. 1978; Cruikshank et al. 1978). The 0.3 μm to 5.0 μm reflectance spectrum of Io is shown in Fig. 20.1 One group, Pollack et al., made these measurements from a high-altitude aircraft, whereas the other (Cruikshank et al.) made them from the 4200-m Mauna Kea Observatory on Hawaii. Both spectra exhibit two striking features: (1) they contain no discernible H_2O band at 3.0 μm, the fundamental O-H stretch frequency, and (2) they show a very deep (30%) absorption at 4.08 μm (see Fig. 20.1). The first observation is important because the absence of the usual bands at 1.4 and 1.9 μm that characterize both ice and even small amounts of bound H_2O had not been regarded as firm evidence of an utterly anhydrous Io surface. However, the fact that Io's spectrum is fairly flat even in the 3 μm region argues strongly against the presence of substantial H_2O ice.

It was suggested by Nelson and Hapke (1978) that the surface was primarily directly sublimed S, which presumably did not involve H_2O as a carrier for its transport. Like Fanale et al., they suggested H_2S as a carrier,

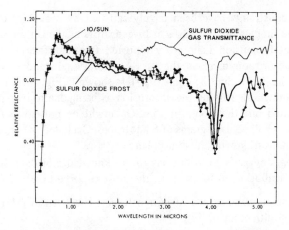

Fig. 20.1. The reflectance spectrum of Io (leading side except the data from 2.9-5.0 μm, which are from Cruikshank et al. [1978] and represent mixed orbital phases) compared with laboratory reflectance spectra. The Io composite spectrum is from Clark and McCord (1980) and contains data from McFadden et al. (1980). The spectra of SO_2 frost are from Fanale et al. (1979). The spectral features near 3.0 μm are primarily due to H_2O contamination. The deep 4.08 μm absorption in Io's spectrum was identified by Fanale et al. (1979) and Smythe et al. (1979) as resulting from the $\nu_1 + \nu_3$ absorption band of SO_2 frost, but whether other forms of SO_2 contribute to the feature is not yet established.

and like Nash and Fanale they cited the work of Meyer et al. (1972) in pointing out the difference between the optical properties of high- and low-temperature phases of S. The Nelson and Hapke model provided rather good matches to the visible and ultraviolet data simply by using various forms of a single substance; it also emphasized that H_2O as a carrier is not required. On a larger scale, Io's overall paucity of H_2O remains a critical issue; this is discussed elsewhere in this book (Chapter 24 by Pollack and Fanale).

We now examine the significance of the deep 4.08 μm band. As shown in Fig. 20.2, vibrational bands of anionic groups in some salts, such as carbonates and nitrates, can occur in this region of the spectrum. (Cruikshank et al. 1978; Pollack et al. 1978). Carbonates exhibit an additional deep band at 3.6 μm, which is also not apparent in Io's spectrum (see Fig. 20.2). The possibility that the "missing" bands are suppressed by unusual physical conditions such as grain size or temperature can be dismissed, since Kieffer et al. (1978) have shown that these parameters have at most a minor effect on relative band depths in the wavelength region of interest. Thus the infrared spectra of Pollack et al. and Cruikshank et al. established both a genetic constraint (the need for exceedingly efficient dehydration of the optical surface, or absence of H_2O deposition) and a compositional requirement (for a component with a deep 4.08 μm absorption).

Both constraints have been largely satisfied by available compositional and genetic models for Io's surface. Investigators at the University of Hawaii and the Jet Propulsion Laboratory reported the identification of the deep 4.08 μm absorption band as the $\nu_1 + \nu_3$ vibrational band of frozen SO_2 (Fanale et al. 1979; Smythe et al. 1979), as shown in Fig. 20.1. These investigators, although identifying the absorption as being due to frozen SO_2 molecules, pointed out that some of the SO_2 might be present as free frost or occluded in small voids in grains of S allotropes. Occluded SO_2 need not be in free communication with the atmosphere.

Among Voyager's contributions to our knowledge of the phase composition and distribution on Io's surface, the most important are:

1. The direct observations of currently active volcanism (Morabito et al. 1979; Smith et al. 1979*b*);
2. The interpretation of spectrophotometry from multispectral images that confirm the presence of various allotropes of sulfur and mixtures of local SO_2 on the surface (Soderblom et al. 1980);
3. The measurement of gaseous SO_2 column abundance equivalent to a basal atmosphere pressure of $\sim 10^{-7}$ bar in the vicinity of one of the large eruptive plumes by the Infrared Interferometer-Spectrometer (Pearl et al. 1979);
4. The measurement of the composition of the surrounding magnetospheric plasma (Bridge et al. 1979; Krimingis et al. 1979).

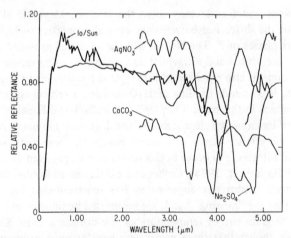

Fig. 20.2. Io reflectance data from Fig. 20.1 reproduced and compared with spectra of carbonate, nitrate, and sulfate phases. The carbonate and nitrate data are from Fanale et al. (1979) and the Na_2SO_4 spectrum was supplied by H.H. Kieffer (personal communication 1979). The deep absorptions in the region 3.0-5.0 μm are due to combinations of vibrational overtones and characterize the three anionic groups present. The absorptions are not prominent in the Io hemispheric spectra.

Shortly before the Voyager encounter, Peale et al. (1979) predicted that tidal heating of Io by Jupiter amounted to perhaps many times the energy input expected from the early "radioactivity only" models of Fanale et al. (1974, 1977a), making the likelihood of any volcanism much greater and allowing derivation of S and Na compounds from Io's interior as suggested by Fanale et al. without any need for H_2O as a carrier. The observation of active volcanism made by Voyager provided qualitative, but not quantitative, support for the hypothesis of Peale et al. However, other studies, based on groundbased observation of Io's eclipse cooling and Earth and spacecraft radiometry, suggest that Io's heat flow exceeds even that predicted by Peale et al. by a large factor (Matson et al. 1981; Morrison and Telesco 1980; Sinton 1981). The importance of this, and the observation of numerous eruptive plumes, for the phase composition is difficult to quantify. However, the ability to expose, recycle, and reexpose volatiles to the atmosphere and surface (see Johnson et al. 1979) would help to explain the degree of dehydration of surface phases, since all atmosphere-magnetosphere models indicate short mean residence times for atmospheric species (e.g., Kumar 1979a). This would be especially true if present suggestions of still greater tidal coupling and heat flow in Io's distant past prove correct (Chapter 3 by Greenberg). The problem of Io's planetary devolatization is discussed in another chapter (Pollack and Fanale, Chapter 24).

The photometric data from the Voyager spacecraft consist of reflectance values for five spectral band-passes with half widths of 40 nm, centered at 410, 480, 560, and 590 nm. These data were obtained for selected regions from what appeared to be five discrete spectral classes of surface areas: white, orange, and red equatorial plains, brownish-yellow polar plains, and dark brown to black caldera floors. As shown in Fig. 20.3a and b, Soderblom et al. (1980) found that the relative reflectances of these typical areas were similar to those exhibited by various allotropes of S. This is not surprising in view of the suggestions that elemental S was responsible for Io's peculiar high visible reflectance and low ultraviolet reflections, as observed in Earthbound wholedisk spectra (Fanale et al. 1974; Wamsteker et al. 1974; Nash and Fanale 1977) and preliminary examinations of the Voyager color images (Carr et al. 1979; Sagan 1979). However, Soderblom et al. also noted that the ultraviolet reflectance of many of the areas was somewhat higher (in the case of the "white" areas, much higher) than could be explained by any S allotrope alone. In the case of the white areas, they conclude that a substance with a very high ultraviolet reflectivity must be substantially admixed ($\sim 50\%$) in order to produce the observed spectrum. They noted (Fig. 20.3b) that a mixture of S_8 (which appears white at Io's surface temperatures) with SO_2 frost, the ultraviolet spectrum of which has been reported by Nash et al. (1980) and is shown in Fig. 20.3c, will satisfy the data.

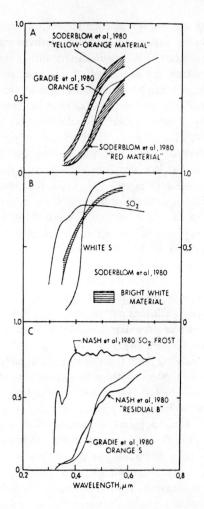

Fig. 20.3. (a) Reflectance data for small areas of red material and yellow-orange material on Io, derived from photometric data in five wide spectral bands from the Voyager photometric imaging data by Soderblom et al. (1980), compared with a spectrum of orange sulfur obtained by Gradie et al. (1980). (b) Spectra derived for bright-white areas on Io and a spectrum for white (− 160°C) sulfur (Soderblom et al. 1980) compared with the spectra of SO_2 frost (Nash et al. 1980). (c) The UV and visible spectrum of SO_2 frost (Nash et al. 1980). When the effect of 20% postulated optically thick SO_2 frost is subtracted from Io's hemispheric spectrum, the residual spectrum (Residual B) is found to generally resemble orange S measured by Gradie et al. (1980) and Nash et al. (1980). This is only an approximation of the anatomy of Io's ultraviolet and visible spectrum, however. It is believed that Io's spectrum is actually a composite of these components plus (at least) white S (S_8), brown and black sulfur, and sulfides of the alkali elements Na and K. The probable existence of SO_2 frost which is not optically thick further complicates the interpretation.

It must be admitted that deducing the phase composition of a possibly complex mixture using only a few wideband reflectances, especially when no specific vibrational bands have been identified in the spectrum, is usually incautious. Nonetheless, the conclusion of Soderblom et al. appears justified, at least for the bright areas. Nash and Fanale (1977) measured ultraviolet and visible spectra of a great many possible surface materials of Io, as have many other investigators; yet the only nonfrost substance other than SO_2 which exhibits the high ultraviolet reflections required by the white area data is $NaNO_3$, the occurrence of which we think is restricted by considerations already discussed. Thus, the firm identification of Io's 4.08 μm band as the $\nu_1 + \nu_3$ absorption band of frozen SO_2, and the detection of gaseous SO_2 by Pearl et al. make the case for abundant SO_2 in the bright areas seem quite strong. Also, as pointed out by Soderblom et al. there are physical-chemical reasons for expecting SO_2 to occur along with the low-temperature form of sulfur, S_8. Finally, white areas will be several degrees colder than adjacent areas that are red, orange, or brown. The vapor pressure curve for SO_2 is steep, and a 10 K temperature difference means more than a factor of 10 change in equilibrium vapor pressure. Thus a white material such as S_8 would have a strong advantage over adjacent dark areas in trapping SO_2.

Nash et al. (1980) performed a similar analysis to that of Soderblom et al. using Earth-based ultraviolet and visible data with much higher spectral resolution but only hemispheric resolution, as shown in Fig. 20.3c. Their results are generally compatible with those of Soderblom et al. They found that an excellent match to Io's ultraviolet and visible spectrum could be obtained with a mathematical mixture of SO_2 and orange sulfur spectra, in a ratio of optically thick areal coverage of 1:5. Since the white and orange areas share domination of Io's surface in the Voyager color images, the results of Nash et al. are qualitatively compatible with those of Soderblom et al.

Another seeming confirmation of the importance of SO_2, shown in Fig 20.4, comes from the identification of an SO_2 frost absorption feature in Io's ultraviolet spectrum (Nelson et al. 1980). Nelson et al. used IUE data to do hemispheric mapping of, and to set up an upper limit on, the SO_2 abundance. The results were generally compatible with the conclusions of Soderblom et al., and Nash et al. IUE data have also been used to set an upper limit on the average disk abundance of SO_2 gas (Butterworth et al. 1980). This limit could become a critical constraint in atmospheric modeling (see Sec. II). However, Kumar and Hunten (Chapter 21) argue that the derivation of this limit is not straightforward.

Nash et al. point out that in all their mixing models, the percentage of area covered by optically thick SO_2 frost must be regarded as an upper limit. This is so because any contributions to the visible and ultraviolet spectral characteristics attributed to optically thick SO_2 may be due instead to solid

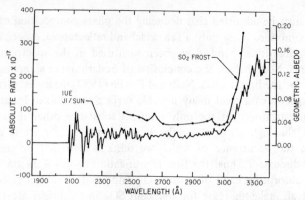

Fig. 20.4. An absorption feature in the reflectance spectrum of Io as obtained from observations from the Earth-orbiting International Ultraviolet Explorer satellite and reported by Nelson et al. (1980), has been tentatively identified by those investigators as corresponding to the absorption features in the ultraviolet spectrum of frozen SO_2 reported by Nash et al. (1980).

SO_2 in another form such as occluded SO_2, less than optically thick frost, dissolved SO_2, or adsorbate. These same points were emphasized by Fanale et al. (1979) and Smythe et al. (1979) in attempting to explain Io's deep 4.1 μm absorption. Smith et al. (1979a) pointed out that as much as 2% by mass of the SO_2 might be dissolved in liquid S. Houck et al. (1973) and Pollack et al. (1978) have shown that a given mass fraction of bound water is typically several orders of magnitude more effective in producing the 3.0 μm absorption due to the O-H stretch fundamental than is the same mass fraction of optically thick ice.

The next key question is: what phases are present on Io's surface other than frozen SO_2 and allotropes of S? Fanale et al. (1974, 1977a) argued that Io must contain on its surface the source of the neutral Na, which is supplied to a cloud by the atomic sputtering mechanism proposed by Matson et al. (1974). The mechanism for supply of the magnetosphere is the subject of several papers in this book, but a few considerations must be dealt with here, since they affect postulates concerning the surface phase composition.

Energetic particle and plasma compositon data suggest that the supply of Na^+ may conceivably be greater than that of Na° (Krimigis et al. 1979; Sullivan and Bagenal 1979), raising the possibility that the rate at which Na (neutral-plus-ionized) is removed from Io could be as high as one-hundredth that at which S and O are removed. Since frosts (such as SO_2) are sputtered orders of magnitude more easily than nonfrosts (Brown et al. 1979), a very abundant surface sputtering source could be required for the Na to compete in flux with the S^+ and O^+. Atmospheric sputtering or entrainment of S^+ and O^+ would probably favor S and O over Na just as strongly because SO_2 is the major constituent of the atmosphere, and it is difficult (but perhaps not impossible; see below) to sustain a high concentration of Na in the atmosphere. Moreover, the possibility that a substance could be supplied directly to the magnetosphere without ever contaminating the surface seems remote, since at least 99% of the plume material simply falls back onto the surface (Johnson et al. 1979; Chapter 17 by Johnson and Soderblom). Therefore we conclude that, even if we do not understand magnetospheric supply mechanisms and the Na flux thoroughly, there is probably a significant Na- (and K-) containing phase at Io's surface. Most of the evidence suggests the sulfide salts Na_2S and K_2S as likely candidates.

The composition of the Na phase(s) is strongly constrained by two sets of observations: (1) the spectral reflectance of Io in the 0.2-5.0 μm region. and (2) the chemical composition of the magnetosphere. The fact that the spectrum of Io can be immediately approximated from 0.3 μm to 5.0 μm by a simple mathematical mixture of the spectra of orange sulfur and SO_2 does not preclude the occurrence of other phases. However, abundant phases having very different spectra from Io's whole disk spectrum are unlikely; it would probably not exhibit vibrational bands in the vast regions where Io's spectrum is utterly bland, or have a very low visible albedo or a very high albedo below 0.3 μm. The problem is illustrated in Fig. 20.2. For example, carbonates and nitrates, once considered as possible major phases on Io's surface, have very deep absorption in the region 3.4-3.7 μm where Io is spectrally bland. Carbonates and nitrates may exist on Io's surface, but their coverage of the optical surface area must be small. An investigation such as the Galileo mission's near infrared mapping experiment with high (204 channel) spectral resolution and $\sim$ 20 km spatial resolution may reveal small domains of spectrally active species which do not strongly affect the whole disk spectra.

The plasma composition not only requires a Na mineral on the surface, but seems to place constraints on surface mineralogy in general. The best evidence on plasma composition comes from the Voyager plasma measurements (Bridge et al. 1979; Sullivan and Bagenal 1980) and the Voyager low-energy particle experiment (Krimigis et al. 1979; see also Chapter 23 by Sullivan and Siscoe). It should be noted, however, that the "sulfur nebula" the "sodium cloud" had been not only detected but carefully mapped from Earth before the Voyager 1 encounter (Brown 1974; Bergstralh et al. 1975; Kupo et al. 1976; Pilcher 1980). The various Voyager charged particle experiments were designed to discriminate various ionic species through energy resolution. The interpretation of these data in compositional terms is complicated in some cases, but the results have produced a fairly clear picture of the major ionic species: the low-energy plasma is mainly S^{++}, S^+, and O^+ in proportion $S^+ + S^{++}:O^+ + O^{++} \sim 0.5$ (Bridge et al. 1979), suggestive of an SO_2 source, and SO_2^+ is a minor constituent. There is however some suggestion of as S excess over the stoichiometric ratio. In addition to the Na^+ mentioned above, the high-energy particles are also mainly S^{++}, S^+ and O^+, in a ratio of $S:O \simeq 0.35$ (Krimigis et al. 1979).

The major constituents of the plasma near Io are thus Na^+, S^{++}, S^+, O^+ and O^{++}. No Cl^- or Fe species were reported, and although C was detected, its abundance relative to He is roughly constant throughout the magnetosphere, indicating a solar wind origin. Calcium has been searched for optically from the Earth, and significant upper limits placed on its abundance (Trafton 1976; see also Fanale et al. 1977*b*). The fact that no Io-derived Cl, N, or C were detected by Voyager strengthens our belief (based on spectral data) that carbonates, nitrates, and chlorides are at most probably minor or local constituents of the surface, implying also that CO_2 as a major gas is absent from Io's atmosphere. We note also that NaCl is very bright in the ultraviolet, whereas Io is very dark. Sulfates, which would contribute only Na, K, and ionized S and O, could conceivably be the host phase for the alkali metals and would not violate the plasma data. Sulfates have a very deep absorption band at 4.4 μm, but the apparent absence of a feature in Io's spectrum does not totally eliminate them from consideration since the observational data are somewhat degraded by CO_2 absorption in the Earth's atmosphere.

The sulfide salts, Na_2S and K_2S, seemed likely phases to Fanale et al. (1979) and Nash and Nelson (1979), in that they exhibit three important spectral properties: (1) they are very dark in the ultraviolet; (2) they are very bright in the visible and infrared; and (3) they exhibit no prominent absorption bands in the region 0.5 μm to 5.0 μm. The spectra of Na_2S and K_2S are shown in Fig. 20.5. Also, sulfides satisfy the plasma data well, since they would contribute only Na and S.

There is yet another possibility that appears attractive to us; Na° may be a significant component of Io's atmosphere, and perhaps its surface. Gooding (1975) pyrolyzed the Allende Meteorite (a Type IV carbonaceous chondrite) and measured the release rates for all volatile species. One view of the Allende pyrolysis data is to ignore all compounds which contain H, C, or other species which plasma or spectroscopic data do not indicate as major constituents of Io's surface. Gooding's results thus leave us with S, SO_2, and metallic Na and K as the most abundant species. Though any agreement with the composition of effusing volatiles on Io may be largely fortuitous, Hapke (1979) proposed that Na and K evaporated from silicate lavas were responsible for the Na and K clouds. Although there are problems with direct escape through this mechanism (see Johnson et al. 1979), this may be a way to obtain Na° and K on the surface or in the atmosphere. Such a process also explains the absence of Ca.

Volatiles supplied to the surface of Io are presumably derived by near-surface volcanic processes from large volatile-rich reservoirs which were in turn derived from a rocky interior by familiar orthomagmatic and hydrothermal differentiation processes, not simply volatilized from an agglomera-

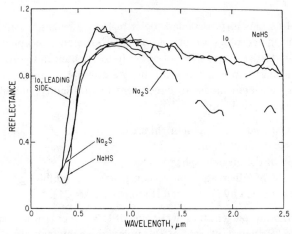

Fig. 20.5. Spectra of Na$_2$S and NaHS compared with a telescopic spectrum of Io from
Fig. 20.1. Host phases for Na and K on Io have been postulated for many years
(Fanale et al. 1974) but most possible phases have strong vibrational absorptive bands
which are not prominent in the Io data, are too bright in the ultraviolet, contain
elements not found in the plasma, or violate some other constraint. The sulfide salts
Na$_2$S and K$_2$S do not violate any of these constraints and therefore are viable Na and
K host phases (Fanale et al. 1979; Nash and Nelson 1979).

tion of silicates and oxides as in the case of the experiment cited. Nonetheless, Na$^{\circ}$ is not known to violate any of the observational constraints and
should therefore be considered more seriously as a phase than those salts
whose anionic groups create problems in explaining either the plasma composition or surface optical properties. Most importantly, it should also be
realized that the existence of Na$^{\circ}$ in the atmosphere could have profound
importance for magnetospheric supply even if the Na$^{\circ}$ could not survive as a
significant constitutent of the optical surface. An alternate mechanism for
atmospheric Na supply might be surface sputtering in areas where the atmosphere of Io was not thick enough to keep out all the sputtering particles but
only to thermalize some of the sputtered particles. Kumar and Hunten (Chapter 21) have suggested a cloud and magnetospheric supply mechanism, a
variant of the sputtering hypothesis of Matson et al., in which the Na sputtered in two stages: first into the atmosphere, then from the atmosphere to
surrounding space.

II. RELATIONSHIP BETWEEN IO'S SURFACE AND ATMOSPHERE

In this section we first discuss the several observations suggesting that Io
may be regarded, as a first approximation, as possessing an atmosphere everywhere buffered by local SO$_2$ frost on the ground. Then we discuss the difficulties of this model that appear to require modifications, for instance the

fact that Io seems to possess dark polar regions, whereas a preliminary analysis of the meteorology of the everywhere buffered atmosphere suggests a flux of SO_2 to the poles sufficient not only to brighten Io's polar regions, but to make a huge cap in geological time. We attempt to modify the simplest, everywhere buffered model to circumvent this difficulty, incorporating several phenomena, such as:

1. Regional cold trapping on bright areas;
2. The role of adsorption;
3. Supply of the magnetospheric;
4. The role of other atmosphere constituents in inhibiting SO_2 migration (first suggested by Kumar and Hunten, Chapter 21).

We argue that, to prevent total whitening of Io's higher latitudes and piling up of tens of kilometers of SO_2 at Io's poles, the combined effect of regional cold trapping, adsorption, and resistance of a permanent atmosphere must be to reduce the SO_2 flux to Io's poles to below $\sim 10^{29}$ atom s^{-1}, which is approximately the total SO_2 destruction rate (including sputtering).

The relationship between Io's surface, atmosphere, and ionosphere is also discussed extensively by Kumar and Hunten (Chapter 21) with a different emphasis. The main difference between our view of Io's atmosphere and theirs is that, while acknowledging the necessity to prevent growth of huge polar caps, they emphasize slowing of SO_2 migration by a permanent O_2 atmosphere. While we do not question the possible occurrence of some permanent O_2 dark-side atmosphere, major retardation of SO_2 migration by O_2 alone requires that the O_2 abundance be equal to or greater than the SO_2 abundance at $> 30° - 60°$ from the subsolar point (Chapter 21). In any event, the possibility of any permanent atmosphere would have many important consequences for Io, including a role in slowing SO_2 migration.

Arguments For and Against an Equilibrium Atmosphere

First, let us consider the evidence suggesting an equilibrium atmosphere everywhere buffered by SO_2 frost. The possibility that Io's surface is buffered by its atmosphere is obvious because frozen SO_2 is a major constituent of the surface material (Fanale et al. 1979; Smythe et al. 1979) and SO_2 gas is thought to be the primary constituent of Io's atmosphere (Pearl et al. 1979; Kumar 1979a). Several observations initially suggested that the surface of Io may be able efficiently to supply unlimited SO_2 to the atmospheric column if the base of the atmosphere is undersaturated at local ground temperature, and also to condense SO_2 from the atmosphere if the atmosphere is saturated or supersaturated at local ground temperature. The SO_2 pressure distribution over the disk for such a hypothetical object "everywhere buffered" is shown in Fig. 20.6. The following observations suggest effective buffering:

1. The IRIS experiment revealed a local atmosphere column abundance of
    0.2 cm - A, equivalent to $\sim 1 \times 10^{-7}$ bar. This is the basal pressure
    which would be in equilibrium with SO_2 frost at the subsolar point
    temperature $\sim 130\,K$ (see Fig. 20.6). The measurement was actually
    made at 13:00 hr Io time and at only $+ 10°$ from the equator—very near
    the subsolar point. Also, although the observation was made rather near
    the volcano Loki and through an active plume (Plume 2), it at first
    appeared that an unreasonably high ratio of gas to particulate matter
    $(> 10^2)$ would be required in order that all the SO_2 molecules in the
    field of view could have been directly and immediately derived from the
    plume. Therefore it was suggested that the SO_2 pressure may have been
    locally buffered by equilibrium with surface SO_2 (Pearl et al. 1979;
    Kumar 1979a,b; Kumar and Hunten, Chapter 21). Also, IRIS observa-
    tions of a dark-side warm caldera failed to reveal any SO_2 gas, suggesting
    effective condensation when saturation was achieved.

2. Pioneer 10 radio occultations of Io (Kliore et al. 1974; Kliore et al. 1975)
    yielded electron density profiles which depended on assumptions con-
    cerning composition and ionization mechanisms (e.g., see McElroy and
    Yung 1975; Johnson et al. 1976; Kumar and Hunten, Chapter 21) may
    be interpreted as equivalent to neutral atmospheric basal pressures of
    $\sim 10^{-9}$ to 10^{-11} bar. Given the assumption that SO_2 dominates the

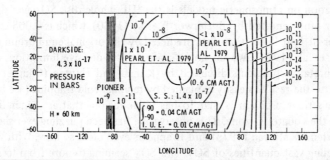

Fig. 20.6. For reference purposes, a contour map of SO_2 isobars on the light side of Io
which would occur if Io's atmosphere were always and everywhere buffered by
equilibrium with SO_2 frost at local ground surface temperatures. Here we derive
local "ground temperatures" by global modeling of Io's temperature distribution
according to a cosine law without regard to regional albedo differences. The SO_2
pressure inferred from Voyager IRIS observation (Pearl et al. 1979) and terminator
pressure inferred from modeling of Pioneer occultation data are shown for
comparison. As shown, the whole disk-integrated value for average SO_2 atmospheric
column abundances would be 0.04 cm-A if the model described were correct, which
can be compared to the upper limit on SO_2 average column abundance , 0.01 cm-A,
tentatively inferred from the IUE data (Butterworth et al. 1980). Kumar and Hunten
(Chapter 21) argue, however, that uncertainties in interpretation cast doubt on the
significance of the apparent discrepancy.

atmosphere and the vapor pressure curve calculated by D.D. Wagman (personal communication, 1979), this range of SO_2 partial pressures is compatible with the assumption of atmosphere-surface equilibrium at terminator temperatures of ~ 100 K-110 K (Kumar 1979*a,b*; Fanale et al. 1979).

3. Some form of frozen SO_2 appears abundant; the spectrophotometric data of Soderblom et al. indicate a high ultraviolet reflectance in many areas, which has been interpreted as evidence for frozen SO_2. This is especially true for the bright areas, where up to 50% SO_2 abundance has been suggested. The results of Fanale et al. (1979) and Smythe et al. (1979) also suggest a high abundance of frozen SO_2.

4. Finally, there is considerable peripheral evidence suggesting the presence of SO_2 free frost; branching flows, apparently of red sulfur, are outlined in white, as are some black calderas. Also there are white regions associated with eroding areas. This has been interpreted to indicate that erosion occurs in some areas, abetted by the oozing and artesian upwelling of possibly liquid SO_2 (McCauley et al. 1979; Schaber, Chapter 15).

Despite all this evidence for widespread SO_2, there are reasons to believe that the atmosphere of Io may not be everywhere buffered by SO_2 at local ground temperatures, and that over much of the surface of Io the basal atmospheric pressure is lower than the local equilibrium value. A potential limitation on atmospheric models is the IUE whole disk SO_2 gas abundance upper limit reported by Butterworth et al. (1980), which is 0.008 cm-A, or a factor of four lower than that predicted by the original equilibrium model of Pearl et al. (1979). We do not consider this a decisive limitation, owing to interpretative problems discussed by Kumar and Hunten (Chapter 21) and because the result is easily explained by regional cold trapping (see below). A more serious problem is that Io has polar regions that are generally darker than the equatorial zone. This observation has been puzzling ever since it was reported by Minton (1973) and it appears even more so now. With volcanoes recycling vast quantities of SO_2, and SO_2 seeming to ooze from Io's surface, why does Io not have huge polar caps? Ganymede's surface is covered with a volatile (H_2O) having a vapor pressure many orders of magnitude lower than that of Io's main surface volatile (SO_2) and lacks Io's nonthermal mechanisms for volatile transport, yet Ganymede has an extended polar cap (see Smith et al. 1979*b*). More importantly, the assumption that each atmospheric column is in equilibrium with surface SO_2 requires the correlative assumption that a pressure gradient of a factor of 10^3 exists between the subsolar point and Io's equatorial terminator. Elementary physics suggests that rapid gas migration would occur in response to such a gradient, but determination of how this migration would interact with sublimation, condensation, and Io's rotation requires quantitative treatment. A preliminary one-dimensional model by

Ingersoll and Summers (1981), which describes quantitatively the migration of SO_2 across Io from a hypothetically saturated subsolar point, suggests migration rates to the poles of $10^{-2}-10^{-4}$ g cm^{-2} yr^{-1}, enough to create a cap tens of kilometers thick in geologic time.

Before listing the factors that we believe contribute to decreasing the SO_2 poleward migration rate and preventing cap development, we should first ask whether the poles are actually dark. There exists very small bright regions near the poles that may be deposits of volatiles or an incipient cap. Also, Soderblom (personal communication) points out that the imaging data in the polar regions have not yet been satisfactorily corrected for limb darkening, so the poles may not really be as dark as the images suggest. Nonetheless it does not appear that Io's near polar regions are generally covered by optically thick SO_2 frost, so a steady relationship of supply and removal must exist.

Effect of Regional Cold Trapping

An important consideration is that cold trapping by bright regions at temperatures up to 10 K lower than their surroundings will sharply decrease both the pressures and the pressure gradients at any point over the disk, owing to the steepness of the vapor pressure curve.

We first note that, as shown in Fig. 20.7, SO_2 frost reflects much more incident solar energy than does orange S. Using the color images from Voyager and the albedo values in typical regions from Soderblom et al., we arbitrarily divide Io into "white" and "orange" domains. Having done this and assuming that we can neglect atmospheric heating of the surface, we calculate that the white regions are 8-9 K cooler than the average regions at any latitude. If we recalculate the model in Fig. 20.6 for this effect alone, allowing white regions to be effective cold traps for adjacent orange regions, we obtain lower SO_2 pressure contours, by a factor of 15. This result is shown in Fig. 20.8. The integrated whole disk gas column abundance is now 0.002 cm-A, vs. 0.04 cm-A in the model of Pearl et al. and well below the Butterworth et al. (1980) upper limit of 0.008 cm-A. We must consider gas migration kinetics in some detail to know whether regional cold trapping on high albedo areas is really effective, but it would be important even if only partly effective. Soderblom et al. deduced an especially high concentration of SO_2 in the bright areas, which they suggested was a preferential stratigraphic association of SO_2 with S_8 as opposed to higher-temperature forms of S. We suggest that regional cold trapping could help to explain this observation. Regional cold trapping is not a bootstrap operation either; Soderblom et al. suggest that the non-SO_2 component in the white areas is itself white. Thus the cold traps would exist even without the volatile that is cold-trapped. The SO_2 distribution in Fig. 20.8 does not contain pressures as high as observed by Pearl et al. (10^{-7} bar), but the IRIS measurements were made over a

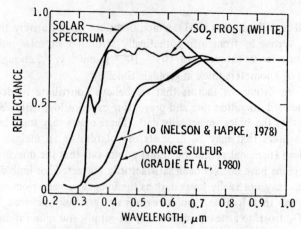

Fig. 20.7. The ultraviolet-visible spectra of SO_2 frost (Nash et al. 1980), orange S (Gradie et al. 1980) and Io (hemispheric) contrasted with the solar radiation spectrum. The difference between the solar energy absorbed per unit area for hypothetical SO_2 frost-covered regions of Io, and areas covered with orange S, is significant and results in $\sim 8°C$ difference between typical whitish and average areas at the same location and an even greater difference between whitish and orange areas at the same general location. Owing to the steepness of the SO_2 vapor pressure curve, such a seemingly trivial difference can greatly effect regional atmospheric pressures and gas transfer patterns.

region much hotter than the subsolar point (Hanel et al. 1979), and estimates of the mass of small particles in the nearby Loki plume (Collins 1981) demonstrate that locally derived volcanic SO_2 could easily account for the mass of SO_2 observed by the infrared spectrometer, as opposed to the original contention that the SO_2 represented an ambient atmosphere (see above). We conclude that regional cold trapping may result in global pressure gradients and local pressures that are both more than an order of magnitude reduced from the equilibrium model of Pearl et al. based on average global temperature. Since the flux of migrating SO_2 is essentially proportional to the product of the pressure and the pressure gradient, it follows that this effect alone can go a long way toward reducing the poleward flux to $\lesssim 10^{-4}$ g cm^{-2} yr^{-1} or $\lesssim 10^{28}$ molecules s^{-1} from all of Io. Since this is $\leqslant$ the sputtered flux and the photolysis rate, it is reasonable to suggest that a steady state between supply to the poles and sputtering could exist and prevent cap build-up. A pressure distribution such as that shown in Fig. 20.8 might also allow some dayside sputtering.

The Role of Adsorption

We shall next consider the effects of adsorption of SO_2, which we believe is a critical influence on atmosphere-surface interaction on Io; observations of

Io and laboratory evidence suggests that the adsorbed phase constitutes an important storage place for SO_2 on Io and a potential target for sputtering of SO_2 into the magnetosphere. We have concluded that there are probably vast regions on Io where the surface pressure is less than the saturation pressure, but still a significant fraction of it. If this is true, then equilibrium can exist locally between an undersaturated atmosphere and a surface soil with large quantities of adsorbed SO_2 covering virtually every accessible surface. A grain size distribution like that of a basaltic ash deposit is a reasonable expectation based on analogy with terrestrial sulfur pyroclastic volcanism, and the individual particles of all sizes are apt to be scoriaceous and have surfaces that are highly involuted and marked by numerous pores (cf. Francis et al. 1980). Thus there may be a great deal of fine grained S available as an adsorbent and even as a site for capillary condensation—both at pressures well below the saturation pressure. Fig. 20.9 shows the general results of E. Laue (personal communication 1980), who measured SO_2 adsorption on S at temperatures near to Io's. Although the temperatures and SO_2 pressures employed were not identical to those on Io owing to experimental problems, adsorption theory suggests that the relationship between coverage and relative pressures in Fig. 20.9 can be extended to apply to Io's range of pressures and temperatures, at least approximately. As shown in Fig. 20.9, it is possible largely to cover any surface of S with SO_2 even though the partial pressure of SO_2 at the base of the atmosphere is well below the condensation pressure. This has several important implications:

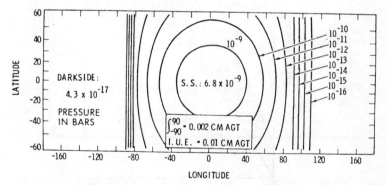

Fig. 20.8. The atmospheric SO_2 pressure contours for a hypothetical model where SO_2 pressure is everywhere buffered by equilibrium with SO_2 frost at local or regional surface temperature. Unlike the case in Fig. 20.6, we assume that temperatures on the ground at any latitude and longitude are $\sim 8°C$ lower on the average than those for "average-Io" (for which the SO_2 pressure contours in Fig. 20.6 are derived) because of the greater amount of solar energy reflected by either condensed SO_2 frost (Nash et al. 1980), or the white sulfur with which it is mainly associated (Soderblom et al. 1980) as opposed to the orange and red regions also indicated by independent evidence to be less SO_2 rich (Soderblom et al. 1980).

1. A great deal of SO_2 may be stored in the adsorbed phase, if the surface particles are fine and convoluted enough, as might be suspected from the low thermal inertia of the surface (cf. Morrison and Cruikshank 1973*b*; Morrison and Telesco 1980);
2. The 1.7-Earth-day diural thermal wave could cause some exchange between the adsorbed layer and the atmosphere;
3. The adsorbed SO_2 might contribute to Io's spectral properties significantly and differently from SO_2 frost;
4. The adsorbed SO_2 could be a target for ion sputtering.

Our point is that although none of the above effects might be important in areas with abundant SO_2 frost, they all might be in areas that lack coverage by optically thick SO_2 frost, perhaps on much of Io. Qualitatively, the above discussion suggests that even areas which are dark in the visible can contain substantial quantities of SO_2 and possibly can supply that SO_2 to the magnetosphere by sputtering. At the same time, ion scrubbing of the adsorbed layer can prevent frost buildup in those SO_2- poor areas.

Sputtering and the adsorbed phase can be related quantitatively as follows. Based on the preceding discussion we expect a good fraction of any

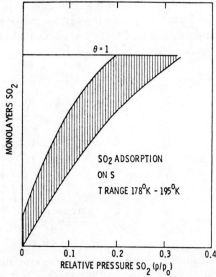

Fig. 20.9. The relation between monolayer coverage (θ) of a sulfur (S_8) surface by adsorbed SO_2 as a function of the relative pressure. p/p_0 of SO_2 is the partial pressure of SO_2 divided by the saturation pressure of SO_2 at the temperature of the gas-solid interface. The shaded area contains all available data. These data were taken for temperatures and SO_2 pressures somewhat higher than those on Io, but experience with analogous adsorbent-adsorbate systems suggests that this relationship will apply, at least approximately, to Io.

exposed surface on Io to be covered with SO_2, including the surface of the regolith that is exposed to space. We also know that for SO_2, one monolayer is equivalent to $\sim 1 \times 10^{-5}$ cm^3 STP cm^{-2}. The supply rate of S and O to the magnetosphere can be estimated in a number of ways to be $\sim 10^{28}$-10^{29} s^{-1} (see next section). Although we believe that atmospheric ionization/entrainment and atmospheric sputtering may be important processes, we will assume for this calculation that the sulfur and oxygen are supplied evenly from the whole disk of Io. This supply rate amounts to $\sim 3 \times 10^{11}$ molecules cm^{-2} s^{-1}. But the monolayer corresponds to 3×10^{14} molecules cm^{-2}. In general, the molecules sputtered from the first monolayer supply a major portion of any sputtered Flux (T.A. Tombrello, personal communication 1970). For adsorbed gas on a solid, this should be especially true considering that energy deposited in a layer by an incoming particle cannot be as efficiently used to launch atoms constrained on all sides by their neighbors as it can be to launch atoms that are unconstrained (only) in the upward direction and also loosely bound. Thus, if the adsorbed layer supplies the entire flux, the total mean residence time on all surfaces could be as short as 10^3 s with respect to sputtering. If, as seems likely, other mechanisms such as atmospheric sputtering also supply the magnetosphere, a longer mean residence time is implied because a lower sputtered flux from any ubiquitous adsorbed layer is needed. Thus we conclude that optically thick SO_2 frost need not exist in order for the near-polar regions to serve as a sputtering target for SO_2 and that a state can exist in which supply of SO_2 to the adsorbed phase is equaled by sputtering so that no buildup of optically thick frost occurs.

Laboratory data on the spectral properties of adsorbed SO_2 have not been reported. However, by analogy to transmission measurements on CO_2 and other comparable spectrally active gases that are adsorbed and chemisorbed, we presume that the coverage of SO_2 we predict would be spectrally significant. For example, Eischens and Pliskin (1958) report that 1% of a monolayer of CO_2 physically adsorbed on Cabosil produced at 40% adsorption in transmission, and they speculate that coverage down to 10^{-7} of a monolayer could probably be detected in this manner. Whether this is true for any $\nu_1 + \nu_3$ absorption band exhibited by adsorbed SO_2 and whether the band center would be the same as for frost or offset, as is SO_2 gas, is not demonstrated as of 1980.

Other Factors Affecting SO_2 Distribution

There are other factors that may slow the flux to Io's poles. One possibility is that a permanent (noncondensible) O_2 atmosphere may exist everywhere on Io (Kumar and Hunten, Chapter 21). According to Kumar and Hunten, if O_2 dominates SO_2 in regions $30°$-$60°$ from the subsolar point, then SO_2 migration would be governed by familiar meteorological processes rather than being driven by the factor-of-one-thousand pressure gradient and would thus be greatly retarded by the presence of the O_2. A severe constraint on such models is provided by the Pioneer occultation results of Kliore et al. (1975), which have been variously interpreted to indicate near-terminator pressures of 10^{-9} to 10^{-11} bar. The terminator pressure gives an upper limit to any O_2 atmospheric pressure since, in the equilibrium model, the SO_2 partial pressure at the terminator must itself be $\sim 10^{-9}$ bar if the equilibrium model is based on average global temperature as depicted in Fig. 20.6. This may suggest that O_2 does not dominate Io's atmosphere except possibly on the dark side, and therefore would not severely inhibit the SO_2 flux to the poles. However, in the regional cold trapping model in Fig. 20.8, a pressure of 10^{-10} bar of SO_2 occurs as close as $60°$ to the subsolar point. In this case, the condition required by the model (O_2 equal to SO_2 within $60°$ of the equator) could be considered to be compatible with the results of the Pioneer occultation experiment (O_2 pressure everywhere $\leq 10^{-9}$ bar). Thus we conclude that resistance from O_2 could slow SO_2 migration to the poles provided that regional cold trapping is effective. The dissociation of SO_2 may itself significantly slow its migration to the poles. Kumar and Hunten (Chapter 21) suggest a rate of dissociation of $\sim 3 \times 10^{11}$ cm^{-2}, which is similar to the magnetospheric supply rate. Thus, while SO_2 dissociation may not entirely account for Io's dark poles, it could slow SO_2 migration rates considerably.

Also, SO_2 distribution may be affected by the rate of volcanic supply. The Ingersoll and Summers model assumes an unlimited availability of SO_2 molecules in the form of solid SO_2. However, the flux of SO_2 across Io's disk may sometimes be at least locally limited by the rate and location of volcanic supply to the frost-atmosphere system. In such a model the poles might be dark because the sources would sometimes provide much less polar flux than expected in a constantly saturated model, thus allowing temporal variability in Io's global properties and the rate at which it supplies material to its surrounding neutral cloud and ionized torus. It can be shown that the global SO_2 flux is currently strongly affected if not dominated by individual volcanic sources, raising the possibility of temporal variations. Let Plume 2 (Loki) be 200 km in diameter and contain 2×10^{11} g of SO_2 (Pearl et al. 1979), and consider that the gas in the plume moves at ~ 100 m s^{-1} (a similar velocity is implied in the model of Ingersoll and Summers). Assume that none of the frost from condensing gas buries itself, but that all flows out across Io. Thus the plume disgorges 2×10^{11} g of SO_2 across the disk of Io every 10^3 s. Note that the 10^3 s lifetime is also consistent with the estimates of Johnson et al. (1979) and Johnson and Soderblom (Chapter 17). The flux to Io's surface is therefore 2×10^8 g s^{-1}. Since the surface area of Io is 4×10^{17} cm^2, the flux is 5×10^{-10} g cm^{-2} s^{-1}, or 1.5×10^{-2} g cm^{-2} yr^{-1}; this value is more than capable of accounting for the poleward flux implied in the model of Ingersoll and Summers (1981) which assumes an infinite available supply of surface SO_2. The actual efficiency with which SO_2 escapes the region of Plume 2 must be much less than 100%, since most of the SO_2 may indeed bury itself in the region of the plume deposits. It has been argued that the plume did not contain as much gas as measured by IRIS and that therefore the SO_2 was more likely to represent a regionally buffered ambient atmosphere (Pearl et al. 1979). If so, then the intensity of regional volcanism would be unlikely to affect SO_2 outflow from the region. However, recent studies by Collins (1981) reveal that the blue color of the Loki plume was not caused by Rayleigh scattering of the gas, but by an enormous load of extremely fine particulate matter, which may be SO_2 snow and may be compositionally unrelated to the throwout closer to the plume source (the latter may be coarser grains of S). A snow of fine SO_2 particles has also been suggested by comparison of SO_2 frost transmission spectra to the Voyager IRIS data (Slobodkin et al. 1980). This discovery has caused the estimate of the amounts of both gas and dust in Loki to be raised by several orders of magnitude, leading to consideration of a direct plume gas contribution in the IRIS field of view, comparable to that observed by the IRIS experiment (Collins 1981; J. Pearl, personal communication, 1980). Moreover, reexamination of the IRIS data has led to the tentative conclusion that Loki may be an atypically SO_2-rich source among the obvious volcanic sources on Io and may even supply a major portion of the SO_2 flux. There is no longer need to hypothesize a regionally buffered SO_2 atmosphere based upon the IRIS data alone (J. Pearl, personal communication 1980). Therefore we conclude that the waxing and waning of individual SO_2 sources could affect the SO_2 gas distribution, frost distribution, and conceivably the torus supply. In an unmodified equilibrium model, the SO_2 gas column abundance integrated over the disk exhibits no long-term changes, and SO_2 ground frost is essentially ubiquitous. The alternative model is shown in Fig. 20.10, which depicts a transient steady state in which the concentration of SO_2 on some regions of Io reflects the competition between supply from more SO_2-rich regions and its destruction and ejection by any of several processes, including sputtering. The current total flux of SO_2 towards higher latitudes is estimated to be equal to the total SO_2 destruction/magnetospheric supply rate, estimated in a variety of ways discussed above to be 1×10^{-3} g yr^{-1}, 3×10^{11} molecules cm^{-2}s^{-1}, or 1×10^{29} molecules s^{-1} for the entire object.

III. CONCLUSIONS

Among the most abundant phases on the surface of Io are allotropes of S formed at a variety of temperatures. SO_2 frost is abundant, and dissolved, occluded and adsorbed SO_2 are presumed to be present as well. Alkali metal containing phases, most likely sulfides, may also be significant components of

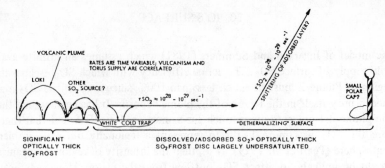

Fig. 20.10. A schematic presentation of a possible variant of the simplest model for Io's SO_2 regime, in which Io's atmosphere is buffered everywhere by local SO_2 frost. In the model shown, much of the disk is sometimes undersaturated with respect to SO_2 because of regional cold trapping on the brighter, colder areas, the slowness with which SO_2 migrates outward from limited regional SO_2 sources, and the existence of possible alternatives for SO_2 molecules other than poleward migration. Even where frost is absent, the adsorbed SO_2 serves as a possible source of SO_2 for the magnetosphere. The global SO_2 regime could be affected by the activity of large individual sources, thus allowing for variability in the global SO_2 regime.

the surface, which is essentially anhydrous. Although the precise mechansim for supply of Io's surrounding neutral cloud and ionized torus is still a matter of debate, Io's surface and/or atmosphere clearly supplies both. Observations of the chemical composition of the neutral cloud and surrounding plasma, together with Earth-based and spacecraft-based spectral reflectance data, provide powerful cross-constraints on Io's surface composition and do not reveal the presence of any major surface components other than those listed above (see Chapter 22 by Pilcher and Strobel and Chapter 23 by Sullivan and Siscoe). Other phases may be important constituents of limited spatial domains, however.

We conclude that a steady state exists between supply of SO_2 to the darker near-polar regions of Io and sputtering of SO_2 from those regions, which prevents SO_2 buildup. If regional cold trapping of SO_2 by the brightest (hence coldest) regions is effective, the flux to the dark regions may be comparable to the present sputtered flux ($\sim 10^{28}$-10^{29} s^{-1}); additionally, the fluxes of migrating SO_2 are then greatly decreased, and the flux to higher latitude regions might be comparable to the present sputtering flux ($\lesssim 10^{29}$ s^{-1}), and the SO_2 pressure may be sufficiently low that SO_2 migration is significantly impacted by the presence of a tenuous but permanent and ubiquitous O_2 atmosphere. Other factors, such as SO_2 dissociation and atmospheric sputtering, will also tend to decrease the SO_2 flux along the SO_2 concentration gradient. In our model with undersaturated areas SO_2 is still abundant as a ubiquitous adsorbed layer, which serves as a sputtering target. Since undersaturated regions may exist and individual volcanic sources can be important for Io's global SO_2 regime, long-term variations in Io's SO_2 gas and frost distribution and correlation with heat flow, post-eclipse brightening, torus brightness, etc. should be sought.

Acknowledgments. We thank D. Hunten, J. Caldwell, R.N. Clark, S. Kumar, A. Ingersoll and D. Morrison for useful technical comments and constructive reviews. This program was carried out in part at the Jet Propulsion Laboratory, California Institute of Technology under the auspices of the NASA Planetology Program.

REFERENCES

Bergstralh, J.T., Matson, D.L., and Johnson, T.V. (1975). Sodium D-line emission from Io: Synoptic observations from Table Mountain Observatory. *Astrophys. J.* 195, L131-L133.

Bridge, H.S., and the Voyager Plasma Science Team (1979). Plasma observations near Jupiter: Initial results from Voyager 1. *Science* 204, 987-990.

Brown, R.A. (1974). Optical line emission from Io. In *Exploration of the Planetary System.* (A. Wozczyk and C. Iwaniszewska, Eds.), pp. 527-531. D. Reidel, Dordrecht, Holland.

Brown, W.L., Lanzerotti, L.J., Poate, J.M., and Augustyniak, W.M. (1979). "Sputtering" of ice by MeV light ions. *Phys. Rev. Letters* 40, 1027-1030.

Butterworth, P.S., Caldwell, J., Moore, V., Owen, T., Rwolo, A.R., and Lane, A.L. (1980). An upper limit to the global SO_2 abundance on Io. *Nature.* In press.

Carr, M.H., Masursky, H., Strom, R.G., and Terrile, R.J. (1979). Volcanic features of Io. *Nature* 280, 729-733.

Clark, R.N., and McCord, T.B. (1980). The Galilean satellites: New near-infrared spectral reflectance measurements (0.65-2.5 μm) and a 0.325-5 μm summary. *Icarus* 41, 323-339.

Collins, S.A. (1981). Spatial color variations in the volcanic plume at Loki, on Io. *J. Geophys. Res.* In press.

Cruikshank, D.P., Jones, T.J., and Pilcher, C.B. (1978). Absorption bands in the spectrum of Io. *Astrophys. J.* 225, L89-L92.

Eischens, R.P., and Pliskin, W.A. (1958). Infrared spectra of adsorbed molecules. *Advances in Catalysis* 1-56. Academic Press, New York.

Fanale, F.P., Brown, R.H., Cruikshank, D.P., and Clark, R.N. (1979). Significance of absorption features in Io's IR reflectance spectrum. *Nature* 280, 760-763.

Fanale, F.P., Johnson, T.V., and Matson, D.L. (1974). Io: A surface evaporite deposit? *Science* 186, 922-924.

Fanale, F.P., Johnson, T.V., and Matson, D.L. (1977a). Io's surface and the histories of the Galilean satellites. In *Planetary Satellites* (J.A. Burns, Ed.), pp. 379-406. Univ. Arizona Press, Tucson.

Fanale, F.P., Johnson, T.V., and Matson, D.L. (1977b). Io's surface composition: Observational constraints and theoretical considerations. *Geophys. Res. Letters* 4, 303-306.

Francis, P.W., Thrope, R.S., Brown, G.C., and Glasscoch, J. (1980). Pyroclastic sulfur eruption and Poas Volcano, Costa Rico. *Nature* 283, 754-755.

Gooding, J. (1975). *A High Temperature Study on the Vaporation of Alkalais from Motton Basutto under High Vacuum: A Model for Lunar Volcanism.* M.S. Thesis, University of Hawaii.

Gradie, J., Thomas, P., and Veverka, J. (1980). The surface composition of Amalthea. *Icarus* 44, 373-387.

Hanel, R., and the Voyager Infrared Spectrometer Team (1979). Infrared observations of the Jovian system from Voyager 1. *Science* 204, 972-976.

Hapke, B. (1979). Io's surface and environs: A magmatic-volatile model. *Geophys. Res. Letters* 6, 799-802.

Houck, J.R., Pollack, J.B., Sagan, C., Schaack, D., and Decker, J.A., Jr. (1973). High-altitude infrared spectroscopic evidence for bound water on Mars. *Icarus* 18, 470-480.

Ingersoll, A., and Summers, M.E. (1981). A dynamically controlled atmosphere on Io. *Icarus.* In press.

Johnson, T.V., Cook, A.F. II, Sagan, C., and Soderblom, L.A. (1979). Volcano resurfacing rate and implications for volatiles on Io. *Nature* 280, 746-750.

Johnson, T.V., Matson, D.L., and Carlson, R.W. (1976). Io's atmosphere and ionosphere: New limits on surface pressure from plasma models. *Geophys. Res. Letters* 3, 293-295.

Kieffer, H.H., Pleskot, L.K., and Smythe, W.D. (1978). Spectra of possible Galilean satellite materials: T and $<$ d $>$ effects. *Bull. Amer. Astron. Soc.* 10, 580 (abstract).

Kliore, A., Cain, D.L., Fjeldbo, G., Seidel, B.L., and Rasool, S.I. (1974). Preliminary results on the atmosphere of Io and Jupiter from Pioneer 10 radio occultation measurements. *Science* 183, 324-329.

Kliore, A.J., Fjeldbo, G., Seidel, B.L., Sweetnam, D.N., Sesplaukis, T.T., Woiceshyn, P.M., and Rasool, S.I. (1975). Atmosphere of Io from Pioneer 10 radio occultation measurements. *Icarus* 24, 407-410.

Krimigis, S.M., and the Voyager LECP Team (1979). Low-energy charged particle environment at Jupiter: A first look. *Science* 204, 998-1003.

Kumar, S. (1979*a*). The stability of an SO_2 atmosphere on Io. *Nature* 280, 758-761.

Kumar, S. (1979*b*). A model of the SO_2 atmosphere and ionosphere of Io. *Geophys. Res. Letters* 7, 9-13.

Kupo, I., Mekler, Y., and Eviatar, A. (1976). Detection of ionized sulfur in the Jovian magnetosphere. *Astrophys. J.* 205, L51-L53.

Matson, D.L., Johnson, T.V., and Fanale, F.P. (1974). Sodium D-line emission from Io: Sputtering and resonant scattering hypothesis. *Astrophys. J.* 192, L43-L46.

Matson, D.L., Ransford, E.A., and Johnson, T.V. (1981). Heat flow from Io (JI). *J. Geophys. Res.* In press.

McCauley, J.F., Smith, B.A., and Soderblom, L.A. (1979). Erosional scarps on Io. *Nature* 280, 736-738.

McElroy, M.B., and Yung, Y.L. (1975). The atmosphere and ionosphere of Io. *Astrophys. J.* 196, 227-231.

McFadden, L.A., Bell, J.F., and McCord, T.B. (1980). Visible spectral reflectance measurements (0.33-1.1 μm) of the Galilean satellites at many orbital phase angles. *Icarus* 44, 410-430.

Meyer, B., Gouterman, M., Jensen, D., and Oomen, T.V. (1972). The spectrum of sulfur and its allotropes. In *Advances in Chemistry Series 110: Sulfur Research Trends.* American Chemical Society, Washington.

Minton, R.B. (1973). The polar caps of Io. *Comm. Lunar Planet. Lab.* 10, 35-39.

Morabito, L.A., Synnott, S.P., Kuperman, P.K., and Collins, S.A. (1979). Discovery of currently active extraterrestrial volcanoes. *Science* 204, 972.

Morrison, D., and Cruikshank, D.P. (1973*a*). Physical properties of the natural satellites. *Space Sci. Rev.* 15, 641-737.

Morrison, D., and Cruikshank, D.P. (1973*b*). Thermal properties of the Galilean satellites. *Icarus* 18, 224-236.

Morrison, D., and Telesco, C. (1980). Io: Observational constraints on internal energy and thermophysics of the surface. *Icarus* 44, 226-233.

Nash, D.B., and Fanale, F.P. (1977). Io's surface composition based on reflectance spectra of sulfur/salt mixtures and proton-irradiation experiments. *Icarus* 31, 40-80.

Nash, D.B., Fanale, F.P., and Nelson, R.M. (1980). SO_2 frost: UV-visible reflectivity and limits on Io surface coverage. *Geophys. Res. Letters* 7, 665-668.

Nash, D.B., and Nelson, R.M. (1979). Spectral evidence for sublimates and adsorbates on Io. *Nature* 280, 763-766.

Nelson, R.M., and Hapke, B.W. (1978). Spectral reflectivities of the Galilean satellites and Titan, 0.32 to 0.86 micrometers. *Icarus* 36, 304-329.

Nelson, R.N., and the JPL International Ultraviolet Explorer Team (1980). Io longitudinal distribution of sulfur dioxide frost. *Science* 210, 784-786.

Peale, S.J., Cassen, P., and Reynolds, R.T. (1979). Melting of Io by tidal dissipation. *Science* 203, 892-894.

Pearl, J., Hanel, R., Kunde, V., Maguire, W., Fox, D., Gupta, S., Ponnaperuma, C., and Raulin, F. (1979). Identification of gaseous SO_2 and new upper limits for other gases on Io. *Nature* 280, 755-758.

Pilcher, C.B. (1980). Images of Jupiter's sulfur ring. *Science* 207, 181-183.

Pollack, J.B., Witteborn, F.C., Erickson, E.F., Strecker, D.W., Baldwin, B.J., and Bunch, T.E. (1978). Near-infrared spectra of the Galilean satellites: Observations and compositional implication. *Icarus* 36, 271-303.

Sagan, C. (1979). Sulfur flows on Io. *Nature* 280, 750-753.

Sinton, W.M. (1981). The thermal emission spectrum of Io and a determination of the heat flux from its hot spots. *J. Geophys. Res.* In press.

Slobodkin, L.S., Buyakov, I.F., Triput, N.S., Cess, R.D., Caldwell, J., and Owen, T. (1980). Spectra of SO_2 frost for application to emission observations of Io. *Nature* 285, 211-212.

Smith, B.A., Shoemaker, E.M., Kieffer, S.W., and Cook, A.F. (1979a). The role of SO_2 in volcanism on Io. *Nature* 280, 738-743.

Smith, B.A., and the Voyager Imaging Team (1979b). The Jupiter system through the eyes of Voyager 1. *Science* 204, 951-971.

Smythe, W.D., Nelson, R.M., and Nash, D.P. (1979). Spectral evidence for SO_2 frost or adsorbate on Io's surface. *Nature* 280, 766.

Soderblom, L., Johnson, T., Morrison, D., Danielson, E., Smith, B., Veverka, J., Cook, A., Sagan, C., Cupferman, P., Pieri, D., Mosher, J., Avis, C., Gradie, J., and Clancy, T. (1980). Spectrophotometry of Io: Preliminary Voyager 1 results. *Geophys. Res. Letters* 7, 963-966.

Sullivan, J.D., and Bagenal, F. (1979). In situ identification of various ionic species in Jupiter's magnetosphere. *Nature* 280, 798-799.

Trafton, L. (1976). A search for emission features in Io's extended cloud. *Icarus* 27, 429-437.

Wamsteker, W., Kroes, R.L., and Fountain, J.A. (1974). On the surface composition of Io. *Icarus* 10, 1-7.

21. THE ATMOSPHERES OF IO AND OTHER SATELLITES

SHAILENDRA KUMAR
University of Southern California

and

DONALD M. HUNTEN
University of Arizona

Voyager measurements of gaseous SO_2 in a hot spot region and of ions of sulfur, oxygen, and SO_2^+ in the plasma torus, combined with the groundbased measurements of SO_2 frost on the surface, indicate that SO_2 is perhaps the dominant constituent of Io's atmosphere. Indeed the ionosphere of Io can be sustained by a pure SO_2 atmosphere in thermal equilibrium with the surface. Upper limits on common volatiles H_2O, CO_2, NH_3, and CH_4 are exceedingly small, although S_2 and Ar may still be present. Photochemistry of SO_2 could lead to O_2 as the major gas on the night side. Reasonable processes for loss to the torus include thermal escape, sputtering, and sweep-up of ions. The symmetry of the sodium cloud between eastern and western elongation is difficult to reconcile with any substantial, diurnally varying atmosphere. Europa, Ganymede, and Callisto may have oxygen atmospheres resulting from photolysis of water vapor. A pressure of $\sim 10^{-6}$ microbar is predicted, consistent with the Voyager upper limit for Ganymede. A much higher pressure might be present if the water vapor pressure is increased by a low albedo (Callisto) or sputtering (Europa).

The year 1973 marked the discovery of an ionosphere (and therefore an atmosphere) on Io by Pioneer 10 (Kliore et al. 1974) and the discovery of the strong sodium-D emissions associated with Io (Brown 1974). The study of the sodium, and later sulfur-II, lines (Kupo et al. 1976) revealed a wealth of unexpected behavior, discussed in detail in Chapter 22 but relevant here for possible insights into the nature of Io's atmosphere. However, it was the Voyager observations that revealed Io as a completely new class of object in the solar system, a highly active planet with a very young surface and a dynamic, tangible atmosphere. The study of the atmosphere is of fundamental importance in the attempt to understand the current structure and state of evolution of Io as well as its relationship to the plasma torus and magnetosphere of Jupiter. For pre-Voyager reviews on this subject, the reader is referred to Brown and Yung (1976), Hunten (1976), Fanale et al. (1977), and Johnson (1978). The entirely different picture of Io and its environment that has evolved from the Voyager observations has required major revision of previous theories. In this chapter we present a review of the current state of observational (primarily Voyager) and theoretical knowledge on satellite atmospheres and their escape and interaction with Jupiter's magnetosphere. The emphasis is naturally on Io, but a brief discussion is included of possible oxygen atmospheres, and the corresponding ionospheres, on the other Galilean satellites.

I. THE ATMOSPHERE OF IO: OBSERVATIONAL EVIDENCE

A number of observations are now available that provide evidence for the presence of a bound atmosphere on Io as well as clues to the composition and structure of this atmosphere. However, at the time of this writing there is considerable debate on this subject, and the point of view presented here reflects our prejudices. Somewhat different and, in some respects, opposing points of view can be found in Chapter 20 by Fanale et al. We begin by summarizing the observational evidence; the constituents observed in the vicinity of Io, namely in its atmosphere, ionosphere and the associated plasma torus, are listed in Table 21.1.

Voyager IRIS Detection of Gaseous SO_2

The most direct evidence of the composition of the atmosphere is the detection of gaseous SO_2 on Io by the Voyager infrared interferometer spectrometer (IRIS) (Pearl et al. 1979). The observed spectrum of Io between 1000 and 1500 cm^{-1} is shown in Fig. 21.1, where it is compared to a synthetic SO_2 spectrum. The feature between 1320 and 1380 cm^{-1} is identified as the ν_3 band of SO_2. This spectrum is the average of seven spectra obtained over a region near the equator (at 13 hr local time) that contained three features: a plume associated with the volcano Loki, a warm region at $\sim$290 K which covered roughly 10% of the instrument's field of view, and a remaining background area at $\sim$130 K (Hanel et al. 1979). The observed SO_2 could originate from any or all of these sources; however, Pearl et al. (1979) concluded that an ambient atmosphere is the most likely source, and derived

TABLE 21.1

Constituents Detected in the Vicinity of Io

Experiments	Measurements Region	Constituent	Density (cm^{-3})	References
Voyager IRIS	Ambient atmosphere above a hot spot	SO_2	5×10^{12} (surface)	Pearl et al. (1979)
Pioneer 10 radio occultation	Pre-sunset ionosphere	e	6×10^4	Kliore et al. (1974, 1975)
	Pre-sunrise ionosphere	e	9×10^3	
Voyager ultraviolet spectrometer	Plasma torus remote sensing	S^{++}	95	Broadfoot et al. (1979, 1981)
		S^{+++}	55	
		O^{++}	850	
		O^{++}		
Voyager plasma instrument	Plasma torus $5.3\ R_J < r < 6.3\ R_J$ in situ	S^{++} or O^+	Total ~ 2000	Bridge et al. (1979) Bagenal and Sullivan (1981)
		S^+ or O_2^+		
		S_2^+ or SO_2^+		
Voyager planetary radio-astronomy	Plasma torus in situ	e	2000–4000	Warwick et al. (1979)
Voyager imaging	Volcanic plumes	Gas and dust		Smith et al. (1979a), Collins (1981)
	Surface of Io	Solid SO_2		Fanale et al. (1979), Hapke (1979) Smythe et al. (1979), Cruikshank (1980)
Groundbased telescopes	Cloud around Io	Na		Brown and Yung (1976)
	Plasma torus near Io	S^+	10	Kupo et al. (1976)
Pioneer 10 ultraviolet experiment instrument	Incomplete torus at Io's orbit	H		Carlson et al. (1975) Judge and Carlson (1974)

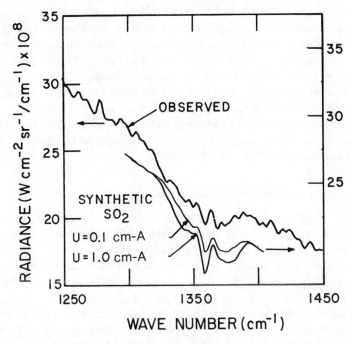

Fig. 21.1. Comparison of the Io spectrum obtained from Voyager IRIS and synthetic spectra in the vicinity of the ν_3 band of SO_2. The synthetic spectra represent absorption by homogeneous layers of 0.1 and 1 cm-A of pure SO_2 gas at 130 K and pressures of 2.6×10^{-8} and 2.6×10^{-7} bar, respectively. The measured and calculated spectra are vertically displaced for clarity. (From Pearl et al. 1979.)

an atmospheric SO_2 abundance of 0.2 cm-A with an estimated probable error of a factor of two. This corresponds to an SO_2 surface density of 5×10^{12} cm^{-3} or a surface pressure of 1×10^{-7} bar, which is in excellent agreement with the vapor pressure of SO_2 at 130 K (1.4×10^{-7} bar) as determined by Wagman (1979).

A recent analysis of spatial color variations in the volcanic plume Loki on Io by Collins (1981) shows that the ultraviolet-bright halo around the plume must consist of $\lesssim 0.1$ μm size particles which are inherently poor scatterers; hence a large number of them are needed to account for the observed brightness. This leads to a factor of 10^4 larger particulate mass in the Loki plume than was estimated by Johnson et al. (1979). If the gas-to-particulate ratio in the plume is close to one, and if the gas expands without condensing quickly (say within 50 m), then the Voyager IRIS observations of gaseous SO_2 would be consistent with the observed gas being transient flow from Loki as well as with an equilibrium atmosphere. However, no SO_2 gas was observed by the IRIS above Pele (J. Pearl, personal communication), which is even hotter (T_s = 700 K) than Loki (T_s = 290 K) and would provide greater thermal contrast for observing SO_2 if it is entirely associated with the plume. The Pele observation is consistent with an equilibrium atmosphere because the equilibrium vapor pressure at the local time (16.4 hr) of this observation is too low to be detected by the Voyager IRIS instrument. In fact, SO_2 was not detected in association with any of the other hot spots or plumes that were observed by IRIS (J. Pearl, personal communication), but all of these observations corresponded to evening or nightside regions where the ambient surface temperature is too small to sustain detectable amounts of SO_2 gas in vapor pressure equilibrium with the surface.

Upper Limits on various Molecular Constituents

The Voyager IRIS also provided upper limits on various other constituents in the "hot spot" region (Table 21.2), indicating the severe lack of other major volatiles on Io. The limits for CH_4, NH_3, N_2S, and N_2O are improvements over previous determinations by Fink et al. (1976) and for CO_2 and H_2O are improvements on values obtained by Bartholdi and Owen (1972). For all of the gases listed in Table 21.2, except H_2O and SO_3, the limits correspond to pressures far below the respective vapor pressures at 130 K. Hence it can be concluded that Io is deficient in these materials. The vapor pressures of H_2O and SO_3 at 130 K place these gases substantially below the IRIS detection threshold, so the present limits only imply that there are not strong nonequilibrium sources of these gases. The absence of near-infrared absorption features of H_2O still provides the best upper limit for this constituent (Pollack et al. 1978).

Pioneer 10 Detection of an Ionosphere

The Pioneer 10 radio occultation experiment provided measurements of a well-developed ionosphere on Io (Kliore et al. 1974, 1975) at both the dawn (5.5 hr) and dusk (17.5 hr) terminators (Fig. 21.2), with peak electron densities of 6×10^4 cm^{-3} and 1×10^3 cm^{-3}, respectively. In comparison, the peak electron densities in the ionospheres of Earth, Venus and Mars are 1×10^6 cm^{-3}, 5×10^5 cm^{-3} and 1×10^5 cm^{-3}, respectively. Significant attention has been given to understanding the source and the maintenance of Io's ionosphere, and a number of models have been proposed (McElroy and Yung 1975; Whitten et al. 1975; Gross and Ramanathan 1976; Johnson et al. 1976; Cloutier et al 1978; Kumar 1980). These models will be discussed in a separate section below; however, it is important to point out that each of these models requires a neutral atmosphere with an atmospheric surface density ranging from 10^9 to 10^{11} cm^{-3} for a molecular atmosphere and $\sim 10^6$ to 10^7 cm^{-3} for an atomic sodium atmosphere. As we discuss below, the

TABLE 21.2

Upper Limits for Various Gases[a]

Gas	Band Used (cm⁻¹)	Vapor Pressure Abundance (cm-A) above Solid at 130 K	Upper Limit (cm-A) for T_{gas} = 130 K	Upper Limit (cm-A) for T_{gas} = 250 K	Mixing Ratio Relative to SO_2
COS	V_1 859	510[b]	3.8×10^{-4}	1.5×10^{-4}	$< 1.9 \times 10^{-3}$
CS_2	V_3 1535	1.4×10^{-1} b	3.5×10^{-5}	2.8×10^{-5}	$< 3.3 \times 10^{-4}$
SO_3	V_2 497	1.6×10^{-6} b	6.5×10^{-5}	2.4×10^{-5}	
H_2S	—	56[c]	7×10^{-2} c		0.35
CO_2	V_2 667	1050	1.5×10^{-4}	5.1×10^{-5}	$< 7.5 \times 10^{-4}$
O_3	V_1 1042	d	1.9×10^{-3}	5.1×10^{-3}	
N_2O	V_1 589	4780	7.4×10^{-3}	2.2×10^{-3}	
H_2O	ROT 254	6.4×10^{-7} b	9.2×10^{-3}	1.9×10^{-4}	$< 4.6 \times 10^{-2}$
CH_4	V_4 1306	d	1.0×10^{-2}	1.7×10^{-3}	0.5
NH_3	V_2 931	11	1.4×10^{-3}	3.4×10^{-4}	
HCl	ROT 206	1.8×10^{-4}	3.8×10^{-3}	3.7×10^{-4}	$< 7 \times 10^{-3}$

[a]From Pearl et al. (1979).
[b]Basic data extrapolated from Hodgman et al. (1956).
[c]From stellar occulation using T_{gas} ñ 110 K (see text).
[d]Above melting point.

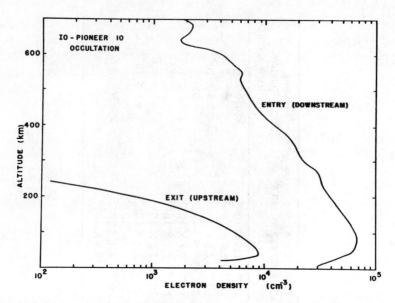

Fig. 21.2. Ionospheric electron density profiles obtained from the Pioneer 10 radio occultation. The entry occultation point was above 17.5 hr local time and exit occultation was above 5.5 hr local time. Upstream site faces Jupiter's corotating plasma at Io's orbit. (From Cloutier et al. 1978 and based on data from Kliore et al. 1974, 1975.)

recent Voyager observations favor a molecular atmosphere with surface densities of 10^{11} cm^{-3} or greater on the day side in order to explain the ionosphere observed from Pioneer 10.

Plasma Torus Observations

The Io plasma torus observations are summarized here with respect to their relevance to an atmosphere on Io; the reader is referred to Chapter 22 by Pilcher and Strobel and Chapter 23 by Sullivan and Siscoe for further details. Recent Voyager observations with the ultraviolet spectrometer (Broadfoot et al. 1979, 1981) and the plasma science instrument (Bridge et al. 1979; Bagenal and Sullivan 1981) have provided substantial information on the morphology and ion composition in the plasma torus that is observed near Io's orbit; important features of these observations are the identification of ions S^{++} (or O^+), S^{+++}, O^{++}, S^+ (or O_2^+), SO_2^+ (or S_2^+) with a total ion density of ~ 2000 cm^{-3}, mean electron energy of 10 eV, and mean ion energy of $30-40$ eV. Ions of nitrogen and carbon were below the detection threshold of ~ 10 cm^{-3} (also see Vogt et al. 1979). S^+ was observed in the vicinity of Io from groundbased observations of S^+ 673nm emission (Kupo et al. 1976) even before the Voyager encounter; it is now known to be a plasma torus ion and is being used as a tracer of the plasma torus for studying temporal variation in plasma torus structure (Pilcher 1980; see also Chapter 23 by Sullivan and Siscoe). O^+ has also been identified in the inner Jovian magnetosphere from groundbased observations (Pilcher and Morgan 1979). The groundbased observations from 1976 to date, the Voyager measurements, and the Pioneer 10 measurements provide sufficient evidence of large fluctuations in the plasma torus density, temperature and composition (Pilcher and Strobel, Chapter 22).

The plasma torus is most likely being populated by gases originating from Io and possibly from Io's atmosphere (Kumar 1979). The observed ion temperature is much lower than the corotation value of ~400 eV, which indicates substantial mass loading of the torus as well as cooling by the ultraviolet emissions. The velocity of the material ejected from volcanoes is $\leqslant 1$ km s^{-1} (Smith et al. 1979b), comfortably less than the escape velocity of 2.56 km s^{-1} on Io. Hence any direct escape is negligible. Since the resurfacing rate due to the dust ejected from the volcanoes is at least 10^{-3} m yr^{-1} (Johnson et al. 1979; also see Chapter 17), comparable amounts of gas should be needed to drive the dust out of the volcanoes. This leads to a gas outflux of 10^{14} molecules cm^{-2} s^{-1} averaged over the globe. In comparison, a supply rate to the plasma torus of only 1.0×10^{11} cm^{-2} s^{-1} (or 2×10^{28} s^{-1} from Io), for ions of S and O, is needed (Broadfoot et al. 1981) if ion corotation energy is the primary source for heating the torus electrons. An upper limit to the total gas supply rate can be estimated by requiring that the observed maximum plasma density of 4000 cm^{-3} (Warwick et al. 1979) be maintained over the diffusion time of ions out of a 2 R$_J$ thick plasma torus. Eviatar et al. (1979) have estimated a diffusion time of 20 days to drift 1 R$_J$ in a radial sense at Io's orbit, leading to a supply rate of ~10^{11} cm^{-2} s^{-1}. The observed mass loading in the torus could require a supply rate as high as 10^{12} cm^{-2} s^{-1} or 2×10^{29} atoms per second released from Io (Sullivan and Siscoe, Chapter 23). On the other hand, Shemansky (1980) argues for long diffusion times of ~100 days in the torus, which imply a much lower supply rate of only ~ 10^{10} cm^{-2} s^{-1} or 2×10^{27} atoms per second from Io. The maximum estimated escape is at least two orders of magnitude less than the gas outflux from the volcanoes. Evidently only a small fraction of what is coming out of Io's volcanoes is escaping; the bulk of it is being retained. The possibility that these gases are retained in the form of an atmosphere is developed below in the section on atmospheric escape.

Sodium Cloud Around Io

Observations of the sodium cloud near Io are reviewed in Chapter 22 by Pilcher and Strobel. For the present purpose we stress four salient results:

1. The velocity of sodium once it is free from Io's gravitational field is typically 2.6 km s^{-1};
2. The observed atoms come mostly from the inside leading quadrant of Io;
3. The source strength is 2 × 10^{25} s^{-1}, or 10^8 cm^{-2} s^{-1} referred to a hemisphere;
4. The cloud is strikingly symmetrical between eastern and western elongations (Brown and Yung 1976; Bergstralh et al. 1977; Smyth and McElroy 1978; Murcray and Goody 1978; Carlson et al. 1978; Matson et al. 1978; Macy and Trafton 1980).

There may be much more sodium close to Io, but, if present, it has so far not been observed. Other unobserved atoms may arise from the outside hemisphere of Io, but be rapidly destroyed by the hot plasma torus.

Pioneer 10 Ultraviolet Detection of an Incomplete Torus

The Pioneer 10 ultraviolet instrument had detected an emission originating from an incomplete torus around Io and at Io's orbit, which was attributed to the Lyman α emission from atomic hydrogen (Judge and Carlson 1974; Carlson and Judge 1975). This prompted the proposal of NH$_3$ as the likely gas in Io's atmosphere to provide the source of H to the torus (McElroy and Yung 1975). However, the Voyager ultraviolet spectrometer, sensitive to the Lyman α emission, did not detect any hydrogen emission associated with the Io torus (Broadfoot et al. 1979). The H$^+$ measurement is outside the mass range to which the Voyager plasma instrument is sensitive, and, therefore, we have no direct measurement of thermal protons in the plasma torus. However, the total ion density (~ 2000 cm^{-3}) measured by the Voyager plasma instrument in the torus is within 10% of the electron densities derived for the same location by the Voyager plasma wave instrument (Warwick et al. 1979), suggesting the presence of a very low, if any, thermal proton density. At present there is no satisfactory explanation of the difference between the Pioneer 10 and Voyager ultraviolet observations concerning the Lyman α emission.

II. IONOSPHERE MODELS AND AERONOMY OF SULFUR DIOXIDE ON IO

The maintenance of an ionosphere on Io has been puzzling due to the relatively strong ram pressure of Jupiter's corotating plasma at Io's orbit, which would sweep away any gravitationally bound ionosphere (Cloutier et al. 1978). To circumvent this problem Cloutier et al. proposed Birkeland currents connecting Jupiter's ionosphere to Io as the source that maintains Io's ionosphere. Alternatively, Kivelson et al. (1979) proposed that if Io has an intrinsic magnetic field, the formation of a magnetosphere could provide the necessary shielding to the atmosphere. This postulate of a magnetosphere has been challenged by N. Ness (personal communication; Acuña et al. 1981) who argues that, since the Alfvén Mach number is only 0.15 for the corotating plasma, making the flow sub-Alfvénic by a factor of six, the field lines of Jupiter and any possible magnetic field of Io would connect easily, and shielding provided by a magnetopause system of currents would be lacking. Kivelson (personal communication) maintains that shielding currents cannot be ruled out by the available information. No such shielding is required, however, if Io has a bound molecular atmosphere thick enough to sustain the observed ionosphere. Such an atmosphere may be maintained by volcanic activity (Kumar 1979). We shall discuss the atmospheric stability and escape in detail in a later section. Here we assume that a bound atmosphere is possible and discuss the models of the ionosphere, which in turn provide necessary information regarding the structure of Io's atmosphere.

Pre-Voyager models of Io's ionosphere have assumed the presence of Na (McElroy and Yung 1975), NH_3 (McElroy and Yung 1975; Gross and Ramanathan 1976; Johnson et al. 1976; Cloutier et al. 1978), N_2 (Gross and Ramanathan 1976), and Ne (Whitten et al. 1975). Most of these models are unacceptable in light of the Voyager data on the atmospheric composition, although certain features of some of them are still valid. The Voyager IRIS upper limit on NH_3 is too low to sustain the observed ionosphere, and ions of nitrogen are conspicuously absent in the plasma torus (Bridge et al. 1979; Broadfoot et al. 1979). An ionosphere of Ne^+ can be ruled out since this ion will charge exchange with SO_2 and be lost rapidly. The sources of ionization have been discussed in detail by McElroy and Yung (1975) who have shown that models based on photoionization require rather large molecular abundances and that a slowly recombining atomic ion such as Na^+ could be a possible candidate. They also put forward a version in which the major source of ionization is soft-electron impact. Johnson et al. (1976) have proposed the latter source to minimize the required atmospheric abundance and permit the escape of sputtered atoms. The difficulty with photoionization as the major source for the ionosphere is readily seen by comparing the ionization and collision cross sections (10^{-17} and 2×10^{-15} cm^2, respectively). Any atmosphere with unit optical depth to ionizing photons has a depth of around 200 mean free paths to sputtered atoms. Johnson et al. (1976) have adopted 100 eV electron fluxes of 3×10^9 and 3×10^7 cm^{-2} s^{-1} for the two sides of Io similar to those observed by Pioneer 10 (Frank et al. 1976). The peak electron densities observed by the occulation experiment (Kliore et al. 1975 and their Figure 2b) were well reproduced. The large flux is for entry (day side or downstream) and the small for exit (night side or upstream).

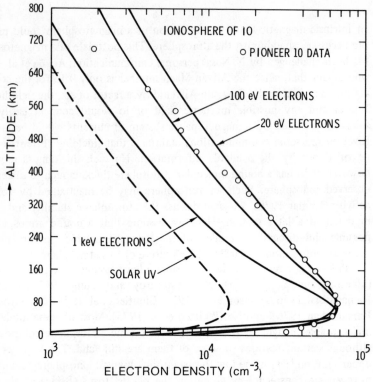

Fig. 21.3. Model calculations for the downstream ionosphere of Io measured by the Pioneer 10 radio occultation (entrance) for a 100% SO_2 atmosphere. Ionization due to solar ultraviolet alone is inadequate. Ionospheric profiles resulting from precipitation of monoenergetic electron beams at 20 eV, 100 eV and 1 keV are shown. (From Kumar 1980.)

The Voyager identification of gaseous SO_2 on Io has prompted Kumar (1980) to build an ionospheric model for a pure SO_2 atmosphere. His model for the ionosphere at the evening terminator is shown in Fig. 21.3 and the corresponding SO_2 atmospheric thermal structure in Fig. 21.4. If atmospheric SO_2 is in thermal equilibrium with SO_2 at the surface, the vapor pressure would vary from $\sim 10^{-6}$ bar at the noon temperature of 130 K to $\sim 10^{-9}$ bar at the terminator temperature of 110 K (Fig. 21.5); these surface temperatures are based on Voyager IRIS measurements (Pearl et al. 1979). The SO_2 density at the surface would therefore be 10^{11} to 10^{13} cm^{-3} on the day side. The solar ultraviolet would heat the SO_2 atmosphere by photodissociation and ionization (rates are taken from Kumar 1980):

$$hv \ (\lambda < 221 \ nm) + SO_2 \rightarrow SO + O \ , \qquad J = 5.6 \times 10^{-6} \ s^{-1}$$

$$hv \ (\lambda < 207 \ nm) + SO_2 \rightarrow S + O_2 \ , \qquad J = 2.9 \times 10^{-6} \ s^{-1}$$

$$hv \ (\lambda < 100 \ nm) + SO_2 \rightarrow SO_2^+ + e \ , \qquad J = 4.8 \times 10^{-8} \ s^{-1}$$

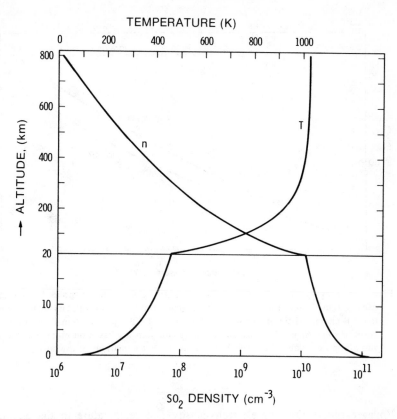

Fig. 21.4. Atmospheric SO_2 density and temperature profiles corresponding to the ionosphere model shown in Fig. 21.3 for 20 eV electrons. (From Kumar 1980.)

Some heating would also result from charged particle precipitation. The SO_2 atmosphere would cool by infrared radiation in the ν_1 (1100–1200 cm^{-1}), ν_2 (480–560 cm^{-1}), and ν_3 (1320–1390 cm^{-1}) bands. Kumar (1980) showed that the solutions of the thermosphere heat equation lead to an exospheric temperature in the 800 - 1200 K range for an SO_2 atmosphere in thermal equilibrium with the surface. If SO_2 is the major gas in the atmosphere, the observed ionospheric scale height for SO_2^+ ions would imply an exospheric temperature of $\sim$ 1000 K, which is consistent with the thermal calculations above. The solar ultraviolet, however, fails to produce the necessary ionization by a factor of four, but ionization by low-energy electrons shows promise. Figure 21.3 shows the calculations for 20-eV, 100-eV and 1-keV electrons; the required electron flux and the SO_2 surface density are shown in Table 21.3. It is evident that the best fit to the ionospheric profile is obtained with the 20 eV electrons. The required flux is consistent with the

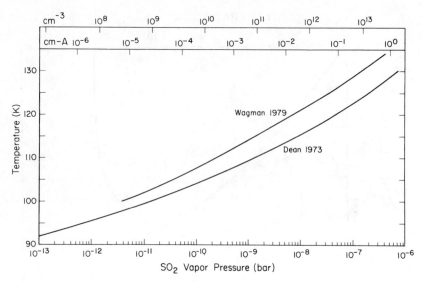

Fig. 21.5. SO_2 vapor pressure versus temperature. Corresponding scales for surface density (cm^{-3}) and atmospheric column (cm-A) are also shown. The temperature range is characteristic of Io's surface. The Dean (1973) curve is from *Lange's Handbook of Chemistry*. The Wagman (1979) calculations are based on more recent data and are more reliable; however, both curves are extrapolations of available measurements.

Voyager observations of electron density and temperature in the plasma torus; the measured density of 2000 cm^{-3} of 10 eV electrons corresponds to a flux of 4 × 10^{11} cm^{-2} s^{-1} in the vicinity of Io. Since the electrons are hot, the electron flux incident on Io is determined by the thermal velocity (~2000 km s^{-1}) rather than the bulk plasma corotation velocity (~57 km s^{-1}). As seen from Io, the flow of electrons would be along the electric field, which is in the **v** × **B** direction where **v** is the corotation velocity and **B** is the corotating magnetic field vector. The exospheric temperature for the 20 eV case is 1030 K. The electron heating is insignificant compared to the solar ultraviolet heating; peak heating rate (at the level of unit optical depth) H_{UV} = 6.1 × 10^{-7} ergs cm^{-3} s^{-1}, and H_e = 4.1 × 10^{-9} ergs cm^{-3} s^{-1}. The large ultraviolet heating is primarily from rapid photodissociation of SO_2. At higher electron energies, the absorption cross section for ionization is lower, hence a greater SO_2 density is needed to match the altitude of the ionospheric peak, but the ionospheric peak itself is narrower and the fit is poorer. The required SO_2 surface density for the 20 eV case is reasonable for an atmosphere controlled by vapor pressure equilibrium, which will be discussed in the following section on global SO_2 distribution.

TABLE 21.3

Ionization by Low-Energy SO_2 Electrons

Electron Energy eV	Electron Flux ϕ (cm^{-2} s^{-1})	Surface Density N_0 (SO_2) (cm^{-3})
20	1.2×10^{10}	1.2×10^{11}
100	2.0×10^9	4.0×10^{11}
1000	1.2×10^8	6.0×10^{12}

The ionosphere above the morning terminator measured during the Pioneer 10 exit occultation (see Fig. 21.2) can also be understood in terms of a pure SO_2 atmosphere. The surface temperature at 5.5 hr, the location where this ionosphere was observed, is <95 K as indicated by Voyager IRIS measurements (J. Pearl, personal communication). The vapor pressure of SO_2 at this temperature is extremely low and would account for a surface density of $< 10^8$ cm^{-3} in thermal equilibrium (Fig. 21.5). The actual surface density, however, must be much higher because on the day side of the morning terminator, where the surface temperature rises sharply, the SO_2 density should be on the order of 10^{11} cm^{-3}, or comparable to that near the evening terminator. Since the shadow height for this ionosphere observation is only 25 km, and the ionospheric peak is just above this altitude, solar ultraviolet alone could produce the necessary ionization, with an SO_2 surface density of 5×10^{10} cm^{-3}. Such high density at 5.5 hr local time could be maintained by transport of SO_2 from the day side of the morning terminator (Yung and Summers 1980). However, the ionospheric scale height of only 25 km is much smaller than that expected from solar ultraviolet heating, possibly resulting from a compression of the ionosphere by Jupiter's corotating plasma as suggested for Mars and Venus (Cloutier et al. 1969; Bauer and Hartle 1974; Taylor et al. 1979) due to the solar wind interaction. This approach was suggested for Io by Cloutier et al. (1978) in the case of an NH_3 atmosphere, but it is equally valid for an SO_2 atmosphere, for which a reduction in the ionospheric scale height by a factor of three will provide a good fit to the data.

If, as seems likely, the atmosphere is rich in sodium atoms, there will be a strong tendency for charge exchange, producing sodium ions. These ions, with their very slow recombination, may help explain the high electron density even in the absence of precipitating electrons (McElroy and Yung 1975). The chemistry of metallic ions is complex (Ferguson 1972; Murad

1978). In the Earth's D region, rapidly recombining ions such as NaO^+, and NaO_2^+ may form, but they are even more rapidly destroyed by reaction with O atoms, with recycling of the Na^+. Photochemical models, described in the next paragraph, predict O densities as high as 10^9 cm^{-3} near the ionospheric peak. Such a density would destroy the oxide ions, as on Earth; a density as low as 10^7 cm^{-3} would permit most of them to recombine. SO and SO_2 present other possibilities for formation of molecular complexes with Na^+, but information is lacking.

Photolysis of SO_2 produces SO, S, O, and O_2. To understand their relative importance in Io's atmosphere, we have examined their chemical reactions; the photochemical scheme including the major reactions is illustrated in Fig. 21.6. The large photolysis rates lead to a maximum SO_2 loss rate of 10^{12} cm^{-2} s^{-1}, which is realized for SO_2 surface density N_0 $(SO_2) \gtrsim 4 \times 10^{11}$ cm^{-3}. For the atmospheric model of Fig. 21.4, the SO_2 loss rate is 2×10^{11} cm^{-2} s^{-1}. Two cases can be examined, corresponding to a chemically passive or active surface. In the passive case, there is no chemical sink of O (except for three-body reactions which are negligible), and S is lost by its reaction with O_2 but is also rapidly produced from recycling of SO, resulting in very little chemical loss of S. However, SO and O_2 recycle quickly to form SO_2 by their reactions with SO; in this case O and S would be major species above ~ 150 km altitude and SO and O_2 would be minor constituents in the atmosphere. In the active case, the surface will provide an efficient sink of S and O; S could form sulfur allotropes but O could be returned to the atmosphere in the form of O_2; this could conceivably lead to an O_2 atmosphere with an abundance comparable to that of SO_2. This case is particularly interesting since the vapor pressure of O_2 is much higher than that of SO_2 at 90-100 K (characteristic of the night side), and therefore the

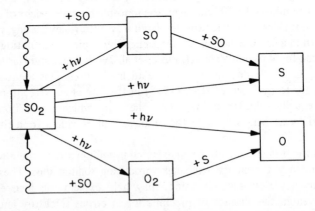

Fig. 21.6. A schematic representation of SO_2 photochemistry.

nightside atmosphere will be dominated by O_2. Io may have a global O_2 atmosphere.

III. SULFUR DIOXIDE IN IO'S ATMOSPHERE AND SURFACE

Chapter 20, by Fanale et al., describes in some detail the evidence for and against the presence of a global layer or film of SO_2 frost, together with an atmosphere in vapor pressure equilibrium. They tend to stress the negative side of the case, whereas we are more impressed by the positive side. The strong 4-μm absorption is most simply explained by SO_2 frost deposits covering 25 - 50% of the surface (Fanale et al. 1979; Smythe et al. 1979; Hapke 1979; Cruikshank 1980). Fanale et al. discuss an alternate hypothesis, that the absorbing SO_2 is dissolved or occluded in the sulfur that covers most of the surface.

The ionospheric model shown in Fig. 21.3 and 21.4 requires a surface density of 1.2×10^{11} cm^{-3}. The expected SO_2 densities in the terminator regions are so close that they suggest a physical connection. The surface temperature quoted by Pearl et al. (1979) based on Voyager IRIS measurements, is $\leqslant 110$ K for the evening terminator and could be as low as 105 K. Vapor pressure equilibrium would predict SO_2 surface densities of 9×10^{10} cm^{-3} and 1×10^{10} cm^{-3} (Fig. 21.5; Dean 1973), respectively, for the two extremes; the higher value is consistent with the ionosphere model. Yung and Summers (1980) have proposed that a flow resulting from the vapor pressure gradient from subsolar to terminator regions will deplete the subsolar densities by a factor of two and will enhance the terminator densities by a factor of ten, but they use 105 K for the terminator temperature giving an increased SO_2 surface density in their model of 3×10^{11} cm^{-3}.

Figure 21.5 shows the vapor density of SO_2 according to a handbook (Dean 1973) and to Wagman (1979). In addition to the obvious sensitivity to temperature, the two extrapolations differ by a factor of five or so. Yung and Summers used the Wagman estimate, and Kumar the Dean version. Thus, both estimates are compatible within their uncertainties and with the requirement of the ionospheric model. It is tempting to regard this as more than a mere coincidence.

J. Pearl (personal communication) has constructed a global model of SO_2 vapor pressure based on the IRIS temperature measurements. The average abundance for the sunlit hemisphere is 0.032 cm-A. Similar models are presented in Fig. 20.6 and 20.8 (Chapter 20); the abundances are 0.04 and 0.002 cm-A. For Fig. 20.8 the temperatures were lowered by 8 K to represent the white areas, whose high albedo reduces the solar energy input. Such colder patches, if large enough, would be expected to attract all the SO_2 frost and control the vapor pressure. These illustrations emphasize the crucial importance of the surface temperature.

An upper limit of 0.008 cm-A has been reported by Butterworth et al. (1980) from the absence of absorption in the 280 - 310 nm range in an IUE spectrum. Using similar data Lane et al. (1979) have reported an upper limit of 0.02 cm-A. The lower of these limits is a factor of four smaller than Pearl's estimate, but is compatible with Fig. 20.8 of the preceeding chapter. The validity of a straightforward interpretation of the spectrum has been questioned by Bertaux and Belton (1979); the data are for the entire disk of Io, but near the limbs the pressure is much lower than the mean and the absorption much weaker. Averaging over the disk, with allowance for this effect, they find that there is no conflict between the results of Pearl and Butterworth et al.

A far more serious argument against an SO_2 atmosphere is the absence of white polar caps, as emphasized in Chapter 20. Flow of SO_2 away from the noon bulge should carry it to the poles at a rate that is not plausibly counteracted by sputtering. A sufficient quantity of a noncondensing background gas could greatly impede this flow, but any such gas must be continuously supplied to the atmosphere if the lifetime against loss to the torus is as short as a few days or even weeks (see the discussion below). The obvious candidate is therefore O_2, which must certainly be present in some abundance as part of the SO_2 photochemical system. If its pressure were comparable to the SO_2 pressure at $30° - 60°$ from the subsolar point, the rapid outflow would be replaced by a more normal meteorological system, and the flux to the poles would drop drastically. Much work is necessary to see if these suggestions can be worked into a plausible model. If not, it may be necessary to abandon the model of an SO_2 atmosphere, but for the present we regard it still viable.

IV. IO: ATMOSPHERIC ESCAPE AND EVOLUTION

Since much of the matter (sodium atoms and ions of sodium, oxygen, and sulfur) in orbit near Io obviously came from that satellite, there must be at least one escape process that acts on neutral atoms. Additional processes have been discussed that are available only for ions, but the possibility remains that most of the actual escape is in the neutral form. Kumar (1979) has suggested that the corotating plasma may be able to ionize and sweep away all matter above a level which crudely represents a magnetopause, if Io has an intrinsic magnetic field (Kivelson et al. 1979), or an ionopause, if there is a cometary or Venus-like situation (Cloutier et al. 1969). In a study of the analogous situation with the solar wind, Michel (1971) has found that the swept flux is limited to a value that translates to $\sim 1/40$ of the incident flux, or 6×10^8 cm^{-2} s^{-1} for Io. This value is negligible compared with the 10^{10} to 10^{12} cm^{-2} s^{-1} required to populate the torus. Kumar (1979) therefore suggested that the Michel limit does not apply to Io. This limit stems from the small momentum carried by the incident plasma. Additional momentum might be available from the Jovian magnetic field, perhaps even to the extent that much of the incident energy could be applied to ionization and pickup. Kumar argued, in effect, that some such process must be operating because clearly the torus is being populated. This suggestion must be viewed with some caution, because, for instance, some other process is ejecting neutral sodium, and may also eject other atoms, molecules, and perhaps ions.

Assuming that the lower boundary is at 200 km, Kumar estimated $1/e$ loss times T_i varying from $1-2$ days for H to 4.9 days for S. For molecular constituents there is an additional delay of a few days, the photodissociation time. If the globally-averaged flux of SO_2, for example, is 10^{11} cm^{-2} s^{-1}, the column amount lost in 10 days is almost 10^{17} cm^{-2}, approximately the amount required to support the ionosphere. Thus, if fluxes of this order actually occur, the atmosphere and torus are replenished on a time scale of weeks.

Whether or not the sweeping mechanism operates for atmospheric gases, it is not applicable to sodium, which is ejected as neutral atoms with the considerable velocity of 2.6 km s^{-1} discussed above. These facts point strongly to the sputtering process of Matson et al. (1974). If this process operates for sodium, it should operate for other constituents as well. If the sputtering is from the surface, the abundance or column density of the atmosphere cannot exceed $1-2$ mean free paths. For a collision cross section of 2×10^{-5} cm^2, this limit is 10^{15} cm^{-2}, far less than the amount advocated here. This difficulty was recognized by Johnson et al. (1976) and led to their advocacy of a thin atmosphere with an ionosphere generated by electron impact. The post-Voyager model of Kumar (1979, 1980), though of this class, requires an abundance of 2×10^{17} cm^{-2}, two orders of magnitude too large to permit the escape of atoms sputtered at the surface.

Three possible solutions are suggested below. The first two have a common difficulty in that they are unsymmetrical between day and night. They should therefore introduce a marked asymmetry in the geometry of the sodium cloud between eastern and western apparitions, whereas the sodium cloud is observed to maintain its form to a remarkable degree. In fact, such a predicted asymmetry would appear to be a necessary consequence of any atmosphere with a large diurnal variation.

The first possiblity is that direct sputtering from the surface occurs only on the night side, where most of the SO_2 has condensed to the surface at the expected low temperature. For the second and third, surface sputtering must be assumed to raise sodium atoms into the atmosphere, where they remain in sufficient density for a separate escape process to operate. There is no doubt that some finite sodium density can be maintained in this way, but its actual value is unknown.

The temperatures near 1000 K (cf. Fig. 21.4) predicted for the SO_2 atmosphere encourage another look at thermal (or Jeans) escape, even though Brown and Yung (1976) stated that it would require an exospheric temperature of order 10,000 K. In contrast, we find that thermal escape should be important at 1000 K, where the escape parameter $\lambda = GMm/kTr$ is 9.0 for a sodium atom (cf. Hunten 1973). This is small enough to support an escape flux of up to 10^8 cm^{-2} s^{-1}. In addition, Jovian tidal forces should considerably enhance the escape through the inner and outer Lagrangian points, as illustrated in Fig. 20 of McElroy and Yung (1975). In contrast to the first mechanism, this one favors the day side. Perhaps its biggest difficulty, however, is that it does not yield the high-velocity atoms required to generate a cloud of the observed size (Murcray and Goody 1978; Smyth and McElroy 1978). Since at 1000 K the escaping atoms after escaping have a modal velocity of 0.6 km s^{-1} instead of 2.6 km s^{-1}, thermal escape may still be important for the innermost part of the cloud, and should give a substantial source of S and O as well.

Sputtering from the top of an atmosphere has been studied in some detail by Haff and Watson (1979). Although their specific results refer to protons and helium ions incident on CO_2, they remark that the heavy ions in the Jovian system would be much more effective. Indeed, their Eq. (9) shows that S^+ would be about 110 times more effective than H^+ at sputtering a sodium atom. For a crude estimate of the actual yield we scale from their 3%, which actually refers to protons incident on CO_2; the result is 3 sputtered atoms per incident S^+ ion. This astonishingly large yield stems from the fact that heavy ions deposit far more of their energy in the exosphere than do protons. Thus, if sputtering at the surface is able to maintain a substantial sodium density in the exosphere, ejection by another sputtering event seems highly plausible. It is well recognized that the anisotropy and high velocity of the emitted sodium can be readily accommodated by a sputtering mechanism. It remains to be seen if a substantial atmosphere can also be accommodated. The east-west symmetry of the sodium cloud could be preserved if there is enough O_2 in the atmosphere to suppress most of the diurnal pressure variation.

Evolution of Io

Io has been in a continuous state of evolution due to large escape of gases from its atmosphere. Strong tidal heating of Io's interior (Peale et al. 1979) implies that fresh material has been brought to Io's surface by convective currents, so that depletion of gases in its atmosphere must represent depletion of material stored in Io's interior. The upper limits on molecular constituents listed in Table 21.2, the conspicuous absence of N (<10% of 0) in the plasma torus (Bridge et al. 1979), and the absence of energetic nuclei of both C (<3% of 0) and N (<7% of 0) in the Jovian magnetosphere near Io (Vogt et al.

1979), require that any evolutionary model must allow depletion of elements H, C and N, but must protect sulfur and specifically SO_2. As a starting point, it is instructive to consider the composition of gases ejected by the Earth's volcanoes. Rubey's (1951) inventory of volcanic gases from Kilauea and Mauna Loa includes the following median values of volume percentages: H_2O (73.5), CO_2 (11.8), N_2 (4.7), SO_2 (6.4), SO_3 (2.3), S_2 (0.2), H_2 (0.4), CO (0.5), Cl_2 (0.05), and A (0.2). Io may have had a similar composition of gases coming out of its volcanoes in early geologic history.

Evolutionary models have been published by Kumar (1979), Hapke (1979), and Consolmagno (1979). The Kumar model, discussed in the previous sections, provides a mechanism for differential loss of gases operating continuously over geologic time. In this model lighter gases escape more rapidly than the heavier ones as a consequence of diffusive separation of gases in the atmosphere. This could explain the complete absence of H_2O and N_2, and the depletion of CO_2 relative to SO_2. Most likely, SO_3, SO_2, and Ar have also been retained, and some CO_2 may still be present as a minor constituent if Io started with Earth-like composition.

Hapke (1979) suggests that, due to the intense tidal heating, the entire body rapidly outgassed and all the volatiles were lost except S, which was protected from loss by being incorporated as iron sulfide in the outer parts of the core. According to this model SO_2 gas is generated in volcanic eruptions by interaction of reactive allotropes of S in the surface deposits with the O that is produced by partial dissociation of oxides in the silicate melts near the mouths of the volcanoes. This model also proposes formation of H_2S by interaction of magnetospheric protons with the surface S and formation of H_2S frost on the surface. No H_2S has been detected; however, since the model does not provide any quantitative estimates of H_2S production, it is difficult to say whether or not it is consistent with the observations.

Consolmagno (1979) proposed that if Io started with the composition of C1 carbonaceous chondrites (20% water, 5% carbon), the entire inventory of H_2O and CO_2 will have escaped due to widespread volcanism if an average eruption rate of 10^8 g s^{-1} was maintained over the age of the solar system. Noting that this rate is of the same order as that from a typical terrestrial volcano, Consolmagno suggests that once the entire planet was outgassed for the first time, only S_2 will remain and serve as the driving mechanism for the present volcanoes. This model was published before the discovery of gaseous SO_2 on Io, and did not consider how SO_2 would remain in the present atmosphere of Io.

V. EUROPA, GANYMEDE, CALLISTO: OXYGEN ATMOSPHERES?

Yung and McElroy (1977) pointed out that any large satellite with a sufficient amount of water vapor would be expected to build up a substantial atmosphere of O_2. They applied this idea specifically to Ganymede, and found O_2 surface pressures in the range 0.4–1.0 microbar. Photolysis of H_2O yields H_2 and O_2, as well as ozone and a number of radicals; the hydrogen escapes rapidly, and the O_2 more slowly (by a nonthermal process). As the O_2 builds up, it shields the H_2O more and more from solar ultraviolet (the "Urey self-shielding process") until the production and loss rates of O_2 are in balance.

A stellar occultation by Ganymede observed by Carlson et al. (1973) gave a suggestion of an atmosphere with a surface pressure in the same range, $\sim$ 1 microbar; theory and observation thus seemed in agreement. To obtain further information, the ultraviolet spectrometer on Voyager 1 was targeted to observe another occultation by Ganymede, this time in the wavelength range 91–170nm (Broadfoot et al. 1979). O_2 has an absorption cross section greater than 10^{-17} cm^2 from 135 to 150 nm. The absence (at the 10% level) of any detected absorption above the limb puts an upper limit of 6×10^8 cm^{-3} on the O_2 number density at the surface; the corresponding pressure would be 10^{-5} microbar. A similar limit applies to H_2O, CO_2, and CH_4. If a gas is really present at the microbar level, it must be one that absorbs strongly only below the Lyman limit; N_2 or any noble gases would qualify.

With minor modifications, the Yung-McElroy model should apply to any Galilean satellite with exposed surface ice. Why then does the model break down for Ganymede, and what are the implications for Europa and Callisto? As discussed below, the system actually has two possible states, differing in O_2 pressure by a factor of $\sim 10^5$. Yung and McElroy chose the high-pressure state, no doubt to fit the Carlson et al. occultation data; they did not mention the other state, except for a statement that their solution requires an average partial pressure of H_2O greater than $\sim 2 \times 10^{-9}$ microbar. Reducing this amount by a mere factor of two puts the system in the low state, with an O_2 number density, 8×10^7 cm^{-3}, a factor of six below the Voyager upper limit. In this regime, the O_2 and H_2O abundances are simply proportional.

We shall now explain briefly the mechanisms at work. Oxygen atoms require 0.63 eV to escape from Ganymede, an amount unlikely to be gained in thermal processes. Energetic atoms are generated in photolysis of O_2 at wavelengths well below threshold. One of the pair will be directed upward, and can escape if created above the exobase. The rate is readily calculated, and has an asymptotic value of 4.2×10^7 atoms cm^{-2} s^{-1} (global mean) if the O_2 column density $N(O_2)$ exceeds 10^{15} cm^{-2}. At $N(O_2) = 3 \times 10^{14}$ cm^{-2} the exobase coincides with the surface, and for smaller amounts the flux of O atoms is approximately $N(O_2)/3 \times 10^{14}$ times the asymptotic value. From Fig. 2 of Yung and McElroy, the generation rate is 3×10^8 atoms cm^{-2} s^{-1} for an H_2O pressure of 1×10^{-8} mbar and with no Urey self-shielding. Since this production exceeds the maximum possible loss, the O_2 must build up until the shielding reduces the source strength by a factor of seven, and the corresponding pressure is in the range of a few tenths of a microbar.

Reduction of the H_2O pressure by only a factor of ten puts the O generation rate (unshielded) at 3×10^7 cm^{-2} s^{-1}, which can be accommodated by the escape mechanism with ~ 0.7 of the full exospheric column density, or $N(O_2) = 2 \times 10^{14}$ cm^{-2}. The corresponding O_2 pressure is 1.5×10^{-6} microbar, which, as mentioned above, is comfortably below the Voyager limit. Yung and McElroy estimated the H_2O pressure from a surface temperature model and an extrapolated vapor pressure curve. The corresponding uncertainties by themselves could probably account for the required factor of ten; additional factors of two to three undoubtedly exist elsewhere in the calculation. One important input for the surface temperature is the albedo which could easily change enough from satellite to satellite to switch the atmosphere to the other regime. Callisto is thus the best candidate to have a microbar atmosphere. Even the same body could conceivably switch from time to time; the time scale is the ratio of the column density (in the dense state) to the escape flux, or nearly 10^5 yr.

Another potential source of H_2O vapor is sputtering by protons and other positive ions (Lanzerotti et al. 1978). A microbar atmosphere would stop protons less energetic than 1-2 MeV, but in the low state the atmosphere is not a significant barrier. Lanzerotti et al. calculated the erosion rate from Ganymede and found that it could be as large as 10^{10} molecules cm^{-2} s^{-1} and could support an H_2O partial pressure almost as large as the value used by Yung and McElroy (1977). The rates are an order of magnitude larger at Europa and a factor of 50 smaller at Callisto. Thus, even if sublimation is much weaker than estimated by Yung and McElroy, a significant oxygen atmosphere could be present on one or more satellites.

As Yung and McElroy mention, an O_2 atmosphere should have an ionosphere with a peak electron density of a few times 10^4 cm^{-3}. The level corresponds to an O_2 density of 4×10^{10} cm^{-3}, some 180 km above the surface for the dense Ganymede model. For the 8×10^7 cm^{-3} version, the atmosphere is optically thin to ionizing radiation, and the electron density near the surface would be $\sim 2 \times 10^3$ cm^{-3}. This value could be lowered significantly by downward flow of ions to the surface, but the magnitude of the effect is difficult to estimate. Similar values should apply to Europa and Callisto.

Soft electrons (~ 400 eV) penetrate to about the same depth as ionizing photons. The density required to match the solar flux in the Jovian system is about 0.5 cm^{-3}, which is met near Ganymede's orbit (Frank et al. 1976; Bridge et al. 1979). Electron impact should be negligible at Callisto, significant at Ganymede, and dominant at Europa, where it could increase the production rate by an order of magnitude.

Acknowledgments. We are pleased to acknowledge helpful discussions with M.J.S. Belton, A.L. Broadfoot, F. Fanale, B. Hapke, D.L. Judge, A.L. Lane, D. Morrison, N.F. Ness, T.C. Owen, J. Pearl, G. Siscoe, and D.F. Strobel. This research was funded by NASA Grants to the University of Southern California and to the University of Arizona.

September 1980

REFERENCES

Acuña, M.H., Neubauer, F.M., and Ness, N.F. (1981). Standing Alfvén wave current system at Io: Voyager 1 observations. Submitted to *J. Geophys. Res.*

Bagenal, F., and Sullivan, J.D. (1981). Direct plasma measurements in the Io torus and inner magnetosphere of Jupiter. Submitted to *J. Geophys. Res.*

Bartholdi, P., and Owen, F. (1972). The occultation of Beta Scorpii by Jupiter and Io. II. Io. *Astron. J.* 77, 60-65.

Bauer, S.J., and Hartle, R.E. (1974). Venus ionosphere: An interpretation of Mariner 10 observations. *Geophys. Res. Letters* 1, 7-9.

Bergstralh, J.T., Young, J.W., Matson, D.L., and Johnson, T.V. (1977). Sodium D-line emission from Io: A second year of synoptic observation from Table Mountain Observatory. *Astrophys. J.* 211, L51-L55.

Bertaux, J.L., and Belton, M.J.S. (1979). Evidence of SO_2 on Io from UV observations. *Nature* 282, 813-815.

Bridge, H.S., and the Voyager Plasma Science Team (1979). Plasma observation near Jupiter: Initial results from Voyager 1. *Science* 204, 987-991.

Broadfoot, A.L., and the Voyager Ultraviolet Team (1979). Extreme ultraviolet observations from Voyager 1 encounter with Jupiter. *Science* 204, 979-982.

Broadfoot, A.L., and the Voyager Ultraviolet Team (1981). Overview of the Voyager ultraviolet spectrometry results through Jupiter encounter. Submitted to *J. Geophys. Res.*

Brown, R.A. (1974). Optical line emission from Io. In *Exploration of the Planetary System.* (A. Woszczyk, and C. Iwaniszewska, Eds.), pp. 527-531. D. Reidel, Dordrecht, Holland.

Brown, R.A., and Yung, Y.L. (1976). Atmosphere and emissions of Io. In *Jupiter* (T. Gehrels, Ed.), pp. 1102-1145. Univ. of Arizona Press, Tucson.

Butterworth, P.S., Caldwell, J., Moore, V., Owen, T., Rivola, A.R., and Lane, A.L. (1980). An upper limit to the global SO_2 abundance on Io. *Nature,* 308-309.

Carlson, R.W., Bhattacharyva, J.C., Smith, B.A., Johnson, T.V., Hidavat, B., Smith, S.A., Taylor, G.E., O'Leary, B.T., and Brinkmann, R.T. (1973). An atmosphere on Ganymede from its occultation of SAO-186800 on 7 June 1972. *Science* 182, 53-55.

Carlson, R.W., and Judge, D.L. (1975). Pioneer 10 ultraviolet photometer observations of the Jovian hydrogen torus: The angular distribution. *Icarus* 24, 395-399.

Carlson, R.W., Matson, D.L., Johnson, T.V., and Bergstralh, J.T. (1978). Sodium D-line emission from Io: Comparison of observed and theoretical line profiles. *Astrophys. J.* 223, 1082-1086.

Cloutier, P.A., Daniell Jr., R.E., Dessler, A.J., and Hill, T.W. (1978). A cometary ionosphere model for Io. *Astrophys. Space Sci.* 55, 93-112.

Cloutier, P.A., McElroy, M.G., and Mitchel, F.C. (1969). Modifications of the Martian ionosphere by the solar wind. *J. Geophys. Res.* 74, 6215-6228.

Collins, S.A. (1981). Spatial color variations in the volcanic plume at Loki, on Io. Submitted to *J. Geophys. Res.*

Consolmagno, G.J. (1979). Sulfur volcanoes on Io. *Science* 205, 397-398.

Cruikshank, D.P. (1980) Infrared spectrum of Io, 2.8-5.2 μm. *Icarus* 41, 240-245.

Dean, J.A., Ed. (1973). *Lange's Handbook of Chemistry.* McGraw-Hill, New York.

Eviatar, A., Siscoe, G.L., and Mekler, Y. (1980). Temperature anisotropy of the Jovian sulfur nebula. *Icarus* 39, 450-458.

Fanale, F.P., Brown, R.H., Cruikshank, D.P., and Clarke, R.N. (1979). Significance of absorption features in Io's IR reflectance spectrum. *Nature* 280, 761-763.

Fanale, F.P., Johnson, T.V., and Matson, D.L. (1977). Io's surface and histories of the Galilean satellites. In *Planetary Satellites* (J.A. Burns, Ed.), pp. 379-405. Univ. of Arizona Press, Tucson.

Ferguson, E.E. (1972). Atmospheric metal ion chemistry. *Radio Sci.* 1, 397-401.

Fink, U., Larson, H.P., and Gautier III, T.N. (1976). New upper limits for atmospheric constituents on Io. *Icarus* 27, 439-446.

Frank, L.A., Ackerson, K.L., Wolfe, J.H., and Milhalon, J.D. (1976). Observations of plasma in the Jovian magnetosphere. *J. Geophys. Res.* 81, 457-468.

Gross, S.H., and Ramanathan, G.V. (1976). The atmosphere of Io. *Icarus* 29, 493-507.

Haff, P.K., and Watson, C.C. (1979). The erosion of planetary and satellite atmospheres by energetic atomic particles. *J. Geophys. Res.* 84, 8436-8442.

Hanel, R., and the Voyager Infrared Team. (1980). Infrared observations of the Jovian system from Voyager 1. *Science* 204, 972-976.

Hapke, B. (1979). Io's surface and environs: A magmatic-volatile model. *Geophys. Res. Letters* 6, 799-802.

Hodgman, C.D., Weast, R.C., and Selby, S.M., Eds. (1956). *Handbook of Chemistry and Physics, 38th Edition.* CRC, Cleveland.

Hunten, D.M. (1973). The escape of light gases from planetary atmospheres. *J. Atmos. Sci.* 30, 1481-1494.

Hunten, D.M. (1976). Atmospheres and ionospheres. In *Jupiter* (T. Gehrels, Ed.), pp. 22-31. Univ. of Arizona Press, Tucson.

Johnson, T.V. (1978). The Galilean satellites of Jupiter: Four worlds. *Ann. Rev. Earth Planet. Sci.* 6, 93-125.

Johnson, T.V., Cook II, A.F., Sagan, C., and Soderblom, L.A. (1979). Volcanic resurfacing rates and implications for volatiles on Io. *Nature* 280, 746-750.

Johnson, T.V., Matson, D.L., and Carlson, R.W. (1976). Io's atmosphere and ionosphere: New limits on surface pressure from plasma models. *Geophys. Res. Letters* 3, 293-296.

Judge, D. and Carlson, R.W. (1974). Pioneer 10 observations of the ultraviolet glow in the vicinity of Jupiter. *Science* 183, 317-318.

Kivelson, M.G., Slavin, J.A., Southwood, D.J. (1979). Magnetospheres of the Galilean satellites. *Science* 205, 491-493.

Kliore, A.J., Cain, D.L., Fjeldbo, G., Seidel, B.L., and Rasool, S.I. (1974). Preliminary results on the atmospheres of Jupiter and Io from the Pioneer 10 S-band occultation experiment. *Science* 183, 323-324.

Kliore, A., Fjeldbo, G., Seidel, B.L., Sweetnam, D.N., Sesplaukis, T.T., and Woiceshyn, P.M. (1975). Atmosphere of Io from Pioneer 10 radio occultation measurements. *Icarus* 24, 407-410.

Kumar, S. (1979). The stability of an SO_2 atmosphere on Io. *Nature* 280, 758-760.

Kumar, S. (1980). A model of the SO_2 atmosphere and ionosphere of Io. *Geophys. Res. Letters* 7, 9-12.

Kupo, I., Mekler, Y., and Eviatar, A. (1976). Detection of ionized sulfur in the Jovian magnetosphere. *Astrophys. J.* 205, L51-L53.

Lane, A.L., Owen, T., Nelson, R.M., and Motteler, F.C. (1979). Ultraviolet spectral variations on Io: An indicator of volcanic activity? *Bull. Amer. Astron. Soc.* 11, 597 (abstract).

Lanzerotti, L.J., Brown, W.L., Poate, J.M., and Augustyniak, W.M. (1978). On the contribution of water products from Galilean satellites to the Jovian magnetosphere. *Geophys. Res. Letters* 5, 155-158.

Macy, W., and Trafton, L. (1980). The distribution of sodium in Io's cloud: Implications. *Icarus* 41, 131-141.

Matson, D.L., Goldberg, B.A., Johnson, T.V., and Carlson, R.W. (1978). Images of Io's sodium cloud. *Science* 199, 531-533.

Matson, D.L., Johnson, T.V., and Fanale, F.P. (1974). Sodium D-line emission from Io: Sputtering and resonant scattering hypothesis. *Astrophys. J.* 192, L43-L46.

McCauley, J.F., Smith, B.A., and Soderblom, L.A. (1979). Erosional scarp on Io. *Nature* 280, 736-738.

McElroy, M.B., and Yung, Y.L. (1975). The atmosphere and ionosphere of Io. *Astrophys. J.*196, 227-250.

Michel, F.C. (1971). Solar wind induced mass loss from magnetic field-free planets. *Planet. Space Sci.* 19, 1580-1583.

Murad, E. (1978). Problems in the chemistry of metallic species in the D and E regions. *J. Geophys. Res.* 83 5525-5530.

Murcray, F.J., and Goody, R.M. (1978). Pictures of the Io sodium cloud. *Astrophys. J* 226, 327-335.

Peale, S.J., Cassen, P., and Reynolds, R.T. (1979). Melting of Io by tidal dissipation. *Science,* 203, 892-894.

Pearl, J., and the Voyager IRIS Team (1979). Identification of gaseous SO_2 and new upper limits for other gases on Io. *Nature* 280, 755-758.

Pilcher, C.B. (1980). Images of Jupiter's sulfur ring. *Science* 207, 181-183.

Pilcher, C.B., and Morgan, J.S. (1979). Detection of singly ionized oxygen around Jupiter. *Science* 205, 297-298.

Pollack, J.B., Witteborn, F.C., Erickson, E.F., Strecker, D.W., Baldwin, B.J., and Bunch, T.E. (1978). Near-infrared spectra of the Galilean satellites: Observations and compositional implications. *Icarus* 36, 271-303.

Rubey, W.W. (1951). Geologic history of sea water. *Bull. Geol. Soc. Am.* 62, 1111-1174.

Shemansky, D. (1980). Mass loading and diffusive loss rate of the Io plasma torus. *Astrophys. J.* 242, 1266-1277.

Smith, B.A., and the Voyager Imaging Team (1979a). The Jupiter system through the eyes of Voyager 1. *Science* 204, 951-972.

Smith, B.A., Shoemaker, E.M., Kieffer, S.W., and Cook II, A.F. (1979b). The role of SO_2 in volcanism on Io. *Nature* 280, 738-743.

Smyth, W.H., and McElroy, M.B. (1978). Io's sodium cloud: Comparison of models and two-dimensional images. *Astrophys. J.* 226, 336-346.

Smythe, W.D., Nelson, R.M., and Nash, D.B. (1979). Spectral evidence for SO_2 frost or adsorbate on Io's surface. *Nature* 280, 766.

Taylor Jr., H.A. and the Pioneer-Venus-Orbiter Ion Mass Spectrometer Team (1979). Ionosphere of Venus: First observation of the dayside ion composition near dawn and dusk. *Science* 203, 752-754.

Vogt, R.E. and the Voyager Cosmic Ray Team (1979). Voyager 1: Energetic ions and electrons in the Jovian magnetosphere. *Science* 204, 1002-1007.

Wagman, D.D. (1979). Data Sheet, Chem. Thermodynamics Data Center, Nat. Bureau of Standards, Washington, D.C.

Warwick, J.W., and the Voyager Planetary Radio Astronomy Team. (1979). Voyager 1 planetary radio astronomy observations near Jupiter. *Science* 204, 995-998.

Whitten, R.C., Reynolds, T.R., and Michelson, P.F. (1975). The ionosphere and atmosphere of Io. *Geophys. Res. Letters* 2, 49-51.

Yung, Y.L., and McElroy, M.B. (1977). Stability of an oxygen atmosphere on Ganymede. *Icarus* 30, 97-103.

Yung, Y.L., and Summers, M.E. (1980). A dynamically controlled atmosphere on Io. IAU Coll. 57. The Satellites of Jupiter (abstract 4-18).

22. EMISSIONS FROM NEUTRALS AND IONS IN THE JOVIAN MAGNETOSPHERE

CARL B. PILCHER
University of Hawaii

and

DARRELL F. STROBEL
Naval Research Laboratory

Jupiter is surrounded by a system of neutrals and plasma that originate at Io. Emission from these particles has been observed in the extreme ultraviolet from Pioneer, Voyager, and the International Ultraviolet Explorer (IUE), and at longer wavelengths from the ground. The emitting species observed as of late 1980 are neutral sodium, potassium, and oxygen, singly- and doubly-ionized sulfur and oxygen, and triply-ionized sulfur. Neutral sodium and potassium have been detected from observations of resonant scattering of sunlight. Neutral oxygen and the ionized species have been detected by means of collisionally excited emissions: forbidden transitions at wavelengths accessible from the ground and allowed transitions at shorter wavelengths. The collisionally excited emissions carry information about the electrons that are the immediate source of the excitation energy. Study of these spectral features thus yields information on the bulk of the plasma in the emitting region. The neutrals, by virtue of their insensitivity to the motion of the rotating Jovian magnetic field, more clearly illustrate the direct effects of the processes that remove material from Io. The study of these emissions in toto thus yields information on a wide range of characteristics and processes concerning Io, the Jovian magnetosphere, and their interaction. We review here the observations as of late 1980 and their interpretation. In addition, we discuss the related processes of ionization, radial diffusion, and dielectronic recombination.

One of the most exciting developments in planetary science during the 1970s has been the discovery of the unique characteristics of Jupiter's innermost Galilean satellite, Io. Among these characteristics are the huge Io-associated neutral and plasma clouds that surround Jupiter. The particles in these clouds appear to originate in the volcanoes that figuratively, if not literally, cover Io's surface. The transfer of these particles to space involves the interaction of Io with its magnetospheric environment, an interaction that may, for example, be strongly affected by volcanism and other geologic processes operative on the satellite. Questions of interest to the geologist such as the rates, surface locations, and velocity characteristics of mass ejection may also be addressed in studies of the clouds associated with Io. To the extent possible, these questions and others are discussed in this chapter.

The species that have been detected around Jupiter as a result of their emissions are listed in Table 22.1 along with the source of the detection, the excitation mode, and the type of transition detected. Our first indication of the existence of this complex system came with the discovery by Brown (1974; Brown and Chaffee 1974) of Io-associated sodium D-line emission; an excellent review of the physics of the emission and the characteristics of the sodium cloud as they were known at the time was presented by Brown and

TABLE 22.1

Species Detected in the Jovian Magnetosphere

Species[a]	Source of Detection[b]	Excitation Mode[c]	Transition Type
Na I	GB	RS	Allowed
K I	GB	RS	Allowed
SII	IUE	Coll	Allowed
	GB	Coll	Forbidden
SIII	Voyager	Coll	Allowed
	IUE	Coll	Allowed
	GB	Coll	Forbidden
SIV	Voyager	Coll	Allowed
OI	GB	Coll	Forbidden
OII	Voyager	Coll	Allowed
	GB	Coll	Forbidden
OIII	Voyager	Coll	Allowed
	IUE	Coll	Allowed

[a]I indicates the neutral, II indicates single ionization, III indicates double ionization, etc.
[b]GB = Groundbased; IUE = International Ultraviolet Explorer.
[c]RS = Resonant scattering; Coll = Collisional excitation.

Yung (1976). An up-dated review including a brief discussion of optical emissions detected from other neutrals around Jupiter is presented here in Sec. I.

Emissions from the plasma torus were first detected by Kupo et al. (1976), who observed the forbidden lines of SII (singly-ionized sulfur) at 6716 and 6731 Å.[a] Brown (1976) pointed out that observations of these collisionally excited emissions could be used to determine some of the characteristics of the ambient plasma. In particular, the line ratio is sensitive to the thermal electron density, n_e. Since Brown's work, the groundbased detections of other emissions, images of the plasma torus, and observations of the response of the neutral sodium cloud to ionizing particles in the plasma have provided a number of means of measuring n_e and the ion and electron temperatures. The results of these measurements are discussed in Secs. I-D and II.

The single biggest advance in our understanding of the plasma torus came as a result of the Voyager mission. Although characteristics of the torus were measured by several instruments, two provided particularly direct information on the plasma composition and distribution. The plasma science investigation (see Chapter 23, by Sullivan and Siscoe) measured the composition and energy distribution of the plasma along the spacecraft trajectory. The ultraviolet spectrometer (UVS) showed the presence of intense extreme ultraviolet (EUV) emission; observations of this emission provide information on the plasma composition and spatial distribution for an extended period of time around encounter. The results of studies of the EUV emissions are discussed in Secs. III and IV. Finally, we discuss in Sec. V the mechanisms and rates of ionization of neutral material to form the plasma torus, in Sec. VI the radial diffusion processes that transport plasma through and out of the torus, and in Sec. VII the competing plasma loss process of dielectronic recombination.

I. THE NEUTRAL CLOUD

A. Notation

Unlike the case of the emitting plasma, it is not strictly correct to refer to the emitting neutrals in the Jovian magnetosphere as a torus or a ring. As detailed below, emission from neutral sodium has been observed from a large volume of circum-Jovian space, sometimes from points tens of Jovian radii (R_J) from Io, the source of the material. Brown et al. (1975) suggested a convenient notation for the sodium cloud that we will adopt here with minor modifications. (The few observations of neutrals other than sodium are

---

[a]The unit of wavelength used in this chapter is the Ångstrom (Å), where 1 Å = 0.1 nm.

discussed in Sec. I–F.) Region A is the portion of the cloud coincident with Io's visible disk and includes all emission from the satellite's gravitationally bound atmosphere. Brown et al. (1975; Brown and Yung 1976) have pointed out that since the mechanism for sodium emission is resonant scattering of sunlight (cf. Sec. I-B), the region A cloud is virtually unobservable from the Earth. This results from the fact that solar D-line photons impinging on region A will be reflected by Io's surface even in the absence of the cloud. A terrestrial observer is unable to distinguish between atmospheric and surface reflection. We will take region B to be that excluding region A in which the sodium D-line emission brightness is at least 1–2 kR.[a] (This is a factor of $\sim$ 10 lower than the defining limit used by Brown et al.; the region of space referred to, however, is essentially the same. Our limit reflects more recent intensity measurements.) An outline of the region B cloud is shown in Fig. 22.1. Region B is largely the region studied in the imaging investigations discussed in the following section. It extends at most $\pm$ 1 R_J from Io in the north-south direction and $\pm$ 2 R_J in the east-west direction as observed projected on the plane of the sky.

We will adopt the division of region C into two regions, C_1 and C_2, as proposed by Wehinger et al. (1976). Region C_1 extends 3–4 R_J from Io in its orbital plane and is filled, according to Wehinger et al., with a disk-like cloud whose D-line emission brightness is between $\sim$ 0.3 and $\sim$ 2 kR. Sodium emission from region C_2, the remainder of the Jovian magnetosphere, is discussed in Sec. I–E.

B. The Region B Cloud

The sodium D-line emission, first detected from Io by Brown (1974; Brown and Chaffee 1974), was shown shortly thereafter to come not just from the satellite but from a substantial volume of surrounding space (Trafton et al. 1974; Mekler and Eviatar 1974; Wehinger and Wyckoff 1974). Early uncertainty as to the relative importance of resonant scattering of sunlight and collisional processes in exciting the emission (McElroy et al. 1974; Matson et al. 1974; Trafton et al. 1974; Parkinson 1975; McElroy and Yung 1975) was decisively resolved in favor of solar excitation by Bergstralh et al. (1975, 1977) and Macy and Trafton (1975a,b), who showed that the D-line brightness is a maximum when Io is at elongation and is at least an order-of-magnitude fainter when Io is near conjunction (aligned with the Earth and Jupiter). This behavior is caused by the presence of deep sodium D

---

[a]The unit of emission brightness or intensity used throughout this chapter is the Rayleigh (R). If the specific intensity I is expressed in the units 10^6 phot cm^{-2} s^{-1} str^{-1}, then $4\pi I$ is in Rayleighs. Physically, a 1 R surface brightness corresponds to an apparent emission rate (equal to the actual emission rate only for an optically thin cloud) of 10^6 phot (cm^2 column)$^{-1}$ s^{-1} (cf. Chamberlain 1961, Appendix II).

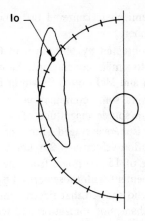

Fig. 22.1. A model of the Io region B sodium cloud as seen from above Jupiter's north pole (from Smyth and McElroy 1978).

absorption lines in the solar spectrum and the sinusoidal modulation of Io's heliocentric velocity as it revolves about Jupiter. At elongation, the Doppler shift of Io relative to the Sun is a maximum; as a result sodium atoms in Io's rest frame scatter the higher intensity solar radiation away from the absorption line center. Near conjunction, Io's heliocentric radial velocity is essentially zero, leaving only the comparatively weak radiation near the absorption line center to be scattered by the Io cloud.

Trafton (1975a) and Trafton and Macy (1977, 1978) found that the combined region A and B line profiles are asymmetric, showing enhanced emission in the blue wing when Io is moving toward the Earth and Sun and in the red wing when it is moving away. They attributed this to preferential sodium expansion in the direction of Io's motion (Trafton and Macy 1978) at velocities up to 18 km s^{-1} (Trafton 1975a). Carlson et al. (1978) found similar line shapes and concluded that both the lines' shapes and Doppler shifts *relative to Io* (shifts also observed by Trafton and Macy) could be accounted for by sodium ejection from the leading hemisphere of Io with a velocity distribution consistent with charged particle sputtering, the ejection mechanism first suggested by Matson et al. (1974). More recent observations by Porco, Trauger, and Carlson (Trauger, personal communication) indicate that the line shapes can be matched by sputtering from the inner hemisphere.

Images of the region B sodium cloud provide information on the velocities and directions of ejection that is consistent with the more recent spectroscopic results. Matson et al. (1978) found the cloud to be banana-shaped, with more atoms preceding Io in its orbit than following. They found that this shape could be produced by sputtering of sodium not from the leading hemisphere, but from a hemisphere centered 60° toward Jupiter from the center of the leading hemisphere.

Analyzing the images obtained by Murcray and Goody (1978), Smyth and McElroy (1978) came to similar conclusions. An outline of the region B cloud as derived by Smyth and McElroy is shown in Fig. 22.1. They found that the source region covers some combination of the inner and trailing hemispheres. They derived a source magnitude of $\sim 2 \times 10^{25}$ atoms s^{-1}, corresponding to a hemisphere-averaged flux of $\sim 1 \times 10^8$ atoms cm^{-2} s^{-1}, and a mean ejection velocity ~ 2.6 km s^{-1}. They also derived a mean sodium atom lifetime of 15–20 hours from the cloud geometry. This value was in reasonable agreement with those derived by Carlson et al. (1975) and Eviatar et al. (1976), who showed that electron impact ionization would occur much more rapidly than photoionization and would limit the sodium lifetime to ~ 20–50 hr. However we now know that the electron densities assumed for these calculations, reflecting the Pioneer 10 results, are greatly exceeded around Io's orbit. Use of the Voyager electron densities (see Sec. II–B) yields a mean sodium lifetime of ~ 1 hr (Shemansky 1980a). The existence of the region B cloud therefore requires a substantially larger sodium source at Io than the $\sim 2 \times 10^{25}$ atoms s^{-1} found by Smyth and McElroy (1978). Brown (1981b), considering the Voyager results, estimated the source magnitude to be $\sim 6 \times 10^{26}$ atoms s^{-1}.

Smyth (personal communication) has suggested that two processes may be involved in the removal of sodium from Io: high-velocity collisional sweeping of sodium atoms by the corotating plasma causing the enhanced wing emission, and sputtering from a hemisphere centered within a few tens of degrees of the sub-Jupiter point accounting for the bulk of the region B cloud. Possible additional evidence for collisional sweeping is presented in Sec. I–C.

In many of the observational studies cited above it was pointed out that the appearance of the region B sodium cloud is primarily a function of Io's orbital phase angle, and that the changes in the cloud's apparent shape and brightness can largely be attributed to changes in the viewing geometry and Io-Sun Doppler shift as the satellite revolves about Jupiter. A clear change in the actual cloud geometry with Io's orbital phase, however, was reported by Goldberg et al. (1978). They found that images of the cloud taken 180° apart were not mirror images as they would be if the cloud shape were constant. This variation, confirmed by later measurements (Goldberg et al. 1980), appears to be linked to the observation by Bergstrahl et al. (1975, 1977) that the combined region A and B cloud is brightest when Io is at eastern elongation. Several explanations for these observations in terms of a correlation between Io's orbital phase and the characteristics of the sodium source were reviewed by Smyth (1979), who offered an additional explanation. He pointed out that solar radiation pressure will tend to

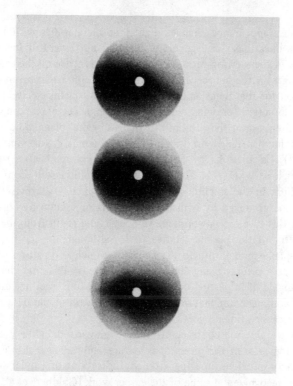

Fig. 22.2. Images of Io's sodium cloud obtained 6 February 1980, UT, showing a directional feature. The images are 5.0 R_J in diameter. Jupiter is to the left (east). Io is centered under the 14 arcsecond (0.6 R_J) occulting disk. The vertical brightness discontinuity at the left of each image is in the background radiation field. North is down and east is to the left. The three 3-minute exposures shown were taken sequentially in a way that emphasizes different portions of the sodium cloud.

decelerate atoms in the cloud when Io is to the east of Jupiter moving toward the Sun, and accelerate them when Io is moving away from the Sun to the west of the planet. The net effect is a distortion of the cloud consistent with the observed east-west asymmetries.

C. Directional Sodium Features

The studies cited above are consistent with sodium ejection over a wide range of direction from regions of hemispheric size on Io. However, D-line images acquired by one of us (CBP) indicate that more directional ejection also occurs. Two of the images have appeared elsewhere (Pilcher 1980a); a set of images showing further evidence is presented in Fig. 22.2. The images are characterized by directional features in the region B cloud, extending into region C_1, whose appearance changes nightly and sometimes within hours. They are sometimes linear and generally directed northward of Io's orbital plane. Changes in the viewing geometry cannot account for all the variations observed; changes in the source and/or sink of sodium atoms are required. One explanation for the directionality is that sodium is being ejected from Io in a particular direction (or small range of directions). The maximum gravitational potential energy of the sodium out of Io's orbital plane indicates that the ejection velocities range up to at least 4–5 km s^{-1} (cf. Smyth and McElroy 1977). Calculations by Smyth (personal communication) suggest velocities as high as 9 km s^{-1}. The directionality could result from a local source of sodium in the satellite's northern hemisphere. Alternatively, the features might be produced by collisional sweeping of sodium by corotating heavy ions. In this case the out-of-plane velocity could result from the north-south asymmetry of the plasma relative to Io, although the mechanism to impart out-of-plane momentum to the neutrals is unclear. Features directed southward of Io's orbital plane have been observed, indicating that some mechanism capable of imparting out-of-plane momentum in either direction operates. The dominance of features directed northward in the data set may be the result of observational selection.

These results are not contrary to those summarized in the preceding section. The brightness of the directional features appears near the detection limit (~ 1 kR) of previous imaging studies. Indications of similar features have now been recognized in some of the earlier work (Goldberg et al. 1980).

D. The Electron Density and Temporal Variations in the Sodium Cloud

A variability in the sodium cloud first noted by Trafton and Macy (1975; Trafton 1977) appears not to be due to time variability in the sodium source. They found that the D-line emission is strongest on the side of Io opposite the Jovian magnetic equator. This effect was also observed by Münch and Bergstralh (1977) and Pilcher and Schempp (1979). Since the magnetic equator is tilted by 10°.6 with respect to the Galilean satellite orbital plane (Smith et al. 1974), the asymmetry (which is in the north-south direction) oscillates with the 13-hr period of Io relative to Jupiter's surface. Trafton and Macy attributed the asymmetry to enhanced charged-particle impact ionization of sodium by the plasma trapped near the magnetic equator. Trafton (1977) calculated a minimum electron flux required of 1×10^{11} cm^{-2} s^{-1} for an assumed electron energy of 15 eV, the energy at which the ionization cross-section is a maximum. The corresponding electron density is ~ 500 cm^{-3}. However, we now know that the characteristic electron energy is probably not more than 3–5 eV (see Sec. II–D). The lower electron velocities and ionization cross-sections at these energies raise the electron density implied by the asymmetry observations to $n_e > 1 \times 10^3$ cm^{-3}, a value in general agreement with those derived from Voyager data and ground-based observations of the plasma emissions (Sec. II–C). Münch and Bergstralh (1977) also concluded that the north-south asymmetry implies an electron density $\sim 1 \times 10^3$ cm^{-3}.

The size of the more distant sodium cloud was used by Mekler and Eviatar (1978) to derive values for n_e. They argued that the size of the region C_1 and C_2 clouds within 10 R$_J$ of Jupiter is inversely dependent upon the electron density. From measurements of the time variability of the extent of this cloud, which they reported to be anti-correlated with the strength of the collisionally excited emissions from singly-ionized sulfur (see Sec. II), they derived n_e values between ~ 50 and ~ 900 cm^{-3}. Most of these values are substantially lower than the Voyager results (Bridge et al. 1979; Warwick et al. 1979). The effect reported by Mekler and Eviatar could be due to variations in the electron density, as they propose, or to variations in the characteristics (ejection velocities and directions) of the sodium source. If the temporal variation in the directional sodium features discussed in Sec. I–C is due to variations in the sodium source, then the implied high ejection velocities suggest that changes in the sodium distribution in the region observed by Mekler and Eviatar would result.

E. Region C_2 Sodium

Wehinger and Wyckoff (1974) first reported D-line emission from region C_2 at distances up to ~ 24 R$_J$ from Jupiter in the Galilean satellite orbital plane. In addition, Wehinger et al. (1976) reported sodium emission over both poles of the planet. Goody and Apt (1977) argued that the population of region C_2 by sodium was implausible on theoretical grounds and that the observations of Wehinger and coworkers were likely due to airglow. The existence of the region C_2 emission, however, is supported by the observations of Trafton and Macy (1978) and Pilcher and Schempp (1979). The former authors reported equatorial emission up to ~ 53 R$_J$ from Jupiter and the latter up to ~ 35 R$_J$. Pilcher and Schempp also showed that the brightness of the emission at 35 R$_J$ (~ 25 R) is consistent with a simple model of sodium ejection from Io, and that the sodium observed at that zenocentric distance may be a combination of recombined atoms on escape trajectories and atoms in high eccentricity, large apoapse orbits.

Brown and Schneider (1981) report high spectral resolution measurements of two types of region C_2 emission, one having the spectral signature of atoms in bound orbits and the other showing a radial velocity distribution peaked near 30 km s^{-1} and extending to velocities in excess of 100 km s^{-1}. This velocity distribution is not that expected for recombined atoms and may reflect a previously unrecognized sodium acceleration mechanism, perhaps related to the directional features in the region B and C_1 cloud discussed above.

F. Other Neutrals

Only two other neutrals have been detected in the Jovian magnetosphere. Trafton (1975b, 1977) reported observations of the resonance lines of potassium at 7665 and 7699 Å. The brightest emission observed was 1 kR at a point ~ 0.3 R_J east of Io. The general characteristics of the potassium cloud appear to be similar to those of the sodium cloud, although the potassium emission intensity seems to be at a minimum when Io is in the magnetic equator. This may reflect the effects of electron-impact ionization on the comparatively small potassium flux from the satellite. Early indications that the potassium emission profiles are more symmetric than those of sodium (Münch et al. 1976) are not supported by more recent observations (Trauger, personal communication). Newer data show no significant differences between the sodium and potassium line shapes.

The forbidden line of neutral oxygen at 6300 Å has also been observed in emission around Jupiter (Brown 1981a). Oxygen is thus the only Jovian magnetospheric species to be observed emitting in both its neutral and ionized forms (cf. Sec. II). Since the plasma ions are created at least partly from the neutral clouds, these observations have important implications for the spatial distribution of the plasma source (cf. Sec. V).

II. GROUNDBASED OBSERVATIONS OF THE PLASMA TORUS

A. General Characteristics of the Torus Emissions

The plasma torus was first observed by Kupo et al. (1976), who detected the forbidden transitions between the ground state and lowest pair of metastable levels of SII at 6716 and 6731 Å. A second ion, OII, was also first detected from the ground by means of its forbidden doublet at 3726, 3729 Å (Pilcher and Morgan 1979). More recently, SIII has been identified by a number of observers from forbidden transitions in the ultraviolet (3722 Å: Morgan and Pilcher 1981), red (6312 Å: Brown 1981a), and near-infrared (9531 Å: Trauger et al. 1979). These emissions are excited by ion-electron collisions that populate both the metastable levels and the higher energy states shown in the Grotrian diagrams of Fig. 22.3–22.5. The EUV transitions shown are discussed in Sec. III. Because of this excitation mechanism, the intensities of the forbidden emissions are particularly sensitive to the characteristics of the electron component of the plasma. This was pointed out for the case of the forbidden SII emissions by Brown (1976), drawing on the extensive treatment of Osterbrock (1974). (We will henceforth use the conventional square bracket notation to denote forbidden transitions, e.g., [SII].) Figure 22.6, taken from Brown (1976), shows for example the sensitivity of the [SII] line ratio I_{6716}/I_{6731} to the electron

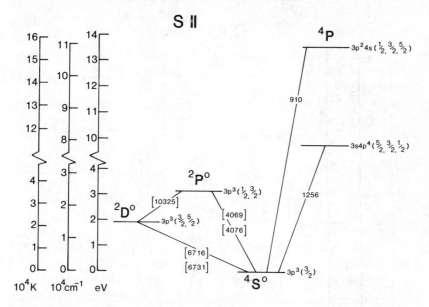

Fig. 22.3. Grotrian diagram for SII. Levels with the same term are shown vertically. The electron configuration and *J*-values are shown for each level.

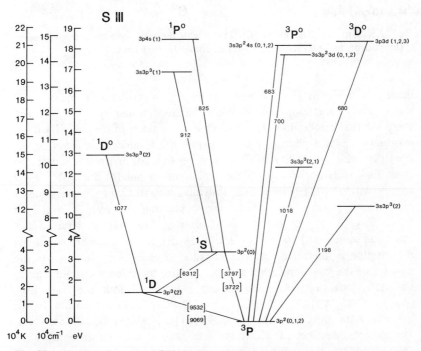

Fig. 22.4. Grotrian diagram for SIII. Levels with the same term are shown vertically. The electron configuration and *J*-values are shown for each level.

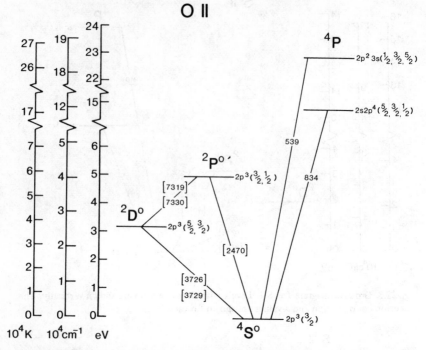

Fig. 22.5. Grotrian diagram for OII. Levels with the same term are shown vertically. The electron configuration and J-values are shown for each level.

density at moderate electron temperature, T_e. A qualitative explanation of this sensitivity is straightforward. When the electron density is low, virtually every ion that reaches the $^2D^\circ$ levels (the upper states of the red [SII] lines) eventually emits a photon; that is, despite the forbidden character of the transition, the probability of collisional deexcitation is small compared to that for radiative decay. Since collisions populate the $^2D^\circ$ levels approximately in the ratio of their statistical weights ($2J + 1$), the value of the line ratio in the low electron density limit is $[I_{6716}/I_{6731}]_{n_e\text{low}} \simeq 6/4 = 1.5$. (Cascading from the $^2P^\circ$ levels affects this value somewhat.) When the electron density is high, i.e., when collisional deexcitation is important, the rate of radiative decay is proportional to the product of the level population and the radiative transition probability. Since the latter quantity differs by almost a factor of four for these lines ($A_{6716} = 4.7 \times 10^{-4}$ s^{-1}, $A_{6731} = 1.8 \times 10^{-3}$; cf. Osterbrock 1974), the line ratio then becomes $[I_{6716}/I_{6731}]_{n_e\text{high}} = 1.5 \ (4.7/18) = 0.4$. The electron densities at which the line ratio is most sensitive to n_e are determined by the collision cross-sections and the radiative transition

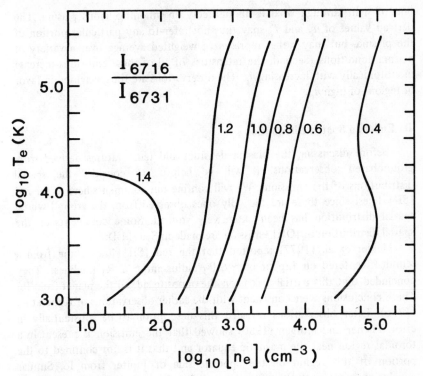

Fig. 22.6. The [SII] line ratio I_{6716}/I_{6731} as a function of n_e and T_e. (From Brown 1976.)

probabilities. As illustrated in Fig. 22.6, the range of maximum sensitivity for the red [SII] lines is $n_e \simeq 10^2-10^4$ cm^{-3}. Since a p^3 outer electron configuration is common to SII and OII, the latter ion has an analogous pair of forbidden lines whose ratio is sensitive to n_e (cf. Fig. 22.5). These lines at 3726 and 3729 Å, first detected from the torus by Pilcher and Morgan (1979), also show maximum sensitivity in the range $n_e \simeq 10^2 - 10^4$ cm^{-3}.

The electron temperature T_e can also be determined if simultaneous measurements of the red [SII] lines and those in the blue at 4069 and 4076 Å are available. The ratio of the intensity of either blue line to that of either red line is particularly sensitive to T_e (for $T_e \lesssim 10$ eV) because of the 1 eV difference between the characteristic excitation energies of each doublet (Fig. 22.3). Although the SII line ratio I_{4069}/I_{4076} is also sensitive to n_e, the primary region of sensitivity lies at densities higher than those characteristic of the plasma torus.

Two properties of plasma measurements of this type should be kept in mind. First, the observable quantity is the total emission intensity along the line of sight through what is undoubtedly an inhomogeneous plasma. The derived values of n_e and T_e may thus not refer to any particular portion of the plasma, but may rather represent a weighted average over a variety of plasma conditions. Second, the intensities of all of these emissions increase monotonically with increasing n_e. The observations are thus weighted in favor of regions of high n_e.

B. Emission Spatial Distribution

Before discussing the plasma densities and temperatures derived from groundbased observations, it will be helpful to consider the spatial distributions of the emissions. We will confine our comments here to the red [SII] lines, since these are the only ones observed from the ground whose spatial distribution has been extensively studied. Some comments on the spatial distribution of [OII] emission are made in Sec. 11-D.

Mekler et al. (1977) reported that the red [SII] lines come from a semidisk centered on Jupiter of ~ 6 R$_J$ radius and ~ 2 R$_J$ thickness. They concluded that this partial disk is on the opposite side of the planet from Io. These conclusions were consistent with the earlier observations of Kupo et al. (1976). Later observations revealed this description to be substantially in error. Pilcher and Morgan (1980) showed that the emission is present in a toroidal region near the magnetic equator and that it is not confined to the portion of this region on the opposite side of Jupiter from Io. Similar conclusions were reached by Trauger et al. (1980) and Brown (1978). Trauger et al. found that the emission was distributed symmetrically not around the magnetic equator, but around the centrifugal equator, the symmetry surface for low-energy ions in the rotating Jovian magnetic field (Hill and Michel 1976; Cummings et al. 1980). Images in the [SII] 6731 Å line acquired by Pilcher (1980b) showed distributions centered on the magnetic equator and the centrifugal equator on successive nights. Owing to the low SII energies that are characteristic of the plasma torus (Sec. II-D), it seems likely that centrifugal rather than magnetic equatorial confinement is normal. It is possible that recently recognized uncertainties in telescope pointing can account for the apparent difference in ring tilt on the two nights discussed by Pilcher (1980b), or that the images showing magnetic equatorial confinement reflect an unusual state of the emitting plasma.

A variation in the red [SII] emission intensity with magnetic longitude has been reported by three sets of observers. Pilcher and Morgan (1980) found that during the 1977-1978 apparition the emission was often strongest over $\sim 180°$ of magnetic longitude centered near $\lambda_{III}(1965) \simeq 250°$. Trafton (1980) reached essentially the same conclusion from data acquired largely a year later. It was pointed out in both of these papers that the longitudes of maximum intensity overlapped those of the active sector — the range of magnetic longitudes over which Io is known to modulate the Jovian decametric radiation (Carr and Desch 1976; Dessler and Hill 1979). The authors suggested that the longitudinal emission variation might be linked to the decametric modulation mechanism. Trauger et al. (1980), observing more than a year before Pilcher and Morgan, found a broad minimum in the emission intensity covering $\sim 90°$ of magnetic longitude centered near $\lambda_{III}(1965) \simeq 280°$, close to the longitude at which the other observers found an intensity maximum. However Trauger et al. found generally high emission intensity at $\lambda_{III}(1965) \simeq 180°$, a region also included in the active sector (Dessler and Vasyliunas 1979). Images obtained by one of us (CBP) in early 1980 showing partial sulfur rings indicate that when longitudinal variation in the emission is present there is generally emission in some portion of the active sector, but that the longitudinal boundaries of the emitting region vary substantially with time. There is as yet no evidence in the groundbased data for an intensity variation with longitude that is stationary with respect to the Jupiter-Sun vector, as has been reported by Sandel (1980) for the UVS plasma torus observations.

The radial distribution of the red [SII] emission has been the subject of considerable discussion. The discovery from Voyager of a strong radial temperature gradient in the plasma torus with SII being the dominant sulfur species only inside of ~ 5.5 R_J (Bridge et al. 1979), coupled with the detection by Trauger et al. (1980) of [SII] emission near 5 R_J but not near 6 R_J, led to the speculation that [SII] emission might be largely confined to the cold inner torus. The inaccuracy of this speculation has been clearly demonstrated by Pilcher and Morgan (1980), Trafton (1980), and Morgan and Pilcher (1981). These investigators showed that the red [SII] lines are often detectable outside of Io's orbit (5.9 R_J). Pilcher and Morgan (1980) reported emission as far as 6.9 R_J from Jupiter. Trafton (1980) found a sharp decrease in intensity beyond 6.7 R_J, measuring some of the largest intensities ever recorded for these lines (500–1500 R) outside of Io's orbit. Morgan and Pilcher (1981) found that the brightest region of emission sometimes lies almost entirely between 6 and 7 R_J. Nonetheless, the observations of Trauger et al. (1980), Pilcher (1980b), Pilcher and Morgan (1980), and Morgan and Pilcher (1981) indicate that the emission is *sometimes* confined to a region inside Io's orbit. These conclusions are consistent with those arrived at by Nash (1979) in an analysis of some of the older groundbased data. This suggests that the spatial distribution of the [SII] emitting plasma varies substantially with time.

One characteristic of the distribution of [SII] emission that seems to be present frequently is the fan-shaped feature shown in the figure (2B) of Pilcher (1980b). More recent images show that this feature often circumscribes a thin ring of emission (similar to that shown in the figure (1) of Pilcher 1980b) centered inside Io's orbit. The origin of the fan and the thin

TABLE 22.2

Forbidden Line Intensity Ratios and Inferred Electron Densities

Reference	I_{6716}/I_{6731}	n_e reported (cm^{-3})[a]	n_e corrected (cm^{-3})[b]
Brown (1976)	0.8 ± 0.2[c]	3×10^3	2×10^3
Brown (1978)	0.68 ± 0.14	5×10^3	3×10^3
Trauger et al. (1980)	0.97 ± 0.22	2×10^3	1×10^3
Trafton (1980)	$0.6 - 0.7$	$1-30 \times 10^3$	$3-5 \times 10^{3}$[e]
Morgan and Pilcher (1981)	0.61 ± 0.07	4×10^{3}[d]	5×10^3
	I_{3729}/I_{3726}		
Morgan and Pilcher (1981)	0.75 ± 0.23	2×10^3	2×10^3

[a]Reported errors generally about a factor of two.

[b]Derived from reported mean line ratios, collision strengths of Pradham (1976 for OII; 1978 for SII) evaluated at $T_e = 2 \times 10^4$ K, and an assumed $T_e = 2.5 \times 10^4$ K. (From calculations of Morgan and Pilcher 1981.)

[c]Assumed on the basis of uncalibrated data of Kupo et al. (1976).

[d]The average of the reported n_e determinations; this value differs from the n_e value corresponding to the average line ratio shown in the adjacent column.

[e]The range in this value corresponds to the range shown for I_{6716}/I_{6731}.

ring and the correspondence to the strong radial gradient in the ion temperature discovered from Voyager (Bridge et al. 1979; Bagenal et al. 1980) are discussed in Sec. II-D.

C. Electron Density

Mekler et al. (1977) derived a value of $n_e \simeq 5 \times 10^2$ cm^{-3} from the intensity of [SII] 6731 Å alone, but in doing so neglected the sensitivity of the I_{6716}/I_{6731} ratio to this parameter. The values of n_e that have been derived from this line ratio and that of [OII] I_{3729}/I_{3726} are listed in Table 22.2. The reported errors associated with these values were generally a factor of two. Trafton (1980) reported a large variation in n_e (1–30 $\times$ 10^3 cm^{-3}), noting that this variation was substantially larger than his measurement uncertainty and inferring a temporal variation. A similar inference was drawn by Morgan and Pilcher (1981). Because the investigators cited in Table 22.2 did not use the same collision strengths and electron temperatures in their calculations, it is not entirely meaningful simply to compare the reported electron densities. To facilitate comparison, we also list in the table the electron densities corresponding to the reported mean line ratios evaluated for a common electron temperature and set of collision strengths (see Table 22.2 footnote) from the calculations of Morgan and Pilcher (1981). These corrected densities yield $n_e \simeq 2.5 \times 10^3$ cm^{-3} with a factor of two uncertainty.

The measurements on which this value of n_e is based refer to an emitting region between ~ 5 and ~ 7 R_J from Jupiter (cf. Sec. 1-B). This n_e value is in excellent agreement with *in situ* measurements of n_e in this region from Voyager (Bridge et al. 1979; Warwick et al. 1979; Bagenal et al. 1980). An average value between 5.5 and 6.0 R_J based on Voyager measurements appears to be $\sim 2 \times 10^3$ cm^{-3}, although the peak value may exceed 4×10^3 cm^{-3}. At the inner boundary the value may be as low as $\sim 5 \times 10^2$ cm^{-3}, and at the outer boundary $\sim 1 \times 10^3$ cm^{-3}.

D. Plasma Temperatures

In earlier studies of the plasma torus (Brown 1976, 1978; Trauger et al. (1980) the electron temperature was assumed. Brown (1981a) reported an electron temperature of $\sim 5 \times 10^4$ K (~ 4 eV) based on the comparison of his groundbased measurements of [OI] 6300 Å and [SIII] 6312 Å with Voyager UVS data. Morgan and Pilcher estimated $T_e \simeq 2 \times 10^4$ (~ 2 eV) from measurements of the SII line ratio I_{4069}/I_{6731} whose mean value they found to be 0.29 ± 0.07. They also found that three spectra showing [SII] emission in the blue lines, but not the red lines, could best be explained by the presence of a dense ($> 4 \times 10^4$ cm^{-3}), cold ($< 4 \times 10^3$ K ~ 0.3 eV) component of the electron plasma. The Voyager plasma science and UVS results show that there is in addition a component of the electron plasma with a characteristic energy of > 50 eV (Scudder et al. 1980; Strobel and Davis 1980), but that the bulk of the electrons in the plasma torus have a temperature of 3–6 eV.

The ion temperatures have been measured by means of three techniques, two of them groundbased. Trauger et al. (1979,1980) measured the Doppler widths of the red [SII] lines and [SIII] 9532 Å, finding characteristic temperatures perpendicular to the field lines of $T_{\text{SII}}^{\text{perp}} \sim 2$–$8 \times 10^4$ K (~ 2–7 eV) and $T_{\text{SIII}}^{\text{perp}} \sim 3 \times 10^5$ K (~ 26 eV). From similar measurements of [SIII] 6312 Å, Brown (1981a) found $T_{\text{SIII}}^{\text{perp}} \sim 1 \times 10^6$ K (~ 86 eV). These measurements are generally consistent with the Voyager results that show the SIII to be present in a plasma outside of 5.5 R_J whose characteristic ion temperature is ~ 50 eV and the SII in a colder plasma closer to Jupiter (Bagenal et al. 1980).

Pilcher (1980b) determined the SII temperature parallel to the field lines from measurements of the ion distribution around the centrifugal equator. Hill and Michel (1976; see also Cummings et al. 1980) showed that the energy required for an ion to travel up a field line against the field-aligned component of the centrifugal force can be expressed in terms of a scale height given by

$$H_{\text{ion}} = \left(2\,kT_{\text{ion}}^{\text{par}}/3\,m\,\omega^2 \right)^{\frac{1}{2}} \tag{1}$$

where $kT_{\mathrm{ion}}^{\mathrm{par}}$ is the ion's thermal energy in the direction of the field, m is its mass, and ω is the angular frequency of rotation. Pilcher (1980b) measured the variation in H_{SII} as a function of radial distance from Jupiter implied by the fan-shaped distribution and derived a radial gradient in $T_{\mathrm{SII}}^{\mathrm{par}}$. He found $T_{\mathrm{SII}}^{\mathrm{par}} \simeq 24$ eV at 5.9 R_J and $\lesssim 2$ eV at 5.1 R_J. These values are in excellent agreement with those obtained from the Voyager plasma science investigation (Bagenal et al. 1980; Bagenal and Sullivan 1981). Pilcher's results are also consistent with the values of $T_{\mathrm{SII}}^{\mathrm{perp}}$ measured by Trauger et al., since the latter refer predominantly to the cold plasma at the inner edge of the radial temperature gradient. Thus there does not appear to be any convincing evidence for a temperature anisotropy in the plasma such as that postulated by Eviatar et al. (1979).

The origin of the fan and the thin ring can be understood in terms of the simultaneous diffusion and cooling of SII (Bagenal et al. 1980). Ions are created near Io's orbit and are first observed with a characteristic energy of several tens of eV. As they diffuse inward they cool, collapsing to the centrifugal equator. The diffusion time increases with decreasing distance to Jupiter (see Chapter 23 by Sullivan and Siscoe), resulting in a buildup of ions on the centrifugal equator that are observed as a thin ring of emission centered well inside Io's orbit.

The [OII] 3726, 3729 Å emission first observed by Pilcher and Morgan (1979) seems to be more extensive, at least in the direction of the field lines, than the red or blue [SII] emission. Morgan and Pilcher (1981) have shown that the [OII] emission is detectable at larger distances from the centrifugal equator than the [SII] lines. This result is consistent with the Voyager plasma science data, which show OII to be generally more extensive than SII (Bagenal and Sullivan 1981). If OII and SII have roughly the same temperature, one would expect the smaller mass of the former to result in a larger OII scale height, in accordance with the observations. Differences in the OII and SII distributions may also arise from the larger ionization potential and hence smaller ionization rate of neutral oxygen.

E. Emission Intensities and Ion Densities

From their original observations of the red [SII] lines, Kupo et al. (1976) reported an order-of-magnitude brightness estimate of 300 R. In analyzing these observations, Brown (1976) assumed that the brightness of the 6716 Å line was in the range 60–300 R. He also assumed that the optical pathlength through the nebula was 4–8 R_J, and derived an SII density of $n_{\mathrm{SII}} = 80^{+170}_{-55}$ cm^{-3} corresponding to a column density along the line of sight of $N_{\mathrm{SII}} \simeq 10^{12} - 10^{13}$ cm^{-2}. Brown (1978) subsequently reported brightnesses below 40 R (averaged over 3 $\times$ 7 arcsec of sky) for each of the red [SII] lines, implying SII densities about an order of magnitude lower than his previous results. However, these low brightnesses are inconsistent with the results of several other observers. Trauger et al. (1980) reported brightnesses in the range 40–60 R for each line averaged over a 34-arcsec (1.5 R_J) diameter field centered near 4.4 R_J. Since the [SII] emission within this field is known to come from a relatively small region (Pilcher 1980b), the actual brightness implied by the Trauger et al. observations is at least several hundred R and may be in excess of 1 kR. Pilcher and Morgan (1979) reported brightnesses for each line ranging up to 500 R averaged over regions of relatively uniform emission; the values of Trafton (1980) are even higher, although most are below 1 kR. Most recently, Morgan and Pilcher (1981) reported red [SII] line brightnesses that average $\sim$ 215 R with a maximum brightness of $\sim$ 370 R. Although there have been some lower values reported on the basis of comparisons with a telluric airglow line (Mekler et al. 1977), it appears that characteristic brightnesses for the red [SII] lines are in the range 200–1000 R per line. Morgan and Pilcher (1981) report characteristic brightness of $\sim$ 60 R for [SII] 4069 Å, 30–40 R per line for [OII] 3726, 3729 Å, and $\sim$ 20 R for [SIII] 3722 Å.

In order to convert these brightnesses into ion spatial number densities, both the plasma conditions and the optical pathlength through the emitting plasma must be known. Figures 22.7 and 22.8 show for [SII] 6731 Å and [OII] 3726 Å, respectively, the relationship between line brightness and ion column density for $T_e = 5 \times 10^4$ K (4 eV; cf. Sec. I-D) and a range of n_e. Assuming $n_e = 2 \times 10^3$ cm^{-3} (cf. Sec. I-C) and brightnesses for [SII] 6731 Å and [OII] 3726 Å of 500 R and 40 R, respectively, we find $N_{SII} = 1.5 \times 10^3$ cm^{-3} R_J (1.1 $\times 10^{13}$ cm^{-2}) and $N_{OII} = 1.0 \times 10^3$ cm^{-3} R_J (7.4 $\times 10^{12}$ cm^{-2}). The optical pathlength through a torus of 1 R_J cross-sectional diameter centered near 6 R_J is $\sim$ 5 R_J. Use of this pathlength yields $n_{SII} \simeq$ 250 cm^{-3} and $n_{OII} \simeq$ 200 cm^{-3}. These values, particularly for SII, are substantially lower than those derived from Voyager plasma science data at 5.3 R_J (Bagenal and Sullivan 1981). The n_{SII} value, however, better agrees with the plasma science result at 6.0 R_J. The n_{OII} value is close to the plasma science result at 6.0 R_J derived assuming a common ion temperature.

Differences between Voyager and groundbased determinations of ion densities may be due to spatial and temporal variations in the plasma. However the values determined from groundbased measurements are quite sensitive to assumed characteristics of the plasma. For example, the derived spatial densities are obviously inversely proportional to the assumed optical pathlength through the torus. The derived densities are also particularly sensitive to T_e. If $T_e = 1 \times 10^4$ K had been assumed, the derived values of N_{SII} and N_{OII} would have been larger by factors of 3 and 9, respectively, than those derived here from Fig. 22.7 and 22.8. Any apparent discrepancies between Voyager and groundbased determinations of ion densities must therefore be viewed with some caution.

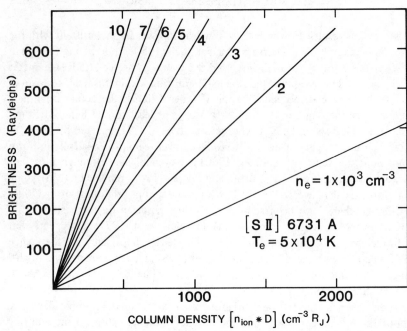

Fig. 22.7. The brightness of [SII] 6731 Å as a function of SII column density for a range of n_e.

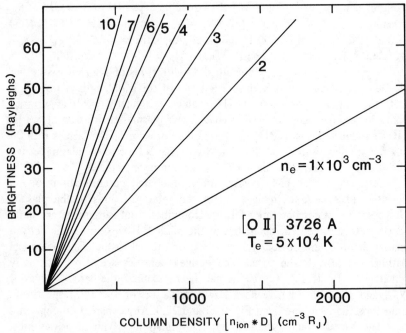

Fig. 22.8. The brightness of [OII] 3726 Å as a function of OII column density for a range of n_e.

F. Plasma Rotational Velocity

Trauger et al. (1980) reported that the wavelengths of the red [SII] lines near 5 R_J correspond to a plasma corotating with Jupiter. Brown (1981a) came to the same conclusion from measurements of [SIII] 6312 Å. This is consistent with the Voyager plasma science results which show plasma corotation inside of $\sim 7.5\ R_J$ (Bridge et al. 1979). However Mekler et al. (1979) have reported evidence for slower than corotational velocities for SII just outside of Io's orbit ($\gtrsim 6\ R_J$). Observations by Pilcher and Morgan (1980) and Morgan and Pilcher (1980) that are similar to those of Mekler et al. (1979) are consistent with corotation. This inconsistency remains to be resolved.

III. SPACECRAFT-BASED OBSERVATIONS

A. Voyager Ultraviolet Spectrometer and Pioneer 10 Ultraviolet Photometer Results

In addition to the visible and near-ultraviolet radiation detected by groundbased instruments, the Io plasma torus can emit substantial radiation in the extreme ultraviolet (EUV). Weak emission from an incomplete torus was first detected by the Pioneer 10 ultraviolet photometer in both its short (~ 40 R for $\lambda \lesssim 800$ Å) and long (~ 300 R for $\lambda \lesssim 1600$ Å) wavelength channels (Judge et al. 1976). It was the ultraviolet spectrometer (UVS) on Voyager 1 in early January 1979 that first measured strong EUV emission ($\sim 2 - 3 \times 10^{12}$ W) from a complete torus (Broadfoot et al. 1979).

Typical EUV spectra from the Voyager UVS, which are displayed in Fig. 22.9, are characterized by two bright features centered at 685 Å and 833 Å. The 685 Å feature is dominated by strong SIII multiplets at 680 Å, 683 Å, 700 Å, and by an OIII multiplet at 703 Å. (The transitions discussed here are illustrated in Figs. 22.3, 22.4, 22.5, 22.10, and 22.11.) The feature is broadened on the short-wavelength side by an SIV multiplet at 657 Å and on the long-wavelength side by SIII multiplets at 724 Å and 729 Å, and an SIV multiplet at 745 Å. Absolute intensities for these multiplets are difficult to determine, since the 30 Å resolution of the UVS results in severe blending. The entire 685 Å feature had a typical brightness of 200–500 R at elongation during the Voyager encounters (Sandel et al. 1979).

The other bright feature with representative intensity of 150–300 R consists of an OII multiplet at 834 Å and an OIII multiplet at 835 Å with smaller contributions from an SIII multiplet at 825 Å and an SIV multiplet at 816 Å. The OII multiplet at 539 Å is absent from the spectra and can be used to infer an upper limit on OII densities in the EUV-emitting torus. At longer wavelengths the spectra contain features that may be identified as SIII

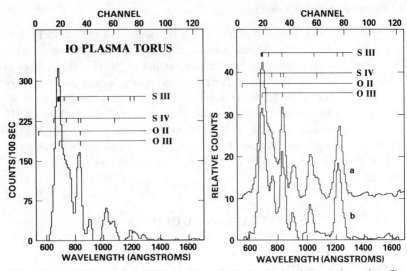

Fig. 22.9. (Left) Voyager 1 EUV spectrum of the Io plasma torus at elongation. Sky background and instrumental scattering have been removed. Multiplets of identified species are indicated. The feature at 685 Å is dominated by SIII and has a brightness of 200 R. (From Broadfoot et al. 1979.)

(Right) Voyager 2 EUV spectra of the Io plasma torus showing differences in spectral content indicating composition and/or electron temperature changes. The spectra have been corrected for instrumental scattering. A large fraction of the feature near 1200 Å arises from interstellar H Lyman α emission at 1216 Å. Spectrum (a) has been displaced upward by ten units. (From Sandel et al. 1979.)

multiplets at 1018 Å and 1198 Å, and an SIV multiplet at 1070 Å. In addition, the SIII multiplet at 1077 Å may contribute to the 1070 Å feature. It is tempting to identify the feature at 900 Å as the SII multiplet at 910 Å, although SIII has a multiplet nearby at 912 Å. However the wavelength of the feature appears to be inconsistent with the SII identification (D. Shemansky, personal communication, 1980). The only other prominent SII multiplet at 1256 Å is, unfortunately, masked by the broad interplanetary H Lyman-α emission at 1216 Å (cf. Fig. 22.9). There are no distinctive features in the Voyager spectra longward of Lyman α (Occasionally a sporadic feature at 1570 Å appears in the torus spectra which also occurs in Jupiter's dayside emission [Sandel et al. 1979].) Noticeably absent are any emission from OI at 1304 Å and 1356 Å and SIV at 1406 Å. From the absence of characteristic multiplets in the Voyager spectra, Shemansky (1980a) has set upper limits on the densities of SiIII, SiIV, ArII, CIII, and HI.

Clearly the Voyager spectra are not dominated by emission from neutral and singly-ionized oxygen and sulfur. Similarly there are no pronounced features in Fig. 22.9 associated with OIV, SV, or more highly ionized species.

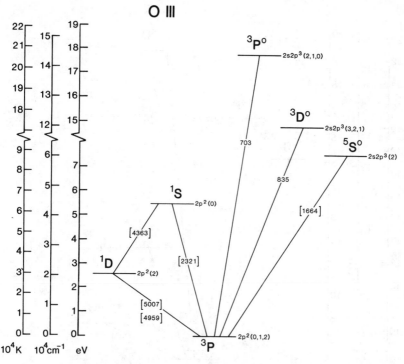

Fig. 22.10. Grotrian diagram for OIII. Levels with the same term are shown vertically. The electron configuration and J-values are shown for each level.

Oxygen IV has three strong multiplets at 789 Å, 609 Å, and 554 Å, whereas SV has a strong multiplet at 786 Å. If these ions were the principal emitters, the EUV radiation would be substantially different from that observed.

The spatial extent of the EUV-emitting plasma torus is illustrated in Fig. 22.12, which shows the spatial intensity variation of the 685 Å feature for scans with the spectrometer slit oriented both parallel and perpendicular to Io's orbital plane. For a torus homogeneous in brightness, a radius of symmetry of 5.9 ± 0.3 R_J and a cross-sectional radius of 1 ± 0.3 R_J give the best fit to the observed intensity. A torus with these dimensions is illustrated above the data in Fig. 22.12. The smooth curve shown superimposed on the data is the brightness distribution corresponding to this model. Clearly the EUV emission extends beyond the outer boundary of this homogeneous torus, but is consistent with a rather well-defined inner edge near 5 R_J. For this homogeneous torus the path length of emitters, as seen by the Voyager UVS inside Io's orbit, varies from 4 R_J to ~ 9 R_J. The intensity exhibits a similar variation, indicating that the plasma is optically thin to the transmission of resonance radiation.

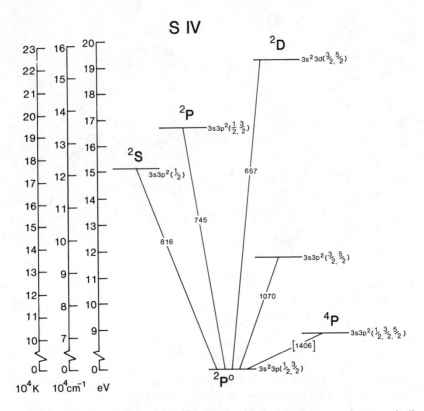

Fig. 22.11. Grotrian diagram for SIV. Levels with the same term are shown vertically. The electron configuration and J-values are shown for each level.

The torus appears to be centered on Jupiter's magnetic equator, which is displaced by the dipole offset angle of $10°.6$ from the rotational equator (Smith et al. 1974). The excursion of Io from the magnetic equator is thus the vertical extent of the torus. However, the analyses to date cannot exclude the centrifugal equator, located at an offset angle of $7°$, as the plane of symmetry (Hill and Michel 1976; Cummings et al. 1980). The incomplete torus observed by the Pioneer 10 ultraviolet photometer had an angular extent of $\sim 1/3$ of Io's orbit, an assumed thickness of $\sim 1 R_J$, and was in the orbital plane (Carlson and Judge 1974). It can be concluded that the spectral characteristics and the spatial distribution of the tori observed by Pioneer 10 on one hand, and Voyagers 1 and 2 on the other, were radically different. If oxygen and sulfur were the principal constituents of the torus observed by Pioneer 10, then the dominant emission features may have been OI multiplets at 1304 and 1356 Å, an OII multiplet at 834 Å, and an SII multiplet at 1256 Å.

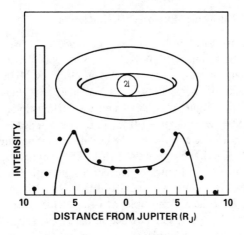

Fig. 22.12. Io's plasma torus. The data points show the measured intensity of the 685Å feature as a function of distance from Jupiter measured in the orbital plane of the satellites. A model torus used to fit the data is shown to scale above the data; the intensity predicted by this model is shown by the solid line. Other observations (not shown) indicate that the intensities at the eastern and western elongation points may differ by a factor of as much as two. (From Broadfoot et al. 1979.)

Smaller differences within the Voyager EUV data are also present. For example, the EUV brightness of the torus increased by a factor of two between the encounters, accompanied by a decrease in the effective electron temperature (Sandel et al. 1979). Broadfoot et al. (1979) reported that the brightness of some EUV features varied by a factor of two during the Voyager 1 encounter. In general the spectral content of the EUV data vary somewhat with both time and position in space (see, for example, Fig. 22.9). These variations should be kept in mind when use is made of results quoted here that are derived from a subset of the data. However, the variations in the EUV emission are not as great as those observed from the ground in the forbidden lines of SII (Trafton 1980).

The qualitative characteristics of the Io plasma torus are consistent with Io being the source of plasma and with plasma diffusing both radially inward and outward. The Voyager UVS observations do not indicate any enhancement of emission or additional spectral features when Io is in the field of view (Shemansky 1980b). This suggests that ionization of neutral material lost from Io, a process that should be accompanied by substantial emission, may occur over an extended region of space around Jupiter or that primarily molecular ions are formed whose ionization-associated emissions are not detectable at wavelengths observed by the Voyager UVS (see Sec. V).

B. International Ultraviolet Explorer (IUE) Results

The Io plasma torus has also been observed from the Earth orbiting satellite IUE in the wavelength interval ~ 1150–1950 Å. With long exposure times (~ 8 hr) and better spectral resolution (~ 11 Å), IUE provides more sensitive detection of weak EUV emission than the Voyager UVS. For example, the SIII multiplet at 1198 Å and the SII multiplet at 1256 Å (cf. Figs. 22.3 and 22.4) were detected from IUE with intensities of 30–60 and 43 R, respectively (Moos and Clarke 1981). The corresponding Voyager UVS observed (SIII 1198 Å) and upper limit (SII 1256 Å) brightnesses, which are masked by interplanetary Lyman-α, are consistent with the IUE values. In addition, marginal detections of ~ 12 R of the OIII multiplet at 1664 Å (Fig. 22.10) and ~ 17 R of the SIV multiplet at 1406 Å (Fig. 22.11) were reported by Moos and Clarke (1981). The absence in the IUE spectra of several multiplets allows upper limits to be placed on the number densities of the following ions: CII, CIII, NIII, NIV, SiII, SiIII, SiIV, AlII, AlIII, CaII, PII, and PIII (Moos and Clarke 1981). In the case of the SIV multiplet at 1406 Å (Fig. 22.11), the observed upper limit does not yield a consistent upper limit on the SIV density for reasons discussed below. As a result of the long-exposure times, the IUE observed brightnesses and inferred densities represent average properties of the torus over approximately one rotation period. It is therefore possible that there are local concentrations of the unobserved ions with densities substantially higher than the reported upper limits.

IV. INTERPRETATION OF THE EUV EMISSIONS

The EUV emissions that have been identified are entirely prompt transitions (radiative decay rate exceeds collisional deexcitation rate) whose intensities depend only on the electron collisional excitation rates. The intensity in Rayleighs corresponding to a particular transition is given by

$$I = \frac{8 \times 10^{-14}}{\omega_\ell} \int \frac{\overline{\Omega}}{T_e^{1/2}} \exp\left(\frac{-\Delta E}{T_e}\right) n_i n_e \, ds \qquad (2)$$

where ω_ℓ is the statistical weight of the ground level, $\overline{\Omega}$ is the thermally-averaged collision strength, T_e is the electron temperature in eV, ΔE is the energy in eV of the emitted photon, n_i is the ion density (cm^{-3}) of the emitter, n_e is the electron density (cm^{-3}), s is the pathlength in cm, and the integration is along the line of sight (Osterbrock 1974). If $\Delta E \lesssim T_e$, then the intensity is most sensitive to the product $n_i n_e$ and is primarily characteristic of regions of plasma density enhancements. This also applies to the observed forbidden transitions of OII and SII as pointed out in Sec. II. If $\Delta E \gg T_e$, then in addition to density enhancements, the observed intensities are also weighted heavily toward regions of high electron temperature. Equation (2) can be extended to a non-Maxwellian energy distribution that is resolvable into a number of isothermal components by the addition of similar terms for each component.

The accuracy of any physical properties inferred from spectroscopy of the plasma torus depends critically on the accuracy of the collision strengths. Unfortunately no experimental values are available, and we must rely totally on theoretical calculations. In the initial interpretation of the Voyager UVS data (Broadfoot et al. 1979) the collision strengths were calculated with standard astrophysical expressions based on the Bethe approximation and the empirical Gaunt factor formula (cf. Van Regemorter 1962). The Bethe approximation is accurate only at high energies and requires that the incident electron lose only a small fraction of its energy. In this approximation the collision strength is proportional to the product of the oscillator strength and the Gaunt factor. The latter quantity is proportional to the probability that the collision will result in excitation of the electronic state in question. For ions, the thermally averaged Gaunt factor is usually assumed to be 0.2 when $T_e < \Delta E$ (Van Regemorter 1962), but this assumption may result in considerable error (Younger and Wiese 1979). In the initial Voyager interpretation (Shemansky 1980a), Gaunt factors were assumed and the Coulomb approximation was used to calculate oscillator strengths when no measurements were available. Strobel and Davis (1980) used the distorted wave approximation to calculate more accurate Gaunt factors. They neglected the effects of configuration interaction, but corrected to first order by multiplying the calculated collision strengths by the ratio of the measured to calculated radiative transition probabilities. Generally the method of distorted waves as used by Strobel and Davis (1980) has consistently predicted collision and oscillator strengths to within a factor of two or better for those cases where comparisons with better theoretical methods can be made. However, higher accuracy is required to interpret the Voyager UVS spectra; currently the more accurate close-coupling calculations of Bhadra and Henry (1980), which include configuration interaction, are available for only a single ion, SIV. Since the bulk of the electrons in the plasma torus have energies below threshold, the major contribution to the collision strength is at threshold where resonance effects may be important. These effects have not been included in any of the collision strengths used to interpret the Voyager UVS data.

In contrast to the groundbased instrumentation capable of resolving the individual lines of each multiplet, the Voyager UVS has poor spectral resolution (30 Å) relative to the spacing of both the multiplets and the lines within each multiplet. (The instrument was designed principally to measure emission at 1216 Å and 584 Å, for which 30 Å resolution was more than sufficient [Broadfoot et al. 1977].) As a consequence the absolute intensities reported are all for blended features. There are no two spectrally distinct and well-separated features of the same ion that can be used to derive an effective electron temperature from the intensity ratio, as has been done for the forbidden emission observed from the ground (Sec. II-D). The effective electron temperature and other properties can be obtained by computing theoretical spectra with Eq. (2) as a function of T_e, n_i, and n_e until a best fit to the observed spectra is achieved. For each spectrum an average value of n_i, n_e, and T_e along the line of sight can, in principle, be deduced (although they may not be unique); successive spectra as a function of radial distance can yield the radial distribution of these plasma properties.

The results of the analyses to date are the average properties of an equivalent homogeneous torus of thickness 2 R_J whose radiative emission equals the observed emission. The electron density is constrained by the average density measured by Warwick et al. (1979; cf. Birmingham et al. 1980). In order to account for the highly ionized oxygen and sulfur species in the torus, which have an ion residence time $\lesssim$ 10 days (Richardson et al. 1980; Froide-vaux 1980; cf. Sec. VI), the electron distribution function must have a high-energy tail with characteristic temperature $>$ 50 eV (Strobel and Davis 1980). Such a distribution was measured in the torus by the Voyager plasma science experiment (Scudder et al. 1980). The high-energy electrons can contribute to the excitation of the EUV emission, but do not affect the longer wave-length forbidden transitions. Average densities (for a homogeneous torus of thickness 2 R_J) derived from the most recent collision strengths for SIV (Bhadra and Henry 1980), the empirically adjusted collision strengths of Strobel and Davis (1980) for other ions, and an assumed cold electron temp-erature (in the hot torus outside of Io's orbit) of $\sim$ 5–7 eV (Scudder et al. 1980), are as follows:

$$n_{SIII} \quad \sim 110 - 180 \text{ cm}^{-3}$$

$$n_{SIV} \quad \sim 130 - 190 \text{ cm}^{-3}$$

$$n_{OII} \quad \sim 70 - 150 \text{ cm}^{-3}$$

$$n_{OIII} \quad \sim 400 - 200 \text{ cm}^{-3}$$

$$n_e(\text{hot}) \sim 60 - 10 \text{ cm}^{-3}$$

For details of the assumptions and techniques used in deriving these values, the reader is referred to Strobel and Davis (1980) and Shemansky (1980a). These values are in reasonable agreement with those derived by Shemansky (1980b). The reversal of the limits for n_{OIII} indicates that this value is inversely coupled to n_{OII}, due to the fact that the only detectable EUV emissions of these ions are coincident at 834 Å. The reversal in the limits of $n_e(\text{hot})$ reflects a similar inverse coupling to n_{OII}, resulting from the absence of the OII feature at 539 Å (see Sec. III-A), i.e., the absence of the 539 Å feature limits the product $n_e(\text{hot})n_{OII}$.

If one-half of the upper limit intensity (after the subtraction of interstellar Lyman-α) at 1256 Å in the spectra of Fig. 22.9 is attributed to SII, then n_{SII} would be $\sim 100 \text{ cm}^{-3}$. If, rather than being distributed in the torus shown in Fig. 22.12, all of the SII were confined to a torus centered at 5.3 R_J with thickness 0.6 R_J, then n_{SII} would be $\sim 1000 \text{ cm}^{-3}$ in agreement with the plasma science measurements (Bridge et al. 1979; Bagenal et al. 1980) and groundbased observations (Sec. II-C).

Note that the O/S ratio is $\sim 1-1.5$. If SO_2 is the parent molecule for these ions (Pearl et al. 1979; Johnson et al. 1979), ionized oxygen may be distributed differently in space than ionized sulfur. A difference in the spatial distribution of these ions was also inferred from the groundbased and plasma science data (Sec. II-D). In addition to the difference in scale height pointed out in Sec. II-D, a difference in spatial distribution may also be caused by a transfer of substantial translational energy to oxygen atoms in the dissociation of SO_2^+ and SO^+ (Strobel and Davis 1980).

The IUE data provide an independent determination of plasma densities. Since only a single multiplet of any ion is observed, the electron temperature cannot be inferred. Emissions from four ions, OIII (1664 Å), SII (1256 Å), SIII (1198 Å), and SIV (1406 Å), have been detected (Moos and Clarke (1981); for an assumed effective electron temperature of 10 eV, their approximate densities are $\sim$ 110, 140, 160–320, and 90 cm^{-3}, respectively, averaged over a torus thickness of $\sim 2 R_J$. Given the strong configuration interaction affecting the SII and SIII multiplets observed from IUE, the reported densities for these ions are in acceptable agreement with those inferred from Voyager UVS data. It should be remembered that SII is not positively identified in the Voyager spectra. Although the OIII density obtained from IUE data is substantially below the Voyager density, it is more consistent with the *in situ* plasma science measurements (Bridge et al. 1979; Chapter 23 by Sullivan and Siscoe). As discussed above, there is a problem in differentiating OIII from OII in the Voyager spectra. One possible resolution of this apparent discrepancy between UVS and IUE data is that OII occupies regions of the torus where the hot electron density is low, with the result that excitation of its 539 Å multiplet is weak in comparison to that of its 834 Å multiplet. This effect would allow a consistent fit to the UVS spectra with larger OII and lower OIII densities than given above. A proper interpretation of the Voyager UVS data may thus require a consideration of plasma inhomogeneities.

Moos and Clarke (1981) also obtained an estimate of the SIV density of 90 cm^{-3} based on a brightness of ~ 17 R for the 1406 Å multiplet (Fig. 22.11). There is however a very serious problem with the Bhadra and Henry (1980) collision strength that is used to infer SIV densities. Straightforward extrapolation of the SIV collision strengths down to threshold for the 1070 Å and 1406 Å multiplets leads to the inference that these multiplets should be of comparable brightness for the electron distribution function including the suprathermal component measured by Scudder et al. (1980) and inferred by Strobel and Davis (1980). The UVS intensity $I_{1070} \sim 100$ R (Broadfoot et al. 1979; Shemansky 1980a) and the IUE intensity $I_{1406} \sim 17$ R are thus in substantial disagreement. (At 1406 Å, the Voyager upper limit is consistent with the IUE data.) The resolution of this discrepancy awaits the extension, in progress, of the Bhadra and Henry (1980) calculations down to threshold with resonance effects included (R. Henry, personal communication). Until it is resolved, the Voyager UVS value for n_{SIV} given above is preferred since it also accounts for emission at 657 and 745 Å observed in the spectra of Fig. 22.9.

V. IONIZATION MECHANISMS AND RATES

The most probable mechanisms for creating new ions in the Io plasma torus are photoionization by sunlight and electron impact ionization, particularly by energetic electrons with large gyration energy (i.e., high temperature perpendicular to the magnetic field lines). Carlson et al. (1975) showed that electron impact ionization was the dominant mechanism for the plasma torus probed by Pioneer 10 even for sodium atoms with the very low ionization potential of 5.14 eV. In the torus probed by Voyagers 1 and 2, the importance of electron impact ionization is clearly indicated by the presence of highly ionized species such as OIII and SIV (Bridge et al. 1979; Broadfoot et al. 1979). Recent experimental measurements of the electron impact ionization cross sections for OI, OII, and OIII are given by Aitken and Harrison (1971), Hamdan et al. (1978), and Brook et al. (1978), whereas only theoretical calculations by Peach (1968, 1970; for corrections see Peach 1971) are available for SI. Ionization rate coefficients can be estimated satisfactorily from formulas given by Chandra et al. (1976).

There is one other ionization mechanism of potential importance in the plasma torus. It is the critical velocity phenomenon proposed by Alfvén (1954) in which ionization can occur if the energy associated with the relative motion of a neutral gas moving through a magnetized plasma exceeds the ionization potential. The relative motion of Io and its extensive neutral clouds in the plasma torus exceeds the critical velocity for all elements (Cloutier et al. 1978). Braking of the relative motion between the neutral gas and magnetized plasma is postulated to generate the necessary energy for ionization through collective plasma processes (cf. Raadu 1978; Varma 1978). Although there is experimental confirmation of this mechanism, the ionization rate coefficients are extremely uncertain, and a large extrapolation from laboratory conditions is necessary.

It is important to determine the site of ionization of neutral material supplied to the torus. In order to explain the departure from corotation of the magnetospheric plasma beyond 10 R_J (McNutt et al. 1979) and the fact that the observed thermal speed of the ions is only one-fourth the expected capture speed (Bridge et al. 1979), Hill (1980) and Goertz (1980) required that most of the ionization occur locally in the vicinity of Io at a rate of $\sim 2 \times 10^{28}$ ions s^{-1}. If OI and SI were the principal neutrals being ionized in the vicinity of Io, then there is a significant probability that some ionization events would be followed by emission of photons detectable by the Voyager UVS (Shemansky 1980b) from the processes

$$e + SI \rightarrow SII\,(^4P) + 2e$$
$$SII\,(^4P) \rightarrow SII\,(^4S^\circ) + h\nu\,(1256\,\text{Å})$$
$$e + OI \rightarrow OII\,(^4P) + 2e$$
$$OII\,(^4P) \rightarrow OII\,(^4S^\circ) + h\nu\,(834\,\text{Å})\,.$$

From UVS data, upper limits of 50 R can be placed on SII (1256 Å) and OII (834 Å) radiation when Io is in the field of view, leading to an upper limit on the local ionization rate of OI and SI of 1×10^{27} ion s^{-1} (Shemansky 1980b), which is at least one order of magnitude less than required by Hill (1980) and Goertz (1980). This limit does not apply to ionization rates of SO_2 and SO, for which no known emission signature lies in the spectral region to which the UVS is sensitive.

If it is assumed that the EUV radiation is powered by the acceleration of newly ionized material (see Chapter 23 by Sullivan and Siscoe), then a lower limit to the total ionization rate can be determined. Each newly created ion will be accelerated by the corotating magnetic field to a velocity no greater than the difference between the neutral's velocity prior to ionization and the corotation velocity. For neutrals in Keplerian orbits similar to Io's this velocity difference is 57 km s^{-1}. The corresponding energies for sulfur and oxygen ions are 8.7×10^{-10} erg (540 eV) and 4.3×10^{-10} erg (270 eV), respectively. In order to supply the ~ 2–3×10^{12} W being emitted by the torus in the EUV (Broadfoot et al. 1979; Shemansky, 1980a), these neutrals must be ionized at a total rate of at least ~ 3–5×10^{28} s^{-1}. (Similar total ionization rates have been derived from the spatial distribution of ions [cf. Sec. VI] and from estimates of the rate of sodium ionization and the mixing ratio of sodium in the plasma [Brown 1981b].) Since this ionization does not appear to be happening in the immediate vicinity of Io, the ion source is probably distributed through space in the form of a torus containing neutral sulfur and oxygen. The recent detection of [OI] 6300 Å radiation from the torus region by Brown (1981a; cf. Sec. I-F) supports this conclusion.

VI. RADIAL DIFFUSION

On the basis of the ionization rates and the dielectronic recombination rates given by Jacobs et al. (1979), the observed ion distribution beyond 5.8 R_J (Shemansky 1980b) is not that expected for collisional ionization equilibrium with the measured electron energy distribution (Scudder et al. 1980). This difference suggests that the plasma is being transported sufficiently rapidly that transport can compete with ionization and prevent the buildup of higher ionization states.

Acceleration and diffusion of magnetically trapped particles require that one or more of the adiabatic invariants be violated (Northrop 1963). Since Jupiter's magnetosphere approximately corotates with the planet, the dominant radial transport mechanism for plasma is probably radial diffusion caused by violation of the third adiabatic invariant (conservation of magnetic flux within a drift orbit). This diffusion is a selective response of each charged particle to that part of a random disturbance in the electromagnetic field that is in resonance with its azimuthal period. Since the azimuthal period is essentially the corotation period, a violation of the third adiabatic invariant requires electric field fluctuations at the corotation frequency (Coroniti 1974).

For Jupiter's magnetosphere the mechanism creating these fluctuations that has gained the most acceptance is global-scale neutral winds in the ionosphere generating fluctuating dynamo electric fields, which are mapped into the magnetosphere. Brice and McDonough (1973) identified solar ultraviolet heating as the energy source driving the global-scale tidal-wind system, whereas Coroniti (1974) preferred a solar wind imposed convection electric field to drive the neutral winds through ion drag forces (Fedder and Banks 1972). Both processes drive electric field fluctuations at the corotation frequency and yield an L^3-dependence for the radial diffusion coefficient, D_{LL}, if the drift period is $\gtrsim 5$ hr (L designates a magnetic flux shell and is given numerically by the planetocentric distance in units of R_J at which the field lines of the shell cross the magnetic equator).

There is an additional mechanism that leads only to outward radial plasma transport from the source: magnetic flux tube interchange by centrifugally driven instability (Ioannidis and Brice 1971). Since the plasma density decreases outward from Io and the principal force acting on the plasma is the centrifugal force, the physical situation is similar to a heavy fluid over a light fluid, an unstable configuration. Ioannidis and Brice (1971) estimated growth times on the order of an hour and calculated that the equatorial plasma density should exhibit an L^{-4}-dependence in regions where this instability dominates.

The equation describing time dependent radial diffusion by either dynamo electric fields or centrifugally driven instability is (Richardson et al. 1980)

$$\frac{\partial(NL^2)}{\partial t} = L^2 \frac{\partial}{\partial L} \left[\frac{D_{LL}}{L^2} \frac{\partial}{\partial L} (NL^2) \right] + L^2 (S-R) \tag{3}$$

where N is the total number of ions (all species in a flux shell per unit L) and S and R are the source and loss strengths in a flux shell per unit L, respectively. The flux may be defined as (Siscoe and Chen 1977)

$$F = -R_J \frac{D_{LL}}{L^2} \frac{\partial}{\partial L} (NL^2). \tag{4}$$

The ordinary number density n is related to N by

$$N = 2\pi R_J^2 f L^2 n \tag{5}$$

where f is the fractional volume filled by plasma in a flux shell per unit L and is equal to L^{-1} for a torus confined within 1 R_J of the equator. It is important to note that NL^2 is the diffusing quantity and that near the source (Io) it, and not the pure number density, should be a maximum. The Voyager plasma density measurements exhibit a single maximum in NL^2, but double maxima in N (which would require two sources). They are thus consistent with radial diffusion described by Eq. (3) rather than with ordinary diffusion driven solely by a gradient in n (Richardson et al. 1980).

There is both theoretical and observational support for a radial diffusion coefficient of the form

$$D_{LL} = k L^m \tag{6}$$

(Brice and McDonough 1973; Coroniti 1974; Goertz and Thomsen 1979) and a corresponding time constant to diffuse ΔL of

$$\tau \sim \frac{(R_J \Delta L)^2}{D_{LL}} \tag{7}$$

where ΔL is unitless. Goertz and Thomsen (1979) deduced from ion densities measured by the plasma instrument on Pioneer 10 that

$$D_{LL} \simeq 3 \times 10^{-9} L^3 \tag{8}$$

in units of R_J^2 s^{-1}, which yields a time constant to diffuse 1 R_J at $L = 6$ of ~ 20 days, about one order of magnitude larger than that calculated by Coroniti (1974) and inferred from the same data by Siscoe and Chen (1977).

Assuming a plasma source localized at Io's L-shell, and using data from several Voyager 1 experiments, Froidevaux (1980) deduced for the region $6 < L < 8$ (diffusion outward from Io) limits of $D_{LL} > 1.5 \times 10^{-10} L^5$ $[R_J^2 \, s^{-1}]$ and $\tau(L = 6) < 10$ days. The density n was inferred to have an L^{-5} dependence, suggestive of centrifugally driven interchange diffusion. In a more detailed analysis of the same data, under the same local source assumption, Richardson et al. (1980) concluded that only a time-dependent description of radial diffusion could adequately describe the variation of NL^2 with L both inside and outside of Io's orbit. In particular they required that the plasma source switch on ~ 10 days prior to Voyager 1's arrival with an injection rate of $2 \times 10^{29 \pm 1}$ ions s^{-1} into the torus. They also required a discontinuity in k (Eq. 6) at $L = 5.8$ of

$$\frac{k(L > 5.8)}{k(L < 5.8)} = 50 \tag{9}$$

but could not discriminate between values of $m = 3-5$. Since $k(L < 5.8)$ in their analysis is determined by dynamo-driven radial diffusion for which Eq. 8 is a representative value, then $k(L > 5.8) \sim 1.5 \times 10^{-7} \, [R_J^2 \, s^{-1}]$ and $\tau(L = 6) \sim 0.4$ day with an uncertainty of an order of magnitude. The discontinuity in k at $L = 5.8$ is consistent with the onset of centrifugally driven interchange diffusion (Richardson et al. 1980).

Diffusion rates of the magnitude deduced from the *in situ* plasma measurements are required to account for the observed distribution of ions in the torus. Transport processes thus play an essential role in the radial dependence of all ion densities and preferentially direct ions outward from the Jovian system. To maintain a steady state ion density of 10^3 cm^{-3} over a torus of thickness $2 R_J$, the time constant for radial diffusion required to balance the total ionization rate (S ions s^{-1}) is

$$\tau \sim \frac{5 \times 10^{29}}{S} \text{ days.} \tag{10}$$

For values of $S = 2 \times 10^{28} \, s^{-1}$ (Hill 1980; Goertz 1980) and $2 \times 10^{29} \, s^{-1}$ (Richardson et al. 1980), the respective time constants are 30 and 3 days. These values are consistent with the diffusion rates derived directly from NL^2 radial profiles.

VII. DIELECTRONIC RECOMBINATION AND COLLISIONAL IONIZATION EQUILIBRIUM

As we have discussed in the previous section, the outward radial transport of torus plasma from Io may be rapid, perhaps exceeding the inward transport by as much as a factor of ~ 50. This suggests that in the inner torus ($L < 5.8$) the time constant for transport is > 30 days and may be > 100 days. For transport time constants in excess of 60 days, dielectronic recombination can begin to compete with radial diffusion, affecting the distribution of ionization states of oxygen and sulfur. In the asymptotic limit where radial diffusion is much slower than dielectronic recombination, the plasma approaches collisional ionization equilibrium with the thermal electrons. Bagenal et al. (1980), in their interpretation of the ion spatial distribution, concluded that inside Io's orbit the plasma cools by radiation, collapses toward the centrifugal equator, and slowly recombines as the plasma diffuses slowly inward towards Jupiter (cf. Sec. II-D). The high plasma temperature outside of Io's orbit is consistent with rapid outward diffusion.

Dielectronic recombination is a process that involves the simultaneous change in the quantum states of two electrons and is the principal path of electron-ion recombination in a high-temperature plasma. It may be viewed as a two-step process: first, the radiationless capture of an electron with simultaneous excitation of an already bound electron,

$$X^{(z)+} + e \rightarrow X^{(z-1)+} \text{ (doubly excited)}$$

followed by a stabilizing radiative transition to a final state below the ionization threshold,

$$X^{(z-1)+} \text{ (doubly excited)} \rightarrow X^{(z-1)+} \text{ (singly excited)} + h\nu,$$

with a subsequent radiative transition. For a sulfur plasma the most accurate dielectronic recombination coefficients currently available are the calculations of Jacobs et al. (1979), whereas for an oxygen plasma the best calculations are those of Jordan (1969) and Jacobs et al. (1979), which are in essential agreement. In addition, the calculations of Jacobs et al. (1978, 1979) yield the distribution of ionization states for isothermal oxygen and sulfur plasmas in collisional ionization equilibrium. These calculations indicate that a $1-2$ eV torus plasma should be composed primarily of OII and SII in agreement with groundbased and Voyager plasma science measurements of the inner torus.

December 1980

Acknowledgments. We thank several of our colleagues for helpful reviews of the manuscript, but particularly D.M. Hunten, W.H. Smyth, and R.A. Brown. We thank J.S. Morgan for assistance with some of the calculations. This work was supported in part by grants from the Planetary Astronomy, Planetary Atmospheres, and Planetary Geology sections of the National Aeronautics and Space Administration.

REFERENCES

Aitken, D.K., Harrison, M.F.A. (1971). Measurement of the cross-sections for electron impact ionization of multi-electron ions. I. O^+ to O^{2+} to O^{3+}. *J. Phys.* B4, 1176–1188.

Alfvén, H. (1954). *On the Origin of the Solar System,* Ch. 2. Oxford, Clarendon Press, London.

Bagenal, F., and Sullivan, J.D. (1981). Direct plasma measurements in the Io torus and inner magnetosphere of Jupiter. *J. Geophys. Res.* 86, 8447-8466.

Bagenal, F., Sullivan, J.D., and Siscoe, G.L. (1980). Spatial distribution of plasma in the Io torus. *Geophys. Res. Letters* 7, 41–44.

Bergstralh, J.T., Matson, D.L., and Johnson, T.V. (1975). Sodium D-line emission from Io: Synoptic observations from Table Mountain Observatory. *Astrophys. J.* 195, L131–L135.

Bergstralh, J.T., Young, J.W., Matson, D.L., and Johnson, T.V. (1977). Sodium D-line emission from Io. A second year of synoptic observations from Table Mountain Observatory. *Astrophys. J.* 211, L51–L55.

Bhadra, K., and Henry, R.J.W. (1980). Oscillator strengths and collision strengths for SIV. *Astrophys. J.* 240, 368-373.

Birmingham, T.J., Alexander, J.K., Desch, M.D., Hubbard, R.F., and Pederson, B.M. (1980). Observations of electron gyroharmonic waves and the structure of the Io torus. *J. Geophys. Res.* 86, 8497-8507.

Brice, N., and McDonough, T.R. (1973). Jupiter's radiation belts. *Icarus* 18, 206–219.

Bridge, H.S., and the Voyager Plasma Science Team (1979). Plasma observations near Jupiter: Initial results from Voyager 1. *Science* 204, 987–991.

Broadfoot, A.L., and the Voyager Ultraviolet Spectrometer Team (1977). Ultraviolet spectrometer experiment for the Voyager mission. *Space Sci. Rev.* 21, 183–205.

Broadfoot, A.L., and the Voyager Ultraviolet Spectrometer Team. (1979). Extreme ultraviolet observations from Voyager 1 encounter with Jupiter. *Science* 204, 979–982.

Brook, E., Harrison, M.F.A., and Smith, A.C.H. (1978). Electron ionization of He, C, N, and O atoms. *J. Phys.* B11, 3115.

Brown, R.A. (1974). Optical line emission from Io. In *Exploration of the Planetary System,* (A. Woszczyk and C. Iwaniszewska, Eds.), pp. 527–531. Reidel, Dordrecht.

Brown, R.A. (1976). A model of Jupiter's sulfur nebula. *Astrophys. J.* 206, L179–L183.

Brown, R.A. (1978). Measurements of SII optical emission from the thermal plasma of Jupiter. *Astrophys. J.* 224, L97–L98.

Brown, R.A. (1981a). The Jupiter hot plasma torus: Observed electron temperature and energy flow. *Astrophys. J.* 244, 1072-1080.

Brown, R.A. (1981b). Heavy ions in Jupiter's environment. In *Space Res. XXI.* Pergamon, Oxford. In press.

Brown, R.A., and Chaffee, F.H., Jr. (1974). High resolution spectra of sodium emission from Io. *Astrophys. J.* 187, L125–L126.

Brown, R.A., Goody, R.M., Murcray, F.J., and Chaffee, F.H., Jr. (1975). Further studies of line emission from Io. *Astrophys. J.* 200, L49–L53.

Brown, R.A., and Schneider, N.M. (1981). Sodium remote from Io. *Icarus* 48, 519-535.

Brown, R.A., and Yung, Y.L. (1976). Io, its atmosphere and optical emissions. In *Jupiter* (T. Gehrels, Ed.), pp. 1102–1145. Univ. Arizona Press, Tucson.

Carlson, R.W., and Judge, D.L. (1974). Pioneer 10 ultraviolet observations at Jupiter encounter. *J. Geophys. Res.* 79, 3623–3633.

Carlson, R.W., Matson, D.L., and Johnson, T.V. (1975). Electron impact ionization of Io's sodium emission cloud. *Geophys. Res. Letters* 2, 469–472.

Carlson, R.W., Matson, D.L., Johnson, T.V., and Bergstralh, J.T. (1978). Sodium D-line emission from Io: Comparison of observed and theoretical profiles. *Astrophys. J.* 223, 1082–1086.

Carr, T.D., and Desch, M.D. (1976). Recent decametric and hectometric observations of Jupiter. In *Jupiter* (T. Gehrels, Ed.), pp. 693–737. Univ. Arizona Press, Tucson.

Chamberlain, J.W. (1961). *Physics of the Aurora and Airglow*. Academic Press, New York.

Chandra, S., Mital, H.P., and Narain, U. (1976). Ionization cross-sections and rate coefficients for atoms, ions, and molecules. *Physica* 83C, 384–388.

Cloutier, P.A., Daniell, R.E., Jr., Dessler, A.J., and Hill, T.W. (1978). A cometary ionosphere model for Io. *Astrophys. Space Sci.* 55, 93–112.

Coroniti, F.V. (1974). Energetic electrons in Jupiter's magnetosphere. *Astrophys. J. Suppl.* 27, 261–281.

Cummings, W.D., Dessler, A.J., and Hill, T.W. (1980). Latitudinal oscillations of plasma within the Io torus. Submitted to *J. Geophys. Res.* 85, 2108-2114.

Dessler, A.J., and Hill, T.W. (1979). Jovian longitudinal control of Io-related radio emissions. *Astrophys. J.* 227, 664–675.

Dessler, A.J., and Vasyliunas, V.M. (1979). The magnetic anomaly model of the Jovian magnetosphere: Predictions for Voyager. *Geophys. Res. Letters* 6, 37–40.

Eviatar, A., Mekler, Y., and Coroniti, F.V. (1976). Jovian sodium plasma. *Astrophys. J.* 205, 622–633.

Eviatar, A., Siscoe, G.L., and Mekler, Y. (1979). Temperature anisotropy of the Jovian sulfur nebula. *Icarus* 39, 450–458.

Fedder, J.A., and Banks, P.M. (1972). Convection electric fields and polar thermospheric winds. *J. Geophys. Res.* 77, 2328–2340.

Froidevaux, L. (1980). Radial diffusion in Io's torus: Some implications from Voyager 1. *Geophys. Res. Letters* 7, 33–35.

Goertz, C.K. (1980). Io's interaction with the plasma torus. *J. Geophys. Res.* 85, 2949–2956.

Goertz, C.K., and Thomsen, M.F. (1979). Radial diffusion of Io-injected plasma. *J. Geophys. Res.* 84, 1499–1504.

Goldberg, B.A., Carlson, R.W., Matson, D.L., and Johnson, T.V. (1978). A new asymmetry in Io's sodium cloud. *Bull. Amer. Astron. Soc.* 10, 579 (abstract).

Goldberg, B.A., Mekler, Y., Carlson, R.W., Johnson, T.V., and Matson, D.L. (1980). Io's sodium emissioncloud and the Voyager 1 encounter. *Icarus* 44, 305–317.

Goody, R.M., and Apt, J. (1977). Observations of the sodium emission from Jupiter, Region C. *Planet. Space Sci.* 25, 603–604.

Hamdan, M., Birkinshaw, K., and Hasted, J.B. (1978). Ionization of positive ions by electrons in the hollow beam trap. *J. Phys.* B11, 331–337.

Hill, T.W. (1980). Corotation lag in Jupiter's magnetosphere: Comparison of observation and theory. *Science* 207, 301–302.

Hill, T.W., and Michel, F.C. (1976). Heavy ions from the Galilean satellites and the centrifugal distortion of the Jovian magnetosphere. *J. Geophys. Res.* 81, 4561–4565.

Ioannidis, G., and Brice, N.M. (1971). Plasma densities in the Jovian magnetosphere: Plasma slingshot or Maxwell demon? *Icarus* 14, 360–373.

Jacobs, V.L., Davis, J., and Rogerson, J.E. (1978). Ionization equilibrium and radiative energy loss rates for C, N, and O ions in low density plasmas. *J. Quant. Spec. Radiat. Transfer* 19, 591–598.

Jacobs, V.L., Davis, J., Rogerson, J.E., and Blaha, M. (1979). Dielectronic recombination rates, ionization equilibrium, and radiative energy-loss rates for neon, magnesium, and sulfur ions in low-density plasmas. *Astrophys. J.* 230, 627–638.

Johnson, T.V., Cook, A.F. II, Sagan, C., and Soderblom, L.A. (1979). Volcanic resurfacing rates and implications for volatiles on Io. *Nature* 280, 746–750.

Jordan, C. (1969). The ionization equilibrium of elements between carbon and nickel. *Mon. Not. Roy. Astron. Soc.* 142, 501–521.

Judge, D.L., Carlson, R.W., Wu, F.M., and Hartmann, U.G. (1976). Pioneer 10 and 11 ultraviolet photometer observations of the Jovian satellites. In *Jupiter* (T. Gehrels, Ed.), pp. 1068–1101. Univ. Arizona Press, Tucson.

Kupo, I., Mekler, Y., and Eviatar, A. (1976). Detection of ionized sulfur in the Jovian magnetosphere. *Astrophys. J.* 205, L51–L53.

Macy, W., and Trafton, L.M. (1975a). Io's sodium emission cloud. *Icarus* 25, 432–438.

Macy, W., and Trafton, L.M. (1975b). A model for Io's atmosphere and sodium cloud. *Astrophys. J.* 200, 510–519.

Matson, D.L., Goldberg, B.A., Johnson, T.V., and Carlson, R.W. (1978). Images of Io's sodium cloud. *Science* 199, 531–533.

Matson, D.L., Johnson, T.V., and Fanale, F.P. (1974). Sodium D-line emission from Io: Sputtering and resonant scattering hypothesis. *Astrophys. J.* 192, L43–L46.

McElroy, M.B., and Yung, Y.L. (1975). The atmosphere and ionosphere of Io. *Astrophys. J.* 196, 227–250.

McElroy, M.B., Yung, Y.L., and Brown, R.A. (1974). Sodium emission from Io: Implications. *Astrophys. J.* 187, L127–L130.

McNutt, R.L., Jr., Belcher, J.W., Sullivan, J.D., Bagenal, F., and Bridge, H.S. (1979). Departure from rigid corotation of plasma in Jupiter's dayside magnetosphere. *Nature* 280, 803.

Mekler, Y., and Eviatar, A. (1974). Spectroscopic observations of Io. *Astrophys. J.* 193, L151–L152.

Mekler, Y., and Eviatar, A. (1978). Thermal electron density in the Jovian magnetosphere. *J. Geophys. Res.* 83, 5679–5684.

Mekler, Y., Eviatar, A., and Kupo, I. (1977). Jovian sulfur nebula. *J. Geophys. Res.* 82, 2809–2814.

Mekler, Y., Eviatar, A., and Siscoe, G.L. (1979). Discontinuities in Jovian sulfur plasma. *Mon. Not. Roy. Astron. Soc.* 189, 15–17.

Moos, H.W., and Clarke, J.T. (1981). Ultraviolet observations of the Io torus from the IUE Observatory. *Astrophys. J.* 247, 354-361.

Morgan, J.S., and Pilcher, C.B., (1981). Plasma characteristics of the Io torus. Submitted to *Astrophys. J.* 253, 406-421.

Münch, G., Trauger, J., and Roesler, F. (1976). Interferometric studies of the emissions associated with Io. *Bull. Amer. Astron. Soc.* 8, 468 (abstract).

Münch, G., and Bergstralh, J.T. (1977). Io: Morphology of its sodium emission region. *Publ. Astron. Soc. Pacific* 89, 232–237.

Murcray, F.J., and Goody, R.M. (1978). Pictures of the Io sodium cloud. *Astrophys. J.* 226, 327–335.

Nash, D.B. (1979). Jupiter sulfur plasma ring. *EOS* 60, 307 (abstract).

Northrop, T.G. (1963). *The Adiabatic Motion of Charged Particles.* Interscience Publ., New York.

Osterbrock, D.E. (1974). *Astrophysics of Gaseous Nebulae.* Freeman, San Francisco.

Parkinson, T.D. (1975). Excitation of the sodium D-line emission observed in the vicinity of Io. *J. Atmos. Sci.* 32, 630–633.

Peach, G. (1968). Ionization of neutral atoms with outer 2p and 3p electrons by electron and proton impact. *J. Phys.* B1, 1088–1108.

Peach, G. (1970). Ionization of neutral atoms with outer 2p, 3s, and 3p electrons by electron and proton impact. *J. Phys.* B3, 328–349.

Peach, G. (1971). Ionization of atoms and positive ions by electron and proton impact. *J. Phys.* B4, 1670–1677.

Pearl, J., Hanel, R., Kunde, V., Maguire, W., Ponnamperuma, C., and Raulin, F. (1979). Identification of gaseous SO_2 and new upper limits for other gases on Io. *Nature* 280, 755–758.

Pilcher, C.B. (1980a). As reported in *Science* 208, 384–386.

Pilcher, C.B. (1980b). Images of Jupiter's sulfur ring. *Science* 207, 181–183.

Pilcher, C.B., and Morgan, J.S. (1979). Detection of singly ionized oxygen around Jupiter. *Science* 205, 297–298.

Pilcher, C.B., and Morgan, J.S. (1980). The distribution of [SII] emission around Jupiter. *Astrophys. J.* 238, 375–380.

Pilcher, C.B., and Schempp, W.V. (1979). Jovian sodium emission from Region C_2. *Icarus* 38, 1–11.

Pradhan, A.K. (1976). Collision strengths for [OII] and [SII]. *Mon. Not. Roy. Astron. Soc.* 177, 31-38.

Pradhan, A.K. (1978). Fine structure transitions by electron impact in singly-ionized sulphur. *Mon. Not. Roy. Astron. Soc.* 184, 89-92.

Raadu, M.A. (1978). The role of electrostatic instabilities in the critical ionization velocity mechanism. *Astrophys. Space Sci.* 55, 125-138.

Richardson, J.D., Siscoe, G.L., Bagenal, F., and Sullivan, J.D. (1980). Time dependent plasma injection by Io. *Geophys. Res. Letters* 7, 37-40.

Sandel, B.R. (1980). Azimuthal asymmetry in Io's hot torus. IAU Coll. 57. *The Satellites of Jupiter* (abstract 5-8).

Sandel, B.R., and the Voyager Ultraviolet Spectrometer Team. (1979). Extreme ultraviolet observations from Voyager 2 encounter with Jupiter. *Science* 206, 962-966.

Scudder, J.D., Sittler, E.C., and Bridge, H.S. (1980). In situ Jovian low energy electron plasma measurement from Voyager 1 and 2. *J. Geophys. Res.* 86, 8157-8180.

Shemansky, D.E. (1980a). Radiative cooling efficiencies and predicted spectra of species of the Io plasma torus. *Astrophys. J.* 236, 1043-1054.

Shemansky, D.E. (1980b). Mass loading and the diffusion loss rate of the Io plasma torus. *Astrophys. J.* 242, 1266-1277.

Siscoe, G.L., and Chen, C.K. 1977. Io: A source for Jupiter's inner plasmasphere. *Icarus* 31, 1-10.

Smith, E.J., Davis, L., Jr., Jones, D.E., Coleman, P.J., Jr., Colburn, D.S., Dyal, P., Sonett, C.P., and Frandsen, A.M.A. (1974). The planetary magnetic field and magnetosphere of Jupiter: Pioneer 10. *J. Geophys. Res.* 79, 3501-3513.

Smyth, W.H. (1979). Io's sodium cloud: Explanation of the east-west asymmetries. *Astrophys. J.* 234, 1148-1153.

Smyth, W.H., and McElroy, M.B. (1977). The sodium and hydrogen gas clouds of Io. *Planet. Space Sci.* 25, 415-431.

Smyth, W.H., and McElroy, M.B. (1978). Io's sodium cloud: Comparison of models and two-dimensional images. *Astrophys. J.* 226, 336-346.

Strobel, D.F., and Davis, J. (1980). Properties of the Io plasma torus inferred from Voyager EUV data. *Astrophys. J.* 238: L49-L52.

Trafton, L. (1975a). High-resolution spectra of Io's sodium emission. *Astrophys. J.* 202, L107-L112.

Trafton, L.M. (1975b). Detection of a potassium cloud near Io. *Nature* 258, 690-692.

Trafton, L.M. (1977). Periodic variations in Io's sodium and potassium clouds. *Astrophys. J.* 215, 960-970.

Trafton, L. (1980). The Jovian SII torus: Its longitudinal asymmetry. *Icarus* 42, 111-124.

Trafton, L., and Macy, W., Jr. (1975). An oscillating asymmetry to Io's sodium emission cloud. *Astrophys. J.* 202, L155-L158.

Trafton, L., and Macy, W., Jr. (1977). Io's sodium emission profiles: Variations due to Io's phase and magnetic latitude. *Astrophys. J.* 215, 971-976.

Trafton, L., and Macy, W., Jr. (1978). On the distribution of sodium in the vicinity of Io. *Icarus* 33, 322-335.

Trafton, L., Parkinson, T., and Macy, W., Jr. (1974). The spatial extent of sodium emission around Io. *Astrophys. J.* 190, L85-L89.

Trauger, J.T., Münch, G., and Roesler, F. (1979). A study of the Jovian [SII] and [SIII] nebulae at high spectral resolution. *Bull. Amer. Astron. Soc.* 11, 591-592 (abstract).

Trauger, J.T., Münch, G., and Roesler, F.L. (1980). A study of the Jovian [SII] nebula at high spectral resolution. *Astrophys. J.* 236, 1035-1042.

Van Regemorter, H. (1962). Rate of collisional excitation in stellar atmospheres. *Astrophys. J.* 136, 906-915.

Varma, R.K. (1978). On Alfvén's critical velocity for the interaction of a neutral gas with a moving magnetized plasma. *Astrophys. Space Sci.* 55, 113-124.

Warwick, J.W., and the Voyager Planetary Radio Astronomy Team. (1979). Voyager 1 planetary radio astronomy observations near Jupiter. *Science* 204, 995-998.

Wehinger, P.A., and Wyckoff, S. (1974). Jupiter I. *IAU Circ.* 2701.

Wehinger, P.A., Wyckoff, S., and Frohlich, A. (1976). Mapping of the sodium emission associated with Io and Jupiter. *Icarus* 27, 425-428.

Younger, S.M., and Wiese, W.L. (1979). An assessment of the effective Gaunt factor approximation. *J. Quant Spec. Rad. Trans.* 22, 161-170.

23. IN SITU OBSERVATIONS OF IO
TORUS PLASMA

JAMES D. SULLIVAN
Massachusetts Institute of Technology

and

GEORGE L. SISCOE
University of California at Los Angeles

Direct measurements of electrons and positive ions in the Io plasma torus were made on 5 March 1979, as the Voyager 1 spacecraft passed through the torus within 4.9 R_J of Jupiter. The torus plasma is characterized by spatially distinct regions with steep gradients in plasma parameters between them. The innermost region has a cool plasma dominated by S^+ ions with a thermal energy of a few eV. In this region, which we refer to as the precipice to describe the sharp decrease in total mass and temperature that were observed there, the plasma corotates rigidly with the planet. The great decrease in temperature in the precipice causes the plasma to collapse toward the centrifugal equator and gives rise to a distinctive localized concentration of plasma well inside of Io's orbit. The next region farther out has a warm plasma consisting chiefly of ions of sulfur and oxygen with a thermal energy of several tens of eV and electrons whose distribution has a two-component structure with a Maxwellian thermal component at low energies and a non-Maxwellian suprathermal component at higher energies. This region includes the L-shell of Io and is the presumed injection region of the plasma. Farther out beyond $\sim 8 R_J$, in the region identified with the inner edge of the radially distended Io plasma disk, the plasma is slightly warmer and departs measurably from rigid corotation. The region between the precipice and the disk containing *the ultraviolet-emitting Io torus is further divided into a plasma ledge extending from 5.8 R_J to ~ 7 R_J in which the density varies slowly with distance compared to the region between 7 and 8 R_J, referred to as the ramp, where the density drops rapidly. To account for such sharply defined changes in plasma characteristics over the radial range 5 to 9 R_J, we invoke centrifugally driven flux tube interchange diffusion to provide radial mass transport. The ramp is shown to result from the impoundment of the Io plasma by the inner edge of the energetic particle population, also known as the ring current. Finally, we show how the power required to excite the ultraviolet emissions of the Io torus and the Jovian aurora determines the rate at which new plasma is fed into the Io torus.*

Io is the origin of one of the most interesting plasma formations existing in the solar system, namely the Io plasma torus with its radially distended plasma disk. The Io torus consists of approximately two million tons of a mixture composed primarily of oxygen and sulfur ions in various charge states. Remote sensing and *in situ* observations of this plasma body have thus yielded important information on the composition of the surface of Io and of its atmosphere. The rate at which Io must replace the material to maintain the torus against dispersal into space can be inferred from the intensities of the observed radiations of the Io torus and the Jovian aurora. This rate is needed to constrain proposed mechanisms for removing material from Io. The Io plasma formation makes up a large part of the spatial environment of the Galilean satellites and possibly of the inner satellites and the ring. Io, Europa and 1979 J2 are virtually continuously immersed in it. Thus satellite and possibly ring surfaces might be modified in important ways as a result of exposure to the magnetospheric "pollution" of Io.

The goal of this chapter is not to explore in any detail the satellite consequences to which we have just alluded. Instead we describe in some detail the physical properties of the Io plasma formation deduced from *in situ* observations. This information together with that contained in Chapter 22 on remote sensing observation of the Io torus should establish the basis for determining the effects that the Io plasma formation has on the satellite-ring system. The first part of the chapter specifies the morphology, density, temperature, motion, composition and other aspects of torus phenomenology; it is a synopsis of Bagenal and Sullivan (1981) augmented by results from the plasma electron analysis of Scudder et al. (1981). The second part (Secs. II, III and IV) concerns the physical principles that govern the behavior in space and time of the torus ions.

I. IN SITU OBSERVATIONS

Since 1956, remote sensing of radio noise from the Jovian system has been used to infer the existence of intense fluxes of relativistic electrons in a strong planetary magnetic field. Further observations have established the control of decametric radio bursts by Io (Bigg 1964). Thus a strong satellite-magnetosphere interaction at Jupiter was known before the start of planetary exploration with spacecraft, and provided a motivation for direct measurements.

The first *in situ* observations of Jupiter were made in December 1973 with the Pioneer 10 flyby and extended a year later by Pioneer 11. Excellent data were obtained on the magnetic field around Jupiter and on the energy spectrum and spatial distribution of charged particles with energies above ~ 0.5 MeV/nucleon. The main satellite-magnetosphere interaction for these charged particles is the sweeping out or absorption of them as they diffuse across the drift shells intercepting various satellites. In fact the existence of a Jovian ring was suggested by Acuña and Ness (1976) on the basis of their analysis of charged-particle absorption on Pioneer 11. Alternative explanations were also advanced (Roederer et al. 1977) so that this interpretation remained ambiguous until the ring was observed optically (Smith et al. 1979).

The Pioneer plasma instrument also provided some evidence concerning the plasma environment. Low-energy positive ions (100 to 4800 eV), interpreted as protons, were detected near Io's orbit (Frank et al. 1976). However, it has subsequently been shown that these ions could not have been protons but were probably heavy ions (Neugebauer and Eviatar 1976; Goertz and Thomsen 1979). The thermal electron measurements were, apparently, seriously compromised by trapped photoelectrons and secondary electrons (Grard et al. 1977; Scudder et al. 1981).

Remote sensing by groundbased optical observations of the region near Io and its orbit have revealed emission from NaI (Brown and Chaffee 1974), from SII (Kupo et al. 1976) and from other atomic and ionic species. The optical radiation observed by Kupo et al. came from a region inside the orbit of Io at 5.95 R_J (1 R_J = 71398 km, the equatorial radius of Jupiter) and was interpreted by Brown (1976) as originating from a dense ring of cold plasma in that region. Although this conclusion provoked little discussion at the time and was not fully appreciated by the energetic particle community, it marks the discovery of the (cold) plasma torus at Io.

The SII emission has been monitored intermittently since its initial discovery. It has shown variabilities on time scales from less than a day (Pilcher 1979) to more than a month (Mekler and Eviatar 1980). Variations with System III longitude are also reported (Morgan and Pilcher 1978; Trafton 1980), so that some of the short-term variations may actually be longitudinal.

Strong SIII, SIV and OIII emissions were observed by the ultraviolet spectrometer on Voyager 1 (Broadfoot et al. 1979; Sandel et al. 1979). These emissions were interpreted as coming from a warm plasma torus centered on

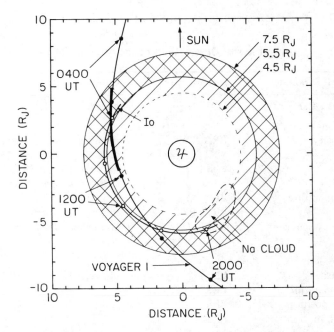

Fig. 23.1. Trajectory of the Voyager 1 spacecraft through the inner magnetosphere projected onto the plane of the ecliptic. Voyager 1 traversed the outer (warm) torus (double hatched) in the mid-afternoon inbound and late evening outbound. The spacecraft moved through inner (cold) torus (single hatched) near closest approach. The extent of the cloud of neutral sodium atoms around Io is illustrated for 2000 UT. All the times are Universal Time (UT) on 5 March 1979. The orbit of Io for the period is also illustrated. (From Bagenal and Sullivan 1981.)

Io's orbit with a cross-sectional radius of 1 R_J that would include most of the optical torus in SII. This warm torus was not observed by Pioneer in 1973 and 1974, but it has been seen since by Voyager 2, International Ultraviolet Explorer, and groundbased observers. Thus it is apparent from remote sensing observations that the Io-magnetosphere interaction is complex and variable and further that it involves a significant cloud of neutral sodium as well as a hot and a cold plasma torus of sulfur. Details on the remote sensing of neutrals and ions may be found in Chapter 22 by Pilcher and Strobel.

Direct measurements of this plasma environment were obtained with the instrumentation on the Voyager 1 spacecraft, which on 5 March 1979 had a periapsis of 4.89 R_J as it passed through the inner magnetosphere of Jupiter, through the region of the plasma torus near Io, and near the Io flux tube (Fig. 23.1). The physical parameters measured *in situ* and available for characterizing the plasma environment are summarized in Table 23.1. Again the magnetic field measurements are definitive and the energetic particle measurements extensive, but it is the *in situ* plasma and plasma wave measurements

TABLE 23.1

Voyager in situ Observations of the Io Torus Region

Physical Parameter	Instrument[a]	Comments
Magnetic field	MAG	high time resolution
Electrons		
density	PRA, PWS	depends on plasma wave mode identification
	PLS	between 10 and 5950 V sensitive to spacecraft
temperature	PLS	potential
	PRA, PWS	indirect argument
energy spectrum	LECP, CRS	compromised by high counting rates
Positive ions		
density	PLS	between 10 and 5950 V
temperature	PLS	
composition	PLS	some model dependence
bulk motion	PLS	
energy spectrum	LECP, CRS	composition above 0.5 MeV compromised by high counting rates
Plasma wave spectrum	PRA PWS	

[a]acronyms: MAG – Magnetometer; PRA – Planetary Radio Astronomy; PWS – Plasma Wave Science; PLS – Plasma Science; LECP – Low Energy Charged Particles; CRS – Cosmic Ray Science (see *Science* 204, 945–1008, 1979).

that provided the primary new data, albeit as a snapshot in time and space, defining the plasma environment. It is these measurements that revealed the unanticipated complexity of structure in the spatially extended plasma body formed out of heavy ions from Io, as well as a distinctive satellite-magnetosphere interaction when Voyager 2 passed through the corotating wake of Ganymede (Bridge et al. 1979*b*; Burlaga et al. 1980).

Morphology of the Plasma Torus

As the spacecraft approached Jupiter, the distended Io plasma disk, which it had sensed at first periodically in the outer magnetosphere and then continually inside of 20 R_J, gave way to a ramp of markedly increasing density near 8 R_J. This led to a more gradually increasing density ledge beginning near 7 R_J.

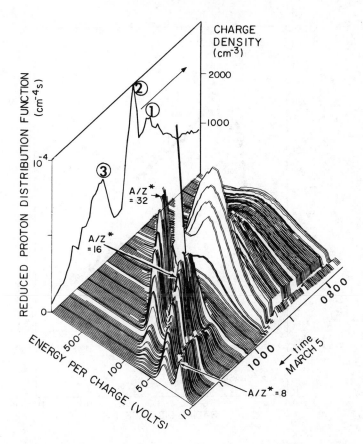

Fig. 23.2. A three-dimensional plot of reduced proton distribution function against energy per charge for spectral measurements made in the C-cup of the main Voyager plasma sensor between 0730 UT (7 R_J) and 1145 UT (4.9 R_J) on 5 March 1979. A total of 160 spectra are shown. Two spectra are omitted every 48 minutes while the instrument was in a different measurement mode. Every tenth spectrum is emphasized with a darker line. The back panel shows the total positive charge density as a function of time determined from fits to the corresponding spectra. (From Bagenal and Sullivan 1981.)

A general survey of positive ion plasma data in the ledge and inner torus (corresponding to the portion in the trajectory shown by the thickened line in Fig. 23.1) is shown in the three-dimensional plot of Fig. 23.2, where the reduced ion distribution function is plotted against (spacecraft event) time and energy per charge. After ~ 1000 UT in the cold inner torus, three density peaks stand out in the energy-per-charge spectra at atomic-mass-to-electronic-charge ratios (A/Z^*) of 32, 16 and 8; the relative amplitudes of these peaks vary systematically, with maximum values for the individual ionic species occurring at 1030, 1050 and 1130 UT, respectively. Before ~ 0930 UT in the

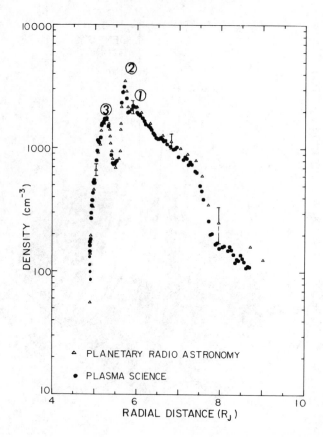

Fig. 23.3. Radial profile of *in situ* measurements of charge density. The plasma science measurements (●) are of the total positive charge derived from fits to positive ion energy-per-charge spectra. The planetary radio astronomy data (△) are of electron density determined from the cutoff frequency of plasma wave modes (the uncertainties in the planetary radio astronomy determinations are shown by vertical bars). (From Bagenal and Sullivan 1981.)

outer torus, the spectra are characterized by a single broad peak at about $A/Z^* = 16$. The back panel shows the positive charge density as a function of time, determined from fits to each spectrum assuming that only ions with A/Z^* between 8 and 64 are present. The local maxima in the charge density profile at 0902, 0924 and 1016 are labeled as peaks 1, 2 and 3, respectively.

Electron number density. A radial profile out to 9 R_J of the charge density determined from fits to the positive ion spectra and from the Planetary Radio Astronomy (PRA) instrument (Warwick et al. 1979) is shown in Fig. 23.3. There is close agreement between the two measurements over the entire region, indicating that most of the ions have $A/Z^* \gtrsim 8$. The three local maxima are labeled as in the previous figure.

We identify the outer edge of the Io plasma torus with the rapid increase in density in the ramp measured by both instruments as the spacecraft moved inside ~ 8 R_J. The density in the ledge built up to the broad maximum around the orbit of Io at 5.95 R_J (peak 1). In the sharp peak at ~ 5.7 R_J (peak 2), well inside Io's L-shell, the density reached ~ 3500 cm^{-3}, the highest value recorded throughout the encounter. However, there were few measurements of such large values, and these occurred within a small radial distance of 5.75 R_J. The bulk of the core of the torus had a charge density from 1000 to 2000 cm^{-3}. In the drop that occurs inward of 5.7 R_J, the charge density plummeted to a value of ~ 740 cm^{-3}, a factor of five decrease within 0.2 R_J. As the spacecraft moved through this cold precipice, the charge density reached the third maximum of 1740 cm^{-3} at 5.3 R_J (peak 3), before its final rapid decrease by more than an order of magnitude as the spacecraft made its closest approach to Jupiter at 4.89 R_J.

In the plasma disk, the values of charge density determined from the positive ion measurements closely match electron densities determined directly from the plasma instrument. Computation of the electron density in the plasma ledge from the plasma science electron data is not yet completed (Scudder et al. 1981).

Temperature. Figure 23.4 shows the radial temperature profile determined from fits to the positive ion data with uncertainty limits due to model dependencies in the analysis. The plasma beyond ~ 5.8 R_J had a fairly constant temperature of $(6 \pm 1.5) \times 10^5$ (~ 50 eV), an order of magnitude less than that expected if the gyro-speed were equal to the full corotational value. Moving inward through the torus, the temperature decreased sharply inside 5.7 R_J, dropping by a factor of ~ 50 to $\lesssim 1$ eV. This sharp transition marking the division between the precipice and the plasma ledge occurs in the same region where the plasma density decreases by a factor of five.

Bagenal and Sullivan noted that throughout the torus significant but small particle fluxes were detected at energies well above the energy of the bulk of the plasma. They listed the alternative interpretations that either very heavy molecular ions were present (Sullivan and Bagenal 1979) or recently created sulfur ions that had not yet thermalized were present.

Scudder et al. (1981) showed that the electron plasma environment in Io's torus is markedly non-Maxwellian with a suprathermal population that becomes more important relative to thermal electrons with increasing distance from Jupiter. Beyond 8 R_J, the Voyager spacecraft reached higher magnetic latitudes off the plasma disk. At these distances, they noted a definite increase with latitude in the number and pressure of the suprathermal

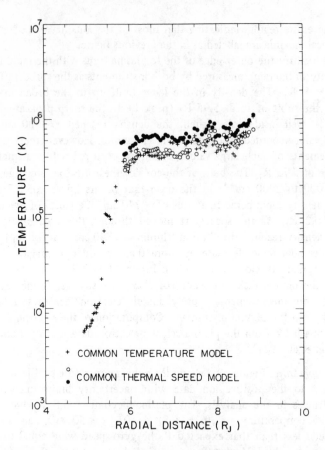

Fig. 23.4. Radial profile of ion temperature derived from fits to the positive ion energy-per-charge spectra. The +'s are from the fits where the ions are assumed to have the same temperature. With the common thermal speed model the average ion temperature has been calculated assuming the $A/Z^* = 16$ spectral peak to be all S^{2+} (●) or all O^+ (o). (From Bagenal and Sullivan 1981.)

electrons relative to the thermal electrons; that is, beyond 8 R_J the effective electron temperature is an increasing function of magnetic latitude. Consequently we might expect a similar latitudinal dependence to hold for the electron temperature inside 8 R_J. Scudder and his colleagues infer distinct electron thermal regimes with mean energies in the disk, ramp, and precipice of the order of 100, 10–40 and < 5 eV, respectively, Figure 23.5 (from Scudder et al. 1981) gives spectra from the three regimes; however, these spectra are not necessarily typical of the regimes. Clearly the plasma torus is not well fit with an isothermal, Maxwellian plasma, but rather it is an inhomogeneous plasma in density and temperature. Scudder et al. (1981) state the following caution: "This experimental fact complicates the interpretation of line-of-site

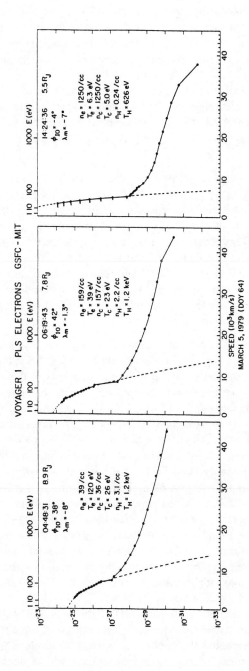

Fig. 23.5. The electron distribution function plotted at different times within Io's plasma tours by the Voyager 1 plasma science instrument. The measurement times, radial distance of spacecraft from Jupiter, System III longitude of spacecraft relative to Io, and dipole magnetic latitude of the spacecraft are indicated. (From Scudder et al. 1981).

measurements; however, in order to extract the maximum information from such integral measurements, the community should be aware of the observed structure in the electron macroscopic and microscopic parameters and that the system is not in local thermal equilibrium." This result from the first *in situ* observations of an emission nubula might well have more general consequences.

The results from analyzing the spectra in Fig. 23.5 are tabulated in each panel. These direct measurements of electron properties are in good agreement with inferences of the electron temperature ratio of the hot and cold components made by Birmingham et al. (1981) based on PRA observations.

Bulk motion. Bagenal and Sullivan report that the plasma in the precipice is corotating to better than 1 %. In the ledge, they find that the plasma flow is also consistent with corotation, but model dependencies of ~ 5% exist. Farther out in the disk, they determine that the plasma flow deviates measurably from strict corotation. Kaiser and Desch (1980), using the PRA instrument as a remote sensor, also find evidence for subcorotation of plasma flow by 3 to 5 % at 8–9 R_J. These results are consistent with the model of Hill (1979), which uses mass loading to cause ionospheric slippage and a consequent progressive lagging behind of the plasma flow compared to strict corotation outside the torus.

Composition. Table 23.2 presents *in situ* densities of various ionic species determined from fits to energy-per-charge spectra at 4.96 and 5.3 R_J (in the precipice); 6.0 R_J (in the ledge); and just outside the torus in the disk (8.6 R_J). Each of the positive ion spectra is representative of the plasma in the respective regions being measured. At 6.0 and 8.6 R_J, the results of fitting the spectra under different thermal models are tabulated. The main difference is seen in the relative abundance of OIII and SIV.

There is often a well-defined spectral peak corresponding to $A/Z^* = 64$ (Fig. 23.3), which is probably due to SO_2^+ (or possibly S_2^+, which has the same A/Z^* ratio). Fitting the $A/Z^* = 64$ peak at 5.3 R_J produces a density of ~ 13 cm^{-3}, which is < 1 % of the ion population. When the plasma becomes too hot for separate spectral peaks to be identified, the density of the ion with $A/Z^* = 64$ can be found from a well-defined shoulder. The density of this heavy ion in the warm torus had a maximum value around Io's orbit of ~ 8 cm^{-3} (< 1% of the ions) for the common thermal speed model and ~ 73 cm^{-3} (~5.5% of the ions) for the isothermal model. Its relative abundance increased to ~ 10% in the disk at 8.6 R_J.

Upper limits have been put on the densities of ions with A/Z^* values that come between two resolved spectral peaks. When fitting a gap in a spectrum between two peaks we choose the minor ionic species for its plausibility. An upper limit for the density of any other ionic species with a similar A/Z^*

TABLE 23.2

Composition of the Plasma in the Dayside Magnetosphere of Jupiter[a]

A/Z*	UT RJ	1120 4.96	1016 5.3	0859 6.0	0527 8.6[e]	
1	H^+					
8	O^{2+}	48	26	160 —	28 26^b	20 9^b
10	S^{3+}	< 5.6	< 3.5	27 170^b	— —	0.5 11^b
16	S^{2+}	14	39	430 560	— 16^d	19 12^d
16	O^+	250	350	140 1100	34 32^c	— 24^c
23	Na^+	< 20	< 72			
32	S^+	91	1100	430 470	8 11^b	28 23^b
48	SO^+	< 2	< 8			
64	SO_2^+	3.5	13 8^b	73 8^b	7 8^b	7
V_c		62	66	75	108	(corotation speed)
V/V_c		1.0	1.0	1.0	1.0	0.9

[a]From Bagenal and Sullivan (1981). All densities are derived from the isothermal model for the ions unless otherwise indicated.

[b]From constant thermal speed model.

[c]From constant thermal speed model when A/Z* = 16 spectral peak is assumed to be all O^+.

[d]From constant thermal speed model when A/Z* = 16 spectral peak is assumed to be all S^{2+}.

[e]The two columns for 8.6 R_J correspond to two different assumptions on the flow speed, as shown in the bottom line, V/V_c.

ratio could be determined in the same way. Finally, as the plasma scan determines only energy per charge, there is an inherent ambiguity in the identification of a given value of A/Z* with any particular species. The identifications given by Bagenal and Sullivan (1981) are adopted here.

Spatial Variations of Plasma Properties

Contour maps of the positive charge density and of the number density of selected ionic species have been constructed for the torus by Bagenal and

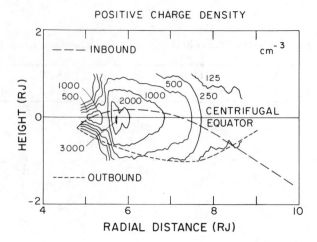

Fig. 23.6. Contour map of positive charge density as a function of radial distance from the center of Jupiter and height from the centrifugal equator. The map has been constructed from plasma measurements made along the inbound spacecraft trajectory (———) by using a theoretical expression for the distribution of plasma along dipolar field lines. (From Bagenal and Sullivan 1981.)

Sullivan (1981). They used the *in situ* density and kinetic temperature of each ionic species and the *in situ* electron temperatures of Scudder et al. (1981) and then applied the equations appropriate to ambipolar diffusion in a rotating, centered, tilted dipole with mirror symmetry about the centrifugal equator. The resulting map of total charge density in the magnetic meridian plane is shown in Fig. 23.6. The apparent asymmetry about the centrifugal equator is due to the geometry of the tilted dipole. While both the inbound and outbound trajectories of the spacecraft are shown, the map is based solely on data obtained on the inbound pass.

Features in the map are interpreted under the dual assumptions that the time scale for intrinsic time variations in the plasma was longer than the time required for Voyager to transit the structure and that the spatial scale of plasma variations in the direction of corotation was large compared to other spatial scales; then the observed variations correspond to structures defined in the magnetic meridian plane. Thus the plasma formation between 5.8 R_J and 9 R_J (where the detailed density and temperature determinations end) has a nearly uniform thickness of $\sim$ 1.5 R_J in the latitudinal direction. The thickness is determined by the balance between thermal pressure and centrifugal and electrical potential along field lines in accordance with the usual theory of ambipolar diffusion. The uniform thickness over this radial stretch reflects the condition of uniform temperature.

In the region inside 5.8 R_J, referred to as the precipice, both the total number of ions and their temperature decline acutely with a sharp temperature gradient of $\sim 7 \times 10^5$ K R_J^{-1}. The drop in temperature to $\lesssim 1$ eV in the precipice reduces the latitudinal thickness to ~ 0.2 R_J. This collapse of plasma toward the centrifugal equator gives rise to a localized knoll in the contour maps of density in the magnetic meridian plane. The spacecraft traversed the crest of this knoll and recorded the innermost peak in the *in situ* density as a result (Bagenal et al. 1980).

The plasma measurements of Pioneer 10 indicate significant amounts of plasma inside the periapsis of Voyager 1 (Frank et al. 1976). Voyager 1 passed under Io at ~ 1500 UT through the region where a neutral cloud of sodium normally surrounds Io; however, there was no direct way of detecting neutral sodium. The precipice around 5.5 R_J is also one of the three regions where whistler mode plasma waves were observed by the Voyager 1 Plasma Wave Science instrument (Gurnett el al. 1979). The other two regions were at ~ 6 R_J on both inbound and outbound legs of the trajectory.

The measured frequency dispersion of the whistlers is largely determined by the density of electrons in the plasma through which these waves have propagated. Gurnett et al. (1981) conclude that the dispersion of these whistlers suggests that there are considerably more electrons between the point of measurement and the source of the whistlers in the ionosphere of Jupiter than are accounted for in the heavy ion torus. If there are additional electrons, they could be associated with light ions such as protons that would not be confined to the centrifugal equator, where they have negative scale heights. Thus, light ions could dominate closer to the planet, while forming a small portion of the plasma near the equator, where the measured total positive-charge density from heavy ions is in fact very close to the electron density.

Bagenal and Sullivan also found that the global density structure on the outbound pass (see Fig. 23.1) was nearly indistinguishable from that on the inbound pass, with similar features at similar L-shells, when the offset of the titled dipole was taken into account. The main effect of including the offset is to change the apparent outbound trajectory shown in Fig. 23.6. When this was done, the electron densities predicted for the outbound structure from inbound data were in good agreement with those measured by Warwick et al. (1979). Consequently, there is no clear evidence that the electron density varies in local time in order to account for the enhanced ultraviolet emission reported from the dusk quadrant of the torus (Sandel 1980). In contrast, radical changes in plasma properties are observed over radial spatial scales as short as $\sim 10^4$ km, e.g., the temperature gradient at 5.5 R_J.

Composition. Contour maps are presented in Fig. 23.7 for the major ionic species observed in the inner magnetosphere. Different assumptions on plasma conditions in the plasma ledge were made for the a-panels and the

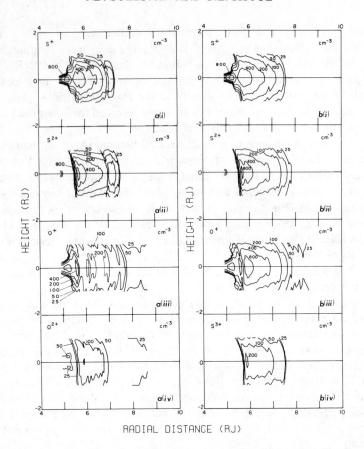

Fig. 23.7. Contour maps of ion number density for the major ionic species found
in the inner magnetosphere of Jupiter. The panels on the left, (a), have been
constructed from fits to the energy-per-charge spectra assuming the ions have
the same temperature. The panels on the right, (b), have been similarly
constructed assuming the ions to be isothermal inside $L = 5.7$ and to have a
common thermal speed at larger L-shells. (From Bagenal and Sullivan 1981.)

b-panels. For the a-panels the ambiguity in A/Z^* between O^+ and S^{2+} is
removed. In both panels several ambiguities remain. For example at $A/Z^* =$
32, S^+ could be O_2^+. This figure is discussed extensively in Bagenal and
Sullivan (1981). Some noteworthy features are:

1. The knoll is predominantly S^+, as seen in panel a(i);
2. In panel a(iv) the O^{2+} is concentrated in two knolls off the equator;
3. The ionic composition is dominated by sulphur unless O_2^+ is a major
    constituent.

The S^+ knoll may well indicate that recombination, which acts quickets on
the highest charge states, is occurring in the precipice. The twin knolls of O^{2+}

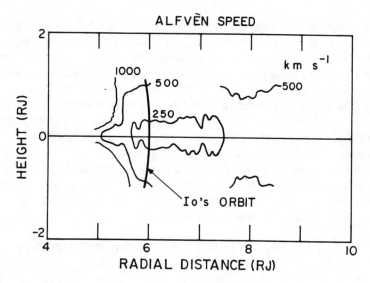

Fig. 23.8. Contour map of local Alfvén speed calculated from the O_4 magnetic field model (Acuña and Ness 1976) and the total ion mass density, measured on the spacecraft trajectory and extrapolated along the field lines. (From Bagenal and Sullivan 1981.)

arise from ambipolar diffusion. Lighter ions of higher charge state have negative scale heights near the centrifugal equator. While the apparent dominance of sulfur is obviously inconsistent with the complete dissociation and ionization of SO_2, we must caution that the actual S/O ratio is not uniquely determined by the plasma science data.

Pilcher et al. (1981) compared the remote images of the torus taken in the red emission line of SII with the *in situ* measurements of S^+ and found excellent agreement in the location of the knoll and precipice, and in the magnitude of the column density.

Alfvén speed. Fig. 23.8 is a contour map of the Alfvén speed determined from the total mass density of the plasma and O_4 magnetic field model of Acuña and Ness (1976). While the calculated local Alfvén speed everywhere exceeds the local corotation speed, it displays a region of uniformly low values in the outer torus with minimum speeds of $\lesssim 250$ km s^{-1} occurring near the centrifugal equator. Above and below the torus the rapidly decreasing density and larger magnetic field produce a rapid increase in Alfvén speed. Io orbits Jupiter in a plane normal to the Jovian rotation axis but at a rate slower than the planetary rotation rate so that the position of the satellite with respect to the centrifugal equator varies as shown in Fig. 23.8. The Alfvén speed of the plasma in the vicinity of Io therefore varies by a factor of two with the System III longitude of the satellite.

The magnetic field of nearly $2 \mu T$ clearly controls the charged particle population in the torus regions. The dynamical β is 0.1 while the thermal β only 0.01. The perturbations of magnetic field and plasma flow measured when the Voyager 1 spacecraft passed beneath Io suggest a standing Alfvén wave is generated near Io making a current system connecting Io and the ionosphere of Jupiter (Acuña et al. 1981; Belcher et al. 1981).

The large Alfvén speeds outside the torus mean that the time for Alfvén waves generated near Io to reach the ionosphere is largely determined by the length of the propagation path in the torus. This transit time will therefore be strongly modulated by the System III position of Io. Similarly, other properties of the propagating waves such as geometry and damping will also vary with longitude. The subsequent changes in the field-aligned current associated with the Alfvén wave may explain the modulation by Io of the decametric radiation (Gurnett and Goertz 1981).

II. FACTORS GOVERNING THE SPATIAL DISTRIBUTION OF IO-DERIVED IONS

Matter comprising the Io torus is transient. New ions enter from a source region localized either in the vicinity of Io or its orbit. They disperse along magnetic field lines as they pendulate with a period of approximately six hours about the centrifugal equator with an amplitude fixed by the latitude of their parent neutrals. In a little more than two such intervals, magnetically enforced corotation smears them in longitude. On a longer time scale, they traverse the torus radially by diffusing outward, except for a small fraction that migrates inward. They exit down a ramp of sharply declining density into a radially distended plasma disk, which conducts them away toward ultimate union with the solar wind. Compared to the term of residence of the outwardly transiting ions, the torus itself is long-lived. It persists in the way a mountain wave cloud outlives its constituting water droplets, which condense, pass through and evaporate as the wind carries them over the crest of a stationary wave.

Such at least is the picture that emerges from the effort to interpret within the framework of mass transport theory the plasma data shown in Fig. 23.9, which was constructed by combining the profiles in Fig. 23.7 (Bagenal and Sullivan 1981). The process responsible for the radial conveyance of the ions was identified as flux tube interchange diffusion after the possibility of removal by means of an organized system of deep circulation was considered and rejected (Brice and Ioannidis 1970; Chen 1977). The operative equation then is (Siscoe 1978):

$$\frac{\partial Y}{\partial t} = L^2 \frac{\partial}{\partial L} \frac{D_{LL}}{L^2} \frac{\partial Y}{\partial L} + S - \mathcal{L} \tag{1}$$

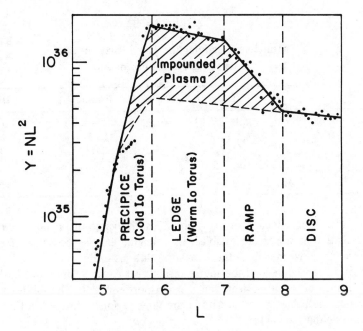

Fig. 23.9. Radial profile of the magnetic flux shell density $Y = NL^2$, where N is the number of ions in a flux shell per unit L. The quantity $Y/2\pi B_J R_J^2$ is the number of ions in a flux shell enclosing unit magnetic flux. (From Siscoe et al. 1981.)

in which $Y/2\pi B_J R_J^2$ is the number of ions in a magnetic flux shell enclosing unit magnetic flux, B_J is the equatorial surface strength of Jupiter's dipole field, R_J denotes one Jovian radius, t is time, L is equatorial radial distance of the flux shell in units of R_J, D_{LL} is the applicable diffusion coefficient, and S and $\mathcal{L}$ are the appropriate source and loss terms. The quantity Y is the proper independent variable to represent the elemental parcels of radial displacement in the process of flux tube interchange, inasmuch as Y is density normalized to magnetic flux, rather than to volume as is usual. It is also often expressed as NL^2, where N is the number of ions in a magnetic flux shell per unit L.

The usefulness of Eq. (1) and the solutions it provides depend in large measure on the accuracy with which the diffusion coefficient can be specified. The coefficient has the generally accepted generic form

$$D_{LL} = k L^m \tag{2}$$

where k and m are constants. The value of m is diagnostic of the physical process that causes the interchange activity. For fixed m, the magnitude of k registers the vigor of that activity. The *in situ* plasma data readily yield the

TABLE 23.3

Parameters Specifying the Three Divisions of the
Outward Io Plasma Formation

	Ledge	Ramp	Disk
Radial range (R_J)	5.8 – 7.12	7.12 – 8.0	8.0 – ?
$Y_i \times 10^{-35}$	17.2	15.5	4.7
L_i	6	7	8
q_i	1.5	9	1
m_i	4.5	12	4

particular L-dependence of D_{LL} and thus the value of m, but additional information is needed to establish its strength, that is, the value of k. To determine m, we merely fit the data with solutions to Eq. (1).

In the space beyond the source region near $L = 5.9$, the terms $\partial Y/\partial t$, S and $\mathcal{L}$ may be safely neglected in the transport equation. The solutions that describe the behavior of the data in the three regions identified in Fig. 23.9 then have the general form

$$Y = Y_i\left(\frac{L_i}{L}\right)^{q_i} \tag{3}$$

where the index i (= L, R, D) denotes the three regions, ledge, ramp, and disk. The parameters defining each domain are given in Table 23.3.

The values for m given in the final row of the table are related to the empirically determined values for q by $q = m - 3$. Thus in Fig. 23.9, the decided shifts in the radial slope q of the ordinate log Y are caused by changes in the exponent m, which in turn disclose the existence of three fundamental transitions over the radial range $5 \leqslant L \leqslant 9$ in the physical mechanism driving diffusion. The nature of these transitions is revealed by considering the expression for the total radially diffusing flux of ions

$$F = -\frac{D_{LL}}{L^2}\frac{dY}{dL} . \tag{4}$$

Using the form for Y given by Eq. (3), we find

$$F = q\, D_{LL}\frac{Y}{L^3} . \tag{5}$$

The requirement that the spatial derivatives in the transport Eq. (1) be defined for all values of L, and in particular at the interfaces between the divisions identified in the figure, necessitates that both Y and F be continuous across the borders. Equation (5) extends the same necessity to the product qD_{LL}. Consequently we see that alterations in the slope of the radial profile of log Y reflect equal but reciprocal variations in the magnitude of the diffusion coefficient. Large slopes characterize zones of weak diffusion, small slopes regions of strong diffusion.

The sharpest break in the slope of the density profile occurs at $L = 5.8$, near the likely location of the inner edge of the source region, which therefore evidently marks the separation between the domains of inwardly and outwardly diffusing ions. The curve in the figure labeled the precipice descends in the direction toward Jupiter by nearly two orders of magnitude in 1 R_J, demonstrating that inward diffusion is comparatively feeble. In a quantitative analysis, Richardson et al. (1980) found that D_{LL} jumped by a factor of at least 50 as one passed outward across this discontinuity. The abrupt change in diffusive properties in virtual coincidence with the location of the source of the ions is naturally explained by the controlling influence of the centrifugal force, which aids outward but inhibits inward transport (Richardson et al. 1980; Richardson and Siscoe 1981; Siscoe and Summers 1981).

The shallow gradients distinguishing the plasma ledge and the plasma disk illustrate the situation expected when centrifugally driven interchange instability is restrained only by ohmic dissipation in Jupiter's ionosphere (Siscoe and Summers 1981). The much steeper decline marking the plasma ramp signifies a substantially reduced level of diffusive activity. The origin of this feature is attributed to pressure gradient inhibition of the interchange motion enforced by a prominent precipitation edge to the ring current that coincides with the ramp (Siscoe et al. 1981). The Io torus is, in effect, impounded by the inner edge of the ring current.

The base of the hatched area in the figure depicts the shape the Io plasma formation would have if there were no ring current. The hatched area itself indicates the amount of Io plasma supported against the outward centrifugal force by the inwardly directed pressure of the inner surface of the ring current. The quantity of matter impounded in this way is very nearly equal to the total bulk of the plasma ledge alone.

We note that the physical processes controlling radial transport have been inferred on the basis of the radial profile of log Y, or in terms of the specification of D_{LL}, on the basis of the exponential parameter m. The diffusion amplitude parameter k is related to a different, important aspect of the phenomenon, namely, the ion production rate S. This can be seen by rewriting the expression for the total diffusing flux of ions Eq. (5), in the form

$$S = (m-3)L^{m-3} \, Y \, k \tag{6}$$

in which we have approximated F by S, as is allowed by the condition that more than 90% of the ions diffuse outward (Richardson and Siscoe 1981). All quantities in Eq. (6) except S and k are listed in Table 23.1 for each of the regions of outflowing plasma. Hence, there remain four unknown quantities S, k_L, k_R and k_D. The above relation and the table may be used to express three unknowns in terms of the fourth, say S. Thus

$$k_L = 2.6 \times 10^{-38} \, S \tag{7a}$$
$$k_R = 1.8 \times 10^{-45} \, S \tag{7b}$$
$$k_D = 2.6 \times 10^{-37} \, S \,. \tag{7c}$$

The source strength or one of the coefficients must be obtained from information originating outside of the *in situ* plasma observations in order to specify the system completely. In principle, external constraints in the form of power requirements and time scales can fix the remaining unknown.

III. POWER REQUIREMENTS AND PLASMA PRODUCTION RATES

There is a basic connection between the amount of fresh material fed into the Io torus each second and the power available for emission from the torus and associated aurora. On the basis of the intensities of the ultraviolet radiation from the Io torus and the Jovian aurora, Broadfoot et al. (1981) found that these luminous features dissipate at least 3×10^{12} W and 1.3×10^{13} W, respectively. The average flux of solar wind kinetic energy incident on the Jovian magnetosphere is 10^{15} W. Earth's magnetosphere can capture 1% of the incident flux of solar wind kinetic energy, which is its main source of energy. The same capture efficiency at Jupiter could supply the power to light the torus and the aurora, as far as quantity is concerned. However, Jupiter's strong magnetic field and rapid rotation combine to shield the deep magnetospheric interior, where Io resides, from the direct influence of the solar wind. For a fuller discussion of this topic see the reviews by Kennel and Coroniti (1975) and Siscoe (1979).

The rotational kinetic energy of Jupiter can power dissipation deep in its own magnetosphere by means of magnetic coupling of the planet's rotation to newly produced plasma and also to radially exiting plasma. Broadfoot et al. (1979) invoked the power released through the coupling of the planet's rotation to fresh plasma to account for the torus emission. Dessler (1981) and Eviatar and Siscoe (1981) extended the principle to radially exiting plasma in order to account for the Jovian aurora. In steady state the production of new plasma is balanced by the radial egress of resident plasma. Thus the two mechanisms provide power in a definite ratio.

The relationship between the amount of rotational energy that can be tapped by the magnetosphere and the rate at which plasma is produced and subsequently exits the magnetosphere is readily derived. The fully general expression for power removed from (or added to) rotation of the planet is

$$P = \Omega T \tag{8}$$

in which the power P is negative if energy is removed from rotation, Ω is the speed of rotation of the body in which rotational energy is stored, and T, the torque on the body, is negative if it acts to slow rotation. By definition, torque is the rate of change of angular momentum. Hence by the principle of conservation of angular momentum, a retarding torque on Jupiter must be matched by an oppositely directed torque elsewhere. Consequently, magnetospheric processes that can support a prograde torque act to drain energy from Jupiter's rotation. It is evident therefore that corotational coupling to newly created ions and to outwardly transported ions are processes that transfer energy to the magnetosphere from rotation.

The torques on Jupiter imposed by plasma production and plasma egress are given by

$$T(\text{production}) = \dot{M} \, \Omega \, R_J^2 \, L_I^2 \tag{9}$$
$$T(\text{egress}) = -\dot{M} \, \Omega \, R_J^2 \, (L_c - L_I^2) \tag{10}$$

in which $\dot{M} = \bar{A} \, m_p \, S$, where $\bar{A}$ is the average atomic mass in AMU of the egressing ions and m_p is the mass of the proton, R_J denotes one Jovian radius, L_I is the value of L at the source of ion production, thus $L_I \simeq 6$, and L_c is the distance at which Jupiter is no longer able to enforce corotation on outflowing plasma (Hill 1979). Voyager plasma data determine the value of L_c to be around 20 (McNutt et al. 1979; Hill 1980). Hence the total energy available from Jupiter's rotation to drive magnetospheric processes is given by

$$P = A \, m_p \, S \, \Omega^2 \, R_J^2 \, L_c^2 \ . \tag{11}$$

Of this total energy output, one half is bound up in the kinetic energy of corotation of the plasma, as can be seen from the expression for the outward flux of rotational kinetic energy

$$F = \frac{1}{2} \left(A \, m_p \, S \right) \left(\Omega^2 \, R_J^2 \, L_c^2 \right) \ . \tag{12}$$

Thus only one half of the total rotational energy released is available for dissipation. As the final result, then, we have for the rotational power available for dissipation

$$P_D = \frac{1}{2} A m_p S \Omega^2 R_J^2 L_c^2 . \qquad (13)$$

Using the numbers given previously for the various quantities in the equation, we arrive at an explicit proportionality between the maximum power (in W) for dissipation and the ion creation rate (in AMU s^{-1}), namely

$$\bar{A} S = 2 \times 10^{16} P_D . \qquad (14)$$

The inferred total power dissipated in producing the aurora and the torus emissions sets a lower limit of approximately 2×10^{13} on the value of P_D, from which we obtain the lower limit $\bar{A} S = 4 \times 10^{29}$ AMU s^{-1}. This production rate accords with that found by Hill (1980) to be necessary to account for the breaking of corotation at $L \simeq 20$.

Equation (14) permits estimates of the power dissipated in radiative processes associated with the Io torus to be used to estimate the source strength for the plasma of the Io torus. Equations (7 a,b,c) then allow the diffusion coefficients to be specified. However, one important dynamical property of the torus can be found directly from S and the total number of ions in the torus, N_L. Equation (3) for Y can be integrated with the parameters given in Table 23.3 to yield $N_L \simeq 5 \times 10^{34}$ ions. The residence time for the plasma in the torus is then $N_L/S \lesssim 1.3 \times 10^5 \bar{A}$ s. If we assume that $\bar{A}$ corresponds to the mean mass of the dissociation products of SO_2, namely $\bar{A} = 21.3$ AMU, then $N_L/S \lesssim 2.8 \times 10^6$ s or ~ 32 days.

IV. THE PROBLEM OF THE THERMAL STATE OF TORUS PLASMA

There is one problem that we have not touched upon, in large part because work on it is in a state of rapid development. This problem is, how does the rotational energy that is released through mass loading of the torus become distributed within the residence time of the plasma in such a way as to produce the observed thermal characteristics of the torus? The torus is an order of magnitude colder than the temperature characterizing a newly created ion that has picked up its full compliment of rotational energy. As noted above in Sec. I, the ion distribution possesses a high energy tail that could signify the presence of new, hot ions; these, however, make up a small fraction of the total ion population. Evidently the temperature of a new ion must drop by an order of magnitude in a time short compared to the residence time in order to explain the observed proportions of hot and cold ions. The ultimate sink of energy that cools the torus is radiation, but the energy that feeds radiation directly is electron thermal energy. Thus, to cool the ions through radiation requires the intermediate step of transferring thermal energy from the ions to the electrons, which is a notoriously slow process. It has not yet been demonstrated how this process can be completed within the available residence time.

A similar apparent difficulty exists in attempting to account for the observed distribution of charge states of the ions in the torus. Oxygen ions exist primarily as OIII and OIV. Shemansky (1980) finds that 100 days are required to reach the observed distribution from the neutral state in an electron population that is characteristic of the torus. Some efforts to resolve the problem of short residence time invoke the electrons and protons from Jupiter's atmosphere that are continuously splashed into the magnetosphere as a result of the impacts of aurora-producing particles. These secondaries have intermediate energy distributions compared to those of newly created ions and electrons from Io. There is the promise that the mix of Io- and Jupiter-derived ions and electrons can significantly shorten time scales for energy transfer.

Acknowledgments. We thank F. Bagenal, H.S. Bridge, A. Eviatar and E.J. Sittler for useful discussions and critical comments. This work was supported by the National Aeronautics and Space Administration.

REFERENCES

Acuña, M.H., and Ness, N.F. (1976). The main magnetic field of Jupiter. *J. Geophys. Res.* 81, 2917–2922.

Acuña, M.H., Neubauer, F.M., and Ness, N.F. (1981). Standing Alfvén wave current system at Io: Voyager 1 observations. *J. Geophys. Res.* In press.

Bagenal, F., and Sullivan, J.D. (1981). Direct plasma measurements in the ion torus and inner magnetosphere of Jupiter. *J. Geophys. Res.* In press.

Bagenal, F., and Sullivan, J.D. (1981). Direct plasma measurements in the ion torus and Io torus. *Geophys. Res. Letters* 7, 41–44.

Belcher, J.W., Goertz, C.K., Sullivan, J.D., and Acuña, M.H. (1981). Plasma observations of the Alfvén wave generated by Io. *J. Geophys. Res.* In press.

Bigg, E.K. (1964). Influence of the satellite Io on Jupiter's decametric emission. *Nature* 203, 1008–1010.

Birmingham, T.J., Alexander, J.K., Desch, M.D., Hubbard, R.F., and Pederson, B.M. (1981). Observations of electron gyroharmonic waves and the structure of the Io torus. *J. Geophys. Res.* In press.

Brice, N.M., and Ioannidis, G.A. (1970). The magnetospheres of Jupiter and Earth. *Icarus* 13, 173–183.

Bridge, H.S., and the Voyager Plasma Team. (1979a). Plasma observations near Jupiter: Initial results from Voyager 1. *Science* 204, 987–991.

Bridge, H.S., and the Voyager Plasma Team. (1979b) Plasma observations near Jupiter: Initial results from Voyager 2. *Science* 206, 972–976.

Broadfoot, A.L., and the Voyager Ultraviolet Spectrometer Team. (1979). Extreme ultraviolet observations from Voyager 1 encounter with Jupiter. *Science* 204, 979–982.

Brown, R.A., (1976). A model of Jupiter's sulfur nebula. *Astrophys. J.* 206, L179–L183.

Brown, R.A., and Chaffee, F.H. (1974). High-resolution spectra of sodium emission from Io. *Astrophys. J.* 187, L125–L126.

Burlaga, L.F., Belcher, J.W., and Ness, N.F. (1980). Disturbances observed near Ganymede by Voyager 2. *Geophys. Res. Letters* 7, 21–24.

Chen, C.K. (1977). Topics in Planetary Plasmaspheres. Ph.D. Dissertation, Univ. California at Los Angeles.

Dessler, A.J. (1981). Mass-injection rate from Io into the Io plasma torus. *Icarus.* In press.

Eviatar, A. and Siscoe, G.L. (1981). Limit on rotational energy available to excite Jovian aurora. *Geophys. Res. Letters.* In press.

Frank, L.A., Ackerson, K.L., Wolfe, J.H., and Mihalov, J.D. (1976). Observations of plasmas in the Jovian magnetosphere. *J. Geophys. Res.* 81, 457–468.

Goertz, C.K., and Thomsen, M.F. (1979). Radial diffusion of Io-injected plasma. *J. Geophys. Res.* 84, 1499–1504.

Grard, R.J., Deforrest, S.E., and Whipple Jr., E.C. (1977). Comment on low energy electron measurements in the Jovian magnetosphere. *Geophys. Res. Letters.* 4, 247.

Gurnett, D.A., and Goertz, C.K. (1981). Multiple Alfvén wave reflections excited by Io: Origin of the Jovian decametric arcs. *J. Geophys. Res.* In press.

Gurnett, D.A., Scarf, F.L., Kurth, W.S., Shaw, R.R. and Poynter, R.L. (1981). Determination of Jupiter's electron density profile from plasma wave observations. *J. Geophys. Res.* In press.

Gurnett, D.A., Shaw, R.R., Anderson, R.R., and Kurth, W.S. (1979). Whistlers observed by Voyager 1: Detection of lightning on Jupiter. *Geophys. Res. Letters* 6, 511–514.

Hill, T.W. (1979). Inertial limit on corotation. *J. Geophys. Res.* 84, 6554–6558.

Hill, T.W. (1980). Corotation lag in Jupiter's magnetosphere: A comparison of observation and theory. *Science* 207, 301–302.

Kaiser, M.L., and Desch, M.D. (1980). Narrow-band Jovian kilometric radiation: A new radio component. *Geophys. Res. Letters* 7, 389–392.

Kennel, C.F., and Coroniti, F.V. (1975). Is Jupiter's magnetosphere like a pulsar's or Earth's. *Space Sci. Rev.* 17, 857–883.

Kupo, I., Mekler, Y., and Eviatar, A. (1976). Detection of ionized sulfur in the Jovian magnetosphere. *Astrophys. J.* 205, L51–L53.

McNutt, R.L., Belcher, J.W., Sullivan, J.D., Bagenal, F., and Bridge, H.S. (1979). Departure from rigid corotation of plasma in Jupiter dayside magnetosphere. *Nature* 280, 803–804.

Mekler, Yu., and Eviatar, A. (1980). Time analysis of volcanic activity on Io by means of plasma observation. *J. Geophys. Res.* 85, 1307–1310.

Morgan, J.S., and Pilcher, C.B. (1978). Longitudinal variations of the Jovian [SII] emission. *Bull. Amer. Astron. Soc.* 10, 579 (abstract).

Neugebauer, M., and Eviatar, A. (1976). An alternative interpretation of Jupiter's "plasmapause." *Geophys. Res. Letters* 3, 708–710.

Pilcher, C.B. (1979). Images of Jupiter's sulphur ring. *Science* 207, 181–183.

Pilcher, C.B., Bagenal, F., and Sullivan, J.D. (1981). A comparison of Voyager and ground-based measurements of the Jovian plasma. MIT preprint.

Richardson, J.D., and Siscoe, G.L. (1981). Factors governing the ratio of inward to outward diffusion of flux of satellite ions. *J. Geophys. Res.* In press.

Richardson, J.D., Siscoe, G.L., Bagenal, F., and Sullivan, J.D. (1980). Time dependent plasma injection by Io. *Geophys. Res. Letters* 7, 37–40.

Roederer, J.G., Acuña, M.H., and Ness, N.F. (1977). Jupiter's internal magnetic field geometry relevant to particle trapping. *J. Geophys. Res.* 82, 5187–5194.

Sandel, B.R. (1980). Azimuthal asymmetry in Io's hot torus. IAU Coll. 57. The Satellites of Jupiter (abstract 5–8).

Sandel, B.R., and the Voyager Ultraviolet Spectrometer Team. (1979). Extreme ultraviolet observations from Voyager 2 encounters with Jupiter. *Science* 206, 962–966.

Scudder, J.D., Sittler, E.C., and Bridge, H.S. (1981). A survey of the plasma electron environment of Jupiter: Direct measurements. *J. Geophys. Res.* In press.

Shemansky, D.E. (1980). Mass loading and the diffusion loss rate of the Io plasma torus. *Astrophys. J.* 242, 1266–1277.

Siscoe, G.L. (1978). Jovian plasmaspheres. *J. Geophys. Res.* 83, 2118–2126.

Siscoe, G.L. (1979). Towards a comparative theory of magnetospheres. In *Solar System Plasma Physics, Vol. II Magnetospheres* (C.F. Kennel, L.J. Lanzerotti, and E.N. Parker, Eds.) North-Holland Publ. Co., New York.

Siscoe, G.L., Eviatar, A., Thorne, R.M., Richardson, J.D., Bagenal, F. and Sullivan, J.D. (1981). Ring current impoundment of the Io plasma torus. *J. Geophys. Res.* In press.

Siscoe, G.L., and Summers, D. (1981). Centrifugally driven diffusion of iogenic plasma. *J. Geophys. Res.* In press.

Smith, B.A., and the Voyager Imaging Team. (1979). The Jupiter system through the eyes of Voyager 1. *Science* 204, 951–971.

Sullivan, J.D., and Bagenal, F. (1979). In situ identification of various ionic species in Jupiter's magnetosphere. *Nature* 280, 798–799.

Trafton, L. (1980). The Jovian SII torus: Its longitudinal asymmetry. *Icarus* 42, 111–124.

Warwick, T.W., and the Voyager Planetary Radio Astronomy Team. (1979). Voyager 1 planetary radio astronomy observations near Jupiter. *Science* 204, 995–998.

24. ORIGIN AND EVOLUTION OF THE JUPITER SATELLITE SYSTEM

JAMES B. POLLACK
NASA Ames Research Center

and

FRASER FANALE
University of Hawaii

According to almost all theoretical models of the origin of the Jupiter system, the gaseous envelope of the protoplanet was initially several orders of magnitude larger than the current dimensions of Jupiter. The most likely, but still far from proven, mechanism for capturing the outer, irregular satellites is by means of gas drag experienced by planetesimals passing through the outer envelope of the protoplanet just prior to the onset of the hydrodynamic collapse. Near the end of the hydrodynamic collapse phase, the outer regions of the protoplanet may have formed a flattened disk of gas and dust from which the regular satellites formed by accretional processes. As a result of the substantial luminosity of the proto-Jupiter, the condensation of low-temperature ices, such as H_2O, was inhibited at close distances and a sharp chemical gradient was created among the Galilean satellites. In order for Europa to be endowed with little H_2O and Ganymede with a great deal of H_2O, satellite formation had to cease $\sim 10^5$ to 10^6 yr after the end of the collapse phase. Under these circumstances, Io may have been constructed, in part, from hydrated minerals; possibly ices other than H_2O were incorporated into Callisto; and Amalthea, 1979 J1, J2, and J3 may have been made only of rather refractory materials.

The Galilean satellites probably formed hot enough to cause some internal differentiation, which was further extended by radioactive heating. Solid state convection is expected to have caused an almost complete freezing of the water mantles of Ganymede, Callisto, and probably Europa during their early history. The surface expression of endogenic activity on Ganymede occurred over a somewhat longer time interval (~ 1 Gyr) than it did for Callisto, although even in Ganymede's case it ceased several billion years ago. This difference is attributed chiefly to Ganymede being more massive and, in particular, having a greater rocky component than Callisto and hence experiencing a larger amount of radioactive and accretional heating. During the late stages of Ganymede's activity, parts of its surface (the so-called grooved terrain) were iced over, thus largely covering up the dark material that was accreted during the first ~ 0.5 Gyr of its history.

Tidal heating is probably the dominant energy source for Io's interior; it has produced important chemical transformations throughout, including the breakdown of primordial minerals into volatiles and refractories, very extensive migration of volatile and underdense phases towards the surface, and the evolution of an S- and SO_2-rich crust and surface. Repeated volcanic outgassing of volatiles, when coupled with their substantial loss to the surrounding Jovian magnetosphere, has resulted in an extensive evolution of the chemical composition of Io's volatiles. Almost all volatile species that may have initially been present, including large amounts of H_2O, N_2, and CO_2, apparently have been totally eliminated from this "open system" during Io's prior history. The remaining volatile species are dominated by sulfur compounds, in part because of the high intrinsic abundance of sulfur in bulk solar system material.

In certain ways, the Jovian system can be considered to be a miniature solar system, with Jupiter playing the role of the Sun and the 16 known satellites playing the role of the planets. This analogy is much more profound than a mere similarity of orbital geometry. The regularity of the orbits of the eight inner Jovian satellites suggests that they were formed within an early Jovian nebula of gas and dust in ways similar to those that led to the formation of the terrestrial planets. The systematic variation of density of the Galilean satellites with distance from Jupiter may reflect a temperature gradient within the early Jovian nebula. Such a gradient is analogous to the one in the primordial solar nebula that apparently led to a chemical zonation within the solar system. Finally, high temperature plasmas – the solar wind and the Jovian magnetosphere – are present in both systems and interact in important ways with the objects in these systems.

These similarities between the Jovian system and the solar system should not obscure some important differences. The outer eight satellites of Jupiter have highly irregular orbits that are usually attributed to their being captured objects. There are no planets identical in bulk composition to Ganymede or Callisto, which probably contain a significant amount of H_2O as well as rock.

The terrestrial planets have only minor amounts of water, while the Jovian planets have massive hydrogen-helium dominated envelopes in addition to cores that may have a certain similarity in composition to Ganymede and Callisto. Although Io and Europa have densities comparable to those of inner solar system objects, they too have unique attributes; Io displays an unprecedented level of active volcanism, and the surface of Europa is almost entirely covered by water ice. Finally, asteroid-sized satellites and a ring exist close to Jupiter.

In this chapter, we assess our current understanding of the history of the Jovian satellites. We pay particular attention to the conditions in the early Jovian nebula that may have led to the formation of the regular satellites and the capture of the irregular satellites, and to the thermal and chemical development of the interiors of the planet-sized Galilean satellites. To lay the foundation for these considerations, we first review relevant observational constraints, dealing separately with groundbased and spacecraft data obtained prior to Voyager and then with the Voyager results. We next discuss theories of the origin of the Jovian system and use these concepts to investigate the origin of the satellite system. We proceed to consider the subsequent evolution of the larger satellites. Finally, we assess our current overall understanding of the Jovian satellites, with particular emphasis on key problem areas, some of which can be addressed by future spacecraft missions, such as Galileo.

I. PRE-VOYAGER OBSERVATIONS

The most significant pre-Voyager data on the Galilean satellites, as far as their cosmogonic implications are concerned, were estimates of their masses and dimensions and hence their mean densities. The latter provide useful constraints on bulk composition. Additional important information about these satellites was provided by remote sensing measurements of their surface compositions and the detection and partial characterization of a gas torus centered about Io's orbit. Finally, the orbital parameters and gross photometric characteristics of the outer eight satellites had important implications for their origin.

Masses, Radii, and Densities

The masses of the Galilean satellites have been known to within $\sim 10\%$ for decades. Values for the masses of Io, Europa and Ganymede deduced by Sampson (1921) from mutual perturbations were within 5 % of currently accepted values (see Chapter 3 by Greenberg). Such accuracy was possible only because of the enhancement of certain mutual perturbations by resonant commensurabilities of the satellites' orbits. For a discussion of pre-Voyager mass estimates, see Aksnes (1977). A more serious problem, as far as bulk characterizations of the Galilean satellites are concerned, was the determination of their radii, since the mean density is inversely proportional to the cube of the radius. Unfortunately, radius measurements by direct visual techniques were uncertain by 10% or more (see Dollfus 1970; Morrison et al. 1977). However, Earth-based observation of the occultation of a bright star by a satellite can allow the determination of the diameter to within only a few kilometers if several competent observations are made along the occultation path. In 1971 Io occulated β Scorpii C (5th mag) yielding a diameter of 3660 ± 4 km (Taylor 1972), and in 1972 Ganymede occulted SAO 186800 (8th mag), yielding a diameter of $5270^{+30}_{-20}.0$ km (Carlson et al. 1973). See Johnson (1969) and Morrison et al. (1977) for reviews of diameter determination. Thus, prior to Voyager, the diameters of Io and Ganymede were fairly accurately known but those of Europa and Callisto were known only to ∼ 7 %.

Considerable improvement in the known masses and radii occurred as the result of the first spacecraft encounters with the system, those of Pioneer 10 and 11. Accurate tracking of these spacecraft allowed determination of the satellites' masses to ∼ 1% (Anderson et al. 1974), while careful analyses of low-resolution pictures permitted radii to be determined to ± 30 km (Smith 1978). Therefore, prior to Voyager, the densities of all the Galilean satellites were known to within ∼ 5%. The best-estimate pre-Voyager density values were 3.41 ± 0.19, 3.06 ± 0.15, 1.90 ± 0.06, and 1.81 ± 0.05 g cm^{-3} for Io, Europa, Ganymede, and Callisto, respectively (Smith 1978). It is interesting to note that the observation with the greatest theoretical impact, i.e., that the inner two Galilean satellites have densities $\gtrsim 3.0$ while the outer two have densities $\lesssim 2.0$, was accomplished from Earth and has stood the test of time. Thus, as discussed below, Io and Europa are made mostly of rock, while Ganymede and Callisto have a substantial ice component. It is interesting to note that the size of the rocky satellites is comparable to the size of the rocky cores of the icy satellites.

In considering the importance of the diameter determination as the greatest error source for measured densities, we did not mention the significance of the magnitude of the diameters. Ganymede and Callisto were known, prior to Voyager, to be as large as or larger than the planet Mercury. This has produced an intuitive consensus that these objects were "planets" or "worlds" in their own right, fully capable of evolving internally rather than simply being four spherical targets for exogenic bombardment (Lewis 1971a,b; Fanale et al. 1974, 1977; Consolmagno and Lewis 1976, 1977; Johnson 1978).

Surface Compositions

We next discuss the spectrometric identification of surface phases. This subject has been reviewed by Johnson and Pilcher (1977) and by Sill and

Clark (see Chapter 7). Pilcher et al. (1972) and Fink et al. (1972) positively identified H_2O frost on the surfaces of Europa, Ganymede, and Callisto (although the form of the H_2O on the last has been the subject of debate); Pilcher et al. suggested H_2O frost coverage of 50 to 100 % for Europa, 25 to 65 % for Ganymede, and 5 to 25 % for Callisto. As far as H_2O was concerned, these workers found no absorption bands in Io's spectrum but suggested that H_2O could conceivably be present and still not provide a spectral signature in the spectral region under investigation (Pilcher et al. 1972; Fink et al. 1972). However, Lee (1972) showed that Io's albedo at 3.0 to 5.0 μm was *generally* very high (these were wideband meausrements), which again argued strongly against H_2O coverage since H_2O has an exceedingly low albedo in this spectral region. It remained for higher spectral resolution measurements by Pollack et al. (1978) and Cruikshank et al. (1978) to establish that there was absolutely no trace of the strong 3.0-μm H_2O band due to O-H stretch. This resulted in upper limits of 10 % areal coverage of pure, optically thick H_2O ice and 0.1 % for a well-mixed bound H_2O on Io (Pollack et al. 1978). As far as the nonicy portion of Europa, Ganymede, and Callisto were concerned, Pilcher et al. pointed out that fractional coverage with a dark (15 % albedo) silicate of 30 %, 60 %, and 90 % for Europa, Ganymede, and Callisto, with the rest being H_2O frost, could approximately explain the data. Pollack et al. (1978) reported 50 % and 65 % H_2O frost coverage on Ganymede's trailing and leading sides, and at least 85 % on Europa's leading side based on spectral data obtained from the Kuiper Airborne Observatory. In the case of Callisto, Pollack et al. reported that although the object also showed a deep 3.1-μm band, it was not necessary to postulate any surface frost; a fine hydrated carbonaceous chondritic material could largely account for this band. Lebofsky (1977), however, interpreted his broadband observations of the 3.1-μm feature to imply the presence of H_2O frost, although the non-H_2O material also seemed to be hydrated as well. The spectra of Pollack et al. also provided marginal evidence for the presence of the 1.5- and 2.0-μm water-ice bands. These features were better defined by Clark and McCord (1980). Thus, some water ice and possibly water of hydration are present on Callisto's surface.

Next we consider what was known of Io's surface other than the fact that it was anhydrous. A detailed review of the spectrometric study of Io's surface phase composition is given in Chapter 20 by Fanale et al. Briefly, prior to Voyager, Io's surface was thought to consist of an anhydrous mixture of various forms of elemental sulfur (Fanale et al. 1974; Wamsteker et al. 1974; Nelson and Hapke 1978) and some Na-containing salts, probably sulfates (Fanale et al. 1974; Nash and Fanale 1977). The investigators who noted the flatness of Io's spectrum in the 3.0-μm region also noted a deep absorption feature at $\sim$ 4.1 μm that was interpreted possibly as being caused by the anionic group of a carbonate (Cruikshank et al. 1978) or nitrate

(Pollack et al. 1978). At approximately the time of the Voyager encounter, three independent groups of investigators identified the feature as the $\nu_1 + \nu_3$ absorption bands of frozen SO_2 (Fanale et al. 1979; Nash and Nelson 1979; Smythe et al. 1979; Hapke 1979).

Io's Torus

Finally, pre-Voyager groundbased studies of the Galilean satellites identified the Io Na° cloud (Brown 1974) and an extensive S^+ torus surrounding Io's orbit (Pilcher 1980). The rate of supply of Na° to Io's torus was generally not considered to be close to that which could affect Io's overall composition (Smythe and McElroy 1977). More importantly, it was difficult to tell if useful constraints could be set on surface composition by examining the cloud and torus emissions since neutral and ionized species varied so much in their oscillator strengths, and the amount of solar energy changed considerably between the regions of the spectrum in which their radiation occurred (see discussion in Fanale et al. 1977).

By the time of Voyager, models for both the bulk composition and the mode of formation of the Galilean satellites were constrained strongly by their densities and, weakly and in a convoluted manner, by their surface optical properties and evidence from their magnetospheric interactions. To illustrate how complex and indirect the constraints on bulk composition imposed by the optical properties were, one need only note that in terms of bulk composition the sequence of increasing ice content is clearly Europa $<$ Ganymede $<$ Callisto, whereas in terms of increasing surface ice coverage the sequence is clearly Callisto $<$ Ganymede $<$ Europa.

Outer Satellites

There appear to be two classes of satellites in terms of their orbital characteristics (Morrison et al. 1977). Since the inner eight satellites, J1–J5 and the newly discovered 1979 J1, 1979 J2, and 1979 J3, have prograde orbits of low eccentricity and inclination, it is reasonable to assume that they were formed within an early Jovian nebula. However, the orbits of the outer eight satellites, J6–J13, are highly eccentric and inclined, with half of them being retrograde. Consequently, it is highly likely that these outer, irregular satellites originated outside the Jovian system and were later captured (see Chapter 5 by Cruikshank et al.). As illustrated in Fig. 5.1 in the Cruikshank et al. chapter, the orbital inclinations and semimajor axes of the irregular satellites tend to cluster into two families, prograde and retrograde.

Limited information exists on the size and composition of the outer satellites. From a combination of visible and infrared photometric observations, Cruikshank (1977) inferred that the largest members of the prograde and retrograde families, Himalia (J6) and Pasiphae (J8), have radii of about 85 and 25 km, respectively. Thus the irregular satellites have asteroidal sizes, in contrast to the Galilean satellites, which have dimensions of several thousand kilometers. The low albedos of Himalia (J6) and Elara (J7) (Cruikshank 1977) and the near neutral U, B, V colors of Himalia (J6), Elara (J7), and Pasiphae (J8) (Degewij et al. 1980) are suggestive of a carbonaceous chondritic composition (see Chapter 5).

II. VOYAGER OBSERVATIONS

Data obtained by experiments aboard the Voyager spacecraft have provided a quantum jump in our knowledge about Jupiter and its satellite system. The new results of particular significance for this chapter include the discovery of a ring system and three new satellites, measurements of the spectral reflectivity properties of Amalthea, improved determinations of the mean density of the Galilean satellites, the discovery of active volcanism on Io, the obtaining of high resolution images of the Galilean satellites, and measurements of the composition of Io's torus. Below, we briefly discuss these data and indicate their relevance for cosmogonic questions.

Images obtained by the Voyager cameras show the presence of a very faint ring system, whose outer edge lies at 1.81 R_J, where R_J is the equatorial radius of Jupiter (Owen et al. 1979). Because the ring particles are only a few microns in size, it is highly unlikely that they have survived over the age of the solar system (Jewitt and Danielson 1980; Burns et al. 1980). Thus they probably require a source, with this source most likely being ejecta from the surfaces of small, nearby satellites, such as Adrastea (1979 J1) and/or numerous kilometer-sized satellites located in the brightest region of the rings (Jewitt and Danielson 1980; Burns et al. 1980); see Chapter 2 by Jewitt. Adrastea is located near the outer edge of the rings (Jewitt et al. 1979), while the brightest portion of the rings occurs at about 1.7 R_J (Owen et al. 1979). Two other small satellites, 1979 J2 and 1979 J3, have been detected at distances of 3.11 R_J and 1.8 R_J, respectively (Synnott 1980).

From the point of view of this chapter, the important aspect of the above results is the occurrence of small satellites quite close to the planet. In view of the regular nature of their orbits and the great difficulty of capturing objects this close to Jupiter, it seems very likely that they represent natural satellites that were formed at about the time that J1–J5 were formed. Thus, proto-Jupiter had to be smaller than 1.8 R_J before the end of the epoch of satellite formation. Furthermore, as illustrated in Table 24.1, there is a tendency among the regular satellites for the smaller ones to lie closer to Jupiter. A similar behavior also appears to be exhibited by the Saturn system (Morrison et al. 1977; Gehrels et al. 1979).

TABLE 24.1

Properties of Jupiter's regular satellites[a]

Satellite		Orbital distance (R_J)	Diameter (km)	Mean density (gm cm^{-3})
1979J1	(Adrastea)	1.8	35	–
1979J3		1.8	–	–
J5	(Amalthea)	2.55	200	–
1979J2		3.1	80	–
J1	(Io)	5.95	3640	3.53
J2	(Europa)	9.47	3130	3.03
J3	(Ganymede)	15.1	5280	1.93
J4	(Callisto)	26.6	4840	1.79

[a]Based on results given in Jewett et al. (1979), Synnott (1980), Thomas and Veverka (1980), and Smith et al. (1979a).

A fundamental constraint on conditions in the Jovian nebula where the satellites were formed is provided by their bulk composition. By combining determinations of satellite sizes from Voyager images with measurements of their masses from Pioneer tracking data, Smith et al. (1979a) have obtained an improved set of values of the mean density of the Galilean satellites. These new values are shown in Table 24.1. They confirm that the mean density monotonically decreases with increasing distance from Jupiter. This result has been widely interpreted as indicating an increasing fractional content of water in the satellites' interior with increasing distance from Jupiter (e.g., Kuiper 1948; Consolmagno and Lewis 1976, 1977).

Very limited information on the bulk composition of Amalthea is provided by spectrophotometric measurements of its surface obtained from Voyager images. Thomas and Veverka (Chapter 6) find that most of Amalthea is very dark (~ 5% albedo) and red with, however, a few isolated places on slopes and ridges having albedos several times larger than the average and exhibiting a green color. They have simulated the dark red areas with mixtures of orange sulfur and carbonaceous material. A sulfur component may have been derived from Io, while the more neutral dark substrate may have been influenced by magnetospheric particle and micrometeoroid bombardment. Jewitt et al. (1979) find that Adrastea is also quite dark. Once again the same cautions about extrinsic contamination need to be noted in making compositional deductions from this result regarding the satellite's interior.

Images of the Galilean satellites permit inferences to be drawn about the level and duration of geological activity on these planet-sized objects. There is an enormous range in the density of preserved craters on these bodies, with Callisto having a crater population comparable to that of the lunar highlands and Io having no recognizable craters. Using admittedly a very crude estimate of the time history of meteoroid bombardment for the Jovian system, based chiefly on lunar analogy, Smith et al. (1979a) estimate that Callisto's surface is $\sim 4 \times 10^9$ yr old, while Io's is $< 1 \times 10^6$ yr old. More detailed calculations in this book (e.g., in Chapters 10, 12, 14, 16) generally support these estimates.

The Voyager photographs show that Ganymede's surface consists of two principal provinces (see Chapter 13 by Shoemaker et al.): a landform that resembles Callisto's in both its low albedo and high crater density, and a terrain of higher albedo that has sets of intersecting grooves. This grooved terrain is thought to be a manifestation of tectonic activity during the early history of the satellite (3 to 4×10^9 yr ago). The morphology of the grooves has been interpreted as indicative of extensional, as opposed to compressional, processes with normal faulting playing a major role.

Images of the surface of Europa do not display any signs of significant topographic relief over length scales of ~ 100 to 1000 km, nor do they show any land masses or features noticeably above or below their surrounding regions (see Chapter 14 by Lucchitta and Soderblom). The almost universal occurrence of terrain having a high albedo and only slight red color suggests that water ice is present over the entire surface (see also Chapter 7 by Sill and Clark). The most dominant features of the topographically mild and slightly colored surface of Europa are sets of enormously long, intersecting, curvilinear features, suggesting cracks, that are somewhat more colored than their surroundings.

Perhaps the biggest surprise to come from the Voyager photographs was the discovery of an unprecedented level of active volcanism on Io (Morabito et al. 1979). This aspect of Io is most dramatically illustrated by the presence of eruptive plumes that attain altitudes ranging from ~ 70 to 280 km above the surface (Strom et al. 1979). In addition there is strong evidence for the occurrence of lava flows of sulfur and possibly silicates across the surface (Carr et al. 1979). From analyses of the Voyager images, Johnson et al. (1979) infer that Io is being resurfaced at the amazing rate of ~ 1 mm yr^{-1} (see Chapter 17 by Johnson and Soderblom). This result implies that over the age of the solar system resurfacing has involved a mass comparable to that of the entire satellite! Thus we can expect that there has been a significant evolution in the composition of Io's interior, particularly for the more volatile species.

Furthermore, as indicated earlier, Io is not a "closed" system, but is losing mass to the surrounding Jovian magnetosphere at a rate of $\sim 10^{10}$ to 10^{11} atoms cm^{-2} s^{-1} (Broadfoot et al. 1979; Chapter 17 by Johnson and Soderblom; Chapter 20 by Fanale et al.; Chapter 22 by Pilcher and Strobel;

Chapter 23 by Sullivan and Siscoe). Not only has Voyager provided useful estimates of this mass-loss rate, but it has also given estimates of the composition of the ionized species in the near-Io environment. This region of enhanced density, the Io torus, is undoubtedly dominated by neutral and ionized gases derived from Io. Sulfur and oxygen ions constitute the dominant species in the Io torus, with oxygen being somewhat more abundant than sulfur (Sullivan and Bagenal 1979). Such a result is consistent with the dominant presence of sulfur and sulfur dioxide on the surface and in the atmosphere of Io. The next most abundant species in the torus is sodium, with sodium being conceivably no more than an order of magnitude less abundant than sulfur and oxygen. Direct evidence for the occurrence of SO_2 gas in Io's atmosphere was provided by IRIS observations made near volcanic Plume 2 (Pearl et al. 1979). Such observations also set severe upper bounds on the amounts of other volatile species.

Tidal heating by Jupiter provides the most likely energy source for the active volcanism on Io (Peale et al. 1979; Chapter 4 by Cassen, Peale, and Reynolds). A necessary condition for the heating is the occurrence of the forced orbital eccentricity, that results from the three-body Laplace resonance among Io, Europa, and Ganymede. Since the Laplace resonance has apparently existed for a significant fraction of Io's lifetime, so has active volcanism. Note, however, that it is not known when the Laplace resonance was established; it may have a primordial origin ($\sim$ 4.6 Gyr ago), coinciding closely with the satellite's origin, or it could have been established as recently as $\sim$ 0.5 Gyr ago (Yoder 1979; Peale and Greenberg 1980; Chapter 3 by Greenberg).

III. EARLY HISTORY OF JUPITER

Recent models of the interior of present day Jupiter are characterized by a two-layer structure (e.g., Slattery 1977; Grossman et al. 1980); almost all of the mass ($\sim$ 95 %) is contained in a gaseous-liquid envelope having approximately solar elemental abundances. Thus H and He are the principal elements in the envelope. At the center of the planet, there is a $\sim$ 15 Earth-mass core made of silicates and perhaps water and other low-temperature condensates. Qualitatively similar models have been constructed for the other Jovian-type planets, with, however, the heavy-element core constituting a larger fraction of the total mass in these cases. For all of these planets, the core is the result of an enrichment of heavy elements for the entire object rather than simply a segregation of heavy elements from an initially solar abundance envelope.

The above structures for Jupiter and the other Jovian-type planets suggest that they may have originated in one of two alternative ways: either the core formed first, from the accretion of smaller bodies, as the terrestrial planets are thought to have formed, and the core then collected a gaseous envelope about itself by effectively concentrating the gas of the surrounding solar nebula (e.g., Perri and Cameron 1974); or the gas envelope formed first, for example, from an instability in the solar nebula (e.g., Bodenheimer 1974), and then obtained a core. As we will discuss later, these two scenarios, core and gas models, represent extreme situations, and it is quite possible that reality corresponds to an intermediate situation. Nevertheless, it is useful to describe the two extreme scenarios in more detail.

Gas Model

The evolutionary history of the gas model is characterized by three main stages:

Stage 1. Early, slow contraction;
Stage 2. Rapid hydrodynamical collapse;
Stage 3. Later, slow contraction (Bodenheimer 1974; Graboske et al. 1975; Bodenheimer et al. 1980; Grossman et al. 1980).

During stage 1, proto-Jupiter contracts from an initial radius of several thousand times its current value to several hundred times on a Kelvin-Helmholtz time scale of $10^5 - 10^6$ yr. At a given instant of time, its structure is quite close to hydrostatic equilibrium, with gravitational energy released during contraction serving to both heat the interior and compensate for radiation emitted to space. By the end of stage 1, temperatures at the center of proto-Jupiter reach a value of 2100 K, at which point molecular hydrogen dissociates into atomic hydrogen.

Stage 2 is initiated by this dissociation, which triggers a gravitational collapse of the central part of proto-Jupiter. The remainder of the object follows the lead of the central portion and experiences a very rapid contraction and heating. On a time scale of only ~ 1 yr, the central density increases by 5 orders of magnitude to ~ 1 gm cm^{-3}, the central temperature rises by an order of magnitude to several tens of thousands of degrees K, and the radius decreases from several hundred R_J to several R_J ($R_J = 71400$ km).

Stage 2 ends and stage 3 begins with proto-Jupiter once more attaining a quasi-hydrostatic equilibrium configuration. By matching their last models near the end of stage 2 with initial hydrostatic models for stage 3, Bodenheimer et al. (1980) found that proto-Jupiter had a size of 1.3 R_J and a luminosity of 2×10^{-6} $L_\odot$ at the start of stage 3, where $L_\odot$ is the current luminosity of the Sun (2×10^{33} ergs s^{-1}). However, a fair amount of extrapolation was involved in performing this matching. Perhaps a better estimate of the size and luminosity of proto-Jupiter at the start of stage 3 has been obtained from calculations of the accretional growth of the planet, which leads to a continuous sequence of models (R. Moore, personal

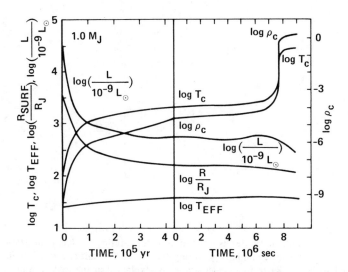

Fig. 24.1. Early evolution of a one-Jupiter-mass gaseous protoplanet. The figure shows the temporal behavior of the central temperature (T_c in units of K), effective temperature (T_{eff}), radius (R_{surf} in units of Jupiter's present radius R_J), intrinsic luminosity (L in units of 10^{-9} times the Sun's present luminosity $L_\odot$), and central density (ρ_c in units of g cm^{-3}). The left side of the figure shows the evolution that occurs during the early hydrostatic phase (stage 1), while the right side shows the evolution that takes place during the hydrodynamical collapse phase (stage 2). (From Bodenheimer et al. 1980.)

communication). Such calculations give a radius of about 3 R_J and a luminosity of about 5×10^{-4} $L_\odot$ at the end of the accretion or, equivalently, the end of the hydrodynamical collapse. Subsequent evolution is characterized by a slow contraction over almost the entire 4.6 Gyr age of the solar system, with the planet's excess luminosity monotonically declining by orders of magnitude to its present value of 8×10^{-10} $L_\odot$.

The above discussion of the gas model of Jupiter's origin is summarized in Figs. 24.1 and 24.2, which illustrate the time history of several parameters of these models for stages 1, 2, and 3. On Fig. 24.2a, we indicate the location of the initial model for stage 3, as given alternatively by Moore and by Bodenheimer et al. (1980). In the discussion below, we will generally use the former. All these calculations were performed for spherically symmetric conditions, i.e., the angular momentum of proto-Jupiter was neglected.

There are two ways in which a heavy element core may have been realized by the gas model. Especially during stage 1, solid bodies that formed within the nearby solar nebula may have passed through the gaseous envelope of proto-Jupiter and been captured by gas drag dissipation (Pollack et al.

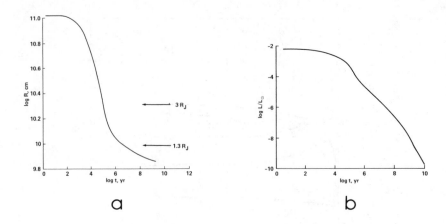

Fig. 24.2. (a) Radius of Jupiter as a function of time during the late hydrostatic phase (stage 3). (b) Intrinsic luminosity during this stage. The initial size of Jupiter was assumed to be 16 R_J in these calculations. As explained in the text, a more realistic initial radius was probably 1.3 to 3 R_J, whose locations are indicated by the arrows on the graphs. (After Graboske et al. 1975.)

1979). The captured bodies would then be subjected to continued gas drag and so evolve rapidly to the center of the protoplanet. In addition, temperature conditions within the interior of the protoplanet close to the end of stage 1 may have been within the liquid stability field of certain species of dust grains (De Campli and Cameron 1979). In this case, rapid particle growth may have ensued, with the particles growing large enough to gravitationally sink rapidly to the object's center. A net gain of heavy elements by the whole planet could have occurred in this way, if most of the protoplanet were subsequently lost or if there were an efficient mixing of material between the protoplanet and the surrounding solar nebula.

Core Model

Perri and Cameron (1974) conducted the first study of the amount of solar nebula gas that could be gravitationally captured by a solid body embedded in the solar nebula. They carried out a series of static calculations for conditions of hydrostatic equilibrium to find the relationship between envelope mass and core mass under the following assumptions: the outer boundary of the envelope was equal to the tidal radius of the composite object, with the pressure and temperature of the envelope at its outer boundary matching those of the surrounding solar nebula, and the temperature gradient throughout the envelope was specified to be the adiabatic value. Perri and Cameron found that the envelope mass constituted an increasing fraction of the total mass as the core mass increased. Ultimately the core mass reached a peak value of about 115 $M_\oplus$ and then decreased as the total mass increased further, where $M_\oplus$ is the mass of the Earth. They interpreted this result as indicating that cores having masses in excess of the peak value would initiate a hydrodynamical instability in the surrounding solar nebula and hence would collect massive envelopes in a relatively brief time. This interpretation was supported by a linear stability analysis of those models having total masses in excess of that for the peak core model. Unfortunately, the peak core mass has a value that is almost an order of magnitude larger than that given by interior models of present-day Jupiter, although lower values could be realized with a cooler solar nebula.

In an effort to overcome the above discrepancy, Harris (1978) and Mizuno et al. (1978) explored similar models with alternative temperature structures. They found that the peak core mass could be lowered by as much as several orders of magnitude by postulating a more isothermal temperature gradient within the envelope. In all cases, the envelope mass approximately equaled that of the core for the peak core model. Thus a hydrodynamical collapse is necessary for this type of model of Jupiter's origin in order to reproduce the planet's total mass (318 $M_\oplus$), and the very large fraction of this mass ($\sim$ 95 %) represented by the envelope.

All of the above calculations of the core model may err in neglecting the evolution of the core-envelope system. In particular, in the same way that the pure envelope cases of Bodenheimer et al. (1980) evolve due to radiation to space and the accompanying contraction of the envelope, a composite system may also evolve. As the envelope contracts, more mass can be added to the composite object from the solar nebula, without the necessity of a hydrodynamical collapse. Preliminary calculations by Bodenheimer, De Campli, and Pollack indicate that even the envelope surrounding a 0.1-$M_\oplus$ core can undergo significant accretional growth. Thus, quite conceivably, both the envelope and the core grew through accretional processes in a quasi-hydrostatic manner before hydrodynamical collapse occurred.

All of the above core models can be characterized by an expanding envelope due to the increase in the total mass and hence the tidal radius of the composite object. Ultimately, however, temperatures at the base of the envelope increase to the point where molecular hydrogen begins to dissociate and a hydrodynamical collapse is initiated. The subsequent stages 2 and 3 are essentially identical to those for the gas model.

IV. FORMATION OF THE JOVIAN SATELLITES

We next relate the above discussion of the early history of the Jovian system to the origin of its satellites. As discussed earlier, the orbital characteristics of the outer eight irregular satellites imply that they originated outside the Jovian system and were later captured. Various theories of satellite capture have recently been reviewed by Pollack et al. (1979). Basically, these fall into three categories: Lagrange point capture, collision of an asteroid and a natural satellite, and gas drag capture.

Lagrange Point Capture

Detailed numerical simulations of the three body problem have shown that an object initially in orbit about the Sun can transfer to an orbit about Jupiter by passing through Jupiter's interior Lagrange point at a very low velocity (Heppenheimer 1975). However, such a capture event is only temporary, with the body eventually escaping back out the Lagrange point. Thus, the operation of some other mechanism is required during the period of temporary capture in order to affect permanent capture. For example, Jupiter's mass may have increased significantly (approximately a few tens of percent) or the Sun's mass may have decreased during the capture phase (Heppenheimer and Porco 1977). Unfortunately, typical time scales of temporary capture are ~ 1 to 10^2 yr, which are much shorter than the time scales over which substantial planetary mass gains or solar mass losses are thought to occur. Plausible alternative processes for affecting permanent capture have not yet been proposed.

Collision of Asteroid and Satellite

Colombo and Franklin (1971) suggested that the two families of irregular satellites originated from a collision of a stray body and a natural satellite of Jupiter, with the retrograde and prograde members being fragments of the stray body and natural satellite, respectively. They showed that this model was consistent with a number of orbital characteristics of the irregular satellites, including the velocity dispersion within each family. However this model does not explain the existence of single irregular satellites about Saturn and Neptune, which do not belong to a group as far as we know. Also, it seems unlikely, based on the discussion of the last section, that natural satellites formed as far out as the prograde group. Furthermore, an icy rather than a carbonaceous chondritic composition would be expected for members of this group, contrary to their observationally inferred composition.

Gas Drag Capture

The first discussion of the possible importance of nebula drag processes for capturing objects was given by See (1910), who envisioned this mechanism as applying quite generally to both planets and satellites, including ones with regular orbits. Kuiper (1951) considered this process within the limited context of explaining the capture of Jupiter's irregular satellites, with Jupiter's outer atmosphere supplying the gas drag. He also pointed out that satellite entry into Jupiter's atmosphere might also cause a breakup of the parent bodies into several pieces, thereby accounting for the occurrence of families of irregular satellites (cf. Fig. 5.1). In its modern guise, this theory has been investigated by Pollack et al. (1979), who performed a detailed analysis of the entry problem, related the capture event to the early history of the Jovian system, and made comparisons between the predictions of their calculations and the observed properties of the irregular satellites. These efforts are detailed below.

As pointed out earlier, normally gas drag continues to operate on a captured object, which quickly spirals into the center of the planet. However, if capture occurred quite close to the end of stage 1 ($\sim$ 10 yr), then only limited orbital evolution could occur before the protoplanet undergoes hydrodynamical collapse and the protoplanet shrinks to a much smaller size on a time scale of about a year. In this event, the captured bodies become outer, irregular satellites.

Pollack et al. (1979) also proposed that the two families of irregular satellites were created in a two-step process; if the strength of the parent bodies was small enough, as might be the case for a carbonaceous chondrite, the gas dynamical pressure experienced by the parent body might have caused it to fragment, much as meteorites entering the Earth's atmosphere are often fragmented. However, the body's self-gravity was probably large enough to. hold the fragments together, in which case a later collision with a meteoroid would be required to separate the fragments into separate satellites.

While the above model for the origin of the irregular satellites is far from proven, several of its predictions are crudely consistent with the observed properties of the irregular satellites. First, the semimajor axes of the prograde and retrograde families are $\sim$ 160 R_J and $\sim$ 310 R_J, respectively, values that are crudely in accord with the size of proto-Jupiter at the end of stage 1.

Second, it is possible to predict very roughly the size of the parent bodies. If the parent body is too large, it experiences too little gas drag to be captured. If it is too small, it will be dragged along by the collapsing protoplanet. The optimum size for capture and survival as a satellite is such that the mass of the protoplanet's envelope encountered by the parent body is a significant fraction ($\sim$ 10 %) of its own mass. Based on Bodenheimer's (1977) model of proto-Jupiter at the end of stage 1, Pollack et al. (1979) obtained an expected radius of the parent bodies ranging from a tenth of a kilometer to several hundred kilometers, with the upper end of this range being consistent with the size of the parent bodies derived from the observed dimensions of the present irregular satellites. Also, the inferred parent body of the outer retrograde satellites is about 3 times smaller than the parent body of the inner prograde group, as crudely expected from the decrease of envelope density with increasing radial distance.

The gas drag model of satellite capture also predicts that the irregular satellites are made of low-strength materials and that family members are quite similar in composition to one another. As discussed earlier, limited photometric information indicates that Himalia (J6), Elara (J7), and Pasiphea (J8) are made of carbonaceous chondritic material, in accord with the former expectation. Furthermore, Elara and Himalia, members of the prograde family, lie closer to each other in a two-color diagram than to Pasiphae, a member of the retrograde family (Degewij et al. 1980).

Finally, Gehrels (1977) has carefully searched for fainter and smaller members of the irregular satellites. His failure to find new members indicates that the size distribution of the irregular satellites does not sharply increase with decreasing size, as occurs for both Trojan and main belt asteroids. This result is consistent with a gas dynamic pressure mechanism fragmenting the parent bodies to produce the irregular satellites, but is inconsistent with mechanisms involving highly catastrophic impact events or metastable capture of stray bodies at the current epoch.

Origin of the Inner Satellites

The inner eight satellites of Jupiter, J1–J5 and the three satellites discovered by Voyager, were probably formed in the outer regions of proto-Jupiter, in perhaps a similar way to the way the terrestrial planets formed in the primordial solar nebula. Formation probably commenced close to the end of stage 2, when the outer boundary of proto-Jupiter shrank to the orbital distances of the Galilean satellites. More precisely, due to the increase in angular velocity in the outermost regions of proto-Jupiter as its size decreased, a flattened disk of gas and dust was left behind as the remainder of the planet continued to shrink to a radius of several R_J. Accretionary processes, occurring in this disk, presumably led to the formation of the regular satellites (e.g., Cameron and Pollack 1976).

Bodenheimer (1977) has carried out a series of two-dimensional calculations of models for proto-Jupiter in order to study the evolution of the angular velocity of the system. He attempted to find initial conditions for the angular velocity distribution such that the system might achieve an end state resembling the present distribution of angular momentum. Thus he required the outermost regions to have the specific angular momentum of Callisto at Callisto's distance and the bulk of the planet to have that of present-day Jupiter. He found that he could not realize these conditions with an initial model having uniform density and angular velocity. However, there was no exchange of angular momentum between various layers of his models during stage 1 because the models were found to be in radiative equilibrium, i.e., stable against convection. These calculations employed values for the grain opacity that now appear to be much too low (Bodenheimer et al. 1980).

Spherically symmetric models having improved opacity values are characterized by extensive convective zones during stage 1, so that angular momentum transport is expected to occur. This transport will have the effect of increasing the angular momentum of the outer zones relative to the inner zones of the protoplanet, thus making it easier to realize the desired end state.

Important constraints on temperature conditions in the Jovian nebula at the time of satellite formation may be contained in the trend of bulk composition of the regular satellites with distance from the planet. In earlier sections, we mentioned the monotonic decrease in the mean density of the Galilean satellites with increasing distance from Jupiter, a result indicative of a larger fractional water content for the outer satellites. Kuiper (1948) presented an early interpretation of these data. He pointed out that Io and Europa probably were mainly silicates and iron to account for their densities and that the occurrence of some silicate or metal with ice was likely for Ganymede and Callisto. He suggested that the apparent monotonic decrease in satellite densities was due to the fact that the inner ones formed at high temperatures owing to the existence of a "red-hot" early Jupiter heated by accretional energy. Finally, he suggested that the high albedos of Io and Europa were caused by a plating of high-temperature oxide smoke. Taken as a whole, this 1948 view of the system must be regarded as a truly remarkable insight.

The next major step in the development of models for satellite formation and explanations of satellite properties *as a set* was undertaken by Pollack and Reynolds (1974). They based their formational model on an early luminosity history for Jupiter calculated by Graboske et al. (1975). Noting the high luminosity of Jupiter during the early part of stage 3, Pollack and Reynolds suggested that the trend in mean density among the Galilean satellites could be attributed to the inhibition of water vapor condensation at close distances to Jupiter during much of the satellite formation period (e.g., Fig. 24.2). In estimating the temperature within the Jovian nebula, they assumed that the nebula was optically thin, so that thermal radiation from the planet directly heated it. Later, Cameron and Pollack (1976) allowed for a wider range of nebula conditions, including optically thick situations, in which case convective energy transport and dissipation of gravitational energy would determine the nebula's temperature structure.

Figures 24.3*a* and *b* illustrate the temporal evolution of nebular temperature at the distances of J1–J5 for these two limiting cases of nebular opacity (Pollack et al. 1976). The vertical axis on the right side of Fig. 24.3 indicates the condensation temperature for various ice species. The time scale for these figures refers to the time from the start of stage 3 and is based on an initial proto-Jovian radius of 16 R_J. As discussed above, the initial radius was probably $\sim$ 3 R_J, and hence $\sim$ 10^5 yr need to be subtracted from the times

Fig. 24.3. (a) Temperature of an ice grain condensing in a low-opacity, primordial Jovian nebula as a function of time from the commencement of the late hydrostatic phase. An initial radius for proto-Jupiter of 16 R_J was assumed. Hence, as explained in the text, $\sim 10^5$ yr must be subtraction from the times given on the horizontal axes. Separate curves are shown for various fixed distances from the planet's center, with each curve corresponding to the present distance of a Galilean satellite or Amalthea. The first letter of the appropriate satellite is shown next to the curve (e.g., C stands for Callisto). The vertical axis on the right-hand side of the graph shows the temperature at which various ice species condense, with the letters E and D indicating the ice species is a member of an equilibrium or disequilibrium condensation sequence. The short vertical line near the top of the figure, labeled A, represents the time, $t(a = r)$, at which proto-Jupiter had the same radius as the orbital distance of Amalthea. (b) Same as (a) except that the nebula is now assumed to be optically thick. (After Pollack et al 1976.)

on Fig. 24.3. Such a correction has been made in the discussion that follows.

A crude estimate of the duration of the satellite formation period can be obtained by noting that much water ice was incorporated into Ganymede, but relatively little into Europa (Cameron and Pollack 1976). Since the nebular disk formed near the end of stage 2 and since the total duration of stage 2 is only ~ 1 yr, the duration of the satellite formation period is essentially equivalent to its longevity from the start of stage 3. Since Jupiter had a radius of ~ 3 R_J at the beginning of stage 3, satellite formation could have commenced then at the orbits of all the Galilean satellites. If we assume satellite formation began then for all four Galilean satellites and ended at the same time for all of them, durations of $\sim 10^5$ and $\sim 10^6$ yr are found for the high- and low-opacity cases, respectively (Cameron and Pollack 1976). Satellite formation presumably ended with the elimination of the disk of gas and dust that resulted from either a strong Jupiter wind driving it away or viscous dissipation causing it to be incorporated into the planet.

Given the above estimates on the duration of the satellite formation period, set by the mean density of Europa and Ganymede and the assumption of coeval formation (or at least coeval termination), we can obtain some interesting bounds on the composition of the other regular satellites. Near the end of the satellite formation period, temperatures decreased to ~ 300 and 400 K at Io's orbit for the low- and high-opacity cases, respectively. As both of these values are within the stability fields of common hydrated minerals (Lewis 1974), it is likely that Io was endowed initially with some water. The corresponding temperatures for Amalthea are 550 and 1200 K so that it, as well as 1979 J1, 1979 J2, and 1979 J3, are predicted to be composed of refractory material. Finally, the corresponding temperatures for Callisto are 150 and 85 K. Especially in the latter case, it is possible that some ices besides water were incorporated into this satellite (see Fig. 24.3).

The smaller size of the innermost regular satellites as compared with that of the Galilean satellites may be due to a combination of two factors. First, the former may have been constructed from a more limited pool of material, refractory minerals, than the latter, due to the much higher temperatures in their region of formation. Second, proto-Jupiter may have extended into the region where the innermost satellites formed during the earliest portions of stage 3, delaying the commencement of satellite formation there. Unfortunately, not enough information is known about the composition of the inner four satellites to assess whether they are made of refractory compounds.

The above model of Jupiter's evolution and its application to the Jovian satellites have been checked in several ways. For instance, the evolutionary calculations for stage 3 predict an excess luminosity for the current epoch that is in good agreement with the observed value (Grossman et al. 1980). Also, the inferred duration of satellite formation is of the same order of magnitude as the formation intervals for some meteorites.

V. EVOLUTION OF THE GALILEAN SATELLITES

The Galilean satellites are large enough bodies to have undergone significant chemical and thermal evolution due to accretional heating and the decay of long-lived radioactive isotopes. Furthermore, they differ in important ways from inner solar system objects. For example, H_2O constitutes a major fraction of Ganymede and Callisto, and tidal dissipation represents an important source of heating for Io and Europa. Below, we review pre-Voyager models of the evolution of the Galilean satellites and then consider the impact of the Voyager results.

In an initial study of the geologic evolution of the individual satellites, Lewis (1971a,b) developed steady-state thermal models for a two-component interior, rock and ice. The only source of heating explicitly modeled was that due to radioactive isotopes, which were contained in the rock fraction. This fraction was modeled after carbonaceous chondrites, while the ice component contained only H_2O. He found that the Galilean satellites were large enough to have undergone almost complete internal differentiation, due to radioactive heating alone. In the case of Ganymede and Callisto, internal differentiation resulted in the occurrence of an inner rock core and a convecting, liquid water mantle of approximately equal mass that was capped by a thin ice crust. Current models of these satellites differ principally in

considering the effect of solid-state convection in the outer solid portion
(Lewis considered only conduction in the solid state) and the role of NH_3
(see below).

Fanale et al. (1977) presented thermal models for Io (radioactivity only),
Europa, Ganymede, and Callisto. Of particular interest is their Europa model.
They postulated that the maximum thickness of any Europa ice crust could
be calculated using an assumed overall density of 3.1 g cm^{-3} together with
the assumption that the nonicy portion of Europa had a density fully as high
as that of Io. (Fanale et al. believed that there was no acceptable scenario
which would allow the density of the nonicy portion of Europa to exceed
that of Io.) Their conclusion was that Europa could possibly possess an H_2O
crust of up to ～ 75 km thick. They also determined that if the crust had to
transport heat to the surface by conduction rather than solid state
convection, there was a chance for liquid H_2O to occur at depths of $\geqslant$ 40 km
even if radioactivity were the only heat source. Thus they suggested the
possibility of a liquid zone at the base of the ice crust. Since that time (see
below) many other effects neglected by Fanale et al. have been introduced as
possibly important factors in the analysis of Europa's near-surface thermal
profile. Some of these (e.g., solid state convection) argue against a liquid
zone, while some others (e.g., marginal tidal heating) tend to support it.
Fanale et al. also argued that the high albedo and exceedingly deep infrared
absorption bands of Europa indicated some effective H_2O redistribution or
resupply process. Another major point argued by Fanale et al. was that
Callisto was much darker than Ganymede because the ratio of silicate (hence
radiogenic isotopes) to ice was significantly greater in the case of Ganymede
and that this resulted in a basically intact initial crust of mixed dark
carbonaceous-chondritic like primordial silicates and ices for Callisto, whereas
Ganymede was more active and able to "purify" its crust by endogenic
activity. Pollack et al. (1978) on the other hand argued that exogenic
contamination by carbonaceous materials (possibly originating from Himalia),
in combination with water ice recoating processes, could explain the albedo
difference. Debris deposition was expected to occur at a greater rate for
Callisto because of its proximity to the outer satellites, and recoating was
expected to be more effective for Ganymede because of its thinner crust.
Each hypothesis encounters some problems explaining the Voyager results
(see below).

The Io model of Fanale et al. (1977) assumed that the bulk composition
of Io was that of a Type II or Type III carbonaceous chondrite based on the
original low-opacity nebula model of Pollack and Reynolds. This model
involved only radioactivity as a heat source and predicted that the basalt
solidus was currently at ～ 400 km depth. As in Fanale et al. (1974), they
postulated an S- and Na-rich surface and assumed that because of the great
depth of the basalt solidus, warm aqueous solutions and H_2O outgassing must

have been the original S and Na "carriers" during an early Io interior devolatization and creation of the S-rich surface. Subsequent H_2O loss to space was recognized to be a necessary postulate. They suggested magneto-spheric ion sputtering as the main cause of surface devolatization, and ioniza-tion followed by magnetic field sweeping as the principal mechanism of re-moval of atmospheric species (Fanale et al. 1977). Shortly prior to the Voyager 1 encounter, Peale et al. (1979) suggested that most of Io's internal heating resulted from tidal interaction with Jupiter, owing to Io's forced eccentricity, and that this heating source dwarfed the nominal contribution from radioactivity. At least from an intuitive point of view, this allowed for a much wider range of possible volatile transport and interior degassing pro-cesses, and even processes of volatile loss from the planet.

Consolmagno and Lewis (1976,1977) provided the most detailed and quantitative models for the internal structure and thermal history of the icy Galilean satellites. These models took into account all phase transitions in the H_2O system, together with the appropriate density changes and latent heat, and agreed completely with the above Europa model of Fanale et al. Their Ganymede and Callisto models were similar, and similar to the much earlier models of Lewis cited above. They emphasized two things: 1) like Fanale et al. (1977) they stressed the differences between the thermal histories of Ganymede and Callisto as the cause of different minimum crustal thicknesses and hence different abilities to preserve the original low-albedo mix in a never-melting crust, and 2) Io- or Europa-sized interior silicate cores of these objects would probably have transformed any possible original carbonaceous-chondritic mineral into a higher temperature assemblage of silicates, such as serpentine or olivine.

Subsequently Reynolds and Cassen (1979) published models for Ganymede and Callisto that departed radically from those of Fanale et al. (1977), Lewis (1971a,b), and Consolmagno and Lewis (1976, 1977). Reynolds and Cassen pointed out that the icy crusts postulated in the earlier models would themselves be unstable with respect to solid-state convection and would therefore transport heat much more rapidly than by solid-state conduction. As a result, the entire liquid mantle would freeze, although subsolidus convection might still occur within much of it. In general, these conclusions were tentatively accepted by most investigators prior to Voyager and, indeed, in 1980. However, it was also recognized that actual laboratory viscosity measurements on ice under appropriate conditions were critically needed and that migration of heat sources, the effect of solutes on the melting-point curves, the effects of tidal heating, and possibly other major effects would have to be taken into account in order to explore satisfactorily parameter space.

Parmentier and Head (1979) provided similar models for the evolution of Ganymede and Callisto. Their models emphasized the role of diaperism as well as solid-state convection in heat transport. Diaperism could arise if the lithosphere is denser than the underlying water mantle, in which case there may be local overturning of the lithosphere and emplacement of water on the surface. They also pointed out that if the icy crusts really contained a high proportion of silicate, they would be unstable and would sink rather than float on any hypothetical mantle, as was the case in Lewis's (1971a) model.

We next consider the impact that the Voyager observations had on our view of Galilean satellite formational conditions and the evolution of individual satellites as geological entities. As indicated earlier, the density of Europa was found to be 3.03, toward the low end of the prior estimates, while Callisto's density (1.79) turned out to be considerably closer to that of Ganymede (1.93) than previously thought (Smith et al. 1979a). In general, the first of these observations had the effect of sustaining the possibility that Europa might have a thick (~ 75 km) ice crust, as suggested by Fanale et al. (1977). Had Europa been found to have a density of ~ 3.5, like that of Io, it would have argued substantially against a thick ice crust. It should be remembered that the low density does not prove that Europa's crust is thick; the estimates are still dependent upon the assumed density of the nonicy material. This could, in fact, turn out to be 3.03 compressed, which would allow for only a veneer of ice. With the new density, the ice layer could be ~ 80 km if the rest of Europa had a density like that of the Moon and ~ 120 km thick if the rest had a density equal to that of Io (Smith et al. 1979b). Additional and perhaps more direct constraints on the crustal thickness of Europa were derived from its surface morphology. The absence of large-scale topography suggested that the crust is not trivially thin (<1 km) and, at least intuitively, the enormous length of the linear features suggested the same. On the other hand, if one interpreted the mottled regions as areas where silicate was closer to the surface, a relatively thin crust might be implied, since enormous local topography on such a small body would be surprising.

Even granting that Europa formed with enough water to form a ~ 100 km ice crust, it is not entirely certain that internal differentiation was complete enough to generate a crust this thick. For example, Ransford and Finnerty (1980) pointed out that the simultaneous postulates that Europa's water was initially all present as water of hydration and that Europa was able to keep its interior fairly cool by subsolidus convection lead to a somewhat different picture of the present planet. In the Ransford and Finnerty model, Europa's ice crust would be $\lesssim 10$ km thick and the fracturing of the crust could be attributed to convective overturn of the underlying silicate (Ransford and Finnerty 1980). They considered accretional heating, but inferred that it was not an important energy source for the interior. However, if Europa was able to retain enough of its gravitational energy of formation, it may have experienced substantial differentiation.

If Europa has a thick ice crust, then a liquid water layer may exist beneath it, as suggested by Fanale et al. (1977). Recent theoretical determinations of the probability of such a liquid layer have involved considerations of the competing effects of solid-state convection and tidal heating. Cassen et al. (1979) found that, if the H_2O had been supplied slowly and incrementally from Europa's interior, the H_2O would never melt, since tidal dissipation in the ice would only slightly augment the heating by interior radioactive decay and the ice crust would probably be convective. (Note that this conclusion differs from that of Fanale et al., who predicted ice melting below 40 km from radioactivity alone because they assumed that conduction would dominate convection in the ice crust.) On the other hand, if Europa's crust *had* ever been largely liquid, and if the current orbital resonance lock among Io, Europa, and Ganymede is ancient, an equilibrium situation could exist in which tidal dissipation in a thin ice crust produced far more heat than decay of radionuclides in the interior and prevented the 10 km crust from thickening. In the latter model, Europa would possess the only liquid H_2O layer among any of the satellites. Moreover, fracturing of a crust overlying liquid H_2O would provide one obvious H_2O redistribution process such as that suggested by Fanale et al. (1977) to explain Europa's optical properties.

Unfortunately the above analysis of Europa's thermal history contains an error in the calculation of the tidal heating rate (Cassen et al. 1980*a*). Allowance for the correct rate considerably weakens, although it does not entirely eliminate, the possibility that liquid water exists today within Europa. The only morphological observation possibly bearing on this particular issue is the global pattern of cracks (see Chapter 14). If it is assumed that the cracks represent genuine crustal expansion due to freezing, much more expansion (5 % to 15 %) would be implied than could be accommodated by the freezing of an initial 100-km H_2O mantle (1 % according to Cassen et al. 1979). On the other hand, piecemeal layer-by-layer emplacement of H_2O at the base of the crust over a period of time could possibly produce the required expansion, providing an argument against a liquid layer, since in the convecting crust models a deep initial ocean is required in order to have a liquid layer now. However, this latter argument is probably fallacious. The large expansion in the case of the piecemeal emplacement results from the assumption that the silicate fraction undergoes no change in mean density as it is dehydrated. When allowance is made for probable concomitant changes in the silicate density, a much smaller expansion takes place (R. Reynolds, personal communication).

The age of the cracks may constitute a somewhat stronger constraint on the phase of the H_2O in the lower portion of Europa's crust at the present time. Preliminary results from a study by Lucchitta and Soderblom (Chapter 14) indicate the possible existence of three families of cracks:

1. Cracks that are roughly concentric about the anti-Jupiter point;
2. Conjugate sets with 30-60° inclinations having a white center strip with dark edges;
3. Concentric cracks about a 70° south pole.

The first set appears to be the youngest and the third the oldest. Both the first and third sets appear to have been episodically produced, yet the second set appears to exhibit a range of ages. The genetic significance of these results is not clear. For Europa, the problem of assigning ages to crater densities is particularly difficult, with Lucchitta and Soderblom assigning ages of several billions of years to the youngest formations and Shoemaker and Wolfe (Chapter 10) assigning ages $< 10^8$ yr. Lucchitta and Soderblom contend that the first set may possibly represent the final (local) freezing of the crust, while the conjugate sets may be the result of tidal stresses.

The Voyager encounter also caused the differences between Ganymede and Callisto to emerge as a more serious problem than before. First, as mentioned above, Voyager revealed the difference in their densities to be only $\sim 8\%$, diminishing somewhat the temptation to attribute dissimilarities in the two objects' geologic evolution to differences in bulk ice: silicate (or radionuclide) ratios. On the other hand, Voyager confirmed, as suspected on the basis of Earth-based measurements (e.g., Fanale et al. 1977; Consolmagno and Lewis 1977), that the higher albedo of Ganymede was indeed associated with a much more active tectonic history for Ganymede than for Callisto.

Cassen et al. (1980b) quantitatively examined several factors that might account for the differences in the longevity of surface activity on Ganymede and Callisto. They found that differences in tidal heating could have extended the period during which Ganymede had a thin lithosphere and hence surface activity by only ~ 0.1 Gyr. While Ganymede may have received a larger amount of accretional heating due to its larger mass, this difference was found to be unimportant in almost all scenarios of their formation and early evolution. The only case where accretional heating was found to make a significant difference was one in which the satellites formed totally inhomogeneously, with a silicate core forming before any water was accreted, and the core of Ganymede was initially molten, but that of Callisto was initially solid. In this extreme case, the period during which Ganymede had a thin crust might have been ~ 0.5 Gyr longer than for Callisto. Finally, Cassen et al. (1980b) examined the ability of differences in radioactive heating to account for the prolonged heating of Ganymede. Despite the close similarity in mean density between the two satellites, they found that this small difference, when coupled with a much larger difference in mass ($\sim 50\%$), implied a significantly more massive silicate core for Ganymede and hence a proportionally larger amount of radionuclides. Furthermore, due to the rapid decay of some of the radionuclides, their heat output decreases by a factor of several over the first ~ 0.5 Gyr. As a result of both of these effects, the time during which the surface heat flux exceeded a "critical" value and hence the lithosphere was "thin" was found to be $\sim 0.5 \times 10^9$ yr longer for Ganymede than for Callisto, with this difference in time being insensitive to the choice of the critical flux. On the basis of these calculations, Cassen et al. (1980b) concluded that differences in radioactive heating constitute the most likely explanation for the differences in the longevity of surface activity on the two satellites.

Parmentier and Head (1980) also considered, in a more qualitative fashion, the effects of differences in various heat sources on the geologic history of the two satellites. They proposed that the differences in the surface evolution between Ganymede and Callisto were due to greater accretional and tidal heating in the case of Ganymede. However, according to Cassen et al. (1980b), the latter is not a major factor and the former is important only in very restricted circumstances.

These theoretical considerations are consistent with Voyager data on the tectonic history of Ganymede. The absence of very young ($\lesssim 3$ Gyr) grooved terrain suggests that the internal forces driving the tectonism died down early in Ganymede's history. If we associate this decrease in tectonism with the thickening of the satellite's lithosphere, then Ganymede's lithosphere became sufficiently thick to suppress the surface expression of its endogenic activity ~ 1 Gyr later than in the case of Callisto, with the precise value of this time interval being highly uncertain. Both this time difference and its occurrence during the early history of the satellite are compatible with the predicted divergence in the thermal histories of the two satellites due to differences in radioactive heating or, less probably, accretional heating (see Chapter 12).

While the Voyager data support the thesis of Fanale et al. (1977) and Consolmagno and Lewis (1977) that the higher albedo of Ganymede is the result of endogenic activity, it is not necessarily consistent with their thesis that the dark material on Ganymede and Callisto is truly primordial. As mentioned in an earlier section, the age of the dark surfaces is probably ~ 4 Gyr and not 4.6 Gyr. Thus, these surfaces are very old, but likely not truly primordial or part of the accretionary flux.

As discussed in Chapter 13 by Shoemaker et al., grooved terrain is produced by extensional, as opposed to compressional, tectonics, largely by normal faulting. Because of the identification with extensional forces and because of the age distribution of various areas within the grooved terrain, Shoemaker et al. also identified the production of the grooved terrain with the freezing of Ganymede's mantle. Shoemaker et al.'s view is supported by the general absence of very large craters on the oldest Ganymede terrain, which suggests that Ganymede started with a liquid mantle and a solid crust < 10 km thick. Note that the "top-down" portion of mantle freezing of an icy object only causes expansion because the familiar Ice I is less dense than water. However, the situation could be complicated because most of the freezing could be from the bottom up, forming the other phases of ice, and all of these are denser than liquid H_2O under appropriate conditions. Thus, overall contraction could result. Internal differentiation provides an alternative mechanism for causing an expansion of Ganymede's crust (Squyres 1980). Both its timing on Ganymede (long after satellite formation) and the absence of grooved terrain on Callisto represent serious problems for this hypothesis.

It was suggested by Schubert et al. (1980) that solid-state convection was so effective for icy bodies that no substantial liquid layer could have persisted for any substantial length of time. Indeed, the model of Schubert et al. is unique in that it envisions the possibility that differentiation of silicate from ice was incomplete, owing to the retention of only a small fraction of the accretional energy and the extreme effectiveness of solid-state convection. We would suggest, however, that models in which steady state is rapidly achieved for both Ganymede and Callisto, and that portray these objects primarily as "dirty ice going around in circles," fail to address the primary datum, i.e., that very small differences in ice-silicate ratio and/or total mass/radius have produced totally different geologic records, totally different surface ages, and totally different optical properties, which other hypotheses discussed herein, such as those of Cassen et al. (1980b) and Parmentier and Head (1980), have been designed to explain.

The Voyager results impacted models of Io's evolution by confirming its high density (3.53 g cm^{-3}), providing additional evidence for the widespread occurrence of sulfur compounds on its surface and atmosphere and, most profoundly, by revealing an unprecedented level of volcanic activity. As discussed in Chapter 17 by Johnson and Soderblom, analysis of imaging data provides a quantitative estimate of the resurfacing rate of ~ 1 mm yr^{-1}.

There appears to be general agreement that tidal heating is responsible for the active volcanism on Io, in which case its solid crust may be no thicker than ~ 20 km (Peale et al. 1979). While much of the crust may be made of silicates, the evidence on surface composition suggests that there is an uppermost layer that is heavily enriched in elemental sulfur and sulfur compounds. The maximum thickness of this top S layer is a few kilometers, this limit being set by the solubility of S in silicate melts and not its cosmic abundance (Smith et al. 1979c). However, if basaltic volcanism is dominant, as argued by Carr et al. (1979), the S layer may be somewhat thinner. The interior of Io presumably is mostly molten or near melting and is believed to be composed of ferromagnesium silicates, with perhaps a core rich in iron sulfide (see below). This inference about the thermal state of the interior is based on the occurrence of active volcanism on Io, the tidal theory of heating, and a large value for the surface heat flow derived from thermal infrared data by Matson et al. (1980), Morrison and Telesco (1980), and Sinton (1981).

Within the rather weak observational constraints provided mainly by knowledge of Io's density and optical surface composition, the years 1979 and 1980 saw Io interior composition and evolution models still striving for reasonableness, not uniqueness (see Chapter 4 by Cassen et al.). Hapke (1979) postulated that the intense tidal heating proposed by Peale et al. (1979) initially resulted in segregation of a Fe° and FeS core. But subsequently, convective transport was so violent that it was able repeatedly to carry up portions of FeS from the core to shallow depths, where the FeS dissociated. This mechanism released Fe, which sank back to the core, and S, which rose, driving the volcanoes and being deposited across the surface. In Hapke's model, the original deposition of S in an FeS core is viewed as an explanation for the preservation of S to the present. Other investigators viewed the potential for forming an FeS core as a problem in that it would drag all the S down permanently, thus forbidding the formation of a sulfur-rich crust.

Some 1979-1980 studies of the chemical history of sulfur on Io suggest that it might have undergone a complex and multistage history influenced largely by the relatively high oxygen fugacity represented by magmas generated from carbonaceous chondritic material. Consolmagno and Lewis (1979, 1980) and Lewis and Consolmagno (1979) have developed a series of Io composition and evolution models that can be described as follows: The bulk initial composition of Io is taken to be that of a C2 carbonaceous chondrite. During the initial differentiation of this object, tidal dissipation causes extensive generation of magmas with moderate oxygen fugacities characterized by the evolution of huge amounts of CO_2 and (later) H_2O. During the process Fe begins to be precipitated into the core as FeS. Later, as H is lost from the planet, there is a change in the geochemical state of mantle iron. Fe is precipitated into the core as pyrite (FeS_2) or magnetite (Fe_3O_4), and some S rises as leftover blobs of native S which are in chemical equilibrium with the species mentioned above. Thus Io starts out fairly oxidized (C2 chondrites have no metallic Fe° and have much of their Fe equally distributed between Fe^{++} and Fe^{+++} in magnetite) and evolves to an even more oxidized state later. The accumulative "fall-in" to the core would include FeS_2, FeS, and Fe_3O_4 as candidate phases. The main solid phase to "float up" to the surface would be S_X, i.e., any of the allotropes of S, and also SO_2 gas. This scenario did not violate anything firmly known about Io in 1980. However, it is important to note that Na_2S, proposed as a major Io surface phase by Fanale et al. (1979) and Nash and Nelson (1979), is even less dense than sulfur! Also, it has been argued by Schubert et al. (1980) that Io might have been able to lose heat by solid-state convection as fast as it gained heat from tidal dissipation, thereby precluding the thin silicate crust "runaway tidal heating" model of Peale et al. (1979). The heat flow estimates (see Chapter 19 by Pearl and Sinton) appear to argue against this model; the bounds on the ratio of the tidal dissipation factor Q to Love number obtained from the heat flow implies a partially molten interior (see also discussion in Chapter 4 by Cassen et al.).

VI. LOSS OF VOLATILES

In this section, we consider the rate at which satellites lose some of their volatile species constituted nontrivial fractions of the lost material, then Earth these losses can cause a substantial evolution of their volatile inventories over the age of the solar system. We center our attention on Io, since it may be the only satellite of the Jovian system for which material has been lost at a high enough rate to result in the complete elimination of some volatile species.

There are several reasons for suspecting that the volatile inventory of Io has been substantially modified with time. First, Io is currently losing material to the surrounding Jovian magnetosphere at a rate of 10^{10} to 10^{11} atoms cm^{-2} s^{-1}, with S and O constituting the dominant species involved in this exchange (see Chapter 23 by Sullivan and Siscoe). If such loss rates were sustained over much of the satellite's lifetime, and if at earlier times other volatile species constituted nontrivial fractions of the lost material, then Earth analogue amounts of N_2, CO_2, A, Ne, and H_2O could have been totally eliminated from the satellite. Second, there is a surprising lack of evidence for the occurrence of volatiles, other than sulfur-containing ones, on Io's surface, in its atmosphere, and in the Io torus. For example, as discussed in an earlier section, very stringent upper bounds have been set on the fractional amount of water ice and bound water on Io's surface from near infrared spectrophotometric measurements (Chapter 7 by Sill and Clark). Upper bounds on the total atmospheric pressure, determined by occultation measurements (Smith and Smith 1972), lie far below the partial pressure expected for gases such as N_2, even when allowance is made for their being buffered by a polar cold trap (Pollack and Witteborn 1980). Finally, severe upper bounds have been set on the N and C content of the magnetosphere near Io (Krimigis et al. 1979; Bridge et al. 1979; Vogt et al. 1979). Thus there are two basic questions that need to be addressed: Could Io have lost essentially all of its original endowment of volatiles such as H_2O, CO_2, and N_2? Why does it still have significant amounts of sulfur-containing volatiles?

In order to answer the above questions, we first need to consider the mechanisms by which Io is presently losing volatiles (Chapter 21 by Kumar and Hunten; Pollack and Witteborn 1980). There are four principal loss mechanisms involving material in the atmosphere: interactions of the magnetospheric plasma with volcanic plume gases, with gases in the background atmosphere, with volcanic plume particles, and Jeans escape (thermal evaporation) of the background atmosphere. The first two of these processes occur at substantial rates when the gases extend into the magnetosphere and are exposed directly to the magnetospheric plasma. In this event, loss occurs principally through charge exchange with ions and electron ionization, with the newly ionized material being swept up by the Jovian magnetic field (Pollack and Witteborn 1980). At present, the highest volcanic plumes reach altitudes that extend into the magnetosphere on the upstream side of the magnetospheric flow past Io, while the high exospheric temperature (~ 1000 K) allows the background atmosphere to extend well into the magnetosphere on the day, upstream side (Pollack and Witteborn 1980; Chapter 21 by Kumar and Hunten). Small solid particles ($\lesssim 0.1$ μm) in the high altitude plumes quickly become charged when exposed to the magnetospheric plasma, and the Lorentz forces acting on them may exceed the force of Io's gravity, in which case they will be lost to the magnetosphere (Johnson et al. 1980). Despite the high exospheric temperature, the dominant atmospheric gas, SO_2, thermally evaporates at a negligible rate. However, one of its dissociation products, O, can escape at a substantial rate (Chapter 21).

There are also several loss processes involving volatiles on Io's surface and in its interior. As first pointed out by Matson et al. (1974), sputtering of surface materials by magnetospheric particles can be an effective loss process. Experimental studies by Brown et al. (1980) and Lanzerotti et al. (1980) showed that very large yields ($\sim 10^3$) can be realized by bombarding ices with ~ 1 MeV heavy ions, where yield refers to the ratio of atoms ejected to incident ions. Thus surface ice deposits of all the major volatile species may be susceptible to large erosion rates due to sputtering. However, in order for the sputtered molecules to escape from Io, they cannot experience more than a few collisions with atmospheric gases. In addition, they must be given speeds in excess of the escape velocity, a condition that has not been adequately demonstrated experimentally. The former condition may not be met on the day side of Io, where surface pressures may range from $\sim 10^{-7}$ to 10^{-10} bar (Pearl et al. 1979), but it may be satisfied on the night side, where the low surface temperatures (< 100 K) result in very low vapor pressures for SO_2. However, atmospheric dynamics may substantially reduce the pressure differences between the day and night sides (Summers and Ingersoll 1980). Also, an appreciable O_2 atmosphere may be present on the night side, arising from the photodissociation of SO_2 on the day side and the subsequent transport of O_2 to the night side (Chapter 21 by Kumar and Hunten).

In the case of water, there may be an additional loss mechanism that is important. Studies of terrestrial volcanoes indicate that water vapor is partially dissociated into hydrogen and oxygen in the magma chambers, with the oxygen being essentially eliminated in the chamber through the oxidation of reduced components of the magma, such as FeO (Pollack and Yung 1980). Such partial dissociation can be expected to occur inside Io in association with silicate volcanism, but not sulfur volcanism, because of the lower temperatures in the latter case. The hydrogen so produced quickly escapes by thermal evaporation at the top of the atmosphere. Because of the extensive recycling of material through the crust of Io (Johnson et al. 1979), much water can be eliminated in this way.

Table 24.2 summarizes the rate at which the above loss mechanisms operate for both the present conditions on Io and for hypothetical early N_2 dominated atmospheres (Pollack and Witteborn 1980). Several of the loss mechanisms are not effective in the latter case. Since the surface pressure on the night side is not negligible for an N_2 atmosphere, sputtered molecules would no longer be able to escape. Also, at the relatively low exospheric temperature (~ 200 K) of these paleoatmospheres, dissociation fragments would not escape rapidly by Jeans escape. Since the exobase extends into the magnetosphere, and the increase in surface pressure has only a very modest impact on the altitudes reached by volcanic plumes, loss mechanisms involving the interaction of the magnetosphere with volcanic plumes and the background atmosphere would operate at rates comparable to those for the current SO_2 dominated atmosphere. Also, silicate volcanism, in contrast to sulfur volcanism, may have been more prevalent at earlier times, in which case a greater loss of water through interior dissociation may have been possible.

Using Table 24.2 as a guide, we now consider the elimination of various volatile species. Thermal dissociation of water offers a way of removing very large amounts of water, amounts far in excess of that derived from a strict terrestrial analog (~ 100 m) and comparable to that expected for some classes of carbonaceous chondrites (kilometers). In addition, once SO_2 became the dominant atmospheric constituent, significant quantities of water might have been removed from Io through the sputtering of surface ice deposits. Finally, as long as water was a major component of plume particles, the charging of these particles by the magnetospheric plasma electrons could have resulted in large additional losses of water.

The above conclusions, based on Pollack and Witteborn (1980), differ from ones drawn by several authors dealing with the loss of water from Io. Hapke (1979) suggested that it was lost simply by being outgassed. But the volcanic plumes have velocities significantly smaller than the escape speed (Strom et al. 1979), and, by the time they reach high altitudes, they are far too cool for Jeans escape to be important (see Chapter 16 by Strom and Schneider; Chapter 18 by Kieffer). Consolmagno (1979) suggested that water in volcanic plumes was lost by both Jeans escape and dissociation and ionization by solar ultraviolet radiation. The former process has been commented upon above. The latter occurs at an even smaller rate than that due to ionization by the magnetospheric plasma, which, according to Table 24.2, is unimportant. Finally, Kumar (1979) has suggested that diffusive separation in Io's atmosphere leads to an enhanced concentration of water vapor in Io's exosphere and hence a preferential loss of it as compared with that of SO_2. But the saturation vapor pressure of H_2O at typical temperatures on Io is many orders of magnitude smaller than that of SO_2. Indeed, so little H_2O can

TABLE 24.2
Loss Rates of Volatiles from Io's Present and Past Atmosphere

Loss Mechanism	Major Atmospheric Constituent						
	SO_2				N_2		
	^{a}Loss Rate (molecules/cm²/s) SO_2	$^a N_2$	$^a H_2O$	$^a H_2$	$^a N_2$	$^a H_2O$	$^a H_2$
Volcanic plume gases/magnetospheric interaction	1×10^8	$1 \times 10^8 \times f_{N_2}$	—	—	$1 \times 10^8 \times f_{N_2}$	—	—
Volcanic plume particles/ magnetospheric interaction	$^b 1 \times 10^{11} - 3 \times 10^{11} \times f_{SO_2}$	$^b 1 \times 10^{11} - 3 \times 10^{11} \times f_{N_2}$	—	—	$^b 1 \times 10^{11} - 3 \times 10^{11} \times f_{N_2}$	$^b 1 \times 10^{11} - 3 \times 10^{11} \times f_{H_2O}$	—
Background atmosphere/ magnetospheric interaction	$^g 1 \times 10^{10}$	$^g 1 \times 10^{10} \times f_{N_2}$	—	—	$^g 2 \times 10^{10}$	—	—
Jeans escape	$^c \sim 1 \times 10^{10}$	$3 \times 10^8 \times f_{N_2}$	$1 \times 10^7 \times f_{H_2O}$	$2 \times 10^{12} \times f_{H_2}$	—	—	$2 \times 10^{12} \times f_{H_2}$
Sputtering of surface ices	$^d 1 \times 10^{11} \times f_{SO_2}$	$^{d,e} 2 \times 10^{10} \times f_{N_2}$	$^d 2 \times 10^{10} \times f_{H_2O}$	—	—	—	—
Thermal dissociation	—	—	—	—	—	$f_2 \times 10^9 - 2 \times 10^{13}$	—

$^a f_{N_2}$, f_{H_2O}, f_{SO_2} and f_{H_2} denote number mixing ratios of the subscripted species at appropriate positions, e.g., surface for sputtering and exosphere for Jeans escape.

^{b}Based on assumption that a non-negligible fraction of the mass in the plumes seen on Voyager images is due to particles whose size is less than 0.1 μm.

^{c}This loss rate refers only to O. Other dissociation products of SO_2 escape at much lower rates. We have assumed that O is the dominant species at the exobase.

^{d}These values are applicable if the exobase lies at the surface on the night side. If it lies at higher altitudes, these loss rates could be negligible.

^{e}The yield values for N_2 ice have been assumed to be equal to those for H_2O ice.

$^f 1$ km of water is assumed here. The numbers scale linearly with the amount of water in or near the crust.

^{g}These numbers are based on ion and electron fluxes calculated on the basis of the corotation of the magnetosphere past Io. However, if electrons are free to move along field lines, as seems reasonable, their flux would be about a factor of 30 higher and the loss rates given in this table should be increased by about a factor of 10.

exist in Io's atmosphere due to the low surface temperature that a few meters at most of water could be eliminated from Io over the satellite's lifetime due to processes involving atmospheric water vapor (Pollack and Yung 1980; Pollack and Witteborn 1980), as also shown in the Jeans escape row of Table 24.2.

Table 24.2 indicates that amounts of N_2 in excess of one bar could be eliminated from Io over the planet's lifetime. This figure may be compared to an amount of ~ 30 mbar expected from a terrestrial volatile analog (Pollack and Yung 1980). If a nitrogen-dominated atmosphere existed during Io's early history, magnetospheric interactions with the background atmosphere would have eroded it at a large rate. Also, much N_2 may have been lost from volcanic plumes then if it constituted a significant fraction of the condensed material in them. Once the SO_2-dominated atmosphere became established and high exospheric temperatures were realized, any N_2 subsequently released to the atmosphere would be lost at an appreciable rate due to Jeans escape (Pollack and Witteborn 1980) and to diffusive separation and magnetospheric interactions (Kumar 1979). Similar statements hold for other volatile species, such as A, Ne, and CO_2, although much CO_2 could still be present in the polar cold trap (Pollack and Witteborn 1980). However, this latter possibility depends on the amount of CO_2 outgassed and the exposure of polar CO_2 ice deposits to magnetospheric high energy ions, since sputtering of surface ices may be an effective loss mechanism.

Finally we consider the factors that have enabled sulfur-containing volatiles to remain abundant on Io. As mentioned earlier, Io is presently losing S-containing material to the surrounding magnetosphere at a rate of $10^{10}-10^{11}$ atoms cm^{-2} s^{-1}. At this rate, Io would lose the amount of S-containing volatiles expected on the basis of the Earth's surface volatile inventory in a time much less than the age of the solar system (Pollack and Yung 1980). However, S is a major element in meteorites ($\sim 10\%$ Si), so that the relative scarcity of S near the Earth's surface is probably the result of the almost complete segregation of S compounds to the Earth's deep interior. Due to the extensive tidal heating of Io, it is reasonable to suggest that a much larger fraction of its total S is contained in its near-surface volatile inventory. In this case, Io will not lose most of the S in this volatile inventory over the satellite's lifetime, as is pointed out by Johnson and Soderblom (Chapter 17). In addition to its large abundance, S-containing volatiles may have lasted until today on Io due to the diffusive separation and the concentration of lower atomic weight gases near the exosphere of Io. As a result, a preferential loss of these gases, rather than SO_2, occurred by magnetospheric interactions (Kumar 1979). In addition, much S may have been initially segregated as FeS to the deep interior of Io and so been protected against extensive early losses (Hapke 1979).

In summary, Io is currently losing and has been losing volatiles at a very large rate due to the combination of its active volcanism and its interactions with the Jovian magnetosphere. Should these rates apply over much of its lifetime, Earth analog amounts of almost all volatile species, including N_2, CO_2, and H_2O, could have been totally eliminated from the satellite. The survival of S-containing volatiles until the present is attributable in part to their high abundance in the *surface* volatile inventory, implying a less effective segregation of S compounds into Io's core, as compared with the situation for the Earth.

VII. PROSPECTUS

As a result of Earth-based observations, laboratory simulations, theoretical investigations, and especially the Voyager flyby mission, much progress has been made in advancing our understanding of the history of the Jovian satellites. Nevertheless, we have achieved so far only a partial answer to most of the fundamental questions about them. Further significant advances can be expected from continued Earth-based activities as well as the forthcoming Galileo orbiter mission. Below, we briefly take stock of our understanding of the origin and evolution of the Jovian satellites and consider the ways in which future investigations can help to resolve areas of uncertainty.

Origin of the Regular Satellites

The regular satellites originated in a primordial Jovian nebula that had a strong temperature gradient, perhaps due to the high luminosity of Jupiter in very early times. The well-measured densities of the Galilean satellites provide useful constraints on the temperature conditions within this nebula and indicate that satellite formation occurred over a time scale of $\sim 10^5 - 10^6$ yr. Important outstanding issues include: the bulk composition of J5 and the new inner satellites, which will further constrain conditions in the nebula and provide a check on the above picture; the abundance of ices other than H_2O in the initial makeup of Callisto and possibly Ganymede; the abundance of hydrated components in the initial makeup of Io and Europa; and a detailed description of the structure and composition (especially the oxidation state) of the primordial nebula.

Origin of the Irregular Satellites

Gas drag capture of stray bodies by the outer regions of a very extended proto-Jupiter, just prior to hydrodynamical collapse, appears to be a promising mechanism for capturing these objects and producing two families through drag-induced fracturing. However, this hypothesis is far from proven. It, as well as alternative hypotheses, need to be further tested against relevant data.

Thermal Evolution of the Galilean Satellites

Surface landforms offer unmistakable evidence that significant endogenic activity has occurred within all four large satellites over a part of their lifetime, with the longevity of this active phase ranging from approximately the first 0.5 Gyr for Callisto to much if not all of Io's history, including the present. In addition to radioactive heating, tidal and accretional heating have played key roles in the thermal histories of the Galilean satellites, with tidal heating playing a dominant part for Io and having some importance for Europa and perhaps Ganymede, and radioactive and possibly accretional heating being most important for the early history of Ganymede and, to a lesser degree, Callisto. Due to the efficient heat transport by subsolidus convection, the interiors of Europa, Ganymede, and Callisto are almost entirely in a solid state at the present epoch, but tidal heating of Io has kept almost all of its interior at temperatures close to and perhaps above the solidus. Key issues include: whether most of Io's interior is partially molten; whether the bottom portion of Europa's water crust is liquid and, if so, its depth below the surface; the reasons for the differences in the duration of tectonic activity on Ganymede and Callisto; and the nature of a number of key surface landforms, such as the global cracks across Europa, and their relationship to endogenic processes.

Chemical Evolution of the Galilean Satellites

Based on our understanding of the thermal history of these satellites, it seems likely that they have undergone at least partial internal differentiation, with the rock fraction being concentrated towards the center and the ice fraction towards the surface. Currently, water ice constitutes almost all of the surface material of Europa, some of Ganymede's surface, and probably a little of Callisto's, with a dark rock component also being present in the latter two cases. Sulfur compounds, including elemental sulfur and sulfur dioxide ice, dominate the surface of Io. As a result of extensive volcanism throughout much of its history and its interactions with the surrounding Jovian magnetosphere, Io has lost and is continuing to lose volatiles at a prodigious rate to the magnetosphere. Thus it may have lost many of the volatile species initially present, with S-containing volatiles surviving until today, in part, because of their high abundance. Key questions include: the degree of internal differentiation of all four satellites; the detailed history of the chemical evolution of Io; the composition of the nonicy components of Ganymede and Callisto's surfaces; and the nature of the processes responsible for the current surface composition of the satellites, including the relative importance of endogenic and external processes.

The 1970s have seen major advances in our understanding of the history of the Jovian satellites. The 1980s promise to be at least as exciting as a result of the Galileo spacecraft mission and complementary Earth-based activities.

Acknowledgments. We are very grateful to R. Reynolds and P. Cassen for their careful reading of this chapter and their many helpful suggestions.

June 1981

REFERENCES

Aksnes, K. (1977). Properties of satellite orbits: Ephemerides, dynamical constants, and satellite phenomena. In *Planetary Satellites* (J. Burns, Ed.), pp. 27-42. Univ. of Arizona Press, Tucson.

Anderson, J. D., Null, G. W., and Wong, S. K. (1974). Gravity results from Pioneer 10 Doppler data. *J. Geophys. Res.* 79, 3361-3364.

Bodenheimer, P. (1974). Calculations of the early evolution of Jupiter. *Icarus* 23, 319-325.

Bodenheimer, P. (1977). Calculations of the effects of angular momentum on the early evolution of Jupiter. *Icarus* 31, 356-368.

Bodenheimer, P., Grossman, A. S., DeCampli, W. M., Marcy, G., and Pollack, J. B. (1980). Calculations of the evolution of the giant planets. *Icarus* 41, 293-308.

Bridge, H. S. and the Voyager Plasma Science Team. (1979). Plasma observations near Jupiter: Initial results from Voyager 1. *Science* 204, 47-51.

Broadfoot, A. L. and Voyager Ultraviolet Spectrometer Team. (1979). Extreme ultraviolet observations from Voyager 1 encounter with Jupiter. *Science* 204, 39-42.

Brown, R. A. (1974). Optical line emission from Io. In *Exploration of the Planetary System* (A. Woszczyk and C. Iwaniszewska, Eds.), pp. 527-531. Reidel, Dordrecht, Holland.

Brown, W. L., Augustyniak, W. M., Brody, E. Cooper, B., Lanzerotti, L. J., Ramirez, A., Evatt, R., and Johnson, R. E. (1980). Energy dependence of the erosion of H_2O ice films by H and He ions. *Bell Laboratory,* preprint.

Burns, J. A., Showalter, M. R., Cuzzi, J. N., and Pollack, J. B. (1980). Physical processes in Jupiter's ring: More than dust by Jove! *Icarus.* In press.

Butterworth, P. S., Caldwell, J., Moore, C. A., Owen, T., Rivolo, A., and Lane, A. L. (1980). An upper limit to the global SO_2 abundance on Io. *Nature.* In press.

Cameron, A. G. W., and Pollack, J. B. (1976). On the origin of the solar system and of Jupiter and its satellites. In *Jupiter* (T. Gehrels, Ed.), pp. 61-84. Univ. Arizona Press, Tucson.

Carlson, R. W., Bhattacharyya, J. C., Smith, B. A., Johnson, T. V., Hidayat, B., Smith, S. A., Taylor, G. E., O'Leary, B. T., and Brinkmann, R. T. (1973). An atmosphere on Ganymede from its occultation of SAO-186800 on 7 June 1972. *Science* 182, 53-55.

Carr, M. H., Masursky, H., Strom, R. G., and Terrile, R. J. (1979). Volcanic features of Io. *Nature* 280, 729-733.

Cassen, P., Peale, S. J., and Reynolds, R. T. (1980*a*). Tidal dissipation in Europa: A correction. *Geophys. Res. Letters.* In press.

Cassen, P., Peale, S. J., and Reynolds, R. T. (1980*b*). On the comparative evolution of Ganymede and Callisto. *Icarus* 41, 232-239.

Cassen, P., Reynolds, R. T., and Peale, S. J. (1979). Is there liquid water on Europa? *Geophys. Res. Letters* 6, 731-734.

Clark, R. N., and McCord, T. B. (1980). The Galilean satellites: New near-infrared spectral reflectance measurements (0.65-2.5 μm) and a 0.325-5 μm summary. *Icarus* 41, 323-339.

Colombo, G., and Franklin, F. A. (1971). On the formation of the outer satellite groups of Jupiter. *Icarus* 15, 186-191.

Consolmagno, G. J. (1979). Sulfur volcanoes on Io. *Science* 205, 397-398.

Consolmagno, G. J., and Lewis, J. S. (1976). Structural and thermal models of icy Galilean satellites. In *Jupiter* (T. Gehrels, Ed.), pp. 1035-1051. Univ. Arizona Press, Tucson.

Consolmagno, G. J., and Lewis, J. S. (1977). Preliminary thermal history models of icy satellites. In *Planetary Satellites* (J. A. Burns, Ed.), pp. 492-500. Univ. Arizona Press, Tucson.

Consolmagno, G. C., and Lewis, J. S. (1979). The evolution of Io. *Bull. Amer. Astron. Soc.* 11, 599 (abstract).

Consolmagno, G. J., and Lewis, J. S. (1980). The chemical thermal evolution of Io. IAU Coll. 57. *The Satellites of Jupiter* (abstract 7-11).

Cruikshank, D. P. (1977). Radii and albedos of four Trojan asteroids and Jovian satellites 6 and 7. *Icarus* 30, 224-230.

Cruikshank, D. P., Jones, T. J., and Pilcher, C. B. (1978). Absorption bands in the spectrum of Io. *Astrophys. J.* 225, L89.

DeCampli, W. M., and Cameron, A. G. W. (1979). Structure and evolution of isolated giant gaseous protoplanets. *Icarus* 38, 367-391.

Degewij, J., Andersson, L. E., and Zellner, B. (1980). Photometric properties of outer planetary satellites. *Icarus* 44, 520-540.

Dollfus, A. (1970). Diamètres des planètes et satellites. In *Surfaces and Interiors of Planets and Satellites* (A. Dollfus, Ed.), pp. 46-139, Academic Press, New York.

Fanale, F. P., Johnson, T. V., and Matson, D. L. (1974). Io: A surface evaporite deposit? *Science* 186, 922-924.

Fanale, F. Johnson, T., and Matson, D. (1977). Io's surface and the histories of the Galilean satellites. In *Planetary Satellites* (J. Burns, Ed.), pp. 379-405. Univ. Arizona Press, Tucson.

Fink, U., Dekkers, N. H., and Larson, H. P. (1972). Infrared spectra of the Galilean satellites of Jupiter. *Astrophys. J.* 179. L155-L159.

Gehrels, T. (1977). Some interrelationships of asteroids, Trojans, and satellites. In *Comets, Asteroids, and Meteorites*. (A. Delsemme, Ed.) pp. 323-325, Univ. Toledo, Toledo, Ohio.

Gehrels, T. and the Pioneer Saturn Photopolarimeter Team. (1979). Imaging photopolarimeter on Pioneer Saturn. *Science* 207. 434-439.

Graboske, H. C., Jr., Pollack. J. B., Grossman, A. S., and Olness, R. J. (1975). The structure and evolution of Jupiter: The fluid contraction stage. *Astrophys. J.* 199, 265-281.

Grossman, A. S., Pollack, J. B., Reynolds, R. T., Summers, A. L., and Graboske, H.C., Jr. (1980). The effects of dense cores on the structure and evolution of Jupiter and Saturn. *Icarus* 42, 358-379.

Hapke, B. (1979). Io's surface and environs: A magmatic-volatile model, *Geophys. Res. Letters* 6, 799-802.

Harris, A. W. (1978). Formation of outer planets. *Lunar Planet. Sci. Conf.* 9, 459-461 (abstract).

Heppenheimer, T. A. (1975). On the presumed capture origin of Jupiter's outer satellites. *Icarus* 24, 172-180.

Heppenheimer, T. A., and Porco, C. (1977). New contributions to the problem of capture. *Icarus* 30, 385-401.

Jewitt, D. C., Danielson, G. E., and Synnott, S. P. (1979). Discovery of a new Jupiter satellite. *Science* 206, 951.

Jewitt, D.C. and Danielson, G.E. (1980). The Jovian ring. Submitted to *J. Geophys. Res.*

Johnson, T. V. (1969). *Albedo and Spectral Reflectivity of the Galilean Satellites of Jupiter*. Ph.D. dissertation, California Institute of Technology, Pasadena, California.

Johnson, T. V. (1978). The Galilean satellites of Jupiter: Four worlds. *Ann. Rev. Earth Planet. Sci.* 6, 93-125.

Johnson, T. V., Cook, A. F., II, Sagan, C., and Soderblom, L. A. (1979). Volcanic resurfacing rates and implications for volatiles on Io. *Nature* 280, 746-750.

Johnson, T. V., Morfill, G., and Grün, E. (1980). Dust in Jupiter's magnetosphere. IAU Coll. 57. *The Satellites of Jupiter* (abstract 5-16).

Johnson, T. V., and Pilcher, C. B. (1977). Satellite spectrophotometry and surface compositions. In *Planetary Satellites* (J. A. Burns, Ed.), pp. 232-268. Univ. Arizona Press, Tucson.

Krimigis, S. M. and the Voyager Low-Energy Charged Particle Team. (1979). Low-energy charged particle environment at Jupiter: A first look. *Science* 204, 998-1003.

Kuiper, G. P., Ed. (1948). Planetary atmospheres and their origin. In *The Atmospheres of the Earth and Planets*, pp. 306-405. Univ. Chicago Press, Chicago.

Kuiper, G. P. (1951). On the origin of the irregular satellites. *Proc. Nat. Acad. Sci.* 37, 717-721.

Kumar, S. (1979). The stability of an SO_2 atmosphere on Io. *Nature* 280, 758-761.

Lanzerotti, L. J., Brown, W. L., and Johnson, R. E. (1980). Sputtering of sulfur from Io. IAU Coll. 57. *The Satellites of Jupiter* (abstract 4-22).

Lebofsky, L. (1977). Callisto: Identification of water frost. *Nature* 269, 785-787.

Lee, T. (1972). Spectral albedos of the Galilean satellites. *Comm. Lunar Planet. Lab.* 9, 179-180.

Lewis, J. S. (1971a). Satellites of the outer planets: Their physical and chemical nature. *Icarus* 15, 174-185.

Lewis, J. S. (1971b). Satellites of the outer planets: Thermal models. *Science* 172, 1127-1128.

Lewis, J. S. (1974). The temperature gradient in the solar nebula. *Science* 186, 440-443.

Lewis, J. S., and Consolmagno, G. C. (1979). Io: Geochemistry and geophysics of sulfur. *Bull. Amer. Astron. Soc.* 11, 599 (abstract).

Matson, D. L., Johnson, T. V., and Fanale, F. P. (1974). Sodium D-line emission from Io: Sputtering and resonant scattering hypothesis. *Astrophys. J.* 192, L43-L46.

Matson, D. L., Ransford, G. A., and Johnson, T. V. (1980). Heat flow from Io (JI). Submitted to *J. Geophys. Res.*

Mizuno, H., Nakazawa, K., and Hayashi, C. (1978). Instability of gaseous envelope surrounding planetary core and formation of giant planets. *Progr. Theor. Phys.* 60, 699-710.

Morabito, J., Synott, S. P., Kupferman, P. N., and Collins, S. A. (1979). Discovery of currently active extraterrestrial volcanism. *Science* 204, 972.

Morrison, D., Cruikshank, D. P., and Burns, J. A. (1977). Introducing the satellites. In *Planetary Satellites* (J. A. Burns, Ed.), pp. 3-17. Univ. Arizona Press, Tucson.

Morrison, D., and Telesco, C.M. (1980). Observational constraints in the internal energy source of Io. *Icarus.*

Nash, D. B., and Fanale, F. P. (1977). Io's surface composition based on reflectance spectra of sulfur/salt mixtures and proton-irradiation experiments. *Icarus* 31, 40-80.

Nash, D. B., and Nelson, R. M. (1979). Spectra evidence for sublimates and absorbates on Io. *Nature* 280, 763-766.

Nelson, R., and Hapke, B. (1978). Spectral reflectivities of the Galilean satellites and Titan, 0.32 to 0.86 micrometer. *Icarus* 36, 304-329.

Owen, T., Danielson, G. E. Cook, A. F., Hansen, C., Hall, V. L., and Duxbury, T. C. (1979). Jupiter's rings. *Nature* 281, 442-446.

Parmentier, E. M., and Head, J. W. (1979). Internal processes affecting surfaces of low-density satellites: Ganymede and Callisto. *J. Geophys. Res.* 84, 6263-6276.

Parmentier, E. M., and Head, J. W. (1980). Some possible effects of solid state deformation on the thermal evolution of ice-silicate planetary bodies. *Proc. Lunar Planet. Sci. Conf.* 11, 2403-2420.

Peale, S. J., Cassen P., Reynolds, R. T. (1979). Melting of Io by tidal dissipation. *Science* 203, 892-894.

Peale, S. J., and Greenberg, R. J. (1980). On the Q of Jupiter. *Proc. Lunar Planet Sci. Conf.* 11, 871-873.

Pearl, J., Hanel, R., Kunde, V., Maguire, W., Fox, K., Gupta, S., Ponnamperuma, C., and Raulin, F. (1979). Identification of gaseous SO_2 and new upper limits for other gases on Io. *Nature* 280, 755-758.

Perri, F., and Cameron, A. G. W. (1974). Hydrodynamic instability of the solar nebula in the presence of a planetary core. *Icarus* 22, 416-425.

Pilcher, C. B. (1980). Images of Jupiter's sulfur ring. *Science* 207, 181-183.

Pilcher, C. B., Ridgway, S. T., and McCord, T. B. (1972). Galilean satellites: Identification of water frost. *Science* 178, 1087-1089.

Pollack, J. B., Burns, J. A., and Tauber, M. E. (1979). Gas drag in primordial circumplanetary envelopes: A mechanism for satellite capture. *Icarus* 37, 587-611.

Pollack, J. B., Grossman, A. S. Moore, R., and Graboske, H. C., Jr. (1976). The formation of Saturn's satellites and rings as influenced by Saturn's contraction history. *Icarus* 29, 35-48.

Pollack, J. B., and Reynolds, R. T. (1974). Implications of Jupiter's early contraction history for the composition of the Galilean satellites. *Icarus* 21, 248-253.

Pollack, J. B., and Witteborn, F. C. (1980). Evolution of Io's volatile inventory. *Icarus.*

Pollack, J. B., Witteborn, F. C., Erickson, E. F., Strecker, D. W., Baldwin, B. J., and Bunch, T. E. (1978). Near-infrared spectra of the Galilean satellites: Observations and compositional implications. *Icarus* 36, 271-303.

Pollack, J. B., and Yung, Y. L. (1980). Origin and evolution of planetary atmospheres. *Ann. Rev. Earth Planet. Sci.* 8, 425-487.

Ransford, G. A., and Finnerty, A. A. (1980). Europa's petrologic thermal history. Submitted to *Nature.*

Reynolds, R. T., and Cassen, P. (1979). On the internal structure of the major satellites of the outer planets. *Geophys. Res. Letters.* 6, 121-124.

Sampson, R. A. (1921). Theory of the four great satellites of Jupiter. *Mem. Roy. Astron. Soc.* 63, 1-270.

Schubert, G., Stevenson, D. J., and Ellsworth, K. (1980). Internal structures of the Galilean satellites. Submitted to *J. Geophys. Res.*

See, T. J. J. (1910). *Researches on the Evolution of the Stellar Systems, Vol. II. The Capture Theory of Cosmical Evolution.* Chapts. 10 and 11. R. P. Nichols and Sons, Lynn, Mass.

Sinton, W. M. (1981). The thermal emission spectrum of Io and a determination of the heat flux from its hot spots. *J. Geophys. Res.* In press.

Slattery, W. L. (1977). The structure of the planets Jupiter and Saturn. *Icarus* 32, 58-72.

Smith, B. A., Shoemaker, E. M. Kieffer, S. W., and Cook, A. F., II (1979c). The role of SO_2 in volcanism on Io. *Nature* 280, 738-743.

Smith, B. A., and Smith, S. A. (1972). Upper limits for an atmosphere on Io. *Icarus* 17, 218-222.

Smith, B. A., and the Voyager Imaging Team (1979a). The Jupiter system through the eyes of Voyager 1. *Science* 204, 951-972.

Smith, B. A., and the Voyager Imaging Team (1979b). The Galilean satellites and Jupiter: Voyager 2 imaging science results. *Science* 206, 927-950.

Smith, P. H. (1978). Diameters of the Galilean satellites from Pioneer data. *Icarus* 35, 167-176.

Smythe, W.D., Nelson, R.M., and Nash, D.B. (1979). Spectral evidence for SO_2 frost or adsorbate on Io's surface. *Nature* 280, 766.

Smythe, W. H., and McElroy, M. B. (1977). *Planet. Space Sci.* 25, 415-431.

Squyres, S. W. (1980). Volume changes in Ganymede and Callisto and the origin of grooved terrain. *Geophys. Res. Letters.* In press.

Strom, R. G., Terrile, R. J., Masursky, H., and Hansen, C. (1979). Volcanic eruption plumes on Io. *Nature* 280, 733-736.

Sullivan, J. D., and Bagenal. F. (1979). In situ identification of various ionic species in Jupiter's magnetosphere. *Nature* 280, 798-799.

Summers, M. E., and Ingersoll, A. P. (1980). Atmospheric dynamics on Io. *Bull. Amer. Astron. Soc.* 12, 695-696 (abstract).

Synnott, S. P. (1980). 1979 J2: Discovery of a previously unknown satellite of Jupiter. *Science.* In press.

Taylor, G. E. (1972). The determination of the diameter of Io from its occultation of β Scorpii C on May 14, 1971. *Icarus* 17, 202-208.

Vogt, R. E., and the Voyager Cosmic Ray Team. (1979). Voyager 1: Energetic ions and electrons in the Jovian magnetosphere. *Science* 204, 63-67.

Wamsteker, W., Kroes, R. L., and Fountain, J. A. (1974). On the surface composition of Io. *Icarus* 23, 417-424.

Witteborn, F. C., Bregman, J. D., and Pollack, J. B. (1979). Io: An intense brightening near 5 micrometers. *Science* 203, 643-646.

Yoder, C. F. (1979). How tidal heating in Io drives the Galilean orbital resonance locks. *Nature* 279, 767-770.

APPENDIX

CARTOGRAPHY AND NOMENCLATURE FOR THE GALILEAN SATELLITES

Merton E. Davies
Rand Corporation

Cartography is an integral part of the program designed to explore the major bodies of the solar system. Even before the Voyager encounters, there was a plan to produce planimetric maps of Io, Europa, Ganymede, and Callisto based on the Voyager 1 and 2 pictures (Batson et al. 1980). Eventually, planetwide maps will be made at the U.S. Geological Survey at scales of 1:25,000,000 and 1:15,000,000 as well as a few sheets at 1:5,000,000 covering regions in which the quality of the pictures merits the larger scale.

The coordinate systems for bodies of the solar system were originally specified in terms of astrometric parameters; e.g., the central meridian for a satellite was defined as the meridian facing the planet at a particular epoch. However, mapping necessitates a redefinition in terms of topographic features observable on the object itself. A working group to recommend coordinate systems for use in mapping has been established by the International Astronomical Union (IAU), as described by Davies et al. (1980) and in the *Transactions* of the 1979 General Assembly of the IAU (1980). These recommended coordinate systems have been used to establish the control networks of the Galilean satellites (Davies and Katayama 1980,1981). These networks contain coordinates of numerous features identified on Voyager pictures; the locations of these control points are measured (in pixels) on many pictures and the results combined to provide optimum values for the size and shape of the satellite and its rotation axis, as well as the planetographic coordinates of the control points. Control nets for each Galilean satellite have been compiled by this photogrammetric procedure (Davies and Katayama 1981; Davies 1981); Table A1 summarizes some of the results. The control net is updated periodically as more pictures and points are incorporated in the computation.

TABLE A1

Satellite Sizes and Rotation Axes[a]

Satellite	Radius (km)	North Pole[b] α_{50}	δ_{50}	Prime Meridian[c] W
Io	1815 ± 5	268°01	64°54	262°7 + 203°4889538d
Europa	1569 ± 10	269°7	64°34	157°6313 + 101°3747235d
Ganymede	2631 ± 10	268°45	64°62	197°8361 + 50°3176081d
Callisto	2400 ± 10	268°25	64°62	157°9790 + 21°5710715d

[a]From Davis and Katayama (1981) and Davies (1981).
[b]Right ascension and declination of north pole (epoch 1950.0).
[c]Prime meridian, where d = JED − 2433282.5.

The astronomical definition of the Galilean satellites' prime meridians is the sub-Jupiter longitude at the first superior conjunction after 1950.0. The cartographic definitions of longitude are based on identifiable surface features for Europa, Ganymede, and Callisto, but on Io, because of the high resurfacing rate due to volcanism, it is not clear which features are permanent. For this reason, the longitude for Io has not been redefined in terms of a surface feature. On Europa, the crater Cilix defines 182° longitude; on Ganymede, the crater Anat defines 128° longitude; and on Callisto, the crater Saga defines 326° longitude.

Astronomical nomenclature is a complex and sometimes emotional topic, and in the past controversies over names often continued for decades until finally resolved by popular usage. Recently, the large numbers of names required for surface features revealed by spacecraft imaging have prompted the IAU to establish a Working Group for Planetary System Nomenclature (WGPSN) to recommend names for approval by the IAU General Assembly. P. Millman (National Research Council of Canada) has been the chairman of the WGPSN since its formation in 1973; he has briefly described its procedures and goals (Millman 1976). T. Owen (State University of New York) is chairman of the Task Group for Outer Solar System Nomenclature, which recommends names for features on the Galilean satellites.

The IAU WGPSN decided several years before the Voyager mission to continue the mythological theme of the satellite names in the nomenclature for surface features, but before the Voyager images were received it was not known what kinds of features would be present or how many would require names. In preparation for the encounters, a "name bank" was compiled from mythologies associated with the personages, places, and objects that figured prominently in the myths from which the satellites' names originated.

The Task Group originally planned to name features on Io from myths of equatorial peoples; Europa was to be assigned names from European temperate-zone myths; Ganymede from non-European myths; and Callisto, the outermost Galilean satellite, from Far-Northern myths.

Images returned from Voyager 1 revealed objects that appeared very different from all known planetary bodies, requiring new generic classes as well as names. Some features are so unusual that new terms such as *macula* (dark spot), *linea* (elongate markings), and *flexus* (low, curvilinear, scalloped ridge) were adopted. The term *regio* (large area of distinctive albedo markings), originally applied to Mars by early astronomers, was revived. In cases in which features of the satellites resemble types of features on the Moon, Mars, or Mercury, terms previously applied to those bodies, like *patera* (irregular depression) and *planum* (high plateau), are used. Whenever possible, these generic terms are descriptive rather than carrying an implication of origin; thus patera could be either volcanic or non-volcanic, and probably refers to features of different geology on different planetary bodies.

For Io, the original plan to use names from equatorial-zone myths was abandoned when prominent volcanic eruptions were discovered. Instead, the features from which the eruptions emanated, and similar features thought to be dormant eruptive centers, were assigned names of fire, sun, and smith gods, and of gods associated with terrestrial volcanoes. For instance, the largest Io volcano was named Pele for the Hawaiian goddess of volcanoes. The original Far-Northern theme was followed for icy Callisto's craters and large ringed basins, such as Valhalla and Asgard. Crater names on Ganymede (e.g. Gilgamesh) were chosen from the most ancient Near-Eastern civilizations. Names on Europa and Amalthea all come from the myths associated with these personages. Myth-related names are also assigned to nonvolcanic features on Io, some craters and a catena on Callisto, and the arrays of linear grooves and a few craters on Ganymede. The astronomers who discovered satellites of Jupiter are commemorated by the names of the ancient regions of low albedo on Ganymede: Galileo, Marius, and Nicholson Regio.

Features on new, high-resolution maps which are still being compiled will be named from the established name banks using schemes already adopted. It may also be necessary to invent new terms and to expand the name banks when the high-resolution pictures are studied in detail.

Tables A2 through A5 contain the IAU-approved names for features on Io, Europa, Ganymede, and Callisto, together with information on the origin of the names. The coordinates are consistent with the November 1980 control nets (Davies and Katayama 1980,1981).

Planetary maps are produced at the U.S. Geological Survey's Branch of Astrogeologic Studies in Flagstaff, Arizona, under the direction of R.M. Batson (see Batson 1978). Preliminary maps of the Galilean satellites at a scale of 1:25,000,000 were produced soon after the arrival of the pictures from the Voyager spacecraft. These maps are shown in Figs. A1, A2, A3, and A4. The airbrush renditions of Io and Callisto were made by P. Bridges, and those of Europa and Ganymede by J. Inge. Because these maps were compiled before the computation of a control net, the positional data may be in error by as much as $10°$ in latitude and an equivalent distance in longitude. The new series of maps will use the control net to position the latitude and longitude grid so that this source of inconsistency will be eliminated.

REFERENCES

Batson, R.M. (1978). Planetary mapping with the airbrush. *Sky Tel.* 55, 109-112.
Batson, R.M., Bridges, P.M., Inge, J.L., Isbell, C., Masursky, H., Strobell, M.E., and Tyner, R.L. (1980). Mapping the Galilean satellites of Jupiter with Voyager data. *Photogrammetric Engineering and Remote Sensing* 46, 1303-1312.
Davies, M.E. (1981). Coordinates of features on the Galilean satellites. *Bull. Amer. Astron. Soc.* 12, 711 (abstract).
Davies, M.E., Abalakin, V.K., Cross, C.A., Duncombe, R.L., Masursky, H., Morando, B., Owen, T.C., Seidelmann, P.K., Sinclair, A.T., Tjuflin, Y.S., and Wilkins, G.A. (1980). Report of the IAU Working Group on cartographic coordinates and rotational elements of the planets and satellites. *Celestial Mech.* 22, 205-230.
Davies, M.E., and Katayama, F.Y. (1980). *Coordinates of Features on the Galilean Satellites*, The Rand Corporation, N-1617-JPL/NASA.
Davies, M.E., and Katayama, F.Y. (1981). Coordinates of features on the Galilean satellites. *J. Geophys. Res.* 86, 8635-8657.
Millman, P.M. (1976). Topographic nomenclature on planetary bodies. *Icarus* 29, 155-157.
Transactions of the International Astronomical Union, Vol. 17B (1980). Proceedings of the Seventeenth General Assembly, Montreal, 1979, D. Reidel Publ. Co., The Netherlands.

TABLE A2

IAU-Approved Names for Features on Io

Name	Information on Origin of Name	Latitude (deg)	Longitude (W) (deg)
Eruptive Centers			
Amirani	Georgian god of fire	28 N - 21 N	118
Loki	Norse smith god; trickster	19 N - 17 N	306 - 301
Marduk	Sumero-Akkadian fire god	27 S	209
Masubi	Japanese fire god	39 S	56
Maui	Hawaiian demigod who sought fire from Mafuike	19 N	122

TABLE A2 (continued)

Name	Information on Origin of Name	Latitude (deg)	Longitude (W) (deg)
Pele	Hawaiian volcano goddess	19 S	256
Prometheus	Greek fire god	2 S	153
Surt	Icelandic volcano god	45 N	338
Volund	Germanic supreme god of smiths	22 N	177
Catena			
Mazda	Babylonian sun god	9 S	313
Mons			
Haemus	Io passed by in her wanderings	70 S	50
Silpium	Io died of grief there	52 S	274
Planum			
Dodona	Io went there after the death of Argus	50 S - 60 S	5 - 335
Nemea	Where Io was turned into a cow by Zeus	75 S - 85 S	210 - 320
Regio			
Bactria	Io passed through in her wanderings	35 S - 50 S	120 - 130
Chalybes	Io passed through in her wanderings	45 N - 70 N	70 - 120
Joppa	Io passed through in her wanderings	0 N - 50 N	130 - 210
Lerna	Meadows of Lyrcea	55 S - 70 S	280 - 310
Media	Io passed through in her wanderings	20 S - 30 N	0 - 100
Mycenae	In some legends, Io transformed there	30 S - 55 S	130 - 180
Tholus			
Apis	Name for Epaphus, son of Io	11 S	349
Inachus	Father of Io	16 S	349
Patera			
Amaterasu	Japanese sun goddess	38 N	307
Asha	Persian spirit of fire	9 S	226

TABLE A2 (continued)

Name	Information on Origin of Name	Latitude (deg)	Longitude (W) (deg)
Atar	Iranian personification of fire	30 N	279
Aten	Egyptian sun god	48 S	311
Babbar	Mongolian sun god	40 S	272
Bochia	Chibcha sky god	61 S	22
Creidne	Celtic smith hero	51 S	344
Culann	Celtic smith hero	18 S	160
Daedalus	Greek smith hero	19 N	275
Dazhbog	Slavonic god of sun	55 N	302
Emaking	Sulca (N. Britain) man who brought fire	3 S	119
Fuchi	Ainu fire goddess	28 N	328
Galai	Mongolian fire god	11 S	288
Gibil	Assyrian fire god	15 S	295
Heno	Iroquois thunder god	57 S	312
Hephaestus	Greek smith god	2 N	290
Hiruko	Japanese sun god	65 S	330
Horus	Egyptian falcon-headed solar god	10 S	340
Inti	Inca sun god	68 S	349
Kane	Hawaiian god of sunlight	48 S	14
Loki	Norse smith god; trickster	13 N	310
Maasaw	Hopi god of fire, death	40 S	341
Mafuike	Hawaiian demigoddess whose fingers contained fire	13 S	260
Malik	Babylonian, Caananite, Syrian sun god	34 S	129
Manua	Hawaiian sun god	35 N	322
Masaya	Nicaraguan sun god	22 S	349
Maui	Hawaiian demigod who sought fire from Mafuike	16 N	124
Mihr	Armenian fire god	16 S	306

TABLE A2 (continued)

Name	Information on Origin of Name	Latitude (deg)	Longitude (W) (deg)
Nina	Inca fire god	38 S	166
Nusku	Assyrian fire god	63 S	7
Nyambe	Zambezi sun god	4 S	342
Ra	Egyptian sun god	8 S	325
Reiden	Japanese thunder god	13 S	236
Ruwa	African (Wachaga) god associated with Mt. Kilimanjaro	1 N	4
Sengen	Deity of Mt. Fuji	33 S	304
Shakuru	Pawnee sun god of the east	24 N	266
Shamash	Assyro-Babylonian sun god	34 S	152
Tohil	Central American god who gave fire to man	25 S	158
Svarog	Russian smith god	48 S	267
Ulgen	Siberian progenitor god who struck first fire	41 S	288
Uta	Early Sumerian sun god	35 S	25
Vahagn	Armenian sun god	26 S	357
Viracocha	Quechua sun god	62 S	281

TABLE A3

IAU-Approved Names for Features on Europa

Name	Information on Origin of Name	Latitude (deg)	Longitude (W) (deg)
Linea			
Adonis	Nephew of Europa	38 S - 60 S	122 - 113
Agenor	Europa's father	44 S - 35 S	221 - 175
Argiope	Europa's mother	20 S - 1 S	204 - 171
Asterius	Europa's husband after Zeus	29 S - 29 N	314 - 251
Belus	Agenor's twin brother	14 N - 26 N	226 - 170

TABLE A3 (continued)

Name	Information on Origin of Name	Latitude (deg)	Longitude (W) (deg)
Cadmus	Europa's brother	29 N - 26 N	251 - 170
Libya	Agenor's mother	66 S - 41 S	263 - 153
Minos	Son of Europa and Zeus	45 N - 31 N	119 - 150
Pelorus	Founder of Thebes	26 S - 90 N	195 - 110
Phineus	Brother of Europa	26 S - 29 S	340 - 314
Sarpedon	Son of Europa and Zeus	23 S - 51 S	94 - 80
Thasus	Brother of Europa	58 S - 76 S	176 - 217
Flexus			
Cilicia	Named for Cilix, who sought Europa	70 S - 58 S	200 - 153
Gortyna	Place where Zeus brought Europa	62 S - 50 S	182 - 146
Sidon	Where Europa was born	70 S - 64 S	215 - 150
Macula			
Thera	Where Cadmus stopped while seeking Europa	48 S	181
Thrace	Where Cadmus stopped while seeking Europa	47 S	172
Tyre	Where Zeus carried Europa away	32 N	147
Crater			
Cilix	Brother of Europa	1 N	182

TABLE A4

IAU-Approved Names for Features on Ganymede

Name	Information on Origin of Name	Latitude (deg)	Longitude (W) (deg)
Craters			
Achelous	River god, father of Ganymede's mother	61 N	14
Adad	Assyrian god of thunder	56 N	0

TABLE A4 (continued)

Name	Information on Origin of Name	Latitude (deg)	Longitude (W) (deg)
Adapa	Babylonian hero; refused immortality when refused to eat the "food life"	72 N	30
Ammura	Phoenician god of the west	31 N	344
Anat	Goddess of dew	3 S	128
Anu	Babylonian-Sumerian-Akkadian heaven god	64 N	346
Asshur	Assyrian warrior god	53 N	335
Aya	Assyro-Babylonian; wife of Shamash	66 N	325
Baal	Caananite god	24 N	332
Danel	Phoenician mythical hero; diviner	4 S	25
Diment	Egyptian goddess of home of dead	23 N	353
Enlil	Babylonian nature god	53 N	315
Eshmun	Phoenician, divinity of Sidon	18 S	192
Etana	Flew high on Eagle's back, then crashed	73 N	342
Gilgamesh	Babylonian hero; sought immortality for friend	62 S	123
Gula	Assyro-Babylonian goddess of health	63 N	13
Hathor	Egyptian goddess of job, love	70 S	267
Isis	Egyptian goddess, wife of Osiris	68 S	197
Keret	Phoenician legendary hero	16 N	38
Khumbam	Elamite creator god	25 S	338
Kishar	Assyro-Babylonian progenitor goddess	71 N	351
Melkart	Phoenician; divinity of Tyre	10 S	186
Mor	Phoenician spirit of harvest	30 N	329

TABLE A4 (continued)

Name	Information on Origin of Name	Latitude (deg)	Longitude (W) (deg)
Nabu	Assyro-Babylonian intellectual god	47 S	9
Namtar	Assyro-Babylonian plague demon	62 S	354
Nigirsu	Assyro-Babylonian war god	63 S	321
Nut	Egyptian sky goddess	61 S	268
Osiris	Egyptian god of the dead	38 S	165
Ruti	Byblos god	10 N	309
Sapas	Torch of the gods	57 N	38
Sebek	Egyptian crocodile god	60 N	358
Sin	Assyrian Moon god	52 N	359
Tanit	Carthaginian goddess	57 N	41
Teshub	Hurrian weather god	7 S	21
Tros	Father of Ganymede	12 N	31
Zagar	Sin's messenger; brought men dreams	58 N	42
Sulci			
Anshar	Sumerian-Akkadian celestial world	32 N - 15 N	211 - 192
Apsu	Sumerian-Akkadian primordial ocean	38 S - 16 S	242 - 218
Aquarius	Zeus set Ganymede among stars as Aquarius	38 N - 53 N	30 - 7
Dardanus	Where Ganymede abducted by eagles	50 S - 4 N	35 - 7
Harpagia	Where Ganymede abducted by eagles	0 N - 2 N	328 - 313
Kishar	Sumerian-Akkadian terrestrial world	20 S - 0 S	230 - 208
Mashu	Mountain where sun rose, set	25 N - 44 N	184 - 221
Mysia	Where Ganymede abducted by eagles	15 S - 9 N	19 - 331
Nun	Chaos; primordial ocean	42 N - 57 N	340 - 321

TABLE A4 (continued)

Name	Information on Origin of Name	Latitude (deg)	Longitude (W) (deg)
Philus	Where Ganymede, Hebe worshipped	36 N - 55 N	218 - 200
Phrygia	Where Ganymede born	4 N - 18 N	25 - 0
Sicyon	Where Ganymede, Hebe worshipped	30 N - 43 N	24 - 359
Tiamat	Sumerian-Akkadian tumultuous sea	19 N - 12 S	223 - 197
Uruk	Babylonian city ruled by Gilgumesh	28 N - 10 S	179 - 143
Regio			
Barnard	Edward Barnard (1857-1923) U.S. Astronomer	5 S - 27 N	25 - 342
Galileo	Galileo Galilei (1564-1642) Italian Astronomer	1 N - 65 N	80 - 180
Marius	Simon Marius (1570-1624) German Astronomer	55 S - 52 N	240 - 180
Nicholson	Seth Nicholson (1891-1963) U.S. Astronomer	42 S - 55 S	61 - 310
Perrine	Charles Perrine (1867-1951) U.S. Astronomer	18 N - 53 N	75 - 340

TABLE A5

IAU-Approved Names for Features on Callisto

Name	Information on Origin of Name	Latitude (deg)	Longitude (W) (deg)
Catena			
Gipul	River; Norse	65 N - 73 N	68 - 36
Large-Ringed Features			
Adlinda	Place in ocean depths where souls go after death; Eskimo	54 S	21

TABLE A5 (continued)

Name	Information on Origin of Name	Latitude (deg)	Longitude (W) (deg)
Asgard	Home of the gods; Norse	28	142
Valhalla	Odinn's hall	17 N	57
Craters			
Adal	Son of Karl and Erna; Norse	76 N	82
Agroi	Finno-Ugric god of twins	44 N	11
Akycha	Alaskan name of the sun	73 N	317
Alfr	Norse dwarf	10 S	224
Ali	Norse strongest ancient man	60 N	56
Anarr	Norse dwarf; Voluspa	44 N	0
Aningan	Greenland Eskimo moon god	53 N	352
Askr	First man	52 N	324
Balkr	Ottar's ancestor; Norse	30 N	12
Bavorr	Norse dwarf	50 N	20
Beli	Norse giant; Jotun	63 N	82
Bragi	Skald-god; Norse	77 N	78
Brami	Ottar's ancestor; Norse	29 N	19
Bran	Celtic ominipotent god	25 S	208
Buga	Tungu (Russian) heaven god	22 N	324
Buri	Norse dwarf; Voluspa	36 S	43
Burr	Giant; sons held up sky	42 N	135
Dag	Ottar's ancestor; Norse	59 N	77
Danr	Son of Karl and Erna; Norse	63 N	78
Dia	Callisto's sister; Greek	72 N	64
Dryops	Son of Dia by Apollo; Greek	78 N	20
Durinn	Norse dwarf	68 N	91

TABLE A5 (continued)

Name	Information on Origin of Name	Latitude (deg)	Longitude (W) (deg)
Egdir	Shepard for giants; Norse	35 N	36
Erlik	Russian first man who became a devil	65 N	349
Fadir	Farmer; Norse	57 N	12
Fili	Norse dwarf	64 N	341
Finnr	Norse dwarf	16 N	4
Freki	Wolf's name meaning insatiable; Norse	80 N	348
Frodi	Hledi's father; Norse	68 N	137
Fulla	Figg's maid; Norse	74 N	111
Fulnir	Son of Thrael and Thyr; Norse	61 N	35
Geri	Wolf's name meaning greedy; Norse	65 N	344
Gisl	Steed ridden by Aesir; Norse	58 N	33
Gloi	Norse dwarf	50 N	243
Goll	Servant to the gods; Norse	57 N	319
Gondul	Valkyrjur maiden	60 N	116
Grimr	A name for Odinn; Norse	41 N	214
Gunnr	Valkyrjur maiden	65 N	106
Gymir	Another name for Norse god Aegir	64 N	53
Habrok	Norse hawk	76 N	136
Haki	Norse giant	25 N	315
Har	Norse dwarf	3 S	358
Hepti	Norse dwarf	65 N	23
Hodr	Norse god; Baldr's blind brother	70 N	91
Hoenir	Norse god; gave blood to first humans	33 S	263
Hogni	Ottar's ancestor	12 S	5
Igaluk	Alaskan name of the moon	7 N	315

TABLE A5 (continued)

Name	Information on Origin of Name	Latitude (deg)	Longitude (W) (deg)
Ivarr	Ottar's ancestor	6 S	322
Jumo	Finno-Ugric heaven god	61 N	9
Kari	Ottar's ancestor	47 N	103
Karl	Son of Rigr and Amma	55 N	329
Lodurr	Norse god; gave color to first humans	49 S	274
Loni	Norse dwarf	4 S	215
Losy	Mongol evil snake; tried to kill all living things	65 N	322
Mera	Another nymph seduced by Zeus; Greek	65 N	76
Mimir	Norse giant	33 N	54
Mitsina	Alaskan old man who perished while hunting	59 N	102
Modi	Son of Thorr and Sif; Norse	67 N	122
Nama	Russian hero; saved family from flood in ark	57 N	330
Nar	Norse dwarf; Voluspa	1 S	46
Nerivik	Alaskan heroine (Sedna)	16 S	56
Nidi	Norse dwarf; Voluspa	67 N	97
Nori	Norse dwarf; Voluspa	45 N	343
Nuada	Irish Chieftan god	64 N	270
Oski	Another name for Odin; Norse	59 N	267
Ottar	Freyja's favorite	62 N	105
Pekko	Finno-Ugric god of barley	19 N	6
Reginn	Norse dwarf	40 N	91
Rigr	Heimdallr in disguise	72 N	238
Saga	Another name for Frigg	1 N	326
Sarakka	Finno-Ugric goddess of childbed	3 S	54
Seginek	Eskimo sun	56 N	25

TABLE A5 (continued)

Name	Information on Origin of Name	Latitude (deg)	Longitude (W) (deg)
Sholmo	Finno-Ugric heaven god	54 N	16
Sigyn	Loki's wife; Norse	36 N	29
Skoll	Norse wolf	56 N	315
Skuld	Norse maiden; governs fate of humans	11 N	38
Sudri	Norse dwarf	56 N	138
Sumbur	Russian (Bariat) world mountain	67 N	324
Tindr	Ottar's ancestor	2 S	356
Tornarsuk	Greenland legendary hero	29 N	129
Tyn	Great god of Germanic peoples	72 N	226
Valfodr	A name for Odinn god of wisdom; Norse	1 S	248
Vali	Ottar's ancestor	10 N	325
Vestri	Norse dwarf	50 N	53
Vitr	Norse dwarf; Voluspa	22 S	350
Ymir	Giant from whom Earth created; Norse Voluspa	52 N	102

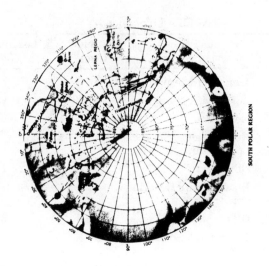

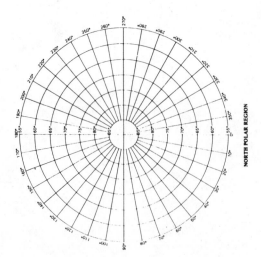

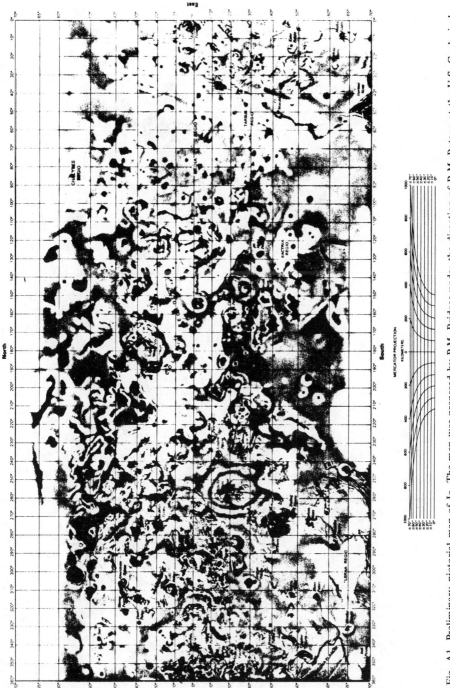

Fig. A1. Preliminary pictorial map of Io. The map was prepared by P.M. Bridges under the direction of R.M. Batson at the U.S. Geological Survey for the Voyager Imaging Science Team.

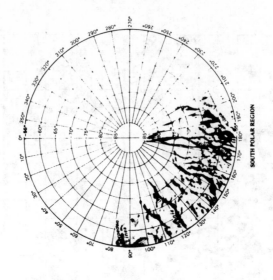

SOUTH POLAR REGION

POLAR STEREOGRAPHIC PROJECTION

KILOMETERS

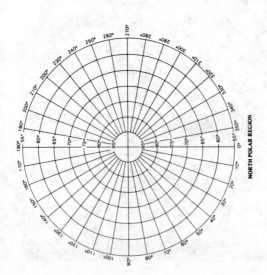

NORTH POLAR REGION

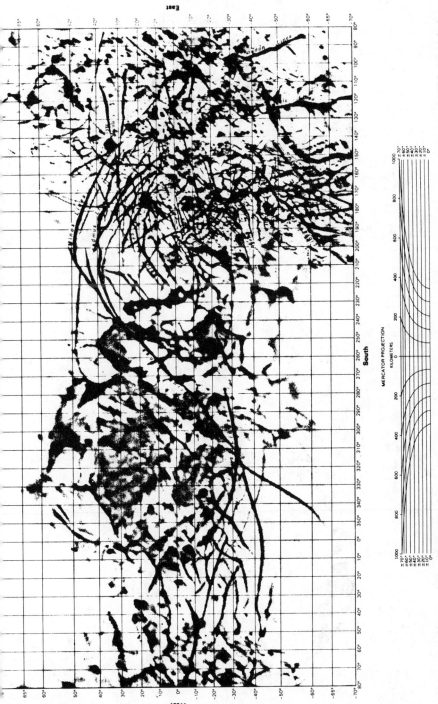

Fig. A2. Preliminary pictorial map of Europa. The map was prepared by J.L. Inge under the direction of R.M. Batson at the U.S. Geological Survey for the Voyager Imaging Science Team.

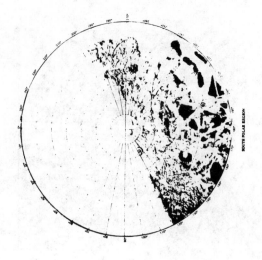

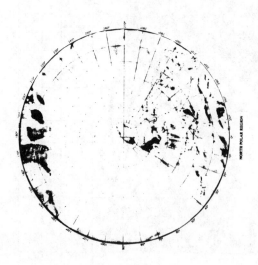

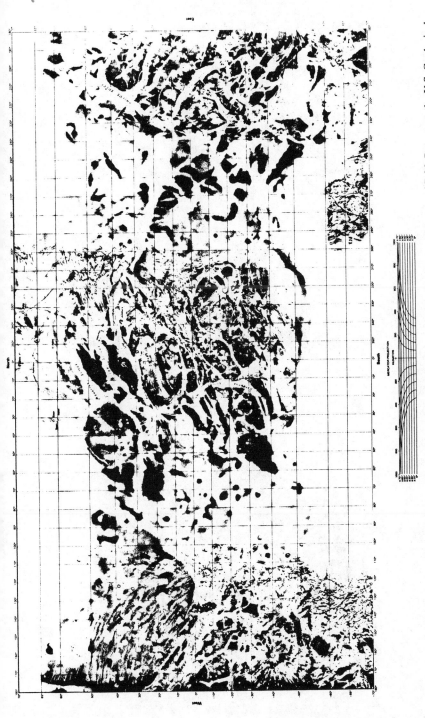

Fig. A3. Preliminary pictorial map of Ganymede. The map was prepared by J.L. Inge under the direction of R.M. Batson at the U.S. Geological Survey for the Voyager Imaging Science Team.

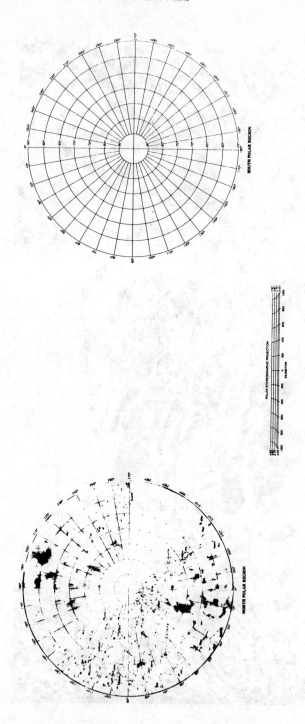

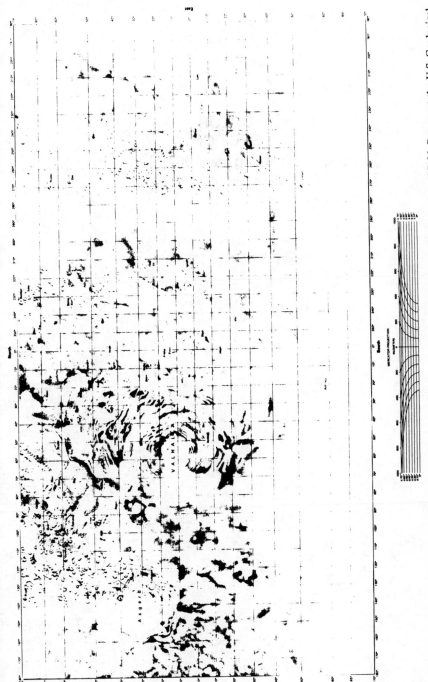

Fig. A4. Preliminary pictorial map of Callisto. The map was prepared by P.M. Bridges under the direction of R.M. Batson at the U.S. Geological Survey for the Voyager Imaging Science Team.

**COLOR
SECTION
(starting page 937)
Follows immediately**